$10.95

ECOLOGY

WITH SPECIAL REFERENCE TO ANIMALS AND MAN

S. CHARLES KENDEIGH

Department of Zoology
University of Illinois at Urbana-Champaign

PRENTICE-HALL, INC., Englewood Cliffs, New Jersey

Library of Congress Cataloging in Publication Data

Kendeigh, S. Charles
    Ecology with special reference to animals and man.

    First published in 1961 under title: Animal
ecology.
    Bibliography: p. 410
    1. Ecology. I. Title.
QH541.K46 1974   574.5   73-14558
ISBN 0-13-222745-2

TO MY WIFE, DOROTHY

10 9 8 7 6 5 4 3 2 1

Ecology with Special Reference to Animals and Man
is a revision of *Animal Ecology* by S. Charles Kendeigh,
© 1961 by Prentice-Hall, Inc., Englewood Cliffs, N.J.

PRENTICE-HALL INTERNATIONAL, INC., *London*
PRENTICE-HALL OF AUSTRALIA, PTY. LTD., *Sydney*
PRENTICE-HALL OF CANADA, LTD., *Toronto*
PRENTICE-HALL OF INDIA PRIVATE LIMITED, *New Delhi*
PRENTICE-HALL OF JAPAN, INC., *Tokyo*

CONTENTS

Part Four   POPULATION ECOLOGY   *187*

Part Five   EVOLUTIONARY ECOLOGY   *241*

Part Six   GEOGRAPHICAL ECOLOGY   *275*

Part Seven   MARINE ECOLOGY   *383*

PREFACE

The science of *ecology*, born at the beginning of the present century after a gestation period of several hundred years, has now not only matured into an honored and respected scholarly discipline and field of research but has also become a household word as it relates to the general condition of the environment. This book is an effort to summarize the basic concepts and principles of the subject, to present the elementary factual information with which a person to be competent in the field should be familiar, and to show how these principles and facts may be applied in a practical way to the interests and welfare of man. Although the book relates especially to animals, enough material is given concerning plants to bring out their essential place in the system of nature and to emphasize the bioecological point of view.

After a background section for orientation, local communities and habitats are discussed in some detail. It is my firm belief that a person beginning the study of ecology should first become thoroughly acquainted with the places where organisms may be found in nature, what kinds of organisms occur in different habitats, the abundance and interrelations of organisms in these habitats, the behavior and the life requirements of the principal species, and the structure and succession of communities as observed in the field.

Only when he is well founded in this knowledge is he fully ready to understand the ecological processes and community dynamics that are presented in the remainder of the book.

I am also convinced that an introductory course in ecology should survey the entire field in a comprehensive and balanced manner. This I have tried to do, as shown in the table of contents, by covering basic concepts and facts in community ecology, ecosystem ecology, population ecology, evolutionary ecology, geographical ecology, and marine ecology. I am a little abashed that I do not include a section on physiological ecology, which is my major research interest at the present time, although it is dealt with sparingly throughout the book. The proper development of physiological ecology takes one extensively into laboratory experimentation, however, which is a different approach. Our emphasis here is kept on the study of the free-living organism in its natural environment. Physiological ecology as well as more advanced and detailed studies in all the various phases of ecology may best be left to advanced courses. Similarly with behavioral ecology, considerable attention is given to the behavioral responses of organisms, even though a separate section on ethology has not been organized. Although the quantitative aspect of ecology is emphasized and continually referred to, I do not believe that in an introductory book it is desirable to approach the subject from a detailed mathematical point of view. This too may well be postponed to advanced courses.

Each chapter is an independent unit, and each major topic progresses from a basic analysis of its natural relations to a discussion of how it applies and is relevant to man. This includes treatment of water, air, and soil pollution; pest control; population explosion; energy yield; cultural evolution; and how man behaves fundamentally as an animal.

This book is designed for a course at the junior-senior-graduate level, and for students who have had at least a year's background in biology. The course is best given in the autumn in order to have early contact in the field with fully developed communities. If given in the spring it would probably be best to reverse the positions of the two sections, Community Ecology and Geographical Ecology. This would provide the student with some knowledge of communities before undertaking the more theoretical sections and leave the discussion of local communities to the end of the course when it can be correlated with field work.

I gave the course during the autumn semester at the University of Illinois for 23 years. There were half-day or full-day trips every Saturday until winter weather set in and two half-day winter trips. Also included in the field work was one weekend camping trip to study communities not found locally. The students got to see at first hand a large variety of plants and animals, and to measure population sizes by quantitative methods that may have been crude but were nevertheless effective in stream riffles and pools, ponds of different ages, bogs, lakes, grassland, deciduous and coniferous forests, and seral stages as they develop on rock, sand, pond, bog, floodplain, and abandoned strip-mine areas. Some experimentation was also done in the field to analyze the manner in which both aquatic and terrestrial species respond to environmental factors. There was a small amount of laboratory work during the winter for learning quantitative methods of counting plankton, examining different kinds of respiratory systems in aquatic organisms, searching and identifying micro-organisms in the soil, and experiments in choice of habitats. Methods for measuring productivity were discussed but actual practice with these methods was left for an advanced class.

Citations to the literature in the text have been done by name and year of publication to encourage students to associate ideas with names of ecologists responsible for them and to enable the student to pursue the subject to greater depth.

Some care has been taken with taxonomic nomenclature. Common names are used throughout the text as far as possible, with the scientific nomenclature restricted to the index. Authorities followed for most scientific names are the following. Mammals: North America: Miller and Kellogg (1955); Eurasia: Ellerman and Morrison-Scott (1951). Birds: A.O.U. Checklist (1957). Reptiles and amphibians: Schmidt (1953). Fish: Bailey (1960). Invertebrates: as given by authors, not standardized. Trees: Dayton *et al.* (1953). Grasses: Hitchcock (1951); and other plants: Fernald (1950), Rydberg (1954). Common names of mammals are mostly from Hall (1957); birds, A.O.U. Checklist (1957); reptiles and amphibians, Conant (1958); and fish, Bailey (1960).

Finally I wish to acknowledge the help of Daniel I. Axelrod of the University of California at Davis; Clarence F. Clark of Ohio State University; Charles E. King of University of South Florida; Peter H. Klopfer of Duke University; James R. Karr of Purdue University; Roland R. Roth of the University of Delaware; and Glenn C. Sanderson, R. F. Labisky, and W. R. Edwards of the Illinois Natural History Survey. I am especially grateful to Robert H. Whittaker of Cornell University and Francis C. Evans of the University of Michigan who read and commented in detail on the whole manuscript. John Riina, Chester C. Lucido Jr., and Zita de Schauensee of Prentice-Hall, Inc., cooperated in every desirable way in obtaining reviews of the manuscript before publication and in efficiently seeing it through the press.

Illustrations come from several sources. I am most grateful to the late Victor E. Shelford for many drawings originally published in his *Animal Communities of Temperate America* (1913); to the Illinois Natural History Survey for original illustrations from several of their publications; to the U.S. Forest Service which allowed me to select what I wanted from their extensive file of photographs; to the Friez Instrument Division; to the University of Wisconsin News Service; to Dr. John W. Aldrich of the Bureau of Sport Fisheries and Wildlife of the U.S. Department of Interior for the detailed map used on the front end papers and several illustrations in the section on Geographical Ecology; to a number of other individuals for supplying photographs or other illustrative material for which acknowledgment is made in the legends of the figures; and to Colleen Nelson, Katherine Little, and Nan Brown for preparing special drawings.

Champaign, Illinois   S. Charles Kendeigh

Part One

BACKGROUND

Chapter 1

SCOPE AND HISTORY
OF ECOLOGY

The word *ecology*, derived from the Greek words *oikos* meaning habitation, and *logos* meaning discourse or study, implies a study of the habitations of organisms.

Ecology was first described as a separate field of knowledge in 1866 by the German zoologist Ernst Haeckel, who invented the word *oekologie* for "the relation of the animal to its organic as well as its inorganic environment, particularly its friendly or hostile relations to those animals or plants with which it comes in contact."

Ecology has been variously defined by other investigators, as "scientific natural history," "the study of biotic communities," or "the science of community populations"; probably the most comprehensive definition is the simple one most often given: *a study of animals and plants in their relations to each other and to their environment.*

OBJECTIVES

Ecology is a distinct science because it is a body of knowledge not similarly organized in any other division of biology; because it uses a special set of techniques and procedures; and because it has a unique point of view. The essence of this science is a comprehensive understanding of the import of these phenomena:

1. The local and geographic distribution and abundance of organisms (habitat, niche, community, biogeography).
2. Temporal changes in the occurrence, abundance, and activities of organisms (seasonal, annual, successional, geological).
3. The interrelations between organisms in populations and communities (population ecology).
4. The structural adaptations and functional adjustments of organisms to their physical environment (physiological ecology).
5. The behavior of organisms under natural conditions (ethology).
6. The evolutionary development of all these interrelations (evolutionary ecology).
7. The biological productivity of nature and how this may best serve mankind (ecosystem ecology).
8. The development of mathematical models to relate interaction of parameters and predict effects (systems analysis).

A study of organisms in the field may bring to light problems which will be most expediently worked out in the laboratory; but field and laboratory investigations must be integrated. The investigator must often study the morphology of dead organisms in the laboratory, and there perform experiments on living animals and plants held under carefully controlled experimental conditions. But unless such studies are perspective to the normal life of an organism, as it is lived under natural conditions, they are not ecology.

The use of exact quantitative techniques is, of course, a general characteristic of all science. But special difficulties arise when such techniques are applied to free-living organisms in natural conditions. For example, size of animal populations has, in the past, often been described in such vague terms as "rare," "common," or "abundant." These are subjective terms, based largely on an impression gained by the observer of the apparent conspicuousness of the species. As James Fisher, an English naturalist, wrote in 1939, a species has usually been indicated as "rare" when actual numbers expressible in one's and two's could be recorded; "common" when the observer began to lose count; and "abundant" when he became bewildered. One of the chief problems of the ecologist is to develop methods by which to measure the absolute size of populations and the productive capacities of different habitats so that the activities of widely varying types of species may be compared. For setting up experiments and organizing and analyzing studies under natural conditions, it is becoming more and more essential that the ecologist be familiar with and employ good statistical procedures (Williams 1954). An objective of ecological research is the establishment of mathematical models or computer simulation programs for the various systems involved. Such models give proper weight to all factors so that the effect of variation of any one or combination of factors can be predicted in advance.

> As a contribution to human knowledge and understanding, ecology is in the fortunate position of being concerned with the most complicated systems of organization, apart from human societies, with which we have to deal. For this very reason it provides a constant challenge to the imagination as well as to experimental ingenuity. It is more difficult to analyze and isolate the relevant factors in a living community than in a simpler system, but the gain in significant understanding of the material world and in comprehending the beauty of its organization is perhaps better in proportion [Macfadyen 1957: 246].

RELATION TO OTHER SCIENCES

Ecology is one of the three main divisions of biology, the other two being morphology and physiology. The emphasis in morphology is on how organisms are made; in physiology, on how they function; and in ecology, on how they live. These divisions overlap broadly. To appreciate fully the structure of an organ, one needs to know how it functions, and the way it functions is clearly related to environmental conditions. The morphologist is concerned with problems of anatomy, histology, cytology, embryology, evolution, and genetics; the physiologist, with interpreting functions in terms of chemistry, physics, and mathematics; and the ecologist, with distribution, behavior, populations, and communities in relation to the environment (ecosystems). The evolution of adaptation and of species is of mutual interest to the ecologist and to the geneticist; biometeorology is a connecting link between ecology and physiology; and systems analysis interrelates ecology and mathematics. All areas, in the final analysis, are simply different approaches to an understanding of the meaning of life.

SUBDIVISIONS OF ECOLOGY

Ecology may be studied with particular reference to animals or to plants, hence *animal ecology* and *plant ecology*. Animal ecology, however, cannot be adequately understood except against a considerable background of plant ecology. When animals and plants are given equal emphasis, the term *bioecology* is often used. Courses in plant ecology usually dismiss animals as but one of many factors in the environment. *Synecology* is the study of communities, and *autecology* the study of species. There is some confusion in these terms since Europeans commonly use "ecology" in a narrower sense—meaning the environmental relations of organisms or of communities. The broader study of communities, including species interrelations and community structure and function as well as environmental relations (synecology), is generally termed "biocenology" or "biosociology" by Europeans.

In this book we shall survey the fundamentals and basic facts of ecology as they relate to animals and have application to man. We will study *community ecology*, the local distribution of animals in various habitats, the recognition and composition of community units, and succession; *ecosystem dynamics*, the processes of soil formation, nutrient cycling, energy flow, and productivity; *population ecology*, the manner of population growth, structure, and regulation; *evolutionary ecology*, the problems of niche segregation and speciation; and *geographic ecology*, concerned with distribution, paleoecology, and biomes. We will be interested throughout the text with how organisms respond and adjust physiologically to the physical factors of their environment, but a full study of *physiological ecology* must be left to another time and place. We will also be concerned throughout with *systems ecology*, that is, the possibility of translating ecological concepts into mathematical models, although we will not go deeply into the actual statistical manipulations involved (Dale 1970, Lieth 1971). This new field is becoming very important in ecological philosophy, changing the emphasis of research from the empirical to the theoretical (Ashby 1956, Margalef 1968). This has potential value in rendering ecology a more exact science so that future events may be predicted when any of several inherent parameters vary. Finally, in several parts of the book we will deal with *human ecology*, involving the population ecology of man and man's relation to the environment, especially man's effects on the biosphere and the implication of these effects for man.

When special consideration of their ecology is given to one or another taxonomic group, we speak of *mammalian ecology, avian ecology, insect ecology, parasitology, human ecology*. When emphasis is placed on habitat, we speak of *oceanography*, the study of marine ecology; *limnology*, the study of fresh-water ecology; *terrestrial ecology*; and so on. *Ethology* is the interpretation of animal behavior under natural conditions; often, detailed life history studies of particular species are amassed. *Sociology* is really the ecology and ethology of Mankind.

Ecological concepts, which may be grouped together as *applied ecology*, have many practical applications; notably *wildlife management, range management, forestry, conservation, insect control, epidemiology, animal husbandry*, and even *agriculture*.

This preview of ecology indicates the great breadth and unique character of the subject material, which justifies the view of ecology as one of the three basic divisions of general biological philosophy.

HISTORY

That certain species of plants and animals ordinarily occur together and are characteristic of certain habitats has doubtless been common knowledge since intelligent man first evolved This knowledge was essential to him for procuring food, avoiding enemies, and finding shelter. However, it was not until the fourth century B.C. that Theophrastus, a friend and associate of

Aristotle, first described interrelations between organisms and between organisms and their environment. He has, therefore, been called the first ecologist (Ramaley 1940).

The modern concept that plants and animals occur in closely integrated communities began with the studies of August Grisebach, a German botanist, in 1838; K. Möbius, a German investigator of oyster banks, in 1877; Stephen A. Forbes, an American, who described the lake community as a microcosm in 1887; and J. E. B. Warming, a Danish botanist, who emphasized the unity of plant communities in 1895 (see Kendeigh 1954 for further details and literature citations). C. C. Adams recognized and described many animal communities in his ecological surveys of northern Michigan and of Isle Royale in Lake Superior, published in 1906 and 1909. V. E. Shelford presented a classic study of animal communities in temperate America in 1913, and Charles Elton published an outstanding analysis of community dynamics in 1927. Although an appreciation that the whole community is one biotic unit, rather than one unit of plants and another of animals, may be discerned in the writings of some early investigators (for example, J. G. Cooper in 1859), the fact has been brought to modern emphasis in the work of F. E. Clements and V. E. Shelford, especially in their *Bio-ecology*, published in 1939. Much interest has been stimulated in recent years by D. Ramon Margalef, Robert MacArthur, and others for analyzing the structure of these biotic communities, particularly in respect to such phenomena as species diversity, niches, and how they came about through evolution.

Succession of plant species after burns and in bogs has been known in a general way since about 1685, and European ecologists have studied succession since the late nineteenth century. The present-day interest

(Left) C. C. Adams, 1873–1955, animal ecologist (courtesy Dorothy Kehaya).

(Right) Victor E. Shelford, 1877–1968, animal ecologist.

(Left) F. E. Clements, 1874–1945, plant ecologist.

(Right) Henry C. Cowles, 1869–1939, plant ecologist (courtesy R. J. Pool).

in succession, however, especially in North America, dates from the plant studies of Henry C. Cowles in 1899 on the sand dunes at the south end of Lake Michigan, and the work of Frederic E. Clements, 1916. C. C. Adams and V. E. Shelford, in the citations noted, were among the first to apply the concept to animals.

Geographic ecology, in the modern sense, dates from the generalizations on the worldwide distribution of animals made by the French naturalist Georges L. L. Buffon (1707–1788), and the explorations of the German naturalist Alexander von Humboldt (1769–1859). There was lively interest and many important contributions in this general field during the nineteenth century; notably, the life-zone concept of C. Hart Merriam (1890–1910) needs special mention. During the present century the concept of biotic provinces is identified with L. R. Dice (1943) and the biome concept with F. E. Clements and V. E. Shelford (1939). The broad survey of ecological animal geography made by R. Hesse in 1924 exerted considerable influence, and this treatise was later translated into English and revised by W. C. Allee and Karl P. Schmidt (1951).

The study of population dynamics, so important in modern ecology, dates back at least to Malthus, who pointed out in 1798 the limitation to population growth exerted by available food. Darwin, in 1859, recognized the importance of competition and predation in developing his theory of evolution. Pearl, 1925, analyzed mathematically the characteristics of population growth, and Lotka, 1925, and Volterra, 1931, developed theoretical mathematical equations to show the manner in which populations of different species interact. These studies led to the classic experiments of Gause, 1935, with interacting populations of predators and prey. Nicholson's publication in 1933 stimulated much thinking concerning the factors that stabilize populations at particular levels. Andrewartha and Birch, 1954, emphasized the importance of climate and other factors on determining the size of populations.

The measurement and analysis of energy use by organisms for existence and growth is now of very great interest in ecology. Attention to biological productivity began in the 1930's in connection with practical pond-fish culturing and the limnological studies of Thienemann in Europe and of Birge and Juday at the University of Wisconsin, but the modern crystallization of the subject came with the fresh-water and marine investigations of Lindeman, Hutchinson, and Riley at Yale University (Ivlev 1945) and of Eugene and Howard Odum. An early study of energy relations within terrestrial communities is that of Stanchinsky (1931).

Physiological ecology had its historical beginnings in the correlation of biological phenomena with variations in temperature stimulated by Galileo's invention of a hermetically sealed thermometer about 1612 A.D. The French naturalist Reaumur summed the mean daily temperatures for April, May, and June in 1734 and again in 1735, and correlated the earlier maturing of fruit and grain during the first year with the greater accumulation of heat. A discovery of parallel significance was of oxygen in 1774 by the English clergyman, Priestley, and the finding by Lavoisier, a Frenchman, in 1777 that it was an essential part of air. Claude Bernard, another French physiologist, enunciated the principle of homeostasis in 1876. This concept originally referred to regulatory mechanisms which maintained the "internal environment" of the body constant in the face of changing external conditions. Later, the concept came to be applied also to maintenance of community interrelations. Van't Hoff, a Dutch scientist, contributed to physiological ecology in 1884 in describing how the speed of chemical reactions increased two- or threefold with each rise of 10°C. K. G. Semper and Charles B. Davenport clearly established physiological ecology in bringing together pertinent information in 1881 and 1897–99, respectively. More recent summaries of knowledge and methods in this general field have been made by V. E.

Shelford in *Laboratory and Field Ecology* (1929) and by Samuel Brody in *Bioenergetics and Growth* (1945).

The development of animal behavior or ethology may be traced back through the natural history of ancient times. More recently the 13 volumes of *Thierleben*, prepared by A. E. Brehm during the period 1911–18, are noteworthy. H. S. Jennings, 1906, and Jacques Loeb, 1918, made valuable contributions to the understanding of the behavior of invertebrates. Precise modern techniques and concepts as applied to vertebrates began to take form about 1920 with the development of banding and marking of individual animals by S. Prentiss Baldwin (1919) and the recognition of territories in the nesting of birds by H. E. Howard (1920). The formulation of the concept of releasers as controlling instinctive behavior by Wallace Craig (1908), K. Lorenz (1935), and N. Tinbergen (1951) has produced a profound effect on present-day thinking.

In regard to other divisions of ecology, the crystallization of studies in oceanography may be credited to Edward Forbes 1843, Maury 1855, Alexander Agassiz 1888, Petersen and his colleagues 1911, and Murray and Hjort 1912; limnology to Forel 1869, Birge 1893, Juday 1896, Ward and Whipple 1918, Thienemann 1913–35, and Naumann 1918–32; and wildlife management to Aldo Leopold 1933.

Ecology, then, is of comparatively recent development as a distinct science, but its roots extend well back into the past. An early comprehensive treatment of the subject is *Principles of Animal Ecology* by Allee, Emerson, Park, Park, and Schmidt, published in 1949 (for citations of historic interest in this chapter, see this reference). Since ecology is a young science, it should be emphasized that its concepts and techniques have not become standardized and that there is opportunity and stimulus here for many new investigators.

The Ecological Society of America was founded in 1915, and in 1971 had a membership of 4700. The British Ecological Society, organized in 1913, has a membership of about 2000. The society in America has given birth to several offspring: The Wildlife Society, Society of Limnologists and Oceanographers, the Nature Conservancy, and the Animal Behavior Society. Each of these daughter organizations has its own journal. Also resulting from promotion by the Society, the Inter-American Institute of Ecology became established in 1971. The Institute, supported by both private and government funds, is designed to conduct and promote research, provide analytical and other services, and relate ecological information to the development of far-reaching public policies and to solving environmental problems.

The Ecological Society of America publishes three periodicals: *Bulletin* for general articles and Society business, *Ecology* for short scientific papers, and *Ecological Monographs* for long papers. The British Ecological Society also publishes three periodicals: *Journal of Ecology* for papers on plants, *Journal of Animal Ecology* for papers on animals, and the *Journal of Applied Ecology*. *Oikos* began publication in 1949 to represent ecologists in Denmark, Finland, Iceland, Norway, and Sweden. The Polish Academy of Science has been publishing *Ekologia Polska* since 1953. *Oecologia*, a German periodical derived from *Zeitschrift für Morphologie und Ökolgie der Tiere*, began in 1968. The U.S.S.R. Academy of Sciences initiated the periodical *Ekologiya* in 1970, and an English translation of it appears under

E. A. Birge, 1851–1950, limnologist.

Aldo Leopold, 1886–1948, wildlife manager.

Some well-known ecologists at the present time (*see* Bibliography), all Ph. D. graduates from the University of Illinois. The photograph was taken on May 26, 1973, on the occasion of Dr. Kendeigh's retirement from teaching (Robert K. O'Daniell, *The News-Gazette*). From left to right: Dr. Eugene P. Odum, University of Georgia; Dr. Robert H. Whittaker, Cornell University; Dr. Robert V. O'Neill, Oak Ridge National Laboratory; Dr. S. Charles Kendeigh, University of Illinois; Dr. Robert M. Chew, University of Southern California; and Dr. James R. Karr, Purdue University.

the title *Soviet Journal of Ecology*. The International Society for Tropical Ecology, including India and adjacent countries, was founded in 1960 and publishes *Tropical Ecology*. The New Zealand Ecological Society came into existence in 1952 and the Ecological Society of Australia in 1960. *Proceedings* of the New Zealand annual conventions are published. *The Japanese Journal of Ecology*, begun in 1954, is the official publication of the Ecological Society of Japan. Many of its articles are in Japanese, but there are summaries in a European language. *Researches on Population Ecology*, started in 1959, is published by the Japanese Society of Population Ecology. In 1968, an International Association for Ecology was formed within the International Union of Biological Sciences to coordinate ecological work in various countries. It began publishing the *Intecol Bulletin* in 1969. No attempt will here be made to list the new journals dealing with pollution and environmental problems of practical or economic importance. Finally, many papers of interest to ecologists appear in biological journals of various sorts that do not carry the word "ecology" in their titles.

Chapter 2

GENERAL NATURE
OF ENVIRONMENTAL RESPONSES

Ecology, by definition, deals with the interrelations of organisms with each other and with their environment. These interrelations become established as organisms respond in various ways to contacts with one another and with the ever-changing environment.

The term *environment* describes the sum total of physical and biotic conditions influencing the responses of organisms. More specifically, the sum of those portions of the hydrosphere, lithosphere, and atmosphere into which life penetrates is the *biosphere*. There are no characteristic or permanent inhabitants of the atmosphere, although the air is traversed by many kinds of animals and plant propagules. Of the hydrosphere, there are two major *biocycles*, marine and fresh water; of the lithosphere there is one, land (Hesse *et al.* 1951).

A *habitat* is a specific set of physical and chemical conditions (for example, space, substratum, climate) that surrounds a single species, a group of species, or a large community (Clements and Shelford 1939). The ultimate division of the biosphere is the *microhabitat*, the most intimately local and immediate set of conditions surrounding an organism: the burrow of a rodent, for instance, or a decaying log. Other individuals or species are considered as part of the community to which the organism belongs and not part of its habitat. The term *biotope* defines a spatial or topo-graphic unit with a characteristic set both of physical and chemical conditions and of plant and animal life.

In order for organisms to exist they must respond or adjust to the conditions of their environment. The first living organisms probably evolved in the sea and must have possessed very generalized adjustments to this relatively uniform and favorable habitat. However, these early organisms had inherent in them the potential for expansion, as they later spread into other and more rigorous habitats, particularly fresh-water and land. As evolution proceeded, organisms became more and more limited in the range of their ability to respond as they became specialized in their adjustments to particular habitats. This led to the great diversification of species that we see at the present time, with each species restricted to its particular microhabitat and place in the community.

Organisms respond to differences or changes in their environment in four principal ways: morphological adaptations, physiological adjustments, behavior patterns, and community relations. Chapters 2 and 3 are a resumé of these responses, the general fundamentals of which must be understood before the subtle relations of an organism to its environment that are the substance of ecology can be appreciated.

Probably the most important of distinctions between organisms in a consideration of their morphological responses to the environment is whether they are sessile or motile (Shelford 1914). Most plants are, of course, sessile; most animals, motile. There are, however, some motile plants among unicellular forms and male gametes, and there are many sessile or slow-moving animals in aquatic habitats. Sessile organisms respond to variations of the environment primarily by changes in form; motile animals, primarily by changes in behavior.

MORPHOLOGICAL ADAPTATIONS

Changes in Form and Structure

Consider a sessile organism, the tree. It is essential to the tree that its foliage be exposed to sunlight. As it grows within a forest, it is usually tall and slender, and little branched except at the top, where the cap of foliage reaches into the full sunlight. Growing on the forest's edge, the tree is shorter, and branching and foliage are dense both at the cap and on that side exposed to full sunlight. The tree which grows solitary in an open place is short, but branching and foliage are dense and uniformly distributed, often starting close to the ground. In similar manner, the variations in form assumed by sessile colonial animals, such as sponges and corals (Fig. 2-1), reflect vicissitudes imposed by habitat (Wells 1954).

Morphological variations induced by peculiarities of habitat do occur in motile animals: thickening of the shells of clams subjected to strenuous wave action; variation in number of vertebrae, scales, and fin rays among fish subjected to different temperatures at critical periods in their growth (Barlow 1961); changes in the number of facets in the bar-eye of the fly *Drosophila* as a correlative of temperature variations during a short critical period in larval growth (Krafka 1920); the many variations in form and size of internal parasites, depending on crowding and other environmental conditions (Baer 1951); pointed tails in certain flatworms crawling over a substratum during growth, contrasted with rounded tails occurring in the same species when individuals are experimentally prevented from crawling (Child 1903).

That individuals of the same species are so much alike attests the great extent to which the course and outcome of morphological development are genetically determined. But that there are variations between individuals, and between groups of individuals, of the same species shows that morphological development is also responsive to environmental influences. Modifications induced by the environment emerge as the individual develops and are not specifically inherited by the succeeding generation. These modifications are called *growth-forms* or *ecophenes*. If the generation following is similar in growth-form to the parent generation, it is a similar morphological response to a similar environment (Schmalhausen 1949). If and when the growth-form becomes inherited as the result of evolutionary processes, it then becomes an *ecotype*. *Life-form* is a general term referring to the shape or appearance of an organism irrespective of how formed (Daubenmire 1947). The prevalence of particular life-forms among the important organisms helps to separate and characterize biotic communities, as we will see repeatedly in later discussions.

Life-forms of Plants

The life-form of a plant is characterized by its vegetative form, its length of life, the arrangement and character of its leaves, whether its stem is herbaceous or woody, its manner of growth, and its means of overwintering. Life-form categories sometimes agree with large taxonomic units, such as ferns or mosses. On the other hand some taxonomic groups contain species exhibiting a variety of life-forms, and some life-forms include species only remotely related taxonomically.

There have been many systems proposed for the classification and terminology of the life-forms of plants. One widely used system (Raunkiaer 1934) is based primarily on position of buds or other meristematic tissues in relation to the soil surface A most useful system for both plant and animal ecologists is one emphasizing differences in plant forms responsible for the structure, or physiognomy, of plant communities,

Fig. 2-1 Form assumed by the coral *Madrepora* as it develops in (a) deep water; (b) barrier pools; (c) rough water (from Wood-Jones 1912).

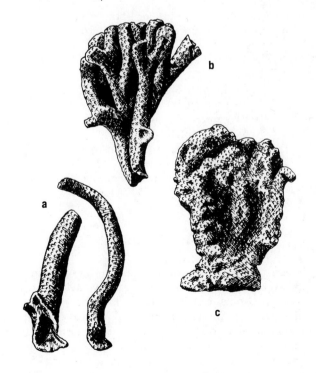

such as the following based on Pound and Clements (1900):

1. Annuals: Passing the winter or dry season in seed or spore form alone, no propagation or accumulation of aerial shoots; living one year.
2. Biennials: Passing one unfavorable season in the seed or spore form and the next in a vegetative stage; no accumulation of aerial shoots; living two years or parts of two years.
3. Herbaceous perennials: Passing each unfavorable season in both seed or spore and vegetative form; no accumulation of aerial shoots; living several to many years.
   a. Broad-leaved herbs: mostly terrestrial
   b. Sod grasses: a continuous turf
   c. Bunch grasses: scattered clumps
   d. Succulents: some broad-stemmed cacti
   e. Water plants:
      (1) Submerged: vegetative body entirely underwater.
      (2) Floating: leaves floating on water surface; water lilies, duckweed.
      (3) Emerging: leaves extending above water surface; cattails, sedges, rushes.
   f. Ferns
   g. Mosses
   h. Liverworts
   i. Lichens
   j. Fungi
   k. Algae
4. Woody perennials: Passing the unfavorable season as aerial shoots or masses, often as seeds also; living many years as a rule.
   a. Lianas: vines
   b. Succulents: some tree or barrel cacti
   c. Bushes: much-branched, low growth, several stems
   d. Shrubs: single stem and tree-like but smaller
   e. Trees:
      (1) Deciduous: shedding leaves during unfavorable season.
      (2) Evergreen: leaves shed irregularly and tree never completely bare.
         (a) Needle-leaved: narrow, elongated or scale-like leaves.
         (b) Broad-leaved: leaves much as on deciduous trees but shed irregularly.

Life-forms of Animals

There have been several attempts to classify the life-forms of animals, but nothing very definitive has resulted (Remane 1952, Krivolutskii 1972). The major life-forms of animals more often agree with their taxonomy than do plants, but some life-forms include representatives from several different taxonomic groups. There can be recognized encrusting forms

such as the fresh-water bryozoan *Plumatella* and some sponges; coral forms, including grass, leaf, or shrub forms; radiate forms, such as coelenterates and echinoderms generally; bivalve forms; snail forms; slug forms; worm forms; crustacean forms; insect forms; fish, snake, bird, and four-footed forms. Each of these major types may be divided into narrower structural or behavioral types, for example among the four-footed form of mammals (Osburn *et al.* 1903):

Aquatic (swimming): Seal, whale, walrus
Fossorial (burrowing): Mole, shrew, pocket gopher
Cursorial (running): Deer, antelope, zebra
Saltatorial (leaping): Rabbit, kangaroo, jumping mouse
Scansorial (climbing): Squirrel, opossum, monkey
Aerial (flying): Bat

Adaptations

Specific life-forms are adaptations of plants and animals to live in particular habitats and to behave in particular ways (Klaauw 1948). The life-forms listed for mammals are largely adaptations to particular strata (water, subterranean, ground, tree, air) within a community rather than to the habitat as a whole; for instance, the subterranean adaptations of mammals living in the Arctic tundra are similar to the subterranean adaptations of mammals in the tropics. In communities lacking one or more strata (for instance, the tree stratum in grassland), animals specifically adapted to the missing strata are also absent. In communities in which all strata are present, a catholic variety of life-forms occurs.

In addition to adaptations to stratum and habitat, there occur ecologically significant adaptations for food-getting and metabolism, protection, and reproduction. The variety of teeth found in mammals and lizards, the variation in shape and size of bills of birds, the different mouth parts of insects, the siphons of clams, the suckers of leeches, the water canal systems of sponges are but a few special anatomical features especially designed for food-getting. Associated with food-getting is a great diversity in structural adaptations for the digestion of the food, for respiration, for circulating food materials and gases through the body, for excreting wastes, for support and movement, and for nervous and hormonal regulation. All these internal organs and structures are necessary to the animal for utilizing the energy resources of the environment.

All animals are subject to predation or competition and must have means of protecting themselves or offsetting losses in the struggle for existence. Such adaptations take a variety of forms such as body armor, concealing coloration, attack weapons, or behavior patterns of escape. High rates of mortality are offset by high rates of reproduction or, in some lower organisms, by considerable power of regenerating whole organisms from fragmented parts.

The manner in which reproduction occurs and the special structures concerned with reproduction vary with each type of animal and often with each individual species. These adaptations are universal and too numerous even to attempt to classify at this point but are certainly obvious to all. The primary objectives in the life of each species are to maintain the existence of the individual and to reproduce its own kind, and all adaptations to live in favorable habitats are designed toward these ends.

Natural Selection

To be heritable, a variation must have been caused either by mutations in the genes or chromosomes of the individual or by new combination of genes. Mutations are produced at random, and mostly independent of natural environmental conditions, although there is some experimental evidence that they may be induced or increased in frequency by cosmic rays, ultraviolet rays, heat, and certain chemicals. There is no reason to believe, however, that environmental factors can ordinarily influence the kind of mutations that occur.

Heritable variations in the structure of organisms, and in their physiology and behavior as well, may be favorable, unfavorable, or of neutral value to the existence of the species. Variations that decrease the efficiency of a species in its struggle for existence against competitors and unfavorable environmental conditions usually disappear, but variations that increase this efficiency give those individuals that possess them a better chance for survival and for giving birth to similar offspring. Thus, there is natural selection of the fitter individuals, and a gradual improvement in the relations between the species and its environment. It is in this way that adaptations are established. A better understanding of the ecological relations between different species and between species and the environment will contribute to a better understanding of the process of evolution. At the same time a thorough understanding of the processes of evolution is necessary to understand how organisms become adapted to live in particular habitats (Simpson 1953).

A close study of differences between individuals shows that within many species convergent evolution occurs under similar environmental conditions. Many of these variations are genetic and apparently due to natural selection. The best established correlations are the following, although even they are subject to frequent exceptions (Mayr 1942, Dobzhansky 1951):

BERGMANN'S RULE: Geographic races of a species possessing smaller body-size are found in the warmer parts of the range, races of large body-size in the cooler parts. This appears true for cold-blooded as well as warm-blooded animals (Ray 1960). Bergmann (1847) originally stated this rule to apply to different species within the genus.

ALLEN'S RULE: Tails, ears, bills, and other extremities of animals are relatively shorter in the cooler parts of a species' range than in the warmer parts.

GLOGER'S RULE: In warm-blooded species, black pigments increase in warm and humid habitats, reds and yellow-browns prevail in arid climates, and pigments become generally reduced in cold regions.

Large body-size and short appendages give less surface area per volume of body and thus minimize heat loss from the body in cold climates. However, the smaller surface area thus attained does not by itself give enough reduction of heat loss to permit adaptation to cold climates in warm-blooded animals, and the advantage would not apply to cold-blooded ones. Rather, the ability of warm-blooded animals to live in cold climates depends on an adaptive strategy that includes better insulation of the body surface, capacity for higher rates of heat production, and greater ability to tolerate a cold tissue temperature, as well as finding adequate food resources in the environment (Scholander 1955, Irving 1957, Kendeigh 1969).

PHYSIOLOGICAL ADJUSTMENTS

Nature of Adjustments

Probably the first response of any organism to a change in the environment is physiological. A physiological response must certainly precede any change in form or structure which requires growth. Even a change in behavior must follow a change in some receptor or sense organ followed by nervous function; a fall in air temperature, for instance, brings a drop in the metabolic rate of cold-blooded organisms but a rise in the rate of warm-blooded organisms. Cold may stimulate nerve endings in the skin of birds or mammals and produce shivering and a search for protective cover. Transference from the dark to light may immediately initiate photosynthesis in resting chloroplastids within a plant cell, or a change in turgescence on opposite sides of a sessile zooid may result in a turning movement, an orientation to or away from the light source. Physiological responses are thus internal responses to factors of the environment. Often they are difficult to detect.

Types of Response

Environmental factors influence organisms physiologically in various ways (Fry 1947). These effects may be classified as follows:

LETHAL: Causing death; for instance, extreme heat or cold, lack of moisture, and so forth.

MASKING: Modifying the effect of some other factor. Low relative humidity increases the rate of evaporation of moisture from body surfaces so that warm-blooded animals are able to survive at otherwise intolerably high air temperatures.

DIRECTIVE: Producing an orienting response in relation to some environmental response so that the organism gets itself into favorable conditions.

CONTROLLING: Influencing the rate at which some process functions, but not entering the reaction. Temperature, pressure, and viscosity, for instance, affect metabolism, secretion, and locomotion.

DEFICIENT: Curtailing an activity because some essential ingredient, such as a salt, oxygen, or the like, is absent or at unfavorably low concentration.

The same environmental factor may produce different effects at different times and under different conditions. Temperature may be lethal, if extreme; masking, as when cold reduces the demand of cold-blooded organisms for food; directive, by inducing a search for more favorable locations; or controlling, as a modifier of the rate of metabolism. Often the distinction between controlling and deficient factors is not made, or they are considered as together constituting *limiting factors*.

Threshold and Rate

Every environmental factor varies through a wider range of intensity than any single organism can tolerate. Characteristically, there is for each individual organism a lower and an upper limit in the range of an environmental factor between which it functions efficiently. For any one factor, different organisms find optimal conditions for existence at different points

along the range; hence their segregation into different habitats.

The *threshold* is the minimum quantity of any factor that produces a perceptible effect on the organism. It may be the lowest temperature at which an animal remains active, the least amount of moisture in the soil that permits growth of a plant, the minimum intensity of light at which a photoreceptor is stimulated, and so forth. Above the threshold, the rate of a function increases more or less rapidly as the quantity of heat, moisture, light, or other environmental factor is augmented, until a maximum rate is attained (Fig. 2-2). Above the maximum, there is usually a decline in the rate of a process either because of some deleterious effect produced, the interference of some other factor, or exhaustion. The curve of decline at high temperatures is usually steeper than the curve of acceleration at low temperatures.

Law of Toleration

For each species there is a range in an environmental factor within which the species functions at or near an optimum (Fig. 2-3). There are extremes, both maximum and minimum, toward which the functions of a species are curtailed, then inhibited. *Upper* and *lower limits of tolerance* are intensity levels of a factor at which only half of the organisms can survive (LD_{50}). These limits are sometimes difficult to determine, as for instance with low temperature, organisms may pass into an inactive, dormant, or hibernating state from which they may again become functional when the temperature rises above a *threshold*. At high temperatures, there may be similar inactivation or aestivation before the lethal level is attained. Even without dormancy occurring, there are normally *zones of physiological stress* before the limits of tolerance are reached.

The species as a whole is limited in its activities more by conditions that produce physiological discomforts or stresses than it is by the limits of toleration themselves. Death verges on the limits of tolerance, and the existence of the species would be seriously jeopardized if it were frequently exposed to these extreme conditions. In retreat before conditions of stress there is a margin of safety, and the species adjusts its activities so that limits of tolerance are avoided. There is variation in hardiness of individuals within a species, so that some hardy individuals find existence possible under conditions that disrupt other individuals. The population level of a species becomes reduced therefore before the limits of its range are actually reached. It is desirable to test by acclimation and breeding experiments whether these differences in physiological adaptiveness between individuals or populations are genetic or phenotypic (Prosser 1955).

Species vary in their limits of tolerance to the same

Fig. 2-2 Interaction between environment and cold-blooded organisms: organism activity as a function of environmental temperature (modified from Fry 1947).

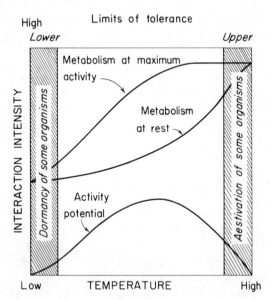

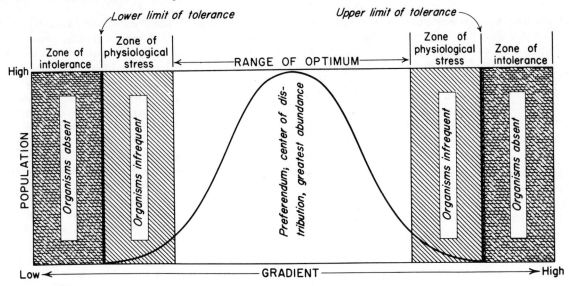

Fig. 2-3 Law of toleration in relation to distribution and population level—often a normal curve (modified from Shelford 1911).

factor. The Atlantic salmon, for instance, spends most of its adult life in the sea, but goes annually into freshwater streams to breed. Most other marine fishes are killed quickly when placed in freshwater, as are freshwater fish when placed in salt water. The following terms are used to indicate the relative extent to which organisms can tolerate variations in environmental factors. The prefix *steno-* means that the species, population, or individual has a narrow range of tolerance and the prefix *eury-* indicates that it has a wide range; thus *stenohaline* and *euryhaline* in respect to salinity, *stenohydric* and *euryhydric* in respect to water, *stenothermal* and *eurythermal* in respect to temperature, *stenophagic* and *euryphagic* in respect to food, *stenoecious* and *euryoecious* in respect to niche or habitat, and so on.

Law of the Minimum

An organism is seldom, if ever, exposed solely to the effect of a single factor in its environment. On the contrary, an organism is subjected to the simultaneous action of all factors in its immediate surroundings. However, some factors exert more influence than do others, and the attempt to evaluate their relative roles has led to the development of the law of the minimum.

The first elaboration of this law was made by the German biochemist, Justus von Liebig, in 1840, who stated:

> If one of the participating nutritive constituents of the soil or atmosphere be deficient or wanting or lacking in assimilability, either the plant does not grow or its organs develop only imperfectly. The deficient or lacking constituent makes those that are present inactive or lessens their activity. If the deficient or lacking constituent be added to the soil or if occurring in insoluble form it be

made soluble, then the other nutrients become active [Browne 1942].

Blackman (1905) developed the more comprehensive concept of *limiting factors*, including both those deficient and controlling, when he listed five factors involved in controlling the rate of photosynthesis: amount of CO_2 available, amount of H_2O available, intensity of solar radiation, amount of chlorophyll present, and temperature of the chloroplast. Any one of these factors will control the rate of the process if the factor is present in least favorable amount, or may actually stop it when insufficient, even though all other factors occur in abundance. The same principle applies to animal functions (Fig. 2-4).

Fig. 2-4 Relation between maximum respiration rate, temperature, and oxygen tension (mm Hg as shown by values in the graph) in young goldfish acclimated to each temperature before measurements were taken (Fry 1947).

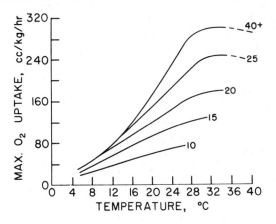

Since the rate of a process may be controlled by too great an amount of a substance, such as heat, as well as by too small an amount, and since the presence or abundance of an organism may be limited by a variety of environmental factors, biotic as well as chemical and physical, and since the limiting effect may be due to two or more interacting factors rather than a single isolated one (Shelford 1952), the *law of the minimum* may be restated in broad ecological terms, as follows:

> The functioning of an organism is controlled or limited by that essential environmental factor or combination of factors present in the least favorable amount. The factors may not be continuously effective but only at some critical period during the year or perhaps only during some critical year in a climatic cycle [Taylor 1934].

BEHAVIOR RESPONSES

Orientation

Behavior responses to changes in environmental factors can usually be detected immediately as turning or locomotor activities on the part of the organisms (Fraenkel and Gunn 1940). These movements tend to take the organism away from points of danger and into more favorable locations, or to perform some task essential to existence, or to reproduction. If the movement involves curvature or a turning movement either toward or away from the source of stimulus, the movement is called a *tropism*. Motile organisms frequently respond by actual locomotion toward or away from the stimulus rather than mere turning, and such guided or directed locomotor movements are called *taxes*. When the movements of the animal are random in direction, and there is no immediate orientation to the source of stimulus, but the frequency of turning or speed of the movements is dependent on the intensity of stimulation, such responses are termed *kineses*. As the result of kineses an animal may arrive by chance in a favorable environment, by which the intensity of the stimulus is reduced or entirely eliminated. To identify the stimulus to which the organism is responding, the following prefixes are employed: *thermo-*, temperature; *photo-*, light; *geo-*, gravity; *hydro-*, moisture; *chemo-*, chemicals; *thigmo-*, contact; *baro-*, pressure; *rheo-*, current; and *galvano-*, electricity.

Jacques Loeb, during the period 1888–1918, vigorously maintained that all tropisms and taxes of organisms were mechanical, automatic, and explainable in simple concepts of physics and chemistry.

> . . . the overwhelming majority of organisms have a bilaterally symmetrical structure. . . . Normally the processes inducing locomotion are equal in both halves of the central nervous system, and the tension of the sym-

metrical muscles being equal, the animal moves in as straight a line as the imperfections of its locomotor apparatus permit. If, however, the velocity of chemical reactions in one side of the body, e.g., in one eye of the insect, is increased, the physiological symmetry of both sides of the brain and as a consequence the equality of tension of the symmetrical muscles no longer exist. The muscles connected with the more strongly illuminated eye are thrown into a stronger tension, and if new impulses for locomotion originate in the central nervous system, they will no longer produce an equal response in the symmetrical muscles, but a stronger one in the muscles turning the head and body of the animal to the source of light. The animal will thus be compelled to change the direction of its motion and to turn to the source of light . . . [Loeb 1918].

The idea that all instinctive activities of organisms were forced and invariable responses to environmental factors met many objections. H. S. Jennings (1906) pointed out that many Protozoa are asymmetrical in body structure and hence could not lend support to Loeb's tonus theory. Furthermore, the movements and responses of many organisms to environmental stimuli were not stereotyped, but random in nature; of a *trial and error* sort. Although much of Loeb's theory has been disproven experimentally and appears untenable on the basis of observations of animal activities under natural conditions, it crystallized the need for objective analysis and interpretation of animal behavior, and the avoidance of teleological and anthropomorphic explanations. The study of orienting responses of organisms is of utmost ecological significance since it is largely by means of such responses that organisms find their proper and favorable habitats.

Preferendum

The behavior responses of animals and their orientation in respect to most environmental factors can be tested experimentally, and results thus obtained correlated with the animal's behavior under natural conditions. There is a variety of procedures and equipment suitable to these purposes (Shelford 1929, Warden, Jenkins, and Warner 1935) and there is distinct value in verifying field observations with experimental analyses.

When the number of favorable responses at each unit intensity of an environmental factor is plotted against the entire range of that environmental factor, the usual result is a normal or Gaussian curve. The maximum number of responses normally occurs near the center of the range, with a progressive reduction in number toward each extreme. An extension in each direction from the peak of the responses to include 50, 25, or some smaller percentage of the total responses is called the *preferendum* for that animal or group of animals.

Innate Behavior

Much of the behavior of organisms is determined by heredity and is characteristic of the species in its proper environment. This behavior may be evident at birth or it may not develop until the nervous system, including both the receptor and effector mechanisms, is fully matured. Such innate behavior is of various degrees of complexity. A *reflex* is a quick, automatic response of a single organ or organ system to a simple stimulus; for instance, the knee jerk in man. Tropisms, taxes, and kineses may involve a series of reflexes and represent a higher level of integration. An *instinct*, or *inherited behavior pattern*, is a complex fixed behavior that is activated, more or less automatically, when the animal is presented with the proper stimulus (Thorpe 1951).

The anatomical basis for these various grades of behavior lies in the structure of the nervous system and especially, in higher types of animals, in the interarrangement of neurones and synapses with each other and in the neural pathways that become established. Hormones and enzymes may also be involved. The genetic basis of behavior is not well known (Dilger 1962), but behavior patterns become elaborated through evolution, are as subject to "mutation" and natural selection as any structural part of the body, and are a means whereby animals respond advantageously to the various factors in their normal environment.

Stimuli

Before an action will take place the nervous mechanism must be released by the reception of a *stimulus*. Stimuli may be either external or internal to the organism. Protoplasm is sensitive to any kind of stimulation, provided it is intense enough. In higher organisms, however, specialized tissues have become particularly sensitive to one kind of stimulus, and these tissues, or sense organs, are called *receptors*. There are several forms of receptors: *photo-receptors, phono-receptors, mechano-receptors, chemo-receptors, thermo-receptors, and stato-receptors*. Not all types of receptors are present in all organisms, and the structure and effectiveness of those present varies from one kind of animal to another. The efficiency of the receptor mechanisms is important, as they largely determine the environmental factors to which the animal will respond and the degree of sensitivity involved.

Internal stimuli are derived either from *hormones* or as *kinesthesia* involving changes in the tension of muscles and tendons or changes in shape or form of muscle fibers. *Motivation* is established when there is an accumulation of internal stimuli potentials as the result of hormone action, kinesthetics, or changes of metabolism (Fig. 2-5). A combination of motivation with proper external conditions and stimuli sets up a *drive*, such as the hunger drive or reproductive drive (Richter 1927).

Once a major drive is initiated, satisfaction of it requires a series of events and stimuli at different levels of integration, so that a hierarchy of drives, actions, and stimuli is established (Fig. 2-6). The significance of this hierarchy is that a major activity in the life cycle of an animal does not take place until the organism is in a proper *physiological state*, which depends, often in large part, on the environment, and then one action leads to another until *consummation* is completed. In the male stickleback, for instance, the reproductive drive is not initiated until hormone stimuli are released as the result of gonad enlargement and response to lengthening daily photoperiods. Once the reproductive stimulus is given, the first secondary drive is the establishment of nesting territories by fighting among male fishes. Then the nest is built. Only after this is completed is the male ready to receive the female.

Even though an animal may have potential capacities in its sense organs with which to respond to the whole environment, a particular action is triggered by

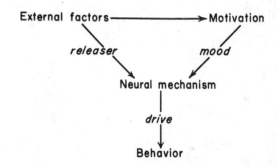

Fig. 2-5 Factors involved in the activation of an instinct.

Fig. 2-6 Hierarchy of drives and actions in the three-spined stickleback (after Tinbergen 1951).

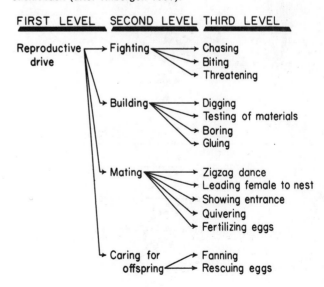

stimuli from only a very small part of the environment. This is a fundamental characteristic of innate behavior, and the discovery of these critical *sign stimuli* or *releasers* is necessary for an appreciation of the interrelation of animals in a community and how they respond to their environment (Lorenz 1935, Tinbergen 1951).

The complete enactment of mating behavior in the stickleback proceeds step by step in an orderly manner, each action a releaser for the next (Fig. 2-7). If any one step is changed, or is interrupted, the behavior subsequent in the sequence does not take place. Releasers are of a variety of sorts in different species, but commonly involve particular colors or

Fig. 2-7 Courtship and mating behavior of the three-spined stickleback (after Tinbergen 1951).

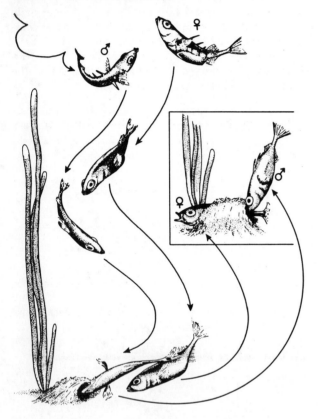

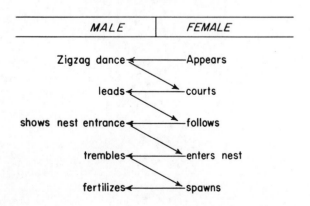

MALE	FEMALE
Zigzag dance	Appears
leads	courts
shows nest entrance	follows
trembles	enters nest
fertilizes	spawns

color patterns, call-notes or songs, shapes, chemicals, or contacts, as well as associated acts, positions, or movements on the part of another animal. If these triggers are not presented, the behavior does not become expressed, even though a specific nervous mechanism is present. The analysis of behavior through observation and experimentation with the objective of understanding how an animal acts under natural conditions constitutes the science of *ethology*, an essential branch of ecology. Ethology differs from psychology in that it is concerned with understanding not only the function and causality of behavior but also the *survival value* of behavior patterns under natural conditions, and the *evolution* of these patterns (Dilger 1962). Psychology is concerned more with analyzing the nervous mechanisms that are involved.

Learning

All behavior is not, of course, automatic and inherited. Much of it represents the adjustment of fixed patterns to changes in and conditions of the animal's surroundings (Thorpe 1956). *Learning* may be defined as the adaptive change in individual behavior as a result of experience.

The simplest form of learning is *habituation*, that is, learning not to respond to stimuli which tend to be without significance in the life of the organism. Young animals, for instance, have an innate tendency to respond to a wide variety of danger stimuli, such as any sudden movement or noise. However, when such stimuli are presented repeatedly without association with further effects, the young animal learns to disregard them. There is some evidence, on the other hand, that instinctive recognition of a specialized predator of a species shows little or no habituation.

Conditioning is a form of learning and consists of the establishment of a connection between a normal reward or punishment and a new stimulus, that is, one that hitherto has had no meaning to the animal.

Imprinting is especially well shown in waterfowl and gallinaceous birds. Grey-lag geese reared from the egg in isolation react to their human keepers, or to the first relatively large moving object that they see, as they would to their parents, by following. This imprinting of the parent companion is confined to a very definite and usually very brief period following shortly on emergence from the egg. Once thoroughly established, the behavior is very stable, if not totally irreversible. Furthermore, this imprinting on a human being as a substitute for its own parents will call forth, a year or more later, sexual reactions to man in the mature bird. There is no innate recognition by birds of parent, species, sex, or home locality, but there is evidence that these are learned through association and contact during the course of development. Imprinting, of course, also occurs in animals other than birds.

Imitation is another form of learning. An individual in a flock or herd may start to feed or run when it observes other individuals feeding or running. A young animal learns much that is traditional of the species by imitating its parents. Vocal imitation is conspicuous in the elaborate songs of some birds.

Trial and error learning involves trial responses to a variety of stimuli with gradual elimination of all responses and stimuli except the relevant ones. A chick pecks at random at all sorts of objects until it accidentally strikes one which is edible, whereafter the chick has a greater tendency to peck at objects that have a similar appearance. Repetition of the same act usually leads to the formation of a *habit*. Habits often appear stereotyped but differ from instincts in that they have to be learned and are not inherited.

Insight learning involves an apprehension of relations and the sudden adoption of an appropriate response without previous trial and error behavior. The mason wasp of India builds a cluster of clay cells. After depositing an egg in each cell, the female fills it with caterpillars and seals it with a lid. Eventually the whole cluster is covered with a layer of clay. While a wasp was away hunting for its prey, an experimenter made a large hole in the side of a cell. On its return, the wasp put in a caterpillar which fell out through the hole. A second caterpillar stuck in the hole with a large part hanging out through it. When the cell was completely provisioned, the wasp appeared to notice the hole for the first time and carefully examined it. With great and prolonged effort she managed to stuff the caterpillar back into place. She then collected a pellet of clay and mended the hole. Such behavior as this involves an apprehension of relations and a sudden adaptive response not preceded by trial and error. Insight learning may be manifested in various ways, as through homing ability, detouring around obstacles, tool-using, discrimination of forms and patterns, and so forth.

Biological Clocks

The behavior of an organism in its natural environment is a mixture of responses to external stimuli and is conditioned by internal physiological readiness and capacities of various sorts. Both the external environment and internal readiness fluctuate rhythmically.

The *circadian* (about a day) rhythm in some species, for instance birds, appears directly induced and controlled by periodically recurrent periods of light and darkness. The *tidal* or *lunar* rhythms of many marine species are clearly influenced by the flow and ebb of the tides. In contrast to these *exogenous* rhythms, *endogenous* ones that persist for at least a period of time under experimentally controlled apparently constant environmental conditions have been demonstrated in a wide variety of animals from unicellular forms to the

higher vertebrates and in some plants. Although in some species the "internal" or "biological clock" may drift out of synchrony with the environment, in other species the rhythm may persist accurately for weeks or months. There is controversy whether this persistence of rhythmicity is due to some fully intrinsic mechanisms or is at least time-regulated by some subtle external force, such as a change in geomagnetism or the earth's electrostatic and electromagnetic fields. Regardless of how regulated, periods of rest or sleep alternating with activity seem to be a fundamental characteristic of life (Brown, Hastings, and Palmer 1970).

Ecological Life Histories

Developmental life histories trace the origin and growth of structures and functions of an animal from the egg stage until maturity is reached. Such studies are largely embryological in nature. *Behavior life histories* attempt to analyze the activities of animals in terms of innate and learned behavior, and the neural mechanisms involved. In order to do this, it is often necessary to trace the origin of each activity to the manner in which it first makes its appearance in the young animal. *Ecological life histories*, on the other hand, are concerned with the activities of a species throughout its life cycle, and in relation to its adjustments to natural conditions. Ecological life histories usually proceed with, first, analysis of the behavior adjustments needed for the survival of the mature animal; then of its reproductive behavior; and, lastly, of the development of behavior and physiological adjustments of the young animal. In general, the proper procedure is

> . . . to discover and establish correlations between the behavior of the organism and the conditions in its environment, and then to test the significance of the correlations by appropriate experiments in nature or in the laboratory. The point should be emphasized that you start with nature, that is, with the organism in its environment. Also it should be noted that morphology and physiology of the organism are entirely subsidiary matters, although most important to the person interested in knowing how the organism behaves as it does . . . [Huntsman 1948].

The behavior of a species in relation to its environment is called its *mores* (Shelford 1913).

The following are important items that should be included in a complete ecological life history of a species:

1. Phylogenetic and geological history.
2. Geographic and habitat distribution with an analysis of adjustments to the physical environment and of biotic interrelations within the community.
3. Variations in population, through time and in space.
4. Changes in seasonal activities and physiological states: breeding, migration, hibernation.

Table 2-1 Comparison of niches of the white-footed and deer mice, both species of the genus *Peromyscus*.

Factor	*P. leucopus noveboracensis*	*P. maniculatus bairdii*
Vegetation, substratum or space occupied	Deciduous forest; subterranean, terrestrial, arboreal; home range 0.12 hectare	Sparse grassland; subterranean and terrestrial only; home range 0.24 hectare
Microclimate	Shade, rich humus, moderate moisture, medium temperature	Sunlit habitat, low moisture, temperature often extreme
Food	Seeds, nuts, insects	Seeds, grass, insects
Enemies	Owls, foxes, weasels, shrews	Owls, foxes, weasels, shrews
Stratum where food found	Surface of ground	Surface of ground
Reproductive site; nesting materials	Nests of leaves in burrows, logs, stumps, or tree cavities	Nests of dried grass in burrows, crannies, or clumps of grass
Diel activity	Nocturnal	Nocturnal
Seasonal activity	Active throughout year	Active throughout year

5. Food, enemies.
6. Parasites, diseases.
7. Reproductive potential, mortality, rate of population turnover.
8. Requirements for reproduction: home range, territory, nest-site, nesting materials, etc.
9. Breeding behavior: mating, nesting, etc.
10. Development of offspring: rate, stages, generations per year, etc.

Useful outlines, methods, and bibliographies for ecological life history studies of different kinds of animals and plants have been published in the scientific periodical *Ecology*, beginning in October, 1949.

Ecological Niche

The *ecological niche* is the particular position in a community and habitat occupied by a species population as the result of its peculiar structural adaptations, its physiological adjustments, and the special behavior patterns that have evolved to make best use of these potentialities. Important factors in the niches occupied by white-footed mice and deer mice are described in Table 2-1. Both mice are equipped with large eyes for nocturnal vision, large external ears for hearing, long vibrissae on the face for aid in running through dark underground burrows, and protective coloration. *P. l. noveboracensis* has a longer tail than *P. m. bairdii*, which appears to be an adaptation for climbing. It is possible that these two species are segregated into different biotopes because *bairdii* is more tolerant of extreme temperatures and low moisture conditions, and hence is more prevalent than *noveboracensis* in the exposed grassland habitat, but is unable to displace *noveboracensis* within the forest because of the latter's tree-climbing ability.

Every species has its own peculiar niche. No two species can permanently occupy exactly the same niche in the same locality. The living together of many species in the same community is possible only because their various niche requirements are different. The analysis of the critical factors in these niche requirements is often very difficult but is one of the main objectives of ecology (Fig. 2-8).

COMMUNITY INTERRELATIONS

The fact that species with similar tolerances and similar or complementary requirements aggregate into similar environments to form communities is a response of special interest. No organism occurs alone. Each must find its place in the community and establish relations with other members of it. The manner in which the response of species to each other is affected is shown in the structure and composition of the community and in its internal dynamics, succession, and distribution. The analysis of the community responses

Fig. 2-8 Graphical representation of the niches of four African bird species in respect to two characteristics of the vegetation (Emlen 1966).

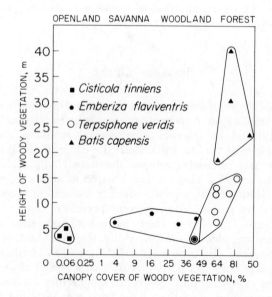

and interrelations of organisms is a major objective of this book.

SUMMARY

The environment, or specifically the habitat, of a species population consists of the range of physical and chemical conditions in which it lives. In order to live in a particular habitat, the species must be *morphologically adapted* to it. This adaptation may be accomplished to a certain extent during growth, especially in sessile forms, but depends mainly on long evolutionary processes of variation and natural selection. Each organism must also be *physiologically adjusted* to the various factors of its environment. Species vary in their limits of tolerance, and those factors in their surroundings that are most immediately unfavorable limit their habitat distribution. In order for an organism to take advantage of its morphological and physiological adjustments, it must have the proper *behavior responses*. These inherited and acquired action patterns involve selective orientation in response to environmental stimuli. Occurrence of different species in the same habitat necessitates the establishment of compatible *community interrelations*. The position of a species population in this system of interrelations in a community and within the confines of a specific microhabitat is its *niche*.

Chapter 3

THE BIOTIC COMMUNITY: STRUCTURE AND DYNAMICS

The concept of the biotic community is basic to an understanding of ecology. We will here be concerned only with laying a foundation of general principles. Details will come in later chapters, but for proper orientation we must know something about how the community is organized, how it functions, and how it may be recognized.

INTERNAL STRUCTURE AND PROCESSES

Community and Ecosystem

A *community*, or *biocenose*, is an aggregate of organisms which form a distinct ecological unit. Such a unit may be defined in terms of flora, of fauna, or both. Community units may be very large, like the continent-wide coniferous forest, or very small, like the community of invertebrates and fungi in a decaying log. The extent of a community is limited only by the requirement of a more or less uniform species composition.

A different community occurs in each different habitat and environmental unit of larger size, and in fact the composition and character of the community is an excellent indicator of the type of environment that is present. Since plants and animals, bacteria and fungi, all occur together in the same habitat and have many interrelations, they can scarcely be considered independently of each other. Together they make up the *biotic community*.

Communities may be distinguished as major or minor. *Major communities* are those which, together with their habitats, form more or less complete and self-sustaining units or ecosystems, except for the indispensable input of solar energy. *Minor communities*, often called *societies*, are secondary aggregations within a major community and are not, therefore, completely independent units as far as circulation of energy is concerned. When in this book communities are spoken of, the reference is to major communities unless otherwise indicated.

The biotic community along with its habitat is called an *ecosystem* (Tansley 1935). The term "ecosystem" has been loosely applied to units of various sizes and characteristics but had best be limited to distinctive combinations of air, soil, and water conditions with vegetative, animal, and microbic life that possess functional unity. It then becomes equivalent to the European *biogeocoenose* (Sukachev and Dylis 1964).

Dominance

When a number of species come together to form communities, each fits into a different niche and plays a different role in the internal dynamics of the community. *Dominance* is the relative control exerted by organisms over the species composition of the community. Species exerting this important control are called *dominants*. Plants are more frequently dominant in terrestrial communities than are animals. In aquatic communities, animals are relatively more important in this role, although dominance is often not developed.

Dominance is most commonly expressed in the *reactions* of an organism on its habitat (Clements and Shelford 1939). Dominants shoulder the full impact of the climate or the environment but modify this effect for other organisms within the community by tempering light, moisture, space, and other conditions. Only those other organisms that find these modified physical conditions tolerable can exist within the community. Furthermore, dominants are ordinarily the most prominent species in the community, make up its greatest mass of living material, and serve as the major source of food, substrate, and shelter for the animals that are present. In a forest community, trees are dominant. They decrease light intensity, increase the relative humidity, intercept precipitation, monopolize most of the moisture and nutrients in the soil, decrease wind velocity, and furnish shelter and food for animals. Grasses play a similar, though less conspicuous, role in prairie communities; sedges, rushes, and cattails in marsh communities; sagebrush in the arid habitat of the Great Basin; mussels and barnacles on a rocky seashore; and so forth.

Sometimes dominance is demonstrated in *coactions*, direct effects of organisms on each other. In some fresh-water ponds, carp and suckers may consume much of the submerged vegetation. This coaction thus prevents the plant constituents from assuming their usual role in the community, and by so much prevents the occurrence of animal species that depend directly upon the plants. These fish also react upon the habitat by stirring up the bottom, from which they derive organic matter, thereby greatly increasing the turbidity of the water. Penetration of light into the water is reduced, greatly handicapping sunfish, bass, and other species which locate food visually.

In primeval days, bison on our western great plains fed on the luxuriant taller grasses more extensively than on the short grasses, with the consequence that, over extensive areas, short grass species replaced tall grasses almost entirely. Thus bison were coactant with and dominant over the composition and character of the community (Larson 1940). In a similar manner, when European meadow voles are numerous, they reduce the vigor and prevalence of the grass dominants in consequence of their feeding and tunnelling in the ground, and flowering plants which are normally absent or scarce appear (Summerhayes 1941). Overpopulations of the European rabbit alter the character of the forest by feeding upon the seedlings of oak, beech, and hornbeam to the exclusion of other species. When introduced into Australia, the European rabbit con-

verted grassy areas into desert-like tracts (Bourliere 1956).

Although animals are more common coactors than plants, plant pathogens may occasionally exert dominance in this way. As a notable example, chestnut blight (a fungus) virtually eliminated the chestnut tree from the deciduous forest of eastern North America during the first few decades of the twentieth century. This fungus infects the cambium, forms pustules under the bark, and causes the bark to fall off and the leaves to wilt. The consequent opening up of the canopy with the death of the trees has allowed the extensive invasion of new species of shrubs, herbs, and animal inhabitants.

Trees are the dominants in a forest community, but species in the lower stratum of shrubs modify the habitat still further and even the herbs exert some control over the physical conditions on the surface of the ground. A *subdominant* species must tolerate the conditions established for it by the dominants; but it in turn is a modifier of the community composition in a secondary manner.

Influence

By *influence* is meant effect upon the abundance, health, and activities of other organisms but not to the extent of directly excluding species from the community. Influence is conspicuously expressed through coactions, but it may be effected through reactions as well. Insects may partially or wholly defoliate a tree; a pack of wolves may diminish a population of deer over winter; squirrels may bury acorns and nuts and thereby aid germination of them; parasitic or poisonous plants may lower the vigor or destroy the life of some other plants or animals; animals may burrow into the soil and thereby increase percolation of water and air, a benefit to plants; and all organisms, upon death, add organic matter to the habitat. These and other actions influence the community, but unless these influences become extreme they do not determine whether or not other species will occur in the community. Influence, then, is of essentially the same nature as dominance but is less vigorous in the modifying role that it plays.

Evaluating and Classifying Animals Ecologically

One of the most important, yet difficult, problems in ecology is determination of *importance-values* of organisms to the structure of communities and of their roles in community dynamics. Various statistical criteria have been used (Whittaker 1970), but we will be concerned only with some of the more general evaluations.

A basis for classifying species is their *fidelity* to the community; that is, a species is *exclusive* when it occurs only in a single area, habitat, or community; *character-istic* (selective, preferential) when it is abundant in one area or community but also occurs in small numbers elsewhere; or *ubiquitous* (indifferent) when it is found more or less evenly distributed in a wide variety of communities. The terms given in parentheses are synonyms used by plant ecologists (Braun-Blanquet 1932). Exclusive species are often rare and of little importance in the dynamics of a community, but when they are conspicuous they often make useful *indicator species* for identifying and recognizing community units.

The recognition of characteristic species presents special difficulties, since one must decide how much more abundant a species needs to be in one community to be sure a definite preference over another is indicated. In a distributional study of breeding bird populations in Ontario (Martin 1960), a species was considered characteristic of one type of vegetation if the species was at least three times more abundant in it than in any other type of vegetation. This was at population levels of from 1 to 9 pairs per 40 hectares (100 acres). For species reaching population densities of from 10 to 100 pairs per 40 hectares, preference was considered demonstrated if the species were twice as abundant in one type of vegetation as in any other. For populations greater than 100 pairs per 40 hectares, differences of 50 per cent are probably significant. It seems logical that a stricter test should be applied to small populations, for errors in measuring the size of populations and random population fluctuations attributable to factors other than choice produce a relatively greater disturbance in the data. An experimental study in measuring foliage insect populations also indicated that populations differing by a ratio of 3:1 could be accepted as statistically significant (Graves 1953). When the bottom fauna of two ponds were sampled, true differences could be detected at minimum ratios between their populations of 1.9 (Hayne and Ball 1956). A species, to be termed characteristic, should have a high *constancy* through a community, this to be indicated by its occurrence in at least 50 per cent of all samples taken (Thorson in Hedgpeth 1957).

Another criterion for evaluating species is by their *density* or numbers of individuals present per unit area. Other things being equal, a species in time of high population affects other organisms to a much greater extent than it does at times of low population. A species that is permanently more abundant than another will consume more food, occupy more reproductive-sites, and demand more space; hence its influence will be greater. *Predominants* are the more numerous constituents of a community, in contrast to *members*, which are species of lesser importance. The dividing line between these two categories is an arbitrary one.

The time and duration of occurrence of a species in a community affect the amount of influence it exerts. Generally, the longer the yearly period during which a species is active, the more important its role becomes.

Species may be classified on a temporal basis into *perennials*, those which are active in a community throughout the year, year after year; *seasonals*, which are present or active only part of the year; and *cyclics*, which may be very important some years but of negligible importance other years, as evidenced by their wide fluctuations in numbers. Even though present, a species is usually considered inactive when it is hibernating or dormant or when it is represented only by eggs, spores, or encysted stages of its life cycle.

The effect produced on the community by individuals and species may be modified by the way they form secondary groupings within the community. These minor aggregations of plants and animals are called *societies* and are of various sorts (Shelford 1932). *Layer societies* occupy different strata, such as the subterranean, ground, herb, shrub, and tree societies in a forest; *local societies* are usually parts of layer societies but are more confined in area, as groups of animals occupying an ant hill, a rotting log or stump, or a restricted but distinctive area of ground; and *seasonal societies* include all the organisms at particular times of the year.

Other factors that affect the influence of a species in the community are the size of individuals, their metabolism, food habits, and general behavior. A moose consumes more food than a mouse, and a warm-blooded mouse more than a cold-blooded salamander of the same size. A carnivore at the top of several food-chains affects the lives of more different species in the community web of life than a herbivore feeding on plants. Burrowing rodents react on the habitat more than do most birds. One factor may cancel another. An individual of a perennial species of carnivorous mammal certainly eats more than an individual cold-blooded herbivorous insect of small size and active only during the warm season. Yet there may be 1000 insects to one mammal, so that in the aggregate a single species of insect may actually produce more disturbance than a single species of mammal. The difficulty of evaluating the relative effects of species is partly alleviated by calculating their respective biomasses and amounts of energy used.

The *biomass* of a species is the average weight or volume of individuals, multiplied by the total number of individuals present. The computation of the biomass of each species thus corrects for differences between species in size and number of individuals. Because of differences between species in body moisture and amount of inert substances such as endoskeleton, chitin, shell, and the like, biomass is expressed with greater accuracy in terms of dry weight than wet weight or volume, and is even more accurate if given in terms of carbon or nitrogen content, or calories.

Attempts have recently been made to compute more significant biomasses of bird populations, using physiological constants. A biomass composed of few but large individuals has a lower metabolic activity than an equal biomass composed of a large number of small individuals. In one study (Turček 1956), the importance of different species in the community was evaluated in terms of the total body surface area rather than total weight presented by each species, using the formula $N \cdot 10 \cdot W^{0.67}$, where N is the number of individuals and W the average weight of the species. In another study (Salt 1957) the number of individuals was multiplied by the mean weight of the species raised to the 0.7 power ($N \cdot W^{0.7}$). One gets the best evaluation of the importance or influence of a species in a community where metabolic activity can be measured directly and expressed in terms of calories per unit of time (Macfadyen 1957, Teal 1957).

Productivity

A characteristic of communities that has become of considerable importance in modern ecological research is productivity. The number of individuals or biomass in a community at any one time is the *standing crop*. At the beginning of the year or reproductive season the standing crop is usually small, but as reproduction and growth take place there is an increase in the amount of organic matter making up the biomass of the community. The production of organic matter per unit of time and area is *productivity*. Productivity is thus a rate and is commonly indicated on a yearly basis, but it is also possible to measure monthly, weekly, or daily production. Small standing crops may have a high productivity and large standing crops a low productivity, hence biomass alone does not give a measure of productivity. The largest standing crop which a habitat can support without deterioration, or the maximum number or biomass of animals that can survive the least favorable yet tolerable environmental conditions during a stated period of time, is the *carrying capacity*. Carrying capacity is determined not just by the amount of food available, but also by shelter, social tolerance, and other factors (Edwards and Fowle 1955). A variety of methods are being used to measure productivity of different kinds of organisms and of different habitats. It is desirable to indicate productivity as accurately as possible in descriptions or analyses of community dynamics.

Species Diversity

Ecologists have long been aware that communities in different environments vary in the number of species that they contain, but only in recent years have quantitative indices been developed to show the relation between community structure not only (1) in number of species but also (2) in the relative number of individuals in each species. Among a number of different indices of species diversity that have been devised that

vary in what they show (Williams 1964, McErlean and Mihursky 1969), the most commonly used one, given below, is based on the information theory of communication engineers (Shannon and Weaver 1949). The application of information theory for the analysis of ecological communities was first made by Margalef for phytoplankton in 1957 and by MacArthur and MacArthur for birds in 1961. The diversity index, H'_i (Pielou 1966), is

$$H'_i = -\sum_i^s p_i \log p_i$$

where s is the total number of species in a sample and p_i is the fraction which the number of individuals of one species (i) is to the total individuals in the population. The higher the value of H'_i, the greater is the species diversity of the community; either there is a greater number of species or there is more even distribution of individuals between the species or both. For example, if we assume (using $\log_e$) a population of 100 individuals:

1. If there is only one species, $H'_i = 0$.
2. If there are 5 species with 20 individuals each, $H'_i = 1.61$.
3. If there are 10 species with 10 individuals each, $H'_i = 2.30$.
4. If there are 100 species with 1 individual each, $H'_i = 4.61$.

However, it is rare, if it ever happens, that each species has the same number of individuals; usually the species may be arranged with a few having large numbers and others having progressively smaller numbers. For instance,

5. If there are 5 species with 50, 25, 15, 8, and 2 individuals each, $H'_i = 1.26$. This gives a lower diversity index than in 2.

However,

6. If there are 10 species with 45, 25, 15, 8, 2, 1, 1, 1, 1, and 1 individuals in each, $H'_i = 1.50$.

In 6 the diversity index is not as large as in 3, because of the less uniform distribution of individuals, but is larger than in 5, both because of the larger number of species and the different arrangement of individuals, even though only five individuals are involved. A paper by Lloyd, Zar, and Karr (1968) will be helpful in calculating measures of diversity. The index (H'_i) has been used thus far primarily to compare the composition, in different communities, of specific taxonomic groups of the same general life-form.

In comparing two or more communities with different diversity indices, the number of species present and the number of individuals in each species are usually apparent, but the degree of evenness in the distribution of individuals between species is not. This evenness, or the lack of it, may be evaluated by an *equitability index*

$(J' = H'/H'_{max})$ (Sheldon 1969). H'_{max} is the maximum diversity possible for the community if all species were equally abundant ($\log_e s$) as in examples 2, 3, and 4 above. The equitability index for community 5 is 1.26/1.61, or 0.78; for community 6 it is 1.50/2.30, or 0.65. The higher species diversity of community 6 is due to its possessing twice the number of species as community 5, which outweighs its lower equitability index. Actually H'_{max} is probably never realized in natural communities, and other means of calculating equitability have been proposed (Lloyd and Ghelardi 1964).

The index of species diversity gives equal value to each individual. This is not entirely realistic ecologically since, even within the same taxonomic unit, there is usually a considerable difference between species in the size of individuals and consequently in their influence within the community. The index is improved by using biomasses instead of number of individuals, that is, by summing the proportions which each species' biomass is of the total biomass of all species (H'_b). Even this is not entirely satisfactory, as the impact of individuals and species upon the energy and other resources of the environment is proportional not to their total size but to an exponential function of their size. An ecologically more significant index of species diversity would be based on the energy demands of each species for existence and reproduction. This index (H'_p) would be calculated from the percentages of the total productivity of the community for which each species is responsible (Karr 1968, Dickman 1968). The indices H'_b and H'_p have not been extensively used to date because of lack of information on species biomasses and productivities (Table 16-4).

SUCCESSION

Communities are in a more or less continual process of change (Clements 1916). These changes result in part from the reactions and coactions of the organisms themselves and in part from such external forces as changing physiography, changing climate, and organic evolution. The habitat is usually affected as well as the community, and as the habitat changes, new species invade it and become established, and old species disappear. These changes are especially noticeable in dominant species, since these species exert a controlling role over the composition and structure of the community as a whole The replacement of one community or ecosystem by another is *succession*, and succession continues until a *climax* or final stage is reached.

Succession is a process. The series of steps or communities comprising a successional sequence leading to the climax is the *sere*. Seres are sometimes classified according to the predominant force that is bringing them about. These forces are *biotic, climatic, physiographic,* and *geologic* and their resultant seres are commonly called *bioseres, cliseres, eoseres,* and *geoseres*. The last three types of succession constitute the subject matter of paleoecology, which will be considered in Chapter 19.

Biotic Succession

Biotic succession is brought about by forces inherent within the community and in the activities of the plants and animals themselves. The most important of these activities are the organismal reactions and coactions that produce modifications in the habitat and interrelations between species. Important reactions involve filling in of ponds with plant and animal remains, the addition of organic nutrients to sterile soil, and the reduction in light intensity by increasing density of plant growth (Horn 1971). With progressive improvement of the soil and changing light and moisture conditions, a series of new dominants come into the area. When invasion of new species occurs, intense competition develops; if the invaders are successful, the old species disappear as a new community replaces the old one.

Contributing factors that may be involved are differences in growth and dispersal rates, which are different for different species. After a forest fire or logging operation, herbaceous plant growth is immediately stimulated; since herbs grow rapidly, they become dominant within a year or two. Shrubs begin to spread, and tree suckers or seedlings also appear quickly, but because they require a longer time to reach maturity, several years may elapse before they gain control of the area.

Succession is considerably influenced by the kinds of propagules available in the vicinity Seeds, spores, and the like are dispersed more or less readily, depending on form. Some kinds of animals roam more widely or spread more readily into new areas than do other kinds. The composition of the community that develops and the rapidity with which it becomes established depend, in large part, on the rates at which different species invade.

When the volcanic island of Krakatoa in the East Indies blew up in 1883, virtually all plant and animal life on it and on two adjacent islands was destroyed. In 1886, it was apparent that the pioneer plant stage consisted of a crust of blue-green algae covering the lava. A few mosses were also present, as were many ferns, and a scattering of some 15 species of flowering plants including 4 species of grass. The vegetation stage following the algae consisted predominantly of ferns, but the grasses had become dominant over most of the island. By 1906, woodland had appeared, which has since developed into an increasingly luxuriant mixed forest. Dispersal of seeds, spores, and other propagules was effected by wind, sea, animals, and man, in that descending order of importance. The order in which the propagules reached the islands greatly influenced the succession that occurred; as vegetation developed it reacted on the soil and habitat, bringing about conditions amenable to the return of the tropical rain forest (Richards 1952).

One year after the explosion, the only sign of animal life on the island was a single spider; six years after the explosion flies, bugs, beetles, butterflies, and lizards were also reported. In 1908, 202 species were found on Krakatoa and 29 on a nearby island. There were no earthworms, snakes, or mammals present, but there were many spiders and centipedes, a number of insects, 2 species each of land snails and lizards, and 16 species of birds. Bats and earthworms were found in 1919. A more thorough survey in 1921 revealed 770 species of animals, including rats, apparently introduced in 1918. In 1933, 1100 species were found, but true forest mammals had not yet appeared and many families of forest birds were unrepresented. As with plants, the invasion by animals depended on wind, sea, and man and other animals, in that descending order of importance. Survival and establishment of the animal species were correlated with the stage of vegetation that was reached. It is of interest, however, that scavenger species appeared first, then omnivores, herbivores, and finally predators and parasites. Succession of animals depended in large part on the speed with which they reached the island and on their finding proper food and shelter (Dammerman 1948).

Bioseres may be broadly grouped as *priseres* and *subseres*, depending on whether they develop on primary or secondary bare areas. A *primary bare area* is a sterile habitat, such as rock, sand, clay, or water (Fig. 3-1). A *secondary bare area* is a denudation resulting from temporary flooding, fire, logging, cultivation, overgrazing, or other phenomenon that does not produce an extreme disturbance of the soil or substratum. Since in the latter the habitat has already supported community life, and since the soil or substratum is already in an advanced stage of development, the resulting subsere progresses rapidly and the early pioneer stages of the prisere are not usually required. A subsere will develop following the destruction of any stage in a prisere, and the species composition of the stages in the subsere will be influenced by the particular priseral community that was destroyed.

The early stages or communities that make up the prisere depend largely on the type of bare area on which the prisere originates. As succession proceeds, however, later stages in the various seres in any area having a relatively uniform and humid climate come to be more and more alike. The successional development from widely diversified communities in initially different habitats to closely similar or identical climax communities in habitats that have also become much alike is called *convergence*.

Biotic succession proceeds rapidly and appears to reach a final, permanent stage in just a few decades or centuries. The climax stage of the biosere is undoubt-

Fig. 3-1 Early stages in the pond sere: open water, floating stage, emergent vegetation, swamp shrubs. Everglades National Park, Florida.

edly more nearly stabilized, self-maintaining, and in steady state in its particular habitat than are the seral stages, yet it also is subject to gradual change over long periods of time. The climax, as well as the seral stages, changes with climate, physiographic forces, and evolutionary processes. However, the clisere usually requires a few thousands of years before changes in the community structure or composition become evident. The progress of an eosere is even slower; that of the geosere, slowest of all. The *climax*, therefore, is best defined as the last obvious stage in the biosere.

The climax may be recognized by the fact that in a uniform climatic area all seres tend to converge into it, and by its steady state in respect to structure, species composition, and productivity. In the climax community, all species, including the dominant species, are continually able to reproduce successfully, and there is no evidence that new and different species are invading. In seral communities, on the other hand, the developing new growth, particularly evident in the dominants, contains many individuals of invading species which will eventually take over and replace species already present.

Bioseres may occur on a small scale in microhabitats as well as in major ones. When hay infusions, prepared in the laboratory, are seeded with representative protozoans, the order of appearance of maximum or peak populations in the various species is bacteria and monads, *Colpoda*, hypotrichates, *Paramecium*, *Vorticella*, and *Amoeba*. Disappearance of species is in the same order, except that *Amoeba* precede *Paramecium* and *Vorticella*. Algae may come in at the final stage, so that a more or less balanced community is established. The succession of species results from the higher reproductive rate of earlier species and from the fact that the excreta of at least some forms, especially the hypotrichates and *Paramecium*, are toxic to themselves (Woodruff 1912, 1913, Eddy 1928).

Another common microsere occurs in the death and decay of trees (Graham 1925, Ingles 1931, Savely 1939, Elton 1966). The sequence of animal species present as decay progresses depends on the species of tree, the community in which the tree occurs, the climate, and the geographic locality The following stages have been recognized: (1) tree dying, but still with leaves and sap; (2) tree recently dead, bark beginning to loosen, termites and other insects boring into wood; (3) wood well seasoned, bark very loose or off, wood borers still predominant; (4) wood softened and permeated with fungus; fungus beetles, elaterids, and passalids common; (5) wood largely disintegrated and crumbly, snails and millipedes occur. Wilson (1959), working in New Guinea rain forests, subdivides stages 2–5 in a different manner, each of which he names after characteristic insects found: (2) scolytid, (3) cucujid, (4) zorapteran, (5) passalid, and (6) staphylinid. Each stage also has a significantly different aggregation of resident ants. Eventually the decaying log becomes a part of the forest floor, and the animal species then present are those in general occurrence on the forest floor.

RECOGNITION OF COMMUNITIES

Community as an Organic Entity

Although the major community or ecosystem is the generally accepted unit of analysis in synecological studies, there is a difference of opinion as to whether the community constitutes a discrete organic entity. Two different points of view are incorporated in the organismic and individualistic concepts which are usually associated with the names F. E. Clements (1916) and H. A. Gleason (1926), respectively, and more recently Phillips (1934–1935), Tischler (1951), and Emerson (1952) on one side and Bodenheimer (1938), Whittaker (1951, 1952, 1956, 1957, 1967), Curtis (Brown and Curtis 1952), and McIntosh (1963) on the other. Ramensky (1926) stated the individualistic concept independently in Russia as early as 1924.

The *organismic concept* considers the community to be a supraorganism, a complex organism, or a social organism. As such, it is the highest stage in the organization of living matter: namely, cell, tissue, organ, organ system, organism, species population, community. There is emergent evolution, so to speak, at each higher stage in this hierarchy; the whole is more than merely the sum of its parts. Tissues have properties, characteristics, and functions over and above those of the individual cells involved; the organ, the organ system, or whole organism functions in a way not to be predicted from a knowledge of the parts of which each consists. The species population has inherent characteristics of density, rate of natality, rate of mortality, and age distribution, while the total community has such unique functions as dominance, cooperation, trophic balance, competition, and succession which are beyond the characteristics of the individual organisms of which the community is composed. The community behaves as a unit in its competitional and successional relations with other communities, in its local and geographic distribution, in its seasonal activities and response to climate, and in its evolution. Although the community varies in its taxonomic composition and structure in different environmental situations, this variation is proportionally no greater than occurs in different cells of the same type or between individuals belonging to the same species.

The *individualistic concept* places emphasis on the species, rather than the community, as the essential unit for analysis of interrelations, activities, distribution, and evolution. Each species responds independently to the integrated influence of the various factors of the physical environment and biotic coactions. The environment may be conceived as a pattern of *gradients* with the intensity of the various factors changing gradually in space from one extreme to the other (Fig. 3-2). The gradient may be a short one, as from the subterranean to the tree stratum in a forest or from the open water of a pond to a nearby swamp or climax forest, or it may be longer, as from the bottom to the top of a mountain or even from the tropical to the arctic zones of a continent. The population density of each species is distributed in a form resembling a normal curve when plotted along the gradient of a given factor, and the curves of many species in relation to various environmental gradients overlap in a heterogeneous, and apparently random, manner. There is seldom agreement between the limits of the distribution curves of any two species; species are not in general bound together into groups of associates which must occur together. Furthermore, the vegetation and its associated animal life very often form a *continuum* of gradually changing composition and complexity from one extreme of the environmental gradient to the other. A vegetational continuum has no sharp boundaries between individual communities of different types, and the ecologist must choose the manner in which he distinguishes these units so as best to suit his interests and objectives. The statistical orientation of species or communities in a continuum along a gradient is called *ordination* (Whittaker 1970).

These two points of view are not necessarily incompatible. There is no doubt that each species is distrib-

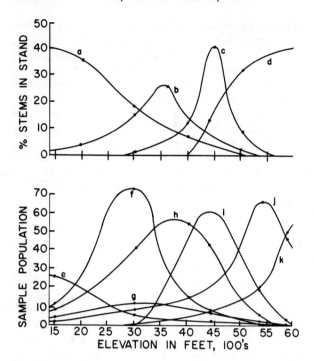

Fig. 3-2 Continuum of tree (a–d) and foliage insect (e–k) species in an elevation gradient in the Smoky Mountains, Tennessee: (a) *Tsuga canadensis;* (b) *Halesia carolina;* (c) *Acer spicatum;* (d) *Fagus grandifolia;* (e) *Graphocephala coccinea;* (f) *Caecilius* sp.; (g) *Agalliopsis novella;* (h) *Polypsocus corruptus;* (i) *Anaspis rufa;* (j) *Cicadella flavoscuta;* (k) *Oncopsis* sp. (after Whittaker 1952).

uted according to its own physiology, its own complex interrelations with other species, and its own tolerances, and that no two species are exactly alike in these various respects and in their responses to the environment. On the other hand, one can be convinced that the community and its habitat, collectively the ecosystem, is a functional system and that every species of necessity occurs in and as a part of such a system so that its distribution is importantly modified by these interactions and community relations.

Kinds of communities are usually recognized and identified by their most important organisms, the dominants and predominants (Shelford 1932). Subdominants and member species, however, are not usually dependent on the dominant species directly; rather, on the environmental conditions that the dominants establish. Different species of dominants in adjacent or related communities may react on the environment in a manner so nearly the same that subordinate species find suitable conditions for existence in each, although they are usually more characteristic of one than the other.

The *community-stand* is an actual aggregation of organisms occurring in a particular locality. In a sense, it is a collection of niches occupied by a particular set of interacting species populations, but it is something that one can see and study in the field. Because

of the great variation in composition and character of community-stands in different habitats and parts of the world, they need to be evaluated and classified in some logical manner for reference purposes. *Community-types* are abstract groupings of individual community-stands which resemble one another and consequently must be defined rather arbitrarily. Different systems of community-types have been proposed, each designed to emphasize a particular point of view. We are, in this book, using the biome system, the various parts and concepts of which will unfold as we proceed. We will consider the community as being at least analogous to an organism in being a functional unit of interacting parts and having some degree of structural integrity. Although community-types are certainly not highly discrete and absolute units, recognition and naming of them is one way of indicating positions in the continua along environmental gradients that are occupied by particular aggregations of plant and animal species.

Physiognomy

The gross structure of a community or its physiognomy is an important basis for its recognition. In terrestrial communities, physiognomy is determined by the life-forms of the dominant plant species and their spacing. The life-forms that prevail in a given area depend on the climate and the substrate or other special features of the habitat and give character to the landscape. The distribution of animal communities is closely correlated with the structure of the vegetation, hence these vegetation-types need to be recognized and defined:

DESERT: Hot, arid habitats with scattered scrubby or thorny vegetation or, in extreme cases, none.

THORN SCRUB: Tropical dry thorny deciduous tall shrubs, succulents, ephemerals.

STEPPE, PLAINS: Semi-arid grassland covered with short grasses.

PRAIRIE: Semi-humid grassland covered with mid and tall grasses.

TEMPERATE SHRUBLAND: Semi-arid areas covered with bushes and shrubs, either deciduous or broad-leaved evergreen (chaparral).

SAVANNA: Grassland with scattered trees or groves of trees or shrubs covering 10 to 25 per cent of the ground (Dansereau 1960).

WOODLAND: Open stand of small deciduous or evergreen trees covering 25 to 60 per cent of the ground and with undergrowth of grassland or desert vegetation.

FOREST-EDGE: Mixture of trees, shrubs, and open country, ordinarily occurring as a narrow belt on the margin of forests.

FOREST: Closed stand of trees forming a continuous canopy over at least 60 per cent of the area.

DECIDUOUS FOREST: Broad leaves fall during cold or dry seasons.

BROAD-LEAVED EVERGREEN FOREST: No regular season of leaf fall, leaves often sclerophyllous, warm climates.

RAIN FOREST: Tall luxuriant forests, often with several strata of trees, foliage retained throughout the year, climate continuously warm and wet.

NEEDLE-LEAVED EVERGREEN OR CONIFEROUS FOREST: Forests of pines, spruces, firs, larches, hemlocks, and the like.

FOREST-TUNDRA: Stunted open growth of coniferous forests in cold climates.

TUNDRA: Extensive flat or gently rolling treeless areas occurring in cold climates.

ALPINE TUNDRA: Treeless areas at higher elevations of mountains.

BOG: Wet areas in cold climates containing sphagnum, heath plants, coniferous trees.

SWAMP: Wet areas in warm climates covered with deciduous trees or shrubs.

MARSH: Wet areas containing sedges, rushes, cattails, and the like.

Inasmuch as animals choose niches in response primarily to the physical structure of the vegetation regardless of its taxonomic composition, it is helpful in describing biotic communities to show the vegetation structure in as much detail as possible (Fig. 3-3). This may be done by semi-realistic diagrams or by a system of symbols (Dansereau 1951, 1956).

The density of the foliage or the percentage of vegetative cover, often of importance to animals, may be conveniently measured at intervals on straight lines through an area. Whether or not leaves are present may be determined by holding a pole upright or sighting through a vertical pipe (MacArthur and MacArthur 1961) or through a viewing tube (Emlen 1967). It is often desirable to obtain the percentage coverage of the foliage of each stratum (Karr 1967).

The 50 per Cent Rule

If the primary basis for community recognition is based on the life-form of the dominants, which on land is expressed in the physiognomy of the vegetation and in some aquatic habitats on the life-form of the predominant animals, then the secondary breakdown of community units must be on the basis of taxonomic units. Here, the species unit is most useful, as the species is

Fig. 3-3 Mixed deciduous-coniferous plant community (after Dansereau 1951). Above, a semi-realistic diagram of the community; below, symbolic structure of the community depicting life-form, size, function, leaf type, and texture.

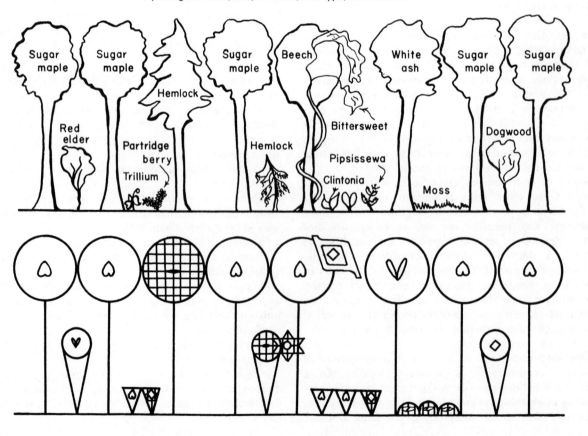

the smallest taxon having objective reality and precise interrelations with its environment.

Two aggregations of species occurring naturally in different areas or in the same area at different times are to be considered as distinct communities when at least 50 per cent of the species of each aggregation are *exclusive* or *characteristic* to the aggregation. This we may call the 50 per cent rule (Mayr 1944). It is necessary to have quantitative information on the size of the populations to evaluate the importance of each species before community classification is attempted (Spärck 1935).

The distinctiveness of communities must work in both directions; that is to say, 50 per cent or more of the species of each aggregation must be different from the other aggregation. This means that the two aggregations are more different than they are alike. If the species composition does not exhibit the 50 per cent distinction, the two aggregations are considered as belonging to the same community. If the difference approaches but does not equal 50 per cent, it is often worthwhile to designate the two aggregations as *facies* of the same community if they are seral, or as *faciations* if they are climax. A number of other statistical procedures for distinguishing communities have been used by different investigators, but no method has yet been suggested that incorporates, as does the 50 per cent rule, not only the presence or absence of a species, but also its relative abundance and ecological significance.

Naming Communities

Since communities are distinguished by differences in life-form and taxonomic composition of the dominant or predominant organisms, these characteristics are usually used also in naming the community. Where the habitat is well defined but vegetation is largely or wholly lacking, as in many aquatic communities, habitat may be used in the terminology. Since names are largely a matter of convenience, they should be short and be derived from some easily recognized feature of the community or habitat. Very often the generic names of two, sometimes three, conspicuous dominants are used to name plant communities. The prevailing type of vegetation or habitat is commonly employed to name animal communities, although two or three predominant characteristic or exclusive animal species may also be used. In case of some large communities, geographic names are more convenient.

Large geographic units, differentiated on the basis of the climax type of vegetation or life-form of animals, are called *biomes*. They are specifically named by the characteristic form of vegetation or life-forms present; tundra biome, grassland biome, or pelecypod-annelid biome, for instance.

Secondary communities within the biome can be distinguished as climax or seral, respectively, by the suffixes *-iation* and *-ies*. An *association* is a climax plant community identified by the combination of dominant species present; an *associes* is an equivalent seral plant community. Thus we may speak, for instance, of the *Fagus-Acer* association, which is a climax deciduous forest community, and of the *Calamogrostis-Andropogon associes*, which is a grass stage in a sand sere (Clements and Shelford 1939).

Animal communities on land are related to different life-forms of plants or types of vegetation, but only seldom to plant communities distinguished by the taxonomic composition of the plant dominants. Thus animal communities must be analyzed and named independently of plant communities. A *biociation* is a climax animal or biotic community identified by the distinctiveness of the predominant animal species; a *biocies* is the seral equivalent. The North American deciduous forest *biociation* is to be contrasted, for instance, with the pond-marsh *biocies*.

When two plant or animal communities merge, either by intermingling of species in the same habitat or by juxtaposition of different communities in the same region, the resultant transitional state is called an *ecotone*. Ecotones occur between consecutive communities in seral development on an area as well as between adjacent existing local or geographic communities.

PRESERVATION OF ENVIRONMENTAL QUALITY

Up to the advent of civilized man, the biosphere of the world consisted of a complex pattern of biotic communities responsive to a variety of environments. The harmony prevailing between the biotic and physical realms was the result of hundreds of millions of years of evolution and adjustment of one organism to another and of each to the habitat in which it lived. The relationship was a resilient one. A sporadic outbreak of a species as the result of some temporary advantage, or a catastrophe to a population as the result of a change in the physical environment, soon became tempered and a balance restored. Primitive man occupied a niche and assumed a role comparable to other animals, and the ecosystem maintained itself in a vigorous healthy condition.

Modern man, however, has initiated a number of abrupt changes during the last few centuries beyond the capacity of the ecosystem to adjust. He has become the dominant organism in the biosphere, forcing changes to fit his needs and desires, and creating ecosystems different from any that have before existed. Emergence of new dominants and evolution of new communities is a characteristic of the geosere, but the replacement of the biosphere by an "anthroposphere" (Marston Bates) or "noosphere" (from the Greek word, *noos*, mind) is unique in the speed and extreme changes produced. This acquisition of dominance is the result of the extraordinary evolution in man of intelligence, ingenuity, and use of tools. He has greatly modified or replaced the natural processes that control stability and balance within the ecosystem with new processes and manufactured products which the environment cannot absorb. The result has been a general deterioration of the environment, a population explosion of the human species, and a disharmony in the world ecosystem that will, unless resolved, lead quickly to catastrophe. Whether man will survive as the dominant organism, be reduced again to the role he occupied as a primitive, or become extinct may well be decided within the next few decades. There is conflict here between the demands of man for expanding population, productivity, power, and pleasures and the functioning of natural laws for maintaining stability and equilibrium between the living and physical environments (Atwood 1970, Wagner 1971).

Immediate problems of concern in the preservation of environmental quality are the elimination of air, water, and soil pollution; the disposal of solid wastes; and climate modification on a planetary scale. Perhaps noise pollution should also be added. *Pollution* is change in environment, caused by man's discharge of matter or energy into it, which alters the normal function of an ecosystem. Fundamental to the attainment of a stabilized and balanced worldwide ecosystem are the control and regulation of the human population at a suitable level and preserving the habitat and the basic community of plant and animal life upon which human life depends. Considerable technical knowledge already exists for solving these problems. The immediate need

is concerted attention for implementation of the necessary remedial measures. Satisfactory resolution of the problems will involve not only science and technology but also law, sociology, politics, and economics (Henning 1970). In the United States, an Environmental Protection Agency became established only in 1970. Emphasis on existing knowledge, however, should not obscure the fact that continuing basic research is required to elevate man's understanding of the environmental system. Scientists are just beginning to study ecosystems on the multidisciplinary basis that is clearly required (Cooke 1969, Turk *et al.* 1972). The various problems involved in the survival of man and in maintaining environmental quality will be identified and analyzed at appropriate places in the text that follows.

MEASUREMENT OF POPULATIONS

Recognition of communities, calculation of species diversity, measurement of productivity, evaluation of species influence, following seasonal and yearly fluctuations, and investigation of a number of other community characteristics and functions depend on a knowledge of species density or biomass. In spite of its fundamental importance, available methods for measuring population size are only moderately satisfactory and are in need of improvement (Southwood 1966, Regier and Robson 1967, Petrusewicz 1967). In most types of ecological research, the aim should be to determine absolute abundance or the actual number or biomass of a species in an area of known size. This is no more difficult than in correcting relative indices of abundance for all the variables that are involved.

Since it is seldom possible to count all the individuals present in a large area, it becomes necessary to take samples over small areas where accurate counting of individuals is practical. *Strip censuses* involve counting all individuals of conspicuous species seen within a prescribed distance on each side of a line of travel over a measured distance (Hayne 1949a, Rasmussen and Doman 1943, Kendeigh 1944, Hickey 1955). *Sample plots* may be distributed at random over an area but care must be taken that they are of proper size (Fig. 3-4), shape, and number adequately to determine both species composition and densities (Cole 1946a, Preston 1948).

Fig. 3-4 Species-area curve for trees to illustrate the method of determining sampling area size adequate for analysis of community species composition (Vestal 1949). Plant ecologists commonly consider a minimum adequate sampling area as represented by a point on the curve where a 10 per cent increase in the sampling area increases the total number of species by only 10 per cent (Cain and Castro 1959).

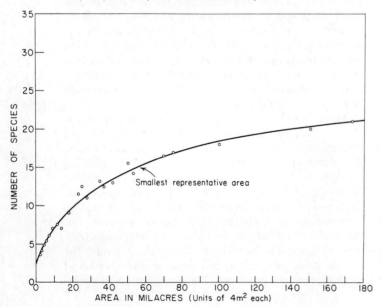

Mammals

Small mammals usually need to be trapped because of their inconspicuousness and nocturnal habits. Killer traps, live traps, and pitfalls may be used, and all animals are captured over areas of known size. Considerable study is underway to standardize procedures. Live trapping has the advantage of allowing the deter-

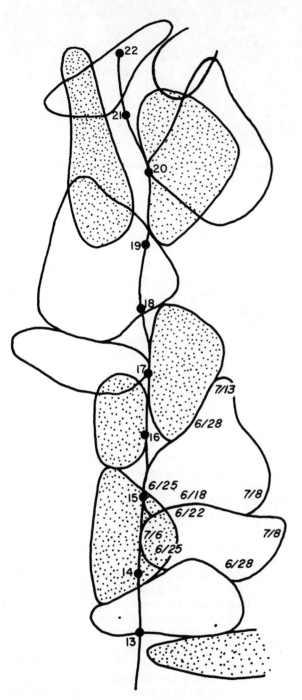

Birds

The spot-map method of censusing (Fig. 3-5) is commonly used for small birds (Kendeigh 1944, Enemar 1959), but a variety of procedures are used in making inventories of waterfowl and upland game species (Hickey 1955).

Foliage Arthropods

In order accurately to determine the insect and spider composition in the foliage, net sweepings, light traps, bait traps, adhesive snares, and the like are necessary (Morris 1960, Menhinick 1963). A series of 48 strokes with a sweep net through the herb and low shrub strata furnishes an approximation of the number of arthropods per square meter (Shelford 1951a, Menhinick 1963). Variations in population estimates obtained by sweep-net samples are not usually significant unless differences of the order of two- or threefold are obtained. Sampling of arthropods in tree foliage is more difficult, but special techniques make this possible (Morris 1960).

Ground Animals

Hand sorting through the ground litter over measured areas permits the obtaining of good estimates of the larger millipedes, centipedes, snails, amphibians, and small reptiles. For the smaller invertebrates, especially insects, spiders, and mites, a common procedure is the use of the *Berlese funnel* (Fig. 3-6). A ground sample of soil and litter, 0.1 m² in size or smaller, is placed bottom side up in an open mesh basket inside a funnel.

Fig. 3-6 Tullgren modification of a Berlese funnel for quantitative sampling of soil animals.

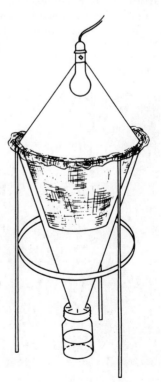

Fig. 3-5 Spot-map trail census of birds showing individual territories of two competing species, wood pewee (strippled) and least flycatcher. Two of the territories show the dates and locations of observations of particular individual birds. The numbered points on the trail are about 50 m (162 ft) apart. Note that territories of individuals of the same species do not overlap, and that territories of the two species are largely but not entirely exclusive (Kendeigh 1956).

mination of home ranges (Stickel 1946, Dice 1952, Calhoun 1956, Grodzinski *et al.* 1966, Petrusewicz 1967, Van Vleck 1969, French *et al.* 1971).

A light bulb above the sample causes the animals to retreat downward and to fall into a bottle of preservative at the bottom (Macfadyen 1961). With the *flotation method*, the litter or soil is placed in a pan and covered with warm water, sugar solution, or water with chemicals added, such as magnesium sulfate, to increase the specific gravity. The material is agitated and the animals come to the surface, where they may be collected (Kajak *et al.* 1968). For the microfauna or for other taxonomic groups, such as the annelids, a variety of special methods are available (Van der Drift 1950, Kevan 1955, Burges and Raw 1967, Wallwork 1970).

Fish

Seines may be used to sample fish populations in shallow water and trammel, gill, or Fyke nets (Fig. 3-7) in deeper water (Regier and Robson 1967). Small bodies of water or representative areas of larger bodies can be blocked off by nets and the fish collected by the use of poison, such as rotenone (powdered derris root). The fish are killed by this method, however, as are most zooplankton and some kinds of larger invertebrates (Brown and Ball 1942). A less drastic method, that of shocking, employs two electrodes inserted into the water a short distance apart (Fig. 3-8). The electric charge temporarily stuns the fish so that they float to the top, where they can be captured (Lagler 1952, Alabster and Hartley 1962).

Plankton

Plankton nets are commonly made of silk bolting cloth arranged as a conical bag and attached to a wire frame. The net may be towed behind a boat either at the surface or submerged to any depth by appro-

Fig. 3-7 Fyke fish trap (courtesy Illinois State Natural History Survey).

Fig. 3-8 Fish-shocker in use (courtesy Illinois State Natural History Survey).

priate weights attached to the tow line. Since the depth at which plankton occurs varies with the time of day, vertical hauls with a Wisconsin plankton net may be preferred. The Kemmerer sampler is used extensively for bringing up known volumes of water from measured depths for plankton and for chemical analyses. Various other types of closing nets or traps are available for sampling at different depths. Net plankton may be counted with the use of a Sedgwick-Rafter cell. For estimating the smaller nannoplankton, filtering or centrifuging is required (Sverdrup *et al.* 1942, Welsh 1948, Ballantine 1953).

Bottom Organisms

Dip-nets are commonly used in shallow water for obtaining macroscopic bottom organisms and those attached to submerged vegetation. Collections may be made quantitative if scoops are taken out of a bottomless cylinder covering a known area. The Surber swiftwater net is standard equipment for sampling rocky stream bottoms (Fig. 3-9). A frame marks out 0.1 m², and a net downstream catches organisms dislodged as the rocks are agitated or removed for closer examination. Dredges of various shapes and sizes may be pulled along the bottom for measured distances to get organisms in deep water, but quantitative determinations obtained in this way are generally too low. Much more reliable are the Ekman bottom sampler on soft bottoms and the heavier Petersen sampler, used also on sand and harder bottoms (Fig. 3-10). Bottom samples must ordinarily be washed through sieves to remove the debris and separate the animals for identification and counting. Mesh of different sizes is used in the sieves to permit sorting out animals of different sizes (Ryther *et al.* 1959, Cummins 1962).

Capture-Recapture Method

K. Dahl in 1917, using methods of marking and recapturing fish developed in 1894 by C. G. J. Petersen of the Danish Biological Station; F. C. Lincoln, of the U.S. Fish and Wildlife Service in 1930, trying to estimate the number of ducks on the North American continent; and C. H. N. Jackson (1933), working with tsetse flies in Africa, all independently derived a formula for determining the population size of various species of animals, much used in recent years (Ricker 1948). The method depends first on capturing a fair sample of individuals in a unit area, marking them in a distinctive manner, releasing them for uniform rediffusion over the area, then, after a short interval, retrapping the area. The ratio of marked individuals recaptured to the total number marked should theoretically be the same as the total marked and unmarked animals captured during the second trapping is to the total population:

$$\text{total population} = \frac{\text{total marked}}{\text{marked recaptured}} \times \text{total captured}$$

Other formulas make use of accumulating totals of marked and unmarked individuals during successive periods of trapping (DeLury 1958). There are several possible errors in arriving at accurate estimates of the total population for which compensation must be made (Buck and Thoits 1965, Eberhardt 1969).

Marking of small mammals, birds, lizards, and amphibians is commonly done by toe clipping, ear or wing notching, tattooing, dyes, or tags (Taber 1956). Birds may be individually identified by means of numbered bands placed around their legs; such a program is sponsored by the U.S. Bureau of Sport Fisheries and Wildlife (Lincoln 1947). Snakes are marked by removing scales from conspicuous locations on the body, frogs may be identified by punctures in the web between the toes, a fin may be clipped in a characteristic manner with fish, and so on. For determining home ranges, a different procedure is labeling individuals with radioactive material and following their movements by means of Geiger counters (Brooks and Banks 1971). Still another method becoming increasingly popular involves attaching miniature radio transmitters to animals and monitoring their movements by means of telemetry (Cochran and Lord 1963).

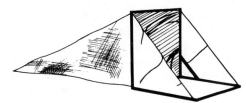

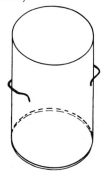

Fig. 3-9 Apparatus for collecting quantitative samples of bottom organisms in streams: above, swift-water net, covers 0.1 m² (Surber 1936); right, sampling cylinder for use in pools, covers 0.2 m², has sharpened lower edge.

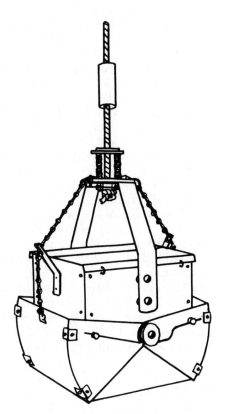

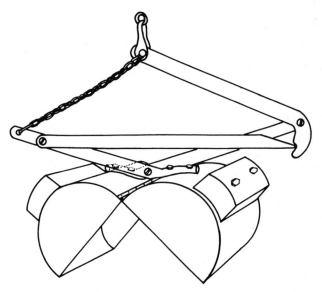

Fig. 3-10 Two types of bottom samplers for measurement of benthic populations. When open, each sampler covers a known area. As each is closed, it scoops up the organisms present. Above, Peterson's bottom sampler for hard or sandy bottoms; left, Ekman's bottom sampler for soft bottoms and deep water (from Welch 1948).

Capture per Unit of Effort

In a closed or stabilized population, when the same time, traps, and effort are employed to capture or count individuals in the same area at different times and there is no loss or increment in the original population, and weather and other conditions remain the same, the number of new individuals captured or discovered with each subsequent effort becomes less and less, and should eventually reach zero (Fig. 3-11). When the number of new individuals captured per unit of effort is plotted against the cumulative number of animals captured, a straight line results. A line thus derived from a few catches may be extended to zero, and the total population of animals in the area determined (DeLury 1947, Zippin 1958). A variation of this method is to use the increasing percentage of marked animals in the total number captured at successive intervals of time, as the increase in these percentages follows a definite trend that would eventually include the total population (Hayne 1949).

SUMMARY

A community is an aggregation of organisms forming a distinctive combination of species. The community and the habitat in which it occurs constitute an ecosystem. Inherent within the community are forces of dominance which control the species composition, and of influence which affects the abundance, health, and activities of organisms. Dominance is exerted primarily through reactions of organisms on the habitat, influence primarily by coactions of organisms on one another.

Fig. 3-11 Total population ($K = 1170$) calculated by extension of a straight line through data on successive catches of new individuals per unit effort, $C(t)$, plotted against the accumulating total catch, $K(t)$ (from DeLury 1947).

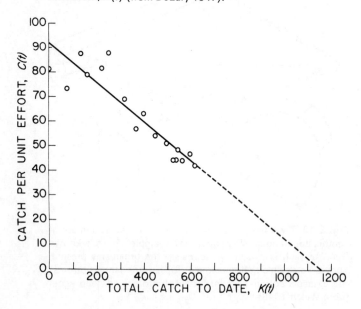

The relative importance of each of the various species within the community is evaluated on the bases of fidelity, constancy, abundance, time of activity, secondary groupings, and influence. Species diversity indices give quantitative expression of community structure in terms of number of species and taxonomic distribution of individuals. Reproduction and growth brings a production of organic matter; the rate of formation of organic matter and energy is called productivity.

Communities are constantly changing, the result of reactions and coactions of the organisms, and of climatic, physiographic, and evolutionary processes. This change is one of succession, an orderly replacement of one community by another until a climax, especially evident in bioseres, is reached.

The community may be considered as a highly integrated self-contained organic unit or as merely an aggregation of independent species whose preferanda coincide in the same habitat. These are extreme points of view; an intermediate one is adopted in this book.

The gross structure of the community is the primary basis for distinguishing and recognizing it. On land, this structure is characterized by type of vegetation; in water, by the life-form of the predominant organisms, which are usually animals. Communities are then subdivided according to their taxonomic composition. Aggregations of species are given community status if at least 50 per cent of the constituent species are exclusive to or characteristic of it. Animal communities are named for the type of vegetation, life-form of the predominant species, or habitat, depending on which is the most conspicuous feature; and secondarily for the predominant two or three exclusive or characteristic species that it contains or for the geographic area in which it occurs. Biomes are major geographic community units. Biociations are subordinate climax communities distinguished by the distinctiveness of their predominant animal species. Biocies are the seral equivalents of biociations.

The rapid evolution of human culture, especially during the last century, has introduced products and effects into the world ecosystem to which the environment has not been able fully to adjust. Control of air, water, and soil pollution, as well as the explosion of the human population, poses problems for the very existence of mankind.

Although determination of relative abundance is sometimes useful in projects of limited scope, the measurement of absolute abundance is generally to be preferred. Measurement of absolute density requires the counting of individual animals or measurement of their biomasses on strip censuses or sample plots. The size, number, shape, and distribution of sample plots and methods of measuring population densities present special problems that must be adjusted for each habitat and group of organisms concerned.

Part Two

COMMUNITY
ECOLOGY

Chapter 4

STREAMS
AND WATER POLLUTION

When rainwater falls on an uneven surface, it collects in depressions. As the water overflows them, the current erodes a narrow channel that deepens with each succeeding shower and may eventually drain the depression. There is usually also a lateral meandering of the stream, by which a valley is formed. The site of the headwaters of such streams is impermanent, and continued erosion forces the headwaters and the channel farther and farther back into the upland. The stream is at first a temporary one, dependent for its waterflow on rainfall runoff, but when its channel is cut below the level of the groundwater table the stream becomes permanent, fed by general, continuous seepage. The headwaters of such a stream are therefore its youngest portions physiographically, and the stream is progressively more aged toward its mouth.

In hilly or mountainous terrain, water may accumulate in large basins until ponds or lakes are formed. In the Great Basin of North America, such lakes have not found an outlet to the sea, and evaporation has left them with a very high salt content. Ordinarily, however, the water level in such a lake will rise until it overflows at the lowest point on the perimeter. Then the waters continue to flow downward until they eventually reach the sea. Streams springing from fixed headwaters (melting snowfields and glaciers, springs) carve valleys that are of essentially the same age throughout. Streams less than 3 m (10 ft) wide are usually called *creeks* or *brooks*; *rivers* are streams 3 m or more wide.

A river system in *youth* is characterized by valleys that are narrow and steepsided; the flow of water is usually fast, there are few tributaries, and there are many waterfalls, ponds, and lakes along its course. As the river system matures, its valleys become wider, its slopes more gentle, and its tributaries more numerous and longer. Many ponds and lakes are drained, and waterfalls are worn down to rapids or riffles. The areas of upland are well dissected, and the land is thoroughly drained. In *old age*, the river system has reached *baselevel*. The upland has been worn down to low ridges between tributary river valleys, and the region as a whole is called a *peneplain*. There are no lakes, ponds, or rapids, and the flow of water is sluggish (Strahler 1951).

HABITATS

Exclusive of its lakes, the principal habitats in a stream are *falls*, *rapids* or *riffles*, *sand-bottom pools*, and *mud-bottom ponds* (Fig. 4-1). The character of the bottom depends primarily on the velocity of the water current, which, along with the volume of stream flow, can be readily measured (Robins and Crawford 1954). Water flowing at the rate of about 50 cm-sec is considered swift-flowing; velocities greater than 300 cm-sec rarely occur. Fast currents roll or slide pebbles and rocks along the bottom, move sand partly by rolling and partly by buoyant transportation, and carry fine materials, such as silt and organic matter, in suspension (Twenhofel 1939).

In places where the topographic gradient is steep, the stream bottom will be composed largely of cobble and boulders too heavy to move, and smaller pebbles which are trapped by obstructions. This habitat is called a *rapids*, if extensive and turbulent; *riffles*, if of a lesser order. Below the riffles there may be a *raceway*, which is intermediate between the riffle and pool in depth of water, rate of flow, and substrate, and its depth is quite uniform (Winn 1958).

When the gradient is less steep and the water current thus slower, gravel (particle size 2–64 mm) is deposited first, then sand (0.06–2 mm), but the finer materials are carried along. Only when the current becomes negligible does the suspended material settle so that silt (0.004–0.062 mm) or mud-bottom *pools* or *ponds* are formed. Clay has a particle size even smaller (Morgans 1956). These mud-bottom pools are the most fertile parts of the stream because of the presence of organic matter entrained in the silt. The rate at which oxygen diffuses into water from the atmosphere increases as the turbulence of the water increases; rapids, therefore, often have the highest oxygen content of a stream's waters. Ordinarily, however, oxygen is near saturation in all parts of a flowing, non-polluted stream. In a general way, riffles and sand- and mud-bottom pools represent three stages in the *erosion cycle* of a stream, and ecological study of them gives a good idea of what the eosere would be over a long period of time.

Trout streams do not normally exceed 24°C maximum summer temperature; streams with higher summer temperatures are more characteristically occupied by species of Centrarchidae and Esocidae (Ricker 1934). Streams have been classified into a variety of different types, using the most characteristic fish present as a basis (Van Deusen 1954) or their pattern of branching (Kuehne 1962). The salt content of stream waters depends both in quantity and in chemical nature on the fertility of the land drained or the rock strata which produce the springs.

STREAM BIOCIES

When quantitative sampling is made of the invertebrate populations of streams, one finds that there is a sharp distinction of species found in riffles and those found in mud-bottom pools (Table 4-1) (Cummins *et al.* 1966). The raceway is here included with the riffles. The sand-bottom pool habitat, shown in the

table, has no characteristic indigenous invertebrate species, but is occupied by small numbers of individuals of species otherwise occurring abundantly in the other two habitats. The unstable bottom apparently prevents the development of a distinctive community. The high index of species diversity is caused by the uniformly low number of individuals in each species, which more than compensates for the smaller number of species present. The unionid clams are the only invertebrate group to become established in this habitat with any degree of success, although they are not exclusive to it (Fig. 4-2). There are, however, several fish species (Table 4-2) that find sandy pools a favorite habitat, although they depend in large part upon riffle organisms for their food. Stream fishes commonly segregate into more or less distinctive societies of species correlated with differences in current, type of bottom, and depth of water (Larimore and Smith 1963). Many fish overwinter in the deeper, more quiescent sand-bottom pools, especially since low water temperature makes them too sluggish to withstand rapid currents.

Mud-bottom pools form in backwaters of the main stream, behind natural or artificial dams in the main channel, or where the current is sluggish. Very often, aquatic vegetation fringes the edges of these pools.

Fig. 4-1 A small stream, showing riffles in foreground and a sand-bottomed pool upstream (courtesy P. W. Smith and the Illinois State Natural History Survey).

Table 4-1 Size and distribution of invertebrate populations in stream habitats of the Vermilion River, Illinois, as determined by class studies through 8 years.

Common name	Classification	Number per square meter		
		Riffles	Sand-bottom pool	Mud-bottom pool
Caddisfly larva	Trichoptera	1006		
Mayfly naiads	Heptageniidae, Baetidae	248		
Hellgrammite	*Corydalis*	46		
Riffle beetle larva	Psephenidae	19		
Riffle beetle adult	Psephenidae	4		
Limpet snail	*Ferrissia tarda*	2		
Bryozoan	*Plumatella*	+		
Fresh-water sponge	Spongillinae	+		
Flatworm	*Planaria*	+		
Broad-shouldered water strider	*Rhagovelia*	+		
Whirligig beetle	*Dineutes discolor*	+		
Stonefly naiad	Plecoptera	61	1	
Snails	*Goniobasis livescens, Pleurocera acuta*	39	2	
White midge fly larva	*Tanypus*	16	0	5
Horse fly larva	Tabanidae	8	1	1
Fingernail clam	*Sphaerium*	8	+	+
Crayfish	*Orconectes propinquus*	6	1	4
Damselfly naiad	Zygoptera	4	2	7
Dragonfly naiad	Anisoptera	2	4	7
Clams (28 species)	Unionidae	++	++	+
Crayfish	*Orconectes virilis*	+	+	+
Snail	*Physa gyrina*	+	0	1
Red midge fly larva	*Tendipes*	+	1	8
Aquatic annelid	Chaetopoda	+	1	134
Burrowing mayfly naiad	*Hexagenia*	+	+	139
Water boatmen	Corixidae			6
Alderfly larva	Sialidae			4
Fishfly larva	*Chauliodes*			2
Crawling water beetle	Haliplidae			1
Amphipod	*Hyalella*			1
Predaceous diving beetle	Dytiscidae			+
Backswimmer	Notonectidae			+
Water scorpion	*Ranatra*			+
Aquatic isopod	Asellidae			+
Whirligig beetle	*Dineutes assimilis*			+
Springtail	*Podura aquatica*			+
Mayfly naiad	*Caenis*			+
Snail	*Gyraulus parvus*			+
Snail	Lymneidae			+
Total taxa		25	12	26
Total individuals		1469+	13+	320+
Percentage taxa characteristic or exclusive		64	0	73
Taxa diversity index (counting + = 0.1, ++ = 0.2 individual) (H')		1.09	2.06	1.37
Equitability index (H'/H'_{max})		0.34	0.83	0.42

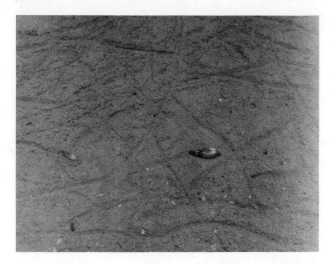

Fig. 4-2 Clam tracks in a sandy pool (courtesy R. E. Rundus, 1956).

These quiet pools are essentially young stages in the development of ponds and support many animal species indigenous to ponds. Such pond animals as aquatic annelids, dragonfly and damselfly naiads, and burrowing mayfly naiads commonly occur also on the muddy margins of streams in which the main channel has a sand, gravel, or rocky bottom.

The *stream biocies* consists most typically, therefore, of the inhabitants of the riffles and sand-bottom pools found throughout the course of the river (Menckley 1963) (Figs. 4-3 to 4-7). The riffle and pool organisms make up two different *facies* in this community. These facies become more fully developed downstream as shown by an increase in the species diversity index, both for benthic organisms (Wilhm 1967) and for fish (Sheldon 1968). Mud-bottom pools and sluggish streams are occupied by the pond-marsh biocies, to be later described.

Plants are not abundant in the stream biocies, although the upper surfaces of rocks in a riffle may be completely covered with branched filamentous algae (particularly *Cladophora*), and a few species of water mosses (Fontinalaceae) may occur. Diatoms, mostly sessile forms, may be numerous in early spring and again in the autumn. Dominance in the true sense, such as occurs in terrestrial communities, does not exist, although the algae and mosses passively provide food and shelter for active forms.

The most characteristic and abundant animal forms of the stream biocies are the caddisfly larvae, mayfly naiads, stonefly naiads, fly larvae, crayfish, snails and clams, sponges and bryozoans, and fish, each occupying its own particular niche (Berg 1948, Hynes 1970, Warren 1971). Plankton is mostly sparse in swift-running water (Carpenter 1928, Coker 1954), but may be abundant in sluggish, pond-like stretches of large rivers. The fishes listed in Table 4-2 and Fig. 4-8 are mostly warm-water fishes. In the colder waters of mountain and northern streams, the fish fauna changes.

Table 4-2 Distribution of predominant fish species in stream habitats of central Illinois (after Thompson and Hunt 1930).

Common name	Riffles	Gravel- and sand- bottom pools	Mud- bottom pools
Suckermouth minnow	+		
Banded darter	+		
Bigeye chub	+		
Log perch	+		
Green-sided darter	+		
Stonecat	+		
Hog sucker	+	+	
Fantail darter	+	+	
Steelcolor minnow	+	+	
Common shiner	+	+	+
Channel catfish	+	+	+
Hornyhead chub	+	+	+
Stoneroller minnow		+	
Silverjaw minnow		+	
River shiner		+	
Redfin shiner		+	
Rainbow darter		+	
Quillback carpsucker		+	
Smallmouth bass		+	
White crappie		+	
Orangespotted sunfish		+	
Longear sunfish		+	+
Green sunfish		+	+
Bluntnose minnow		+	+
White sucker		+	+
Northern redhorse		+	+
Shorthead redhorse		+	+
Creek chub		+	+
Johnny darter		+	+
Golden shiner			+
Creek chubsucker			+
Grass pickerel			+
Blackstripe topminnow			+
Pirateperch			+
Freshwater drum			+
Gizzard shad			+
Highfin carpsucker			+
Largemouth bass			+
Bigmouth buffalo			+
Carp			+
Black crappie			+
Black bullhead			+
Total species	12	23	24

Trout, sculpins, and sticklebacks become the most conspicuous species. Streams that empty into the ocean

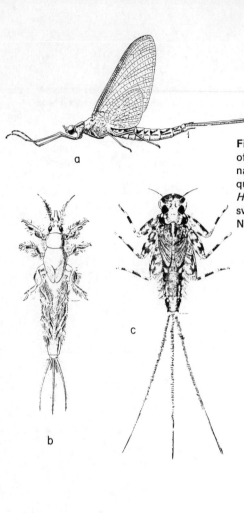

Fig. 4-3 Mayflies: (a) adult of *Hexagenia limbata;* (b) naiad of *H. limbata* from quiet water; (c) naiad of *Heptagenia flavescens* from swift water (courtesy Illinois Natural History Survey).

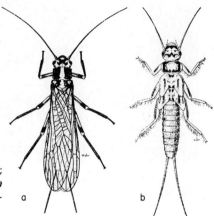

Fig. 4-4 Stonefly: (a) adult; (b) naiad, *Isoperla confusa* (courtesy Illinois Natural History Survey).

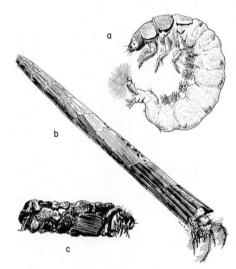

Fig. 4-5 (a) External features of a caddisfly larva; (b) larva and case from a weedy lake; (c) larva and case from a spring-fed brook (courtesy Illinois Natural History Survey).

Fig. 4-6 Immature stages of the black fly: (a) larva; (b) pupa; (c) pupa case (Shelford 1913 after Lugger); (d) enlarged detail of arrangement of hooks on the posterior end of the larva (after Nielson 1950).

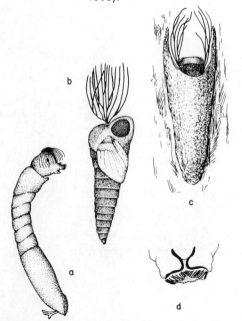

Fig. 4-7 Water pennies, larva of the psephenid beetle: (a) dorsal and (b) ventral views (Shelford 1913); (c) larva of the net-veined midge, showing the central row of six suckers (after Hora 1930).

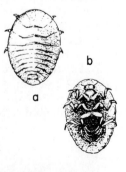

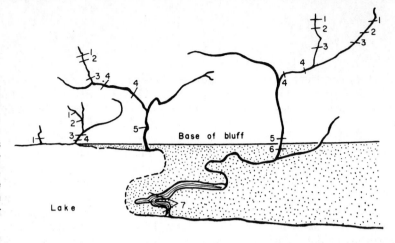

Fig. 4-8 Diagrammatic arrangement of streams of different physiographic age on the south shore of Lake Michigan. Each number shows the location nearest a head-waters of that pool which first contains these fish: (1) creek chub; (2) redbelly dace; (3) blacknose dace; (4) suckers, minnows; (5) grass pickerel, bluntnose minnow; (6) sunfish, bass; (7) northern pike, lake chubsucker, and others (after Shelford 1913).

may have a special fauna of *anadromous* ("upstream") fish, such as salmon, shad, striped bass, and some trout, that spend much of their lives in the sea but migrate into fresh-water streams to spawn, and *catadromous* fish, such as the eel, which migrate "downstream" into the sea to reproduce. There are a few vertebrates other than fish commonly found in streams. Some salamander species occur only in fast mountain streams; other species are more typical of pond-like pools. In North America, the belted kingfisher feeds on stream fishes, and nests in adjacent clay banks. In the western mountains, the water ouzel feeds underwater on the insect larvae and naiads of the riffles. Muskrats make their burrows in the stream banks and feed on vegetation and clams. Mink patrol the streams for the muskrats and fish that serve them as food. The once-abundant otter is now absent from most localities. Beaver dam streams to enlarge the pools in which they build their lodges and find shelter. Beaver feed on the bark and cambium of aspen, willow, and other trees and shrubs occurring on the shores of the stream.

ADJUSTMENTS TO CURRENT

Probably the characteristic of a stream most critical to the life therein is the current. All organisms that occur in streams must adjust to it to maintain constant position. Torrential floods scour the stream bed, move rocks and sand, cut new channels, and destroy entire populations. Recovery after such catastrophes may take place within a few weeks, especially by those species possessing short life-cycles, but sometimes it may require years (Moffett 1936, Surber 1936). Position is ordinarily maintained by clinging to the substratum, avoidance of the current, or vigorous swimming, and requires a good development of orientation behavior.

Clinging Mechanisms

The growth form of fresh-water sponges is affected by a number of factors, but in riffles sponges are usually simple encrustations. In quieter water, long, slender, finger-like processes may form. The distribution of species depends both on current and organic content of the water (Jewell 1935).

Plumatella is a common bryozoan that forms an encrusting, plant-like, branching colony on the underside of rocks or fallen trees in swift water. *Pectinatella*, on the other hand, forms a gelatinous spherical ball, and is more commonly found in ponds or slow-flowing portions of streams.

Turbellarians, such as *Planaria*, and swift-water snails, such as *Goniobasis* and *Pleurocera*, and the limpet *Ferrissia*, cling to the substratum by means of flat, slimy, adherent body or foot surfaces, and are most common on the protected lower surfaces of rocks.

Mayfly naiads have efficient adaptations which enable some of them to tolerate currents up to 300 cm-sec (Dodds and Hisaw 1924). The animals cling to the smooth undersurfaces of the rocks, keeping their heads toward the current and their bodies parallel with it as they move sideways, forward, and back. The head is flattened, and when pressed firmly against the substratum the water current exerts a downward pressure which helps to hold the animal in position. Compared with forms found in quieter water, they show a larger thorax and legs, a smaller abdomen, absence of hair on the caudal cerci, shorter middle cercus, and smaller gill lamellae. These modifications enhance body streamlining and reduce the drag of the water. Furthermore, the legs are articulated in a way which allows the current to press them firmly against the substratum. The body itself swings freely in the current.

Mayfly naiads that occur in quiet waters do not have these modifications. They commonly spend most of their time in burrows, dug into the mud. They come out at night to swim around and search for food. The abodomen of the mud-inhabiting forms is thick, with little taper, sometimes bowed ventrally, and the three terminal cerci are provided with long stiff hairs that overlap and make an excellent oar for swimming.

Stonefly naiads are not limited to stony habitats. Some species occur in the masses of leaves that lodge against rocks or along the banks, in the algae growing on the rocks, on sand bottoms, and in small mud-bottom streams rich in organic matter. The general form of the body is similar to that of swift-water mayfly naiads, although the gills are filamentous and located at the base of the legs.

Caddisfly larvae occur most abundantly in streams with medium to swift currents, but some species occur only in sluggish rivers, in lakes, or in pond vegetation. Caddisfly larvae are of especial interest because of the cases they construct, in which the pupae also occur later. In some species these cases are portable. They are made of pieces of leaves, twigs, sand grains, or stones which are cemented or tied together with silk that the animals secrete. In standing or sluggish water, the cases are often large and made of buoyant plant material, or they may be made of sand grains, more fragile and slender. In swift water, the cases are stout, cylindrical, tapered posteriorly, and are usually smaller and more solidly constructed of sand, small pebbles, or rock fragments (Dodds and Hisaw 1925). The Hydropsychidae, Philopotamidae, and Psycho-myiidae are unique in spinning fixed abodes in the form of a finger, a trumpet, or a tube. The Hydropsychidae erect a net at the front end of the tube to catch particles of food washed down with the current. Some psychomyiid larvae, particularly *Phylocentropus*, burrow into sand and cement the burrow walls into fairly rigid cases. Some larvae belonging to the Rhyacophili-dae are free-living. Found in algal growth, they crawl around seeking food, and are provided with large abdominal hooks as devices to supplement the legs for clinging. However, they form a stone case, or cocoon, for pupation (Ross 1944).

The black fly larvae, Simuliidae, are often very abundant in the swift waters of mountain brooks and northern streams. The larvae secrete from their salivary glands a delicate silken thread by which they attach to the rocky substratum, and by manipulation of which they can move short distances. At the posterior end of the semi-erect body is a circlet of rows of outwardly directed hooks which, when the muscles of the disk are relaxed, move outward and catch on to a silk web placed there previously by the larva; the anterior end of the body then swings freely in the current. There is a fan-like food-gathering organ on each side of the mouth. Before pupation, the larvae spin a sedentary cocoon. The pointed end faces the current and the other end, open, faces downstream. Out of it, the peculiar gills of the pupa float in the water (Hora 1930, Nielsen 1950).

The net-veined midge larvae, Blepharoceridae, are unique in possessing six unpaired suckers on the ventral side, by means of which they fasten to the substratum. The original segmentation of the body is almost obliterated; it has been replaced by a secondary segmentation correspondent with the number of suckers.

Adult riffle beetles (Psephenidae, Dryopidae, Elmidae) are small in size and are the only coleopterans that live in or near running water. The legs are not fitted for swimming, but rather possess hooked claws for clutching the substratum. The body is covered with silken hairs that hold a thin film of air about it when the beetle is submerged. The larvae are disc-shaped and pressed close upon the substratum, to which they cling with their legs and backward-directed spines. They are sometimes called water pennies. When ready to pupate the larvae crawl out of the water.

Avoiding the Current

Diminuitive body and appendage sizes and assumption of a stream-line shape keep the amount of surface exposed to the full impact of the current at a minimum. The conical shape of the limpet *Ferrissia* and the flatter cone of water pennies offer little resistance to water flow. Flat bodies, such as are found in many swift-water animals, appear to be not only an adaptation lowering resistance to current (Madsen 1969) but also to escape it by enabling the animals to seek shelter in crevices and underneath stones (Dodds and Hisaw 1924, Nielsen 1950). Most species, even those with specialized means of clinging to the bottom, are more abundant on the undersides of rocks in riffles than they are on the uppersides. Some species, however, such as the free-living caddisfly larvae, rotifers, tardigrades, water mites, and protozoans, find shelter within the mass of algae that may cover the top of the rocks. The hellgrammite, tabanid fly larvae, and stream crayfishes possess no special structures for withstanding currents and only occur in riffles providing protection or lodgement underneath and between rocks. Even swift-water fishes, strong swimmers, take maximum advantage of whatever protection is available.

The clams avoid the full force of stream current, and at the same time retain position, by lodging their bodies between stones. In pools, they bury themselves in an oblique position in the gravel, sand, or mud. Their posterior ends are directed upstream (according to Max Matteson), and their siphons usually maintain contact with open water so there can be circulation through the mantle cavity, for gaining food and oxygen. Clams occurring in pools $\frac{1}{2}$ to 1 m in depth may remain more or less sedentary, but those occurring in shallower waters move around considerably, especially in response to changes in water level and temperature.

Swimming

Locomotion of swift-water invertebrates is, in the main, restricted to short-distance crawling. Mayfly naiads that occur in riffles, such as *Baetis*, do some swimming for short distances but not nearly as much as related species frequenting quiet waters. Only the more vigorous fishes can maintain position in swift currents by swimming, and many of them do so only when feeding. At other times they congregate in the pools that occur between riffles. Salmon and trout are well known for their ability to swim against strong currents, an accomplishment of sheer force of powerful, muscular, tails. The subfamily of darters, Etheostominae, which contains a variety of brightly colored small fish, are especially adapted to live in the riffles. The air bladder of the darters has become very degenerate, even absent, so that the specific gravity of the body is increased. Their fan-shaped pectoral fins are enlarged, and project at right angles from the lower side of the body. When at rest they maintain position by contact with the bottom, fins lodged between pebbles or the body partly buried. They never float suspended in the water, as do other fish; when disturbed, they dart swiftly from one anchorage to another. The Etheostominae are confined to North America east of the Rocky Mountains.

Stream fishes are in general quite sensitive to current, and the discontinuous distribution of a species within the same stream may be closely correlated with gradient (Trautman 1942, Burton and Odum 1945). The smallmouth bass, for instance, is mostly absent in southern streams of gradient less than 40 cm/km (2 ft/mi); is of moderate abundance in gradients up to 135 cm/km (7 ft/mi); is very abundant in gradients of 135–380 cm/km (7–20 ft/mi); and becomes less common again, until it disappears altogether, in gradients above 475 cm/km (25 ft/mi). Perhaps streams with very slow current do not provide suitable gravel nest-sites for spawning, and in streams with very fast currents they are unable to maintain position.

Salamanders that live in swift mountain streams generally have short limbs and toes, reduced size of fins, smaller lungs in the adult and shorter gills in the larvae, and relatively few large eggs, which they fasten to the underside of flat rocks (Noble 1931).

Structural adaptations for withstanding or avoiding current are of no avail without appropriate behavior responses to make use of them. The rheotactic responses of animals may be tested either in the field or in the laboratory by means of special apparatus (Fig. 4-9).

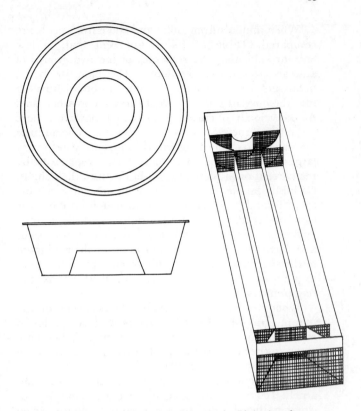

Fig. 4-9 Apparatus for studies of rheotaxis. Right, box for use in streams where water enters at upper end, flows through center trough, and out lower end. Controls may be run in side troughs filled with still water. Left, rheotaxis pan in which current is produced artificially with a rod or finger. An organism's response is positive when it turns to confront the current; negative, when it faces downstreams; indifferent, when it orients crossways.

Table 4-3 Rheotactic responses of invertebrates from riffles and pools (from Shelford 1914).

Velocity of current:	4–6 cm/sec				10–12 cm/sec				16–20 cm/sec			
Response in percentages:	Positive	Indifferent	Negative	Inactive	Positive	Indifferent	Negative	Inactive	Positive	Indifferent	Negative	Inactive
Riffles animals												
Crayfish, *Orconectes virilis*	30	40	28	2	54	8	16	22	78	2	6	14
Snail, *Goniobasis livescens*	45	27	28	0	65	22	0	13	76	7	0	17
Caddisfly larva, *Hydropsyche* sp.	23	26	16	35	18	9	6	67	26	2	4	68
Damselfly naiad, *Argia* sp.	79	0	17	4	63	18	4	15	63	4	0	33
Stonefly naiad, *Perla* sp.	31	24	3	42	65	6	15	14	61	3	3	33
Mayfly naiad, Heptageninae	25	12	14	49	52	3	0	45	52	3	0	45
Water penny, *Psephenus* sp.	26	32	36	6	67	26	0	7	74	15	11	0
Averages	37	23	20	20	55	13	6	26	62	5	3	30
Pool animals												
Damselfly naiad, *Calopteryx maculata*	78	0	22	0	59	8	0	33	63	0	0	37
Snail, *Campeloma subsolidum*	51	32	6	11	80	0	0	20	10	0	0	90
Burrowing dragonfly naiad, *Macromia* sp.	17	36	41	6	12	72	10	6	0	0	0	100
Clam, *Anodontoides ferussacianus*	16	66	18	0	17	67	16	0	0	0	0	100
Fingernail clam, *Spaerium* sp.	17	66	17	0	16	67	17	0	0	0	0	100
Averages	36	40	21	3	37	43	9	12	15	0	0	85

When animals from riffles and those from pools are compared (Table 4-3), it is apparent that, at low current velocities, the responses of the two groups of animals are nearly the same. The elongate body, notably of stream animals, brings an automatic turning into the current much as wind directs a weathervane. As the velocity of current is increased, however, there is a marked increase in the percentage of riffle animals that face into or move against the current, while a very large percentage of pool animals are swept away by the current or are forced to withdraw into their shells. Caddisfly larvae, free of their cases, are not very able to withstand a strong current, although within their case they readily maintain position.

Blackfly larvae can tolerate water currents as swift as 180 cm/sec, and studies indicate that their clinging to the substratum is a response to current rather than to any associated factor, such as food or oxygen requirement (Wu 1931).

When tested experimentally, 80 per cent of the stream crayfish *Orconectes propinquus* were able to maintain position in currents of 50 cm/sec, but only about 20 per cent of the pond crayfish *O. fodiens* were able to do so (Bovbjerg 1952).

Fishes generally respond to current by showing nearly 100 per cent positive response, regardless of whether they be taken from streams or ponds. Since the response involves a tendency to swim upstream, other factors must be involved for the fish to maintain a constant location in the stream; otherwise they would all move to its headwaters.

Some stream fishes, such as the blacknose dace and the common shiner, can be shown experimentally to respond visually to landmarks on stream bank and bottom to maintain their location. Some pool fishes, such as the sunfish and topminnow, likewise respond visually, but much more sluggishly and irregularly. Darters are less unresponsive to visual stimuli, depending on the tactile stimulus of contact with the bottom for maintaining position (Lyon 1905, Clausen 1931). Smell may be important to some fish for orientation. The backswimmer *Notonecta* (Schulz 1931) and whirligig beetle *Dineutus* (Brown and Hatch 1929) have also been shown to use visual orientation in running water.

Stream Drift

In spite of various mechanisms for maintaining position against the current, there is continuous drift of living invertebrates downstream (Needham 1928, Denham 1938). This drift consists of many immature forms, insects in the process of emerging from the water as adults, terrestrial animals that fall into the water, and invertebrates of various sorts entering the stream from lakes and ponds. Very common are mayfly naiads; black fly, midge fly, and caddisfly larvae; and amphipods, but representatives from all types of bottom fauna occur. The total drift over 24 hours may at times be greater than the standing crop in the immediate vicinity.

Drifting in most forms is more pronounced at night than during the daytime. A negative phototaxis keeps the animals close to the substratum while there is light, but at night the organisms loosen their hold, move about, and may be caught by the current (Hughes 1966, Holt and Waters 1967, Bishop 1969, Thomas 1969, Elliott 1969). Drift is also more common during late spring and summer when the animals are active at high temperatures. At times of flood and swift current, more animals become dislodged (Anderson and Lehmkuhl 1968). This is often called "catastrophic drift" in contrast to "behavioral drift," which results from the normal activities of the organisms as above described, and of "constant drift," which is of low intensity and due largely to accident or death (Waters 1966).

If the drift animals are carried into a lake, most of them quickly perish (Dendy 1944). Commonly, however, the organisms are carried only a short distance before they become lodged or attached to the substratum (Waters 1965). This moving into and out of the current may occur repeatedly. By such means, areas of stream bottom made bare by the scouring of floods or by insecticides become recolonized. This period of re-establishment may sometimes take as long as three years, during which there is little drift, but as soon as the community becomes re-established, the rate of drift increases rapidly (Dimond 1967). Although the volume of drift may not always be proportional to the standing crop (Elliott and Minshall 1968), it is related to the productivity of the bottom fauna (Waters 1966) and is a means of regulating population size to the carrying capacity of the habitat. Overcrowding leads to increased jostling between individuals and risk of dislodgement from the substratum.

The continual drift of organisms downstream is compensated for by the tendency of adult insects to lay their eggs in shallow rather than deep water. Thus adult insects emerging downstream commonly fly upstream for reproduction. However, immature stages of several kinds of insects as well as strictly aquatic forms, such as amphipods, turbellarians, and isopods, regularly move upstream underwater. These upstream movements occur more extensively at night than during the daytime and are usually more prevalent close to the bank than in mid-stream. Such underwater upstream movements, however, are always only a small percentage of the drift downstream (Hultin *et al.* 1969, Hughes 1970, Elliott 1971).

RESPONSES TO BOTTOM

The segregation of stream animals between riffles and sand- and mud-bottom pools may be, in part, a response to type of bottom. With no current flowing, the species listed in Table 4-3 were, in another experiment, given a choice between a hard bottom and a sand bottom. Eighty-five per cent of the riffles animals selected the hard bottom, but only 10 per cent of the pool animals did so. Of the pool animals, all species made 100 per cent response to sand, except the damselfly naiad, *Calopteryx maculata*, which divided equally between the two types of bottom. When the riffles animals were given a choice between loose stones and a bare bottom, nearly all individuals selected the stones, and they distributed themselves among the stones or on top or underneath in the manner one would expect of them under natural conditions (Shelford 1914). Stream crayfish, when given a choice between mud and cinders, oriented 88 per cent to the cinders, while the pond crayfish responded 40 per cent to cinders and 60 per cent to mud (Bovbjerg 1952). The fingernail clam, *Sphaerium tranversum*, found in pools in the Mississippi River, when given a choice in a gradient experiment, showed preference first for mud, then sandy mud, and then sand (Gale 1971).

Type of bottom is important to invertebrates for support and locomotion. Sand bottoms are noteworthy as unstable and shifting. Insect larvae and naiads find footing very uncertain; planarians, sponges, and bryozoans find no stable anchorage; and rock-inhabiting snails and limpets are quickly buried. Clams, however, find a sandy bottom suitable, if it is firmly packed, as they are adapted to burrowing and plowing their way through a loose substratum. They are able also to move through a mud bottom, but where silting is heavy they close their valves to avoid an accumulation of silt within the mantle cavity and on the gills. The anodontas seem to be the most tolerant of mud bottoms.

Some of the mayfly naiads, such as *Hexagenia*, are adapted to burrowing in mud, and the surface of the bottom in shallow water is often closely dotted with the openings of their burrows (Eriksen 1968). These burrows are relatively permanent in compact mud but would quickly collapse in loose sand. The genus *Caenis* is peculiar in possessing covers at the anterior end of the abdomen; they protect the gills from becoming clogged with silt. Midge fly larvae and aquatic annelids exist in mud bottoms; they would be ground to bits among moving sand particles. The pond crayfish will burrow into mud down to water level as a pond dries up, but stream crayfish will not do so and consequently suffer high mortality (Bovbjerg 1952).

The bottom is important to invertebrates and vertebrates alike for placement of eggs. Some caddisfly eggs are fastened to smooth rock surfaces in long strings by a cement-like substance. The eggs of other species occur in jelly-like masses and may be secured to plant stems or other submerged objects. Jelly-like masses of snail eggs are often quite common on the undersides of rocks in riffles. Some fish, such as the fantail darter (Lake 1936), make nests in small cavities under stones, but other species, for instance the rainbow darter (Reeves 1907), creek chub (Reighard 1908), and river chub (Reighard 1943), build nests in gravel bottoms in the upper parts of riffles. Some of the suckers (Reighard 1920) spawn in shallow water; their eggs scatter downstream, finding lodgment in various riffles.

RESPIRATION AND OXYGEN REQUIREMENTS

Oxygen is usually ample in streams, often saturating the water in turbulent riffles. The oxygen concentration is sometimes low, however, in sluggish streams and standing pools. The difference in oxygen tension of the two habitats is reflected in the respiratory adaptations of the organisms that inhabit them.

The lamelliform gills of the mayfly naiads inhabiting mud-bottom pools are larger in size than those of species inhabiting streams, are doubled in number on the anterior abdominal segments of some species, and are almost continuously flicked back and forth for better aeration. The gills of naiads living in riffles, or in waters in which the oxygen content is high, may have the surface area of the gills reduced by two-thirds in proportion to body weight, compared with mud-dwelling forms. They are never flicked, since the water movement continually brings oxygen to them (Dodds and Hisaw 1924). Other species do not flick their gills at high oxygen tensions, but will do so when tension is reduced. In some swift-water species, there appears to be sufficient oxygen diffusion through the general body surface to make gills inessential equipment (Wingfield 1939).

Caddisfly larvae have filamentous gills, and there is some evidence that they increase in number as body size increases and oxygen content of the water decreases. It is probable that oxygen also diffuses readily through the thin skin. A constant current of water is maintained through their cases by undulations of the abdomen. Stonefly naiads have poorly developed filamentous gills, located on the thorax, or have none at all. As a result, they are more sensitive to variation in oxygen supply than are the other forms mentioned.

The respiratory equipment of pond-inhabiting animals permits them not only to live in habitats with lower oxygen tensions but also to survive longer at high water temperatures. Often, these animals display relatively low rates of general body metabolism and oxygen requirement. Such relations between riffle and pond animals have been observed for mayfly naiads, caddisfly larvae, isopods, crayfish, and fishes (Allee 1912–13, Wells 1918, Fox et al. 1935, Clausen 1936, Whitney 1939, Bovbjerg 1952), and to some extent for limpet snails (Berg 1951).

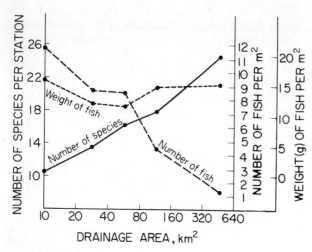

Fig. 4-10 Relationships of weight, number of individuals, and number of species to size of the drainage area (from Larimore and Smith 1963.)

RESPONSES TO STREAM SIZE

Of the species of clams indigenous to Michigan, the 3 commonest are largely limited to creeks, 14 others to medium-sized rivers, and 5 to large rivers (Van der Schalie 1941). In central Illinois, the number of species of fish per collection increases progressively downstream while the number of individuals per collection decreases. However, the average weight of fish increases downstream so that the total biomass per unit area remains about the same (Larimore and Smith 1963). Large species of fish can occur only in streams with sufficient volume of water to permit freedom of movement; small fish may find orientation difficult in large rivers. The preference of fish for streams of specific size is evident in the tendency for some species to travel upstream in times of flood and downstream in times of drought.

An increased number of species downstream correlates with greater variety of available niches and moderate environmental conditions (Fig. 4-10). In many instances the correlation between distribution of species and stream size, or volume, is not direct but dependent on associated changes in temperature, type of bottom, fertility, silting, pollution, and other factors. In the Patuxent River, Maryland, primary productivity per unit volume of water was only one-third as great downstream as upstream, but productivity per surface area was greater. This is correlated with the thicker euphotic layer downstream (Stross and Stottlemeyer 1965).

Headwaters

The headwaters of drainage streams present a highly variable habitat. During dry periods, pools shrink and may disappear; temperature may be very high in summer and the water largely converted to ice in winter; there may be a lack of oxygen, an excess of carbon dioxide, and a high acidity; fishes and other organisms may become greatly overcrowded. In times of heavy rain, on the other hand, the stream is swollen, there is considerable erosion of materials into the stream, and animals are washed downstream. At all times food is likely to be scarce.

Only the hardiest species can exist under these conditions. The creek chub is such a hardy fish; it may be found in large numbers in shrunken pools, stirring up the water with tail action and gaping for air at the water surface. Crayfish burrow into the bottom when the pool dries up. Small snails may survive desiccation of habitat by crawling under rocks or into crevices, secreting a mucous membrane across the aperture of their shells, and remaining dormant until water returns. The occurrence of insect larvae and naiads is hazardous, for if the aquatic stages of their life-cycles are characteristically prolonged, they perish at times of low water or drought.

Temperature and Altitude

In drainage streams the temperature of the headwaters is variable, but as the water volume increases downstream and becomes more constant, the range of temperature variation decreases. The headwaters of spring-fed streams, or of streams arising at high elevations, usually have a progressive increase in temperature downstream.

Some species of stonefly and mayfly naiads and caddisfly larvae are absent from the headwaters of Ontario streams because the temperature never gets high enough to permit them to complete their life-cycle. Downstream, the headwaters species tend to emerge early in the season while the waters are still cold. Still farther downstream, the headwaters species disappear altogether. Species that are limited to the lower portions of the stream emerge late in the season, when the waters are the warmest. Closely related species are thus segregated to different positions in the stream by temperature tolerances. Headwaters species have generally a northerly distribution over the continent and the downstream species a southerly distribution (Ide 1935, Sprules 1947). Linear distribution of fish in streams may be, in part, a result of differences in temperature preferenda. Brook trout, for instance, do best in waters cooler than 19°C in Virginia, while some varieties of introduced rainbow trout prefer waters above 19°C (Burton and Odum 1945). The altitudinal zonation of various species of invertebrates and fish in mountain streams is well defined, and is in large part contingent on differences in temperature (Dodds and Hisaw 1925).

Shape and Size of Individuals

In the Tennessee River, riffles snails of the genus *Io* show a progressive change in shape from the headwaters on downstream. There is a decrease in shell diameter, a decrease in globosity, and an increase in number and length of spines (Adams 1915). However, the riffles snail *Pleurocera* was found to increase in globosity downstream in Michigan (Goodrich 1937). Some pond snails, such as *Lymnaea stagnalis* and *Galba palustris*, develop a larger foot and shell aperture when exposed to wave action (Baker 1919). Primitive types of clams, on the other hand, such as *Fusconaia*, *Amblema*, *Quadrula*, *Pleurobema*, and others, change progressively downstream from a large, compressed, smooth shell to one that is shorter, more obese, and sculptured with tubercles (Ortman 1920). Some species of clams show no such changes in shape. In some fish of central Asia (Nikolski 1933), the body changes downstream from a torpedo-shape to a flatter, longer form. These changes are probably a result of downstream reduction of water current, increase in amount of calcium in the water, and higher temperatures. The formation of spines and tubercles, for instance, would require an abundance of calcium and quiet water.

EVOLUTION

In all probability species inhabiting quiet waters are ancestral to those occurring in running waters (Dodds and Hisaw 1925, Hora 1930). Invasion of stream habitats requires mechanisms for contending with the force of current, and orientation behavior for maintaining position. Convergent evolution has occurred in many kinds of animals under the influence of current, as shown by similarities in structure and habits (Shelford 1914a). Inducements to the invasion of swift waters have doubtless been new sources of food, escape from enemies, and avoidance of competition with the abundant life of lakes and ponds. As adaptations to stream habitats evolved, animals have largely lost their ability to occupy quiet waters. They no longer can tolerate the lower oxygen tension, silt bottoms, and the absence of current which brings them food and oxygen, and, in some forms, such as the Hydropsychidae, helps build their shelters and nests.

LIFE HISTORIES

The life cycle of stream insects is remarkable for the long duration of the immature stage in many species and the brief life of the adult. The naiads of mayflies pass through a number of molts (20–40), and this immature stage may last from 6 weeks to 2 years.

When ready to emerge, the naiad comes to the water surface or crawls out onto a stone, molts into a subimago, and flies away. Within a few minutes, or a period of 1 to 2 days at the longest, the subimago undergoes another molt, unique in insects, into the fully mature adult. The adult insect does not eat and lives only a few hours or days; during this time reproduction takes place. Mating occurs in flight, hundreds or thousands of individuals swarming in flight together. The females lay their eggs almost immediately after mating. In some species, deposition is made upon the water surface, the eggs sinking to the bottom; in other species the female crawls down into the water and attaches the eggs, as they are laid, to a rock surface. The eggs have a viscid surface or filaments and quickly become attached to submerged objects. Embryonic life may last 11 to 23 days, at the end of which time the naiad is fully formed (Needham *et al.* 1935, Elliott 1967).

The life cycle of stoneflies is also 1, 2, or possibly 3 years long in different species, of which time all but a brief interval is spent in the water (Frison 1935). Molting into the adult occurs after the naiad crawls out of the water onto a rock or other projecting object, and there is no subsequent molt in the adult stage. Adult diurnal stoneflies may feed, although the adults of nocturnal species apparently do not. It is of great interest that many species emerge, mate, feed, and carry on all essential activities during the coldest months of the year (Frison 1935). At all seasons, the eggs may be dropped into the water while the female is in flight over the water, or as she alights on its surface. The eggs are mucilaginous and may contain surface filaments or hooks.

Caddisfly larvae pupate submerged in cases. As the pupa approaches the adult form, it leaves the case; and, after crawling and swimming, emerges either upon the water surface or on some protruding object. Larval life in different species may be as short as 25 to 80 days, but since overwintering occurs in this stage it may be greatly prolonged. The pupation period is ordinarily shorter than the larval period, and the adults, which probably feed, may live from several days to a few weeks. Species living in temperate climates have either one or two generations per year. Some females drop their eggs while in flight but others crawl under the water to deposit them. The eggs are laid in masses in either a single-layered, cement-like encrusting form or in a jelly or gelatinous matrix that swells in water. Eggs are sometimes deposited on objects above water. Usually, 10 to 24 days are required for their hatching (Balduf 1939).

The common hellgrammite of North America appears to require 3 years to complete its life cycle, of which it spends 2 years and 11 months as an aquatic larva. When ready to pupate, the larva crawls out of the water and underneath some loose stone or piece of wood. The adults do not eat and live only a few days. The female lays her eggs in masses attached to supports situated near water or to the upper surface of leaves. Upon hatching, the larvae make their way back into the water (Balduf 1939).

In the crayfish *Orconectes propinquus* copulation occurs in cool climates from July to November. Farther southward, copulation is delayed until September, continues until cold winter weather, and is renewed again during March and April. Eggs are laid beginning in late March or early April and are carried around by the female, attached to her pleopods, or swimmerets. The eggs hatch in 4 to 6 weeks, and the young are carried for another week or two before they become free-swimming. The majority of the young become sexually mature at the end of the first growing season in early October (Van Deventer 1937).

The female adult black fly deposits her eggs in a mass or string on a stone or other object at water level during late afternoon, usually with only the tip of the abdomen submerged. If the eggs become exposed to the air they do not hatch; normally, the larvae appear in 4 or 5 days at medium water temperature of 20°–22°C. The larval stage persists 13 to 17 days, the pupal period a little more than 4 days, and the adult stage a little over a week when the adults feed, or only 5 or 6 days when they do not (Wu 1931).

Stream snails attach their eggs in a jelly mass to the sides of stones during late spring and summer, and development leads directly to the adult. Clams of the family Unionidae, however, have a peculiar mode of reproduction. The sexes are separate, and fertilization of the eggs takes place in the suprabranchial chambers of the female. Development continues through several weeks in these marsupial gills, and each egg grows into a minute glochidium. These larvae are later shed into the water, where further development requires that the glochidia become attached to the gills, skin, or fins of fish. The larvae may be parasitic, feeding on nutrients absorbed from the fish; this stage may last from 9 to 24 days. Later, the cyst formed by the fish around the glochidium weakens, and the young animal escapes to take up a free-living existence. Breeding occurs from May to August in *Quadrula* and *Unio*, while in some species (*Anodonta*, *Lampsilis*) breeding does not occur until late in the summer, and the glochidia are retained in the female over winter (Purchon 1968).

The life history of clams is of special significance in showing that dispersal depends, to a large extent, on the movements of the fish to which the glochidia are attached. There is evidence that some species of glochidia cling to particular species of fish only, so that distribution of the two forms in the stream is closely correlated. The fingernail clams Sphaeriidae, on the other hand, are hermaphroditic and lack the glochidial stage. The fingernail clams are annuals; the larger unionid

clams may live 10 to 15 years (Coker *et al.* 1922, Boycott 1936, Matteson 1948).

Some sponges, and perhaps also bryozoans, are perennial, although they may become fragmented as a result of floods or freezing during the winter; they may die during times of low water. Both kinds of animals have vegetative buds, gemmules in sponges and statoblasts in bryozoans, that become free of the parent body. The buds are adapted to withstand unfavorable drought or winter periods, and to germinate and form new colonies when favorable conditions return.

The nesting habits of some stream fishes have already been mentioned. Some of the darters and dace defend their nests, or small territories around their nests, against intruders; other species appear not to do so. Individuals of territorial species do little wandering, and it is possible that a darter may persist through several generations in the same riffles. There is increasing evidence that some larger species of stream and pond fishes have definite home areas, and that the fish population of a small stream with riffle-pool development may be considered as a series of discrete, natural units. This has been demonstrated with tagged individuals for species of bass, sunfish, suckers, and bullheads (Gerking 1953). Homing tendencies, however, are developed to varying degrees, and some species appear to move around in a quite random manner (Thompson 1933).

FOOD COACTIONS

The basic food substances for stream animals are detritus, diatoms, and filamentous algae. Detritus consists of dead fragments of plants; partially decomposed and finely divided and derived mostly from the adjacent land, and a certain amount of dead animal matter. In small streams, nearly the entire energy input is detritus and dissolved organic matter derived from the stream basin (Fisher and Likens 1972). Plankton, either plant or animal, is not normally a common source of food, except in outlets from the lakes and ponds from which it is derived and in the sluggish waters near the mouth of the stream. Larger aquatic plants are not characteristic of swift-flowing streams; they occur in sluggish pools. Filamentous algae, however, may be abundant in riffles, and a rich microflora of diatoms, with scattered protozoans, may furnish a thin slimy film over the surface of rocks. Animals are adapted to these food resources as filter feeders, microflora eaters, or carnivores (Nielsen 1950).

The caddisfly larva *Hydropsyche* is a fine example of a *filter feeder*. This species and related forms construct silken nets at the entrances of their shelters and strain out food particles brought down by the current. The anterior legs of some caddisfly larvae and mayfly naiads are furnished with brushes of hair-like setae which catch and transfer the detritus to the mouth as the animal faces the current. Black fly larvae have a pair of fans at the anterior end of the body. These fans are of long, curved setae. The larva folds them periodically, and the mandibles comb or brush off the detritus that collects. Clams siphon water through the mantle cavity, and detritus material and plankton are carried to the mouth through the activity of the cilia of the mantle, gills, and labial palps. Sponges and bryozoans also take detritus into body cavities for feeding purposes.

Feeding on the microflora and filamentous algae are planaria, snails, and various insects. Some caddisfly larvae have mouthparts specially adapted to scrape the thin film of microflora from the surface of rocks. The maxillae of mayfly naiads serve as a comb or brush with which diatoms are swept up into the mouth.

Carnivorous species may also be partly herbivorous (Table 4-4). Too, there is apt to be seasonal variation in food habits and there are differences of habit between closely related species. Fall and winter stonefly naiads are largely herbivorous, but spring and summer forms comprise genera that are either carnivorous, herbivorous, or omnivorous. Hellgrammites are largely carnivorous, feeding on immature insects. Crayfish are omnivorous; they appear to prefer dead and decaying material. The smaller fish, including the darters, are largely insectivorous, but also consume some plant material. Suckers, carp, and catfish feed on bottom debris as well as small living animals and plants. Young bass and trout are largely dependent on insects for food, but as they grow larger they turn also to young crayfish and small fish. A good proportion of the food of insectivorous fish is derived from the stream drift but attached forms are also consumed. The population density of fishes is ultimately determined, therefore, by the abundance of invertebrates and, when fishes rely on vision for finding their food, also on the turbidity of the water.

The average weight of food in the stomach of fantail darters of all sizes, sampled from October to May in New York State, was found to be 0.01354 g (Daiber 1956). If the average biomass of the living food averages 2.83 g/m² of bottom, then one individual of this species

Table 4-4 Food habits of immature stream insects in Yellowstone National Park, Wyoming (Muttkowski and Smith 1929).

Insect	Number of specimens examined	Food types consumed (%)		
		Animal	Plant	Detritus
Stonefly naiads	80	54	22	24
Mayfly naiads	109	4	30	66
Caddisfly larvae	115	28	54	18
Diptera larvae	20	0	77	23

Table 4-5 Relation between numbers per square meter and biomass of insect groups in a riffles of a California coastal stream during February and March (after Needham 1934).

Insect	Number of individuals	Per cent	Wet weight	
			Grams	Per cent
Caddisfly larvae and pupae	742	22.2	5.66	43.9
Mayfly naiads	1,853	55.5	3.61	28.0
Fly larvae and pupae	343	10.3	1.02	7.9
Stonefly naiads	260	7.8	1.58	12.2
Miscellaneous	137	4.1	1.02	7.9
Totals	3,335		12.89	

could get 209 full meals from 1 m² if it captured everything that was there. Similarly, mottled sculpins could obtain 130 meals from a square meter. It would be interesting to know what actual percentage of the invertebrate population can be readily captured by fish and how frequently the fish feed, for correlation with the density of the fish population. Fish, however, also depend to a considerable extent, especially in summer, on small terrestrial organisms that fall, or are washed, into the stream.

BIOMASS AND PRODUCTIVITY

Of the kinds of animals present in one short coastal stream in California, the caddisfly larvae were found to be not the most populous. But when size was considered, they constituted more bulk than any other invertebrate group (Table 4-5). The invertebrate biomass per unit area of riffles is invariably much greater than in sand-bottom pools, whether biomass be computed in terms of wet weight, dry weight, or volume. However, the abundance of species within the riffles depends on whether the stones are loose or are fastened to the bottom, and on whether or not they are covered with algae, moss, or other vegetation (Percival and Whitehead 1929, Minckley 1963). The biomass of mud-bottom pools may sometimes exceed that of the riffles, especially if it contains the burrowing mayfly naiad *Hexagenia* (Behney 1937, Forbes 1928, Lyman 1943, Needham 1932, O'Connell and Campbell 1953, Pennak and Van Gerpen 1947, Smith and Moyle 1944). In the mud-bottom Silver Springs stream in Florida, the dry weight biomass of plants averaged 809 g/m², herbivores 37 g/m², small carnivores 11 g/m², and large carnivores 1.5 g/m² (Odum 1957a).

Insect populations in streams vary with the season (Table 4-6). Peak populations commonly occur during late spring and again in autumn (Daiber 1956, Lyman 1943, Needham 1934, 1938, Stehr and Branson 1938).

Table 4-6 Seasonal variation in invertebrate populations per square meter in a California coastal stream (Needham 1934).

Month	Number of individuals	Wet weight (g)	Predominant species
February	2,862	7.89	Mayfly naiads
March–April	2,324	9.76	Mayfly naiads
May	18,254	52.94	Blackfly larvae and pupae
August	4,524	19.37	Caddisfly larvae and pupae
November	6,531	23.03	Mayfly naiads

Populations become reduced in summer because of low water; in winter, because of low temperature and ice.

Small streams tend to have greater densities of insect populations per unit area than do large streams. In New York State, streams up to width 2 m have biomasses that average 22.2 g/m² wet weight; from 2 to 4 m, 18.0 g/m²; from 4 to 6 m, 10.1 g/m²; and over 6 m, 7.7 g/m² (Needham 1934). In small streams, the distribution of organisms is nearly uniform from one side to the other, but in large streams there is a decrease in density from the sides toward midstream (Behney 1937). Larger streams actually contain more organisms, however, in spite of lower densities per unit area, because they have a much larger total bottom surface. The reason for this variation in density per unit area is not clear, but it may be that per given population of sexually mature adult insects in the surrounding region, small streams offer less area than large streams, over which the females can spread their egg-laying.

The standing crop of fish in Indiana streams varies from 5.2 to 106 g/m² (46–939 lb/acre) wet weight for minnows, suckers, centrarchids, darters, and bullheads (Gerking 1949) to 2.7–4.2 g/m² (24–37 lb/acre) for rock bass (Scott 1949). The fish crop in warm-water streams is generally higher than in cool trout streams, a relation that also holds for the biomass of invertebrates (Pennak and Van Gerpen 1947). Fish are usually more abundant in relatively deep streams than in shallower ones. Brook trout and three other species in one stream in New York State averaged 10.9 g/m² (97.5 lb/acre), a ratio of 1:2.1 to the invertebrate food supply (Moore *et al.* 1934).

Of a stream, the richness of a fauna and the size of the biomass that develops depend largely on the fertility and chemical composition of the water. Hard-water streams, with an abundance of salts in solution, tend to have a large and more varied fauna than do softwater streams. Calcium salts, in particular, are required by mollusks for building their shells, and by crayfish for the exoskeleton. The salts and organic matter which are basic substances in all aquatic food chains depend directly on the fertility of the soil over

which the water drains. Streams draining areas of fertile soil usually have an abundance of stream organisms; biomasses of both invertebrate organisms and fish in streams occurring in areas of poor soil are low.

The productivity of insects in Algonquin Provincial Park, Ontario, was periodically measured during one summer by collecting, in cages a yard square, all insects as they emerged from the water and transformed into adults. The count varied over different kinds of bottom between June 1 and August 31, 1940, as follows: rubble 6603, gravel 1636, sand 1079, mud 2618 individuals per m². Various mountain streams in different parts of the country have shown an annual productivity of trout taken by fishermen of 2.2 to 3.9 g/m² (20 to 35 lb/acre) wet weight (Surber 1937). Fish productivity in streams, lakes, and ponds is obviously of much interest and importance and has been much studied (Gerking 1967).

PRESERVATION OF THE AQUATIC ENVIRONMENT

The chief problems in applied ecology related to streams and other aquatic bodies are those of maintenance of a water supply for man, flood control, recreation, fish management, and control of pollution. Pollution takes four forms: *silting*, resulting from excessive erosion of the surrounding upland; *industrial*, produced by inorganic chemical wastes from lead and zinc works, tanneries, breweries, paper mills, gas plants, mines, etc. (Parsons 1957); *thermal*, as hot-water effluents from atomic plants; and *organic*, principally municipal sewage and drainage from agricultural land.

Erosion and Silting

Stream erosion becomes considerable when upland vegetation is so reduced that there is little or no retardation of runoff from heavy rains. Continuous erosion throws a heavy load of fine silt into the stream. This is detrimental. It makes the water opaque; reduces or prevents photosynthesis in algae, water moss, and other plant life; handicaps those fish and other animals that depend on sight for finding and capturing food; and clogs the filtering mechanism of various invertebrates. Clams are ordinarily closed less than 50 per cent of the time, but in silted waters they may stay closed up to 95 per cent of the time. Clams secrete mucus to keep the mantle cavity cleansed, but when silting is heavy this may not be sufficient and mortality will result (Ellis 1936). Deposition of silt on rock or sand bottom may bring a considerable change in species composition of animals present. During the last several decades, greatly increased soil erosion in agricultural areas has reduced pan and game fishes in our streams, and rough fish, such as carp, have taken their place.

Chronically muddy streams may often be cleared by reforesting the watershed, and by practicing modern erosion control in cultivated areas. With slower runoff, more rainwater soaks into the ground, and the water table is raised. It is also desirable to maintain vegetation on the immediate stream banks to slow up undercutting. Streambank vegetation is also beneficial for shading the water and keeping it cool enough for such fish as trout. If artificial dams are necessary, they should be small, and located where the drainage begins in the numerous headwaters of the streams. Contour plowing, strip planting, and sod ditches also slow up water movement in hilly areas and should be practiced.

Water Supply, Flood Control, and Recreation

As the human population has expanded during the last century, man has exerted more and more control over stream flow to conserve it for drinking and home use, irrigation of arid lands, industrial and municipal use, navigation, swimming and boating, to produce electric power, and to prevent flood damage. This extensive use commonly involves construction of dams and reservoirs, dredging of river channels, raising of levees along the shores to confine the course of water flow, and manipulation of flow rates. As a result, the riffle habitats are eliminated, water levels are altered, and floodplains are permanently inundated. The stream biota disappears to be replaced by that of lake or pond, mud-flats are created with drawdowns or utilization of water in excess of the inflow, and consequently the environment is drastically changed. Unfortunately, sedimentation and filling of the basin with silt eroded from the upland often limit the life of such reservoirs to 50 to 100 years. There is a limit in height to which a dam can be repeatedly raised and to the number of reservoirs that can be built along the main channel of a river. Alternative solutions are the tapping of underground aquifers for water supply, constructing of multiple dams on headwaters of tributary streams for flood control, and restricting water use to what can be readily supplied. Compounding the difficulties is the rapid depletion of the groundwater reserves. These depend on precipitation for recharging, but man in Europe and America is extracting water in some localities at a rate two to three times what the hydrologic cycle is returning.

In 1968, the U.S. Congress enacted into law a National Wild and Scenic Rivers System so that selected rivers and adjacent shoreline areas possessing special values for scenery, recreation, geology, fish and wildlife, history, and culture may be preserved in a free-flowing condition for the benefit of present and future generations. Several states have set up similar systems to include rivers of local rather than national significance. Important to the selection of rivers to be put into these systems is for them to be free of impoundments, unpolluted, and relatively primitive or undisturbed by man. By this Act (Public Law 90-542), Congress recognizes the need of a declared national policy to complement that of dam and reservoir construction, whereby appropriate rivers may be maintained instead for their scientific, aesthetic, and recreational values.

Fish Management

Basic to fish management in streams is the control of soil erosion and pollution. Pollution is reported to have killed over 15 million fish in 42 states in 1968 (Federal Water Pollution Administration). In clean, clear streams, both invertebrates and fish can attain high populations through normal reproduction. Artificial propagation and release of reared fish into streams to improve fishing is not necessary except where habitats have been depleted of breeding stock or where the fishing pressure is excessive. The artificial raising and releasing of fish of suitable size for quick recapture in sport fishing is expensive but sometimes justified in highly populated areas. In most regions the fish manager is better concerned with improving habitats and letting the fish repopulate them to full carrying capacity on their own accord (Fig. 4-11).

Stream fishes suitable for sport and food are primarily those inhabiting the pools rather than the riffles. The carrying capacity of streams can sometimes be raised by artificially increasing the number of pools without destroying too many of the riffles, the main source of fish food. The formation of pools may often be done inexpensively by making simple log or rock dams or deflectors. Occasionally, it may be desirable to haul in gravel from elsewhere to make spawning beds and to provide artificial log or brush shelters (Needham 1938, Lagler 1952, White and Brynildson 1967).

The composition of the fish fauna of streams has changed since primitive times. Three extensive fish surveys of the streams in east-central Illinois during 1899, 1928-29, and 1959-60 (Larimore and Smith 1963) depict the effects of intensive agricultural, industrial, and urban development. Since 1899, east-central Illinois has been converted from wet prairie to well-drained farmland with scattered human settlements. Drainage, dredging, and elimination of the native vegetation have lowered the water table and increased soil erosion. Canalization has altered water courses and produced uniformity in stream environments. Organic and chemical pollution and increase in suspended silt have become chronic in some streams. Ninety species of fish were recorded in 1959-60, of which 7 were introduced by man. Of the remaining 83 species, 16 showed a decided increase in numbers over

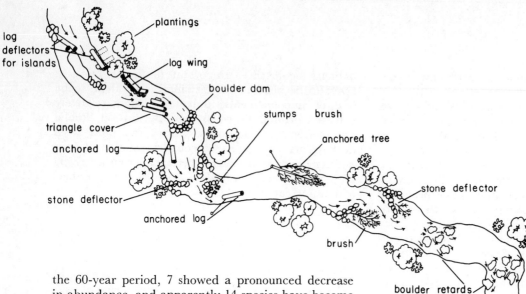

the 60-year period, 7 showed a pronounced decrease in abundance, and apparently 14 species have become extirpated. All other species have maintained nearly normal populations. Changes in fish fauna, such as this, have occurred in many parts of the country, and even more drastic changes may be expected in the future with acceleration in the building of dams and reservoirs and other manipulations in the use of water. Hopefully, with the establishment of Wild and Scenic Rivers Systems, some rivers may be preserved in near-primitive condition and thus remain useful for scientific study, for preventing extinction of aquatic species, and as a historical record of the past. The U.S. National Park Service has also attempted to preserve a certain number of natural streams in their original condition, including the prohibition of fishing therein (Kendeigh 1942a).

Industrial Pollution

Industrial and mine wastes are often acid, and extreme acidity will kill fish and other organisms. Clams are greatly reduced or disappear altogether in acid waters. Industrial wastes contain a great variety of chemical compounds, including salts of the heavy metals, and many of them are very toxic to fish. Young fish and species of small fish appear especially sensitive, and the polluting materials may cause physical or chemical injury to the gills without actually being absorbed into the body (Doudoroff and Katz 1953). Only such invertebrates as the alderfly larvae *Sialis*, midge fly larvae, particularly *Tendipes plumosus*, dytiscid beetles, caddisfly larvae *Ptilostomis*, and some periphytic algae were found to tolerate acid mine wastes in a stream in Ohio (Warner 1971).

Mercury pollution began to be a problem in Sweden and Japan in the mid-1950s and has since become of worldwide concern. Trace amounts of mercury occur naturally in the environment, but the difference between tolerable natural levels and levels harmful to man and animals is very small. Mercury is used in agriculture and paper mills as a fungicide. Seeds so treated have caused mortality in terrestrial birds. Mercury from these sources, from its use in industry,

Fig. 4-11 Schematic diagram of possible stream modifications affording improved protection and spawning facilities to fish (after Lagler 1952).

and from the combustion of some coal and petroleum becomes washed into fresh-water streams and lakes and eventually into coastal waters of the ocean. Metallic or inorganic mercury does some damage to animals, but ordinarily it does not accumulate in the body to serious levels. When acted upon by microorganisms, however, methylation occurs and menthyl mercury may become so highly concentrated, via the food chain, in fresh- and saltwater fishes as to be dangerous to man and fish-eating birds. In fact, mercury poisoning in Sweden first came to notice as a result of drastic declines in bird populations. Banning the use of mercury in fungicides has recently brought improvement in fish and bird populations (Ackefors *et al.* 1970, Hammond 1971, Peakall and Lovett 1972).

The control of radioactive wastes from uranium mills and other atomic energy plants and from fallout has become an especially serious problem in modern times. No stream or other aquatic body can purify itself of these wastes through biological means. However, they become dilated downstream, undergo physical decay, settle out in the mud bottom, and are taken up by organisms. Organisms assimilate elements at equal rates whether the elements are radioactive or not. Some radioactive elements have a half-life extending over several years and may thus accumulate and become concentrated in organisms to an extent many thousands of times greater than their concentration in water. This is of potential harm to man when he consumes these organisms. Mammals are most sensitive to radioactive material, then, in order, seed plants and lower vertebrates, insects, and microorganisms (Odum 1971).

Thermal Pollution

Water is required in atomic reactors for absorbing the waste heat that is generated, and the heated water is then returned to the stream, lake, or ocean. Engineering devices are available to recool the effluent and either recirculate it through the reactor, return it to the stream, or use it for other purposes, but to do this has been considered uneconomic (Smith 1972). A challenge to engineering is to use the heat generated for useful purposes, such as heating homes, in irrigation of crops, in industry, and so on.

Some bacteria, blue-green algae, rotifers, nematodes, and protozoans can survive in waters up to 50° to 60°C, but very few higher organisms can tolerate aquatic temperatures above 35° to 40°C. In fact, most animals come under stress at significantly lower water temperatures (McLean *et al.* 1969). With the rise in water temperature, the solubility of oxygen decreases whereas that of salts increases. Various physicochemical changes occur in the functioning of cellular membranes, hydration of enzymes, and so on (Drost-Hansen 1969). Chemical reactions such as those involved in decay of organic matter and other oxidation processes proceed more rapidly in warm than in cool water. Likewise, metabolism of organisms and hence their demand for oxygen increases in warm weather. The augmentation of metabolism and oxygen requirements of organisms may well exceed the capacity of photosynthesis for producing oxygen. The resulting change in habitat brings a drastic change in the species composition of the biota or, when extreme, its elimination (Clark 1969).

The heat produced by atomic reactors, as well as heat generated by man in industry and modern living generally, eventually reaches the atmosphere. The total heat now being produced by man is estimated at 1 per cent of the total radiated by the earth. This is too small to have a significant effect on the average temperature of the earth's surface. However, man's output of heat is increasing about 7 per cent per year. At this rate, in 108 years the mean global temperature would rise 3°C. This may not seem so great until we realize that there is only about 5°C difference between a glacial period and an ice-free world, and that we are now somewhere between these two extremes. If the icecaps of Antarctica and Greenland and other arctic regions were melted, the sea-level would rise 100 meters. All of Florida would be put underwater, and most of the world's major cities which are located along coastlines would be drowned (Cole 1969).

Organic Pollution

The introduction of small quantities of organic wastes may increase the size and productivity of plant and animal populations by adding to the basic nitrogen and phosphorus supply (Flemer 1970). The limit of the sewage load that a stream can carry without harm is, however, low and soon reached. As fresh organic material oxidizes, carbon dioxide and toxic gases are released into the stream, and there is a drastic reduction in the oxygen content. Fermentation is more

rapid in summer than in winter, and may begin in wastes before they are discharged into the stream. The decomposing organic material continues to be oxidized through bacterial and other action as it is carried downstream, and when this action is completed the stream is again pure (Fig. 4-12).

There have been many attempts by means of chemical analyses of the water to determine the degree to which a stream is polluted and the rapidity at which it purifies itself. There is difficulty, however, in evaluating the extent to which each of the many chemical compounds to be found is harmful to the various kinds of organisms. There is considerable variation in this respect, even between different stages in the life-cycle of the same species. Furthermore, the sewage load may vary from time to time, and infrequent heavy loads may wipe out the animal life in localities where chemical measurements made at other times do not indicate harmful pollution.

Various investigators (Hynes 1960, Tarzwell 1965) have attempted to use invertebrate animals as indicators of pollution. The presence of midge fly larvae *Tendipes riparius, Glyptotendipes,* mosquito larva *Culex pipiens,* rattail maggot, and sludge fly delimit zones of septic pollution. There are relatively few species that can tolerate septic conditions, but those that do may become very abundant. The tubificid worms *Tubifex* and *Limnodrilus,* and certain midge fly larvae, such as *Tendipes plumosus,* indicate low oxygen when they

Fig. 4-12 Diagrammatic presentation of the effects of an organic effluent on a river and the changes as one passes downstream from the outfall: (a) and (b), physical and chemical changes; (c), changes in microorganisms; (d), changes in larger animals (from Hynes 1960).

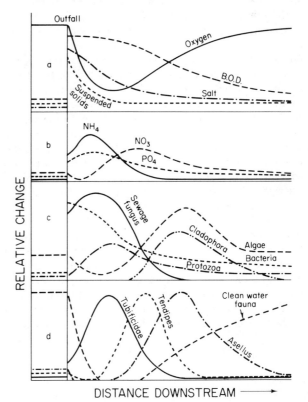

predominate in the fauna. In general, pond invertebrates are much more tolerant of low oxygen concentration than are those belonging to stream habitats. Species especially tolerant of pollution are those that have adaptations for obtaining oxygen at the water surface, such as the dipteran larvae of Culicidae, Syrphidae, and Stratiomyidae, aquatic Coleoptera and Hemiptera, and pulmonate snails. Gill-bearing species generally require clean water of high oxygen content. Among fish, pond species such as carp, bullhead, perch, and crappie are relatively more tolerant than stream species.

The amount of dissolved oxygen required to degrade organic and chemical wastes until the waters are again purified of foreign material is its *biochemical oxygen demand* (BOD). Municipal sewage plants are designed to remove the BOD, but this is done with varying success, depending on the extent and manner in which the wastes are treated. *Primary treatment* commonly starts with removal of solids, grease, and scum by screening and sedimentation. The clarified waste water then goes to *secondary treatment* (if it is used), where microorganisms carry out the assimilation and degradation process that breaks down the organic matter. Often during secondary treatment the water is aerated to supply oxygen to the microorganisms. The solids formed are again removed by sedimentation and the clarified effluent from this treatment is returned to the stream. Primary treatment has an overall removal efficiency for the BOD of about 35 per cent; primary combined with secondary treatment has an efficiency of about 90 per cent. However, this efficiency varies with different elements; for instance, primary and secondary treatment removes only 20 and 50 per cent of the nitrogen and 10 and 30 per cent of the phosphorus (Cooke 1969).

Considerable research is now underway, and practical applications have been installed at some localities, for advanced waste treatment, *tertiary treatment*, which will remove 99 per cent or more of the BOD. It is technically possible, although expensive, to return water to the stream in a purer form than it was received. Such water could be used for recreational purposes or even recirculated through the municipal water system. Tertiary treatment involves such processes as coagulation-sedimentation to remove suspended solids and colloids, carbon adsorption to remove residual dissolved organics, electrodialysis to reduce the dissolved mineral content, and chlorination (Bylinsky 1970).

The Federal Water Quality Act of 1965 requires that states maintain certain standards for interstate waters. These standards include a minimum of 4 mg of dissolved oxygen per liter and specific limits in hydrogen-ion concentration (pH), temperature, dissolved solids, toxic substances, and bacterial count.

To meet acceptable minimal limits, low-flow augmentation has been practiced in some places. Water is stored in reservoirs during rains and is released at rates sufficient to dilute the industrial and municipal sewage effluents to attain the standard at times when the natural stream flow is reduced. However, this procedure is not legally permitted "as a substitute for adequate treatment or other methods of controlling waste at the source." As techniques for advanced waste or tertiary treatment become perfected and economically practical, water quality control by low flow augmentation may disappear.

A big expense in waste treatment is the concentration and disposal of the sludge resulting from the treatments. It is often used for landfills, incinerated, or dumped into the sea. Efforts are now being made to use it as fertilizer on agricultural land. Pilot experiments of piping the liquid effluent with its suspended solids to irrigation rigs for spraying over the land appear to be giving promising results (Sheaffer 1970).

A most difficult problem in maintaining water quality in streams running through intensive agricultural areas is the control of drainage of nitrogenous and phosphoric compounds and pesticides (Taylor 1967). These materials, derived from heavy use of commercial fertilizers, livestock feedlots, and other farming practices, may be transported both by surface runoff or underground seepage. Farm animals in the United States Midwest alone provide excrement equivalent to that from a population of 150 million people (Hasler 1965), but doubtless some of this material is widely dispersed over the land by grazing animals and presents no special problem. Because of the diffuse distribution of the effluents in rural areas, treatment similar to that given municipal and industrial wastes is generally impractical. Nitrogen is especially important since it is readily soluble and transported by water drainage, is often the critical element releasing excessive growth of aquatic algae and other plant growth, and when in excess in drinking water constitutes a hazard to the health of babies (Commoner 1971). Phosphorus and pesticides are less soluble and become adsorbed instead on fine soil particles. Erosion and water control over watersheds would reduce the amount of these chemicals, as well as silt, reaching streams and other bodies of water. In evaluating the twelvefold increase of artificial fertilization of farmland in mid-America since 1945, recognition needs to be made that this application compensates to some degree for the loss of fertility that the soils have experienced since the beginning of intensive farming in the last century. Increasing the yield of crops per unit area presents the possibility of using only the best-situated areas for food production and taking marginal lands and lands subject to high erosion out of cultivation, thereby decreasing the loss of nutrients (Aldrich 1972). The virgin prairie, from which our most productive farmland is derived, was certainly very fertile. The high nutrient content was held in the soils by the dense cover of vegetation, so the loss through runoff from rains was very low. The loss of nutrients at the present time is often excessive because cultivated lands lack this protective cover for much of the year.

SUMMARY

Streams contain riffles and sand- and mud-bottom pools. Inhabitants of the riffles and sand-bottom pools constitute a distinct stream biocies. Mud-bottom pools are inhabited by species from the pond-marsh biocies. Animals adjust to the action of water current by clinging mechanisms, avoidance, or vigorous swimming. They are generally positively rheotactic, and several forms maintain orientation to a particular position in the stream by means of visual landmarks. In spite of these mechanisms, organisms continually lose their footholds and are carried by the current as stream drift. Segregation of species to different habitats depends largely on differential response to the substratum; that is, preference for rock, sand, or mud. Animals occurring in mud-bottom pools are usually negatively rheotactic, or become helpless in strong current. They also have adaptations tolerant of lower oxygen concentrations in the water. Changes in the size of the stream, occasioned by various physical factors, also affect the responses of animals. Stream animals have apparently evolved from ancestral types that occupied the quiet waters of lakes and ponds. The life-cycles of many stream insects are remarkable for the long duration of immature stages and the brief life of adults. Animals have various adaptations to feed on detritus in the water, on diatoms, on filamentous algae, or for being carnivorous. Density of individuals, biomass, and productivity of invertebrates are ordinarily less in sand-bottom pools than in either riffles or mud-bottom pools. Clams and game fish, however, inhabit sand-bottom pools.

Preservation of the stream environment requires control of water use by man, evaluation of flood control, proper forms of recreation, and avoidance of pollution. Maintenance or improvement of fish productivity may require management. Pollution comes from silt erosion in the surrounding uplands, inorganic chemical wastes from industry, heated water from atomic plants, and organic sewage from municipalities and agricultural land. Maintenance of pure water in streams requires control of erosion and of industrial, thermal, and organic pollution at its source.

Chapter 5

LAKES
AND EUTROPHICATION

Lakes are large bodies of fresh water, usually deep enough to have a pronounced thermal stratification for part of the year. Typically, shores are barren and wave-swept (Muttkowski 1918).

Lakes are formed in youthful stages of river system development. Water from upland runoff, groundwater seepage, springs, and melting snow-fields and glaciers collects in basins. As the basins overflow, erosion of outlets continues and the water level of the lake drops. Products of erosion from the surrounding upland are carried into the basin by wind and water, and along with the products of animal and plant decay fill in the basin in the course of time.

Morphometry aside, the essential distinction between lake and stream habits is the characteristic of water movement; continuous, rapid flow is characteristic of the stream, the *lotic* habitat. The lake is a *lentic* habitat; the water is essentially a standing, quiescent body, although at times wind action stirs surface layer and margins into great turbulence. Lentic habitats and biota have received considerable study to date (Welch 1948, Hutchinson 1957, 1967, Macan 1963, Frey 1963).

HABITAT

The aquatic habitat offers a variety of unique conditions to which organisms must be adapted before they can occur. We must understand these conditions in order to interpret how they are distributed.

Pressure, Density, and Buoyancy

The pressure imposed on a lake-dwelling organism is the weight of the column of water above it plus the weight of the atmosphere. Most lakes have a maximum depth of less than 30 m; the Great Lakes of North America vary from 64 to 393 m in depth, Crater Lake in Oregon is the deepest on the continent, 608 m (Welch 1952). Maximum pressures are much less than in the ocean, and organisms appear to adjust to them readily. The absence of animal life from deep water is ordinarily a consequence of low oxygen supply, or low temperature, rather than pressure.

The density of water varies inversely with temperature and directly with the concentration of dissolved substances. Water is most dense at approximately 4°C and becomes progressively less dense as it is cooled below +4°C. Ice also expands markedly (that is, becomes less dense) the colder it gets. It is because the coldest water is at the surface in winter that ice forms there, rather than at the bottom. In summer, the coldest waters of deep lakes are at the bottom. Dissolved salts increase the density of water; the density of most inland water-bodies is much less than that of the ocean.

When great evaporation occurs in a lake having no outlet, as in the Great Basin, the lake may come to contain a higher percentage of salts than the ocean. The few species capable of living in these very salty lakes include some algae and Protozoa, the brine shrimp *Artemia gracilis*, and the immature stages of two brine flies, *Ephydra gracilis* and *E. hians*. There are no fish in the Great Salt Lake of Utah (Woodbury 1936).

By the law of Archimedes, the buoyancy of an object is equal to the weight of the water it displaces. Buoyancy varies with the density of water, and is influenced by the factors that affect density. Viscosity, the measure of the internal friction of water, varies inversely with temperature and also influences buoyancy.

An organism will sink unless it keeps station by swimming movements, or unless it has special adaptations to decrease the specific gravity of the body and take advantage of any turbulence in the water. Such adaptations take several forms: absorption of large amounts of water to form jelly-like tissues; storage of gas or air bubbles within the body; formation of light-weight fat deposits within the body, or oil droplets within the cell; increase of surface area in proportion to body mass, which increases frictional resistance (Davis 1955). When an organism so equipped dies, the special mechanisms quickly cease to function, and it sinks to the bottom. If dead organisms did not sink to the bottom, living organisms, with the exception of some bacteria, could not exist in an aquatic habitat.

An interesting phenomenon is *cyclomorphosis*, a seasonal change in body form that develops in many plankton organisms, both plant and animal, including protozoans, cladocerans, and rotifers. In general, the summer generations have higher crests, longer spines, longer beaks, or longer stalks than do the winter generations (Fig. 5-1). It is believed that the increased surface area provided in the summer forms is induced by the higher water temperatures obtaining then, and may be an adaptation to the decreased buoyancy of the water at this season, although this is uncertain (Brooks 1946).

Light

The daily alternation of light and darkness establishes a rhythm in the activities of many aquatic organisms. Light is essential to plant photosynthesis; some fish require light by which to feed. Many organisms orient to light, and some are sensitive to light of particular wavelengths, notably ultraviolet. Small, soft-bodied, bottom-dwelling organisms are particularly sensitive to light, and it is thought that the evolution of pigmentation, chitinous exoskeletons, shells, cases, and similar structures may have helped certain otherwise photosensitive species to survive in shallow, well-lighted areas (Welch 1952).

A common way to measure the relative transparency of water is to lower a *Secchi disc*, a white plate 20 cm in diameter attached to a cord marked off in linear units, marking the depth at which the disc disappears from sight. The disc is lowered a bit farther, then raised until it reappears, and that depth marked. The two depths are averaged. The light intensity at the depth of disappearance of the disc is usually about 5 per cent of that at the surface (Hutchinson 1957). Other more exact procedures employ photographic methods, pyrlimnometers, or photoelectric cells (Shelford 1929).

The depth to which light penetrates into water is affected by intensity of the light, angle of ray incidence, reflection at the surface, scattering within the water, and absorption. Penetration anywhere is reduced when the sun is away from the zenith, is less in waters at high latitudes, and is much less in winter compared with summer. About 10 per cent of the light falling on Lake Mendota, Wisconsin, during the spring and summer is reflected; about 15 per cent during the autumn (Juday 1940). In the unusually clear waters of Crystal Lake, Wisconsin, measurements with a pyrlimnometer indicated only a small surface-reflection light loss, a penetration of 67 per cent of full intensity to a depth of 1 m, and 10.5 per cent of full intensity at 10 m (Birge and Juday 1929). In pure water, red light is absorbed most rapidly, at a rate of 64.5 per cent per meter; orange, at 23.5 per cent per meter; yellow, at 3.9 per cent; green, at 1.1 per cent; blue at only 0.52 per cent. Blue penetrates the farthest. Violet is absorbed at 1.63 per cent per meter. Very little ultraviolet penetrates the water, and nearly all the infrared is absorbed in the first meter (Clarke 1939, Ruttner 1953).

Suspended material in water produces *turbidity*, and reduces light penetration. In western Lake Erie, the depth to which 1.0 per cent of surface light penetrates varies from 9.7 m, when turbidity is 5 ppm, to

Fig. 5-1 Cyclomorphosis of *Daphnia retrocurva* in a Connecticut lake (from Brooks 1946).

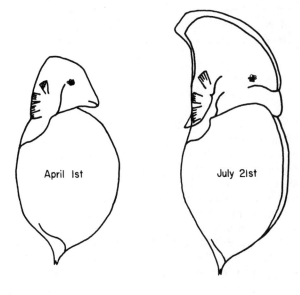

0.8 m, when turbidity is 115 ppm (Chandler 1942). Since phytoplankton requires light for photosynthesis, abundance varies inversely with turbidity. Light penetration is also affected by the abundance of organisms themselves, both phyto- and zooplankton.

An appreciable amount of light passes through ice in the winter. This enables phytoplankton photosynthesis to continue. In eutrophic lakes many fish may suffocate when snow overlies surface ice, preventing photosynthesis and, thus, the generation of oxygen (Greenbank 1945).

The *apparent color* of water bodies may be the result variously of sky reflections, the color of the bottom, suspended materials, or plants and animals. But apart from these extraneous factors, water often has an intrinsic color deriving from its chemical contents. The blue color of pure water is a result of blue-light scattering by water molecules. Iron gives water a yellow hue. A green color is usually associated with high concentrations of calcium carbonate. Water from bogs or swamps contains humic materials and is often dark brown. Many waters are essentially colorless. In a Wisconsin lake showing practically no color, maximum photosynthesis of algae occurred at a depth of 1 m on bright days: some photosynthesis occurred down as far as 15 m. In a highly colored lake, maximum photosynthesis occurred at 0.25 m, none at 2 m (Schomer 1934).

Photosynthesis releases oxygen into the water; respiration and decomposition absorb it. The upper layer of a lake, where photosynthesis predominates, is called the *trophogenic zone*. Below this zone there may still be considerable photosynthesis, but oxygen absorption is greater than oxygen release. The deeper portion of a lake is called the *tropholytic zone*. The two zones are separated by a thin layer where the oxygen gains from photosynthesis during the daylight hours are balanced by the respiratory and decomposition losses during the day and night. This is the *compensation depth*, to which generally about 1 per cent of the full sunlight at the water's surface penetrates. The compensation level in a dark-colored bog may lie less than 1 m below the surface; in a deep, clear lake it may be 100 m down.

Wind and Currents

Wind is an important environmental factor of lakes because of the water currents it generates. The effect of wind action depends largely on the extent of the exposed water surface, the presence or absence of protecting upland, and the configuration of the lake relative to the prevailing wind direction.

Waves may become sizable in large lakes, but the forward motion of a wave does not involve any great mass of water. The rate of movement of surface water is usually less than 5 per cent of the velocity of the wind. The wave form moves on while the water beneath

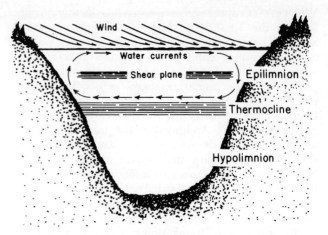

Fig. 5-2 Water currents and thermal stratification in a deep lake.

undergoes a more nearly cycloidal motion, except along the shore, where the wave mass progresses forward and breaks as surf. The water washes back off the beach as an undertow, only to be carried forward again by the incoming waves. The problem of maintaining position here is similar to the problem of maintaining position in streams. The depth of wave action in the open lake and along the shore depends largely on the strength of the wind (Ruttner 1953).

In summer, surface water is warmed by solar radiation and its density, weight, and viscosity decrease. In deep lakes the warm water piles up on the exposed shore until, moving down along the bottom, it encounters colder and denser waters, which resist mixing. The warm water is then diverted horizontally to the opposite shore. Thus the lake becomes stratified horizontally into an upper *epilimnion*, where the water circulates and is fairly turbulent, and a lower *hypolimnion*, which is relatively undisturbed. This difference in circulation in deep lakes is closely correlated with differences in temperature and oxygen characteristics; it is of considerable importance in the distribution of the biota (Fig. 5-2).

Temperature

The thermal conductivity of water is very low; but because of the thorough mixing of the waters in the epilimnion during the summer by wind action, the temperature is nearly uniform down to the thermocline. The *thermocline* is the zone of most rapid temperature decrease, generally involving a drop of at least 1°C per meter of depth (Birge 1904) and occasionally as much as 7°C/m. The thermocline, as here defined, equals the *metalimnion* of some authorities, who have a different conception of the thermocline (Hutchinson 1957). When the thermocline forms, early in the season, it is close to the surface (Fig. 5-3). As the season progresses, it sinks lower, increasing the volume of the epilimnion and decreasing the volume of the hypolimnion. The temperature of the hypolimnion is fairly uniform, although it declines gradually from the lower edge of the thermocline to the bottom, where it is seldom below 4°C.

During the autumn, the surface water cools and the thermocline sinks. The epilimnion increases in thickness until it includes the entire lake. The waters are then uniform in temperature and density, at all depths. Even slight winds produce complete circulation. This is the *autumn overturn*, which may last for several weeks, or until ice forms.

As surface waters cool below 4°C they no longer sink, and ice may form. Less dense than the underlying water, the ice floats. Immediately below the ice, the temperature of the water is very close to 0°C, but in

1 or 2 m of additional depth it usually rises rapidly to 4°C, although in some lakes temperatures below 4°C occur even at considerable depths.

As the ice melts during the spring and the surface waters warm up, a *spring overturn* occurs when the water at all depths is at the same temperature. The time and duration of the spring overturn depends on weather conditions; it may last several weeks. It often occurs intermittently, however, corresponding with changes in weather and water temperatures.

When a lake has two overturns during the year, it is called a *dimictic lake*. Such lakes are characteristic of, but not limited to, temperate climates. In warm, oceanic climates and in the tropics, the surface waters may not cool sufficiently to permit complete circulation, except during the coldest period of winter. Lakes under-

Fig. 5-3 Vertical temperature distribution throughout a year in a dimictic lake of the second order—Convict Lake, California; elevation 2308 m (from Reimers and Combs 1956).

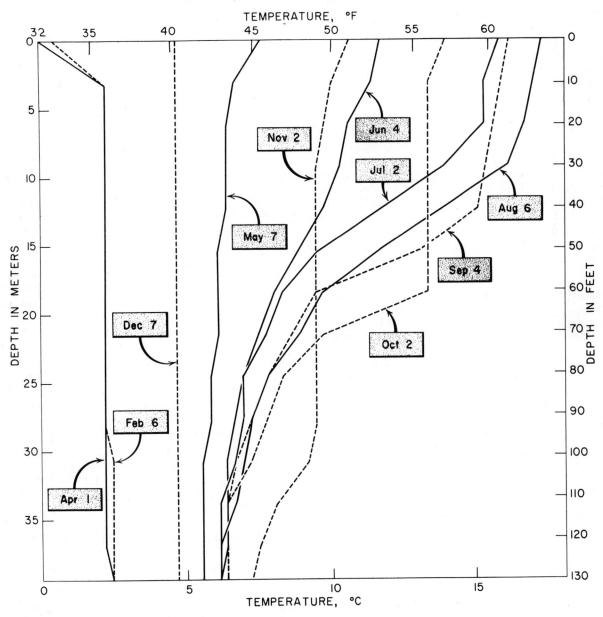

going a single overturn are called *warm monomictic*. The temperature of the water in the hypolimnion of such a lake is never lower, of course, than the mean air temperature during the period of the last complete circulation; in warm climates this may be several degrees above 4°C. On the other hand, lakes in polar or alpine regions may never warm above 4°C, and complete circulation occurs only in the middle of the summer. These are *cold monomictic* lakes (Hutchinson 1957). The three types of lakes were formerly called temperate, tropical, and polar, but this terminology is undesirable since their geographical segregation is not precise.

Lakes of the first order are those in which the bottom water remains at or near 4°C throughout the year, and while one or two circulation periods are possible, there is often none. In *lakes of the second order*, the temperature of the bottom water rises above 4°C during the summer, and there are one or two regular circulation periods during the year. *Lakes of the third order* do not develop extensive thermal stratification, and circulation of water is more or less continuous (Whipple 1927). These are essentially large ponds. In general, lakes over 90 m in depth belong to the first order; those between about 8 and 90 m, to the second order; and those less than 8 m, to the third order.

The specific heat of water is greater than most other substances; accordingly, a vast amount of heat must be absorbed to cause a temperature change. Temperature change is, in any event, slow. Much of the energy of solar radiation is lost by reflection from the water surface. The rest of the radiation is absorbed by the water, the solutes, and the suspended material. But much of the diurnal energy increment may be dissipated by re-radiation at night, by evaporation, and by convectional cooling. The amount of heat actually retained by a lake to melt its winter ice and warm it from the winter

minimum up to the summer maximum is its *annual heat budget* (Table 5-1). For many dimictic lakes this is between 20,000 and 40,000 g-cal/m² of surface; there is wide variation in different kinds of lakes. The annual heat budget is important in determining a lake's productivity.

Oxygen

The distribution of oxygen at various depths depends upon the presence or absence of a thermocline, the amount of vegetation, and the organic nature of the bottom (Fig. 5-4). The amount of oxygen in water is only one-fortieth to one-twentieth of that present in an equal volume of air when the two are at equilibrium, although their partial pressures are the same. Diffusion of oxygen from the air into comparatively sedentary water occurs very slowly; agitation of the water increases the surface area and promotes a faster rate of equilibration.

The amount of oxygen released by plants varies with their abundance and time of day; photosynthesis takes place only in light. Phytoplankton and the rooted vegetation restricted near the shore are important sources of oxygen to the water. With rapid photosynthesis in relatively small volumes of water, the water may be supersaturated with oxygen for short periods of time.

The oxygen supply of lakes is reduced in various ways, most notably through the respiration of animals and plants and the decomposition of organic matter. As lake waters warm up during the summer, their capacity to hold oxygen is reduced and oxygen may be released into the atmosphere. The saturation capacity of water at 0°C is 10.2 cc/liter, but at 25°C it is only 5.8 cc/liter. In some lakes, decomposition of organic material at the bottom may deplete the hypolimnion of its oxygen content for several weeks during the summer, lower than the level minimal to the support of many forms of life. This is called the *summer stagnation period*. During the winter, if the lake is covered with ice and snow, there may be a *winter stagnation period*. The oxygen supply of the deep waters is renewed with the autumn and spring overturns. Before decomposition can proceed very far there must be calcium in the water. Hence, decomposition is slow in soft or acid waters.

At temperatures of 15°–26°C oxygen concentrations of less than 2.4 cc/liter (3.5 ppm) are fatal within 24 hours to several species of fish. From 0°–4°C, oxygen concentrations can decline through 48 hours to 1.4 cc/liter (2.0 ppm), or even to 0.7 cc/liter (1.0 ppm), before the same mortality results (Moore 1942). This is related to the reduced metabolic requirements of cold-blooded organisms at low temperatures. Some planktonic invertebrates can tolerate oxygen concentrations as low as 0.2 cc/liter (0.3 ppm) and, for short periods,

Table 5-1 Monthly change in cumulative heat budget and solar radiation in the Bass Islands region (depth 7.5 m) of western Lake Erie, in 1941. A 20.3-cm ice covering formed in mid-January melted in late March. The maximum heat budget, reached on July 30, was 19,575 g-cal/cm². The heat budget was about 15 per cent of the total solar radiation received during the year (after Chandler 1944).

Month	g-cal/cm² Heat budget	g-cal/cm² Solar radiation	Month	g-cal/cm² Heat budget	g-cal/cm² Solar radiation
January	105	3,364	July	18,765	17,291
February	206	5,849	August	18,112	16,375
March	581	10,201	September	16,331	12,737
April	6,405	13,952	October	11,115	6,829
May	12,581	17,156	November	4,350	4,147
June	16,369	15,960	December	1,369	2,599

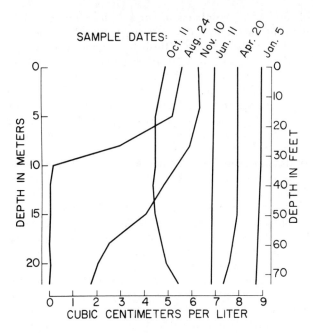

SAMPLE DATES: Oct. 11 Aug. 24 Nov. 10 Jun. 11 Apr. 20 Jan. 5

Fig. 5-4 Changes in the vertical distribution of oxygen throughout a year in a dimictic eutrophic lake—Lake Mendota, Wisconsin (after Birge and Juday 1911).

even 0.1 cc/liter (0.1 ppm). Some bottom-dwelling protozoans, annelids, mollusks, and insect larvae may survive actual anaerobic conditions for periods of days, even weeks. Organisms that tolerate a lack of oxygen do so by creating an *oxygen debt;* that is, the lactic acid and other breakdown products produced in consequence of muscular activity simply accumulate until conditions permit oxidation of them. In true anaerobes these acid waste products are eliminated from the body; no oxidation debt is established.

European workers, principally Thienemann and Naumann, have devised a classification of lake habitats into three main categories on the basis of fertility and the amount of oxygen in the hypolimnion during the summer. The oxygen concentration in the hypolimnion is, of course, a reflection of the fertility of the lake, since it is inversely proportional to the amount of decaying organic matter. Dystrophic lakes contain considerable organic matter but are infertile because the organic matter does not completely decompose and there is release of organic acids.

Oligotrophic lakes are usually deep (over 18 m) with very little shallow water, and little vegetation around margins. Bottom contours are V-shaped; they are low in fertility, rich in oxygen in the hypolimnion (*orthograde* distribution), low in CO_2, and the color of the water varies from blue to green. The volume of the epilimnion is usually less than the volume of the hypolimnion. The fish population is not large. Characteristic species are lake trout, whitefish, and cisco. The midge fly larva, *Tanytarsus*, predominates. Plankton is not abundant. The Finger Lakes of New York are of this type.

Eutrophic lakes are usually less than 18 m deep, the bottom contour is U-shaped, water color varies from green to yellow or brownish green, and there are larger areas of shallow waters and more marsh vegetation. Fertility is high, and because of rich bottom humus the oxygen content of the hypolimnion is greatly reduced during the summer (*clinograde* distribution). The CO_2 content is accordingly high. The volume of the epilimnion is usually greater than that of the hypolimnion. Plankton is abundant and consists of different species from those in oligotrophic lakes (Hutchinson 1967). The midge fly larva *Tendipes* is very numerous and the culicid larvae *Chaoborus* is usually present. The bottom fauna is rich, and there is a large fish population in the epilimnion. Characteristic fish species are the largemouth bass, perch, sunfish, and pike. These lakes occur in relatively mature river systems; many lakes in Minnesota and Wisconsin are of this type.

Dystrophic lakes are bog-like, very rich in marginal vegetation and organic content. Oxygen is likely to be scarce at all depths. The water is usually conspicuously colored, yellow to brown, and may be acidic because of organic acids and incompletely oxidized decomposition products. Plankton, bottom organisms, and fish are usually scarce, but blue-green algae are sometimes abundant. *Tendipes* may predominate among the bottom forms, but at times only *Chaoborus* is present. Characteristic fish are sticklebacks and mud minnows. Many lakes of northern latitudes are dystrophic in type.

All gradations exist between these three types of lakes, and individual lakes are often difficult to classify. *Oligotrophy* is indicated if the loss of oxygen in the hypolimnion during the summer is not over 0.025 $mg/cm_2/$ day; *eutrophy*, if it is over 0.055; *mesotrophy*, if it is between the two (Hutchinson 1957). A lake may change from one type to another as succession proceeds (Lindeman 1942). Probably all lakes start as oligotrophic, but as they accumulate vegetation and decaying organic matter, they change into eutrophic lakes; or, if the organic matter does not completely decompose, into dystrophic lakes. Eutrophic lakes may later develop into ponds and marshes; dystrophic lakes, into bogs. These changes take a very long time, usually centuries.

Carbon Dioxide and Other Gases

Carbon dioxide is required by plants for photosynthesis. Its presence in lake waters tends to vary inversely with oxygen. Carbon dioxide is derived from the atmosphere, the respiration of both animals and plants, decaying organic matter, groundwater, and bicarbonate salts. It may occur in the *free* state (dissolved CO_2), *half-bound* state (HCO_3), or *fixed* state (CO_3). These three states are associated respectively with pH values 7, 7 to 10, and above 10. Algae and some rooted aquatic vegetation are able to obtain the half-bound CO_2 from the soluble bicarbonate salts, thereby converting them into the less soluble carbonates:

$$Ca(HCO_3)_2 \rightleftharpoons CaCO_3 + CO_2 + H_2O$$

Mollusks, a few insects, and some bacteria are also able to precipitate carbonates. Carbonates may accumulate as to make conspicuous marl deposits on the bottom of some lakes. When marl formation becomes considerable, there is a decrease in lake fertility and a consequent decrease in animal life present, including bottom-inhabiting organisms.

When there is sufficient free carbon dioxide in the water derived from other sources, carbonates are converted back into bicarbonates and marl does not form. The degree of *alkalinity* of a lake is measured by the amount of carbon dioxide or acid required to convert the excess carbonates into bicarbonates, yielding neutral water. *Soft-water lakes* contain not over 5 cc/liter of fixed carbon dioxide; *medium-class lakes* contain 5 to 22 cc/liter; *hard-water lakes* may have from 22 to as high as 50 cc/liter (Birge and Juday 1911).

Marsh gas (methane) evolves from organic matter decomposing at the bottom. It rises in bubbles to the surface of the water. Methane formation may be extensive during the summer stagnation period. Methane does not appear to be particularly toxic to organisms until it is generated in very large amounts.

Hydrogen sulfide results from anaerobic decomposition of sulfurous organic matter. It may be conspicuous in sewage-polluted waters. It is inherently very poisonous.

Nitrogen occurs in water by reason of diffusion from the atmosphere. When present in excessive amount it has been known to form bubbles in the circulatory systems of fish causing death, but this does not commonly occur in natural waters.

Ammonia may occur naturally in water, a result of decomposition of organic matter. Ammonia may also be dumped into streams and lakes from industrial plants, often in concentrations toxic to fish. Fish are apparently unable to detect the presence of ammonia in water.

Dissolved Solids

Falling rain may contain as much as 30 to 40 ppm of solids, and the runoff dissolves more as it drains over the upland into streams and lakes. Water draining off siliceous or sandy soils may contain 50 to 80 ppm of dissolved minerals; off more fertile calcareous soils, 300 to 660 ppm. Lake waters commonly vary from about 15 to 350 ppm of dissoved minerals, although in some lakes of the Great Basin, the total dissolved salts exceed 100,000 ppm. The ocean contains only 33,000 to 37,370 ppm.

Inorganic salts especially important for plants include ammonium salts, nitrites, and nitrates as sources of nitrogen; phosphates to supply phosphorus, which, with nitrogen and sulfur, are raw materials for protein synthesis; silicates, which furnish silicon to diatoms and sponges; and salts of calcium, magnesium, manganese, iron, copper, sodium, and potassium for proper development of chlorophyll and growth of plants and, indirectly, of animals. Mollusks require calcium salts for shells. Crayfish and other arthropods require calcium for the carapace; vertebrates, for their skeleton. Absence of these necessary salts in lake waters limits the kinds and abundance of animals that can live there. Phosphorus and nitrogen are the most likely to be deficient. Nutrient salts tend to accumulate in the deeper waters and at the lake bottom, but they are brought to the surface at the autumn and spring overturns. Lakes in prairie regions tend to have more salts than those in deciduous or hardwood forests, which, in turn, have more salts than lakes in coniferous forest areas (Moyle 1956). The total dissolved content of a lake is important in determining its general level of productivity (Northcote and Larkin 1956). Standard methods for the chemical analysis of waters have been given by the American Public Health Association (1955).

Information is accumulating on the amount of amino acids, fats, and carbohydrates occurring in natural bodies of water and how this nutrient material may be absorbed by organisms (Fontaine and Chia 1968). Dissolved organic matter is derived chiefly from plankton remains, other dead plants and animals, feces, as well as from bottom mud and external sources. Some living organisms may excrete amino acids directly into the water (Johannes and Webb 1965). In Wisconsin lakes, there is about 15 mg/liter, of which crude protein constitutes 15 per cent, fats or ether extract 1 per cent, and carbohydrates about 83 per cent. Dissolved organic material becomes higher, of course, in dystrophic lakes and peat bogs (Birge and Juday 1934).

Hydrogen-ion Concentration

The acidity or alkalinity of water depends on the ratio between the H^+ (or hydronium, H_3O^+) and OH^- ions. The amount of acidity or alkalinity is commonly expressed in terms of potential hydrogen ions in a pH scale. The values on this scale represent the logarithm of the reciprocal of the normality of free hydrogen ions. When the number of H^+ ions is equal to the number of OH^- ions, the pH value is 7, the value which represents absolute neutrality. All pH values less than 7 indicate a greater number of H^+ ions than OH^- ions, which is to say the closer the pH value approaches 0, the more acid the water. Above pH 7, there is a preponderance of OH^- ions; the higher the pH value, up to 14, the more alkaline is the water.

The hydrogen-ion concentration of most unpolluted lakes and streams is normally between pH 6.0 and 9.0, but extreme values of pH 1.7 and pH 12.0 occasionally occur (Hutchinson 1957). In some bodies of water, the pH value fluctuates considerably. Hydrogen-ion concentration increases (low pH values) with active decomposition of organic matter.

In general, aquatic animals can tolerate great changes in pH, although the range of toleration varies between species. Mollusks are not ordinarily found in acid lakes, but some snails can survive pH as low as 6, and the fingernail clam *Pisidium* down to pH 5.7. At the lower pH values, the shells of mollusks become thin, fragile, and transparent, but it is believed that the cuticular covering is partially protective and prevents complete dissolution of the calcium carbonate by the acid (Jewell and Brown 1929). In *Campeloma* snails, the apex of the shell may completely dissolve, exposing the apex of the visceral mass. Most fish can tolerate pH 4.5 to 9.5, provided there is plenty of oxygen (Brown and Jewell 1926, Wiebe 1931), and many invertebrates will tolerate even greater extremes. Fish as individuals become acclimated to certain pH values, and will select those values when given choice in a gradient. Such acclimation of individuals may have an effect on their choice of natural habitats, although when forced into a habitat with a different hydrogen-ion concentration, they change their acclimation. Although the direct ecological importance of differences in hydrogen-ion concentrations is doubtful, the measurement of pH may serve as an index of other environmental conditions, such as the amount of available carbon dioxide (with which it varies inversely), dissolved oxygen (with which it varies directly), dissolved salts, and so on. Sometimes the difference in species of plankton found in bodies of water with permanently different pH values, for example in granite and limestone, is very striking (Reed and Klugh 1924).

LAKE BIOCIES

If we reserve ponds and peat bogs to separate consideration, there remain two major lake communities. They differ in species composition, abundance of organisms, distribution of niches, productivity, and physical characteristics. Inasmuch as these two communities correspond fairly well to the oligotrophic and eutrophic types of lakes, we may name them simply the *oligotrophic* and *eutrophic lake biocies*. Various facies of each community, or intermediate types (Deevey 1941) are affected by variations in the abundance of component species and correspond to differences in temperature, depth, fertility, and other features of the habitat. The communities that occur in dystrophic lakes, for instance, are an impoverished facies of the eutrophic lake biocies. In spite of taxonomic differences in constituent species, each lake biocies contains organisms belonging to the same life-forms and with similar mores, so they may be discussed together.

Depending largely on their morphological adap-

tations and behavior, aquatic organisms are, for convenience, divided into plankton, neuston, nekton, and benthos, although the differences between the groups are not precise. *Seston* is a collective term that includes all small particulate matter, both living and non-living, that floats or swims in the water. *Plankton* are free-floating or barely motile organisms, either plant (*phytoplankton*) or animal (*zooplankton*), that are readily transported by water currents. Most plankton are microscopically small, although some forms are visible to the unaided eye. Species that can be caught with a net are called *net plankton* to distinguish them from the minute varieties that pass through No. 20 silk bolting cloth meshes. The latter, less than 0.06 mm in length, include most protozoan, bacterial, and fungal forms, collectively called *nannoplankton*. Organisms that depend on the surface film for a substratum are called *neuston* and are more important in the quiet waters of ponds than in lakes. *Nekton* are larger animals that are capable of locomotion independent of water currents. Aquatic birds that swim and dive are included in this group. *Benthos* organisms are attached to or dependent on the bottom for support; there are sessile, creeping, and burrowing forms. *Periphyton* is the assemblage of organisms attached to surfaces, either living or non-living, submerged in water above the bottom (Hanson 1962). Organic *detritus*, important in terrestrial as well as aquatic ecosystems, is "particulate material originating from disintegrating and decomposing biomass" (Odum and Cruz 1963). All these major categories may be subdivided into smaller units (Hutchinson 1967).

PLANKTON

Fresh-water plankton (Welch 1952, Pennak 1946, Davis 1955) includes representatives from the photosynthetic algae, Bacillariaceae (diatoms), Myxophyceae (blue-green), Chorophyceae (green), and occasional other form such as *Wolffia* among the higher plants; the non-photosynthetic bacteria and other fungi; and among the zooplankton, all classes of Protozoa except Sporozoa, Rotatoria, Entomostraca (especially Cladocera, Copepoda, and Ostracoda), some immature Diptera, the statoblasts and gemmules of bryozoans and sponges, the rare fresh-water jellyfish, *Craspedacusta*, and occasional aquatic mites, gastrotrichs, and others (Fig. 5-5). Fresh-water plankton lack many forms common in the plankton of the ocean. On the other hand, the rotifers, aquatic insects, and water-mites are mostly absent from the sea, and the Cladocera are only poorly represented. It is probable that plankton evolved from benthonic forms occurring near the shore (Ruttner 1953), and many species of groups listed above, notably Ostracoda and Rotatoria, are still largely benthonic in behavior.

The algae in fresh water may vary in numbers from hundreds of thousands to tens of millions of cells per liter; Protozoa, from thousands to hundreds of thou-

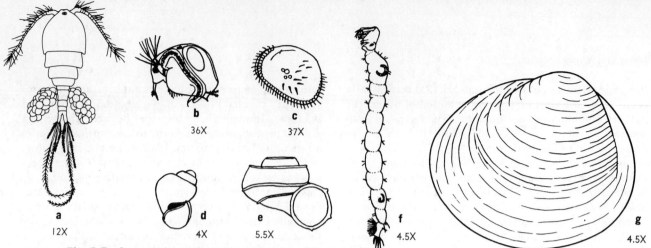

Fig. 5-5 Common invertebrates found in lakes: (a) copepod; (b) cladoceran; (c) ostracod; (d) the snail *Amnicola limosa*; (e) the snail *Valvata tricarinata*; (f) the ghost larva *Chaoborus albipes*; (g) the fingernail clam *Pisidium* (modified from various sources, Pennak 1953).

Fig. 5-6 Vertical distribution of net plankton in an oligotrophic lake (left) and in a eutrophic lake (right), Wisconsin. Note that the horizontal scale is different for the two lakes, and for the algae as compared with the zooplankton. The crosshatched horizontal belts show the region of the thermocline (from Birge and Juday 1911).

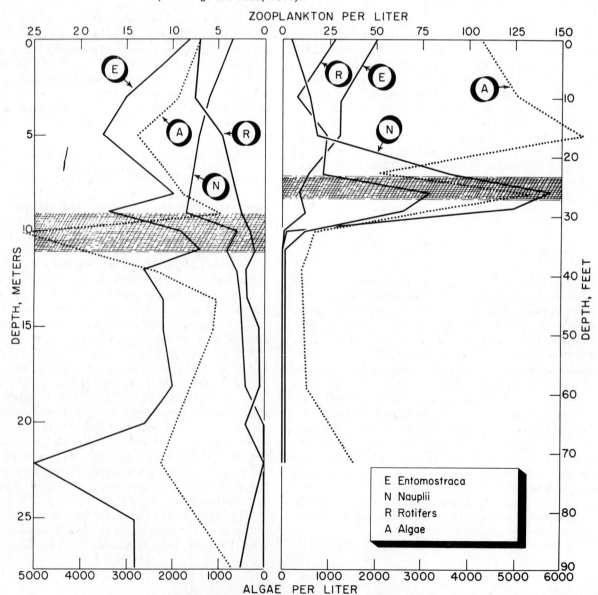

sands of individuals per liter; and the rotifers and entomostracans, from less than ten to hundreds per liter.

Distribution

Many species of plankton are nearly worldwide in distribution, particularly those that occur in the larger lakes. Cosmopolitan distribution and the many primitive types of the plankton community indicate that its origin is ancient. Some plankton, however, such as the copepod *Pseudodiaptomus* and the rotifers *Brachionus* and *Keratella*, have a very limited distribution.

The plankton found in the open water of small to medium-sized lakes is seldom more than one to three species of copepods, two to four species of cladocerans, and three to seven species of rotifers, although the species change from one time of the year to another. It is also unusual to find more than one species of the same genus at the same time. When two do occur, one of them is usually much more abundant than the other. It is commonplace to find that 80 per cent or more of all limnetic copepods present belong to a single species, 78 per cent of all cladocerans to a single species, and 64 per cent of all rotifers to a single species (Pennak 1957).

In any one lake the horizontal distribution of the plankton may be irregular because of water currents, inflowing streams, irregularity of shoreline, or *swarming* of a particular species in local areas. The vertical variations in the composition and abundance of species is even more striking. The chlorophyll-bearing algae require light and are most numerous in the upper stratum, although diatoms commonly occur at greater depths (Fritsch 1931). The vertical distribution of zooplankton varies widely with the species, but it is strikingly affected by light, food, gravity, dissolved gases (particularly oxygen), and thermal stratification. Few zooplankton occur in the hypolimnion of eutrophic lakes during the summer stagnation period, but they occur at all depths during the spring and autumn overturns (Fig. 5-6).

Diel Movements

Several species of net zooplankton exhibit pronounced vertical migrations, moving upward into surface strata during the night and returning to greater depths during the day (Hutchinson 1967). In some instances this daily shifting of position may extend to 60 or more meters, in other instances it may be only a fraction of a meter, and some species do not exhibit the phenomenon at all (Langford 1938). An old explanation of these movements is that the animals are negatively geotactic by nature, but that during the day this drive is suppressed by a negative phototaxis and can be expressed only at

night (Parker 1902). Probably the zooplankton actively orient to a band of optimum light intensity and move up and down at different times to avoid light of too great or too little intensity (Cushing 1951, Hardy and Bainbridge 1954). In certain crustaceans, as the amphipod *Nototropis*, the rhythm may persist for 4 or more days in continuous darkness (Enright and Hamner 1967).

These diel movements are most widespread among Cladocera and Copepoda (Fig. 5-7), but other species are also involved. One of the most interesting cases is the dipteran larva *Chaoborus punctipennis*, which rests on the lake bottom during the daylight hours but is often teeming in the surface waters at night. It appears that the buoyancy of this larva varies with the size of its two pairs of air-sacs (Damant 1924). There are a few rotifers, *Mysis* among the Malacostraca, and *Ceratium* among the Mastigophora, in which vertical day and night movements have been demonstrated (Pennak 1944).

Seasonal Distribution

The different species of plankton vary in their response to seasonal changes in the physical and chemical nature of the water, in number of generations per year, and in time of occurrence. Accordingly, there is a marked seasonal variation in total numbers during the year. In larger and deeper lakes, a maximum population usually occurs between April and July, a minimum in August, a second maximum in late September or October, and the yearly minimum in late winter, February or March (Fig. 5-8). However, not all species follow this schedule; some species have a maximum in the spring and not in the autumn, or vice versa; and some species reach greatest abundance during the general summer or winter minimum.

A species can also exhibit alternate increases and decreases in population at other times; these, as well as fluctuations in total plankton, are called *pulses*. At times, especially during the summer when the water is warm,

Fig. 5-7 Vertical distribution of three species of copepods in the daytime (stippled) and at night (black) in the oligotrophic Lake Nipissing, Ontario, on a July day, when the thermocline occurred between 12 and 15 m (from Langford 1938).

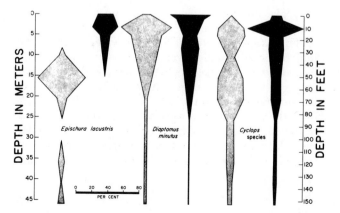

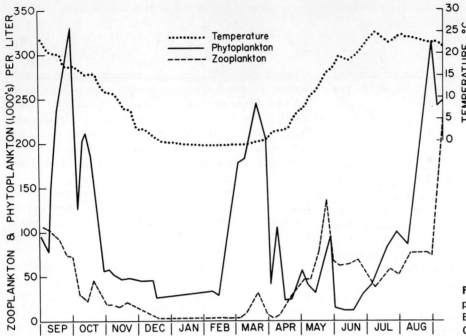

Fig. 5-8 Seasonal plankton populations in western Lake Erie through a year (after Chandler 1940).

an algal form, most commonly a blue-green species, may become so abundant that it discolors the water; these irruptions are known as *blooms*. The death and decay of such masses of vegetation may so deplete the oxygen supply that great mortality of fish and other animals results. In some cases the algae produce chemicals toxic to animals.

The ways in which environmental factors control seasonal and other changes in population are not all clearly understood, but it is significant that the maxima in total plankton of deep lakes often come at the times of the two annual overturns, times when food and oxygen are abundantly distributed at all depths. But the bimodal curve may also be found in shallow lakes and ponds that do not possess thermoclines. In small lakes, however, there is greater irregularity, and one, two, three, or no maxima may occur at various times of the year (Pennak 1946). Periods of high rainfall, which means increased drainage of nutrients into a lake, may be a factor of importance in producing maxima; seasonal changes in light, water temperature, and oxygen tension certainly are important.

BENTHOS

Divisions

The lake bottom can be divided into a littoral zone and a profundal zone.

The *littoral zone* extends from the water's edge to the limit of rooted aquatic vegetation. It may be subdivided into the *eulittoral zone*, between high and low water marks at the water's edge, where the beating of waves is most effective, and the *sublittoral zone*, which extends from the lower limit of wave action to the lower limit of rooted vegetation. Where such vegetation is absent, the sublittoral zone may be considered the bottom of the epilimnion down through the thermocline.

The *profundal zone* is the entire bottom below the rooted vegetation, or commonly the bottom of the hypolimnion. The boundary lines between the zones are variable and change with the depth of the thermocline. The open water of the lake above the bottom is known as the *limnetic zone*.

Littoral Zone

The bottom of the littoral zone may be rock, cobble, gravel, sand, or mud. The muddy shallows of protected bays may have considerable rooted vegetation; they are essentially pond habitats. Differentiation of species distribution is primarily between the hard-bottom and mud-bottom habitats; sand-bottom habitats are transitional (Table 5-2). Sand bottoms ordinarily have the lowest population of most species except clams because they are unstable habitats at best; indeed, they are often destructive by reason of the action of sand grains grinding on each other (Rawson 1930, Krecker and Lancaster 1933, Lyman 1956). A lake-bottom and a streambed of similar composition will contain many of the same kinds of organisms because of the similarity in the physical conditions of existence. The respective species compositions, however, are often different. The higher taxa diversity index for the sand bottom of a lake, in contrast to the sand-bottom pools in a stream (Table 4-1), is correlated with both a larger number of taxa and a more uniform distribution of individuals in the different taxa.

Oneida Lake in New York has an unusually high mollusk population. Baker (1918) recorded 59 species and varieties. It is interesting that most of them occurred on mud and sand bottoms. The highest populations were 1890 individuals per m² on mud at depths less than 2 m, and 1573 individuals per m² on sand. On rocks and gravels there were only 656 individuals per m².

Table 5-2 Size and distribution of invertebrate populations on different types of bottom in the littoral zone of western Lake Erie (after Shelford and Boesel 1942).

Common name	Classification	Number per square meter		
		Cobble and gravel	Sand	Mud
Midge fly larva	*Cricotopus exilis* and others	1000		
Caddisfly larva	*Hydropsyche*	70		
Mayfly naiads	*Stenonema tripunctatum, S. pulchellum, S. interpunctatum*	15		
Snail	*Physa* sp.	13		
Water penny	*Psephenus contei*	7		
Sponge colonies	Spongillinae	3		
Snail	*Planorbula crissilabris*	2		
Leech	*Glossiphonia*	2		
Snail	*Amnicola limosa porata*	1		
Mayfly naiads	*Baetis, Centroptilum*	+		
Clam	*Elliptio dilatatus sterkii*	+		
Damselfly naiad	*Argia moesta*	+		
Midge fly larva	*Tendipes pallidus*	+		
Bryozoan colonies	*Plumatella*	32	+	
Parnid beetle and larva	*Stenelmis crenata*	17	+	
Caddisfly larva	Trichoptera	2	1	
Clam	*Leptodea fragilis*	+	1	
Flatworm	*Planaria*	+	+	
Snail	*Goniobasis livescens*	12	1	2
Midge fly larva	Chironomidae	1	2	1
Clam	*Amblema costata*	+	0	+
Clam	*Lampsilis siliuoidea rosacea*	+	2	1
Clam	*Obovaria subrotunda*		1	
Clam	*Lampsilis ventricosa*		+	
Mayfly naiad	*Ephemera*		+	
Alderfly larva	Sialidae		+	
Midge fly larva	*Tendipes flavus*		+	
Parnid beetle larva	*Stenelmis bicarinatus*		+	
Clam	*Anodonta subglobosa*		+	
Clam	*Micromya fabilis*		+	
Snail	*Pleurocera acuta*		7	+
Clam	*Fusconaia flava parvula*		1	+
Mayfly naiads	*Hexagenia occulata, H. rigida*		+	33
Midge fly larva	*Tendipes digitatus*		1	6
Midge fly larva	*Procladius culiciformis*		+	3
Amphipod	*Gammarus limnaeus*		+	+
Water boatman	*Arctocorixa lineata*		+	+
Midge fly larva	*Tendipes decorus*			1
Snail	*Valvata tricarinata*			1
Leech	*Herpobdella punctata*			1
Amphipod	*Gammarus fasciatus*			+
Leech	*Glossiphonia stagnalis*			+
Crayfish	*Cambarus argillicola*			+
Mite	*Lunnesia undulata*			+
Midge fly larva	*Cricotopus trifasciatus*			+
Clam	*Protera alata*			+
Clam	*Ligumia nasuta*			+
Clam	*Truncilia donaciformis*			+
	Total taxa	22	23	22
	Total individuals	1177	17	49
Percentage taxa characteristic or exclusive		73	48	64
Taxa diversity index (H') (counting + = 0.1 individual)		0.70	1.57	1.30
Equitability index (H'/H'_{max})		0.23	0.50	0.42

In eutrophic Douglas Lake, Michigan, bottom deposits in the littoral areas show zonation down to a depth of about 18 m. Beginning at the shoreline, there are belts of barren, wave-washed sand, muddy sand, sandy mud, and deep-water soft black ooze, in that order. The average number of macroscopic benthic animals is large, varying in the different types of bottom from 369 to 1178 to 3822 to 1713 per m², respectively. The abundance of animals is related not only to the nature of the bottom but also to depth and vegetation present. Where vegetation was scarce there were only 162 animals per m², but with increasing density of plants from sparse to common to abundant the population of animals rose to 1531, 2525, and 4407 per m², respectively. Vegetation was most dense at depths of 7 to 14 m in mixtures of sand and mud. Most abundant animal species, in decreasing order, were the amphipod *Hyalella azteca*, the dipteran larvae *Tendipes* and *Protenthes*, the snail *Amnicola*, tubificid worms, and the sphaerid *Pisidium* (Eggleton 1952). For comparison, the depth distribution of animals in an oligotrophic lake is shown in Fig. 5-9.

There is also a fauna of microscopic animals inhabiting the bottom. This consists of Protozoa (especially Ciliophora, Webb 1961), *Hydra*, Rhabdocoela (flatworm), Nematoda (Prejs 1970), Rotatoria, Gastrotricha, Oligochaeta. Cladocera, Copepoda, Ostracoda, Acarina (mites), and Tardigrada. These organisms are often very numerous in the thin organic ooze-film that covers mud bottoms (Bigelow 1928), but may penetrate underlying deposits to depths of 20 cm. Sand bottoms also support a varied and abundant microfauna (Pennak 1940, Cole 1955). In addition to the bottom, microscopic animals and plants, including especially bacteria and diatoms (Fox *et al.* 1969), occur abundantly as periphyton on the surface of larger organisms and structures in clear water. In general, number of

microfauna species and individuals varies inversely as the depth of water; only a few species remain active in the profundal zone during the summer stagnation period (Moore 1939). Bacteria, a source of food, are abundant in the bottom at all depths.

Much of the bottom fauna of the littoral zone consists of immature stages of otherwise terrestrial insects. The pulmonate snails and water mites have evolved from terrestrial species. Other aquatic species, however, have related forms in the sea, and this may indicate their evolutionary origin. The fundamental problem involved in dispersal from the sea into fresh water is that of osmoregulation, and the ability to live in fresh water has doubtless constituted a selection factor in the origin of this community.

Profundal Zone

In oligotrophic lakes (Fig. 5-9), species characteristic of the littoral zone are found at much greater depths than they are in eutrophic lakes, in which the oxygen supply during the summer stagnation period is reduced. The amphipod *Pontoporeia* occurs only in the deeper oxygenated cold waters of some northern lakes (Adamstone 1924). It is a relic from the glacial period, when it was probably more widely distributed. The profundal benthos of one oligotrophic lake in British Columbia increased from 470 individuals per m² in January to 1270 in August (Ricker 1952).

The most common bottom organisms are the annelids *Tubifex* and *Limnodrilus*, and the insect larvae *Tendipes*, *Chaoborus*, and *Protenthes*. There may be a few mollusks, such as *Pisidium* and *Musculium*, nematodes, and other forms, including a microscopic fauna (Eggleton 1931, Webb 1961).

The midge larvae represent a variety of species.

Fig. 5-9 Variation with depth in abundances of various organisms in oligotrophic Great Slave Lake (Rawson 1953).

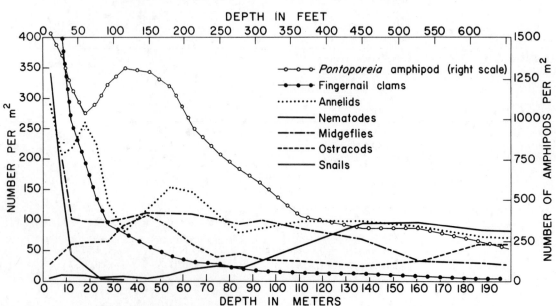

Fifty species occur in one small lake in Algonquin Park, Ontario, that has a pH range of 4.6 to 6.6 and thermal stratification in summer with ample oxygen in the hypolimnion. Of this number, 33 species are confined to the littoral and sublittoral zones, 7 to the profundal zone, and 10 occur throughout (Miller 1941).

The bottom mud of eutrophic lakes commonly consists of a thin, upper, brown, detritus layer of newly deposited organic matter that has drifted down from above; a relatively thick gray layer containing many fecal pellets and much organic matter, as well as diatoms; and a relatively barren bottom layer. In England's Lake Windemere, 85 per cent of all bottom organisms occur within 6 m of the surface, and 100 per cent within 12 m of the surface (Humphries 1936).

Maximum populations of insect larvae in eutrophic lakes are ordinarily reached during the winter. Minimum populations occur during late spring and summer, both in the littoral and profundal zones, because many immature insects have completed their development and emerged, as adults (Eggleton 1931, Ball and Hayne 1952). Although relatively few genera make up the bottom fauna, populations may at times be enormous. *Chaoborus* larvae alone have been recorded in populations of 97,000 individuals per m², *Tendipes* larvae at 26,000 individuals per m² (Deevey 1941), and the fingernail clam, *Sphaeruim*, at over 12,000 individuals per m² (Carlson 1968).

Summer Stagnation Period

The low concentrations or complete disappearance of oxygen in the hypolimnion of eutrophic lakes for periods of several days or weeks in the summer requires special adjustments by organisms. Some bacteria are truly anaerobic, and perhaps some animals are, too, but most forms simply accumulate an oxygen debt that is repaid when the autumnal overturn takes place. It is of interest that the annelid worms and those midge fly larvae that tolerate the lowest oxygen concentrations possess hemoglobin in the blood, the pigment which has the greatest capacity and efficiency in transporting oxygen at low tensions. Tubificid worms extrude farther from their tubes and wave their tails more vigorously for a time as the oxygen content becomes reduced. The nightly excursions of *Chaoborus* larvae into the oxygenated epilimnion certainly provide opportunity for replenishing of their oxygen needs. A considerable proportion of the larvae migrate out of the profundal zone during the spring, and do not return until autumn or early winter (Wood 1956). The copepods *Cyclops bicuspidata*, *Canthocamptus staphylinoides*, and perhaps others, encyst and lie on the bottom during the summer period (Moore 1939), although this action has not definitely been related to any particular environmental factors (Cole 1953). Some midge fly larvae also form inactive cocoons.

NEKTON

The nekton of lakes consists principally of fish. There is an interesting small shrimp, *Mysis relicta*, found in the deeper waters of many northern lakes of North America that is often included with the nekton. This species is believed to be a relic of a marine fauna that happened to get cut off from the sea in some past geological period, yet was able to survive as the water became fresh.

Numbers and species of fish are more concentrated in the littoral zone of lakes than in the open, deeper waters of the *limnetic* zone. Limnetic species also invade shallow waters for spawning. In deep waters, fish tend to remain close to the bottom, where their food supply is located, unless there is a deficiency in oxygen there. Caged fish, lowered to various depths in a eutrophic lake, did not survive long below the thermocline (Smith 1925). Fish may, however, make short excursions into the hypolimnion.

In a study of fishes in six Wisconsin lakes, Pearse (1934) found that

> Usually most fishes per unit area occur in muddy, vegetation-filled, shallow ponds, but the characteristic fishes (carp, crappie, sunfish, dogfish) are not the most desirable for food. Rich eutrophic lakes produce considerable quantities of desirable fishes (perch, largemouth bass, white bass, rock bass). Oligotrophic lakes produce littoral game fish of good quality and size (smallmouth bass, wall-eyed pike, pickerel) and ciscoes in deep water.

The average catch with gill nets in two oligotrophic lakes was 3.5 per hour; in two eutrophic lakes, 4.2 per hour; and in two shallow lakes or ponds, 5.1 per hour.

In the littoral zone, fish species are segregated according to the composition of the bottom, as are the invertebrates. The species living over rock and gravel bottoms in lakes are mostly different from those inhabiting similar bottoms in streams, but the mud-bottom forms are nearly the same as in ponds (Shelford and Boesel 1942, Nash 1950).

Amphibians and reptiles do not commonly occur in lakes except around margins supporting attached aquatic vegetation, and here pond species occur. Such pond mammals as the muskrat, mink, and otter are not typical of lakes as such, although they are frequently found in shallow littoral waters. There are a number of bird species, however, that occur most commonly in lakes: American and red-breasted mergansers, loons, pelicans, cormorants, terns, gulls, ospreys, bald eagles, and swallows. These species get their living from the lake, but nest on neighboring shores or islands. In addition, there are many pond and marsh birds that occur along vegetated lake margins.

FOOD CYCLE

The lake is a closely knit ecosystem whose inhabitants are largely independent of the rest of the world but very much dependent on each other for existence. It is almost a microcosm in itself (Forbes 1887), but it depends on the insolation of the sun for energy, rain and snow for water supply, and on minerals dissolved out of the surrounding uplands for the basic nutrient salts essential to the formation and functioning of protoplasm.

Basic to this food cycle are the bacteria. A few bacteria occur free-floating in the water. For the most part, however, they are either attached to algae, to other plankton organisms, to submerged objects, or occur on the bottom as part of the benthos (Henrici 1939). Their number varies from one place and time to another, as do the numbers of other organisms; they are more abundant in eutrophic than oligotrophic lakes. Their action is to transform the dead organic matter into nutrients, especially nitrates, that the green plants then absorb.

The phytoplankton is also basic in the food cycle because of its ability to manufacture carbohydrates with the aid of sunlight and to anabolize proteins after absorbing nitrogen and other compounds liberated by the bacteria into the water. Rooted vegetation around the lake margin is important in this respect, although in large lakes the proportion of food substances formed by marginal vegetation is small as compared to the

amount manufactured by phytoplankton. In Wisconsin lakes, the daily production of glucose during clear days in August varies from 14 to 44 kg/hectare (12 to 39 lb/acre) (Manning and Juday 1941).

Zooplankton feed upon phytoplankton, Protozoa, bacteria, detritus, and each other (Gliwicz 1969). Some species appear to discriminate in their choice of food, but most species filter out and ingest all particulate matter, within size limits, non-living as well as living, with which they come in contact. The ratio of number of entomostraca to number of phytoplankton cells has been found to vary from 1 : 1800 to 1 : 63,000. Ratios of rotifers to phytoplankton vary from 1 : 50 to 1 : 37,500 (Pennak 1946). The plant cells, however, are much smaller than individual animals. The mean ratio of zooplankton to phytoplankton by volume is commonly about 1 : 4 (Davis 1958), but in alpine and northern oligotrophic lakes, the ratio may be reversed (Pennak 1955, Rawson 1956). In the nannoplankton, Protozoa depend largely upon bacteria, although some forms feed also on algae and detritus; a few species prey chiefly upon other protozoans (Picken 1937).

When the plankton dies, it settles to the bottom and furnishes food for the benthos. The accumulation of dead plankton and other aquatic organisms on the bottom may be extensive enough to form a distinctive brownish layer. The benthic midge fly and other insect larvae, annelids, clams including the sphaeriids, snails, and bottom-dwelling entomostraca feed on this detritus layer, on organic matter held in suspension, and on algal plankton and attached forms.

The variety of food habits in fish is reflected in their anatomical adaptations. Fish feeding on bottom matter have soft-lipped sucking mouths; fish feeding on plankton have numerous slender gill-rakers; fish feeding on other fish have large mouths and sharp teeth. The adults of some fish, such as the cisco, gizzard shad, paddlefish, and sunfish, consume large quantities of plankton. The gizzard shad also feeds on bottom mud, straining organic particles out of it and grinding them up in a stomach that resembles the gizzard of a chicken. Sturgeon, whitefish, buffalo fish, carp, catfish, bullheads, suckers, sunfish, and many others feed largely on bottom annelids, insect larvae, mollusks, and vegetation in shallow waters. As many as 354 midge fly larvae have been found in a single whitefish stomach; 331 were found in a sturgeon stomach (Adamstone and Harkness 1923). Bass, crappies, perch, pike, gar, and lake trout feed principally on other fish. The bottom feeders scoop up the bottom ooze indiscriminately (Darnell 1964). Several forms maintain contact with the bottom by means of sensitive barbels hanging from the chin, but plankton-feeders and carnivorous species depend largely on sight for seizing individual prey.

Young fish of many species live largely on plankton, even though as adults they feed on something quite different (Fig. 5-10). A 10-cm perch requires 150 mg dry weight of food per day during the summer, the

Fig. 5-10 Change in food habits of perch as they increase in age and size (after Allen 1935).

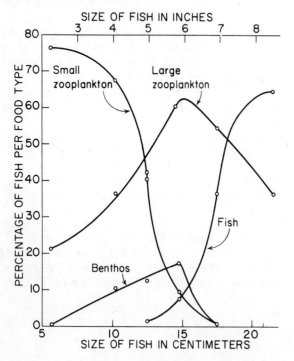

equivalent of about 37,500 *Cyclops*. The perch would have to consume *Cyclops* at a rate of 26 per minute throughout the day in order to ingest such a total. A 20-cm perch would require 600,000 *Cyclops* per day, ingested at a rate of 417 per minute, which is doubtless beyond its efficiency of intake. By consuming only four small fish 0.3 g dry weight each, the perch could obtain the same energy intake (Allen 1935).

Most lake-inhabiting birds subsist mainly on fish, diving for their food. Gulls take only dead fish, which they find floating on the surface or washed up on the shore. Swallows skimming over the water surface consume enormous numbers of emerging adult midge flies and other insects.

BIOMASS AND PRODUCTIVITY

The dry weight of total organic matter of seston in 529 fresh-water lakes in Wisconsin was found to range from 0.23 to 12.0 mg/liter with an average of about 1.36 mg/liter (Birge and Juday 1934). Of this, living plankton organisms constituted an amount ranging from 20 to 80 per cent. The biomass of green phytoplankton is usually, but not always, greater than the zooplankton. The biomass of net plankton may be only one-third to one-tenth of the total net and nannoplankton. Net plankton is generally more abundant in hard water than in soft water, more abundant in eutrophic lakes than oligotrophic lakes (Rawson 1953). The dry weight of net plankton during the summer in 18 lakes of western Canada and in 2 lakes of Wisconsin varied from 0.9 to 17.7 mg/liter, and averaged 5.0 mg/m² of water surface area (Rawson 1955).

The biomass of benthos varies with the nature of the bottom, amount of vegetation, and depth. When computed for the total bottom of 10 Canadian lakes exceeding 11 m in depth, it was found to vary from 0.07 to 2.47 g/m², and average 0.63 g/m² dry weight, not counting the shells of mollusks (Rawson 1955). The mean of 36 lakes in Connecticut ranging in depth from 1.1 to 11.1 m varied from 1.09 to 34.8 g/m², and averaged 7.5 g/m² (Deevey 1941).

When lakes of different depths are analyzed, it is found that the mean biomass per unit area of both net plankton and benthos exist in inverse relation with mean lake depth (Table 5-3). Likewise the biomass of benthos, especially midge fly larvae and annelids, varies inversely with the size of the lake (Pieczynska *et al.* 1963). These correlations indicate that the morphometric characteristics of a lake influence its carrying capacity, although perhaps indirectly by affecting the dissolved salt content, oxygen content, temperature, or some other factor.

The biomass of plankton is generally greater than the biomass of benthos. In addition to the five Canadian lakes listed in Table 5-3, Deevey (1940) found the ratio between plankton and benthos in five other lakes to

Table 5-3 Interrelations among depth, biomass of plankton, and biomass of benthos in five Canadian lakes (from Rawson 1955).

Average depth (m)	Average dry weight of net plankton (g/m²)	Average dry weight of benthos[a] (g/m²)	Ratio total biomass (plankton/ benthos)
11	9.05	2.47	2.3
26	3.65	0.41	4.5
38	3.2	0.45	5.7
69.5	2.6	0.20	7.2
120	0.9	0.07	9.6

[a]Minus weight of shells in mollusks.

vary from 3.8:1 to 10.0:1. In one eutrophic lake in Michigan, the standing biomass of fish to benthos was in the ratio 2.7:1 (Ball 1948).

Primary net productivity in lakes is correlated with the biomass of phytoplankton present, transparency of the water, and other factors. In 20 Danish lakes and 14 Indiana lakes, it varied between a low of 62 and a high of 1691 g of carbon per m² per year (Whiteside and Harmsworth 1967). Nannoplankton may often be responsible for the greater share of the primary production of the phytoplankton (Hillbricht-Ilkowska and Spodniewska 1969).

Secondary productivity may be measured in various ways. If, throughout the year, the plankton population of Lake Mendota, Wisconsin, should replace itself every 2 weeks, then the annual productivity would be 624 g of ash-free, dry, organic matter per m² of water surface. Of this amount, 585 g would come from phytoplankton and 39 g from zooplankton. The benthos reproduces less rapidly, nekton, still less so. The annual productions of bottom fauna and fish in Lake Mendota is estimated at 4.5 and 0.5 g/m², respectively, and the large aquatic vegetation at 51.2 g/m² (Juday 1940). Disregarding the large aquatic plants, the ratio of productivity between plankton and benthos is approximately 139:1; between benthos and nekton, 9:1; and between plankton, benthos, and nekton taken together, 1248:9:1. The ratio of annual productivity between plankton and benthos is much higher, therefore, than is the ratio of their biomasses or standing crops. No attempt was made in this study to determine the standing crop of fish.

In a detailed study of the net productivity of the benthos in the Russian Lake Beloie (Borutsky 1939). it was found that the standing crop increased during the year by 125 per cent. Of this total biomass, 55 per cent died without being eaten by other organisms or was replaced by the small biomass of new eggs being laid; 14 per cent was consumed by other organisms, chiefly fish; and 6 per cent emerged as adults that subsequently left the lake ecosystem. The remaining 25 per cent constituted the standing crop of the fol-

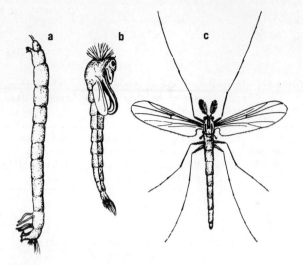

Fig. 5-11 (a) Larva, (b) pupa, and (c) adult of a midge fly (from Shelford 1913 after Johannsen).

lowing year. However, this standing crop was only 56 per cent of what it was the year before, so these percentages are not representative of stabilized populations.

It was estimated that in Costello Lake, Ontario, the standing population of midge fly larvae was replaced during the 135 days of summer eight or nine times in the epilimnion, and two or three times in the hypolimnion. Consumption of larvae by fish was small in shallow waters but amounted to 50 per cent of the standing crop in deep water (Miller 1941).

LIFE HISTORIES

Although most *midge fly larvae*, Chironomidae or Tendipedidae, are aquatic, some forms live in decaying organic matter, under bark, or in the ground. The earliest larval stage is a wiggler, which may be carried by the current or may squirm about from place to place (Fig. 5-11). Later, this wiggler larva becomes sluggish and builds a case or tube, open at both ends, by cementing particles of sand, debris, or silt about itself with mucus from its salivary glands. Construction is accomplished in about 3 hours. The larvae extend themselves from these cases for feeding, and in some species may even move the cases to better feeding areas. The larval period is the longest part of the life cycle; it lasts at least 2 months (Macdonald 1956). Most of the pupation period, which is probably less than a week, is spent in the larval case, but toward the end the pupa swims to the surface of the water. At this time it is preyed upon extensively by fish. The adult imago struggles out of the case and flies off. The adult lifespan is probably short, as there is no evidence that they feed. They may occur in immense swarms in the evening. Eggs are laid in masses of several hundred, in sticky gelatinous strings that float attached to some object at the surface, or sink to the bottom. The eggs hatch in a few days, and the cycle is repeated (Cavanaugh and Tilden 1930, Johnson and Munger 1930). Some larvae (for example, *Procladius, Tanypus*) are carnivorous, but most are herbivorous, feeding on algae, or sapropha-

gous, feeding on detritus. They do not build cases, but roam over the bottom. The number of generations varies in different species from two per year, to one per year, or one in two years, and depends in part on the depth and temperature of their habitats (Miller 1941).

The *tubificid worms* do not leave the lake bottom. Many of them occur in tubes or cases, similar to the habit of midge fly larvae. It is evident by the presence of sexually mature adults and the reproductive cocoons that reproduction occurs principally during the periods of autumn and, especially, spring overturns. There follows a large increase in numbers of small immature worms (Eggleton 1931).

The cisco or lake herring has been selected to illustrate the life history of a lake fish (Cahn 1927, Fry 1937). Life history data on other fresh-water fishes may be found in Carlander (1961).

The *cisco* spends much of the year in deep water, feeding very largely on plankton. Because its food habits require a large volume of water to be strained through its gill-rakers, the fish swim almost continuously, usually in a constant and definite direction, in schools of from 20 to several hundred individuals. During hot summers the cisco may leave the cool, deep, but oxygen-poor carbon dioxide-rich waters and ascend into the epilimnion. There they are sometimes killed in large numbers by temperatures higher than they are able to tolerate. The fish spawn in November or December, when the water temperature drops to 4°C. For this purpose the fish move into water only 1 to 3 m deep, or even up into rivers. The males precede the females by 2 to 5 days. When the females arrive, several males consort with each. When she is ready to spawn, the female descends to within 20 cm of the bottom and sheds about 15,000 eggs. At the same time, the accompanying males discharge sperm, and fertilization is completed. The eggs are viscous and become attached to rocks or bottom debris. No nest is made and no further attention is paid to the eggs. Incubation may last 10 to 12 weeks; hatching normally occurs in late March. The young fish later return to deep water and reach breeding condition in 3 years. Doubtless the slow rate of development in this species is related to the low temperature of the habitat. After spawning is completed, the adults may remain in shallow water until water temperatures reach 20°C. This temperature is above their preferendum, although they can tolerate temperatures up to at least 25°C.

APPLIED ECOLOGY

Lake and Fish Management

Applied ecology involves the management of lakes and the control of their resources for man's benefit. Aside from their use in transportation, in industry, and as sources of drinking water, lakes are of importance to man for fishing, swimming, sight-seeing, and boating. For swimming, clean, clear water with a sand bottom is desirable. Sewage and industrial wastes must be diverted or eliminated for reasons of health and the appearance of the water. Algal growth, when excessive, can sometimes be controlled with copper sulfate; and rooted vegetation can be reduced by sodium arsenite treatment. When chemical treatment of water is limited to low concentrations administered with discretion, there is generally no great harm to fish; some invertebrates, such as midge fly larvae, mayfly naiads, and fresh-water shrimp, are adversely affected (Machenthun 1958).

Where there is excessive erosion of the surrounding upland, silting may render the waters of small lakes turbid, decreasing the growth of algae, a basic food substance for lake organisms. The rapid accumulation of silt on lake bottoms covers up bottom organisms, clogs the gills of mollusks, and generally reduces the lake's productivity. The obvious remedy is the control of erosion at the source.

The management of large lakes to the end of increasing fish productivity is difficult because of the area and depth of water involved. Where commercial fishing is commonly practiced in large lakes, a careful yearly catch record for each species should be maintained. This will suggest regulations such that annual cropping will not exceed annual production. To maintain good fishing and productivity there must be good chemical and physical characteristics of the water, an abundance of food, plenty of breeding areas, and exclusion of exotic predators. The drastic decline in the annual yield of lake trout in the Great Lakes is attributed in part to invasion by the predaceous marine sea lamprey (Fig. 5-12) and other foreign species as well as overexploitation (Smith 1968). Artificial fertilization of lakes presents problems but may be practicable for smaller lakes (Hasler and Einsele 1948).

The smaller the lake, the easier is the management of the habitat. The water level may be manipulated by damming to increase the area of shallow water available for spawning at certain seasons, or lowered at other times to permit growth of marginal vegetation or prevent spawning of undesirable species. Artificial shelters or spawning areas may sometimes be created (Hubbs and Eschmeyer 1938). In general, rearing small fish in hatcheries for later release has not proven economically practicable. Any proposed introduction of exotic species should be investigated with considerable skepticism.

Fig. 5-12 Commercial production of lake trout (1,000,000 pounds = 486 metric tons), trends in abundance of the sea lamprey, time of sea lamprey establishment (E), start of chemical lamprey control (S), and completion of treatment of all lamprey-producing streams (C) for Lakes Superior, Michigan, and Huron. Lines showing 10 and 20 per cent of the average of the peak years of sea lamprey abundance are shown for each lake. The commercial fishery for lake trout in Lake Superior was closed in mid-1966 (Smith 1972; reproduced by permission of Information Canada).

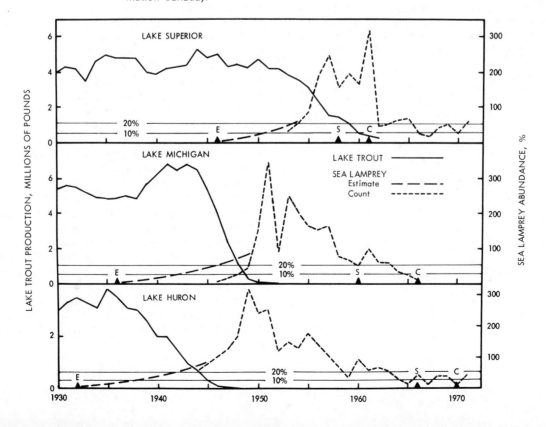

Eutrophication

Eutrophication is a natural process, as oligotrophic lakes change to eutrophic ones in the course of centuries. Eutrophication becomes of concern when its rate is greatly accelerated by products of human activities that bring equivalent changes in decades or years. Modern agricultural practices cause increased erosion of uplands and correspondingly siltation of lakes. Excessive organic wastes from cities and farmlands become destructive pollutants even in large bodies of water. What has happened to Lake Erie during the last few decades illustrates what is going on.

One hundred years ago, the Maumee River filtered through 7770 km² (3000 mi²) of marsh and swamp before it emptied its relatively clean waters into the western end of Lake Erie. During the latter part of the nineteenth century, the wetlands were drained and converted, along with surrounding forests and prairies, into farmland. The Maumee River then began to carry increasingly heavy loads of silt into the lake, and the water became so turbid that less than 5 per cent of sunlight was able to penetrate even to a depth of 1 meter.

Meanwhile the Detroit River began to empty progressively heavier loads of organic and industrial wastes into the western basin of Lake Erie, and other rivers carried in similar loads into other parts of the Lake. In the 14 years between 1948 and 1962, the pH was raised from 8.7 to 9.2, the oxygen content near the bottom was lowered from 6 cc/liter to nearly zero, the nitrogen content of the water was increased from 260 to 330 μg/liter, and the soluble phosphorus from 7.5 to 36 μg/liter. From 1964 to 1968, the increase in

Fig. 5-13 Contribution of major fish species (percentage) and total commercial catch (pounds) of Lake Michigan for various periods 1898–1966. Shaded portion indicates exotic species (Smith 1968; reproduced by permission of Information Canada).

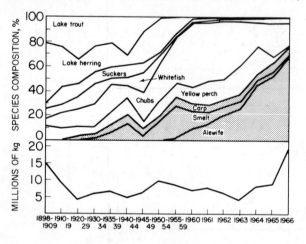

organic nitrogen accelerated 30 per cent and soluble phosphorus 50 per cent. The alarming increase in phosphorus is attributed to the widespread substitution of phosphate detergents for ordinary soaps and, to a lesser extent, its use as a rust inhibitor, in the salt applied to city streets in the winter, and in other ways. With increased availability of nitrogen and phosphorus, phytoplankton and filamentous algae have become abundant, and as this material dies and decays, it depletes the oxygen in the water faster than it can be replenished. Much of the nitrogen and phosphorus remains in the water and sinks to the bottom with the decaying mass of algae to form a thick layer of muck. The filamentous alga, *Cladophora glomerata*, has become so excessive that it fouls fishermen's nets and gets washed up on the beaches in large masses. There it decays, forming a disagreeable stench, and has to be hauled away (Verduin 1969). Photosynthesis in the western basin of Lake Erie has varied since the late 1940s from 98 to 7340 and in Sandusky Bay up to 36,000 mg C/m²-day. This is to be compared with measurements from 50 to 1260 in Lake Superior and Lake Michigan and up to 2560 mg C/m²-day in Lake Ontario (Saunders 1964).

Correlated with changes in the physical and chemical characteristics of the water, there has been a change in the predominant species of phytoplankton, and the mayfly naiad, *Hexagenia libata*, has been superseded in the benthic fauna by the midge fly larva, *Tendipes plumosus* (Verduin 1969). In the early nineteenth century, muskellunge, northern pike, and lake trout were commonly taken in the commercial fisheries, but the harvest of these species began to decline rapidly toward the end of the century. Cisco, whitefish, sauger, blue pike, walleye, ling, fresh-water cod, and channel catfish were among the principal commercial fish during the first half of the twentieth century but have decreased considerably since the 1940s and 1950s. On the other hand, increases in recent years have occurred in catches of yellow perch, smelt, carp (Fig. 5-13), freshwater drum, alewife, and gizzard shad (Clarence F. Clark, personal communication).

It is obvious that Lake Erie is not a "dead lake," but drastic changes have occurred in its physical and chemical characteristics and in its biota. Actually the fish productivity of the lake may be higher now than it ever was, but it consists of species of fish that man considers much less desirable as food.

The remedy for situations leading toward accelerated eutrophication, particularly as it results from the use of detergents, would seem to be either the reduction of the phosphate content in the product, replacement of phosphorus by some other chemical to produce the same cleansing effect, return to the use of soap, development of special sewage-treatment methods to remove the excess chemicals, or a combination of alternatives. There are difficulties with each option (Rukeyser 1972). The minimum amount of phosphorus recommended to make an acceptable detergent is still more than the environment can absorb without harm. A suitable substitute for phosphorus has not yet been found. The return to soap will create residue problems with present automatic washing equipment and will require considerable new sources of animal, or possibly plant, fats for its manufacture and will increase the biochemical oxygen demand of its effluents. Water-treatment technology is improving and phosphorus removals up to 90 per cent or more have been achieved, but conversion of existing sewage-treatment plants to these new techniques will take time and money. Some combination of these different procedures may be required for quick improvement of lake habitats.

Reduction of phosphorus from inflows into bodies of water may not always bring a corresponding decrease in excess algae present, since at times the organic growth is controlled by the nitrogen supply rather than the phosphorus (Ryther and Dunstan 1971). It is even more difficult to remove nitrogen from water effluents than to remove phosphorus (Amer. Chem. Soc. 1969). There may be need, as with phosphorus, for regulation of the use of nitrogen at its source, which is often fertilizer. Certainly cultural eutrophication can be arrested, but this will require ingenuity, time, and money (Hasler 1969, Rohlich 1969).

SUMMARY

Important factors in aquatic habitats are pressure, density, light, current, temperature, oxygen, carbon dioxide and other gases, dissolved solids, and hydrogen-ion concentration. Of special importance in most lakes is the occurrence of a thermocline that divides the water into an epilimnion and a hypolimnion. The hypolimnion retains a low temperature throughout the year and in some lakes becomes deficient in oxygen in late summer. These differences in temperature and oxygen greatly affect local and seasonal occurrence of organisms. Lakes are classified several ways on the basis of physical characteristics; biologically, only two distinct communities, the oligotrophic and eutrophic lake biocies, are distinguishable.

The life-forms of lake organisms are chiefly plankton, benthos, and nekton. Zooplankton exhibits diel movements to greater depths in the daytime and general dispersal, including movements toward the surface, at night. Peak populations are commonly reached in late spring and again in autumn; low points occur in summer and winter. Benthos decreases in abundance from the littoral to the profundal zone. Profundal animals in eutrophic lakes are adjusted in various ways to survive the low-oxygen late summer stagnation period. Nekton includes aquatic birds as well as fish.

The lake is a closely knit ecosystem, largely independent of the rest of the world except for its solar energy, inflowing water, and mineral salts. The base of food chains is composed of detritus, bacteria, and phytoplankton, then zooplankton and small benthic organisms, and finally fish and birds. All dead organisms, as well as their excreta during life, decompose so that their nutrient substances start the food cycle over again. The biomass and productivity of the three life-forms usually rank, from high to low: plankton, benthos, and nekton.

The life cycle and behavior of lake organisms are closely adjusted to the various environmental situations available. Control or management of fish production by man is difficult, except in lakes of small size. Erosion of the watershed is making many lakes turbid, and organic and industrial wastes from cities and agricultural lands are increasing the process of eutrophication so rapidly as to render many lakes of greatly decreased usefulness to man.

Chapter 6

PONDS, MARSHES, SWAMPS, AND BOGS

Pond is a popular term for lakes of the third order that are small, shallow, and, when mature, have rooted vegetation over most of the bottom. There is no clear distinction between ponds and lakes of the first and second orders. The littoral zone of eutrophic lakes, for instance, is pond-like in habitat and organisms. Floating and emergent vegetation commonly occurs around the margin of ponds to form extensive tracts of marsh. The pond habitat may originate as a shallow basin, as a large pool in a stream, as the result of the filling in of a lake, or from a stream dammed by beavers, man, or landslide. Slow-flowing rivers are essentially elongated ponds, and have a similar fauna (Kofoid 1908, Richardson 1928). Because of the slight water movement in ponds, the surface film becomes an important microhabitat for some species. Pondwater temperature is often uniform at all depths, but during warm sunny weather, ponds well protected from the wind may show considerable stratification, not only in temperature, but also in oxygen content and other characteristics (Wallen 1955). Daily and seasonal variations in temperature may be great because of the small volume of water. Ice forms earlier and lasts longer in ponds than in lakes, freezing shallow ponds to the bottom in severe winters. Light penetrates to all depths, encouraging growth of vegetation except in high turbidity. Young ponds may have rocky, sandy, clay, or mud bottoms; in mature ponds, there is ordinarily an accumulation of organic matter and silt.

The dissolved oxygen content of ponds varies widely from temporary supersaturation when there is excessive photosynthesis of plants to near depletion when decomposition predominates. Oxygen content is often highest in the spring; very low in late summer; and sometimes low again under the winter ice cover. Oxygen content is usually higher during daylight hours than during the night because of the daytime photosynthetic cycle of plants. Oxygen may become so low at night as to become critical, especially for fish. Decomposition of organic matter evolves carbon dioxide and, at times, considerable methane, hydrogen sulfide, and other gases. In ponds, as in lakes, there is wide variation in hydrogen-ion concentration. As ponds mature and accumulate humus, pH value decreases.

PLANT SERE

The plant hydrosere, or pond sere, typically contains the following stages and characteristic species:

Submerged vegetation: Water weed, pondweed, milfoil, hornwort, naiads, buttercup, bladderwort, eelgrass, and the herb-like alga *Chara*.

Floating vegetation (Fig. 3-1): Water lily, pond lily, pondweed, smartweed, duckweed, and water hyacinth. All except the last two are rooted in the mud, often at depths of 2 to 3 m, and may have rhizomes from which long petioles extend to the leaves floating on the surface. Duckweeds and water hyacinths are unattached floaters, and cover the surface extensively in some localities.

Emergent vegetation (marsh): The dominant species are: cattail, reed, bulrush, bur-reed, swamp loosestrife, wild rice, and sawgrass. They invade waters of over 1 m depth, but in shallower waters or in secondary succession they are replaced by a sedge meadow composed of sedge, rush, and spike rush.

Swamp shrubs: Buttonbush, alder, dogwood, swamp rose, and sometimes shrubby willow and cottonwood.

Swamp forest: Red and silver maples, elm, ash, swamp white oak, and pin oak.

Succeeding stages depend on the climate of the region. In arid regions the swamp forest may be poorly developed, and grassland or desert vegetation may come in quickly. In the mesic climate of the Eastern states, an oak-hickory associes follows the swamp forest, replaced in turn by a climax of sugar maple-beech or mixed mesophytic forest. The hydrosere in the broad-leaved evergreen climax area of southeastern North America brings in cypress and a number of other unique species.

Vegetative debris and animal remains, together with inwashed silt, fill the basin gradually, reducing the depth of water and allowing vegetation to encroach on the periphery. As this process continues, the succession is effected. Ultimately, open water entirely disappears as the ground stratum is built up above the water table, and climax vegetation replaces all other types.

ANIMAL SERE

Animals characteristic of marshes and ponds constitute a distinct *pond-marsh biocies* which extends into sluggish or base-leveled streams and the littoral zone of eutrophic lakes (Fig. 6-1). Most animal species are not restricted to a single stage or community of the plant sere, but commonly occur in several stages in varying abundance and for various activities. Fish, for instance, feed in open water but spawn in shallow water among the emergent vegetation. Submerged, floating, and emergent vegetation represent different levels or strata in a single biotic community, and each stratum has about the same degree of distinctiveness as forest community strata.

With the invasion of swamp shrubs and with the ground level well above the water table most of the year, pond and marsh species largely disappear,

replaced by many characteristic new species. This animal community represents the swamp facies of the *deciduous forest-edge biocies*, which will be discussed later. The swamp forest is often quite open at its outer margin, and forest-edge species remain common. But as this forest develops a closed canopy and drier ground stratum, it is invaded by the swamp facies of the *deciduous forest biociation*. In the pond sere there is a succession of animal adaptations from aquatic to amphibious to terrestrial.

POND-MARSH BIOCIES

Neuston

The *supraneuston*, organisms which move on top of the surface film in pursuit of most of their life activities, consists of the water striders Gerridae, Veliidae, Mesoveliidae; the water measurers Hydrometridae; the whirligig beetles Gyrinidae; the springtails Collembola; some spiders; and occasional other forms. The gyrinids of several species commonly occur in social groups (Robert 1955). They are remarkable for having each eye divided so that the upper portion looks into the air and the lower portion into the water. Several of the forms listed have long legs that distribute the weight of the body over a large area of surface film. The portions of the legs or body that contact the surface film are water-repellant.

The undersurface of the water film supports an *infraneuston* of Hydra, planarians, ostracods, cladocerans, snails, and insect eggs, larvae, and pupae (mosquitoes, certain kinds of midge flies, and so forth). For all except the insects, however, the use of the surface film in this manner is usually transitory. Some cladocerans, such as *Bosmina* and *Daphnia*, occasionally break through the surface film from below, fall over onto their sides, and cannot return.

Plankton

The species of plankton found in ponds differ somewhat from those in lakes (Klugh 1927), but the transition from lake species to pond species is a gradual and progressive one. Protozoa and Rotatoria are usually more abundant in ponds than in lakes.

Although plankton distribution does not vary with depth as much in ponds as in lakes, seasonal fluctuations are as extensive and similar in nature. Eddy (1934) lists 15 perennial species of zoo- and phytoplankton that may be found in ponds throughout the year, 2 seasonal species which reach their peak of abundance between December and April, 4 between February and June, 12 between March and December, and 5 between July and September.

Representative animals of the emerging association: a, the common newt; b, the common pond snail; c, a predaceous diving beetle.

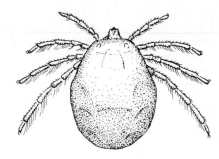

Red mite

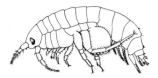

An amphipod

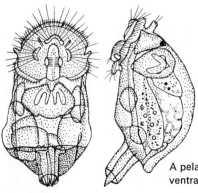

A pelagic rotifer, ventral (left) and side (right) views

A garter snake feeding on the dead fish left in a dry-season pond.

Representative animals of the submerged vegetation: a, a viviparous snail; b, a green sun-fish above a yellow perch, both juvenile; c, a shrimp; d, a winter body, or statoblast, of the gelatin-secreting polyzoan.

Fig. 6-1 Pond animals (from Shelford 1913).

78

Benthos

Subaquatic animals dwell not only on the bottom but also on the stems and leaves of submerged plants. Aquatic plant species have a diversified animal periphyton of insects, amphipods, mites, and snails, using them for the food, shelter, or reproductive sites denied them in the mud bottom below. Other kinds of insect larvae and oligochaetes are more abundant in the mud than on the plants. The undersurface of lily pads often contains many small organisms, including Protozoa, *Hydra*, flatworms, rotifers, and snails. The biomass

Table 6-1 Development (ecesis) of invertebrate bottom populations (calculated in number per m²) in strip-mine ponds of different ages, as determined by studies conducted in October, through 3 years. The ponds 1 and 8 years old had no rooted vegetation; the ponds 21 and 30 years old had a little vegetation in protected coves; the pond 80 years old had submerged, floating, and emergent vegetation. Range of pH: 7.1 to 8.5; locality, near Danville, Illinois.

Common name	Classification	Age of pond				
		1 yr	8 yr	21 yr	30 yr	80 yr
Red midge fly larva	*Tendipes*	16	7	6	27	25
Damselfly naiad	Zygoptera	2	23	4	22	51
Caddisfly larva	Trichoptera	1	—	—	2	3
Backswimmer	Notonectidae	+	—	—	+	3
Whirligig beetle	*Dineutes assimilis*	+	+	—	+	
Alderfly	*Sialis*		23	3	2	
Water boatman	Corixidae		13	+	1	
Burrowing mayfly naiad	*Hexagenia*		21	262	5	1
Dragonfly naiad	Anisoptera		9	4	11	27
Crawling water beetle	Haliplidae		5	+	+	+
Clam	Unionidae		5	+	—	+
White midge fly larva	*Tanypus*		3	1	35	10
Aquatic annelids	Tubificidae, Lumbriculidae		4	+	3	2
Ghost larva	*Chaoborus*		+	+	+	17
Springtails	*Podura aquatica*		+	+	+	+
Crayfish	*Orconectes propinquus*			1	1	
Fly larva	Diptera			3	+	
Water spider	Arachnida			+	+	3
Fingernail clam	*Sphaerium*			+	—	2
Mayfly naiad	*Caenis*				+	
Snail	*Physa gyrina*				14	58
Amphipod	*Hyallela*				5	105
Other mayfly naiads	Ephemerida				1	21
Aquatic isopod	*Asellus communis*				1	5
Flatworm	*Planaria*				1	1
Limpet snail	*Laevapex*				1	1
Snail	*Gyraulus parvus*				+	17
Snail	*Helisoma trivolvis*				+	9
Water scorpion	*Ranatra*				+	1
Leech	Hirudinea					5
Snail	Lymnaeidae					2
Fingernail clam	*Pisidium*					1
Shrimp	*Palaemonetes*					1
Water strider	*Gerris*					+
Total taxa		5	13	16	27	28
Total individuals		19+	113+	284+	132+	371+
Percentage taxa characteristic or exclusive		0	31	12	4	57
Taxa diversity index (counting + = 0.1 individual (H')		0.60	2.09	0.44	2.40	2.33
Equitability index (H'/H'_{max})		0.37	0.81	0.16	0.73	0.70

Table 6-2 Succession of dragonfly and damselfly naiads in western Lake Erie (after Kennedy 1922).

Dragonfly and damselfly naiads	Lake biocies		Pond-marsh biocies			Forest-edge biocies
	Open lake	Lake margin with submerged vegetation	Lake margin with floating and emergent vegetation	Young pond	Mature pond	Old marsh with invading shrubs and trees
Gomphus plagiatus	++	++				
Gomphus vastus	++	++	+			
Neurocordulia yamaskinensis	++	++	+			
Macromia illinoiensis	++	++	+			
Argia moesta	++	++	++	+		
Enallagma carunculatum		++	++	++	++	
Enallagma exsulans		+	+	++	++	
Enallagma ebrium		+	+	++	++	
Ischnura verticalis		+	++	++	++	++
Tramea lacerata			+	++	++	
Anax junius			+	++	++	+
Enallagma signatum				++	++	
Libellula luctuosa				++	++	
Libellula pulchella				++	++	+
Lestes rectangularis				+	++	+
Leucorrhinia intacta					++	
Erythemis simplicicollis					++	
Plathemis lydia					++	
Nehalennia irene					+	
Pachydiplax longipennis					++	+
Lestes forcipatus					++	++
Sympetrum obtrusum					++	++
Sympetrum vicinum					++	++
Sympetrum rubicundulum					+	+
Enallagma hageni						+
Lestes uncatus						+
Lestes unguiculatus						+

of animals varies directly with the biomass of vegetation, and the quantity of invertebrates is especially great on those plants possessing finely dissected leaves (Gerking 1957). Very few species found in lake bottoms are not found in ponds, but the pond-marsh biocies contains many species not found in lakes.

The number of species and individuals found in the bottom fauna increases with the age of the pond from the time the pond is formed until attached vegetation becomes excessive (Tables 6-1, 6-2). Coincident with the development of the bottom fauna is an increase in variety and abundance of plankton (Eddy 1934) and fish. Different-aged ponds clearly do not have distinct communities. This is characteristic of ecesis, since relatively few species drop out and more and more species are added as the community matures. In spite of irregularity in the values, there is a trend for the taxa diversity index to increase as the community matures. This is reflected both in an increase in the number of species and in higher equitability indices.

The construction of a beaver dam in a small Ontario river changed the riffle habitat into that of a pond and brought a reduction in mayfly naiads, stonefly naiads, and caddisfly larvae within 2 years. Other stream animals fell from 68.7 to 15.6 per cent of the total population while midge fly larvae increased from 31.3 to 84.4 per cent (Sprules 1940). Shallow ponds develop more rapidly than deep ones, and mud-bottom ponds develop more rapidly than sand- (Shelford 1911) or rock-bottom ponds (Krecker 1919). The increase in number of species and individuals in ponds depends on an increase in the variety of microhabitats, types and amount of food, and vegetation. With the development of the pond into a marsh there is generally an increase in humus and an increase in bacteria effecting its decomposition, carbon dioxide, and marsh gases. Oxygen and pH decrease.

Two predominantly terrestrial orders of insects, Coleoptera and Hemiptera, have invaded the pond community but are not found in lakes except those which have pond-like margins. The Coleoptera are represented by three families of diving beetles, Haliplidae, Dytiscidae, and Hydrophilidae, and by the whirligig beetles, Gyrinidae. The haliplids are herbivorous; the dytiscids are predacious; some hydrophilids and gyrinids are predators, others are scavengers. The aquatic bugs or Hemiptera are the Corixidae, which feed on the bottom ooze; the Notonectidae, which prey upon small Entomostraca; the Nepidae, the Belostomatidae, and the Naucoridae, which are all carnivorous; and the Veliidae, Mesoveliidae, Gerridae, and Hydrometridae, which are probably both carnivores and scavengers. Some of these species, as already noted, are usually found on the surface film, but they may occasionally dive and cling to submerged vegetation. The true diving forms, especially the beetles and some of the hemipterans, have evolved oar-like legs for rapid propulsion.

Respiratory Adaptations Air-breathing aquatic insects, as well as the pulmonate aquatic snails *Lymnea*, *Helisoma*, *Gyraulus*, *Physa*, and *Laevapex*, have evolved special mechanisms and behavior for respiration. Most species rise to the surface of the water at intervals to replenish their supply of air. Insects are so buoyant that they must cling to the vegetation or some other object to maintain a submerged position. As soon as they let go of the substratum, they float to the surface and must return by swimming. Snails commonly creep to the surface along plant stems or other submerged objects, or suddenly emit mucous threads that float them to the surface (Max Matteson, personal communication). They find their way to the surface, at times of oxygen need, by negatively geotactic behavior. After they have obtained a fresh supply of oxygen, they become positively geotactic (Walter 1906). Pulmonate snails probably also absorb some oxygen from the water; indeed, some species appear never to come to the surface. The gill-bearing or branchiferous species of snails are seen to be segregated into rather distinct niches when their habitat relations are analyzed in detail (Baker 1919).

Diving beetles carry a bubble of air beneath the elytra, and the entire body of *Dryops* is enclosed in air (Fig. 6-2). The hemipteran notonectids and corixids carry a bubble over the ventral surface of the body, trapped there by hair-like setae. The spiracles of the tracheal system open into these bubbles. The body surface and the setae holding the bubble are water-repellent, or *hydrofugous*. The fresh air bubble contains 21 per cent oxygen and 78 per cent nitrogen, the same proportion as the atmosphere. The nitrogen dissolves into the water very slowly. The carbon dioxide given off by the insect passes first into the bubble, replacing

the oxygen absorbed, and then into the water. If the water contains ample oxygen, it will diffuse into the bubble as rapidly as it is used, and perhaps three times as fast as the nitrogen diffuses out. Under these conditions backswimmers, *Notonecta*, have survived for nearly 7 hours without coming to the surface. The bubble is really a physical gill mechanism, but functions only as long as the nitrogen present provides an adequate surface for oxygen diffusion. The insect's trip to the surface is as much to get a fresh supply of nitrogen as it is to get a fresh supply of oxygen. If, however, the water contains little or no oxygen, the oxygen content of the bubble is quickly reduced and the insect may be forced to the surface for a fresh supply every 3 or 4 minutes (Wolvekamp 1955).

The air-breathing respiratory mechanisms of other aquatic insects are equally remarkable. In many larvae, *Dytiscus*, Culicidae, and other Diptera, and in the aquatic Hemiptera, only the terminal abdominal spiracles are functional. The tracheal trunks of mosquito larvae and *Dytiscus* larvae, among others, store considerable air so that the animal may remain submerged for long periods. *Ranatra* and other Nepidae have long respiratory tubes extending from the tip of the abdomen so that they can cling to vegetation well below the water surface, yet respire directly into the air.

Dragonfly naiads pump water through the anus, in and out of an enlarged rectum. The walls of the rectum are abundantly supplied with a network of tracheae for interchange of gases directly with the water. In the larvae of midge flies, black flies, and corixid beetles, the general body surface is richly

Fig. 6-2 Beetle *Dryops* freshly submerged, crawling along a stem, encased in a bubble of air (after Thorpe 1950).

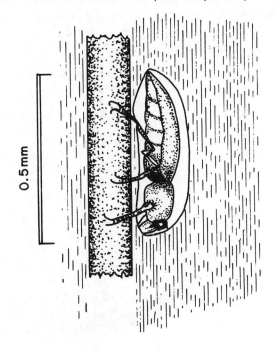

0.5 mm

supplied with fine tracheae for exchange of gases directly with the water. The anal papillae of midge flies and mosquito larvae are not respiratory in function, as formerly supposed; rather they serve for osmoregulation. Tracheal gills, plates, or filaments are found on many immature insects, Ephemeridae, Plecoptera, Zygoptera, Trichoptera, Neuroptera, and some Diptera, and effectively increase the area of surface available for oxygen absorption. The larvae of the beetle *Donacia* and certain Diptera including mosquitoes have a unique ability to puncture the walls of submerged plants and collect air from the intercellular spaces (Miall 1934).

Terrestrial Invertebrates

The terrestrial insects found in marsh vegetation are in the main adult mosquitoes, midges, dragonflies, damselflies, mayflies, and alderflies, whose immature stages live submerged (Fig. 6-3). On bare ground around ponds may be found toad bugs, shore bugs, springtails, tiger beetles, and sometimes ground beetles and pigmy locusts. Spiders become numerous throughout the vegetation, and the snail *Succinea* appears. In addition to these true marsh and pond species, invertebrates belonging to the forest-edge biocies may occasionally be found.

Fish

Fish are often very abundant (Table 6-3). Included among the species that occur are several bottom-feeders —suckers, bullheads, buffalo, and carp—the last, a species introduced into North America from Europe in 1877. By feeding on the submerged vegetation and stirring up the bottom, they may control the habitat and the composition of both animal and plant species

Table 6-3 Differences in species composition and number of individuals of fish present in two similar-sized Wisconsin ponds (Cahn 1929).

Species	Pond with carp	Pond without carp
Carp	5891	0
Shorthead redhorse	66	0
White crappie	17	0
Bigmouth buffalo	1	0
Northern redhorse	14	10
Walleye pike	4	20
Bowfin	7	340
Northern pike	3	380
Rock bass	1	940
Bluegill	2	1220
Longnose gar	0	30
Pumpkinseed	0	610
Yellow perch	0	680
Black crappie	0	730
Largemouth bass	0	1120
Total species	10	11
Total individuals	6006	6080

present in the community. This condition, however, does not last indefinitely. Vegetation encroaches on the margins of ponds, and the fish are gradually eliminated because of the disappearance of suitable breeding sites. The mudminnow, bowfin, and bullhead are usually the last to disappear before the pond becomes a dry marsh (Shelford 1911).

Amphibians and Reptiles

Salamanders and frogs are basically aquatic animals, although they show varying degrees of adaptation to terrestrial life. *Siren* and *Necturus* have permanent

Fig. 6-3 Dragonfly niches (after Needham 1949): (1) on sand, *Macromia*; (2) in sand, *Gomphus*; (3) in muck, *Libellula*, *Neurocordulia*; (4) on massed *Nitella*, damselflies; (5) on tips of *Websteria Enallagma laurenti*; (6) in open tangles of bladderwort, *Erythemis* and damselflies; (7) in fallen brown leafage, *Pachydiplax longipennis*; (8) at sides of ditch, *Tetragoneuria*, *Celithemis*, *Erythrodiplax*; (9) on invading roots of woody plants, *Argia fumipennis*; (10) at water-line rooted green plants, damselflies; (11) in rafts of fallen pine needles, aquatic Hemiptera that are enemies of dragonfly naiads.

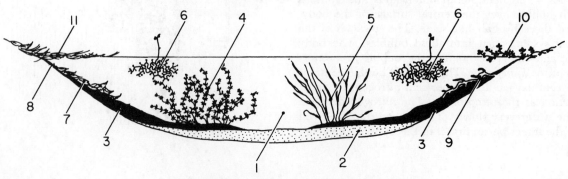

external gills and spend all their lives in the water. Most other forms lay their eggs and pass through their early development in water, but the adults are air-breathing and wander over the land. Since their skins must be kept moist, they are confined to the vicinity of water, to humid climates, or to damp humus. A few species, such as *Plethodon cinereus* and *P. glutinosus,* lay their eggs in the cavities of well-rotted logs and seem largely independent of standing water. The ability of salamanders and frogs to live temporarily away from water appears positively correlated with thickness, cornification, and relative impermeability of the skin. The aquatic tadpoles and larvae are scavengers or herbivorous in their food habits, the adults feed on insects, earthworms, or other animal matter that they catch on land (Noble 1931).

Reptiles are terrestrial. Desert reptiles never go to water. Painted, geographic, and snapping turtles bask in the sun on the shore or on protruding logs, but quickly plunge into the water to escape danger, to cool off, or to feed. The alligator, musk, and soft-shelled turtles spend nearly all their time in water. The soft-shelled turtle is able to utilize dissolved oxygen in the water and hence has evolved special readaptation to water. Like other turtles, however, they lay their eggs on land, placing them in holes excavated in sand, loose soil, muck, or decaying stumps or logs. Water snakes give birth to living young that enter the water immediately. Water snakes feed on insects, small fish and amphibians, crayfish, or whatever other animal food they can find. The food of turtles is similar to that of snakes; some species are also scavengers. The cottonmouth moccasin is a prominent poisonous snake in southern marshes and swamps. The massasauga rattlesnake occurs in wet areas in the north.

Birds

Bird populations are high, and the pond-marsh, swamp shrub or forest-edge, and forest communities are especially clearly defined (Table 6-4). Both the number of species and the equitability index increases with succession. There is an abundance of nest-sites and food, but the aquatic and terrestrial species exploit different niches to avoid competition as much as possible. The aquatic species feed in all stages of the plant sere, beginning with the open water, but nest for the most part in the emergent vegetation (Beecher 1942). Herons commonly feed in shallow water but nest in tree-top colonies. Grebes, cormorants, and terns feed on fish in the open water; the herons, egrets, and bitterns get fish in water shallow enough for them to wade in; cranes and coots are omnivorous; ibises, stilts, snipes, and rails probe around in the mud for invertebrates; avocets sweep their curved bills back and forth through the water, catching aquatic insects; gallinules eat seeds, roots, and soft parts of succulent plants as well as some invertebrates; most ducks feed on submerged and floating vegetation and attached animal

Table 6-4 Populations of breeding birds in units of pairs per 40 hectares (100 acres) in marshes of northern Ohio (after Aldrich 1943).

Bird species	Marsh	Swamp shrubs	Swamp forest
Virginia rail	22		
Least bittern	12		
Short-billed marsh wren	10		
Florida gallinule	5		
Sora	3		
Mallard	3		
Killdeer	1		
Long-billed marsh wren	78	3	
Red-winged blackbird	113	144	
Swamp sparrow	68	49	4
Song sparrow	8	49	12
Yellow warbler		80	
Traill's flycatcher		80	
Eastern kingbird		24	
American goldfinch		21	
Tree swallow		7	
Catbird		31	4
Green heron		7	1
Yellowthroat		28	13
Robin		3	4
Red-eyed vireo			9
Black-capped chickadee			9
Northern waterthrush			9
House wren			7
Ovenbird			7
Downy woodpecker			7
Eastern wood pewee			5
Tufted titmouse			4
White-breasted nuthatch			4
Blue jay			3
Rose-breasted grosbeak			2
Crow			2
Scarlet tanager			2
Yellow-shafted flicker			2
Crested flycatcher			2
Cardinal			1
Wood thrush			1
Veery			1
Hairy woodpecker			1
Black-billed cuckoo			1
Red-shouldered hawk			1
Eastern bluebird			1
Brown-headed cowbird			1
Prothonotary warbler			1
Total species	11	13	30
Total individuals	323	526	121
Percentage taxa characteristic or exclusive	73	62	80
Species diversity index (H')	1.71	2.14	3.40
Equitability index (H'/H'_{max})	0.71	0.83	0.90

organisms; song birds inhabiting the marsh feed chiefly on insects captured outside the water.

Mammals

One of the most characteristic mammals of the marsh is the muskrat. A well-developed marsh may contain one of their haycock-shaped lodges on each acre (2.5 per hectare), with perhaps five animals per lodge during the autumn. The diet of the muskrat is largely the leaves and roots of marsh vegetation, although they also feed to some extent on crayfish, clams, snails, and sluggish fish. Overpopulations of two or three lodges per acre (5.0 to 7.5 per hectare) may lead to "eat-outs" or local destruction of the marsh vegetation (Dozier 1953).

The mink is probably the most common mammalian predator of the marsh; it is an enemy of the muskrat. Foxes, raccoons, and coyotes may invade the marsh when the water level is low. The otter preys on fish and crayfish of the marsh; it has now been exterminated from most of its former range.

The beaver makes its own marsh habitat by damming small streams, flooding the surrounding lowland. Here it builds its large lodge and feeds on the bark and twigs of adjacent aspen, willow, and cottonwood, and on the roots of aquatic plants. When the supply of aspen and other food is exhausted, the colony disappears, the dam decays, the water level subsides, and marsh vegetation invades. After some years the pond is converted into a beaver meadow.

In the southern Atlantic seaboard and Gulf states, the herbivorous rice rat is common to marsh vegetation. In northeastern Ohio, the meadow vole attains populations averaging 58 per hectare (23 per acre) in the marsh and persists in smaller numbers in the swamp shrub and swamp forest (Aldrich 1943) and in old fields in Michigan (Getz 1960). The smoky shrew averages 8 per hectare (3 per acre) in the marsh. The cinereous shrew and short-tailed shrew are found in marsh vegetation and are common in the swamp-shrub stage (30 and 52 per hectare, 12 per acre and 21 per acre, respectively). All three of the insectivorous shrews are also found in the swamp forest. The white-footed mouse increases in numbers from the marsh through the swamp-shrub into the forest stages (2.5-22-32 per hectare, 1-9-13 per acre). Moles, chipmunks, and squirrels also occur in small numbers where the ground is drier.

FOOD CHAINS IN PONDS

Many animals in ponds depend for food on floating phytoplankton, bacteria, and bottom detritus (Fig. 6-4). Filamentous algae are more abundant in ponds than in lakes and are a source of food to various imma-

ture insects and frog tadpoles. Ponds, unlike lakes, have additional producers of organic matter in the rooted pondweeds. Pondweeds are consumed by insects, ducks, and herbivorous mammals. The periphyton of bacteria, diatoms, green and blue-green algae, and microscopic animals is important as food to many small crustaceans, immature insects, oligochaete worms, and snails (Frohne 1956). Creeping predators among the pondweeds and algae are leeches, dragonfly and damselfly naiads, and water mites. Small swimming predators are the dytiscid beetles, most of the hemipterans, and the swimming leeches *Erpobdella* and *Macrobdella*. At the top of the food chains, feeding on all these small animals, and often on plants as well, are the fish and other vertebrate groups (Lindeman 1941).

SEASONAL CHANGES AND TEMPORARY PONDS

Seasonal changes are greater in a pond than in a lake, a consequence of the smaller volume of water. Because of decreased rainfall, increased evaporation, and continuous seepage or drainage, ponds often become greatly diminished during the summer, or the open water may entirely disappear. As the volume of water shrinks, water temperature rises, and the oxygen content and pH decline. Animals must either adjust to these conditions of the pond or disappear altogether. The limit of tolerance of several invertebrates to decreased oxygen content is in the range 0.1–0.4 ppm (Moore and Burn 1968).

Under the winter ice, active pond life is slight because of the low oxygen and pH, but all groups increase in numbers as the temperature rises during the spring. In one small *Chara-cattail* pond near the south end of Lake Michigan, the snails *Amnicola* and *Helisoma deflectus*, the amphipod *Hyalella azteca*, the isopod *Lirceus danielsi*, the back swimmer *Plea striola*, and diving beetles Haliplidae attained maximum populations during April and May, but then declined in numbers through August, increasing again in the autumn. On the other hand, the water strider *Gerris*, mayfly naiads Ephemeridae, damselfly naiads Agrionidae, dragonfly naiads Libellulidae, and the snails *Physa* and *Helisoma parvus* had highest populations during June, July, and August (Petersen 1926). Maximum populations of Protozoa are also attained during the warm period of the year (Wang 1928).

Fish sometimes suffer from lack of oxygen in the winter when the ice cover lasts a long time. During the summer thermal stratification often forces them out of the deeper stagnant water. When photosynthesis is curtailed at night or in cloudy weather, mortality may become high.

Bird nesting is ordinarily completed by the time marshes dry up in late summer. At this time ducks

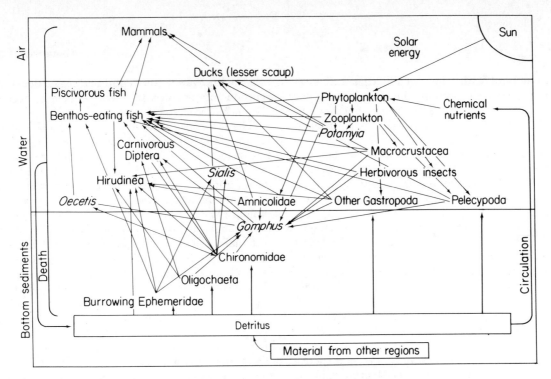

Fig. 6-4 Diagrammatic representation of nutrient and energy flow in the biological community of mud-flats along the eastern shore of the Mississippi River above Dam 19, Keokuk, Iowa (Carlson 1968).

concentrate in the remaining deeper bodies of water, and other marsh species start their southward migration.

Periods of drought are times of stress and increased mortality for muskrats, since the lack of water interferes with their normal locomotion and predation upon them by invading terrestrial species increases. Intraspecific competition becomes intensified as animals become crowded together in the shrinking habitat. Many individuals undertake overland journeys to new areas (Errington 1939).

Many invertebrates have spores or eggs resistant to the effects of desiccation, which enables them to pass over the period during which the pond is dried up. These include representatives of the protozoans, sponges, hydras, turbellarians, nematodes, annelids, bryozoans, rotifers, mollusks, crustaceans, and insects (Mozley 1932, Kenk 1949). It is an interesting experiment to collect top soil from a dried-out pool, place it in an aquarium with fresh water, and see what comes out of it (Dexter 1946). When water returns to the pond the hatching of plankton organisms and their growth to reproductive maturity is very rapid. The life cycle of some species appears shorter than required for a normal year; perhaps this is an adaptation to survive extreme years (Table 6-5). During years when the

Table 6-5 Monthly changes in the ostracod fauna of temporary ponds in central Illinois. The ponds dried up in mid-July (after Hoff 1943).

Ostracod species	March 9	March 17–20	April 1	April 15	May 4	May 18	June 9	June 22	July 10
Cypricercus reticulatus	+	+	+	+	+	+	+		
Cypria turneri		+	−	−	+	+	+	+	+
Candona simpsoni		+	+	−	+	+	+	+	
Candona fossulensis			+						
Candona distincta			+	−	+				
Candona indigena				+					
Candona biangulata				+					
Cypria maculata					+	+			
Cypria opthalmica					+	+	+		
Cypridopsis vidua					+	+	+	+	+
Candona suburbana					+	+			
Cypria obesa						+			

pond does not fill with water at all, eggs and cysts remain in a dormant condition; they may hatch after lying dormant for several years. Some, but not all, species of crayfish survive dry periods by burrowing down to the water table.

Among the more interesting inhabitants of temporary ponds are the phyllopods. The fairy shrimp is not found in permanent ponds, except those with a wide, shallow shore that dries out during the summer. Shrimp nauplii develop quickly from the egg, usually in January or February after the ice melts, but sometimes as early as November if the pond becomes filled with water after an autumn dry period. In the spring, adults mature in 3 or 4 weeks, egg-laying takes place forthwith, and the species may be gone by late May. The period when the pond is dry is passed through in the egg stage, and either the drying or freezing of the eggs facilitates their hatching (Weaver 1943).

Some toads, *Bufo* spp., *Microhyla olivacea*, and spade-foot frogs, lay their eggs, after a warm spring rain, in temporary pools rather than permanent ponds. Development is rapid, and metamorphosis of the tadpoles may be completed in a month's time, before the pool evaporates (Bragg et al. 1950).

LIFE HISTORIES

On hatching from the egg, *copepods* first pass through six free-swimming *nauplius* stages by a series of molts, during which the small compact animal possesses only three pairs of appendages; then through five *copepodid* stages, when additional appendages are added; and, finally, into the adult form. Both sexes occur regularly.

Ostracod eggs also hatch into nauplii, but these already possess a shell like that of the adult. Several molts are required, however, before maturity. Some species always reproduce sexually; others are partially or always parthenogenetic.

The reproduction of the *cladocerans* is of special interest. Most of the time only females are present, and eggs develop parthenogenetically during the summer into more females. The thin-shelled eggs are held in a brood pouch on the dorsal side of the body, and the young are well grown before they are set free. There are no free-swimming larvae. After a number of generations, the number varying with the species, and as the pond begins to dry up in the summer or winter, conditions reach a point where there is a crowding of females, an accumulation of excretory products, and a decrease in available food. Parthenogenetic male as well as female eggs are then produced. The resulting males are usually smaller than the females, but subsequent eggs are fertilized and a thick shell is formed around them. These ephippial eggs are produced in smaller numbers and are very resistant to drying and freezing. When the pond again becomes

filled with water, the ephippial eggs develop into reproducing females to start the cycle over again (Pennak 1953).

The life cycle of the *rotifers* bears some resemblance to that of cladocerans. A few species are viviparous, but in most forms development of the egg takes place outside the body and is direct into the adult form. Two kinds of females are not distinguishable by external characters. One kind, amictic, produces large diploid eggs that are never fertilized and only develop parthenogenetically into more females. The other kind of female, mictic, occurs only at critical times of the year and produces smaller haploid eggs. If not fertilized, these small eggs develop into males; if fertilized, they form the thick-walled winter eggs which, under subsequent favorable conditions, develop into females. The males are usually small compared with the females; they lack an alimentary tract, and consequently live only 2 or 3 days. Females live 1 to 3 weeks or longer. The production of males appears to be periodic and is often correlated with a change in type or amount of food, or degree of crowding. Males have never been seen, and may not occur, in some groups (Pennak 1953).

The amphipod *Hyalella azteca* breeds only during the warmer months of the year. The male carries the female on his back for 1 to 7 days before copulation occurs. Oviposition follows copulation by 12 to 24 hours. The incubation period is 21 days, and the female may carry the young another 1 to 3 days in her brood pouch. A period of 24 to 36 days elapses between successive broods, and each brood is larger than the last. The females may live into a second summer and reproduce again. The young on hatching in the spring have all the adult appendages and can reproduce later in the summer (Gaylor 1921).

In the spring, aquatic Hemiptera commonly glue their eggs to submerged vegetation. Some species insert their eggs into incisions made in leaves or stems. The young emerge directly into the water and resemble the adults except that they do not acquire wings until after several molts. In the Sialidae of the Megaloptera, eggs are deposited in masses on leaves or bare ground near water; on hatching, the larvae proceed into the water. Here they stay for a full year, after which they leave the water and pupate for several months in a hollow that they scoop out of moist earth. The adult does not live over winter. Dragonflies may fasten their eggs to plants below or above the water surface, puncture leaves or stems for egg insertion, oviposit eggs in the bottom, or may scatter them through the water and over the bottom. The naiads hatch out in about 3 weeks and are of varied forms and sizes. Dragonfly naiads may be divided into three groups on the basis of their habits: the climbers that crawl through the vegetation; the sprawlers that lie half buried in the mud with legs extended and backs covered with silt; and the burrowers. They all undergo several molts under water, some forms living 11 months in this stage. For their last molt they crawl up the stem of some plant or onto a rock on the shore, molt, and emerge as adults. They live for a few weeks only. Adult dragonflies commonly feed on adult mosquitoes, and the naiads feed to some extent on the mosquito larvae (Needham and Westfall 1955).

Aquatic beetles commonly attach their eggs to water plants or bore holes into plant tissues to hold them. Some hydrophilid beetles make floating silk cocoons containing many eggs, anchoring these cocoons to surface plants. Beetle larvae live only a few weeks before they leave the water and pupate in characteristic mud cells that they build for themselves. Pupation varies from a few weeks to several months, depending on the temperature, before emergence of the adult occurs. The adults are the chief survivors of the winter but sometimes eggs or larvae live through it, too (Miall 1934, Balduf 1935, Rice 1954).

Mosquitoes reproduce abundantly in marshes, ponds, or even in small pools, tree holes, or other water-holding depressions. Some species of mosquitoes lay hard-shelled chitinous-covered eggs on the ground which are capable of withstanding freezing, extreme heat, and drought but hatch very quickly after being covered with warm water. In water, eggs may be laid singly or in rafts. The adult female *Culex vexans* stands at the margin of the pool or on some floating object and deposits as many as 300 eggs. The individual eggs are cigar-shaped and are placed vertically to form a floating raft. They fit together so snugly that the surface film of water does not penetrate between the eggs, and the surface of the raft is dry. The larvae hatch in 12 to 28 hours and hang head down from the surface film. Vibrating vibrissae continually sweep food particles into the larval mouth. The respiratory tube at the posterior end of the body penetrates the surface film and also prevents the body from sinking. At other times the larvae may suspend themselves from the surface film, dorsal side uppermost, and feed on floating materials (Renn 1941). After 3 or 4 molts (5 to 8 days), the larva changes into a quite differently shaped pupa, which hangs from the surface film by two respiratory tubes proceeding from the thorax. The winged adult may emerge in 2 days. Some species may have seven broods per year.

Sexual and other behavior of mosquitoes varies considerably among species (Horsfall 1955). Ordinarily only the female mosquito bites, this to obtain the blood nourishment necessary for egg-laying. The male feeds only on plant juices. Studies made on marked individuals of *Aedes vexans* showed that 73 per cent of the individuals confined their activities to within a radius of 5 miles (8 km), but that 19 per cent traveled 5 to 10 miles (8–16 km) and some even to 16 miles (26 km) away from the point of marking (Clarke 1937). Other species, however, appear not to have such wide ranges.

The pulmonate snail *Physa gyrina* lays its eggs in the spring when water temperatures reach 10°–12°C; thereafter the adult population dies. Snails born in the spring may reach sexual maturity by autumn, but oviposition is normally delayed until spring because of cold weather. The life span is usually 12–13 months, but may be prolonged if development is interrupted by aestivation resulting from the drying up of the pond during the summer (DeWitt 1955).

Many warm-water pond fish, such as the black bass and sunfish (Breder 1936), spawn in nests or redds prepared in shallow water by removing all debris and vegetation over circular areas of $\frac{1}{2}$ to 1 m diameter. There is some preference for gravel and sand bottoms when they are available. The male remains to guard the several thousand eggs during the few days required for their hatching, and the fanning movements of his tail and fins doubtless help to aerate them. He may also guard the young until they can take care of themselves. Both bullhead parents guard the egg masses and keep them continually agitated for aeration; they may even suck the eggs into their mouths and expel them forcibly. The adults keep the young in compact groups by swimming about them. The European carp may spawn promiscuously a half-million or a million eggs during the early spring. The eggs settle in the water and adhere to the roots and stems of vegetation there. The eggs are not guarded, and the young are left to care for themselves.

Salamanders commonly hibernate in humus, under logs, or in other nooks or crevices on land. They usually emerge during the first warm rains of early spring and proceed to the nearest pond, there to lay their eggs. The males deposit their spermatophores on submerged leaves or twigs from whence the female picks them up for fertilizing the eggs. The eggs are laid in jelly-like masses and require several days to hatch if the temperature is low. The eggs of *Ambystoma maculatum* (Gilbert 1944) and *A. texanum* (Burger 1950) hatch more successfully and at a faster rate if they contain unicellular green algae within the capsule. These algae apparently create a symbiotic relationship for oxygen and carbon dioxide. Larval salamanders possess gills, but in all but a few forms these are later absorbed and the adult returns to land. *A. tigrinum* sometimes breeds while still retaining the larval gills, and never leaves the water.

Frogs commonly hibernate in the mud at the bottom of ponds, although some forms, including toads, hibernate in the soil on land. In the spring the males go to small bodies of water where their loud choruses attract the females for mating purposes. The jelly-like masses or strings of eggs require only a few days to hatch, but the tadpole stage lasts longer. Metamorphosis in toads that lay their eggs in temporary ponds takes place rapidly, but in other species, such as the bullfrog, adults do not occur until two years after the eggs are laid (Wright and Wright 1933).

Practically all species of birds characteristic of northern latitudes that nest in the marsh are migratory, as the freezing of the water and drying of the vegetation eliminate their food supply. Nests are located in a variety of situations: on floating masses of plant debris built above the water level, typical of grebes, terns, gulls, black-necked stilt, and ducks; in plant material, placed in tufts of vegetation or formed into platforms, or nests attached to cattails and other emergent plants well above the water level, typical of cranes, gallinules, rails, avocets, snipes, bitterns, ibises, marsh wrens, swamp sparrows, and blackbirds; in swamp shrubs, typical of flycatchers and yellow warblers; in holes in trees, typical of tree swallow, prothonotary warbler, and wood duck; in the tops of trees in adjacent forests, typical of herons, cormorants, egrets, and wood ibis.

The muskrat is one of the most conspicuous and important mammals of both salt- and fresh-water marshes as well as river banks. Along rivers, the animal lives in burrows that it excavates well back in the bank. In marshes, it constructs a dome-shaped lodge, as high as 1 m, by heaping up freshly cut marsh vegetation. The lodge is hollow and dry within, the floor is placed well above the water level. The lodge has several underwater entrances and exits. In it the animal cares for its young and finds protection from enemies and weather in both winter and summer. In addition to lodges, the muskrat constructs shelters, where it may feed out of sight of enemies, and breathing holes, called push-ups, through the winter ice.

BIOMASS AND PRODUCTIVITY

The benthic biomass of ponds in Europe averages 3.6 times greater than in eutrophic lakes (Pieczynska et al. 1963). In a pond in Iowa, the average summer population of bottom invertebrates in water 0.5 m deep averaged 3819 individuals, 1334 mg/m^2; in water 1.5 m deep, 1540 individuals, 1370 mg/m^2. In the shallow water the most important components of the biomass were, in descending order: snails (shells removed), midge fly larvae, annelid worms, and the amphipod *Hyalella*. In the deeper water the biomass was mostly midge fly larvae (Tebo 1955). Productivity of the midge fly *Tanytarsus*, one generation per year, averaged 7.5 g/m^2 in a Michigan lake (Anderson and Hooper 1956). By mooring a floating cage over open water throughout the season in an English pond, a total of 8988 midge flies and other insects per square meter were caught as they emerged from the bottom mud. In shallow water, where the vegetation was thicker, a total of 5979 individuals per square meter were captured, a total which included fewer midge flies and more dragonflies and caddisflies (Macan and Worthington 1951).

Average standing crops of fish in backwaters and oxbows may be almost 500 lb/acre (57 mg/m^2), in midwestern North American reservoirs almost 400 lb/acre (45 mg/m^2), in other reservoirs and ponds 200–300 lb/acre (23–24 mg/m^2), in warm-water lakes 125–150 lb/acre (14–17 mg/m^2), and in trout lakes less than 50 lb/acre (5.7 mg/m^2). There is no tendency for the standing crop to decrease with increase in size of the body of water (Carlander 1955). Biomass varies with the fertility of the pond and the food supply. Ponds and lakes receiving water that drains over fertile soil will have more basic food substances than water draining over poor soils. The presence of certain species depends also on suitable breeding sites (Shelford 1911).

Biomass is further affected by the food habits of the fish species present. In fertile ponds in Alabama containing species feeding largely on phytoplankton, the median biomass of fish was 925 lb/acre (105 mg/m^2); in ponds with fish feeding largely on insects, 550 lb/acre (62 mg/m^2); and in ponds with fish feeding largely on other fish, 175 lb/acre (20 mg/m^2) (Swingle and Smith 1941). It was estimated that about 5 lb of food (2.26 kg) are required to produce 1 lb of fish (0.45 kg). The same ratio has been found characteristic of ponds in Michigan (Hayne and Ball 1956). Hence, the biomass of animal life decreases with each additional link in the food chain.

In two small Michigan ponds where it was possible to tabulate the entire fish population, the benthic production (at least, that portion used as fish food) during one growing season was calculated at about 17 times the standing crop when fish were present. This equaled 811 lb/acre (92.0 g/m^2). The productivity of the fish during the same period was 181 lb/acre (20.5 g/m^2), giving á ratio of 4.5: 1 (Hayne and Ball 1956).

When the standing crop of fish remains the same year after year, its productivity is indicated by the number or biomass harvested. In northern Wisconsin the maximum annual yield of desirable food fishes is about 21 per cent of the mean standing crop; in central Illinois it is about 50 per cent; in southern Louisiana, 118 per cent (Thompson 1941).

The productivity of ponds and marshes for vertebrates other than fish has been measured in a few localities. In northwest Iowa, redhead ducks annually produce about 0.8 young per hectare (33/100 acres); ruddy ducks, 0.6 young (24/100 acres); (Low 1941, 1945). Nine species of ducks in the Bear River marshes of Utah average 16 young per hectare (640/100 acres) (Williams and Marshall 1938). In Idaho, nine species of ducks produce over 22 young per hectare (880/100 acres) and Canada geese about 0.1 young (Steel et al. 1956, 1957). On a well-developed marsh, it is generally possible to remove two-thirds of the muskrats each year and still reserve sufficient brood stock for a sustained annual crop. This is about 2.5 muskrats per lodge (Dozier 1953).

POND AND MARSH MANAGEMENT

Aquaculture, in both fresh and brackish waters, of oysters, shrimp, carp, rainbow trout, channel catfish, the milkfish of southeastern Asia, and other species yields 2 million metric tons of animal flesh per year (Bardach 1968). The maintenance and control, throughout the year, of the water level of ponds and marshes is important for increased productivity. This may often be accomplished by damming the outlet. It is also important to retard the plant succession which, if left alone, will eventually bring about the total disappearance of the habitat. Aquatic vegetation may be reduced by cutting, burning, use of chemical sprays, flooding, and ditching. Artificial fertilization may bring excessive growth of pondweeds (Holm *et al.* 1969). Small ponds, called pot holes, with a good margin of marsh vegetation, or a marsh interspersed with numerous small areas of open water, give the highest yield of waterfowl and other birds, and muskrats. The abundance of waterfowl is often proportional to the extent of the pond margin rather than the acreage of emergent vegetation. In the Louisiana coastal marshes, the highest sustained yield of muskrats (14.5 per hectare/yr or 580 per 100 acres/yr) is in areas with *Scirpus americanus* (O'Neil 1949).

Artificial ponds are easily constructed (Anderson 1950, Musser 1948) and are an asset to farms as a source of food and recreation as well as water for domestic animals. Such ponds are commonly stocked with bluegill and largemouth bass, although other combinations may be used. High rates of reproduction bring the fish population up to full carrying capacity within 1 or 2 years. If the pond were stocked with an herbivorous fish only, such as bluegills, normal reproduction would soon become so excessive that a dense population of stunted fish would be present. Using a prey-predator combination in proper proportions, the predator (largemouth bass, for instance) will consume the excess offspring of the prey species, and the average size of the remaining fish will be increased. The development of aquatic vegetation in these farm ponds is discouraged since it allows too many prey individuals to escape the predator. The fertility of poor ponds can be increased by applying fertilizer encouraging the abundant growth of bacteria, plankton, and bottom organisms providing fish food (Howell 1941). The control of turbidity is also important. Clear ponds with less than 25 ppm turbidity may have 12.8 times more plankton and 5.5 times more fish than ponds with a turbidity exceeding 100 ppm (Buck 1956).

Repeated stocking of ponds with artificially propagated fish is undesirable as there is more trouble in controlling overpopulation than underpopulation (Fig. 6-5). The productivity of a pond is determined not by the number of fish introduced but by available food supply. The available food supply is divided between

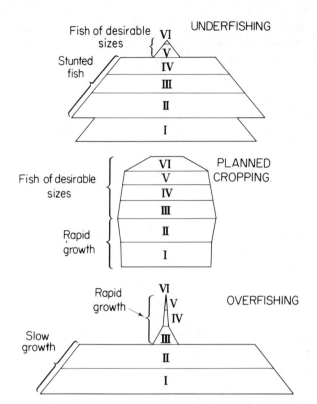

Fig. 6-5 An *underfished* population is characterized by high survival of all 6-year classes and stunting, with few fish attaining desirable sizes for harvesting. *Planned cropping* can produce a population the members of which grow rapidly to large average sizes. Overpopulation is prevented, and therefore an abundance of food is usually available. An *overfished* population is characterized by the first 2-year classes of very young fish being over-abundant and showing slow growth until they reach sizes large enough to interest anglers. At this point, many are taken and the few that escape grow rapidly to large average sizes (from *Management of Artificial Lakes and Ponds* by George W. Bennett, 1971, by Litton Educational Publishing, Inc.; reprinted by permission of Van Nostrand Reinhold Company).

the individuals present. One study showed that 6500 bluegills per acre (16,250 per hectare) averaged 25.5 grams each, 3200 per acre (8000 per hectare) averaged 51.0 g each, and 1300 per acre (3250 per hectare) averaged 104.9 g each (Swingle and Smith 1942). When the available food supply must be apportioned to a relatively large population, growth of individuals is retarded, but with small populations there is more food available per individual, and growth is surprisingly rapid (Fig. 6-6). Ponds are frequently underfished, resulting in large stunted populations. To obtain maximum yield, the harvest must be regulated.

A problem involved in the management of ponds and marshes is the control of mosquitoes. Oiling the water surface will kill mosquitoes, but it also renders the habitat unsuitable for other organisms. Mosquito larvae and pupae are good food for such minnows as *Fundulus* and *Gambusia* that regularly feed at the surface.

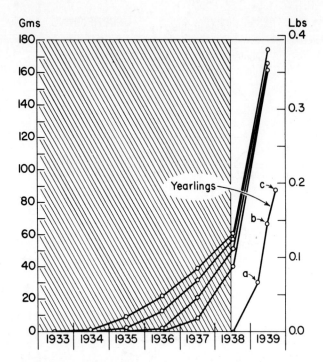

Fig. 6-6 Increase in weight of bluegills upon removal from an overpopulated pond to a pond of lower fish populations (from Bennett, Thompson, and Parr 1940). The cross-hatched portion represents the circumstance of overpopulation; a, spring; b, summer; c, autumn.

Stocking of these fish species will often keep mosquitoes under control. Elimination of aquatic vegetation in the shallow marginal areas will do away with hiding places and leave the larvae more exposed to fish predators.

Because ponds and marshes produce great numbers of fish, muskrats, and waterfowl, and are a source of recreation for hunting, fishing, boating, and swimming as well, the actual economic value of maintaining such areas is often greater than it would be if they were drained and planted to crops (Bellrose and Rollings 1949, Bardach 1968). Proper management of them is therefore a challenge to applied ecologists.

BOGS

Characteristics

Bogs or moors typically develop in the hydroseres of cold northern regions; while marshes and swamps, which are markedly different from bogs (Dansereau and Segadas-Vianna 1952), are characteristically southern in their location. Bogs commonly develop into a coniferous forest climax; swamps succeed to deciduous forest or other southern climax types. Several thousand years ago, when glacial climates gripped the northern states, extensive bogs developed and have persisted as relic communities in spite of the warming of the climate. These bogs are slowly being replaced by pond-marsh species at equivalent seral stages by cliseral succession (Table 6-6).

Bogs occurring in the Great Lakes region are ordinarily small in area and have little or no drainage (Fig. 6-7). There may be oxygen present in the open water of the larger bogs, but it is characteristically in very low concentration, at most, in small bogs or in the marginal zones. Bog water has a distinct brown color; a low nitrogen content; a low temperature beneath the surface; a low pH, at least in the marginal vegetated zones; and a low dissolved salt content.

A false bottom is characteristic of bogs. It consists of finely divided plant material of a light brown color, held suspended in the water at varying depths below the surface. This false bottom may extend downward several meters before a true solid bottom is reached. The material disperses on slight disturbance and may render all the open water turbid. Ordinarily, the surface waters are quiet and clear. Dead vegetation does not completely decompose; as it accumulates, it becomes compressed to form peat.

Plant Bog Sere

In the early stages of development of the bog, organic detritus may accumulate mostly in the deepest portions (Potzger 1956). As time goes on, however, a definite concentric-circle zonation of vegetation is

Table 6-6 Relation of biotic succession to climatic succession in ponds and bogs. Vertical succession from open water to climax forest is taking place in both the pond and the bog, but as the climate gets warmer, there is simultaneously a horizontal succession from the various stages in the bog sere to equivalent stages in the pond sere.

	Stage	Clisere	
		Cold climate bog	Warm climate pond
Biosere	Floating vegetation	Pond and water lilies, or absent	Pond and water lilies
	Emergent vegetation	Sedge mat	Marsh: cattail, reeds, bulrushes
	Low shrubs or heath	Leatherleaf, labrador tea, bog rosemary	Absent
	High shrubs	Mountain holly, chokeberry	Buttonbush, alders
	Swamp or bog forest	Tamarack, black spruce	Soft maple, elm, ash
	Climax forest	Hemlock, pine, white cedar or spruce, fir	Oak, hickory or beech, sugar maple

Fig. 6-7 Plant sere at Bryant's Bog, Michigan, from open water through a narrow broken mat stage of sedge, a low shrub stage of leather-leaf, a high shrub stage (in middle rear) of holly, to tama-rack and black spruce (cour-tesy R. E. Rundus).

Fig. 6-8 Profile of a bog plant sere (from Dansereau and Segadas-Vianna 1952).

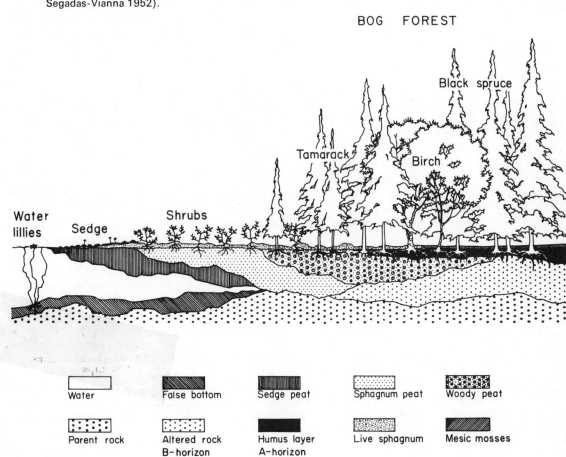

BOG FOREST

Black spruce

Tamarack

Birch

Water lillies

Sedge

Shrubs

| Water | False bottom | Sedge peat | Sphagnum peat | Woody peat |
| Parent rock | Altered rock B-horizon | Humus layer A-horizon | Live sphagnum | Mesic mosses |

established around the margin (Fig. 6-8). As peat accumulates, each zone encroaches on the next inner; the inmost shrinks until all open water disappears. The area becomes finally covered with climax forest (Dachnowski 1912).

In some bogs (Gates 1942, Dansereau and Segadas-Vianna 1952) the first plant stage may be composed of floating vegetation (*Nuphar, Nymphaea, Potamogeton, Sparganium*), but floating vegetation is often absent and the first stage is a *sedge-mat* composed of sedges, cotton-grass, and buckbean. The rhizomes of the sedges grow out into the water and become so interlaced that they form a floating mat. At the water edge the mat may be very thin, but toward shore it may become as much as 1 m thick. Since the mat floats on open water it jars easily, hence the name quaking bog—one must watch his step that he does not break through. *Sphagnum* moss is not essential for the formation of a mat, but it invades the mat quickly and helps bind it together. *Sphagnum* species replace one another and persist into the following shrub and bog-forest stages. Interesting insectivorous species such as the pitcher plant and sundew are common, as are various members of the orchid family.

The next plant stage is dominated by *low shrubs*, which encroach on the floating mat. The leatherleaf, bog rosemary, laurels, labrador tea, sweet gale, and cranberries are important species.

A *high shrub* stage commonly follows the low shrubs at such time as the mat becomes thicker or grounded. Common shrub species are holly, willow, chokeberry, alders, and dwarf birch.

The first tree of the *bog forest* to invade the shrubs is commonly the tamarack, but this species is now less common than formerly because of fire, logging, and the depredations of the sawfly larvae *Lygaeonematus erichsonii*. Black spruce may either invade the shrubs directly or follow the tamarack. Later, the northern white-cedar may become dominant and persist for a very long time, but the ultimate fate of the bog, upon addition of upland soil or lowering of the water table, is to be covered with the *climax* forest of the region.

Animal Life

In bogs that have a large body of open water, or an inflow of water entraining oxygen, and in which the pH is not extreme, invertebrate life comparable to that found in ponds and marshes occurs. True bogs, however, have little oxygen and a low pH, and many pond species do not appear. Mollusks are characteristically absent; sphaeriids may persist but their shells become very thin. Bottom organisms in general are poorly represented because of the tenuous physical nature of the substratum.

Desmids predominate among the phytoplankton,

although dinoflagellates, Chlorophyceae, and Myxophyceae are common. Rotifers and a variety of Protozoa are the principal zooplankters (Graaf 1957).

The chief fish found in acid waters in Michigan are the brown bullhead, northern pike, bluegill, yellow perch, and mudminnow (Jewell and Brown 1929). The mudminnow may be found in waters almost devoid of oxygen since it is one of the few species that can live indefinitely by gulping air at the surface.

In a northern Illinois bog, 11 species of *Sphagnum* moss-inhabiting pselaphid beetles occur. One species is predominant on the sedge mat, another species takes over under the tamarack (Reichle 1969).

Amphibians and reptiles are not characteristic of bogs, although the leopard frog is sometimes numerous on the sedge mat of bogs in Minnesota (Marshall and Buell 1955). Proceeding northward, boreal species of birds become progressively more prominent in the bog avifauna (Brewer 1967). The muskrat and beaver persist into northern Canada.

In general, the productivity and economic value of bogs is very low compared with ponds and marshes. Liming experiments, calculated to improve productivity, are being made. Calcium combines with the humic colloids, which then flocculate and fall to the bottom. This clears the water, light penetrates deeper, pH is raised, and greater algal, zooplankton, and fish growth is induced (Hasler *et al.* 1951). Peat is a special product of bogs of importance in northern Europe. It is cut out in blocks, dried, and used as fuel and for mixing in the soil of gardens and lawns.

It is doubtful if the aquatic fauna of bogs is sufficiently distinct or unique to constitute more than a facies of the pond-marsh biocies. It is succeeded, however, by a distinct *shrub biocies* that differs from the deciduous forest-edge community. The shrub biocies is replaced by *coniferous forest biociations*.

SUMMARY

Ponds differ from lakes in that they are generally small and shallow, and, when mature, have rooted vegetation over most of the bottom. Bogs are limited to northern regions, contain a northern type of vegetation, and are generally acid and deficient in oxygen. As the climate of northern regions slowly warms, stages in the bog plant sere are replaced by corresponding stages in the pond plant sere. The pond sere consists of six or more plant stages but only three animal stages: pond-marsh biocies, deciduous forest-edge biocies, and deciduous forest biociation. These animal communities correspond with the *types of vegetation* in the plant sere, but not with the plant communities identified by *taxonomic composition* of the plant dominants. The animal community in bogs is an impoverished facies of the pond-marsh biocies.

The pond-marsh biocies contains plankton, benthos,

and nekton, as do lakes; in addition, neuston is present. Species that constitute these life-forms are mostly different from those in lakes. In ponds pulmonate snails replace the gilled snails of lakes, and clams are of less importance. Air-breathing adult beetles and bugs, mostly absent from the lake biocies, are often abundant. Adult stages of aquatic insects and terrestrial forms occur in the surrounding marsh. Fish spend most of their lives in the ponds, but go into the marshes to reproduce. Amphibians, reptiles, birds, and mammals are usually numerous.

Food chains in ponds and marshes are based in part on detritus, bacteria, and phytoplankton, true also of lakes; and, in part, on rooted plants, the periphyton that covers them as well as other objects in the water, and filamentous algae. Biomass and productivity are usually greater in ponds than in lakes. Ponds, however, may become stagnant during dry periods, especially in late summer, with great adverse affect upon their carrying capacity. Pond and marsh management for high economic yield of fish, waterfowl, and muskrats requires control of the water level and control over plant succession; an incidental problem is mosquito control. The unique adaptations and behavioral adjustments of animals to meet the critical periods of summer stagnation and winter freezing, characteristic of ponds, are most interesting. Ponds and marshes are available to all ecologists for the study of the life-cycles and adjustments of animals in the pond-marsh biocies.

Chapter 7

ROCK, SAND, AND CLAY

The origin of life was undoubtedly in the sea. Physiological adjustments were necessary before organisms could occupy fresh water. Although some organisms may have become air-breathers and invaded land habitats directly from the sea, most evolution of terrestrial forms has doubtless come from fresh water. Relatively few major groups of animals have been successful in this invasion of land, the most notable being oligochaete worms; gastropod mollusks; many arthropods, especially the insects and spiders; reptiles and their evolutionary offspring, the birds and mammals.

ADJUSTMENTS TO THE TERRESTRIAL HABITAT

Living on land presents many problems. Our present concern in to analyze the ways in which animals have met these problems and to trace the succession of communities in the extreme terrestrial habitats of rock, sand, and clay. A variety of techniques have been developed for measuring environmental factors (Wadsworth 1969).

Gravity

In water, organisms counteract gravity by means of various flotation and swimming mechanisms. Fluid buoyancy permits water-dwellers to attain huge size; consider the whale. A land animal, on the other hand, must support its entire weight. Some terrestrial animals gain a modicum of support by burrowing into the soil; others drag their bodies over the ground surface. But the supportive advantage they thus gain is costly in other directions, for they are slow moving and relatively helpless before predators. The animals best adapted to terrestrial life have evolved appendages in the form of legs or wings that not only raise the body above the ground but are also the means of more or less rapid locomotion and adroit movements over the surface or through the air. Terrestrial adaptation has involved the development of a tough body covering to hold fluids and internal organs in place; a skeletal framework to give permanent shape to the body and, as a system of levers, to furnish means of locomotion; and powerful muscles to lift and move the heavy body. Gravity thus limits the mass of land animals; dinosaurs, mastodons, and elephants approach the maximum practicable size.

94

Moisture

In sharp contrast to aquatic forms, terrestrial animals are not constantly enveloped with a continuous watery medium, with the limited exceptions of protozoans, nematodes, and other small organisms living in moist soil. Land animals, lacking constant contact with the water medium, are faced with the problems of obtaining water and preventing excessive water losses from the body.

Water becomes available to animals in varying amounts in the forms of rain, snow, hail, frost, and fog. Whatever the form it arrives in, the significant things are the amount of free, liquid water added to the substratum, accessible to plants and animals, and the humidity of the air. Considerable amounts of moisture are lost to organisms as runoff water flowing into streams, by evaporation back into the air, and as water bound in snow and ice.

Evaporation of water from the earth's surface or from the bodies of organisms increases as temperature rises, air movement (wind) accelerates, and the amount of moisture already in the air decreases. When measured as grains per cubic foot or as millimeters mercury pressure, the actual amount of moisture vapor in the air is known as *absolute humidity*. This measurement is of less ecological importance than is *relative humidity*, the ratio of amount of water vapor actually in it to the quantity required to saturate the air at existing temperature and barometric pressure. Relative humidity is easily determined by means of sling or cog psychrometers, and may be continuously recorded with temperature by hygrothermographs (Fig. 7-1).

The evaporation rate of water is more closely related to *saturation deficit* than it is to relative humidity. Saturation deficit is a quantity which cannot be directly measured, it must be calculated. It is that additional amount of moisture required to saturate air under prevailing temperature, relative humidity, and barometric pressure conditions, commonly expressed as grains per cubic foot or as millimeters of mercury pressure.

The most exact and desirable measurement affecting water evaporation is the *vapor-pressure gradient* obtaining between the organism and the surrounding air. The gradient is positive if water molecules leave the organism at a rate faster than the rate at which the organism is absorbing them from the air, and negative if the reverse is true (Table 7-1). The determination of gradient magnitude involves the measurements of body temperature of the organism and air, permeability of body membranes, and rate of air movement over the body surface (Thornthwaite 1940).

Water is obtained by a land animal by various devices. There may be some direct *absorption* through body surfaces such as occurs in the toad in moist soil and in some beetle larvae in moist air; this device is important in only a few species. Large mammals frequently travel several miles each day to water holes to imbibe *drinking* water. Many, but not all, birds require drinking water; some species, for instances quail and partridge, get it as morning dew on vegetation. Butterflies may frequently be observed drinking water from small pools. An important source of water is the *free water in food*, particularly in succulent vegetation and in the blood and body fluids of animals. Desert animals depend

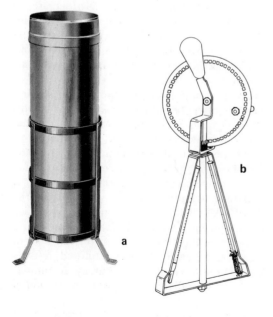

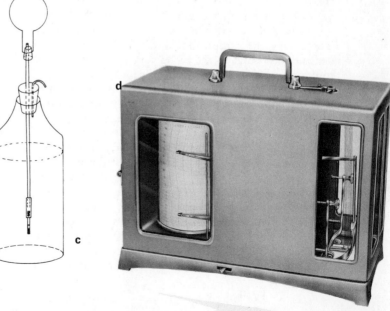

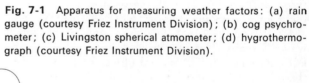

Fig. 7-1 Apparatus for measuring weather factors: (a) rain gauge (courtesy Friez Instrument Division); (b) cog psychrometer; (c) Livingston spherical atmometer; (d) hygrothermograph (courtesy Friez Instrument Division).

Table 7-1 Evaporation and condensation on a free water surface in relation to temperature, relative humidity, saturation deficit, and vapor-pressure gradient (after Thornthwaite 1940).

Factor	Water evaporates	Water condenses
Water temperature (°C)	16	16
Air temperature (°C)	16	27
Relative humidity (%)	70	70
Vapor pressure of saturated air (mm Hg)	133	263
Vapor pressure of air 70% saturated (mm Hg)	93	184
Saturation deficit (mm Hg)	40	79
Vapor pressure of the water (mm Hg)	133	133
Vapor-pressure gradient (mm Hg)	+40	−51

almost entirely for water on that contained in their food and on *metabolic water*, liberated when fats and carbohydrates, and to a lesser extent, proteins, are oxidized in their bodies.

Water is lost from the body through the skin and lungs as insensible moisture and perspiration. Rapid and largely uncontrolled loss of moisture through the skin of amphibians, snails, annelids, and insect larvae is a limiting factor confining these animals to moist habitats or to activity only in times of high humidity. Adult insects and other arthropods, reptiles, birds, and mammals have evolved body surfaces of chitin and waxes, scales, or cornification of the surface layers of the skin that largely prevent uncontrolled loss of moisture. Moisture loss through the respiratory surfaces in these forms remains considerable, however. Water is also lost with the *feces*, although in some species much water is reabsorbed by the large intestine before the feces are ejected. The amount of water removed from the body by excretory organs, particularly the *kidneys*, varies directly with the amount of water intake and inversely with the amount lost through other devices. The kidneys are critical to maintenance of proper concentration of salts in the blood and body fluids. It is important that water intake balance water loss. Organisms are very sensitive to disturbances in body water balance, and this factor is significant in determining the type of niche which a species comes to occupy.

In order that animals could exist in terrestrial habitats, they had to acquire the ability to carry on reproductive activities in the absence of water. The chief reproductive adaptations involve the following (Pearse 1950): internal fertilization; a shell covering the egg to conserve moisture and salts; food provision to the embryo and young, by yolk in the egg cell, placenta in the uterus of the mother, or direct feeding by the adults; reduction in number of young with more efficient parental care; reduction or elimination

of free-swimming larval stages; and greater segregation of species into different niches to avoid interspecific disturbances.

Temperature

Aquatic animals are not ordinarily subjected to temperatures below freezing and are in a relatively stable temperature environment, but terrestrial species are exposed to highly variable temperatures that may reach extremes of about −68°C and +55°C. No single species is required to withstand such a wide range of temperatures, however. Optimum and tolerance limits vary from one species to another.

No aquatic species has evolved control over its body temperature. Aquatic warm-blooded mammals and birds are derived from terrestrial forms. An ability to maintain a constant body temperature has survival value in terrestrial habitats, however, and consequently physiological mechanisms for *homoiothermism* developed independently in birds and mammals. All other land organisms are *poikilothermal*; that is, they have no *physiological mechanism* for maintaining a constant body temperature. Some poikilotherms, such as bees, have developed *special behavior patterns* that enable them to maintain fairly constant conditions in the hive by cooperative efforts; some moths and other insects with hair-like body coverings maintain a degree of constancy in body temperature by shivering their wings; some lizards, snakes, and turtles are able to exert some control over their body temperatures by moving into and out of sunlit areas (Heath 1968).

Rates of activity, food consumption, metabolism, growth, and other physiological functions increase, to a certain limit, with rise of body temperature. Homoiotherms maintain a continuous high rate of functioning because of their constant high body temperatures, but the rate at which poikilotherms function varies with the temperature of their habitats.

Latitudinal distribution of both poikilotherms and homoiotherms is often limited northward and southward by the extremes of temperature that they can tolerate. The rate of energy exchange in poikilotherms is so directly dependent on the amount of heat in the habitat that the total growth and reproduction of a species may be determined by the extent to which it can accumulate developmental heat units during the year (Shelford 1929: Chap. 7); this process is similar to accumulation of heat budgets in lakes. The closer a region lies toward either Pole, the shorter the growth season is, and distribution may be limited not by extreme low temperatures as such, but by accumulation of heat energy insufficient to permit completion of life-cycles.

The relation of energy balance in homoiotherms to air temperature is even more complicated (Kendeigh 1969). In cold regions an animal may require all the energy its food provides simply to maintain its own existence; no surplus is available to meet the high demands of reproduction. Under such conditions, a species cannot become permanently established in a region. A warm-blooded animal requires a range of temperature that is comfortable and in which it can ingest and metabolize food at a rate sufficient to maintain normal body temperature, sustain physical existence, and carry on reproductive activities, too. We can thus speak of *existence energy* and *productive energy*, concepts essential to understanding the relation of an organism to the temperature of its environment.

All organisms outside the tropics must adjust to meet the critical winter season. In those species active throughout the year, there is an increase in physiological resistance to cold and in the body insulation of plumage or fur. Those species incapable of maintaining activity *in situ* over winter either migrate to more favorable regions or hibernate, or the adults die. Many invertebrates survive the winter in resistant egg or larval stages.

Oxygen

Dry air at 760 mm Hg pressure contains approximately 21 per cent oxygen, 0.03 per cent carbon dioxide, 78 per cent nitrogen, and traces of other gases. Oxygen is thus much more abundant, constant, and available at all times in air than it is in water. Oxygen availability seldom becomes a critical factor for land animals, with the occasional exception of forms that live in the soil or invade high altitudes.

Although terrestrial organisms have evolved simple moist chambers, branched tracheal systems, or complicated lungs to replace the gills found in many aquatic forms, the fundamental requirement of moist membranes for the exchange of oxygen and carbon dioxide between body fluids, tissues, and the surrounding medium remains the same. The skin still serves this purpose in some terrestrial forms—annelids and some amphibians—but in most forms the moist membranes are within the body. Internal placement decreases the loss of water through evaporation. The evolution of an ability to take oxygen directly out of the air apparently preceded the actual invasion of land, and may have been induced in the pond and marsh habitat when oxygen dissolved in the water became reduced or absent during summer stagnant periods (Pearse 1950). The evolution of internal air-breathing organs was probaby concurrent with the evolution of mechanisms to prevent excessive water loss from the exposed surfaces of the body.

Solar Radiation

Solar radiation furnishes the energy for circulation of the oceans and the atmosphere, for the water cycle between oceans and lakes and the land, and for photosynthesis. It takes the form of an endless procession of waves. The length of a light wave from crest to crest, or trough to trough, determines its character in respect to energy and color; the height of the wave determines its intensity. Wavelength is commonly expressed in millimicrons   (1 mμ = 0.000001 mm = 10 Ångstrom units). All wavelengths have a velocity of 299,340 kilometers per second. The solar spectrum varies from 51 mμ, which is the shortest ultraviolet radiation, to 5300 mμ, which is the longest infrared radiation. The spectrum visible to man is between 390 and 810 mμ. Considerable ultraviolet is absorbed by the atmosphere, and the ultraviolet wavelengths reaching the earth's surface are mostly between 292 and 390 mμ. Color perception by man is as follows: violet, 390–422 mμ; blue, 422–492 mμ; green, 492–535 mμ; yellow, 535–586 mμ; orange, 586–647 mμ; and red, 647–810 mμ. The longer waves are rich in heat energy; the shorter, in actinic energy.

Solar radiation may be measured with a pyrheliometer, by which readings are given in terms of heat energy (g-cal/cm²/sec). Results are not accurate for the shorter wavelengths. Photoelectric cells accurately measure intensities in the shorter wavelengths; readings

Fig. 7-2 Ultraviolet-hydrogram for February populations of pronghorn antelope in Yellowstone Park for the years indicated in italic numerals. It appears that the number of young produced in any year has been determined in a sensitive period two Septembers earlier; hence, the ultraviolet data given for each year are for the second preceding September, and rainfall is for that September through the August following, inclusive (modified from Shelford 1954).

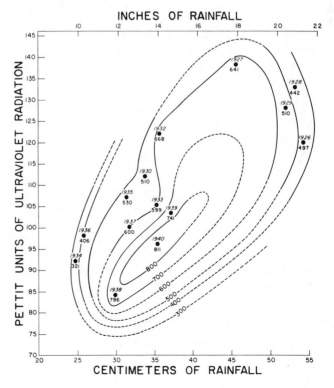

may be given in foot-candles. With any photometric instrument, the measurement of intensity of any portion of the spectrum requires the use of calibrated color filters that screen out everything but the desired wavelengths.

Different wavelengths have different effects on organisms. Green light is reflected by plants; little is used in photosynthesis. Some early experiments on tadpoles, fish, snails, and other forms (Davenport 1908) indicate that there is an increasing growth rate in different wavelengths, in the following order: green, red, white, yellow, blue, violet. The physiological basis of this phenomenon is not known. An excess of infrared may produce overheating of the animal. Ultraviolet in large concentrations is harmful to most animals, but in lower intensities is beneficial to elaboration of vitamin D. Evidence indicates that ultraviolet combined with rainfall is important in controlling numbers of some mammals, forest-edge birds, and insects (Fig. 7-2). In general, terrestrial organisms are exposed to much higher intensities of solar radiation than are aquatic organisms and have evolved horny or chitinous body coverings, hair, or feathers that function in part to protect internal structures from lethal concentrations. Long-range vision has developed only in land animals and is correlated with the high light intensities characteristic of terrestrial habitats.

Diurnation

Animals may be divided into *diurnal* (daytime), *crepuscular* (late evening and early morning), *nocturnal* (night), and *arhythmic* (irregular) species. Animals that occupy microhabitats where temperature and light changes are negligible tend to be arhythmic—cave crayfish, log-inhabiting beetles, moles, shrews, and some ants, for instance. The microfauna of the soil is probably arhythmic. About two-thirds of the mammal species occurring in both temperate deciduous and tropical rain forests are nocturnal. Birds are predominantly diurnal, except that owls are nocturnal and goatsuckers crepuscular. The majority of amphibian and reptile species are nocturnal; some frogs and lizards are diurnal. The drosophilid flies are crepuscular, having pronounced peaks of activity at dawn and dusk (Taylor and Kalmus 1954). The major period of activity of many nocturnal animals occurs during the first half of the night period, although they often possess a secondary pre-dawn period of activity. With diurnal animals, the major period of activity usually comes during the first portion of the day, although there may be a secondary pre-dusk period of activity (Calhoun 1944–46).

It is of interest that nocturnal animals are less frequently gregarious and social than the diurnal forms. Adjustments for night activity involve development of

luminescent organs such as fireflies (Lampyridae) possess; infrared-sensitive vision, suggested for some insects and birds but not proven for owls (Dice 1945); increase in visual acuity by modification of eye structures (Walls 1942); keenness of smell displayed by some mammals; and increased sensitivity to sound, remarkably developed in bats. Bats have evolved a radar system, called *echolocation*, whereby the animals emit ultra-high-frequency sound waves, which are reflected from objects back to the ears (Griffin 1953). Color vision, well developed in some diurnal insects, fish, reptiles, amphibians, birds, and mammals, is largely lost in those nocturnal species active at such low intensities of light that colors would be indistinguishable anyway. Correlated with loss of color vision is restriction of body coloration to blacks and whites or intermediate shades. Nocturnal animals escape such diurnal predators as reptiles, birds, and hymenopterous insects. There are nocturnal predators, to be sure, but predation pressure at night appears to be less intense than during the day.

There is decreased competition for food and shelter when some species are active by night, others by day, over the same range. Butterflies are predominantly diurnal; moths, nocturnal. Animals with moist skin, like snails and amphibians, suffer less evaporation of water from their bodies at night, when the relative humidity is higher and the temperature is lower. There is some belief (Clark 1914) that nocturnal forms are derived from originally diurnal forms; an adjustment, perhaps, to avoid competition from aggressive diurnal species occupying the same niches.

Seasonal Variations (Aspection)

In tropical rain forests there is very little seasonal variation in the number of species active and size of populations attributable to length of day, temperature, and humidity, for these factors are nearly uniform throughout the year. In other parts of the tropics, however, there are definite wet and dry seasons, and a considerable change in numbers and activities of animals correlates with the seasonal variations in vegetation and food supply. In temperate and arctic regions, seasonal differences in length of day and temperature become increasingly great the closer the region lies toward a pole (Fig. 7-3).

Correlated with seasonal changes in climate are adaptive adjustments of metabolism and energy balances, regulation of breeding time, change in food habits, and migration or hibernation. Birds breed in the spring and early summer, since lengthening daily photoperiods stimulate maturing of the gonads (Burger 1949). *Photoperiodism* also controls the breeding time of some mammals, fish, and invertebrates as well as plants. However, in some species, say trout and deer,

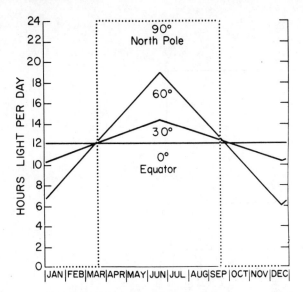

Fig. 7-3 Monthly variation in daily photoperiods at cardinal latitudes of the northern hemisphere (after Boggs 1931).

shortening rather than lengthening photoperiods are stimulating, and such species regularly breed during the autumn.

In deciduous forests, seasonal differences in the development of the foliage greatly affect animals. When trees are bare, sunlight penetrates to the forest floor more readily than when foliage is in full development. Foliage is important because it is protective cover from weather and offers refuge and concealment from predators; to many species it is a direct source of food.

Six main ecological seasons, or aspects, may be recognized:

Hiemal aspect (middle November to early March). Deciduous trees bare of leaves; herbs mostly dead; most invertebrates, amphibians and reptiles, and some mammals in hibernation or dormancy; only permanent resident and winter visitor species of birds present.

Prevernal aspect (middle March to middle April). First flowers appear, animals come out of hibernation, early migrating birds arrive, reproductive activities begin.

Vernal aspect (late April to late May). Deciduous trees become fully foliated, early flowers replaced by shade-tolerant species, bird migration reaches its peak.

Aestival aspect (early June to late July). Vegetative growth reaches maximum and animals at peak of reproductive activities.

Serotinal aspect (early August to middle September). Vegetative growth declines, many ground plants dry up, fruits and nuts begin to ripen, birds become quiet and undergo molt, mollusks and some other invertebrates and some mammals aestivate, insects attain maximum populations.

Autumnal aspect (late September to early November). Deciduous tree foliage changes color and falls, autumn flowers appear, insect populations decline, many animals enter hibernation or dormancy, many birds migrate to the south.

These periods are best developed in the temperate deciduous forest but occur in modified form in all temperate communities. The beginning and end of any aspect cannot be set with exactness, since aspects even in the same locality vary from year to year, and they vary with latitude and type of community. A more detailed classification of aspects is given by Macnab (1958).

Substratum

The substratum greatly influences the kind of plants and animals that occur in the pioneer stages of succession. Bare rock presents one extreme physical habitat, sand another, and clay yet another. The substratum affects animals indirectly in terms of the kinds of plants it supports and the variety of niches it affords. Differences between early sere stages notwithstanding, later ones tend to be more and more alike so that *convergence* occurs. In temperate humid regions, where the seres pass through several stages, the climax communities of all seres are very much alike, regardless of the type of bare area on which they originated (Fig. 7-4).

ROCK SERE

Plant Communities

Stages in the plant sere on bare rock are lichens, mosses, herbs and grasses, shrubs, and forest. The species composition of each stage varies with the chemical nature of the rock, the prevailing climate, and the locality.

In the first stage, various kinds of lichens compete for a foothold, but crustose types usually precede foliaceous types. Mosses and such fruticose lichens as *Cladonia* follow foliaceous lichens; or may initiate the sere, telescoping the earlier lichen stages (Keever, Oosting, and Anderson 1951).

Lichens and mosses soak up moisture in wet weather. They derive mineral nutrients from the underlying rock. Carbon dioxide secreted from the rhizoids forms a weak acid with water and dissolves the binding material of the small rock particles. Rhizoids may penetrate rock for several millimeters. These plants trap windblown dust and obtain nitrogen from organic compounds in it. When the plants die, they become an addition to the accumulation of organic matter. Herbs, grasses, ferns and later stages invade to continue the crumbling of the rock and buildup of soil. Freezing and thawing of water may crack the rock, and in these

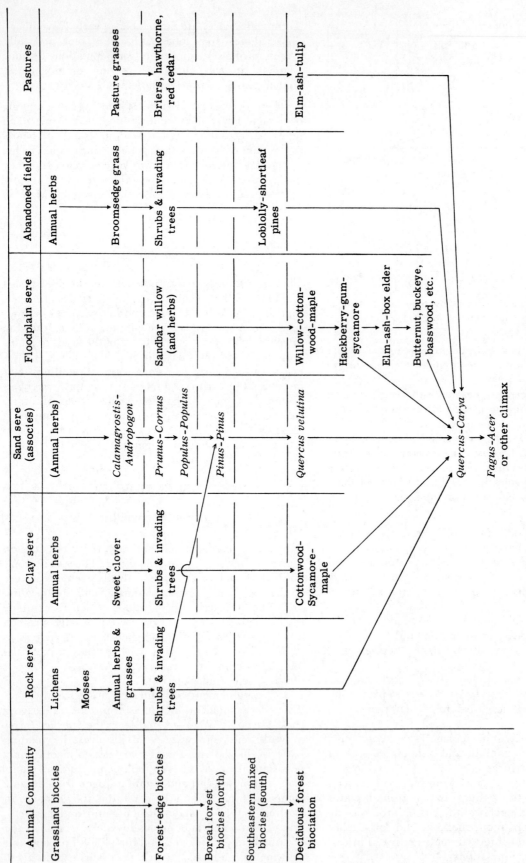

Fig. 7-4 Correlation of animal communities with converging stages of plant seres in the eastern United States.

cracks wind- and water-borne soil lodges and supports plants. Once shrub and tree roots get started in crevices, their growth exerts a powerful force, further splitting and crumbling the rock.

Animal Life

Animal life in the pioneer plant stages on rock is scanty. Ants and spiders roam over the bare rock, and insects of various sorts may stop there, temporarily. Spiders may construct webs and nests in rock crannies or among the lichens. Some tardigrades find preferred niches in lichens. Mosses offer a somewhat more substantial microhabitat, but only those animals that can tolerate great extremes of flooding, dessication, heat, and cold can survive. Such forms are found in the rhizopod protozoans, nematodes, bdelloidid rotifers, tardigrades, copepods, small insects, and mites (Heinis 1910). They often have special means of attachment to keep them from being blown away by the wind, such as strong claws or cement glands on the feet, long bristle-like threads to entangle among the moss filaments; stickers or spines covering the eggs. Since wet periods are often too short to permit complete development, all stages must be tolerant of desiccation, at which time activities and growth are largely suspended.

Animal life in general and land snails in particular are usually more abundant in vegetation (grassland and forests) established on calcareous soils derived from limestone than in the vegetation established on soils derived from sandstone, granitic, or volcanic rock. Calcium carbonate is a mineral essential to the metabolism of most animals and for building such skeletal structures as bones and shells. Snails are less numerous in the grass stage than in the later, moister forest communities that develop in the succession.

SAND SERE

Plant Communities

Sand is the product of mechanical pulverization of various rocks. It is deposited by wind and water. Where extensive areas of sand occur, strong winds pile the sand into shifting dunes. These dunes have a characteristic shape as the sand grains are blown up a long, rather gentle windward slope and swept over the crest onto a steep lee slope. Moving dunes may engulf whole forests; they eventually move on, leaving the denuded trunks of trees that they have smothered. The dunes continue to move until they reach the shelter of some other dune, get beyond the full force of the wind, or until invading vegetation covers the surface and anchors them down. The most successful sand-binding plants are the grasses *Ammophila*, *Calamovilfa*, and *Agropyron*, willows, sand cherry, and cottonwoods. Willows

and cottonwoods will survive even when almost buried. Each succeeding stage ties the sand down more firmly, but any break in the vegetation occasioned by a blow-down of trees or disturbance by man may invite the wind to start moving the exposed sand, and change the partially anchored dune again into a moving one. Only when the pine stage or the black oak stage is reached does the dune become relatively secure from the wind.

The plant sere on the south shore of Lake Michigan consists essentially of the following stages (Cowles 1899):

Lower beach: Washed by summer storms and devoid of vegetation.

Middle beach: Washed only by severe winter storms; comparatively dry in summer; upper limit marked by driftwood and debris. Scattered annual plants present (Fig. 7-5).

Calamagrostis-Andropogon associes (upper beach): This is where the dunes begin to form. In this early developmental stage (associes) grasses are dominant, particularly *Calamagrostis longifolia*, *Andropogon scoparius*, *Agropyrum dasystachyum*, *Ammophila arenaria*, *Elymus canadensis;* various biennial and perennial herbs make their appearance. The sandbur grass occurs extensively in some areas.

Prunus-Cornus associes: The commoner shrubs are sand cherry, chokecherry, red-osier dogwood, creeping juniper, and the frost grape vine. Shrubs may invade the grass directly but become more common in the following tree stages.

Populus-Populus associes: The first tree stage in the southern portion of Michigan is made up principally of the eastern cottonwood and, in the northern portion, of the balsam poplar. The trees commonly occur in open stands with grasses and shrubs forming the lower strata. The habitat is essentially forest-edge. The shrub and cottonwood stages are often missing so that the sere progresses from the grass directly to the pine or black oak stage.

Pinus-Pinus associes: Jack pine, red pine, and eastern white pine may occur in relatively pure or mixed stands. Northern white-cedar and eastern redcedar also occur; the former, more commonly northward. Succession to this stage is mainly contingent on stabilization of sand in dunes, and more efficient utilization of water resources. For succeeding stages to emerge, soil must develop by deposition of humus. The floor of pine forest is covered with a carpet of needles, although patches of bare sand still occur. As the sere advances, all bare areas become covered with a layer of humus.

Quercus velutina consocies: Black oak often forms a nearly homogeneous stand that may persist for a long time.

Quercus-Carya associes: Black, white, and, to a lesser extent, red oaks are commonly mixed with shagbark and bitternut hickories and, in moist habitats, American basswood.

Fig. 7-5 Sand sere at Ludington State Park, Michigan. (a) The lower beach (at right center) is washed by ordinary waves; the middle beach (in center) contains driftwood left by heavy storm waves; the upper beach (at left) has a sand dune well anchored by grass and frost grape (light areas), shrubs (dark areas), and cottonwood trees. (b) Grass stage, showing blowouts devoid of vegetation; a mixed pine stage is shown in the distant background (courtesy R. E. Rundus).

Fagus-Acer association: When soil humus and moisture become sufficient, American beech and sugar maple invade the sand to form the final climax stage.

In other localities, the taxonomic composition of the communities, especially the later stages, differs considerably. The character of the climax varies according to climate and geography, but perhaps the sere is as complete and as complex in the Lake Michigan region as it would be anywhere.

Habitat

The sand dune habitat is characterized by extreme fluctuations in physical conditions, generally resembling those of a desert (Chapman *et al.* 1926). Temperatures, especially at the ground surface, are very high during bright sunny days; relative humidity is very low. Evaporation from spherical atmometers is 2.5–3 times higher than in forest habitats at the same time of day. At night the ground surface temperature

may be even lower than that of the air since there is little or no surface covering to prevent rapid heat radiation.

Correlated with the diurnal changes of temperature, relative humidity, and light, the kinds of animals active on the sand during sunny days are quite different from those active on cloudy or rainy days and at night. When the temperature of the sand nears 50°C, all insects leave the surface (Fig. 7-6). Some climb grasses to get off the ground, others enter their burrows. Insects flying above the sand can select an optimum temperature from widely different temperatures merely by changing their elevation a few inches. They make hurried landings when entering their ground burrows. The female velvet-ants are usually among the last to retreat into their burrows in the morning and the first to leave them in the evening. Experiments show that they of all insects in this habitat are the most tolerant of the high temperatures. Animals living here must either be physiologically tolerant of extreme heat or possess behavior patterns that enable them to avoid it.

Grasshoppers and Other Orthoptera

There have been detailed studies of a few special groups of animals occupying the Lake Michigan sand dunes. Three species of wood roach, 2 species of walking-stick, 20 species of short-horned grasshopper, 13 species of long-horned grasshopper, and 6 species of field cricket occur in various stages of the sand sere in the Chicago area (Strohecker 1937). A breakdown of this list shows that 7 species, all short-horned grasshoppers, occur in the grass and cottonwood stages; of these, one species is not found in the pine stage, and the other 6 species disappear by the time the black oak stage is reached. Eight new species of orthopterans, including 4 short-horned grasshoppers, enter the sere at the pine stage, but only 5 species persist into the black oak stage. Altogether there are 23 species of orthopterans in the black oak forest, an increase of 18 new species. There are only 25 species of orthopterans listed for the climax, but this includes 4 species of camel crickets which for the first time can find their proper niches under logs, and a katydid that appears in the trees. The greatest change in species composition within the sere occurs at the black oak stage upon the disappearance of 67 per cent of the species present in the earlier stages and the appearance of 78 per cent of the species as new forms. Of that 78 per cent, 61 per cent persist through all later stages. The change in species composition at this stage can be correlated with the development of a canopy of foliage and the resulting reduction in light intensity and soil temperatures.

The community or niche restriction of the short-horned grasshoppers appears to be determined either by soil conditions or by the vegetation (Isely 1938a).

Fig. 7-6 Temperature gradient on a sand dune, on a rainy day and on a sunny day (after Chapman *et al,* 1926).

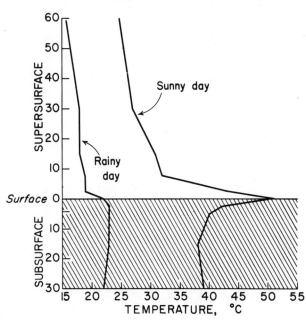

Before short-horned grasshoppers lay their eggs in the ground, the female tests the soil with her ovipositor until she finds soil of proper conditions. Experimental studies show that in certain cases soil texture is the critical factor in the choice of the egg-laying site, while in other cases soil structure or degree of compaction is most important. Soil conditions appear particularly important for the subfamily of band-winged grasshoppers; for other groups, vegetation is of greater significance.

In an experimental study of how an available choice between foods may affect distribution (Isely 1938a), one-half of 40 species of short-horned grasshoppers showed a feeding preference for grasses and one-half for broad-leaved herbs. The latter group included the spur-throated grasshoppers. Four species were restricted to feeding on a single plant species; 30 species confined themselves to a few plant species only, and usually of a single plant family at that; only 2 species fed on a wide variety of plants. In several instances grasshoppers starved in cages, when there was an abundance of fresh plant materials present that were palatable for other species, because their own preferred food species were absent.

All five species of false katydids studied in Texas (Isely 1941) confined their choice of food to related species of broad-leaved herbs or forbs, refusing grasses; adults showed a marked preference for the flower parts and tender fruit pods. Two species of shield-backed grasshoppers were wholly carnivorous. The flower-feeding false katydids disappeared from the prairie in late spring and early summer as the flowing plants passed their peak, but the insect-feeding grasshoppers persisted to the end of July or until temperatures became too high for their comfort.

Ants

Ants cannot get established on the beach because of its unstable character and are scarce even in the grass and cottonwood stages because of the shifting character of the dunes (Talbot 1934). Species found are crater-formers *Lasius niger neoniger* and *Pheidole bicarinata*, and two species of *Camponotus* that find protection under the occasional log that occurs. On hot dry days these ants withdraw to several inches below the surface and emerge only in the cool of the evening.

In the pine stage, the slight mixture of humus in the sand is decidedly favorable, food is more abundant and varied. Of 18 ant species found, 9 live in patches of open sand with no shelter, 6 require sand with some protection above it (logs, bark, needles), and 3 are strictly log-inhabiting forms. *Monomorium minimum* and *Paretrechina parvula* are characteristic species.

In the black oak community, 29 species occur of which only 6 live in scattered open areas of sand. These 6 species are quickly crowded out when there is development of a complete leaf covering over the ground. *Formica pallide-fulva*, which was becoming important in the pine community, is the predominant ant in the black oak stage. Its nests are invariably found under pieces of bark or branches lying on rather open ground.

As the sere advances into the white and red oaks stage, open areas of sand disappear, humus and moisture increase, logs in all stages of decay occur, the whole area becomes shaded, and the daily extremes in temperature and humidity typical of the open dunes are considerably curtailed. Species of ants characteristic of the early stages disappear, and forms that are found in mesic deciduous forests generally predominate, although there are only six species found here that do not also occur in the black oak community. *Formica truncicola obscuriventris* is the most numerous species. The number of colonies and variety of species reach maximum in the oak stages.

In the climax beech-maple community, the number of soil-dwelling forms is reduced, perhaps because of the thick rich humus, although log-inhabiting forms are numerous. *Lasius niger alienus americanus* and *Aphoenogaster fulva aquia picea* are the only ants abundant in the deep woods; ants are more numerous in the forest-edge than in the forest-interior.

Although different species reach peaks of abundance at different points in the habitat gradient proceeding from open sand to dense forest, the nature of the substratum divides the species into two major groups: those that tolerate and reach their greatest abundance in the sandy areas where vegetation is scattered, and those that are limited by sand and require humus in the soil or the microhabitat of decaying logs. The transition or ecotone between these two ant communities comes at the pine and black oak stages. Experimental studies of six species in the genus *Formica* indicate that physiological differences occur, and that some species are able to invade places of low relative humidity that others cannot.

Spiders

In the sand dunes on the south shore of Lake Michigan and in adjacent areas, 228 species of spiders are to be found (Lowrie 1948). In England, 188 species were recorded from the water's edge through the grass stage alone (Duffey 1968). The number of families represented in the Lake Michigan area, the number of

Table 7-2 Number of species of each spider family found in each sand sere plant stage except the oak-hickory (after Lowrie 1948).

Spider family	Middle beach	Grass stage	Cottonwood	Pine	Black oak	Beech-maple
Web-builders						
Ariopidae	2	5	3	5	26	20
Micryphantidae	2	2	0	2	8	3
Theridiidae	1	1	2	5	10	13
Dictynidae		1	1	4	3	2
Linyphiidae		2	0	0	5	7
Agelenidae				1	3	8
Ciniflonidae					1	1
Hahniidae					2	1
Mimetidae					1	1
Uloboridae						2
Total	5	11	6	17	59	58
Per cent all species in stage	29	35	34	40	35	48
Non-web-builders						
Lycosidae	9	4	2	1	24	11
Gnaphosidae	2	1	0	3	11	6
Salticidae	1	7	5	10	29	17
Thomisidae		7	4	10	22	14
Clubionidae		1	1	1	14	7
Anyphaenidae				1	1	3
Dysderidae					1	0
Oxyopidae					1	1
Pisauridae					6	4
Total	12	20	12	26	109	63
Per cent all species in stage	71	65	66	60	65	52
Number individuals of all species in herb stratum per 50 sweeps		8	6	10	18	24

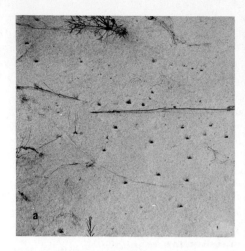

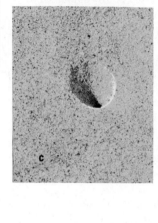

species involved, and the abundance of individuals per unit area increase as plant stages in the sere succeed one another (Table 7-2). Probably because of the greater diversification of the vegetation, the availability of logs, the increase in number of strata, and the consequent greater variety of niches, spiders, like ants, are represented by a larger number of species in the oak communities than in the earlier stages of the sere or in the climax. It is of significance that up through the black oak stage new species appear in each succeeding stage with very few dropping out. In the beech-maple climax, however, 51 per cent of the spider fauna occurring in preceding stages are no longer found, while 79 per cent of the species are either new with this stage or came in at the black oak stage and remained. Up to the black oak stage the species composition of the spider population shows ecesis, but with the advent of deciduous forest, the change in the fauna composition is sufficiently extensive to indicate succession of distinct communities.

There is also a change in the mores of spiders as the sere advances. Small lycosids that hide during the day under driftwood or other debris and run over the sand at night hunting for insect prey washed up by the waves are most characteristic of the beach. The permanent population is small. A burrowing spider, *Geolycosa wrightii*, is usually common. The burrows in which the spiders stay during the day may be easily spotted on the beach and through the grass and cottonwood stages. Web-building species are at a disadvantage in the early stages of the sere, however, because of the general lack of vegetation to which their webs may be anchored and because of the destructive effect of unchecked wind. With the appearance of grasses, a substratum in which spiders can build webs becomes available. In later stages, the percentage of web-builders increases considerably as stratification progresses and the forest furnishes a scaffold.

Other Animal Life

Strong offshore winds often blow insects out over the water where they are forced down onto the surface and washed ashore. Windrows of such insects, many thousands of individuals representing a wide variety of species, are sometimes to be seen. Dead fish washed up on shore are fed upon by flesh-flies and histerid,

Fig. 7-7 Burrows made in sand by arthropods: (a) burrows of a digger wasp, *Microbembex monodonta*; (b) a digger wasp, *Bembex spinolae*, and a cross-section sketch of its burrow (Shelford 1913); (c) the white tiger beetle and its burrow (Shelford 1913); (d) excavated burrow of a sand spider (courtesy R. E. Rundus). The upper portion, shown with a stick in it, is intact; the lower portion, in the shadow, is broken open.

dermestid, and rove beetles. The tiger beetles *Cicindela hirticollis* and *C. cuprascens*, a white ground beetle and other carabids, shore bugs, digger-wasps, robber flies, and other insects and spiders come down from higher ground to feed on the scavenger species and those washed up by the waves (Shelford 1913, Park 1930). The tiger beetles, ground beetles, digger-wasps, and sand spiders build their burrows and larval stages far enough back to escape the summer waves. Termites feed on buried wood that is decaying or on the undersides of logs that have drifted ashore. The piping plover and spotted sandpiper place their nests in the middle and upper beaches. At night, the toad, opossum, raccoon, and the deer mouse come down to scavenge whatever is available. The light coloration of many of the insects and spiders that occur on sand is doubtless an adaptation for concealment (Hart 1907).

The kinds of animals occurring in the grass, shrub, and cottonwood communities are similar except that new species invade with each successive plant stage. The white tiger beetle *Cicindela lepida* first appears on the upper beach and reaches maximum populations in the cottonwood stage, as do the digger-wasps, robber flies, and sand spiders. Another tiger beetle, *Cicindela formosa*, occurs in the ecotone between the cottonwood and the pine stages (Fig. 7-7). Snout beetles, spittle bugs, and miscellaneous other insects are occasionally very numerous. Some 592 species and varieties of beetles have been taken from various stages of this sere (Park 1930). Fifty species were found to occur in the cottonwood stage, 23 in the conifer stage, and about 200 in each succeeding forest stage. The occurrence of bees is dependent to a large extent on the variety and abundance of flowers, but the number of species in each plant stage increases up to the black oak and then declines to the climax (Pearson 1933).

Vertebrates are not usually numerous on sandy flats or dunes away from the water's edge. The vesper and lark sparrows occur among the grasses, the prairie warbler and chipping sparrow are found among the shrubs, and the kingbird is conspicuous in the trees. Tracks of the prairie deer mouse are frequently to be seen on the sand. Fowler's toad and the hognose snake are the only amphibian and reptile that regularly occur. The grass, shrub, and cottonwood stages ordinarily occupy relatively narrow belts parallel to the lake shore. Extensive sandy areas inland may have a larger variety of species present (Vestal 1913).

The pine community in the sere is not so well developed around the south end of Lake Michigan as it is northward. The coniferous forest penetrates southward from the north, and some northern animals move with it. Nesting birds are represented by the slate-colored junco, red-breasted nuthatch, black-throated green warbler, blackburnian warbler, and myrtle warbler, all belonging to the boreal forest biociation. Forest-edge and deciduous forest birds also occur. The red squirrel is a characteristic boreal mammal that occupies this stage, and the white-tailed deer browses on conifer foliage, especially the white cedar. The six-lined racerunner and blue racer snake appear.

Among the invertebrates are the bronze tiger beetle *C. scutellaris* and the ant-lion.

With the advent of the black oak and later forest stages, most species requiring open areas or depending on patches of bare sand disappear. Although the bronze tiger beetle remains abundant in the black oak community, it, as well as the other tiger beetles, disappears in the higher plant stages. Only the green tiger beetle *C. sexguttata* is in the climax, a species that requires bare spots on the forest floor, but not sand.

Reptiles are not common in the sand sere around Lake Michigan, but elsewhere around the world lizards and snakes are quite characteristic of sandy habitats. They are remarkable in showing a variety of structural and behavioral adaptations specific to locomotion in sand (Fig. 7-8) and for protection of their sense organs and body openings from sand (Mosauer 1932). The sidewinder rattlesnake, for instance, has evolved, in addition to the usual undulatory lateral movement of snakes, a rolling sidewise movement that involves spiral contractions of the body and applies vertical rather than lateral pressure to the sand. Sand offers the snake an unstable footing—lateral undulations alone do less to propel the snake forward than to merely push sand aside. Sand provides firm footing only if it is pushed down upon, hence the effective, if singular, action of the sidewinder.

Life History of Tiger Beetles, *Cicindelidae*

The intimate adjustments of a species to its habitat and the manner in which it selects a particular stage in the

Fig. 7-8 Sidewinder rattlesnake and the track it makes (Mosauer 1935).

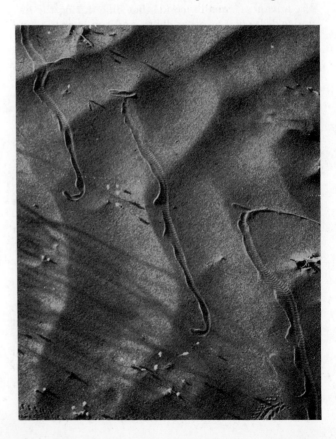

Table 7-3 Percentage location of ovipositor holes and larvae in different soils under experimental conditions (from Shelford 1911, 1915).

Tiger beetle species	Holes or larvae	Number	Sand (%)	Sand and humus (%)	Humus (%)	Clay (%)		Niche under natural conditions
Cicindela	Holes	69	40	50	7	3		Wet sandy beaches
hirticollis	Larvae	50	56	42	2	0		
Cicindela	Holes	141	34	48	2	16		Adults on sandy ridges
tranquebarica	Larvae	129	7	75	2	16		covered with vegetation, larvae on sand or clay
Cicindela	Holes	117	8	22	0	70		Clay soils, oak-hickory forests,
sexguttata	Larvae	93	15	53	0	32		prefer leaves on ground
						On level	*On slope*	
Cicindela purpurea	Holes	51	0	2	0	23	75	Adults on sand or clay, larvae
limbalis	Larvae	47	0	2	0	23	74	entirely on clay banks

sere may be illustrated by briefly describing the life histories of tiger beetles.

Adult tiger beetles are bright-colored, alert, swift fliers. They are frequenters of bare ground. Both adults and larvae feed predatorily on ants, sowbugs, centipedes, spiders, beetles, flies, dragonflies, butterflies, and larvae of various forms. Tiger beetles commonly dig shallow burrows in the soil for shelter. They reach sexual maturity after several warm days in spring or early summer after they have emerged from hibernation. They copulate on warm, humid days when there is an abundance of food and sunlight. After laying their eggs, they die. The female deposits one egg at a time, and lays up to 50 in all, in small vertical holes, 7–10 mm deep, which she makes with her ovipositor. The female tests the soil with her ovipositor until she locates soil of the required characteristics. Hatching occurs in about 2 weeks.

The larvae are elongated, yellowish, and grub-like. Anteriorly directed hooks, spines, and bristles on the dorsal side of the body prevent the larvae from being pulled out of their burrows by the larger prey on which they feed. At the site of the ovipositor hole the larva excavates a vertical cylindrical burrow 8–50 cm deep in temperate climates, much deeper in colder northern regions. Most of the time the larva stations itself at the top of its burrow with its mandibles extended, and with its head and prothorax just closing the round opening. It grabs passing prey and carries it to the bottom of its burrow to devour; larger prey are eaten at the entrance. Inedible parts are cast out on the surface of the ground around the burrow entrance. After feeding 3–4 weeks, the larva closes the mouth of its burrow with soil and goes to the bottom to molt. The second larval stage lasts 5 weeks or longer, after which there is another molt. The last of the larval stages closes the entrance to its burrow in late August or September and goes to the bottom to hibernate over winter (some species hiberate in the second larval stage). The larva comes out of hibernation in late spring

and feeds until summer. Then it closes the entrance of its burrow and constructs a side chamber in which it pupates. The adult emerges in late summer and feeds until October. It then digs a hole in which to hibernate over winter. Two years are commonly required to complete a generation, although in various species the interval between successive generations may be 1 to 4 years, depending in part on regional temperatures.

The niche requirements or seral stage preferred by different species are rigid and appear determined, in large part, by the character of the type of soil a species finds suitable for deposition of eggs and larval growth. Studies performed under experimental conditions demonstrate the nature of these requirements (Table 7-3) but suggest no physiological explanation (Shelford 1908, 1911, 1915, Balduf 1935).

CLAY SERE

Plant Communities

Erosion or calculated removal of overlying material in strip mining for coal may leave bare areas of clay exposed and impound small bodies of water, often with low pH values. These water bodies undergo a modified pond succession (Riley 1960). In clay above pH 4.5 annual plants, of which smartweed is particularly important, appear within a few weeks to 2 years; the higher the clay pH, the quicker the appearance of vegetation. Within 2 to 5 years thereafter sweet clover invades and develops nearly complete dominance over large areas (Fig. 7-9). Sweet clover is a biennial, and an exotic species unimportant in the sere in some parts of the country (Bramble and Ashley 1955). Prior to its introduction, this stage in the sere on bare clay may have consisted of the perennial grasses still found in small scattered patches, or it may not have been well developed. A shrub stage seldom takes dominance over extensive areas, but thickets of raspberries and black-

Fig. 7-9 Stages in the strip-mine plant sere in east-central Illinois: (a) sweet clover, aster, ragweed, after 7 years; (b) same area 10 years later, trees invading; (c) silver maple-cottonwood-sycamore floodplain forest after about 40 years (Wetzel 1958).

berries, smooth sumac, trumpet creeper, and various other species succeed the sweet clover and grass stage. The first trees begin to invade early in the sere, but they are scattered and slow of growth, and do not attain dominance for 25 to 30 years. The tree stage is commonly made up of eastern cottonwood, American sycamore, silver maple, and American elm. Willows occur in wet spots. Herb species of the first two plant stages disappear, for the most part, in the shrub stage. The herb stratum now consists largely of wood nettle. Advanced forest stages of oaks, hickories, basswood, and sugar maple will likely invade in the future, as they occur now in adjacent areas.

Animal Life

The number of invertebrate species tends to increase as the sere advances, although not always regularly. In a study of a formerly strip-mined area (Smith 1928), 18 species were found to be important in the annual stage, 41 species in sweet clover, 40 species in shrubs, 32 species in the early forest stage, and 67 species in the upland climax. More species would be found in advanced stages because of the greater variety of niches available. Thus, in the initial bare area there is only the ground stratum; in the annuals and sweet

clover stages there are the ground and herb strata; in the shrub stage there are ground and shrub strata. The herb stratum is poorly developed or absent altogether. In the forest there are the ground, herb, shrub, and tree strata. Since the early forest is on a floodplain, the ground is frequently swept by floods, and the shrub stratum is poorly represented. The climax forest has all strata, richly developed, and possesses the greatest number of animal species. There is an increase in the abundance of individuals per m^2 with the progression of the sere: annuals, 268; clover, 531; shrubs, 532; early forest, 748; climax, 2445 (Davidson 1932).

Beetles, spiders, ants, and mites are the most abundant animals in the annuals stage, and along with aphids remain most abundant also in the sweet clover community. Grasshoppers are fewer in number but especially characteristic of the first three stages; they practically disappear in the forest. Earthworms are absent in the annuals and scarce in the sweet clover, as are the springtails; as the amount of soil humus increases with the development of the sere, both groups become more and more numerous. Snails first appear in the sweet clover stage and increase in importance in the forest stages.

Starting with bare ground and continuing through early and late shrub stages to bottomland forest, there is a progressive increase in bird species diversity, density, and biomass. A nearby climax forest on upland showed a decrease, however, in all these characteristics (Table 16-4) (Karr 1968).

The first small mammal (Wetzel 1958) to invade the annuals and sweet clover stages is the prairie deer mouse. It attains populations as high as 22 per hectare (9 per acre). It persists until the shrubs and trees have become well established. Its place is taken in advanced stages by the woodland white-footed mouse. The prairie vole prefers the grassy areas and is found under briars and other shrubs. Peak populations are about 18 per hectare (7 per acre). The short-tailed shrew invades the sweet clover stage but does not establish a stable population until the shrubs come in; it persists into the climax forest. Woodchucks commonly occur throughout the early stages of the sere, but mostly disappear in the forest. The cottontail rabbit is common in the early stages, and the fox squirrel invades with the first trees.

FLOODPLAIN SERE

A stream continuously deepens its channel, thus lowering the water table of the surrounding land. At times of heavy rain, the channel may not be deep enough to carry the discharge, and the river overflows its banks. In mesic climates this commonly happens three or four times per year, usually during late winter and spring, and each flood may last up to several days. The more extensive floods, persisting for longer periods, come at less frequent intervals. The lowland area between the river and the bluffs on each side is called the *floodplain*. In the course of time, the river in its erosion meanders back and forth across the floodplain, cutting new channels and abandoning old ones, and frequently leaves a sequence of terraces between its present channel and the surrounding upland. The higher the terrace or the farther an area is from the channel, the less frequently and for shorter periods is it flooded, so that there is commonly a sequence of plant communities that constitute the plant sere.

Plant Communities

Gravel, sand, or silt is deposited on the inner side of river bends. Attached aquatic vegetation may occur in the water. On land, such herbs as smartweed, cocklebur, ragweed, beggar's ticks occur. At some bends small sand dunes may occur, displaying their characteristic plants and animals; usually this stage is narrow at most, and may be entirely absent. On sandy islands in the river or on sandy shores, the sandbar willow often forms dense, shrubby thickets. The first tree stage is frequently black willow mixed with eastern cottonwood, and sometimes silver maple. On the floodplain of the Canadian River in Oklahoma (Hefley 1937),

the sere proceeds next to an edaphic subclimax of either tall grass prairie or elm-oak. The climatic climax on the surrounding upland is mixed prairie. The normal sequence of stages in this region has become considerably modified by the extensive ecesis of the exotic tamarisk tree, introduced from Asia.

On the Mississippi floodplain in western Tennessee (Shelford 1954b) the mature cottonwood-willow associes contains an abundance of vines of several species that form such tangled masses as to be almost impenetrable. The next stage is one in which sugarberry, sweetgum, American elm, and American sycamore predominate; several other species are present in small numbers. This leads to an oak-hickory stage that includes a complex variety of species, and eventually to the regional climax of western mesophytic forest. Cypress becomes part of the composition of the early floodplain forest around the edge of small oxbow ponds or other standing water. The schedule for this sere, the time from the start to the beginning of dominance by each successive plant community, has been estimated as follows: sandbar willow, 3 years; cottonwood-willow, 35 years; sugarberry-sweetgum, 82 years; early species of oaks and hickories, 260 years; intermediate species of oaks and hickories, 350 years; early climax of oaks and tuliptree, 440 years; full development of the climax, 600 years.

Elsewhere in the eastern United States, the cottonwood-willow stage gives way to a narrow zone of sycamore. Two or three species of elm, white ash, and boxelder follow; then a mixed forest that includes black walnut, butternut, black maple, Ohio buckeye, red mulberry, American basswood, tuliptree, and hackberry; next an oak-hickory stage; and finally the beech-sugar maple climax. The herb and shrub strata are usually well developed in mature floodplain forests. Telescoping or skipping of stages is not uncommon in this sere, since variation in ground level or in height of terraces is considerable and the transition between heights is often abrupt. The later stages occur only on the very oldest terraces and may be hard to find at all.

Animal Life

In the bare areas, in the herbs, and among the invading trees occur such beetles as *Heterocerus pallidus* and *Bembidion laevigatum* that feed on the algae and detritus present on the shore. They make their burrows in sand. Fly larvae, a cocklebur weevil, a cocklebur mirid, and a cocklebur fly also occur. The tiger beetles *Cicindela hirticollis*, *C. cuprascens*, and on slightly higher ground *C. punctulata*, prey on the ground species and may even dig them out of their burrows. Spiders, ground beetles, and rove beetles invade from higher stages. In the herb stratum and in the shrubby growth of willows, adult midge flies and other flies are sometimes very abundant. Tarnished plant bugs, 12-spotted cucumber beetles,

Table 7-4 Distribution of annelid worm species in the Sangamon River floodplain forest of central Illinois (Goff 1952). Original nomenclature revised by W. J. Harman.

Annelid species	Family	Ruderals	Willow-silver maple	Silver maple-elm	Elm-bur oak	Elm-shingle oak	Oak-hickory upland
Lumbricus terrestris	Lumbricidae	++	++	++	++	++	
Allolobophora iowana	Lumbricidae	+	++	++	++	+	
Bimastos tumidus	Lumbricidae		++				
Octolasium lacteum	Lumbricidae			++			
Henlea urbanensis	Enchytraeidae				++	+	
Henlea moderata	Enchytraeidae				+	++	
Diplocardia singularis	Megascolecidae					+	
Fridericia agilis	Enchytraeidae					++	
Fridericia sima	Enchytraeidae					+	++
Fridericia tenera	Enchytraeidae						++

and other insects of open-area habitats are present, and there is invasion of various species from the forest itself (Hefley 1937, Shelford 1954b).

The animal life of the floodplain forest is much the same as that of the deciduous forest in general (Chapter 8) and does not need to be discussed here except for its unique features. Annelid worms make their appearance in the ruderal stage, become very abundant in the moist soils of the elm-ash and mixed floodplain forests, then decrease in numbers in the drier soils of the late seral stages. They occur mostly in the first 5 or 10 cm below the surface in moist soil, but up to 30 cm or more in dry soil. During the winter they keep below the frostline, and in very dry weather they roll up in small knots and aestivate. Ten species occur in the floodplain of the Sangamon River in central Illinois, and each species has its particular range of moisture requirements between the river's edge and the upland forest (Table 7-4).

Snails and slugs are moisture-loving animals and occur in large numbers and great variety in floodplain forests; it is not hard to find 15 to 20 species with a little searching. *Mesodon thyroidus* is a common snail, and on a floodplain in central Illinois an average population of 6.3 individuals per m² was found during the autumn (Foster 1937). This amounts to a biomass of living flesh (shell excluded) of 15.8 g/m² (141 lb/acre). *Succinea ovalis* on another old Illinois floodplain (Strandine 1941) averaged 6.5 individuals per m² in September with a biomass of only 0.878 g/m² (7.84 lb/acre). Snail flesh is an important source of food for such small mammals as the short-tailed shrew.

Effects of Flooding

Animals living on floodplains must tolerate flooding of their habitats—in periods with heavy precipitation, often several times annually. Leaves are swept up from the forest floor and piled with other debris against shrubs and the bases of trees. Herbs and shrubs may be damaged, sometimes killed. Silt is deposited as the flood waters recede, covering the foliage and filling in surface cavities and tunnels made by small animals.

Observations during time of flood (Stickel 1948) showed that emergent brush, the bases of trees, and debris rafts supported masses of insects, spiders, millipedes, snails, and amphibians. Debris rafts were refuges for box turtles and pine-mice as well. Snakes, turtles, and amphibians were also seen swimming or floating in the water. No white-footed mice were found in the flood, but the size of the population per unit area, determined by live-trapping immediately after the flood, was the same as it was immediately before the flood, and a number of tagged individuals were found surviving. This species readily climbs trees and may well have passed the danger period arboreally. Tagged box turtles were found on the identical home ranges they had occupied before the flood. This flood lasted only a few days. Severe flooding persisting for long periods is known to have virtually exterminated species of small mammals from wooded floodplains in grassland areas (Blair 1939). Larger mammals, such as rabbits, oppossums, and foxes, quickly leave flooded areas and may be temporarily concentrated around their margins. Squirrels and raccoons easily obtain refuge in the trees, but if the flood persists for some time, they may have trouble finding food. Woodchucks normally spend considerable time in underground burrows and may be trapped there by floodwaters (Yeager and Anderson 1944). Ground-nesting birds are few or absent on floodplains.

Invertebrates in the soil are also affected by flooding, since a gradually rising water table may eventually displace all the air from the soil so that the arthropods are killed. Many earthworms leave their tunnels when these are inundated and are killed. Some species that regularly exist in areas subjected to frequent flooding, however, are not injured. Crane fly larvae are flood-resisting. In normal flooding, bubbles of air trapped in the soil provide sufficient oxygen for at least the smaller arthropods (Kevan 1955). The eggs of some floodplain mosquitoes are laid just above the water level of pools during late summer or autumn and must be flooded the following spring before they will hatch. Hibernating insects and other ground animals are more tolerant of flooding during the winter, when their metabolism and oxygen requirements are low, than they are during the active summer season. The washing away of surface litter, essential for shelter, food, and egg-laying sites, may prevent full recovery of the ground populations for weeks or months.

SUBSERES

Abandoned Fields

When farmland is abandoned, succession back to natural vegetation and ultimately to the climax is rapid, since the soil is already relatively fertile and does not need a great deal of conditioning. On the Great Plains the subsere proceeds rapidly through stages of annual herbs, several of which may be exotics naturalized from other continents; mixed annual and perennial herbs; a short-lived perennial grass; dense stands of triple-awned grass; finally, the climax of short grasses. This last stage may be attained in 10 to 20 years.

Small mammals and grazing domestic animals retard the succession by feeding on grasses while avoiding the herbs. Sheep have the opposite effect, preferring the herbs. Harvester ants denude the vegetation in a circle around their mounds and consume a considerable amount of the available seed supply. Ant coactions

may be of very great importance when we consider that the population of a mound may average 10,000 individuals and the number of mounds per hectare range from 0 to 10 (0 to 4 per acre) in the annuals stage, 7 to 28 (3 to 11 per acre) in the mixed annual and perennial herbs, 12 to 52 (5 to 21 per acre) in the first grass stage, 40 to 142 (16 to 57 per acre) in the triple-awned grass, and 0 to 32 (0 to 13 per acre) in the final stage (Costello 1944).

Six plant stages are recognized in the sere that develops in the mixed prairie region of Oklahoma: an initial stage of mixed herbs; three intermediate stages involving different proportions of triple-awned grass; a subclimax; and the climax of *Andropogon* and *Bouteloua* grasses. The insect population consists of 293 species representing the following orders, ranked in decreasing abundance: Coleoptera, Hemiptera, Homoptera, Diptera, and Orthoptera. There was a greater variety of species and a greater abundance of individuals in the intermediate stages than in either the early or the climax stages, probably because of the greater variety of plant species present in the intermediate stages. Of the 144 species of insects in the climax, 58 per cent entered the sere in its initial stage, 15 per cent in the second stage, 12 in the third, 5 in the fourth, 2 in the fifth, and only 8 per cent were limited to the climax itself. The ecesis of the mature animal community was therefore a gradual and progressive one. On the other hand, many species that were present in the early stages did not persist into the climax community (Smith 1940).

Succession in abandoned fields of the southern Atlantic and Gulf states is of special interest (Fig. 7-10). During the first year, crabgrass and horse-weed,

Fig. 7-10 Plant succession in abandoned fields in Virginia: (a) annual herbs invading old cornfield; (b) broom-sedge grass; (c) invasion of young pine trees.

Table 7-5 Breeding bird pairs per 40 hectares (100 acres) in sere developing on abandoned fields, Georgia Piedmont region, averaged from two stations in herb-shrub (1, 3 years old), three stations in grass-shrub-tree (15, 20, 25 years old), four stations in pine forest (25, 35, 60, 100 years old), and one station in oak-hickory (over 150 years old) (condensed from Johnston and Odum 1956).

Bird species	Herb-grass	Grass-shrub-tree (forest-edge)	Pine forest	Oak-hickory climax
Grasshopper sparrow	20	8		
Eastern meadowlark	8	6		
Yellowthroat		11		
Yellow-breasted chat		7		
Prairie warbler		4		
Catbird		1		
Indigo bunting		1		
American goldfinch		+		
Bobwhite		+		
Field sparrow	36	4		
Bachman's sparrow		5	1	
Rufous-sided towhee		9	13	
White-eyed vireo		3	3	
Mourning dove		+	+	
Pine warbler		5	43	
Cardinal		6	15	23
Summer tanager		2	14	10
Chuck-wills-widow		+	+	+
Brown-headed nuthatch			2	
Brown thrasher			1	
Solitary vireo			1	
Yellow-throated warbler			1	
Pileated woodpecker			+	
Hooded warbler			11	11
Carolina wren			10	10
Ruby-throated hummingbird			6	10
Blue-gray gnatcatcher			5	13
Tufted titmouse			5	15
Eastern wood pewee			4	3
Blue jay			4	5
Carolina chickadee			4	5
Crested flycatcher			4	6
Red-eyed vireo			4	43
Yellow-throated vireo			3	7
Wood thrush			2	23
Yellow-shafted flicker			1	3
Hairy woodpecker			1	5
Downy woodpecker			1	5
Yellow-billed cuckoo			+	9
Black and white warbler				8
Acadian flycatcher				5
Kentucky warbler				5
Total species	2	18	30	22
Total pairs	28	104	163	224
Percentage taxa characteristic or exclusive		(61)		(82)
Species diversity index (counting + = 0.1) (H')	0.60	2.23	2.70	2.78
Equitability index (H'/H'max)	0.86	0.77	0.80	0.90

annuals, predominate. During the second year an aster and a ragweed, and in the third year the perennial broomsedge grass, become dominant. The grass is invaded quickly by loblolly and shortleaf pines which form closed stands in some areas in as little as 10 to 15 years. These pines do not reproduce in their own shade. They mature in 70 to 80 years, and are replaced by the climax oaks, hickories, beech, and sugar maple which take complete dominance by the time the area is 150 to 200 years old (Oosting 1942).

The bird succession (Table 7-5) shows an increase in number of species up to the pine stage, with a decrease in the climax, the same as noted previously with the clay sere. However, in this case the climax has the greatest density of individuals. The herb-grass and forest-edge may best be considered as making up one distinct bird community. Percentagewise, there is little distinctiveness between the pine forest and the climax deciduous forest in species composition, but when taken together there is a clear separation between forest-edge and forest bird communities. Species diversity increases progressively with succession, but there is no consistent trend in the equitability index. The pine stage may be more distinct from deciduous forest with invertebrates than it appears to be with birds or with mammals.

Small mammals are most numerous in the grass stage of the sere. Principal species are little shrew, eastern harvest mouse, oldfield mouse, hispid cotton rat, pine mouse, and house mouse. This community largely disappears in the pine stage. The final stage of deciduous forest is occupied principally by the rice rat, cotton mouse, and golden mouse (Golley et al. 1965).

In Michigan, the sere in old fields passes through the following stages: annuals-biennials; perennial grasses; mixed herbaceous perennials; shrubs; and finally three tree stages, the first reached in 21 to 25 years. Prairie deer mice are at their most abundance during the early stages, meadow voles in the intermediate grassy stages, and the woodland white-footed mouse and short-tailed shrew in the shrub and tree stages. Such game species as ring-necked pheasants, bobwhite, and cottontail rabbits are common on abandoned farmlands but give way to another group of game species, including white-tailed deer, ruffed grouse, and gray squirrels, when the forest stages become established (Beckwith 1954).

Pastures

Pastures in northern Ohio contain a sod principally of blue grass. With light grazing, this sod will resist invasion of other species for a long time, but with heavy grazing, resistance is weakened and unpalatable herbs, briars, and hawthorn come in. The latter two species are armed with prickles or thorns discouraging animal browsing. When they become dense enough they kill the grass beneath them. Eastern redcedar may establish itself in horse pastures, but not in cattle pastures; cattle browse it but horses will not. In the middle of protecting thickets of briars, hawthorns, and redcedar such deciduous trees as elm, ash, tuliptree, sycamore, and oak come in. After a few years they grow beyond the reach of animals, shade out the briars, hawthorns, and red cedar, and establish a forest dominance. Where left undisturbed by man, the succession of native vegetation will thus bring about the elimination of domestic animals from the area and replacement with the biotic climax natural for the region.

In western areas too dry for deciduous forest, overgrazing reduces the vigor and abundance of the taller climax grasses, and the short grasses that are less easily grazed are favored. Unpalatable herbs, sagebrush, cacti, and mesquite may also replace grasses over extensive areas. Although native animals such as the bison and pronghorn may have heavily grazed the original prairie in locally arid regions, the result was less drastic than that produced by the heavy concentrations of grazing stock on our farms and ranches at the present time. When the most favored vegetation was reduced, native animals commonly dispersed into other areas so that the carrying capacity of the land was not critically reduced.

Burns

Prairie fires, frequently started by lightning or by Indians, were doubtless important in preventing deciduous forest from succeeding grassland in parts of the Midwest. More lately, fires are started by careless campers or travelers. Fires are especially destructive in coniferous forests, as the clinging dry needles encourage crown as well as ground fires to develop. Many thousands of square miles of forests are burned over annually.

The extensive pure stands of longleaf pine on the coastal plain of the southeastern states are probably a consequence of ground fires that regularly occurred at intervals of 3 to 10 years before white men came. The terminal bud of the longleaf pine is well protected by a thick covering of green leaves, one of several characteristics that make the species extremely fire resistant (Chapman 1932). Fire destroys all seedling hardwood trees as well as other species of conifers. In forest management, fire may be used deliberately for maintenance of pine stands for pulp and lumber.

When northern and western coniferous forest is destroyed by fire, the first trees to invade are usually quaking aspen, paper birch, and sometimes balsam

Fig. 7-11 Tillamook Burn in coniferous forest in Oregon, 1944 (courtesy U.S. Forest Service).

Fig. 7-12 This aspen forest in New Mexico will be replaced by one of spruce as the young trees now forming in the undergrowth reach maturity (courtesy U.S. Forest Service).

poplar (Figs. 7-11 and 7-12). These forests cover extensive areas in Canada and southward on the Rocky Mountains. Jack pine in the north and lodgepole pine in the western mountains either come in with the deciduous trees or succeed them. The cones of these two trees take several years to open and shed the seeds held within, and may not do so at all unless heated by forest fires. Aspen and pine are eventually replaced by the climax forest. In many western areas, particularly in the Sierra Nevada, chaparral may occur in dense stands after fires and persist for a long time.

Since the burn subsere in coniferous forest commonly includes the aspen-birch associes, many of the typical animals in this stage are deciduous forest and forest-edge species, although there is a penetration of coniferous forest species as well. The birds and mammals are not generally very numerous in the aspen-birch community, but ground invertebrates may be more abundant here than in the poorly decomposed acidic ground duff found in the coniferous climax.

ANIMAL COMMUNITIES

Although the plant communities that make up the stages of the different land seres and subseres we have described are numerous and varied, the number of distinct animal communities that can be clearly recognized are few (Fig. 7-4). Actually, we can distinguish in eastern North America only the animal communities of *grassland, forest-edge, deciduous forest, south-eastern evergreen forest*, and *coniferous forest*. Each of these communities varies in the different habitats of rock, sand, and clay, and in the various subseres, but the variations are of minor significance and are best treated as facies of the larger communities.

SUMMARY

For terrestrial living, animals must actively support themselves against gravity, obtain water, and prevent excessive water losses from the body. They must be equipped to endure a wide range of fluctuating temperatures, to secure oxygen, to endure intense solar radiation, adjust to diurnation (day and night) and aspection (seasonal changes), and often maintain close contact with particular kinds of substratum.

Succession occurs on all primary bare areas, such as rock, sand, clay, and floodplains, and in such secondary bare areas as abandoned fields, pastures, and burns. In a given area, best shown in humid regions, all seres converge toward the same broadly defined climax community. There are normally more plant than animal stages in any sere. The succession of animal communities correlates with the succession of vegetation-types or life-form of the plant dominants, not with plant communities identified by the taxonomic composition of the plant dominants. In eastern North America, we can distinguish only the grassland, forest-edge, deciduous forest, southeastern evergreen forest, and coniferous forest terrestrial animal communities.

Chapter 8

GRASSLAND,
FORESTS, AND FOREST-EDGES

We have seen that succession of animal communities in humid climates passes through three terrestrial stages before attaining climax: grass, shrubs and scattered trees (forest-edge), and forest. In arid climates, the climax may be reached at the first or second stage. It is important for us to examine each community in more detail, therefore, if we are to gain an understanding of the ecology of animals prevailing locally in different parts of the world.

VEGETATION

Grassland vegetation differs from forests in that the above-ground vegetation is completely renewed each year. Grasses may be divided into three categories on the basis of height: *tall grasses* (1.5–3 m tall), such as big bluestem and slough grass; *mid-grasses* (0.5–1.5 m tall), such as little bluestem and needle grass; and *short grasses* (less than 0.5 m tall), such as buffalo grass and grama grass. The taller grasses grow in wet habitats, the short grasses in arid habitats. Most native grasses are *bunch grasses* in that they grow in clumps with the areas between the clumps either bare ground or occupied by other species. Broad-leaved herbs occurring between the dominant grasses are called *forbs*. A few species are *sod formers* in that their growth

is continuous over the ground surface. The leafy aerial parts of perennial grasses die in the winter or in dry season, leaving the underground stems or rhizomes to propagate the plant the following year (Weaver and Fitzpatrick 1934).

Forests are composed of trees growing sufficiently close together to dominate the entire area of ground surface. In cold climates most forests are needle-leaved evergreen; in seasonally warm, moist climates, they are mostly broad-leaved deciduous; and in continuously warm, moist climates, they are broad-leaved evergreen. In spite of these secondary differences in life-form, the structure and internal dynamics of all forest communities are quite similar. Useful methods for measuring the density of trees per unit area are described by Cottam and Curtis (1956).

Between forests and open country, scattered trees are interspersed with shrubs and grasses. This transition area is usually narrow around the margins of a stabilized forest (Fig. 8-3), but where succession is occurring, large areas of mixed grass, shrubs, and trees may provide in close proximity a variety of habitats which animals take advantage of in a variety of unique ways. Likewise, in agricultural areas, hedge and fence rows, or narrow strips of trees and shrubs along streams, are really edges without the adjacent forest and have the same characteristic animal life.

115

Deciduous trees shed their foliage in the autumn, are bare over winter, and obtain new foliage in the spring. Evergreen coniferous trees, on the other hand, retain their foliage throughout the year, although old dried leaves fall a few at a time at all seasons. Differences in the size, shape, and structure of the leaves are important to many animals. The lack of foliage in deciduous forests during the winter permits a greater light penetration to the forest floor, more wind circulation, and relatively lower temperatures than in coniferous forests. During the summer, deciduous forests generally have higher but more variable temperatures and lower relative humidities than do coniferous spruce and fir forests (Blake 1926, Dirks-Edmunds 1947). Pine forests, however, commonly develop in habitats that are warm and dry.

As shade producers, the deciduous and coniferous trees do not vary as groups, but only as individual species (Weaver and Clements 1938):

Deciduous trees	Coniferous trees
Heavy shade producers	
Sugar maple	Yew
Beech	Spruce
Basswood	Hemlock
	Firs
	Thujas
Medium shade producers	
Elms	Eastern white pine
White oak	Douglas-fir
Northern red oak	
Ash	
Black oak	
Light shade producers	
Silver maple	Ponderosa pine
Bur oak	Tamarack
Birches	Lodgepole pine
Poplars	
Willows	

It is interesting that in forest climates light shade producers are species found in the early stages of succession while the heavy shade producers are mostly climax species.

There is an important difference between deciduous and coniferous forests in the nature of the decomposing dead leaves that fall from the trees. Decomposition of broad leaves is rapid and relatively complete to form a rich humus that mixes gradually with the mineral soil beneath. Needle leaves decompose slowly and form a somewhat acid humus sharply defined from the underlying mineral soil. Humus formed in humid grasslands is similar to but richer than that of deciduous forest; in arid grasslands it is poorly developed. The nature of the humus and litter affects the number and kinds of animals that occur in the soil.

The grassland community typically contains *subterranean*, *surface*, and *herb* strata of vegetation. The forest in addition has *shrub* and *tree* strata. Correlated

Table 8-1 Dry weight (kilograms per hectare) of plant biomass (April to November) in tall grass prairie and oak-pine forest in central Minnesota (from Ovington *et al.* 1963).

Stratum	Prairie	Forest
Subterranean (roots and subterranean stems)	4,824	14,997
Herb	449	88
Shrub	10	512
Tree	0	163,076
Total biomass living vegetation	5,283	178,673
Dead plant material:		
Ground litter	2,788	36,735
Standing	0	21,837
Total biomass	8,071	237,245

with the differences in microclimate, both plant and animal species show segregation of their activities and presence to one or more of these strata.

The herb stratum in mature grassland is much better developed in biomass than in the forest, but the woody vegetation of the forest is so great that the total biomass of the community is considerably greater. In one study (Table 8-1), it was 29 times greater. Good comparative data on the biomass of all animals are not available, but class studies in central Illinois during early autumn indicate no great difference in the biomass of the macroscopic invertebrates. Seral grassland communities may, however, in contrast to climax ones, be relatively impoverished (Chapter 7).

HABITAT

Grassland, Forest-edge, and Forest-interior Compared

In studying the distribution of animals in relation to climate, it is obviously not sufficient to consider only the macroclimate. Animals respond to the microclimate of their particular niches, and the relation between these microclimates and the prevailing macroclimates of the region must be demonstrated.

At the University of Illinois, no significant difference in mean monthly temperatures, calculated bihourly day and night, has been found between the interior of a virgin oak-maple forest and an adjacent open grassland. In the forest, however, the daily extremes are not so great; that is, the maximum midafternoon temperature is not as high, nor the minimum night temperature so low, as in the grassland (Fig. 8-1).

Relative humidity during a summer day in Iowa was found (Aikman and Smelser 1938) to average 20 per cent lower in grassland than in a shrubby forest-edge, and 5 to 8 per cent lower in the forest-edge than in the forest-interior. There is less difference between the three habitats, however, at night. Rate of evaporation, as measured with Livingston atmome-

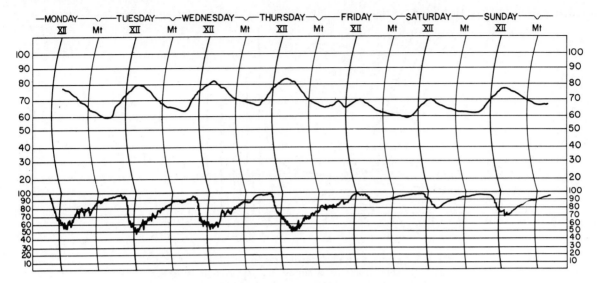

Fig. 8-1 Weekly chart from a hygrothermograph placed at shrub level in a deciduous forest in central Illinois; temperature above, relative humidity below.

ters, is inversely correlated with humidity, being greatest in grassland and least in the forest-interior. Daily changes in relative humidity between day and night tend to vary inversely with the temperature, except when there is rain.

During 4 years at the University of Illinois woods, rain gauges recorded 88.8 cm (35.5 in.) per year in the adjacent grassland, upon which full precipitation fell, and 70.1 cm (28.0 in.) *throughfall* (the amount reaching the ground) under the tree canopy of the forest. There was variation of throughfall from spot to spot in the forest, depending on the location of openings in the canopy and drip-points from the leaves and stems. *Stem-flow* of water down the tree trunks was not measured. Throughfall and stem-flow together make up the *net rainfall*. In a shortleaf pine plantation in southern Illinois (Boggess 1956), the net rainfall over 3 years averaged 91.2 per cent of the total rainfall. *Interception*, the amount of rainfall presumably evaporated back into the air, was 100 per cent in very light rainfalls but less than 5 per cent in rainfalls exceeding 5 cm (2 in.).

In a beech-maple forest in northern Ohio, which bordered on an open field, wind velocity at a distance 245 meters (about 800 ft) inside the west margin was reduced to a minimum of 10 per cent when the trees were in leaf and 25 per cent when not (Williams 1936). With a protective edge of shrubs, the wind velocity would doubtless have been decelerated more quickly.

Summer light intensities are much less under foliage than out in the open. Noontime illumination under shrubs in Iowa averaged 26 per cent of full sunlight; within the forest interior, 6 per cent (Aikman and Smelser 1938). The forest floor is not uniformly illuminated because small openings in the canopy admit sun-flecks of varying intensity. In the cottonwood,

pine, black oak, and sugar maple stages of the sand sere at the lower end of Lake Michigan, the percentages of the forest floor shaded during the midday hours were 68, 87, 75, and 90 per cent, respectively (Park 1931). There may be some change in the quality of light that filters through the forest canopy, as there is of intensity, as some wavelengths are used more than others in photosynthesis; green is transmitted or reflected and not absorbed. Where a stand of trees abruptly confronts an open field, light penetrates laterally under the forest canopy and, the typical edge configuration reversed, the light permits shrubs to extend 40 m or more into the interior.

Vertical Gradient

There is a gradient in microhabitat factors from above the grasses down to the ground. In one study of virgin prairie (Weaver and Flory 1934), light intensity varied from 100 per cent in full sunlight to 25 per cent at one-half the pile depth of the grasses to 5 per cent at the base of the stems, and 0 per cent in the subterranean stratum. The relative humidity above the grass was 20 per cent; in the grass, 31 per cent. The wind velocity above the prairie grasses was 14.5 km/hr (9 mph); at the top level of the grasses, 6.0 km/hr (3.7 mph); at the soil surface, zero. The rate of water evaporation from white spherical atmometers was 55.3 cc/day above the grasses, 33.3 cc at top surface of the grasses, 15.1 cc at one-half the pile depth of the grasses, and only 13.4 cc just above the soil surface. The temperature gradient varies with the height of the grass and between day and night (Fig. 8-2).

The vertical gradient of temperature in a deciduous forest in central Ohio varies with the season and with the height of macroclimatic temperature (Table 8-2).

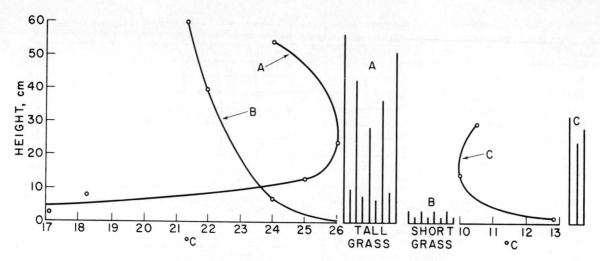

Fig. 8-2 Gradient of air temperatures in and above (A) tall grass, (B) short grass on a sunny day, and (C) in grass of medium height at night (after Waterhouse 1955).

In the summer, the greatest extremes of temperature occur in the canopy, but at other levels, both above and below the ground, summer daily mean temperature is more stable than at any other season. Because the canopy largely controls the air temperature beneath it, there is little or no thermal stratification between it and the ground. Summer soil temperatures are always lower at 1.2 m below the surface than at the surface. For comparison, air temperatures in a coniferous forest in Wyoming during July and August averaged 12.3°C at 0.1 m above the ground and 7.6°C 0.1 m below the surface litter (Fichter 1939).

During the winter, temperatures in deciduous forests are lowest near the ground and more uniform at all higher levels than during the summer, since the absence of a canopy permits greater turbulence, hence less stratification, of the air. Soil temperature at 1.2 m depth is generally higher than surface temperature; beneath the litter in the central Ohio area temperatures do not usually go below freezing (Table 8-2; Christy 1952). A covering of snow gives added protection against freezing of the leaf litter. Another study (Wolfe et al. 1949) revealed differences between temperatures above and below a snow covering 2 to 10 cm deep during a period of 2 months that averaged 8.9°C, and on one occasion reached 15.5°C.

Relative humidity decreases from the ground stratum upward. In a young elm-maple forest in Tennessee, the relative humidity from mid-February to mid-August averaged 77.9 per cent at the surface of the leaf litter, 75.2 per cent in the herb stratum 0.5 m above the ground, 72.5 per cent in the shrub stratum at 0.9 m above ground, and 67.4 per cent in the trees at 7.6 m above ground (Adams 1941). In this same forest, the rate of evaporation between May and November in the four strata respectively averaged 29.4, 60.7, 72.8, and 99.2 cc/week. In a spruce-fir forest in Wyoming, the average weekly evaporation

at 0.1 m was 50.5 cc, at 1 m 75.2 cc, and at 3 m 103.9 cc (Fichter 1939).

In the Tennessee elm-maple forest (Adams 1941), the average daily mid-summer light intensities measured with a MacBeth illuminometer for ground, herb, shrub, and tree (beneath the canopy) levels were respectively 52.3, 60.3, 60.4 to 76.2 foot-candles; in early May, before the foliage was fully developed, intensities of 65.8, 78.3, 104.4, 119.1 foot-candles were measured. Under the leaf litter and in the soil, the light intensity was, of course, zero. Above the trees it was doubtless several thousands of foot-candles. Maximum light intensities from the sun occasionally reach 15,000 foot-candles. There are, therefore, three distinct sections in the vertical gradient: below ground surface, between ground and tree canopy, and above the canopy.

Ground insects, millipedes and isopods, when placed in experimental gradients, show a preferendum for lower light intensity, higher humidity, and lower temperature than do insects taken from the herb or shrub stratum (Table 8-3). In grassland, motile organisms can quickly vary the microhabitat to which they are exposed by changing their vertical position only a few centimeters, and they do shift in position as the gradient varies at different times of day or from day to day. To obtain an equivalent change in microhabitat in the forest gradient requires a shift of several meters in vertical position. Experiments show, however, that each forest animal species occupies a stratum approximating its preferendum for a particular microhabitat (Todd 1949).

Slope Exposure

Microclimatic differences between north- and south-facing slopes are great. South-facing slopes receive a greater amount of solar radiation and are commonly exposed to the prevailing winds in the summer. As a consequence, both air and soil temperatures are higher on south-facing slopes than on north-facing slopes; relative humidity is lower, soil moisture lower, and the rate of evaporation is higher. The differences between the two slopes are most marked close to the ground, increasingly less so at higher levels (Cantlon 1953).

The vegetation on the protected north-facing slopes is usually more mesic in type and more luxuriant than on the exposed south-facing slopes, and there is a deeper organic leaf litter on the ground. Types of vegetation characteristic of arid habitats penetrate humid climates on south-facing slopes; mesic vegetation penetrates relatively arid climates on north-facing slopes. In mountain areas, vegetation characteristic of lower altitudes ascends higher on south-facing slopes, and vegetation of the upper altitudes descends farthest on north-facing slopes. Animals are locally distributed in a similar manner, in response partly to the climate, partly to the differences in vegetation. Many other differences in microclimate occur in various parts of the forest and forest-edge (Wolfe et al. 1949), and in grassland.

Fire

Although often overlooked, fire is a significant environmental factor in determining the composition and

distribution of vegetation and in affecting animal populations both directly and indirectly. Fire is most important in grassland, chaparral, and other temperate shrublands, and coniferous forest. Although many fires are now started by man, before man lightning was probably responsible for periodic fires in these communities. The occurrence of fire at intervals retards the invasion of forest into grassland and deciduous into coniferous forest. Grasses quickly recover by growth from underground parts, but seedling trees, especially if deciduous, are usually killed. Fire-dependent plant communities tend to have inherent flammable properties and burn more readily and frequently than non-fire-dependent communities (Daubenmire 1968, Mutch 1970).

After dry-season fires in tropical climates, the new shoots that sprout from perennial grass bases provide a sparse forage for grazing animals which would not be available otherwise until the rainy season (Vesey-FitzGerald 1960). Burning off the surface litter allows increased temperature and light exposure at the ground level and indirectly an increased availability of nitrogen in the soil in response to increased microbial activity. The result is greatly increased productivity and flowering during the growing season (Old 1969). Fire rotated on a 3-year schedule is used at the University of Wisconsin Arboretum to maintain prairie in a semblance of natural condition and to prevent invasion of shrubs and trees.

Coniferous forests are more susceptible than decidu-

Table 8-2 Vertical gradient of temperature (°C) in a beech forest in central Ohio (Christy 1952).

Elevation	May-September			November-December		
	Minimum	Maximum	Range	Minimum	Maximum	Range
Macroclimate	10.0	34.4	24.4	−19.4	18.3	37.7
+25.0 m	7.8	31.7	23.9	−17.8	13.3	31.1
+18.9 m	8.9	30.0	21.1	−17.2	13.3	30.5
+6.1 m	8.9	30.0	21.1	−16.1	13.9	30.0
+1.5 m	8.9	28.9	20.0	−21.1	12.2	33.3
Surface of leaf litter	9.4	28.3	18.9	−8.3	8.3	16.6
Under leaf litter	12.2	22.2	10.0	0	10.0	10.0
−0.15 m	12.2	18.9	6.7	2.8	12.8	10.0
−1.2 m	10.6	16.1	5.5	4.4	13.3	8.9

Table 8-3 Results of experiments conducted in the field to establish light orientation of arthropods taken from different strata.

Mixed Species	Number of Experiments	Light-Intensity Gradient (%)			Control— No Gradient (%)		
		Strong	Intermediate	Weak	Left	Middle	Right
Grassland animals from herb stratum	5	47	34	19	40	31	29
Forest animals from herb and shrub strata	7	46	21	33	33	30	37
Grassland animals from ground stratum	5	24	27	49	34	23	42
Forest animals from ground stratum	8	18	23	59	34	29	37

ous forest, especially to crown fires, because of the many dry needles adhering to the branches. Crown fires may kill the trees over large areas. Ground fires may become crown fires if there is an excessive accumulation of dry decaying litter, logs, and stumps present. Ground fires coming at frequent intervals may prevent such dangerous amounts of combustible material from accumulating. Foresters are beginning to consider controlled burning as a desirable tool in maintaining certain forest types no longer subject to natural fires (Oberle 1969).

GRASSLAND COMMUNITY

Invertebrates

Snails, earthworms, and myriapods are not numerous in most grasslands because of the dry habitat. Insects, however, are abundant; 1584 species were recorded over a period of 10 years in the herb stratum of a 5-hectare (12 acres) oldfield in Michigan, of which hymenopterans (especially Ichneumonidae), flies, and lepidopterans were exceptionally numerous, making up 75 per cent of the total composition. In addition, there were 112 species in the ground surface stratum, chiefly carabid beetles and ants. There was a continual replacement of species in the herb stratum from April into October, with the largest number at any one time coming in late July. Eighty-five per cent of the species were herbivorous as adults, these being mainly flower-feeders, while 12 per cent were carnivorous. Of the larvae, 41 per cent of the species were herbivorous and 52 per cent fed upon other insects, chiefly as parasitoids. The remaining species had different food habits altogether. The stability of the trophic structure of the community was indicated, in spite of the considerable replacement of species, in the persistent, nearly constant ratio throughout the year in the number of species with herbivorous and carnivorous larvae (Evans and Murdoch 1968). In this same oldfield, soil arthropods consisted chiefly of mites, pauropods, symphylids, proturans, japygids, and three families of springtails (Fig. 8-7). All these groups consist of small-sized individuals. Although their numbers per square meter may be in the thousands or tens of thousands, in biomass they amount to only a few milligrams (Engelmann 1961).

Spiders make up about 7 per cent of the total arthropod population in grassland. In one study made in Nebraska (Muma 1949), 111 species were collected from 128 hectares (320 acres) of mixed high and low prairie containing some shrubs. Less than a dozen species were web-builders; there is a lack of suitable web-building sites in grasslands. The vast majority were wandering cursorial forms. In regard to strata in this prairie, 45 species were restricted to the soil and litter, 30 to the herbs, 1 to the shrubs. Thirty-five

species occurred in two or more strata. The total population for the area was least in the spring and greatest in the autumn. Peak populations in the ground stratum were reached during the winter, however, because of the presence of many hibernating immature forms. Similar seasonal fluctuations occur with other invertebrates, although the peak populations of insects are usually attained during the summer (Shackleford 1939, Fichter 1954).

In grazed pastures and in grassy meadows in New York State, invertebrates average 777 individuals per m^2 (Wolcott 1937). Of this population, ants make up 26 per cent, leafhoppers 15 per cent, other insects 34 per cent, spiders 9 per cent, millipedes 9 per cent, sowbugs 2 per cent, snails and slugs 2 per cent, earthworms 2 per cent, and large nematodes 1 per cent.

Some insects show structural adaptations for living in grassland (Hayes 1927). May beetles in forested regions commonly feed at night on the foliage of trees and have well-developed wings, but closely related species in grassland areas feed on low-growing plants during the day and are flightless. The development of pilosity and thick integuments in some insects appears to be an adaptation to prevent evaporation. Prairie May beetles pupate in the spring rather than autumn, probably in correlation with their change in food habits, and adults appear in mid-summer rather than late spring.

An insect microhabitat of special interest is the dung of the larger mammals. Bison formerly occurred in frequency one to 10–20 hectares. Inasmuch as the output of each animal is about 25 droppings per day, the number of these microhabitats available was considerable. Some 83 species of arthropods have been collected from cow dung, mainly beetles and flies, but including annelids, nematodes, and protozoans. There is a regular succession of insect species breeding and maturing. The microsere is completed in about 8 days, the length of time required for the droppings to dry. The first species that arrive are the obligatory breeders on dung. They have the shortest life histories, and remain for the shortest time. Predaceous and parasitoid species prey on the coprophagous ones. The greatest

Table 8-4 Population of small mammals per hectare (2.5 acres) in mixed prairie of western Kansas (after Wooster 1939).

Mammal species		Number
Prairie vole		9.6
13-lined ground squirrel		7.6
Prairie deer mouse		6.8
Harvest mouse		2.8
Little shrew		2.4
Short-tailed shrew		1.3
Black-tailed jack rabbit		0.7
Cottontail rabbit		0.1
	Total	31.3

variety of species is present at the middle of the microsere, but the composition of species varies with the season. Species disappear as the dung disintegrates into the general surroundings (Mohr 1943, Laurence 1954). A comparable microhabitat and succession occurs in carcasses of dead animals (Chapman and Sankey 1955, Payne 1965).

Vertebrates

Table 8-4 gives a representative sampling of small mammal populations found in grassland, although it is to be expected that the species composition and size of populations will vary locally and from year to year. The mores of grassland mammals, which show how they are adjusted in behavior to live in this community, are tabulated in Table 8-5.

Birds are not numerous in grassland. In northwestern Iowa (Kendeigh 1941b), grasshopper sparrows, western meadowlarks, bobolinks, ring-necked pheasants, marsh hawks, and short-eared owls averaged less than one pair per hectare (2.5 acres). Prairie chickens and sharp-tailed grouse formerly occurred where now is to be found only the introduced pheasant. The eastern meadowlark predominates over the western meadowlark in the wetter and smaller pastures east of the Mississippi River. Vesper sparrows and horned larks occur in short grasses, but usually not in climax areas with dense tall grasses. Upland plovers, Henslow sparrows, lark buntings, and longspurs are common locally.

Some 14 species of snakes are generally distributed over the prairies (Carpenter 1940). To the east, the blue racer, massasauga, bullsnake, and garter snakes are frequently found. The prairie rattlesnake is increas-

Table 8-5 A tabulation of certain grassland mammal mores (Carpenter 1940).

Mammal species	Solitary or family groups	Herds or packs	Fossorial, locally cursorial	Fossorial	Cursorial	Herbivorous	Omnivorous	Carnivorous	Migratory	Hibernating	2 litters/yr	1 litter/yr	Diurnal	All 24 hours	Crepuscular	Nocturnal
	Social life		Stratum			Food habits			Seasonal activity		Production of young		Daily period of activity			
Bison		×			×	×			×			×	×			
Pronghorn antelope	×				×	×						×	×			
Wapiti		×			×	×						×		×		
White-tailed deer	×				×	×						×			×	×
Mule deer	×				×	×			×			×			×	
Cottontail rabbit	×				×	×					×				×	
White-tailed jack rabbit	×				×	×						×			×	
Prairie dog		×	×			×				×		×	×		×	
Prairie deer mouse	×				×		×				×				×	×
Prairie vole	×				×		×		?		×				×	×
Jumping mouse	×		×			×				×		×	×			
Pocket mouse	×				×		×				×				×	×
Harvest mouse	×				×		×				×				×	×
Franklin ground squirrel	×		×				×			×		×	×			
13-lined ground squirrel	×		×				×			×		×	×			
Pocket gopher	×			×		×						×			×	×
Richardson ground squirrel		×	×				×			×		×	×			
Wolf		×			×			×	×			×		×		
Coyote	×		×				×					×			×	×
Badger	×			×				×				×			×	×
Bobcat	×				×			×				×			×	×
Skunk	×		×					×				×			×	×
Weasel	×				×			×				×			×	×
Red fox	×		×				×					×	×			
Swift fox	×		×					×				×			×	×
Shrew	×				×			×			×				×	×

ingly common westward. The lizards *Cnemidophorus sexlineatus*, *Sceloporus undulatus*, and *Holbrookia maculata*, commonly occur in grassy areas at forest-edges. The horned toad is found in arid habitats.

The most characteristic amphibian of grassy areas is the toad. All species breed in the ephemeral bodies of water resulting from the rains of spring and summer. One species, *Bufo cognatus*, will not breed unless it rains, even though bodies of water are present. During the hot, dry weather of late summer, the toads retreat to burrows in the earth or to other shelter until favorable conditions again return (Bragg and Smith 1943).

Grazing Food Coactions and Range Management

Since the vegetative productivity of grasses is very high, herbivorous animals, especially large mammals, are favored in the grassland community (Renner 1938). Unlike trees and shrubs, the terminal bud on grasses lies close to the ground and is not ordinarily injured by grazing. Meristematic tissue lies at the base of the leaves so that when the terminal portion of the leaf is eaten off, the leaf keeps on growing. Actually, lateral branching at the base of the grass stem is stimulated by grazing, and a thicker and more succulent growth with less fiber is produced. Productivity of grass is reduced if the herbage is removed more than two or three times during the growing season. Light to moderate grazing can usually be carried with full or even increased productivity. Heavy grazing, however, should not be permitted. In addition to reducing herbage production, heavy grazing may destroy seed stalks prior to the dropping of the seed or so weaken the plants physiologically that seed is not even produced. The growth of underground rhizomes and vegetative reproduction is retarded when photosynthetic activity is reduced. The best pastures are those in which grazing animals do not consume more than 70 to 80 per cent of the total herbage productivity of the grasses (Stoddart and Smith 1943). Removal of the herbage overwinter permits earlier growth the following spring. Overgrazing always brings about a reduction in abundance of the more palatable species and an increase in the less desirable ones with the consequent deterioration of the range and the productivity of the community (Weaver and Tomanek 1951, Kucera 1956). The carrying capacity of grassland or the largest number of animals that can be supported without deterioration of the range varies with the type of grasses involved, the climate, and the soil (Table 8-6).

Although often overlooked, invertebrates constitute one of the three important groups of grazing animals. The total biomass of insects in a New York pasture amounted to 3.2 g dry weight per m². This is to be compared to 14.5 g for the dry weight of cows per m² that the pasture was supporting. Feeding experiments showed that in one pasture where grazing by the cattle was moderate and the vegetation was ample, the insects ate more of the grasses and clovers than the cows did, but in another pasture which was being overgrazed and in which the vegetation was short, the cattle ate more than the insects did (Wolcott 1937). Grasshoppers and Mormon crickets are sometimes very destructive in the arid west. In one area in Montana, a population of 25 grasshoppers per m² destroyed enough forage on 3 acres during 1 month to support one cow for a month (Stoddart and Smith 1943).

Rodents and rabbits consume very considerable amounts of grasses and other herbs and cause great damage at times of high populations. In a study performed in Arizona (Taylor 1930), grazing by Gunnison's prairie dogs alone consumed 87 per cent of the total grass production and grazing by cattle and rodents combined, 95 per cent. In California, Beechey's ground squirrels eliminated 35 per cent of the green forage by the end of the season, pocket gophers 25 per cent, and kangaroo rats 16 per cent (Fitch and Bentley 1949). Since these various rodents have food preferences of grass species similar to those of cattle, there is obviously severe competition between them, especially in times of drought. When rodents are not overly abundant, they have some beneficial effects in fertilizing, aerating, and mixing the soil.

Among big-game mammals, bison and wapiti are largely grass-eaters. Food consumption of bison is about equal to that of cattle, but wapiti eat only about half as much per individual. In Yellowstone Park it

Table 8-6 Carrying capacity of natural grasslands for big game and livestock (from various sources, compiled by Petrides 1956, Petrides and Swank 1965).

Location	Game or livestock	Number/km²	Biomass (kg/km²)
Oregon	Antelope (64%), mule deer (36%)	3.5	175
Tanganyika, Africa	Bush country game	3.8	575
Montana	Bison (50%), mule deer, elk, bighorn	8.1	2,450
Arizona	Bison	6.5	3,000
Western United States	Cattle, average all grassland types	7.7	3,500
Western United States	Cattle, tall grass prairie	11	4,900
Nairobi Nat. Pk., Africa	(1) Herbivorous big game	33	4,900
Nairobi Nat. Pk., Africa	(2) Herbivorous big game	52	8,300
Queen Elizabeth Nat. Pk., Africa	Herbivorous big game	–	17,500

Table 8-7 Relative abundance (per cent of total specimens collected) of various orders of arthropods in normal and overgrazed grasslands in Oklahoma (after Smith 1940).

Order		Normal prairie	Properly grazed	Slightly overgrazed	Heavily overgrazed	Severely overgrazed and eroded
Coleoptera		29	27	19	14	11
Hemiptera		17	11	22	36	14
Homoptera		21	24	22	26	8
Hymenoptera		9	11	6	30	45
Diptera		19	22	23	30	6
Orthoptera		15	16	34	20	15
Lepidoptera		11	13	22	17	38
Arachnida		25	21	25	21	9
	Total	19	19	25	24	13

has been estimated that wapiti may utilize 67 per cent of the available grass forage, 47 per cent of forbs, and 30 per cent of browse. Browse and forbs are used more than grasses by pronghorn antelope and deer (Stoddart and Smith 1943). Because of difference in food preferences, competition between deer and cattle, although significant, is not as great as between wapiti and cattle (Mackie 1970). Deer and wapiti are able to graze steep slopes and other areas which cattle ordinarily do not (Stoddart and Rasmussen 1945). Competition between deer and wapiti on the one hand, and sheep and goats on the other, is more direct, however, because sheep and goats also feed largely on forbs rather than on grass.

Overgrazing produces a change both in the kinds and numbers of animals present (Table 8-7). This is correlated with the change from mid grasses to short grasses to weedy perennials. The short-horned grasshoppers increase in variety of species with this change, but in other orders of insects, the number of species present in overgrazed pastures either remains the same or declines. There is generally an increase with overgrazing in population level of all groups of arthropods, except beetles, until the pasture deteriorates to such an extent that erosion becomes severe; then there is a decline in abundance of all groups except the Hymenoptera and Lepidoptera.

Meadow voles, cotton rats, and cottontails are less numerous in overgrazed than in undisturbed grassland, but other rodents and lagomorphs increase in abundance. Tall grass in ungrazed pastures hinders the vision of jack rabbits, kangaroo rats, prairie dogs, and ground squirrels. Some rodents are benefited by the larger and more numerous seeds of the annual weedy species, and pocket gophers find more tap- and bulbous-rooted plants in deteriorated range (Bond 1945). Increased populations of insects and rodents are a result, not a cause, of overgrazing. If grazing by larger mammals is eliminated, succession back to thick grassland will occur in spite of the smaller animals, and prairie dogs and ground squirrels may actually be eliminated from the area (Osborn and Allan 1949).

In the luxuriant native prairie of early days, there was seldom overgrazing by such large mammals as bison, antelope, and wapiti, although this sometimes occurred in the more arid Great Plains. Insects and rodents occurred in populations that were in equilibrium with their food supply, and overpopulations of the species were held in check partly by the vegetation itself and partly by predatory birds, mammals, and reptiles. The most important of the larger predators were the hawks and owls, coyotes, foxes, badgers, black-footed ferrets, bullsnakes, and rattlesnakes (Shelford 1942). In California, it has been estimated that these predators eliminate about half of the annual increase of ground squirrels (Fitch 1948). Because coyotes and wolves occasionally took calves and lambs, they were systematically killed by ranchers; many other predators suffered with them. With the elimination of these predators, one of the checks on the rodent population was removed at a time when increased grazing by livestock rendered this control even more desirable. Damage done to the range by increased populations of rodents and rabbits has undoubtedly been much greater than the monetary value of an occasional killed lamb, calf, or chicken. In the great grasslands of the West, where human populations are low, there would be advantage not only in reducing the amount of grazing by livestock to the carrying capacity of the land but in restoring balanced populations of herbivorous and carnivorous species.

FOREST-EDGE COMMUNITY

Grassland animals are usually restrained from penetrating forests in the same way that true forest animals are restrained from penetrating grassland, although the home ranges of these species may overlap at the forest margin and in shrubby areas. Since shrubs are especially numerous at the forest-edge and animals have an opportunity to make use of these as well as both grassland and forest, the forest-edge biocies is well developed for some groups of animals (Fig. 8-3).

Fig. 8-3 Forest-edge at William Trelease Woods, University of Illinois: prairie grasses and herbs in foreground, briers and shrubs in middle, forest in background.

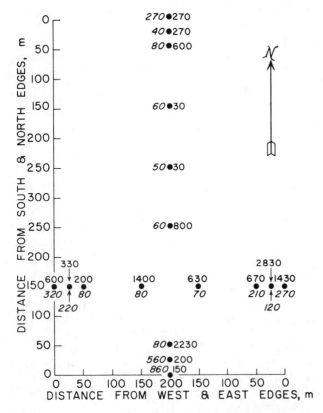

Fig. 8-4 Population (number per m²) of hibernating insects in the soil during the winter at different locations in a 22-hectare (55-acre) rectangular deciduous forest tract (Trelease Woods) in central Illinois. The forest is in contact with grassland on the south side and with farmland on the other sides. The italicized numbers indicate field and crop insects that have invaded the forest to hibernate; the other numbers indicate true forest species (modified from Kennedy 1958).

There are probably no soil or small ground animals characteristic of the forest-edge. There are some foliage insects that find their preferred niches here. Many insects of grassland and agricultural crops that over-winter as adults migrate into the forest-edge to hiber-

nate (Fig. 8-4). Since many game species of interest to man reach their greatest abundance on the forest-edge, he has become impressed by this *edge effect*. When total populations of all species are measured, however, the density of birds (Table 8-12) or mammals is not always higher than in the forest. When two forest types come in contact, for instance different deciduous forest types or deciduous and coniferous forests, there is no consistent change in the density of animal species (Barick 1950). The forest-edge is the preferred nesting site of many birds (Johnston 1947).

FOREST COMMUNITY

Since the censusing of each group of animals furnishes special problems, there have been no studies of total animal populations in single forest communities (Fig. 8-5, 8-6). By Table 8-8, however, it appears that the ratio in numbers of individuals per hectare between different animal groups is of the order: 1 bird, 3 mammals; 13,000 snails and slugs; 20,000 centipedes, millipedes, and sowbugs; 35,000 arachnids; and 225,000 large insects. The total mesofauna would number in the tens of millions and the microfauna in numbers so large as to be scarcely conceivable. The biomass of bacteria would probably be somewhat less than that of fungi but exceeding that of algae, protozoans, and nematodes combined (Burger and Rau 1967). In general, the number of individuals representing a species varies inversely with the body size characteristic of the species. There is, however, considerable variation in population levels both geographically and temporally. We must give special consideration to each of these various groups of animals.

Ground Animals

On the basis of size, ground animals (Figs. 8-7, 8-8) are grouped into *megafauna*: large millipedes, centipedes, and snails, amphibians, reptiles, small mammals;

124

Fig. 8-5 Interior of a temperate deciduous forest of sugar maple, basswood, and American elm in Wisconsin (courtesy U.S. Forest Service).

Table 8-8 Size of animal populations in forest and forest-edges, May to September, exclusive of some mesofauna and microfauna in the ground.

Taxonomic group	Deciduous forest, central Ill.[a]	Coniferous forest, Utah[b]	Chaparral, Utah[c]
Vertebrates (number per hectare)			
Shrews, mice, chipmunks	62	31	87
Squirrels, cottontails, raccoons, etc.	1	20	+
Birds	12	24	25
Snakes, lizards	+	+	+
Frogs, toads, salamanders	+	0	+
Invertebrates (number per square meter)			
Snails, slugs	79	+	1
Spiders	158	16	10
Harvestmen	12	0	0
Pseudoscorpions	10	0	4
Sawflies, wasps, bees, etc.	22	5	20
Ants	141	17	142
Flies	100	20	14
Moths, butterflies	8	0	+
Beetles	165	5	20
Leafhoppers, aphids	82	27	27
True bugs	40	3	10
Thrips	131	0	1
Psocids	1	0	0
Lacewings	1	0	+
Crickets, roaches, etc.	9	0	1
Insect larvae	307	4	20
Centipedes	67	5	3
Millipedes	31	1	2
Sowbug	24	0	0

[a]Including and extending data by Shelford (1951a, b).
[b]Hayward 1945.
[c]Hayward 1948.

Fig. 8-6 Interior of a virgin coniferous forest of Engelmann spruce in Colorado (courtesy U.S. Forest Service).

macrofauna: large insects and spiders, small millipedes, centipedes, and snails, earthworms; *mesofauna*: enchytraeid or potworms, small arthropods such as springtails, symphylans, pauropods, proturans, mites, insect larvae; and *microfauna*: protozoans, rotifera, nematodes, tardigrades, turbellarians.

Some animals, *geobionts* (Table 8-9), spend all their lives in the ground; certain protozoans, flatworms, nematodes, annelids, tardigrades, snails, amphipods,

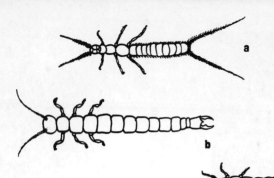
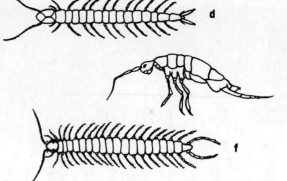

Fig. 8-7 Soil fauna: (a) campodeid; (b) japygid; (c) proturan; (d) symphylid; (e) springtail; (f) centipede (from Kevan 1955).

millipedes, centipedes, some spiders, mites, pseudoscorpions, true scorpions, many small wingless insects, some beetles and other winged insects, and a few mammals are examples. Other animals, *geophils*, live in the ground only as eggs, larvae, or pupae, such as do many flies and beetles; in cocoons, as do some moths; or for hibernation, as do many beetles and bugs. There is much of interest to learn about the natural history and biology of ground animals (Kuhnelt 1961, Burges and Roll 1967, Schaller 1968, Wallwork 1970, Ghilarov 1971).

Ground animals are most abundant in undisturbed virgin areas (Table 8-9). In a longleaf pine forest suffering frequent burning, the number of small animals in the humus layer was reduced to one-fifth and the number in the top 5 cm of the mineral soil was reduced to one-eleventh of the number in unburned areas (Heyward and Tissot 1936).

Of the microfauna, some 250 species of flagellate, amoeboid (including testaceans), and ciliate protozoans have been recorded in the soil (Sandon 1927), but only a few species are limited in distribution exclusively to the soil since they also occur in fresh-water habitats. Many species occur in practically worldwide distribution. Flagellates commonly range from 100,000 to 1,000,000 or more individuals per gram of soil; amoeboid forms, from 50,000 to 500,000; and ciliates, from 50 to 1,000 (Waksman 1952). The wet weight biomass of protozoans commonly varies from 5 to 20 g/m². Over 150 species of rotifers are known as ground inhabitants, and about one-third of these species have

Table 8-9 Numbers of soil animals per square meter (mostly mesofauna) in three different communities.

Locality: Community:	England[a] Disturbed grassland	Trinidad[b] Tropical rain forest[c]	Tennessee[d] Southeastern pine forest
Season of censusing:	Nov.	July–Sept.	June–Aug.
Depth of sampling:	30 cm	23 cm	3–4 cm
Sowbugs	0	12	0
Pseudoscorpions	56	138	203
Spiders	142	138	537
Mites	164,363	20,022	83,920
Millipedes	401		27
Pauropods	629	366	183
Centipedes	648	366	77
Symphylids	3,867		267
Telson-tails	1,363	0	930
Japygids	6,605	42	73
Springtails	61,269	354	12,400
Termites	0	5,394	127
Thrips	1,129	42	443
Ants	141	2,736	357
Miscellaneous adult insects	23,047	992	1,030
Immature insects	*	1,578	660

[a]Salt *et al.* 1948.
[b]Strickland 1945.
[c]From 5 samples from 3 forest reserves.
[d]Crossley and Bohnsack 1960.
*Larvae classified with adults.

Fig. 8-8 Some inhabitants of the ground stratum in a temperate deciduous forest: (a) camel cricket; (b) yellow-margined millipede; (c) round red millipede, *Narceus*; (d) *Mesodon pennsylvanicus*; (e) *Allogona profunda*; (f) *Anguispira alternata*; (g) *Anguispira kochi*; (h) *Haplotrema concava* (Shelford 1913).

been found only in the soil. They feed on organic material and, to a lesser extent, on nematodes and protozoans. Nematodes of many species occur to the extent of 1,000 to 10,000 individuals per cubic centimeter. Most of these forms belong to the Anguilluliformes and are more or less worldwide in distribution. They commonly possess mucous glands in the skin, the secretions of which aid locomotion. These nematodes are very resistant to desiccation and will quickly become active when moisture is added to soil that has been dried out for years. They feed on both macro- and microflora, on bacteria, and some are predators on the microfauna (Wasilewska 1970). Tardigrades occur regularly, sometimes abundantly; they too are very tolerant of desiccation. Land planarians are not common except in moist tropical regions. Some of these soil animals are detritus-eaters, some bacterial and algal feeders, some partly carnivorous, and some partly parasitic on plant roots.

The majority of these small organisms are active only in soil water, present as a thin film lining the surfaces of the soil particles. Swimming forms are necessarily very small; often, they appear dwarfed compared to the size they have been brought to in cultures. Nematodes are somewhat less restricted in their movements. They can distort the surface of the water film by means of muscular movements, and thereby bridge intervening air spaces to the next soil particle. Amoeboid organisms and hypotrichous ciliates usually accommodate their shapes to irregularities of the solid surfaces over which they crawl and can become larger in size but still remain in the water film. The variety of microhabitats in the soil accommodating the large number of species that occur includes spaces between surface litter, caverns walled off by soil aggregates, root channels, fissures, and pore spaces between individual soil particles. These microhabitats vary in size, temperature, and moisture conditions (Birch and Clarke 1953).

Most of the insects, as well as the myriapod and arachnid groups that belong to geobiontic meso- and macrofauna, are wingless or nearly so (Lawrence 1953); many species are also eyeless. Some arthropods burrow into the ground (digger wasps, crayfish), some

are primarily inhabitants of the litter (millipedes, fly larvae), others may follow cracks or other microcaverns down into the soil (mites, springtails), and several small forms are directly dependent upon the presence of water in the soil (copepods, tardigrades). Amphipods of the family Talitridae occur in moist leaf litter and have evolved an entirely terrestrial mode of life. They occur in countries and islands adjacent to the Indian Ocean and south Pacific (Hurley 1959). The springtails jump around by means of a special springing apparatus. Millipedes and centipedes, of course, have numerous legs. Many of these animals feed on plant litter and fungus, but the pseudoscorpions, spiders, some of the mites, and centipedes are carnivorous. Most of these species are annuals or have even shorter life cycles. Favorable soil moisture and food are most important in maintaining their numbers; temperature and hydrogen-ion concentration are secondary factors. Differences in the character of soil, whether sand or clay, do not appear to affect the size of populations greatly; however, the amount of decaying humus present is important. In Denmark the biomass of soil organisms decreases from oak to beech to spruce forests (Table 8-10), but there is an increase in number of individuals in beech and spruce over the oaks, attributable to increased numbers of mites and springtails, which are so small that they do not greatly affect the biomass. Springtails also increase in abundance from oak to spruce to beech in the forests of Yugoslavia (Stevanovic 1956). Wet-weight biomasses of mites and springtails in an English grassland area varied from less than 0.1 to 1.4 g/m^2; they were generally at peak during the winter months (Macfadyen 1952). In a hemlock-yellow birch forest in Michigan, mites and springtails were over twice as numerous in winter as in summer (Wallwork 1959). In deciduous upland forests of Indiana, however, mites showed a peak of numbers in the spring and autumn with lower numbers in both winter and summer. Pseudoscorpions were most abundant during June and July (Gasdorf and Goodnight 1963). In general, mites predominate in the mesofauna in all types of soil with springtails second in abundance in all except perhaps tropical soils (Table 8-9).

Table 8-10 Number, biomass, and metabolism of ground invertebrates per square meter in forests of Denmark (after Bornebusch 1930).

Type of forest: Number of stations:	Oak 1		Beech 6		Spruce 3	
Invertebrate group	Number	Biomass (g)	Number	Biomass (g)	Number	Biomass (g)
Lumbricid worms	122	61.0	79	15.8	50	2.5
Other humus-eating animals	2,675	15.0	9,338	10.6	10,807	7.0
Predaceous animals	181	0.8	264	1.5	290	1.5
Totals	2,978	76.8	9,681	27.9	11,147	11.0

There are two main groups of annelids in the soil, the large red earthworms, Lumbricidae and Megascolecidae, and the small, whitish potworms, Enchytraeidae. In New York State, 10 species of earthworms were recorded in the top 15 cm of the soil, and they varied in abundance from 37 per m² in oldfields to 34 in deciduous forest to 20 in white pine forest to none in spruce forest (Stegman 1960). In rich, moist humus soil, the red annelids may reach populations of 500 individuals and a biomass of 150 g/m²; potworms sometimes occur in hundreds of thousands per m². Earthworms ingest particles of mixed humus and mineral soil, absorb the organic matter out of them, and defecate around the entrances and along the length of their burrows. The earthworm is remarkable in its capacity to withstand desiccation—up to 75 per cent of the water content of its body. The native North American annelid fauna has been extensively modified by the widespread invasion of introduced *Lumbricus terrestris* and *Allolobophora caliginosa*. These species are found in forested areas, especially along rivers to which they have been carried by fishermen.

Potworms feed more on plant and animal detritus, but may ingest some mineral particles. Potworms may also exert some control over parasitic nematodes of plant roots. Minimum numbers in Wales occur in late winter, maximum numbers in the early summer, and the biomass varies from 2.7 to 13.2 g/m² (O'Connor 1957). In other localities the biomass of potworms may at times exceed 50 g/m².

The gastropod fauna is rich in moist, humus soil, but becomes scarce when the soil dries out. It is more abundant in deciduous than coniferous forests, because

Fig. 8-9 Wood-eating beetle, *Passalus cornutus*. Top left, adult; top right, pupa; bottom, larva (Shelford 1913).

coniferous forest soils tend to be acid. In eastern North America there are three common genera of slugs, *Philomycus*, *Deroceras*, *Pallifera*, and a variety of snails belonging principally to the Polygyridae, Zonitidae, Entodontidae, Haplotrematidae, Pupillidae, and Succineidae. Fifty species of snails were collected in 74 hours of searching in the Great Smoky Mountains (Glenn Webb). The haplotremes are carnivorous, feeding on other snails, but otherwise the gastropods feed chiefly on detritus, algae, lichens, and fungus.

The assemblage of small animals that dwells in the soil surface litter and under stones, rotting logs, and the bark of trees is sometimes called *cryptozoa*. Many species occurring in this microhabitat also commonly occur through the litter and soil generally, especially in moist climates with rich soil humus. In temperate forests, however, some snails, sowbugs, some spiders, lithobiid centipedes, julid and polydesmid millipedes, entomobryid springtails, roaches, earwigs, staphylinid, carabid and histerid beetles, and some ants reach maximum populations as cryptozoa (Cole 1946). The cryptozoan habitat is a favorite of salamanders. Many common soil animals are found as well in the special tree-hole forest microhabitat (Park *et al.* 1950); indeed, some species are specifically limited to tree-holes. Certain species of sowbugs *Porcellio scaber* commonly inhabit high elevations in trees during the summer but migrate to the base of the trees to pass the winter (Bereton 1957).

Decaying logs and stumps are preferred by many species (Fig. 8-9). During the first three years following the cutting of the pine trees in a North Carolina stand, 130 species of insects, myriapods, annelids, mites, and mollusks were found (Savely 1939). Coleoptera was by far the most numerous order of insects. Of all species, approximately 7 per cent were phloem-feeders, 15 per cent sapwood-feeders, 44 per cent rotten wood- and fungus-feeders, 30 per cent predaceous, and 4 per cent parasitic. The phloem-eaters were most numerous during the first year. Their mode of feeding prepared the way for the later entrance of fungi, fungi-eating species, and predaceous forms.

In the decay of logs of such northern trees as pine, spruce, and fir, the character of the food available is important to the succession that occurs (Graham 1925, Ingles 1931). Bark beetles require fresh green tissues of the inner bark and cambium, and hence occupy the tree only for the few weeks that these tissues remain. The long-horned beetles and wood borers require green tissue for their younger stages. As they mature, they are able to digest the solid wood. The horntail larva can digest solid wood as soon as it hatches. The outer bark is most difficult to digest but it furnishes food for some species of Lepidoptera and Diptera.

In addition to invertebrates, there are several vertebrates, particularly mice, shrews, moles, amphibians, and reptiles, that may be mentioned as part of the ground fauna. These animals often have extensive underground runways and feed on the invertebrates in the soil, as well as on each other (Fig. 8-10).

Foliage Arthropods

Foliage insects and spiders are represented by large numbers both of species and individuals (Fig. 8-11)

Fig. 8-10 Relation of small mammals to the forest floor (Hamilton and Cook 1940).

(Graham 1952). Spiders, ants, flies, beetles, leafhoppers, bugs, and larvae ordinarily predominate (Table 8-8); population depends directly on the amount of green foliage present. The species present depend on the type of vegetation, stratum, season, time of day, locality, and climate. Outbreaks of particular species may occur irregularly or periodically, and several hundreds or even thousands of individuals may occur in each tree. Spruce budworms and walkingsticks are sometimes so abundant that their excrement or eggs dropping to the ground sounds like the patter of raindrops.

Birds

Breeding-bird populations (Fig. 8-12) in forest communities vary with the fertility of the forest, but are commonly between 100 and 400 pairs per 40 hectares (100 acres), which would be equivalent to 5 to 20 individual birds per hectare (2 to 4 per acre) (Table 8-12). In addition, there is often a large non-breeding population present. On a 16-hectare (40-acre) spruce-fir forest in Maine there were 154 territorial males present prior to June 13. Between June 21 and July 5 a determined

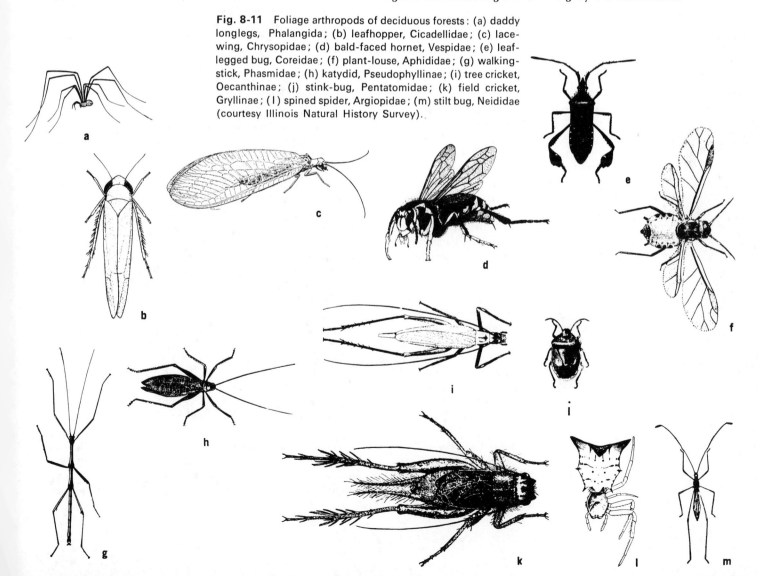

Fig. 8-11 Foliage arthropods of deciduous forests: (a) daddy longlegs, Phalangida; (b) leafhopper, Cicadellidae; (c) lacewing, Chrysopidae; (d) bald-faced hornet, Vespidae; (e) leaf-legged bug, Coreidae; (f) plant-louse, Aphididae; (g) walkingstick, Phasmidae; (h) katydid, Pseudophyllinae; (i) tree cricket, Oecanthinae; (j) stink-bug, Pentatomidae; (k) field cricket, Gryllinae; (l) spined spider, Argiopidae; (m) stilt bug, Neididae (courtesy Illinois Natural History Survey).

Fig. 8-12 Adult red-shafted flicker and young at nest in an aspen forest, Oregon (courtesy U.S. Forest Service).

effort was made to reduce the population of songbirds, and a total of 352 males was taken from the area, more than were actually nesting at the start of the operations. Yet on July 11 there were still 40 males present and proclaiming territories. One hundred twenty-six females and 49 individuals of undetermined sex were also removed (Hensley and Cope 1951). This floating population is more numerous than the number of nest sites available in the community, but it functions as a pool from which individuals may replace any nesting birds that die, or take over any suitable niche that becomes available for one or another reason. This population pressure doubtless keeps the community saturated with breeding birds, tends to maintain the nesting population at a high peak of efficiency, and is a challenge to individuals to exploit new adaptations and to occupy new niches. It is therefore a potent factor in evolution.

In a mixed deciduous-coniferous forest in Europe, a bird population of 662 individuals per 40 hectares (100 acres) was estimated to have a biomass of 47 kg (103 lb). In the same area, there were 528 individual mammals of biomass 264 kg (580 lb) (Turček 1952). The number and biomass of birds is generally less than that of mammals (Hamilton and Cook 1940).

Mammals

Rodents (mice, chipmunks, squirrels) and insectivores (shrews, moles) are the most abundant small mammals

of forest communities. Resident summer populations commonly vary from 25 to 100 per hectare (10 to 40 per acre). In rich, moist, undisturbed forests, populations may sometimes attain temporary levels of up to 500 per hectare (200 per acre). Considerable data on population sizes and biomasses of individual species have been compiled by Mohr (1940, 1947).

In forests of eastern North America (Hamilton and Cook 1940:469), small mammals fall into several categories.

The deer [white-footed] mice and the flying squirrels are adept climbers and often have their homes thirty feet or more [10 meters] from the ground in some hollow snag, deserted nest, or abandoned woodpecker hole. Flying squirrels feed among the trees and descend to the ground to forage about old logs and brush piles. They also dig down into the litter from the surface. Chipmunks forage in much the same manner, although they climb less frequently. Deer mice occupy several levels, from the trees to the burrows of moles and shrews. The red-backed mice, the lemming mice, and probably the jumping mice dig fairly permanent tunnels and runways through the soil and the litter. These they use as bases for food-gathering in both the litter and the upper layers of the mineral soil. These runways are often used by the shrews and the deer mice. The short-tailed shrews dig substantial tunnels. The diminutive long-tailed shrews thread their way through the easily parted litter and top-soil, and make intricate temporary labyrinths daily in search of food. The moles remain in their . . . tunnels during the daylight hours, but often come to the surface at night, no doubt attracted by the countless invertebrates that swarm over the ground with darkness. In winter they remain safe under the snow.

In mixed coniferous forests in the Sierra Nevadas the total number of rodents varied from 150 per hectare (60 per acre) in July–August to 52 per hectare (21 per acre) in December; their biomass varied from 27 to 4 kg/hectare (24 to 3.3 lb/acre) (Storer *et al.* 1944). The home ranges of individuals of these small rodents and insectivores are commonly only a fraction of a hectare (Blair 1953).

Of forest species, the larger the mammal, the fewer its numbers; and, usually, the wider an individual's range. The home ranges of the weasel, raccoon, and bear are more extensive than those of rodents and insectivores. In the aggregate, their biomass does not exceed that of the more numerous smaller species. A population density of one deer per 20 hectares, for instance, translates into a deer biomass of about 2.8 kg/hectare (2.5 lb/acre).

Stratification

Animal life in the community is separated into different niches in relation to strata (Elliot 1930). Food, shelter, and microclimatic differences are the chief limiting factors. Because microclimate in each stratum varies from hour to hour, day to day, and season to season, classification of a species by stratum must be in terms of the stratum it is observed to frequent for the major portion of a relatively long period of time. The inhabitants of the five strata divide into two major groupings, or societies. The soil invertebrates and some mammals move freely back and forth between subterranean and surface strata, and may be considered a *ground society*, distinct from *a foliage society* which occupies the combined herb, shrub, and tree strata. The two societies are not mutually exclusive, for ground animals such as millipedes and snails climb up onto herbs and tree trunks during humid weather, and foliage animals often rest on the ground, and search for food, hibernate, and lay eggs, there.

The majority of arthropod species carry on their main activities within a single stratum for their major activities (Table 8-11). The tree stratum spans a greater vertical distance than any other. Within this broad stratum, arthropods often show segregation to particular heights above the ground (Davidson 1930). There is a striking difference in species composition of spiders between the shrub, herb, and surface strata (Luczak 1960). Likewise, several species of homopterans in grassland communities show segregation between the litter and different heights in the grass (Andrzejewska 1965). Ants and beetles appear to move more freely between different strata than do other species.

Population density of invertebrates ranks highest in the surface stratum, followed by subterranean and herb strata, while shrub and tree strata rank lowest. The largest mammal populations occur in the subterranean and ground strata.

In a European oak-hornbeam forest (Turček 1951), 15 per cent of the bird species nested on the ground, 25 per cent in the herb and shrub strata, 31 per cent in or on the trunks of the trees, and 29 per cent in the tree canopy. The largest number of individuals (32 per cent) occurred in the forest canopy, although the biomass of these birds constituted a smaller percentage of the total (16 per cent) than did the ground and herb population (67 per cent). In respect to feeding, however, the distribution was different: 52 per cent found their food on the ground, 9 per cent in the herbs and shrubs, 10 per cent on the tree trunks, 23 per cent in the tree foliage, and 6 per cent in the open spaces between the canopy, trees, and shrubs.

Seasonal Changes

Outside of the tropics, the forest community changes drastically with the seasons such that six aspects may be recognized (p. 99). The total population of the soil macrofauna in the temperate deciduous forest is highest during the winter because of the migration of many foliage insects into this stratum to hibernate. However,

Table 8-11 Stratal distribution (%) of arthropod species in Missouri (Dowdy 1951).

Distribution of species:	Oak-hickory forest	Redcedar forest
Number of species:	161	96
Confined to one stratum	69	78
Confined to two strata	19	15
Confined to three strata	11	5
Confined to four strata	1	2
Found in all five strata	0	0

many species overwinter only as eggs or other immature stage. Forest species hibernate in densities that vary randomly throughout the forest, except where there are differences in topography or substratum. Forest tracts adjacent to grass- or farmland, however, receive an influx of non-forest species that hibernate principally on the forest-edge generally, and along the south edge in particular, where exposure to solar radiation and protection from cold northerly winds produces warmer temperatures (Fig. 8-4) (Weese 1924).

During the vernal aspect, insects and other invertebrates come out of hibernation, and the adults of forms variously frequenting the herb, shrub, and tree foliage return to their characteristic stratum (Fig. 8-13). The population of ground animals remains relatively high throughout the year, however, which can be attributed to the reproduction of the geobionts and to the fact that the immature stages of many foliage arthropods, particularly Diptera, Coleoptera, and Lepidoptera, occur in the ground. Various groups of these geobionts and geophils reach peak numbers of adults at different times during the year (Pearse 1946).

An insect species may show more than one population peak during the year (Shelford 1951), depending on the number of generations produced and the specific life span. A species may appear, attain to very large numbers, and disappear, all in a matter of a few days, or a few weeks at most. Considerable variation also occurs from year to year in the population fluctuations of individual species and of whole groups, correlative with differences in weather, particularly temperature and moisture (Leopold and Jones 1947).

Bird populations in temperate zones reach peak populations with the passage of transients during the seasons of migration (Fig. 8-14). These migration peaks are inconspicuous or absent in northern coniferous forests, since most birds that arrive stay to nest. In the tropics, birds are most abundant during the winter period of the north temperate zone, since the fauna then contains many migrant species from the north.

Mammal populations in temperate regions commonly reach their maximum numbers in the autumn,

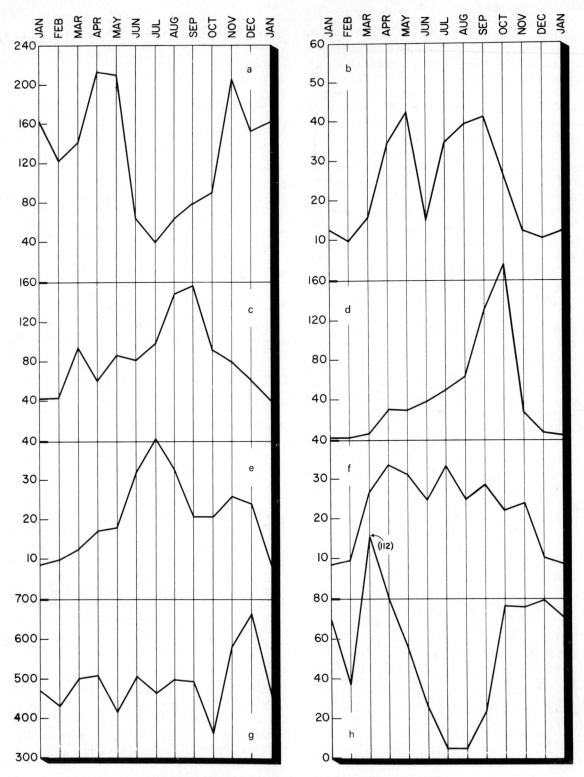

Fig. 8-13 Monthly variations in total population size of different animal groups in a temperate deciduous forest (Trelease Woods) of central Illinois. The data for each month are averages of 10–14 year records. (a) Diptera larvae; (b) Lepidoptera larvae; (c) spiders; (d) Homoptera; (e) snails; (f) centipedes; (g) invertebrates (macrofauna) in ground; (h) non-forest species.

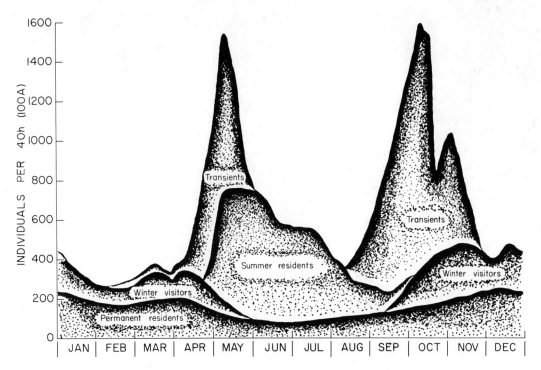

Fig. 8-14 Seasonal fluctuations in the bird populations of a deciduous forest area in Ohio (modified from Williams 1936).

at the end of the breeding season. Populations decline progressively as winter wanes, the result of mortalities from severe weather, lack of food, and predation.

Comparison of Animals in Different Forest Types

There are more niches and microhabitats available in forest and forest-edge communities than in any other type of terrestrial community. The stratification and the diversification of plant forms are responsible for this. Many forest niches are much the same, regardless of whether they occur in deciduous or evergreen forest, and regardless of geographic location (Blake 1926, Dirks-Edmunds 1947). Species occupying these niches differ, but are often taxonomically related; related or not, they have similar mores. Thus a predominant, shrub-inhabiting, plant-juice-sucking leafhopper in an Illinois oak-maple forest is *Erythoneura obliqua*, but in a Maine pine-hemlock forest it is *Graphocephala coccinea*. The common herbivorous woods mouse in Illinois is *Peromyscus leucopus*, but in the Douglas fir-hemlock forests of Oregon it is *Clethrionomys occidentalis*. An insectivorous hole-nesting chickadee in Maine is *Parus hudsonicus*; in Illinois, *P. atricapillus*; and in Oregon, *P. rufescens*.

Since coniferous forests have needle leaves and deciduous forests, broad-leaves, special niche adjustments are required which may not permit a species to

occupy successfully both kinds of forest (Fig. 8-15). The tube-building moth attaches its eggs to the pine leaf, and, when the larva hatches, it makes a nest of 6–15 needles, bound together with silk-like threads. The larva, and later the pupa, is protected within this tube but can come out and feed on the end of the leaves of which the tube is constructed. Deciduous leaves are clearly incompatible with these behavioral patterns, so nicely adapted to the peculiarities of pine needles. On the other hand, the red-eyed vireo experiences difficult feeding in coniferous trees because of the arrangement and close position of the needle leaves on all sides of the twig (Kendeigh 1945). While the vireo can feed in coniferous forests, it is considerably more profitable for the bird to confine itself to the deciduous forest, for which it is better adapted. Some animal kinds inhabit the soil and litter of both deciduous and coniferous forests, but a kind may be more abundant in one forest type than the other because coniferous and deciduous leaves form two distinct types of humus. Because of differences in foliage character, persistence through the year, climate, and the considerable difference in the taxonomic composition of the plants and animals involved, coniferous and deciduous forests are separated as distinct *biomes*, each with a number of *biociations* in different parts of the world (Chapters 20 and 21).

Bird populations are not necessarily consistently higher in one type of forest than in the other (Table 8-12); rather, population varies with the luxuriance

Fig. 8-15 Foliage insects of coniferous forests: (above) sawfly larvae, *Neodiprion lectonei*, on jack pine; (right) spruce budworm eggs, larva, chrysalis, and moth (courtesy U.S. Forest Service).

Table 8-12 Average densities of total breeding-bird populations in forests and forest-edges of different types in eastern North America (compiled from Fawver 1950).

Type of vegetation	Number of areas censused	Number of species	Number of territorial males per 40 hectares (100 acres)
Spruce-fir (coniferous) forest	5	30	311
Mixed coniferous forest	6	33	207
Mixed hemlock-deciduous forest	5	28	224
Beech-maple-hemlock forest	5	31	190
Mixed deciduous forest	17	26	255
Deciduous floodplain forest	7	28	229
Oak-hickory forest	5	24	181
Mixed deciduous and southern pine forest	5	19	157
Southern pine forest	1	23	163
Broad-leaved evergreen forest	1	23	162
Coniferous forest-edge	7	30	241
Deciduous forest-edge	6	27	265

of the vegetation. Animal populations in both coniferous and deciduous forests are generally highest in regions where ample rainfall brings rich development of vegetation as the basic food supply (Odum 1950). Population densities of both birds and mammals decrease progressively westward from the Appalachian Mountains to the eastern edge of the prairie as the climate becomes progressively drier (Wetzel 1949, Fauver 1950). The variety of snail species decreases in a similar manner from moist to dry forest types (Shimek 1930).

Forest and Game Management

The forest productive capacity of greatest economic interest is the timber yield. Forests are also of great importance to man for the protection of watersheds against erosion; for such recreational purposes as hunting, camping, and hiking; and for the inspirational values of unspoiled scenery and primitive nature. Complete logging, as practiced universally in colonial and even in modern times, destroys the forest, most of its animal life, and its usefulness for these purposes. Logging on a sustained-yield basis converts the forest into a forest-edge community and allows the forest-edge animals to increase in abundance, while the animals dependent on dense forest decline.

Such game animals as the gray squirrel, black and grizzly bears, moose, the fur-bearing marten, and the ruffed grouse and wild turkey belong primarily to the forest proper, although they often feed in the forest-edge, or brushland, and openings scattered through the forest. Populations of these species may be maintained simply by preserving large tracts of virgin or dense forests.

Most game animals of interest to the ordinary sportsman, however, belong to the forest-edge. These species are the cottontail, fox squirrel, deer, bobwhite, pheasant, and dove. Increase in populations of these species requires interspersing the forest with open areas, development of shrubby forest margins, or creation of artificial cover along fence rows, uncultivated field corners, around ponds, along drainage ditches or streams, on steep slopes subject to erosion, and on waste lands (Trippensee 1948). Intelligent management may involve control of plant succession to prevent its proceeding to a normal closed forest, and harvesting the forest for both timber and game. Procedures for managing timber on a sustained-yield basis are fundamentally the same as for managing populations of game animals on a permanent basis. Soil conservation and erosion control can also be readily combined with wildlife management, especially when trees and shrubs selected for planting to regulate soil erosion are species useful to game as cover and food.

The farmer can encourage establishment of small game species on his land by practices that do not interfere with the raising of crops (Fig. 8-16). The maintenance of brushy fence rows does not increase the number of insect or other crop pests (Dambach 1948), as has sometimes been maintained. A knowledge of the fundamentals of life history and ecology is essential to wildlife management, as wildlife management is applied ecology and involves the management of the total community, not merely the game species in it (Leopold 1933).

LIFE HISTORIES

We here choose four species to show life history adjustments to the habitat and community: two mammals for the grassland, a bird for the forest-edge, and a millipede for the forest.

Voles (*Microtus, Pedomys*)

The meadow vole, *M. pennsylvanicus*, and prairie vole, *P. ochrogaster*, are small, dark gray or brownish, compactly built mice with short legs and tail, small eyes, and partly hidden ears. They spend most of their time in an elaborate system of tunnels, partly underground and partly as almost hidden galleries in dense grass. The food of these species consists mostly of grasses but also includes legumes, composites, fruits, and occasionally insects. Grass stems grow close together. The voles thrust them aside to form paths on the ground surface. These runways are heavily trafficked networks, and feeding is done in them. Runways formed through clover or alfalfa are less permanent. Underground

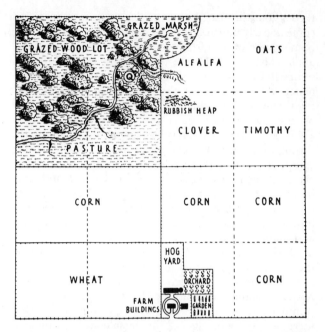

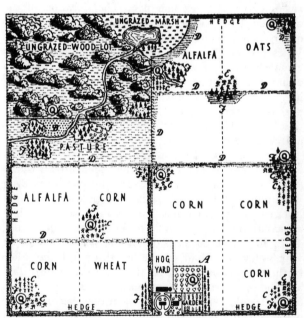

Fig. 8-16 Sketch of a 100-acre farm before and after improvements to encourage small game: (*A*) windbreak, of some value to game as cover; (*B*) hardwood planting, perhaps black locust usable later for fenceposts; (*C*) a portion of the crop (corn) left, near cover, for wildlife; (*D*) a field or fence border; (*E*) emergency food, a few shocks of grain placed near cover; (*F*) cover planting of coniferous trees; (*Q*) a quail habitat with food and cover (Phelps 1954 in *Virginia Wildlife*).

passages lead to nests and chambers where food is stored. Nest cavities are round, lie 7 to 45 cm below the surface, and often have two tunnels leading up to the surface. The cavities are lined with dried grass and leaves (Jameson 1947, Martin 1956).

The voles may be active at any hour of day or night

throughout the year, but the periods of greatest activity come in early morning and evening. The mean monthly home range of an individual is very small, commonly about 364 m² (0.09 acre), although males wander somewhat farther than do females. There is no defense of territory. Because of their small home ranges and high rates of reproduction, vole populations in years of abundance may reach 366 or more per hectare (146 per acre). The level of population fluctuates through the years, however, in response to rainfall and competition, and there is some evidence of a 4-year cycle. Populations regularly decline each winter and increase to an annual peak in the autumn. Predators on the two voles are numerous and varied, and include hawks, owls, crows, weasels, foxes, coyotes, badgers, and snakes (Hamilton 1937, 1940, Martin 1956).

During the peak of a population cycle, or when there is a good protective ground cover of snow, breeding may continue throughout the year, but it is ordinarily curtailed during the winter and periods of summer drought. The number of young in a litter increases with the age of the female but commonly varies between three and five. The duration of the estrus cycle in the female, if such a cycle actually occurs, is not known, but is likely only a few days long. Voles are promiscuous, and the female may accept a male within a few minutes or hours after the birth of her young and be capable of ovulating and conceiving a new litter within 5 or 6 days. The gestation period is short (about 21 days), and it is estimated that 8 to 10 *Microtus* litters may be produced in a favorable year (Hamilton 1941). *P. ochrogaster* is less prolific.

The young are born pink-skinned, hairless, blind, and with ear pinnae closed. They soon attach themselves to the teats of the mother, who may even drag them along as she forages. They weigh 2 to 3 g at first but grow rapidly, gaining $\frac{1}{2}$ to 1 g/day. Meadow voles at sexual maturity weigh 25 to 30 g; when fully adult, 40 to 50 g. The backs of the young voles are covered with soft velvety hair on the fourth or fifth day after birth, the incisors erupt on the sixth or seventh day, and the eyes open and the pinnae unfold on the eighth or ninth day. With their eyes open they become more active and may take short trips away from the nest to nibble on succulent vegetation. The young voles are weaned at 2 to 3 weeks, but may remain with their mother for several days longer. Males may become sexually mature when 5 weeks old, and females may mate successfully when only 4 weeks old. The mortality rate in young mice is high, but the normal life span of adult meadow voles under natural conditions is 10 to 16 months. Prairie voles appear not to live as long (Hamilton 1941, Fitch 1957).

Bobwhite (*Colinus virginianus*)

The bobwhite is found over the eastern part of the United States and south through Central America (Stoddard 1931). It occupies open woodlands, shrubby fields, fence rows, and forest-edges bordering on grassy fields or farmlands. Nests are usually located within 16 m of roads, paths, or cultivated fields. This vegetation serves as cover against both weather and enemies. The bird feeds primarily on seeds, occasionally on fruit and herbage, and, during the summer and autumn, on insects. It also ingests a quantity of mineral matter in the form of grit. The young chicks feed largely on insects the first 3 weeks after hatching, but then become grammivorous like the adults. With ample cover and food, the species may reach a maximum population of 2.5 per hectare (1 per acre), but populations of one bird to 2–5 hectares are more common. In the South, populations are fairly stable year after year, but in the North they fluctuate widely (Kendeigh 1944).

The birds have a number of call notes for communication between individuals of a pair or covey. These notes serve to attract mates, express alarm or distress, indicate that the individual is lost, for feeding, to reassemble the covey, for battle cries, and so on. Pairing of male and female usually begins in April as the winter covey breaks up, and the males give their *bob-bob-white* calls. During this period, there may be song competition between males, fighting, chasing, plumage displays, and bluffing. Competition is intense because there are more males than females. Two to 4 weeks may elapse before the pair begins to nest, during which time the two birds stay close together.

Nesting may start in April in the South and extend from May to August in the North. Nests are placed in good cover where the herb stratum is open enough so the birds can run around over the ground near the nest. A slight hollow is scratched in the ground, and the nest is commonly constructed of grasses, pine needles, mosses, or whatever is immediately available. A grassy arch is made overhead to serve as a roof and to conceal the nest from predators. It is ordinarily located on well-drained high ground.

One egg is laid daily until the full clutch of 14 or so is attained. An occasional day may be skipped, and clutches laid early in the year are larger than those laid later. The incubation period lasts 23 days, and incubation may be performed either by the male or female; three out of four times it is the latter. During this period the incubating bird usually leaves the nest for a time to feed early in the morning and often again in late afternoon. The incubating bird joins its mate at a distance from the nest and they feed and rest together from 1 to occasionally 9 hours, depending on the weather. The birds do not need surface water for drinking, but get what water they require from their food or from dew. About 86 per cent of eggs hatch, and all of these within about an hour. The young chicks quickly leave the nest and are cared for and brooded against cold, wetness, and too much sun by both parents for another 2 weeks. By that time, juvenal plumage is replacing the natal down, and the birds will flush and fly a short distance when disturbed. The young birds become similar in plumage to the adults at the end of 15 weeks. There is some feather molt about the head in the spring and a complete molt from August to October each year.

The winter covey forms in the autumn, and comprises one to three pairs of adults, their surviving young, and a few birds that were unmated. As birds die, small coveys unite and maintain an average size of about 14 birds. A covey may confine its activities within a range of 16 to 60 hectares (6 to 24 acres), and the ranges of adjacent coveys may overlap. The birds commonly freeze when approached by enemies, relying on their protective coloration for escape. If too closely approached they burst forth in rapid flight that carries them in all directions for 400–500 m, whence they then drop down into other cover. When the enemy disappears the covey call reunites them again. Coveys feed together and roost together. They roost on the ground in compact circles with heads pointing outward.

Species predaceous on eggs, chicks, and adults include skunks, rats, foxes, weasels, opossums, raccoons, dogs, snakes, red ants (eggs), cats, shrikes (chicks), Cooper's and sharp-shinned hawks, and great horned owls. Numbers of parasites and disease organisms potentially dangerous to it are harbored by the species. Heavy rains may be destructive to nests and young birds in the summer, while extreme cold combined with long periods when snow covers the ground may kill adults during the winter. The population turnover during a year is therefore large.

Millipede (*Pseudopolydesmus serratus*)

This species feeds on decaying leaves and other organic material. Adults occur in populations up to 5 per m², and immature stages may be present up to several hundred per m². High populations, however, occur only in poorly drained, moist forests. During periods of low precipitation, individuals migrate and become concentrated in wet depressions. The dependency of the species on moisture is further indicated by higher reproduction during wet than dry years.

Copulation occurs March to December, but there are two principal peaks of egg-laying; one in April, the other, during the first half of July. This results in two generations per year, but these are not genetically distinguishable.

There are 7 larval or instar stages. At hatching, the larva has only 7 post-cephalic somites and 3 pairs of legs. At each molt, more somites and legs are added until in the adult there are 20 post-cephalic somites and 30 or 31 pairs of legs. The first and second instars are whitish in color, but later instars develop a reddish-brown pigment. The April generation reaches the morphological adult stage in the autumn but sexual maturity not until the next spring. The July generation

overwinters in the fifth to seventh instars, reaches the adult stage the following spring, and sexual maturity in June or July. Molting takes place in small chambers similar to those in which the eggs are laid.

The egg chamber is unique (Fig. 8-17). It is made of fecal pellets containing ingested soil and organic material. These pellets are placed in a ring of diameter about 6 mm. More and more pellets are piled on until the ring reaches a height of about 3 mm. Some 200 to 400 eggs are then deposited inside the ring, after which the ring is closed at the top to form a capsule. The whole process requires 6 to 12 hours. After breeding is completed, the adults die.

PRESERVATION OF NATURAL AREAS

It is of utmost importance for the future of ecological *research* and *education* that adequate samples of virgin primitive areas—forest communities, tundra, grassland, desert, tropical and rain forest, and all seral as well as all climax types of communities in all parts of the world—be preserved intact. Balanced primitive communities are the result of processes at work through eons of time. Primary communities once destroyed, there is never assurance that the secondary communities which develop can ever exactly duplicate them. This involves not only the replacement of all species in the original fauna, but also their replacement in the same relative numbers so that an integrated balanced community is fully re-established. The preservation of such natural

Fig. 8-17 Formation of egg-chambers by the millipede *Pseudopolydesmus serratus*. Top and bottom, base of two egg-chambers being formed; middle, two egg-chambers filled with eggs and partially capped over (Hanson 1948).

areas is of historical value to future generations as a record of natural conditions over the country in pre-colonial days. Natural areas serve as controls for the agricultural development of the country, for the evaluation of various farming practices and uses of the land, and to show the potentialities of vegetative development of various parts of the continent. Preserves serve to *safeguard germ-plasm* represented in the diversity of species. No one can know the potential value for food, medicine, biological control, or domestication of any organism that makes up primitive communities. Finally, natural areas have *scenic, recreational, aesthetic,* and *inspirational* values of direct value to all persons with any interest in the out-of-doors.

Large primitive areas in the United States are preserved in some of the National Parks, National Monuments, and in some of the larger of our state parks (Kendeigh 1951). Natural, wild, and wilderness areas have been set aside in several of the National Forests. Congress passed the Wilderness Act in 1964 for the preservation of these and other large roadless areas. Smaller areas of ecological value are being preserved in state, city, and private preserves. Not all community types are represented; more areas need to be set aside in other parts of the country, and constant vigilance must be exercised to keep them undisturbed. These projects are being sponsored by the Nature Conservancy, the National Parks Association, the Wilderness Society, and other organizations which deserve the support of all ecologists.

SUMMARY

Daily fluctuations of temperature, precipitation reaching the ground, light intensity, and wind velocity are greater in grassland than in forests, but relative humidity is usually less. Lightning-set fires was an environmental factor of importance in primitive time. A gradient in habitat conditions extends from above the vegetation to the ground in both grassland and forest. Segregation of animal species into subterranean, surface, herb, shrub, and tree strata is partly explained by differences in response to this gradient. North-facing slopes are generally cooler, moister, and with lower light intensities than south-facing slopes.

The species composition of animals differs between grassland, forest-edge, and forest. Within each community there is a vertical division into a ground (subterranean-surface) society and a foliage (herb-shrub-tree) society. Animal density and biomass are generally greater in the former. Food, shelter, and microclimate are the chief limiting factors. Outside the tropics, there is considerable seasonal variation in the abundance of animals.

Many niches are similar in forests of different types; say, coniferous and deciduous. The species occupying these niches are often different, however, although they may have similar mores.

Grasses tolerate considerable grazing, and grassland productivity may provide a high carrying capacity for large grazing animals. Overgrazing by large populations of insects, rodents, or domestic stock, however, may bring deterioration of the range. Economic utilization of grassland requires proper balancing of grazing pressure against vegetative productivity throughout the year.

Forests are of great interest to man for timber, protection of soil against erosion, and recreation. Game species are usually more varied and abundant in the forest-edge than in the forest-interior. Game management is concerned with controlling the vegetation and habitat to produce the highest yield of the desired species and to regulate the number taken. It is necessary to know the intimate life histories of the species concerned before this can be accomplished intelligently. Finally, it is of utmost importance to ecological study and to man generally that adequate samples of primitive areas be preserved in an undisturbed condition.

Part Three

ECOSYSTEM ECOLOGY

Chapter 9

SOIL FORMATION,
CHEMICAL CYCLES,
SOIL AND AIR POLLUTION

Communities cannot exist without a habitat, nor is a habitat likely to remain long without a community developing in it. The functional interrelations between community and habitat are many and complex, constituting an *ecosystem*. Most important are soil formation, nutrient cycling, and energy flow. Human interference in these processes often causes pollution, and exploitation often brings exhaustion of natural resources.

We have already considered many of the reactions of plants upon the habitat, such as reduction of light and wind intensities, mitigation of temperature extremes, interception of rainfall, and increase in relative humidity. Plants also exert important effects on the formation, structure, and characteristics of the soil or substratum produced by accumulation of dead plant remains: they further the weathering of rock through acid excretion and the mechanical action of roots; they offer obstruction to wind- and water-borne materials; they help stabilize moving sand and talus slopes and help prevent erosion generally; they variously increase or decrease the water content of soil; they foster decomposition of raw humus into usable nutrients; and so forth. Water plants form marl. It is by these reactions that plants exert dominance in terrestrial communities and establish the physical conditions of the habitat, which must be acceptable to all minor plants and animals that dwell there. Succes-

sion of plant stages eventually brings the interactions between habitat and community into equilibrium upon the establishment of the climax (Weaver and Clements 1938). In this chapter we will be primarily concerned with soils as dynamic components of ecosystems, including the cycling of nutrients between the soil and the biota (Witkamp 1971).

SOIL FORMATION

Texture, porosity, consistency, arrangement of particles, chemical nature, and organic content of soils are determined by three sets of factors: the parental rock material, the biota, and the climate. Differences in *topography* modify the relative effects of these three factors, and *time* is required before their full effects are realized.

Parental Rock

The basic rock from which the mineral portion of a soil is derived determines, to a large extent, not only its chemical composition but also its structure. For instance, soils derived from limestone are highly calcareous and more alkaline than soils derived from sandstone. Clay soils are derived from feldspar; sandy soils, from quartzite. Clay forms a finely textured, compact, water-retaining soil. Sand is coarse-textured and porous. Loam is a mixture of sand and clay and makes the best soil. The presence of iron oxides and silicates produces the red and yellow colors of some soils. Humus produces black soils. Soil from swamps or bogs and very rich in organic material is called muck.

Residual soils are formed *in situ* from underlying bedrock. Soils may, however, be formed in one locality and moved considerable distances. Soils transported and deposited by wind are called *loess*; by water, *alluvium*; by glaciers, *till*.

Biota

Plants and animals have a highly important role in the formation of soil, both as they affect its structure (Jacot 1936, 1940) and as they aid in the production of humus. Plants contribute to the mechanical and chemical weathering of rock. Plant roots, especially those of trees, can split large rocks. Lichens, mosses, and even bacteria and fungi excrete acids in the course of metabolism, which dissolve the substances that cement rock granules together. When plant roots die, fungi convert them to dry, soft, spongy material (punk), used as food by saprophytic microarthropods. Usually the bark of the root remains intact the longest. Hollow tubes are thus formed that permit water and air to penetrate considerable depths into the soil. These channels gradually become filled with silt and animal excreta.

The addition of plant and animal organic matter to heavy compact soils or clay tends to open the soil, making it more porous. Addition of organic matter to sandy soils binds the particles closer together, making the soil less porous.

Earthworms may be divided into deep- and shallow-working species (Kevan 1955). Deep-working species dig narrow tube-like channels which may reach 2–3 m down through overlying soil to parent rock. Earthworms ingest soil while burrowing, digest and absorb organic matter from it, and egest the residue in a semi-liquid form which is used to cement the walls of the burrow or else is deposited at the surface as castings. Earthworms prefer easily digested succulent vegetation and dung for the purpose, but in the autumn may pull the freshly fallen leaves down into their burrows to use as food or nest linings. Ejected petioles may form midden piles around burrow entrances. In an undisturbed virgin prairie in Texas, earthworm casts made a layer 2–3 mm thick over the entire ground surface and when air-dried weighed about 2400 g/m^2 (10.7 tons/acre) (Dyksterhuis and Schmutz 1947). Earthworms are not, however, important soil builders in disturbed grassland; they may be absent altogether in arid regions. In other studies (Evans and Guild 1947), the dry weight of casts brought to the surface annually by earthworms varied from 475 g/m^2 (2.1 tons/acre) in a moderately hot dry climate, to 24,000 g/m^2 (107 tons/acre) in the White River valley of the Sudan, during the rainy season. Earthworm casts compared with the surrounding soil show higher total nitrogen, organic carbon, exchangeable calcium, exchangeable magnesium, available phosphorus, exchangeable potassium, organic matter, base capacity, pH, and moisture equivalent (Lunt and Jacobson 1944). Only certain species make these surface castings; other species void the ingested soil into subterranean spaces.

The ant *Lasius niger neoniger* spends most of its time in its underground burrows and deposits excavated soil upon the ground surface around burrow entrances. In an oldfield community in Michigan such deposits amounted, at one sampling, to 85.5 g/m^2 (750 lb/acre). However, entrances are abandoned and new ones made, so that in the course of a few weeks a much larger quantity of soil is brought up (Talbot 1953).

In the semi-arid Great Plains of western North America there is at least one species of ant that excavates extensively underground and builds a conical mound of this excavated material above the surface. A single such mound weighs approximately 77 kg (170 lb); there are as many as 50 such mounds per hectare (20 per acre) in some localities. Plainly, these little excavators move prodigious amounts of soil. The relatively sterile subsoil is gradually mixed with organic material and spread over the surface of the ground, thus increasing the depth of the fertile topsoil. Scarabeid beetles, bees, wasps, and in tropical regions mound-building termites also move considerable subsoil to the surface (Thorp 1949).

The crayfish *Cambarus diogenes* often occurs in poorly drained fields; it burrows down to the water table, sometimes a depth of 3 m. Excavated material is brought to the surface and built into chimney-like affairs which may be 20 cm high and almost that much in diameter. Where crayfish are abundant, as much as 600 to 2000 g/m^2 (2.7 to 8.9 tons/acre) of soil per year may thus be moved (Thorp 1949).

The burrows of prairie dogs and badgers may extend 2 to 3 m below the surface, and a single mound of excavated dirt weigh from 100 to 10,000 kg. Mounds made by pocket gophers and ground squirrels weigh from 7 to 180 kg each; it is not unusual to find 42 such mounds per hectare (17 per acre). These animals thus move from 7 to 9 kg of subsoil for each m^2 of surface (30–40 tons/acre) in a period of several months (Taylor 1935, Thorp 1949).

Large terrestrial animals trample the soil into greater compaction and destroy vegetation at sites where numerous individuals foregather; around water holes in grassland where bison and antelope come to drink, for instance, or winter yards of deer and moose, trails on hillsides, wallowing places, and so on. These reactions are usually very local, however.

As animals burrow and bring large quantities of loose soil to ground surface exposure, the likelihood of water and wind erosion destruction is greatly increased, especially true if the burrowing is done on hillsides where the flow of water is faster and where the animals always tend to deposit the soil on the downslope side of burrow entrances. On the other hand, the very same activities may decrease erosion where a soil is, in consequence, made more porous so that there is less water runoff.

Humus In soil, organic matter that is partly or entirely decomposed is called *humus*. The amount of humus varies from less than 1 per cent to as much as 20 per cent of the soil; peat soil may be largely organic material, but much of it resists decomposition, and hence is not true humus. Decomposition breaks down complex organic compounds into simpler ones that are washed back into the soil, thus becoming available again as nutrients.

On virgin prairie in Texas the ground litter of dead grasses and herbs amounted to over 300 g/m² when measured in April (Dyksterhuis and Schmutz 1947). The annual dry weight of leaves that fall to the ground in deciduous and coniferous forests varies from site to site and with the density of the trees, but is commonly in the range of 50 to 400 g/m² (Olson 1963). In mature climax forests the rate of decomposition of the litter and re-absorption by plants of the nutrients thus yielded keeps pace with the annual accumulation so that an equilibrium is established. In seral stages, decomposition and utilization do not keep up with the annual accumulation, so that the organic content of the forest floor increases with time (Fig. 9-1). Litter production is lower in arctic than in tropical regions (Bray and Gorham 1964):

REGION	METRIC TONS/HECTARE-YEAR		
	Leaves	Other	Total
Arctic-alpine	0.7	0.4	1.0
Cool temperate	2.5	0.9	3.5
Warm temperate	3.6	1.9	5.5
Equatorial	6.8	3.5	10.9

However, in warm tropical regions the amount of humus that accumulates on the forest floor is low because of the high rate of decomposition, runoff, and leaching (Fig. 9-2) (Olson 1963).

The thick organic layer on the ground moderates extremes in the daily and seasonal rhythms of soil temperature, retards freezing of the ground in the autumn and thawing in the spring, and retains soil moisture. Because of humus formation (involving oxidation) and the respiration of plant parts and animals underground, soil air contains little oxygen but much carbon dioxide, and it possesses a higher moisture

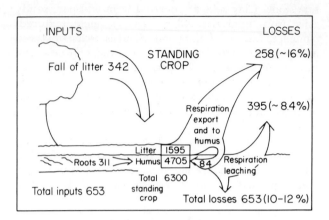

Fig. 9-1 Annual budget of organic matter in soil of an oak-pine forest (g dry weight/m²) (Woodwell and Marples 1968).

content than does the general atmosphere above ground. This is especially marked in warm summer months when these processes go on more rapidly. The decay of organic matter usually makes the top soil somewhat acid (most commonly pH 5 to 7), but in the mineral subsoil, the acids are often neutralized by the basic salts commonly present.

The mineral content of leaf fall varies according to the species of tree, but in the northern United States it averages about as follows (in g/m²) (Chandler 1941, 1944):

Element	Hardwood forests	Coniferous forests
calcium	7.3	3.0
nitrogen	1.8	2.6
potassium	1.5	0.7
magnesium	1.0	0.5
phosphorus	0.4	0.2

Silicon, copper, manganese, carbon, and zinc are also present in the leaves of hardwood trees. Carbon is relatively more abundant and nitrogen less abundant in coniferous than in deciduous leaves, but commonly the carbon/nitrogen ratio is 55:20 (Ovington 1954).

Both plants and animals are important agents effecting the decomposition of organic matter and the formation of humus. An animal digests and metabolizes plant foods, much of which is returned to the soil, in part as the excreta of the living animal, in part as the body of the dead animal, in part as gas. Fully formed humus is, in fact, derived mostly of fecal material. The larger herbivorous and carnivorous animals pass urine and feces containing simple nitrogenous compounds and compounds of phosphorus, potassium, and traces of calcium, magnesium, sulfur, and other elements. Humus is but one point in a continuous cycle of decomposition of plant and animal organic matter, absorption of decomposition products by plants; ingestion and metabolization of plant matter

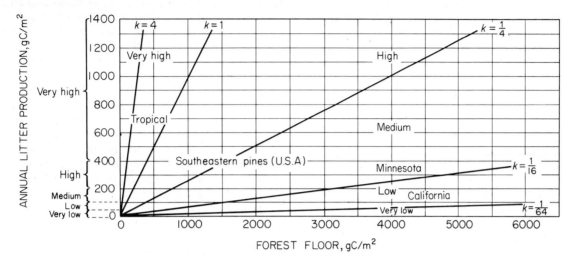

Fig. 9-2 Relation between annual production of litter (g C/m²) and amount stored on forest floor (k = fraction of stored carbon lost annually) (modified from Olson 1963).

by animals, decomposition of plant and animal organic matter—ad infinitum. The consumption by saprovores and herbivores of living and dead plant matter and the consumption of herbivores by carnivores neither add nor subtract from the total nutrient supply of an ecosystem. The chemical elements available in the air, water, and soil of an ecosystem pass, in one compound or another, from one organism to another, and from one stage in the cycle to another. They continue thus to circulate within the ecosystem unless and until they are physically withdrawn from it. To remove plant and animal crops from an ecosystem is to withdraw nutrients from it, and thus to reduce the fertility of the system. Fertility can then be maintained only if the nutrient supply is kept replenished by artificial fertilization.

Kangaroo rats defecate promiscuously throughout their underground burrow systems. The soluble nitrate content of the soil in the region of one burrow system averaged 221 ppm and in another one 570 ppm, compared with a maximum of 15 ppm in the surrounding desert soil generally (Greene and Reynard 1932). It is a reasonable estimate that the total bird population in a deciduous forest would deposit 0.1 g dry weight of organic excrement per m² in a year's time; the mammal population, perhaps 0.5 g; and the total invertebrate fauna, possibly 2–3 g. The accumulation of excrement under the roosts of birds is sometimes enough to kill the ground vegetation and even the trees (Young 1936). The guano deposits on the coast of and islands off Peru and elsewhere in the world were originally several meters thick, as the result of centuries of occupancy by nesting colonies of marine birds, but have now been largely depleted by man for use as crop fertilizer (Hutchinson 1950). Bat excrement, deposited in caves, was exploited in years past as a source of saltpeter for gunpowder.

The conversion of raw organic matter into materials suitable for re-absorption and utilization by plants is a complicated process and depends almost entirely on the reactions of plants and animals (Lutz and Chandler 1946, Waksman 1952, Edwards *et al.* 1970). Especially important soil animals are the mites, earthworms and enchytraeids, woodlice, millipedes, springtails, and other saprophagous insects. Decomposition of litter proceeds more rapidly with these soil animals present than in their absence (Fig. 9-3). The role of beetles and flies in the decomposition of trees (p. 128) and of carrion (pp.120–121) on the ground surface have been mentioned in connection with the microseres that are formed. The digestion of animals produces both mechanical and chemical changes in raw humus that can be measured quantitatively (Franz and Leitenberger 1948). The non-nitrogenous substances in fresh litter are sugars, starches, pectins, pentosans, celluloses, cutins, tannins, lignins, oils, fats, waxes, and

Fig. 9-3 Rate of litter decomposition with and without larger invertebrates present (Ghilarov 1967).

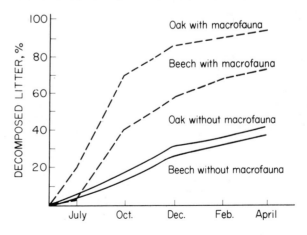

resins. Most of these substances are readily broken down in the soil by fungi, actinomycetes, bacteria, and protozoans, but tannins, lignins, waxes, and resins decompose very slowly. The end products of complete decomposition are H_2O and CO_2, but sometimes decomposition is incomplete and organic acids are formed instead.

The most important soil organisms concerned in the decomposition of the litter are the bacteria, both aerobic and anaerobic forms. They are commonly divided into two types. *Heterotrophic* bacteria obtain their energy from the oxidation of the carbohydrates and fatty substances as described above. They use this energy for the synthesis of cell substances and the production of enzymes that break down complex compounds in the litter into simpler compounds, including proteins into ammonia compounds. They then use part of the ammonia compounds in synthesizing the amino acids they need in building their own proteins. *Autotrophic* bacteria, in turn, are of two types: *chemosynthetic* species, which obtain their energy from the oxidation of inorganic compounds (hydrogen, sulfur, hydrogen sulfide, iron, ammonia) and *photosynthetic* species, which include purple and green sulfur bacteria, possess a form of chlorophyll, and utilize the energy of sunlight. Chemosynthetic bacteria convert ammonia compounds into nitrites and nitrates, part of which they use in their own anabolism, the rest becoming available for plants to absorb. Photosynthetic bacteria use the ammonia compounds in their own anabolism and do not render them directly available to plants. Chemosynthetic bacteria are more abundant than photosynthetic bacteria in soil; photosynthetic bacteria are the more abundant in water.

The decomposition products available to plants are reabsorbed by the roots and built up into plant tissue.

Plants are eaten by animals, and as plants and animals die the minerals are returned to the soil. All essential minerals cycle repeatedly through the ecosystem, the cycle of each mineral differing in various ways from the cycle of every other mineral (Duvigneaud and Denaeyer-DeSmet 1970).

Nitrogen Cycle In the nitrogen cycle proteins are broken down, yielding ammonia (NH_3) compounds in the course of the metabolic processes of all animals and by the activities of heterotrophic bacteria, filamentous fungi, and actinomycetes (Fig. 9-4). The process is called *ammonification*. Some of the ammonia is oxidized to form nitrites (NO_2) and nitrates (NO_3) through the action of autotrophic bacteria; the process is called *nitrification*. Other types of bacteria act on ammonia in the process of *denitrification*, by which nitrogen (N_2) is liberated into the atmosphere. Nitrogen is removed from the air by the *nitrogen-fixing* bacteria which live either freely in the soil or as symbionts in the root and leaf nodules of many legumes and some non-legumes, *Ceanothus, Elaeagnus, Alnus,* and *Myrica,* among others (Stewart 1967). Blue-green algae also fix nitrogen in moist soil and are more important than bacteria in this regard in aquatic habitats (Howare *et al.* 1970). Ammonia compounds, nitrates, and other substances are added to the soil in small amounts with rainfall; sources of these nitrogen compounds are volcanic eruptions, terrestrial decomposition, and atmosphere nitrogen fixed by lightning. An attempt to estimate the quantities of nitrogen involved in the different parts of the cycle has been made by Hutchinson (1944). Man has developed an important chemical fertilizer by artificially fixing nitrogen from the air, and much research is underway for the better understanding of the enzymatic mechanisms of biological fixation.

Fig. 9-4 Steps and processes in the nitrogen cycle.

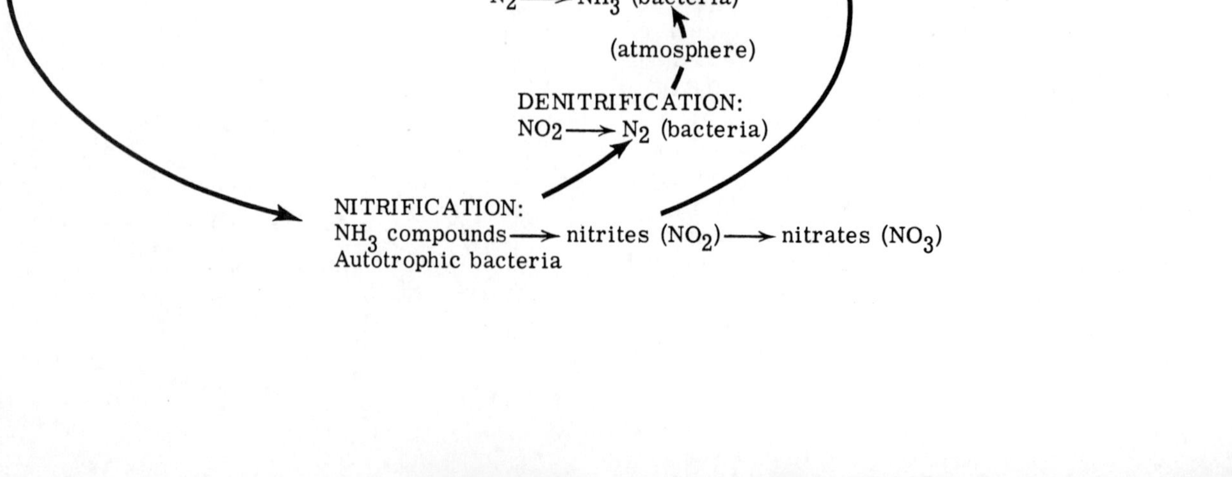

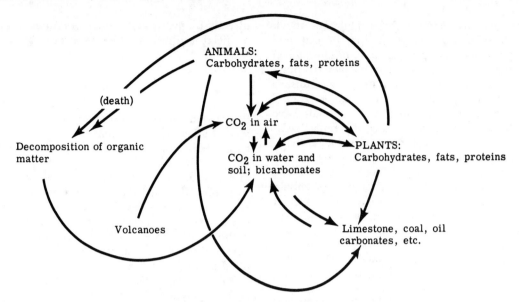

Fig. 9-5 The carbon cycle.

Nitrates and perhaps also the simpler nitrogen compounds are absorbed and used by plants for the synthesis of amino acids and proteins. In a comprehensive review, Rodin and Bagilevich (1965) state that nitrogen uptake by plants varies from 1.8 in cold shrub deserts, 2.1 in the Arctic tundra, to 42.7 g N/m^2-yr in wet tropical forests. The corresponding amounts of nitrogen in the standing biomasses are 6.1, 8.1, and 294 g/m^2, respectively. When animals digest plants, plant proteins are broken down into amino acids and re-combined into animal proteins and body tissue.

In a balanced or stabilized ecosystem, nitrogen molecules continue indefinitely to cycle among the soil, plants, and animals with nitrogen-fixation replacing that lost in denitrification, leaching, and so on. In seral communities, increase in soil fertility depends in large part on nitrogen-fixation exceeding denitrification so that there is a progressive increase of nitrogen molecules in the ecosystem as succession proceeds.

Carbon Cycle Animals obtain much of their carbon, as well as nitrogen, from plants, although some forms are also able to fix carbon directly from salts dissolved in water (Fig. 9-5) (Hammen and Osborne 1959). In photosynthesis, carbon dioxide obtained from the air and from dissolved bicarbonates in the substratum is combined with water to form carbohydrates, a portion of which may be converted to fats. Plants combine carbon with oxygen, nitrogen, hydrogen, and sulfur to form proteins. Carbon dioxide in the air comes chiefly from the respiration of animals, but small amounts arise from the respiration of plants, the decay and fermentation of organic matter, springs, volcanic action, and solution of sedimentary rock. Volcanoes were probably the original providers of carbon dioxide to the biosphere. Organisms tie up carbon dioxide as carbonates in skeletons and shells. Carbon is also tied up in the formation of peat, oil, shale, and coal. When limestone

and other carbonaceous sediments are exposed to water erosion, the carbonates may be hydrolyzed to bicarbonates and thus become a source of CO_2. The concentration of carbon dioxide in the air is stabilized at 0.03 per cent by the buffering action of bicarbonates and carbonates in the oceans and fresh-water bodies (Hutchinson 1948):

$$CO_2 + H_2O \rightleftharpoons H_2CO_3 \rightleftharpoons HCO_3^- \rightleftharpoons CO_3^{-2}$$

On the other hand, oxygen in the air (21 per cent) is derived almost entirely from the photosynthesis of plants. Studies are underway to quantify the various steps in the carbon cycle (Olson 1970) as well as other mineral cycles (Duvigneaud and Denaeyer-DeSmet 1970).

Other Elements In addition to oxygen, carbon, hydrogen, and nitrogen, animals require at least 13 other elements that are all derived from the soil: calcium, phosphorus (Hutchinson 1948), potassium, sodium, chlorine, sulfur, magnesium, iron, copper, manganese, iodine, cobalt, and zinc. Only traces of some of these elements are required, but calcium is used in large amounts for skeletons, shells, antlers, and other organs, and in metabolism generally. Phosphorus is a constituent of nucleoproteins, phospholipids, and skeleton. Goiter occurs in mankind and some animals in regions deficient in iodine. These elements are obtained from food, drinking water, salt licks, and grit taken into the stomach. A salt lick is a local, usually clayey, area characterized by a high concentration of salts where deer and other animals foregather to lick the soil for the salt. Soils deficient in or lacking these various elements support sparse animal populations; individuals are in more or less poor health; reproduction rates are low (Albrecht 1944, Crawford 1950).

An excess of some elements is harmful. Too much fluorine in drinking water causes mottling of teeth and possibly pathological changes. Selenium in soils of arid plains becomes dangerous when it reaches 0.5 ppm, and some grasses, asters, and certain legumes absorb and retain it in concentrations that can be highly injurious to herbivorous animals. Wild animals have apparently learned to avoid eating these particular plants, but domestic stock blunder into them, eat them, and die (Knight 1937). Certain plants concentrate specific elements, a factor which may affect the food habits of animals. Black tupelo concentrates cobalt, and inkberry concentrates zinc to a much greater extent than do other species growing in the same areas (Beeson et al. 1955). Likewise, the feathers of birds reflect in their chemical composition the mineral content of the soils of the region they inhabit during the period of feather growth (Hanson and Jones 1968).

Soil Fertility and Agricultural Practices Soil fertility depends on the prevalence of various minerals, especially nitrogen and phosphorus. Surface runoff resulting from rains and leaching through the soil to subsurface drainage brings removal of both particulate and dissolved nutrient material. A study in New Hampshire showed a loss of particulate matter amounting to 2.5 metric tons/km^2-yr and dissolved matter of 14 metric tons. Losses were proportional both to rate and volume of water runoff and erodability of the soil. Cutting off of the forest greatly increased the runoff (Bormann et al. 1969). Leaching is especially important in tropical regions during the rainy season, and soil fertility is inherently very low. The land quickly becomes sterile after removal of the vegetation.

In agricultural practice, man removes nutrients from ecosystems in a form that can be used as food and replaces the nutrients as fertilizers that are not usable as food. Without such replacement, the agricultural ecosystem would quickly become sterile and unproductive.

Agricultural technology is placing increasing reliance on chemical nitrogen fertilizers. Rotation of crops with legumes is becoming less and less common. Using heavy loads of chemical fertilizers brings the risk of destroying the bacterial soil flora essential to the nitrogen cycle and of destroying the capability of the soil to maintain its own fertility. The application of organic humus, on the other hand, helps to maintain the soil bacteria, provides other essential organic compounds, and is important for persistence of the soil structure. It is possible that organic sludges, coming as sewage wastes in large quantities from our cities, could be used advantageously as organic fertilizers (Albrecht 1956, Bylinsky 1970). Still another approach suggested is extending the symbiotic nitrogen-fixing bacteria to other non-leguminous species. If means could be devised to associate these bacteria as root nodules to crop plants generally, the great source of nitrogen in the atmosphere could be more extensively tapped

directly by plants and reduce or eliminate the need for artificial fertilization (Phillips et al. 1971).

Climate

Water, temperature, and wind are important weather factors affecting soil formation. Water is an agent of rock erosion and transportation, sorting, and deposition of soil-building erosion products. Water freezes and expands in cracks and crevices of massive rock structures, breaking them into fragments and particles. Daily and seasonal heating and cooling cycles produce cracking because of different coefficients of expansion of the minerals in the rock. Wind erosion is particularly devastating in arid regions; fine soil particles may be lifted and transported many miles.

Weathering of rock is a chemical as well as physical process. Hydrolysis of some rock materials brings absorption of carbon dioxide and the formation of soluble bicarbonates. Hydration softens and increases the mass of some minerals, so that physical weathering of the rock bearing them is facilitated. Oxidation discoloring of many rocks, especially those containing iron, is symptomatic of chemical changes in progress; binding materials are weakened and crumbling occurs easily. Finally, many substances simply go into solution and are carried away. Where precipitation is frequent, water percolating through the soil carries soil nutrients to greater depths than where precipitation is light. In hot dry climates, organic matter may oxidize completely and so quickly its nutrients are lost to plants and microfauna.

The prevailing climate is a determinant of the kinds and luxuriousness of plant and animal life in an area. The biota has much less effect on soil formation in arid climates than in humid climates. Desert vegetation is usually quite as locally distinctive as are the soil parent materials on which it develops, but in humid regions, where many plant stages succeed one another, climax vegetations may be much the same (or may differ in species composition but not structure) regardless of whether the sere originally started on limestone, sandstone, or in a pond. Because of the interactions of parental rock, biota, and climate, different soil profiles are formed, each characteristic of a specific climatic region and type of climax vegetation. An understanding of soil profiles is prerequisite to understanding vagaries of animal distribution.

SOIL PROFILES

As a result of the specific circumstances of weathering, biotic reactions, and climatic influences it has experienced, a mature soil has a definite structure characteristic of its different environment (Lutz and Chandler 1946). The living plant draws nutrient materials from

L	Mostly fresh litter
F	Fermentation layer
H	Amorphous humus
A_1	Humus and mineral soil mixed; light texture, dark color
A_2	Leached of soluble salts and organic matter; coarser texture, light color
B	Enriched with precipitated salts and humus; sometimes forms hardpan, often brownish or yellowish in color and columnar in structure
C	Weathered parent rock materials
D	Parent rock or sediment too deep to be affected by weathering or animals, and may be of different origin from parental soil in C

Fig. 9-6 Letter designation of horizons in a mature soil profile. The L horizon is also designated the A_{00} horizon; and the combined F and H horizons, the A_0 horizon. The A horizons make up the topsoil, the lower horizons the subsoil.

the deeper layers of the soil, but these materials are deposited on the surface soil when the dead plant decays. Rain falling on the ground surface carries the nutrients and salts back down into the soil, at least as far as the water percolates. Soil animals are doubtlessly also involved, as the depth of the humus layer is correlated with the depth to which ground insects penetrate (Ghilarov 1964). This sequence of events produces a definite layering of the soil. Each layer is called a *soil horizon*, and the series of horizons characteristic of a soil is called the *soil profile* (Fig. 9-6).

A *mature* or fully developed soil profile is characteristic of climax or late seral stages of a succession. Horizons are not fully expressed in early seral stages, so these profiles are called *immature* or undeveloped. Mature profiles are found best under virgin vegetation, for erosion and cultivation disturb horizons. Profiles are best developed in humid climates, where abundant precipitation carries humus and salts well into the soil.

Hardpan

In arid regions, evaporation may be in excess of rainfall. Moisture in the soil has no opportunity to percolate downward; rather, it rises to the surface of the ground and is lost. Where rainfall is inadequate for efficient leaching, *hardpan* may form in the B horizon as the result of deposition here of ferric oxide, alumina, colloidal clay, or calcium salts (Fig. 9-7). This layer becomes so compact and hard that it is impervious to root penetration and the burrowing of animals, although during periods of wet weather it disappears temporarily. In arid regions, the hardpan may be at

Fig. 9-7 Relation of lime hardpan to types of prairie vegetation extending from west to east in central North America (Shantz 1923).

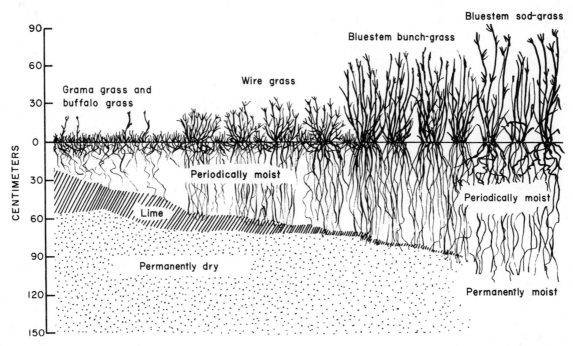

or close to the surface; but in more humid climates it occurs at progressively greater depths until it disappears altogether.

Mull and Mor Humus

Distinction between mull and mor humus is made primarily for forest soils. Both humus types of soil may be subdivided, but these subdivisions need not concern us here (Lutz and Chandler 1946, Wilde 1966).

Mull (endorganic layer) is a porous, friable humus layer of crumbly or granular structure and only slightly matted, if at all. The A_1 horizon is well developed; bacteria are abundant, annelids numerous, and nitrification occurs.

Mor (ectorganic layer) is a strongly matted or compacted humus layer. There is no A_1 horizon; the transition from humus layer to mineral soil is abrupt. The underlying soil is more acid, bacteria are much less numerous, and nitrification is usually reduced if not absent. Abundant fungi reduce the raw humus to punky material; thereafter it is worked on by the small arthropods, particularly springtails, mites, flies, and ants (Thomas 1968). Decomposition of mor is much less rapid than of mull. Annelids are few or absent, and moles are less common than in mull soils. Snails are scarce in acid soils. In general, the biomass of organisms inhabiting a mor soil is smaller, species are less diverse, and individuals are smaller than in mull soils.

Mor humus is common in cold regions and at high elevations, but is not limited to such climatic zones. It occurs especially under coniferous forests, even in warm climates, and under ericaceous vegetation. It is also found in very wet or very dry habitats, where there is accumulation of poorly decomposed or sharply delimited humus layers on top of the mineral soil, sand, or rock underlying. Mull humus commonly develops in warm, humid climates and is found especially under hardwood or deciduous forests. Patterns of animal and plant distribution correlate closely with these two humus types; indeed, the formation of each is a result, in the main, of unique combinations of biota reactions, climate, and mineral soil characteristics (Romell 1935, Fenton 1947).

Depth Distribution of Organisms

The small animals in the soil, including protozoans and nematodes, are most abundant in the L, F, and H horizons, becoming rapidly less abundant in the mineral soil (Table 9-1). Bacteria, actinomycetes, and fungi are also most abundant in these top layers, especially in F and H, although they occur well down into the B and C horizons.

The depth distribution of soil animals depends on temperature and varies with season. When the top layers freeze during the winter months, much of the fauna keeps well below the frost line, although many species are tolerant of freezing. During the cold months the depth at which most soil insects, mollusks, and annelids occur varies from about 9 cm in silty clay-loam to 38 cm in gravelly clay-soil (Dowdy 1944). With the return of warm weather, the fauna ascends to the top horizons.

Soil Erosion

As rain water drains over the surface of the ground, it picks up and transports particles of soil more or less proportional to the rate of flow. This is a continuous process that has gone on throughout geological time, resulting in base-leveling of mountains, the filling in of valleys, and the formation of river deltas. Normally this is a slow process, with changes becoming apparent only after thousands of years. With the ground covered with vegetation, the flow of water is retarded and more of it percolates into the ground or is evaporated. In fertile areas, new soil is formed faster than it is eroded or at least there is stabilization.

With removal of the native vegetation and cultiva-

Table 9-1   Depth of distribution of soil arthropods in the Adirondack Mountains. The figures indicate approximate number of individuals per square meter in the total thickness of each layer (computed from data given by Eaton and Chandler 1942).

Soil horizon	Depth (cm)	Mites	Springtails	All others
Mor humus under red spruce and balsam fir				
L and F	0–5	150,000	19,800	1,400
H	5–25	62,000	19,200	400
A_2	25–33	1,500	0	0
B	33–58	1,900	100	0
Mull humus under beech, sugar maple, and yellow birch				
L and F	0–5	62,000	17,800	2,800
A_1	5–15	15,000	6,200	600
B	15–55	6,200	5,800	200

tion of the soil in agriculture, surface particles are more easily moved and surface water drainage is more rapid. As a result, topsoil is being lost at an alarming rate. Hugh H. Bennett, former chief of the U.S. Soil Conservation Service, estimates that of the original 23 cm, the average depth of the topsoil over the country, one-third has already been washed away. Much land has been so badly damaged that there is little likelihood that it will ever again be productive of crops. Conservation of the fertile topsoil is one of the major problems facing modern agriculture. Putting in underground drainage tiles in level farmlands helps to draw rainwater into the soil and to decrease surface runoff and evaporation. Contour plowing of slopes decreases the rate of surface runoff and erosion. *Soil conservation* procedures should be strictly followed.

MAJOR SOIL GROUPS

The interrelations between the basic mineral content of the parent substrate, biotic reactions, and climate can be seen through an analysis of the development of the great soil groups of the world. We will give only a simple classification of them. A detailed classification would include many subdivisions and intermediate categories (Lutz and Chandler 1946, Simonson 1957).

Podzolic soils are formed in humid temperate climates, under forest vegetation. The A_2 horizon is moderately well developed, for there is sustained leaching. Soils are more or less acid and only moderately fertile. *Podzols* develop under coniferous forest and have a mor type of humus. *Gray-brown* and *brown* podzolic soils are found under hardwood forests and have a mull humus.

Latosolic soils develop in humid tropical or semitropical forested regions. Humus is quickly oxidized by action of microorganisms and hence does not accumulate. The soil fauna is depauperate. Chemical weathering of the parental material is intense. Water drainage through the porous soil is rapid, so leaching is extensive. In early stages of its formation, the soil is neutral or slightly alkaline, but as leaching continues, it becomes acidic. The soil has a thin organic layer (A_0 and A_1 horizons) on a reddish, leached soil (A_2 horizon) that extends to great depths below the surface.

Chernozemic soils occur in humid to semiarid temperate climates under grass vegetation. The grasses on dying return considerable organic matter to the soil. The A_1 horizon is consequently dark in color and of great thickness. The soil contains more bases and hence is less acid than in the two types above. The B horizon in humid regions is indistinct, but where there is less rainfall, calcium salts may accumulate to form a hardpan. Prairie soils in temperate climates are among the most fertile soils of the world, but fertility decreases in the tropical and desert climates.

Desertic soils are characteristic of arid climates and contain very little organic matter. A profile is poorly developed. The surface soil is brownish gray, and grades quickly into the calcium carbonate horizon which usually forms a hardpan just below the surface. Wind erosion removes the finer soil particles, leaving the coarser material to form a hard pavement. The soils are but slightly weathered and leached; lacking nitrogen, they are infertile.

Mountain and *mountain valley* soils vary from shallow layers on eroding rocks to deep organic soils of valleys and swampy areas.

Tundra soils occur in cold northern areas where the substratum remains continuously frozen and the vegetation of lichens, mosses, herbs, and shrubs makes a peaty surface layer. The region is poorly drained and characterized by many scattered shallow ponds.

Alluvial soils may be important locally. These soils are mostly without a developed profile and are the result of deposition by streams. They are usually very fertile and support luxuriant vegetation.

Saline soils are found in dry climates where rapid evaporation of water results in surface deposition and accumulation of salts leached from surrounding upland areas.

This classification shows clearly the correlation that exists among soil groups, climate, and vegetation. The composition and particularly the abundance of animals is also affected. The average biomass of soil animals under different categories of vegetation-soil-climate systems is as follows (Ghilarov 1967):

Moss tundra: 3 g/m^2
Coniferous forest: 20 g/m^2
Broad-leaved deciduous forest: 100 g/m^2
Short grass plains: 25 g/m^2
Semi-desert: 1 g/m^2

REACTIONS IN WATER

Considerable attention has already been paid to the reactions of animals and plants in streams (Chapter 4), lakes (Chapter 5), and ponds (Chapter 6). These involve changes both in the chemical and physical characteristics of the habitat and are fundamentally the same as occur on land.

Water plants, especially those, such as water lilies and water hyacinths, that float, and surface concentrations of both zoo- and phytoplankton reduce light intensities like forest canopies. There is accumulation of plant and animal remains and fecal material on the bottom of the water bodies just as on land, and this material is worked over by bacteria and a large variety of microorganisms which differ only in taxonomic composition, not in activities, from those on land (Henrici 1939). Water plants obstruct the flow of water and cause deposition of suspended materials, simulating the reduction of wind velocities inside forests. An important physical reaction is the damming of streams by beavers so that ponds are formed. Such beaver activity can sometimes be usefully co-ordinated with waterflow management (Beard 1953). These ponds eventually fill with silt and organic matter, succession occurs, and so-called beaver meadows are produced (Van Dersal 1937).

The nitrogen cycle (Cooper 1937) and carbon cycle in aquatic ecosystems are essentially the same as on land. The absorption of oxygen by organisms and by the decomposition of organic matter in some lakes and ponds causes in the habitat a seasonal change of profound importance.

SOIL AND AIR POLLUTION

Soil, water, and air pollution results from contaminants added to ecosystems beyond what can be readily absorbed through normal nutrient cycling. In undisturbed ecosystems, dead organisms and excreta quickly decompose and their nutrient elements recycle for reuse. The existence of plants and animals depends on the proper utilization of these "natural wastes," and there is no harmful accumulation of surpluses. Man, however, introduces into ecosystems excessive amounts of organic and other wastes and has created through chemistry an expanding quantity and variety of complex substances which are highly resistant to existing enzyme action and normal cycling processes. Consequently, there is a general alteration and deterioration of the environment. Included are the chemicals that man uses for maintaining his dominance over the "anthroposphere." The application of herbicides, fungicides, insecticides, fumigants, and rodenticides often produces far-reaching and unanticipated disturbances between predators and prey, producers and consumers, and in the various nutrient cycles. We need a worldwide system of careful monitoring of environmental quality as a basis for evaluating and controlling all kinds of disturbances to environmental quality (Wilson and Matthews 1970). Man's existence may depend on it.

Noise pollution results when undesirable sounds become irritating to the general public. Excessive noise may not only be annoying but may also cause permanent hearing loss, loss of efficiency in performing tasks, and other less-well-defined organic, sensory, or physiological disturbances. The loudness or acoustical effect of sound is nearly proportional to the logarithm of its intensity. An acoustical scale in decibels (dB = 0.1 Bel) is commonly used, which is 10 times the measured intensity of a sound relative to the intensity of sound just perceptible to the average human ear, or approximately 10^{-16} watt/cm^2. Thus an intensity 10 times the threshold, 10^{-15} watt/cm^2, gives a decibel rating of $10[10 \cdot (10^{-15}/10^{-16})]$; an intensity 100 times the threshold, 10^{-14} watt/cm^2, gives a decibel rating of $20[10 \cdot (10^{-14}/10^{-16})]$; and so on. Ordinary conversation on the decibel scale is about 60, heavy auto-

mobile traffic or a rock and roll band about 100, and a jet aircraft taking off, 120.

The amount of continuous undesirable noise that a person can tolerate without discomfort or annoyance, or to which he can become habituated, is a subjective matter and varies with the individual and the situation but is probably below that of ordinary conversation. The U.S. Air Force recommends wearing ear "defenders" when noise levels exceed 85 dB. The threshold of pain is about 140 dB. More research is required to establish a noise index consistent with human comfort and well-being. The remedy for excess noise, chiefly from use of power tools and machinery, road traffic, and aircraft, is better engineering to reduce the noise at its source, the use of sound barriers of various sorts, and isolation (Rhodda 1967, Burns 1969, Anthrop 1969, Ehrlich and Ehrlich 1970, Detwyler 1971: pp. 175–189).

Solid Wastes

The United States generates 4.5 kg (10 lb) per person per day of municipal, agricultural, commercial, and industrial wastes in the form of paper, plastics, garbage, textiles, wood, rubber, glass, ceramics, metallics, and so on. This figure does not include junked automobiles. Annual per capita increase is 2 per cent per year, and with the population increasing 2 per cent per year, there is a compounding increase of 4 per cent per year (Cronin 1971). Many of these wastes are odorous and unsightly and are, in mere mass, excessive. Furthermore, natural materials, formerly used and easily decomposable, are being extensively replaced by synthetic ones that are resistant (nylon hose for cotton hose, for instance). Several methods are in use or under study for disposing of solid wastes.

Sanitary Landfill   The material is spread in depressions in thin compressed layers, alternating with dirt, to bring the ground level up to the general topography. Aerobic decomposition takes place at first and then is replaced by anaerobic decomposition. There is some danger from gases that are formed and of contamination of ground waters (McGauhey 1968). Suitable landfill sites may not be available locally.

Incineration   Much of the solid wastes may be reduced by incineration at temperatures of 760°–980°C. The residue may contain iron, aluminum, zinc, copper, lead, and tin, which are recoverable, but has only limited marketing feasibility at the present time. Combustion gases consist of carbon dioxide, water, oxides of nitrogen and sulfur, and other gases in lesser amounts. Air pollution by these gases is usually not serious, but emission of particulate fly-ash may be important. In addition, considerable heat escapes into the atmosphere. New techniques are being developed to increase the percentage of combustion, prevent the escape of particulate matter, and recover the heat for useful purposes.

Composting The nitrogen, phosphorus, and potassium in the wastes, if recovered, could be used as fertilizer on agricultural land. Present procedures involve first removing noncompostible substances and those items that have salvage value, grinding the remainder for faster decomposition, and then subjecting the matter to bacterial action until it becomes humus. The material spread over fields in windrows commonly requires 6 to 7 weeks to decompose, but some mechanical systems require only 5 or 6 days. The humus is then bagged and offered for sale. There are some plants in operation, but the procedure requires more development to become generally economic.

Recycling Junked automobiles present a special problem since there are 9 to 16 million cars abandoned yearly. Useful parts are commonly salvaged. Of the remainder, combustible parts may be stripped off and burned and the useful metals recovered. A difficulty lies in obtaining these metals in pure enough form for their reuse to be practical. The same problem exists in recycling industrial and mining wastes. Doubtless this can be solved if sufficient research and effort are made to do so.

Reuse A different attack on the problem is the elimination of the waste, as far as possible, at its source. For instance, containers for holding liquids and foods could be standardized in a few different shapes and sizes and made in a form that could be returnable and used repeatedly, rather than being discarded after the first use. Containers and items that cannot be reused could in many instances be made of degradable materials. Recycling glass and metal containers bears an energy cost that often offsets the value of the materials that are saved.

Certainly we must face up to the fact that space for storing solid wastes in the environment is limited. There is continuing pressure to use the deep sea as a natural sink for solid wastes, but this could soon mean the loss of the marine ecosystem for other purposes, such as a source of food. The more shallow waters of estuaries and on the continental shelf would be especially vulnerable. There is no choice but to find alternatives to prevailing practices at the present time (Cooke 1969, Breidenbach and Eldredge 1969).

Herbicides

Herbicides are used for the control of weeds, shrubs, and young trees. There are also fungicides, including mercury compounds (p. 52). The herbicides 2,4-D (2,4-dichlorophenoxyacetic acid) and 2,4,5-T (2,4,5-trichlorophenoxyacetic acid) began to be used extensively in the mid-1940s. They are derived from hormone-like compounds which activate growth in minute concentration but kill plants in high concentrations. They may be sprayed on the plants either in the dormant condition or when in full leaf. Residues disintegrate by bacterial action in 6 to 8 weeks and are not very toxic to animals or man. Even after heavy applications, it is unusual to find phytotoxic residues in the soil after 1 year. These two are the most common among some 40 herbicides that are available, and their use is becoming more and more prevalent among farmers for the holding down of weeds among crop plants. Some herbicides, such as simazin and monuron, have a different mode of action, interfering with photosynthesis until the plant dies for lack of energy. Although present-day herbicides in normal applications are not known to harm soil microorganisms, there is potential danger here with new kinds of herbicides that may destroy bacteria essential in the cycling of nutrients (Egler 1968, Cole 1968, Ehrlich and Ehrlich 1970).

Insecticides

Rachel Carson's *Silent Spring* aroused the general public to the problem of pesticides in 1962 and perhaps may be credited in large part for initiating the modern awareness of ecology. DDT (2,2-bis[*p*-chlorophenyl]-1,1-dichloroethane) is one of the first and most extensively used of the insecticides, although several other chlorinated hydrocarbons have been commercially developed and used (aldrin, dieldrin, endrin, heptachlor, toxaphene, BHC, etc.) (O'Brien 1967). Other chemicals, such as lead arsenate, are also important. DDT is relatively inexpensive, highly toxic, and very stable—characteristics attractive to the agriculturist. The resistance of DDT to degradation is high; it or its derivatives, especially DDE and DDD, may persist for months or years, dependent upon type of soil, soil moisture, soil temperature, cover crop, cultivation, and method of application. The organophosphate and carbamate insecticides, on the other hand, mostly disappear after several days or weeks, but they are more expensive.

The evaluation of pesticides as to their ecological effects involves the degree to which their action is limited to the undesirable species. Indiscriminate spraying of insecticides and herbicides is often destructive to non-target organisms. Soil and aquatic organisms are especially exposed because of the accumulation and persistence of the toxins there. In Florida an estimated 1,175,000 fishes of at least 30 species were killed by dieldrin, when sand fly larvae (*Culicoides*) were the target species. Crustaceans were virtually eliminated (Harrington and Bidlingmayer 1958). The unintentional killing of insect pollinators with insecticides often remains unnoticed until the desired crop fails to set (Wilson and Matthews 1970).

Different insecticides vary in their toxicity to organisms. DDT is considerably less effective on wireworm larvae, for instance, than is BHC (1,2,3,4,5,6-hexachlorocyclohexane). Likewise, a particular insecticide is not equally toxic to all species. Repeated applications of DDT in an orchard killed many of the insects but not the red spider (Tetranychidae). Earthworms and potworms are not killed by many insecticides at the usual rates of application. This differential toxic effect often upsets the "balance of nature." When DDT was applied to one of two mowed grasslands, there was a decided mortality of predaceous forms of centipedes, mites, and adult and larval insects. This loss of predators was followed by tremendous increases in phytophagous aphids and coccids and in fungivorous and saprophytic mites and springtails (Menhinick 1962).

Populations subjected to repeated exposures of an insecticide may build up a resistance to it. Fly and mosquito populations formerly sensitive to mere traces of DDT now require much heavier dosages to produce the same mortality or are entirely resistant. Apparently this resistance has come about through a selective process, similar to natural selection, in that only individuals possessing an immunity survive to reproduce. The continuance of artificial control of such insects requires use of progressively higher dosages or switch to a different kind of chemical. If this phenomenon proves to be general, and over 100 important insect pests now show definite resistance to chemicals, it would appear that the use of chemicals for controlling insect populations will become more and more difficult in the future (Conway 1971). Outbreaks may then occur that may be devastating and uncontrollable. Not only that, but resistant strains may contain so much of the insecticide residue in their bodies as to kill the predators that normally keep their populations at reduced levels. This has been shown especially with fish (Rosato and Ferguson 1968).

Persistent insecticides become concentrated in the higher links of food chains. On the University of Illinois campus, elm trees were sprayed regularly from 1949 to 1953 in an attempt to destroy a leafhopper responsible for spreading phloem necrosis, which was killing the trees. The DDT spray that settled on the ground, especially with the falling of the leaves in the autumn, was consumed by earthworms in sublethal amounts. However, earthworms are a favorite food of robins and a consumption of fewer than 100 earthworms by a robin was sufficient to accumulate a dosage that was intolerable. As a consequence, the robin population on the campus was virtually eliminated (Barker 1958).

An even more striking case comes from Clear Lake, California (Hunt and Bischoff 1960). This lake was sprayed with DDD, a derivative of DDT, to control gnats, *Chaoborus astictopus*. Concentration of the insecticide in plankton was found to be 265 times more than the concentration applied, in frogs 2000 times more, in sunfish 12,500 times, and in grebes up to 80,000 times more. Previous to spraying, a thousand pairs of grebes nested on the lake; after the spraying no young grebes were produced for the next 12 years, when a single young bird hatched.

Recent declines in numbers of eagles, falcons, hawks, gulls, and brown pelicans are linked with unsuccessful reproduction. The predators accumulate DDE in their tissues concentrated from the fish, rodents, and small birds that they eat. The DDE is correlated with disturbances in liver enzyme action and interference in calcium metabolism. Eggs are produced with little or no hard shell, are easily broken, and do not hatch (Ratcliffe 1970, Cade *et al.* 1971). With the concentration of toxins to vulnerable levels in the higher links of food chains, predator pressure that acts as one on the checks limiting populations of species in the lower links becomes reduced. Thus the pesticides favor the herbivores, the very organisms they were invented to control (Woodwell 1970).

Spraying of insecticides from the air may leave particles airborne for several hours and spread them over considerable distances. Attached to soil particles, they may also be swept aloft by the wind. Pesticide residues found in the fat of polar bears in the far north and in penguins in the Antarctic (Tatton and Ruzicka 1967) must have been carried there by wind or ocean currents. Although pesticide contamination of water and air poses a potential threat, the most likely danger to man is from the food that he eats. Many insecticides are fat-soluble and hence may be carried in a wide variety of foods, including milk. Even though foods may contain low concentrations of the chemicals, man may over the course of years concentrate them, as do predators, to a level that becomes harmful.

The persistence and accumulation of chlorinated carbons in the substratum may be a continuous health hazard to man. DDE in the waters of Lake Michigan, as a result of leaching and runoff from the surrounding upland, has become concentrated in the introduced coho salmon beyond the levels set by the Food and Drug Administration for its use as food. Likewise, contaminated fresh waters flowing into the ocean build up concentrations of toxic chemicals in the highly food-productive coastal waters. Burrowing marine organisms and detritus feeders keep these chemicals circulating through the ecosystem for long periods of time. Some of the destruction formerly attributed to DDE may actually be due to polychlorinated biphenyls, but experimental evidence is conflicting as to the importance of the polychlorinated biphenyls (PCBs).

Balanced against the actual and potential harm from pesticides is their value for agriculture, for control over the vectors of disease, and for the control of pests that merely cause discomfort to man. The use of DDT, for instance, in the control of mosquito and other vector populations has been most important in the worldwide suppression of malaria, plague, and typhus. Likewise, the use of herbicides and insecticides has permitted, along with heavy application of fertilizers, great increases in crop productivity. With human populations expanding in size all over the world at rates challenging the ability to feed them adequately, increased yields from food-producing areas should certainly be encouraged. This increased crop productivity, however, cannot be permitted at the expense of harmful declines in environmental quality. Humanity faces the alternative of extinction from starvation or from pollution unless drastic control becomes established over the basic cause of it all, the explosive population increase now underway.

If chemical pesticides must continue in use, as is likely, research is required to develop compounds that decompose rapidly, are more selective or limited in their biological effects, and can be applied more efficiently on the organisms of concern. Modern research, for instance, has developed a new form of DDT that is highly toxic but at the same time is quickly degradable, so that it does not accumulate in food chains (Kapor *et al.* 1970). There is question, however, whether chemical methods of pest control need to be as widely used as they are at present. A number of non-chemical procedures are in use, and new ones are being developed. The problem of pest control will be considered more thoroughly in Chapter 14 (Cook 1969, Cottam 1965, Edwards 1969, Niering 1968).

Gases

Clean dry air is composed of nitrogen, 78 per cent; oxygen, 21 per cent; argon, 0.93 per cent; carbon dioxide, 0.03 per cent; and traces of other elements. The five most important air pollutants in the United States are carbon monoxide, sulfur oxides, hydrocarbons, nitrogen oxides, and particles, both liquid and solid, of industrial dusts, ash, and spray (Fig. 9-8). The major sources from which these are derived are

Fig. 9-8 Energy cycle involved in the combustion of fossil fuels begins with solar energy employed in photosynthesis millions of years ago. A small fraction of the plants is buried under conditions that prevent complete oxidation. The material undergoes chemical changes that transform it into coal, oil, and other fuels. When they are burned to release their stored energy, only part of the enegy goes into useful work. Much of the energy is returned to the atmosphere as heat, together with such by-products of combustion as carbon dioxide and water vapor. Other emissions in fossil-fuel combustion are listed at right (Singer 1970; from *Scientific American*).

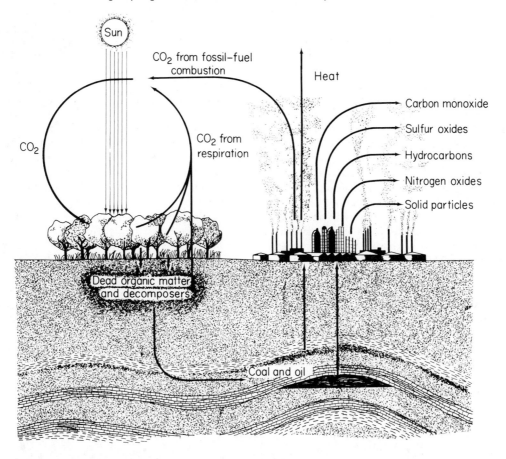

automobiles, industry, electric power plants, space heating, and refuse. Lead aerosol is a common contaminant in urban areas; in a few years the rising concentration level at which it occurs may render it as important as some of the other chemicals mentioned. Very important here was the shift during the first half of the twentieth century from animal power to mechanical power in transportation and agriculture.

Complex reactions occur between these various contaminants and water vapor. The acute toxicological effects of most air contaminants is well known, but the continuous or repetitive exposures to lower concentrations of these contaminants and to their heterogeneous mixtures is just beginning to be understood. Sulfuric, nitric, and other acids are commonly formed (Kellogg *et al.* 1972). As they and other toxic, irritant, and corrosive substances in the air precipitate, they come in contact directly or indirectly with all living matter and exposed non-living surfaces.

Smog, a word coined from *smoke* and *fog*, is sometimes used loosely, but true smog is a mixture of pollutants, some of which are generated photochemically. The reactions that take place are complex and obscure but involve ultraviolet light from the sun, nitrogen oxides, hydrocarbons, and a number of secondary products. Atomic oxygen and ozone are also involved. Smog readily forms when the lack of wind, surrounding mountains, or other conditions prevent the diffusion and dispersal of the aerial pollutants.

Although *carbon dioxide* is a normal constituent of air, it has increased about 7 per cent since 1900 and is continuing to rise. This is the result of emissions of the gas from electric power plants, cement factories, and internal combustion engines. Fossil fuels in the form of coal, petroleum, and gas are being burned in far less time than was required to form them. Under natural conditions the carbon dioxide content of the air is constant because of the exchange of carbon dioxide and oxygen between plants and animals in the processes of photosynthesis and respiration and the buffering action of bicarbonates in the sea and fresh water. This buffering action takes place slowly, however, requiring hundreds or thousands of years, and has not been able to absorb the rapid input of carbon dioxide into the air by man. Increasing the carbon dioxide in the air may accelerate the rate of photosynthesis, but this is offset by the decrease in volume of green plants undergoing photosynthesis and adverse effects of other pollutants. Measurements of the total biomass of photosynthesizing material are not available, but with the logging of forests, plowing up of grasslands, covering of the ground surface with concrete in our cities and highways, and pollution of our waters, it is doubtlessly less now than it was even 100 years ago.

This blanket of carbon dioxide also affects the heat balance of the earth. The *heat balance* is the ratio between the incoming solar radiation and outgoing re-radiation. Turbidity in the air, produced by dust particles from agriculture and industry, and solid and liquid aerosols of all sorts scatter the incoming solar radiation of all wavelengths, thus decreasing the amount reaching the earth's surface. The albedo is a fraction of incoming light that is reflected from the earth's surface as short wavelengths. This may be only 5–15 per cent from forest vegetation and the sea or up to 90 per cent or more from snow and ice. Considerable radiation is also reflected from clouds. The remaining portion of the incoming solar radiation is absorbed as heat by the land, water, and organisms exposed to it and is captured as potential energy in the products of photosynthesis. Organisms, as well as land and water, however, maintain an average temperature so that as much heat is lost as is absorbed. Even the potential energy of photosynthesis not used in work is degraded into heat and lost. The outgoing radiation is in the form of long wavelengths and may be so great around cities as to produce heat islands having air temperatures 1°–2°C higher than the surrounding country. The mean temperature of the earth's surface, as the result of re-radiation and albedo, would be 25°–30°C lower than it is were it not for the water vapor and carbon dioxide in the air. These normal constituents of the atmosphere are impervious to long wavelengths and redirect the radiation back again to the earth. This is called the *greenhouse effect* (Ehrlich and Ehrlich 1970, Landsberg 1970, Wilson and Matthews 1970).

The global temperature depends on the balance among all these factors, and their interaction is not fully understood (MacDonald 1971). Between 1880 and 1940 the mean global temperature appears to have risen about 0.4°C, but between 1940 and 1967 it fell nearly 0.2°C. An increase in the blanket of carbon dioxide and water vapor during the first period of years may have been responsible for the rise in the earth's temperature. An accelerating increase in air turbidity during the more recent period apparently overcame the greenhouse effect to bring the decline in temperature. It is estimated that carbon dioxide is unlikely to increase more than 8 times even in the next several thousands of years; dust concentration in the atmosphere could, on the other hand, increase 4 times in the next century and bring a decline of 3.5°C in the global temperature (Rasool and Schneider 1971).

The free *oxygen* content of the air is the result of the photosynthesis of plants. However, as much oxygen is consumed in the decay of plants and in the metabolism of animals that feed on the plants as is freed in photosynthesis. That there is surplus free oxygen available at all is due to plants becoming fossilized before they could be consumed or oxidized. Combustion of these fossil fuels utilizes and depletes the oxygen surplus that has accumulated through geological time. Fortunately, much of the photosynthesized carbon products are widely dispersed in shales and rocks not readily used as fuel. It is unlikely that they will ever become completely oxidized.

Maintenance of a large oxygen-consuming population of animals and man, however, requires continual replenishment of the free oxygen supply in the air. When a point is reached that the rate of combustion exceeds the rate of photosynthesis, the supply of free oxygen will decrease. With the continuing reduction of green vegetation on land and the pouring of pesticides, radioisotopes, and detergents from our rivers into the sea, the capacity for oxygen production could conceivably decline rather rapidly. A decline in oxygen supply below a critical level could destroy all living organisms except for anaerobes. There is, however, an enormous reserve supply of oxygen in the atmosphere and in the ocean, enough to supply requirements for hundreds, if not thousands, of years. Man may well succumb to some form of pollution long before he feels a lack of oxygen (Broecker 1970, Ryther 1970).

In recognition of all these dangers from pollution of the atmosphere, the U.S. Congress passed the Air Quality Act of 1967, and efforts are being made on a national scale, as must also be done on a worldwide basis, to preserve the air fit to live in. The cleaning up of the contaminants already introduced into the environment and keeping the environment clean thereafter will require basic adjustments at the very roots of our socioeconomic system (Miller 1971). Modern civilization is, however, at stake and the costs must be met. Man, for all his ingenuity and arrogance, is still a part of nature and subject to natural laws (Cole 1968, Cook 1969).

SUMMARY

The characteristics of soil are determined by the parent rock material, the reactions of plants and animals, and climate. The burrowing of earthworms, ants and other ground insects, crayfish, and rodents brings subsoil to the surface, where it becomes mixed with humus. Animal metabolic processes aid in the formation of humus by breaking down complex organic matter into simpler compounds which the animals then excrete. Bacteria, actinomycetes, and fungi are doubtless even more important in this respect. Plants then reabsorb the nutrients from the soil. Animals require nitrogen, carbon, oxygen, hydrogen as well as some 13 other elements, and hence are also involved in nutrient cycles of these elements in the ecosystem. Nutrient cycles occur in aquatic as well as terrestrial ecosystems.

Climate is directly involved in the weathering of soil particles and as rainwater percolating into the ground carrying nutrients into the soil, and indirectly in determining the kind and luxuriance of the vegetation and animal life that occurs in the area. As a result, mature soils of climax-stabilized ecosystems have profiles characteristic both of types of vegetation and of climatic regions. The species composition and density of ground animals vary with the profile horizon and with the various soil types found in various parts of the world.

Soil, water, and air pollution results from contaminants added to ecosystems beyond the capacity for their absorption into normal nutrient cycling. Common contaminants are solid wastes, herbicides, insecticides, and gases. Some of these pollutants are resistant to decomposition and accumulate in the environment or in food chains to be harmful to man. Excessive output from industry into the atmosphere of carbon dioxide, dusts, and vapors threatens the world's heat balance and global temperature. Humanity faces the alternative of extinction from starvation or from pollution unless drastic action is taken, including curtailment of the population explosion.

Chapter 10

FOOD AND
FEEDING RELATIONSHIPS

Food-getting necessarily involves interrelations between organisms and between species; these interrelations are among the most important coactions in any community. Animals are adapted variously to capture and utilize certain types of food, and to avoid being captured by other animals. One must understand these adaptations and interrelations to appreciate properly the role that food-getting plays in the dynamics of the community.

FEEDING BEHAVIOR

Free-living animals are commonly classified on the basis of normal feeding behavior, thus:

Herbivores: Feed principally on living plants
Carnivores: Feed principally on animals that they kill
Omnivores: Feed nearly equally on plants and animals
Saprovores: Feed on dead plants and animals, and excreta

The various categories are capable of further subdivision. Thus, herbivores include (a) those that feed chiefly on the foliage, as the large cursorial grazers, such as bison, antelope, muskox, caribou, sheep; small surface-living grazers, such as rabbits, mice, grasshoppers; subterranean-living grazers, such as woodchucks, prairie dogs, kangaroo rats, ground squirrels; browsers, which feed on foliage buds and twigs of trees and shrubs rather than strictly on grass or ground herbs, such as wapiti, deer, moose, grouse, and defoliating types of insects such as the hemlock looper, spruce budworm, and larch sawfly; (b) those that feed chiefly on the products or non-photosynthetic parts of the plant, as seed-, nut-, and fruit-eaters, such as squirrels, chipmunks, gallinaceous birds, sparrows; plant-juice suckers, such as aphids, leafhoppers, mosquitoes, chinch bugs; and cambium feeders, such as bark beetles, gall flies, cynipids; and (c) those that feed primarily on fungi and bacteria—*fungivores*. Some forms that appear to be saprovores are actually fungivores, feeding on the fungi or bacteria that they obtain from decaying organic material. A useful guide to the literature on insect-plant relations is that of Kingsolver and Sanderson (1967).

Carnivores are also called predators. Carnivores restricting their food chiefly to insects are called insectivores; those limiting themselves largely to fish are called piscivores; and so on. Parasitoids eventually consume their hosts, and hence are a special type of carnivore. Some plants are carnivorous. The pitcher-plant, Venus's-flytrap, and sundew, which grow in bogs or wet places, and bladderwort, which occurs in ponds, depend for their nitrogen supply largely on animals that they capture and consume. Some soil fungi trap and consume nematodes (Pramer 1964). Perhaps the bacteria fungi, and viruses that cause disease in animals also belong to this classification.

Many species eat both plant and animal matter, on occasion, or at particular seasons, but animals are considered to be truly omnivorous only if they feed on plants and animals in nearly equal amounts or indiscriminately. Omnivores occasionally also consume dead organic matter. Some aquatic organisms are filter-feeders and may consume everything within a particular size range that passes through their feeding apparatus (Jorgensen 1955). However, filter-feeders may demonstrate selectivity by feeding in neighborhoods where certain species predominate. Some copepods select particles of a particular size, rejecting larger ones, by regulating the distance between the maxillae in the filter mechanism (Hutchinson 1951).

Probably most, if not all, animals have chemoreceptors of some sort, either to discriminate chemical substances dissolved in drinking water or food (taste), or chemical substances that are water- or air-borne (smell). Essential oils and alkaloids in plants are important as conditions of acceptability to insects. Hairiness, other surface features, or the visual stimuli that plants present are also conditions of attractiveness or acceptability of a food item.

The food preference of any species depends on chromosomal inheritance, parental training, and personal experience of that species, but the relative significance of each of these factors has not been evaluated for most animals. Young birds and mammals, in their first experiences at independent feeding, may pick up a variety of material but reject those items that are distasteful or indigestible; they soon learn to distinguish acceptable substances. This process is established as the parents feed offspring only those things traditional to the species, or so direct the feeding movements that untraditional food is excluded. Some adult insects lay their eggs on material that will serve the larvae as food. The larvae acquire the habit of feeding on that material, and do not readily change to something else as adults (Brues 1924, Thorpe 1939).

FEEDING ADAPTATIONS

Among kinds of mammals, teeth show considerable adaptive radiation, correlated with type of food consumed. The molar and premolar teeth of insectivorous species, such as shrews and bats, are low and have sharp-pointed cusps for crushing weak-bodied prey. The piscivorous toothed whales have largely lost all differentiation in their teeth, which are simple, conical, grasping structures. The teeth of the carnivorous dogs and cats are high-crowned and tubercular, well fitted for shearing flesh. Herbivorous ungulates and rodents have teeth that are flat-crowned, suited to grinding

harsh grasses and other vegetation. Their jaws are capable of considerable lateral motion. Omnivores may have both grinding and pointed teeth. Saprovores have rather blunt teeth. The ant-eating sloths and their relatives have no teeth, and the mouth is almost tubular in shape. The tongue has become long and prehensile for lapping up tiny ants.

The bills of birds display great variety in shape and size, adaptations to feeding in numerous quite specialized niches in the environment (Fig. 10-1). The tongues of birds are variously modified to serve as long probes or spears (woodpeckers and nuthatches), as a strainer (ducks), as a long capillary tube for obtaining nectar from flowers (hummingbirds), as a rasp (hawks and owls), as a finger for manipulating the food in the mouth (parrots and sparrows), and as a tactile organ (sandpipers and herons) (Gardner 1925).

The mouth parts of insects are adapted primarily either for biting and chewing or for piercing and sucking. Among marine invertebrates, adaptations for feeding on detritus (Blegvad 1914) include pseudopodia (Foraminifera); ciliated epithelium that maintains a flow of water through the animal (sponges, clams); prehensile, often ciliated arms (various polychaetes; holothuroideans); and soft eversible gullets (various polychaetes, sipunculids). Those that are herbivorous or carnivorous, as well as detritus feeders, have prehensile tentacles armed with nematocysts (hydroids, actinians), radulae in the mouth (mollusks), eversible stomachs (starfish), and masticatory structures in the mouth or stomach (crustacea, diptera larvae).

In addition to mouth-part adaptations, there are many modifications in other parts of the digestive tract for handling particular types of food. These adaptations occur throughout the animal kingdom, but are especially evident in birds. A crop is present in some species but not in others. The walls of the stomach are more muscular in seed-eating birds than in flesh-eaters. Owls and some other species form and regurgitate from the stomach pellets of indigestible matter. Gallinaceous birds are able to retain or shift the supply of gravel or grit in the stomach as an aid to

Fig. 10-1 Adaptations in the bills of birds: (a) seed-eating sparrow; (b) insect-eating warbler; (c) plant-eating duck; (d) fish-eating heron; (e) predaceous hawk; (f) aerial insect-eating whippoorwill.

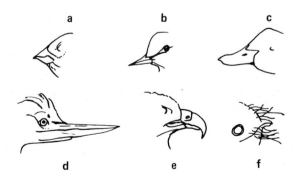

grinding seeds. The length of the intestine varies with the type of food consumed; the caeca are longer in browsing and seed-eating gallinaceous birds for digesting cellulose.

Associated with these anatomical and histological adaptations are adjustments in function and behavior. Obviously, if an animal has morphological adaptations for ingesting and digesting flesh, it must behave as a carnivore and not as an herbivore. The possession of these adaptations and adjustments means that animals are generally restricted to the particular types of food that they can use most efficiently. The kinds of food eaten by animals is of fundamental ecological and economic importance.

METHODS OF STUDY

A common procedure for analyzing kinds of food consumed by organisms is to identify the contents of the crop, stomach, cheek pouches, or other parts of the digestive tract (Hartley 1948).

The diet may be described in terms of number of items of each kind of food found in one specimen, or percentage of specimens containing a particular item, but it is usually more satisfactory to measure in one specimen the percentage volume of each food item against the total contents (McAtee 1912).

This procedure has the advantage of showing accurately what an animal has actually ingested, but has the disadvantage that the animal must usually be killed; thus, the information obtained is on only one meal, or portion of a meal. It also gives no information on where or how the food was obtained. Nevertheless, considerable information on the food habits of animals has been obtained in this manner (Henderson 1927, McAtee 1932, Davison 1940, Martin et al. 1951). Improved techniques make it possible to secure the stomach or crop contents without killing the animal. This is done by manual manipulation of the crop (Errington 1932), or by use of flushing tubes (Vogtman 1945, Robertson 1945). Artificial beaks with open gapes placed as decoys among nestlings have been used to collect food brought by the parents (Betts 1954).

There are usually indigestible parts in all kinds of food, and these indigestible or undigested parts are eliminated from the body. The contents of fecal droppings or regurgitated pellets can often be identified by differences in shape, size, color and texture, or by histological techniques (Dusi 1949). Collection of droppings is not practicable for aquatic animals, or for more than a few of the terrestrial invertebrates. Moreover, the droppings must be relatively fresh, as they quickly disintegrate in wet weather. The analysis of owl pellets is very fruitful, for owls swallow all parts of their mammalian or avian prey, and then regurgitate the hair, feathers, and skeleton. Hawks, gulls, and shrikes also produce pellets. The considerable advan-

tage of pellet analysis is the possibility of continuous diet analysis on the same individual, or species, through long periods of time, without disturbance to its normal behavior (Dalke 1935, Errington 1932).

Whatever method is used, field observation of the feeding behavior of animals in the natural environment is desirable. For instance, one series of stomachs of the house sparrow contained a large number of May beetles, which would suggest that the bird was important for the control of this insect pest. Observations disclosed, however, that the sparrows were picking up dead beetles littering the pavement under street lights. Field observation itself often furnishes considerable information concerning the kinds of food consumed, but the results are usually not quantitative, do not disclose the less conspicuous kinds of food taken, and may be misleading. A hawk visiting a game farm may take not game animals but undesirable rodents that are also present (Kalmbach 1934).

Food chains are commonly determined by correlating the food eaten by different species in the community. Tracing the movement from one species to another of radioactive elements introduced into the ecosystem gives promise of more direct determination of how matter flows (p. 174).

CHOICE OF FOOD

The kinds of food eaten by animals depend on their genetic heritage and parental training or conditioning while young. Involved in the evolution of the food habits of a species and the "profitability" of consuming one prey species rather than another (Royama 1970) are the animal's physical adaptations for ingesting and digesting particular types of food, the nutritional values of the food, its palatability, the size of it, its availability or abundance, and its ease of procurement, which depends in large part on the various protective devices that it possesses.

Nutritional Values

Animals generally require proteins, carbohydrates, fats, vitamins, minerals, and water. Proteins are used as the basic substance in the composition of protoplasm; carbohydrates and fats are oxidized to furnish energy for the body; vitamins serve as catalysts for specific metabolic processes; minerals are needed to regulate osmotic pressure and as constituent elements of various body organs; and water is used as a general solvent, lubricant, and circulatory medium. Species differ, however, in their needs for particular substances. The beetles *Tribolium*, *Lasioderma*, and *Ptinus*, for instance, grow slowly but nonetheless satisfactorily on diets lacking carbohydrates. Hence, they may be distributed more widely than are species which require carbohydrates in their diet (Fraenkel and Blewett 1943).

Foods differ in composition. Foods *staple* to an organism's diet are those easily digested, and of high caloric and protein content. They are adequate for sustenance of the weight and vigor of the animal, but usually need to be supplemented with vitamins and minerals. The bobwhite and ring-necked pheasant, for example, eat certain cultivated grains and weed seeds, such as corn, sorghum, barley, wheat, rye, soy beans, pigeon grass, and lesser ragweed, as staple foods, at least on a mixed diet.

Non-staple or *emergency* foods are not in themselves sustentative, and animals limited to them gradually lose weight and die. Such foods are, however, often abundant and easily procured in emergencies, when staple foods are covered with snow or ice, and furnish sufficient energy to tide the animal over the critical period. In emergencies, the bobwhite and ring-necked pheasant eat black locust beans, fruits of the bittersweet and sumac, rose hips, dried wild grapes, and sweet clover seeds (Errington 1937).

During good acorn years, squirrels, deer, and raccoons feed extensively on the acorns of white and black oak, but almost completely ignore northern red oak. Experiments with fox squirrels show that the animals gain weight on an exclusive diet of white oak acorns, scarcely maintain weight on acorns of the black oak, and lose weight rapidly on acorns of the red oak. The percentage of tannin in red oak acorns is twice that in white oak acorns, and animals are probably able to distinguish red oak acorns by a bitter taste (Baumgras 1944).

Vitamins are necessary for the maintenance of good health in wild animals, just as in domestic animals or man. The symptoms of vitamin deficiency, induced experimentally, are similar. Evidence has been difficult to secure, however, that animals suffer from vitamin deficiencies in their natural environments (Nestler 1949, House and Barlow 1958).

Animals obtain most of their required minerals from their food and water. Additional salt must sometimes be given caged animals to prevent cannibalism. The gnawing of castoff deer antlers by rodents is apparently for additional salts. The use of certain soil deposits and springs as natural "licking sites" by deer and other ruminants is apparently for sodium salts lacking in their general diet (Stockstad *et al.* 1953). Some birds, such as the evening grosbeak, are also attracted to sources of salt supply. Grit in the form of gravel, sand, or small stones in the stomach of gallinaceous and other birds may not only aid in the grinding of seeds and hard vegetable matter but also be a source of minerals (Nestler 1949).

Animals appear to become aware of nutritional deficiencies in their diet through physiological and neurological mechanisms. Experiments with rats show that when the body lacks some necessary element, such as sugar, salt, or a vitamin, the animal consumes more of that particular substance than usual. Discrimination and selection are apparently made by taste, and a special need for a particular substance sharpens the taste for that substance so that it can be detected even when present in food or water in very small quantities (Richter 1942). Nutritional needs are neither the sole nor necessarily the most important factor involved when animals show preference for one type of food over another. Many other factors condition the choice (Dethier 1954).

Digestibility

Different species of animals vary considerably in efficiency of digestion and utilization of particular food substances. Thus, clothes-moth larvae can digest cloth and bird lice can digest feathers, because among other things they have an exceptionally high hydrogen-ion concentration in their intestines.

Digestive enzymes occur generally throughout the animal kingdom, although less is known about them in the Protozoa. In the lower phyla the enzymes are generalized in respect to the kinds of foods on which they act; in the higher phyla, they become highly specialized (Prosser *et al.* 1950). A specific enzyme, however, does not differ greatly from one animal group to another. Carnivores have strong proteases and weak carbohydrases, correlated with their meat diet. Herbivores, on the other hand, have weak protein, but active carbohydrate, enzymes. Herbivorous mammals and birds possess a bacterial flora in their digestive tracts that makes possible digestion of cellulose. Omnivores have a full complement of enzymes and can utilize a wide variety of foods.

Practically all food contains some indigestible matter, which is ordinarily passed through the digestive tract and eliminated in the feces. If the indigestible material is excessive, or if it contains toxic substances, regurgitation or vomiting may result.

Size of Food Item

The size of the food in relation to the animal is not of major importance to many herbivores or saprovores as they normally feed in or on the organism or substance. With carnivores, the size of the prey must be within their power of conquest. Ordinarily the size of prey is less than that of the carnivore that feeds on it, but a high degree of ferocity and audacity, or pack hunting in the manner of wolves, often enables the carnivore to take prey larger than itself. On the other hand, a predator cannot profitably prey on species so small that the energy derived from its consumption does not equal the energy expended in its capture. Evolution of prey species subject to heavy predation tends to proceed either to very large size or very small size in relation to the size of their most important predators (Valverde 1964). Some very large aquatic predators, however, have become adapted to feed with a minimum of effort on very small organisms occurring in dense concentrations through the evolution of a filtering apparatus in their mouth parts. A good example is the feeding of baleen whales on plankton.

Availability

In order to determine if a species is fed upon in proportion to its abundance (McAtee 1932), it is necessary to find out what animals have been eating of that which is available in a habitat. The relation between the two may be shown graphically (Hamilton 1940a) and expressed as *forage ratio*: per cent of species in animals' food divided by per cent of species in habitat (Hess and Swartz 1941). A value of unity indicates that the food item is taken in proportion to its abundance; a value greater than unity indicates that it is taken more frequently; values of less than unity indicate that the item is either inaccessible, of the wrong size, too difficult to obtain, or is actually avoided. Table 10-1 is an example of such a study. It is apparent that while there is a relationship between the relative abundance of various species and the degree to which they were taken as food, there are also several discrepancies. Weed-inhabiting organisms, more accessible to fish than organisms buried in the mud, are accordingly fed upon heavily, in disproportion to their rela-

tive abundance in the total fauna. The fingernail clam is fed upon heavily in spring, since it lies on the surface of the mud. But there is no explanation of why it is fed upon less heavily during the summer. Water boatmen and water mites, however, are generally not acceptable. Although the alderfly larva is a mud dweller, it is fed upon in large numbers, suggesting that there may be something in its behavior that makes it especially vulnerable, or that brown trout have evolved special methods for securing it. The caddisfly larva *Leptocerus* also appears to be easily taken, as it is devoured in numbers greatly disproportionate to its relative abundance. The mayfly naiad *Caenis* is not much fed upon in spring, at which time it is buried in the mud, but it is fed upon in large numbers in summer, when it comes to the surface of the water to emerge. Midge flies also become more vulnerable during the process of emergence. Zooplankton is of very minor importance in the diet of the fish. Forage fish were not present.

The vulnerability to a predator of a prey species is directly proportional to its relative abundance among the other species available. Voles are preferred by the red fox over white-footed mice, but both mice and voles are preferred over moles, shrews, and snakes. Predator preferences among them, however, occur only when all are abundant. When prey species are reduced in numbers and difficult to get, little or no predator preference occurs (Scott and Klimstra 1955).

There is some evidence that predators—particularly insectivorous birds—when searching for prey concentrate on one or a few species at a time. By a kind

Table 10-1 Quantitative comparison of food organisms eaten by the brown trout, a carnivore, with those present in the fauna of an English fishpond (data from Frost and Smyly 1952).

Common name	Classification	Spring			Summer		
		Per cent eaten	Per cent in fauna	Forage ratio	Per cent eaten	Per cent in fauna	Forage ratio
Mud-living organisms							
Midge fly larva and pupa	Chironomidae	36	66	0.5	36	48	0.8
Alderfly larvae	*Sialis lutaria*	10	1	10.0	4	1	4.0
Mayfly naiads	*Caenis* sp.	+	6	+	20	+	20.0
Fingernail clam	*Pisidium* sp.	17	16	1.1	5	27	0.2
Worms	Oligochaeta	0	1	0	0	2	0
	Totals	(63)	(90)	(avg. 0.7)	(65)	(79)	(avg. 0.8)
Weed-living organisms							
Caddisfly larvae	*Leptocerus* sp.	21	+	21.0	21	1	21.0
Caddisfly larvae	Limnophilidae	3	+	3.0	1	1	1.0
Caddisfly larvae	Polycentropidae	1	1	1.0	2	1	2.0
Mayfly naiad	*Leptophlebia* sp.	7	4	1.8	0	+	0
Damselfly and dragonfly naiads	Odonata	3	1	3.0	2	5	0.4
Beetle adults	Coleoptera	+	+	1.0	+	+	1.0
Water boatmen	Corixidae	1	3	0.3	2	9	0.2
Snail	*Lymnaea pereger*	1	+	1.0	5	+	5.0
Water mites	Hydracarina	+	1	+	+	3	+
	Totals	(37)	(10)	(avg. 3.7)	(35)	(21)	(avg. 1.7)

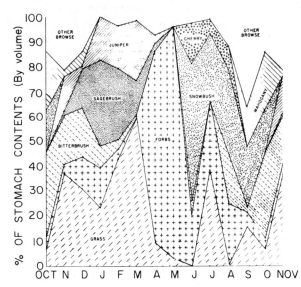

Fig. 10-2 Monthly variation in kinds of food consumed by a herbivore, the mule deer, in California and Oregon (Interstate Deer Herd Committee 1951, courtesy California Division of Fish and Game).

of learning, they acquire what may be called specific *searching images* for these species and thereby mostly disregard other species. When a new species becomes numerous in an area, they feed on it at first only as the result of chance encounters. To obtain a preference for the new species, they must become conditioned to it gradually and must learn the proper cues of where and how to search it out (Tinbergen 1960).

The importance of availability is illustrated by seasonal variations in the kinds and amounts of food consumed by animals. Predator species that remain in one habitat throughout the year must adjust their feeding to the kinds of food available in each season (Fig. 10-2). Species that are unable to do this are compelled to migrate, hibernate, or make other adjustments to survive the unfavorable seasons. Thus birds that are strictly insectivorous may occur in a given area only during the warm part of the year, leaving it before insects disappear. Omnivorous species commonly change from an insect diet in summer to a diet of seeds and fruits in the autumn and are often nonmigratory.

Seeds and fruits are most abundant and easily obtained in the autumn. During the winter they decompose, become buried in the snow or softened ground, or are consumed. In Michigan, the weed seeds available on agricultural lands in March are only 9 per cent of those available in October (Baumgras 1943). The most critical time of the year, as far as food supplies are concerned, is early spring, the time before new vegetation and hibernating prey animals appear. Abundance of seeds also varies with fertility of soil, which thus influences survival, density, and distribution of animal populations.

Animals are subject to considerable variation in the abundance and kinds of food available to them from year to year. In a four-year study of the yield of fruit and seeds from 27 species of trees and shrubs in West Virginia, only 33 per cent of the species produced a crop every year, 29 to 33 per cent failed to produce a crop in 3 of the 4 years, and 22 per cent failed twice within the 4 years (Park 1942). When herbivorous species vary in abundance, because of variation in food supply or other factors, carnivorous species that prey upon them often vary in direct proportion.

Protective Devices

Most, if not all, plants are equipped with defense mechanisms against the grazing and browsing of animals. The most widespread defenses are chemical—various kinds of bitter or toxic materials in plant tissues that make them unpalatable to many animals. Animals evolve tolerances and neutralizing devices, however, so that most plants are fed on; their defense is relative, not complete. New leaves often contain less of the defensive chemicals, and are thus more vulnerable. Trees and shrubs may be deformed or killed by excessive browsing that destroys terminal twigs and buds. Grasses are not so affected, for they grow from the base of the leaves; they may even be benefited by moderate cropping. Grasses also are protected by grains of silica embedded in their tissues; only by evolution of high-crowned teeth adapted to abrasion by the silica have some ungulates become able to graze on grass. Some species of trees, shrubs, and cacti are protected from browsing by prickles or thorns. This protection is important to the plants for survival in deserts and in grazing subseres of humid regions. Some plants are particularly noxious or toxic, and animals quickly learn to avoid them. Seeds may develop alkaloids or other disagreeable substances that give them protection (Janzen 1969).

Various kinds of arthropods have defensive secretions in the form of sprays that they emit when attacked. These occur particularly in insects, millipedes, and phalangids (Eisner and Meinwald 1966). The odoriferous spray of the skunk is well known. There is increasing evidence that attractant and repellant secretions may be important in both predator-prey and social relation of animals (Watkins *et al.* 1969).

When coloration renders it inconspicuous in its normal environment, an animal is said to have *concealing* or *cryptic coloration* (Cott 1940, Portmann 1959). When the coloration or markings reproduce the general tone or characteristics of background, it is called *protective resemblance* (Fig. 10-3); *disruptive* (Fig. 10-4), when the markings break up the outlines of an animal and replace it by some irregular configuration so that the animal is not recognized as prey. The white collar

Fig. 10-3 Protective resemblance of an incubating ruffed grouse to her surroundings (courtesy U.S. Forest Service).

Fig. 10-4 Disruptive coloration of a killdeer. The white color obscures the connection between head and body.

of the killdeer, observed casually from a distance, tends to the human eye to separate the head as a distinct object from the rest of the body. *Obliterative coloration*, or *countershading* (Fig. 10-5), describes the condition where the upper side of the body, exposed to the brighter illumination, is heavily pigmented and the lower side of the body, which is in the shadow, is lighter in color. This coloration obliterates the effect shadows have of making a body stand out from its surroundings (Thayer 1910). *Aggressive resemblance* is where the animal closely resembles some particular object rather than the general environment. The *Kallima* butterfly of the Orient and the preying mantis of Central America match the shape, markings, and color of leaves when the insects repose with wings folded. The familiar walkingstick resembles a twig. Several insect species look like bird-droppings. Such resemblances doubtless serve the animals to escape the attention of predators only as long as they remain motionless. Even slight movements quickly call attention to animals, regardless of any concealing coloration that they may have. Behavioral orientation is well shown by those caterpillars that are lightly colored dorsally and darkly colored ventrally. They bring their countershading into proper position by coming to rest upside-down along plant stems (Ruiter 1955). The tell-tale shadows cast by animals may be eliminated when the resting animal lies lengthwise, rather than crosswise, to the sun, or when they lie pressed close to the ground.

Some animals, on the other hand, are vividly marked with strikingly conspicuous patterns or bright colors, and this *aposematic* or *warning coloration* is accompanied by unpalatableness in certain butterflies, bugs, beetles, ants, and birds; stings, in wasps and bees; a disagreeable odor in skunks; or some other offensive feature (Poulton 1887). There is experimental evidence that such animals are actually avoided by predators

Fig. 10-5 Countershading: Left: Caterpillar, last instar of *Dicranura vinula*, in normal upside-down position on a willow twig in natural diffuse light. The back of this animal is lighter than the underparts, annulling the shadow. Right: The same caterpillar, inverted, is much more conspicuous, for the countershading effect is lost. This caterpillar has second and third lines of defense: when touched, it turns a kind of grotesque "face" toward you; when pressed, it squirts acid (courtesy N. Tinbergen).

(Finn 1895–97, Jones 1932, Cott 1947). However, a hungry predator is less selective in its choice of food than one that has recently fed. Apparently each individual carnivore must have a personal experience with an animal so marked before it learns to avoid it by associating the coloration with the disagreeable feature. Aposematic coloration has an advantage over cryptic coloration since it permits the animal to move freely and safely during the daytime.

There are many examples of *mimicry* among insects (Goldschmidt 1945). *Batesian mimicry* is the resemblance of a palatable species in external features to an unpalatable one that in turn possesses warning coloration and is the more abundant of the two species (Fig. 10-6). The palatable species derives benefit from the relation, since predators, especially birds, avoid them as well as the unpalatable ones. The effectiveness of the mimicry decreases as the number of mimics equals or exceeds the number of models (Huheey 1964). The viceroy butterfly mimics the disagreeable monarch butterfly and differs strikingly from other members of its own genus. The monarch butterfly is inedible because its larvae feeds on milkweed, from which it derives cardiac glycosides (Brower 1969). There is considerable controversy about mimicry, however, and even the classical example of the butterflies is disputed (Urquhart 1957a). In *Mullerian mimicry*, both model and mimic are unpalatable. Pooling of numbers between the two species gives more chances for inexperienced birds to learn to avoid them and reduces the losses per species during the learning process. The yellow jacket and polistine wasps are examples and many more can be found among butterflies (Wickler 1968).

Some animals possess bright spots or colors so placed on the body as to be *deflective*. The attention of a pursuing predator is drawn to less vulnerable parts of the body; for instance, eyespots on the fins or tail of a fish. Eyespots on the wings of some butterflies and moths are concealed at rest, but when flashed out by spread wings may frighten away an attacking bird or other predator.

Concealing coloration and resemblance to other objects are apparently also useful to animals aggressively. A carnivore that matches its background can approach its prey undetected more easily than can a conspicuously marked one. Some predators have *directive markings* to confuse their prey as to the location of their mouths or to allure them in various ways.

There has been considerable controversy about the value of concealing coloration. McAtee (1932) minimized its importance because he found both protectively colored and conspicuously colored species in the stomach contents of the birds he examined. Probably few species are entirely immune to predation, but if coloration to match the surroundings, mimicry of some other avoided species, or peculiarities of form or structure render predation even slightly less frequent than it would otherwise be, it can well have survival value and evolve as characteristic of a species.

A number of experiments in regard to concealing coloration have been performed by exposing different kinds of insects (Carrick 1936, Isely 1938), fish (Sumner 1935), and mice (Dice 1947) to bird predators, with the result that those individuals that most closely matched the color of their background were taken less frequently than those that did not do so. In a black aquarium in which equal numbers of black and white mosquitofish were exposed to the predation of a penguin, 27 per cent of the fish eaten were black and 73 per cent were white, but in a white aquarium 62 per cent of the fish eaten were black and only 38 per cent were white.

RANGE OF FOOD SELECTION

There is considerable range in the variety of foods eaten by most species. Herbivorous species are often more specific in their feeding habits than are carnivorous forms. However, even herbivorous forms have various degrees of restriction, as shown in the following analysis of 240 species of plant-juice sucking aphids (Clements and Shelford 1939):

	Percentage
Species restricted to a single plant species	27
Species restricted to a single genus but feeding on more than one species	40
Species feeding on several different genera	33

One may argue with considerable justification that animals such as aphids, gall wasps, some bugs, and others that show host specificity are really parasitic rather than herbivorous in their feeding behavior, and hence are not good examples of free-living animals on a restricted diet. Actually it is very difficult to find proven cases of animals that confine themselves to a single species of food. It is more common to have an animal feeding on a small group of related species, as do aphids. The potato beetle originally fed chiefly

Fig. 10-6 Mimicry of (a) the monarch by (b) the viceroy butterfly.

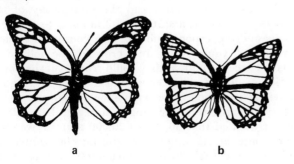

a   b

on the sand-bur *Solanum rostratum* in the Rocky Mountains until about the year 1859, when it began to infest the potato *Solanum tuberosum* and spread across the country.

Herbivorous species may be classified in respect to the diversity of their food into

Monophagous: Restricted to a single food plant
Oligophagous: Restricted to a few very definite food plants
Polyphagous: Feed on many species

The restriction of animals to particular foods may be the result of chemicals affecting odor or taste, or to structural adaptations. Chemical stimuli are especially important with insects (Dethier 1947). The crossed bills of certain birds (*Loxia* sp.) are well fitted for prying seeds from between the bracts of coniferous tree cones. The Siberian nutcracker has special structures in its bill for cracking the nuts of the Siberian cedar, on which it depends almost exclusively for food (Formosof 1933).

In spite of the monophagy exhibited by some species, many herbivorous species have a wide choice of food; the bobwhite quail in Georgia is known to feed on 927 different food species, 107 of them regularly. These are mostly seeds and fruits, although about 14 per cent of the food of this species consists of insects and spiders, taken chiefly during the summer months (Stoddard 1931). Restriction of feeding to a single or a few species is a specialized behavior. Feeding on a wide variety of substances or prey usually represents the more generalized primitive condition (Dethier 1954).

FOOD CHAINS

A single *food chain* should have at least three links to be complete: plant→herbivore→carnivore. Very often, however, a small carnivore or omnivore may be preyed on by a larger carnivore, and so on until four or five links are involved. Rarely are food chains longer than five links. An example of a three-link chain occurring on the North American Great Plains is: grass→pronghorn→coyote. A four-link chain common to deciduous forest communities might include tree foliage→leafhopper→vireo→hawk. A five-link chain would have to involve a number of small species, as bacteria→protozoan→rotifer→small fish→large fish. A food chain does not need to start with a living plant; consider, for instance: detritus→snail→shrew→owl. A strict predator need not necessarily be the last link, it could be an omnivore: flowers→bees→bear. Saprovores do not fit into food chain diagrams very well because they feed on all links of the chain. Food chains occur in all kinds of habitats and communities, even with the microorganisms of the soil: detritus→nematodes→mites→pseudoscorpion.

The feeding coactions between the many species that constitute the community are seldom as simple as the food chains just described. The rotifer feeds not only on protozoans but on bacteria. The small fish feeds on insect larvae and many other plankton species besides the rotifer. The shrew in the forest feeds not only on snails but on a variety of insects, and is fed upon in turn not just by owls but also by hawks, foxes, weasels, and others. If all these feeding relations between species in a community were diagrammed, a complicated web would be formed—the *food web* (Fig. 10-7).

BALANCE OF NATURE

Charles Darwin explained, a hundred years ago, that there was a balance in nature between the abundances of plants, herbivores, and carnivores. Were carnivores for some reason to increase unduly in numbers, they would soon exhaust their food supply and die of starvation. On the other hand, were food plants or herbivores to fluctuate excessively, then predators would vary in a similar manner. There is no doubt that marked variations in abundance in one link of a food chain will cause variations in the other links. Rodent plagues in a local area will bring an influx of foxes and hawks; a spruce budworm outbreak will result in an increase in the bird population; increases and decreases in soil bacteria are correlated inversely with decreases and increases in soil amoebae that feed upon them (Russell 1923).

As a case in point, the relation between the mule deer population and its predators on the Kaibab Plateau in northern Arizona is worth citing in detail (Rasmussen 1941). When this area was made a game preserve in 1906, killing of deer for sport was prohibited. During the period from 1906 to 1939, there were 816 mountain lions, 863 bobcats, 7388 coyotes, and 30 wolves trapped or shot, mostly by government hunters. The wolves were exterminated, and the other predators were markedly reduced. At the same time, there was a marked decrease in grazing by domestic sheep that would have favored an increase in the numbers of deer (Caughley 1970).

The deer increased from about 4000 in 1906 to an enormous herd variously estimated up to 100,000 in 1924. The deer's habits and the topography of the country prevented a scattering of animals to adjacent ranges. The overpopulations of deer consumed all new growth of young trees and browsed the foliage of mature trees as high as they could reach, yet the population could not secure enough food to keep it in good physical condition. The population far surpassed the carrying capacity of the range. In September 1923, it was estimated that 30,000 to 40,000 animals were on the verge of starvation and during the winters of 1924–25 and 1925–26, an estimated 60 per cent of the population died. Plainly the balance of nature was upset in this instance, with dire results. Hunting was again permitted in 1924, and the herd was reduced to about 10,000 by 1936. This history of the Kaibab deer herd is not unique; similar cases have been reported in many other areas (Leopold, Sowls, and Spencer 1947). Overpopulations in all areas frequently follow reductions in the number of predators, although other factors may also be involved.

When populations of species with differing food habits are in equilibrium, the surplus of prey species resulting from reproduction is destroyed and consumed by the predators. If predation does not destroy the total yearly surplus, the prey population increases in size; if predation takes more than the surplus, the population of the prey decreases in size. Even in entirely natural communities undisturbed by man, a strict balance of nature is probably never maintained for any appreciable period of time. It is characteristic for populations to vary in size, but these fluctuations tend to vary rather closely around a certain constant population mean. Depending on the length of its life cycle, a species population in an area may fluctuate from day to day, from season to season, or from year to year.

Many factors other than food coactions cause fluctuations in the abundance of animals, and predation is not always the most important (see Chapters 14 and 15). Close interdependency between populations of prey and predator species occurs most commonly when a prey species has only one or two important species preying upon it, while the predator species is largely restricted to that one species of prey (Pennington 1941). It is obvious, however, that an equilibrium of a sort exists between different species, and this is what is referred to by the concept of balance of nature.

TROPHIC LEVELS

In order to analyze the intricate coactions involved in the food web and balance of nature, it is desirable to simplify the relationships into nutritional or *trophic levels* (n) (Thienemann 1926, Lindeman 1942, Allee *et al.* 1949, Odum 1953). The lowest level (P) is composed of photosynthetic plants that are able to use solar energy for the manufacture of food, and certain types of bacteria that use either the free energy of unstable inorganic compounds or are activated by light to synthesize a limited amount of new organic matter. These are the *producers*. At the second level (C_1) come the herbivores, or *primary consumers*; at the third level, the smaller carnivores, or *secondary consumers* (C_2); and at the fourth level, the larger carnivores, or *tertiary consumers* (C_3). Occasionally there may be *quaternary consumers* (C_4). The terms "producer" for plants and "consumer" for animals were used, and the essential relationship understood, by Dumas in

1841. These two groups are also distinguished as *autotrophic* and *heterotrophic*, respectively.

The levels in subdivision of consumers are not sharply defined, as the feeding behavior of species often involves them simultaneously in several levels. Actually, the more remote an organism is from the initial source of energy (plants), the more likely it is that it will prey on two or more levels. This need not confuse the essential relationships involved (Lindeman 1942). Omnivores overlap between levels C_1 and any of the higher levels. Large saprovores, the heterotrophic bacteria, and fungi derive their nourishment from the excreta and dead bodies of organisms from all trophic levels. Since they are reducers or decomposers, they may for simplification be grouped with the autotrophic bacteria, and be called *decomposers* (Dec). Their total effect is to convert dead organic matter into nutrients that green plants can again absorb.

Figure 10-7 illustrates how a complicated food web may be simplified, somewhat arbitrarily, into trophic levels. Detritus, derived from the disintegration of dead organisms and excreta from organisms, is worked over by the bacterial decomposers, and the detritus and bacteria represent an independent base of the food web separate from the green plants.

A characteristic of trophic levels in most communities is that the nearer a level is to the source of energy, the greater the diversity of species involved. Thus in Fig. 10-7, the primary consumers (C_1) include some 12 taxa; the secondary consumers (C_2), 4; the tertiary consumers (C_3), 3; and the quaternary consumers (C_4), but 1. Species in the lower trophic levels have a higher rate of reproduction than those in the higher levels, to compensate for the greater predation. Evident also in Fig. 10-7, the higher the trophic level occupied by a predator, the more diverse is the choice of prey. This relationship has also been noted with predatory gastropods (Paine 1963).

PYRAMID OF NUMBERS AND BIOMASS

When the total animals in a community are grouped according to an arbitrary series of size ranges (Elton 1927, Allee *et al.* 1949), there are always a larger number of small individuals present than large ones. Plotting these data gives a pyramid of numbers (Fig. 10-8).

Pyramids of numbers arranged by trophic levels, rather than size, have special interest in respect to food coactions. In a bluegrass field, the number of green plants at the producer level (P) was over 8 times the number of herbivorous invertebrates in the level of primary consumers (C_1); the number of primary consumers was 2 times the number of spiders, ants, and predatory beetles at the secondary consumer level (C_2); and the number of secondary consumers was over 100,000 times the number of birds and moles

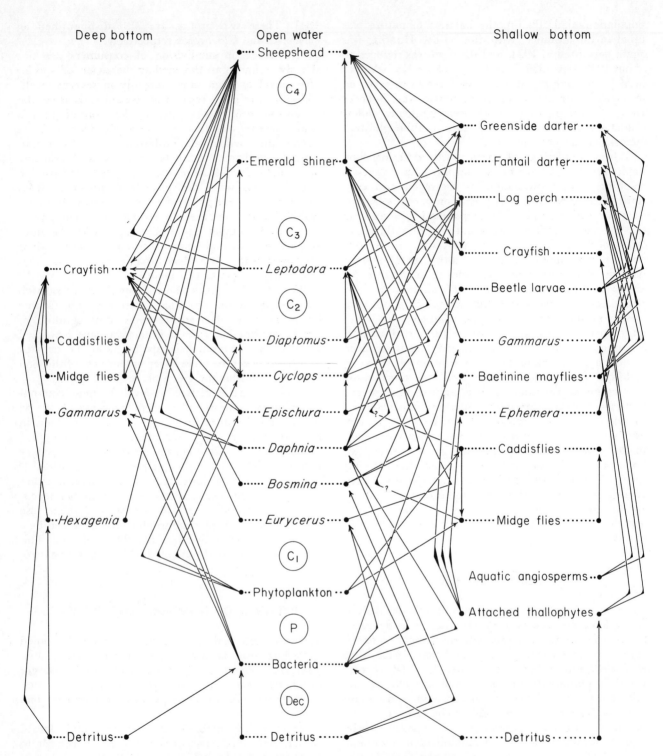

Fig. 10-7 Food web in western Lake Erie, leading to the sheepshead fish. Species are separated into their different trophic levels. The diagram would be even more complicated if other fish species in the C_4 trophic level of the community had been included. Although adult sheepshead are not preyed on by other animals, young during their first year serve as food for three or four other fish species (considerably modified from Daiber 1952).

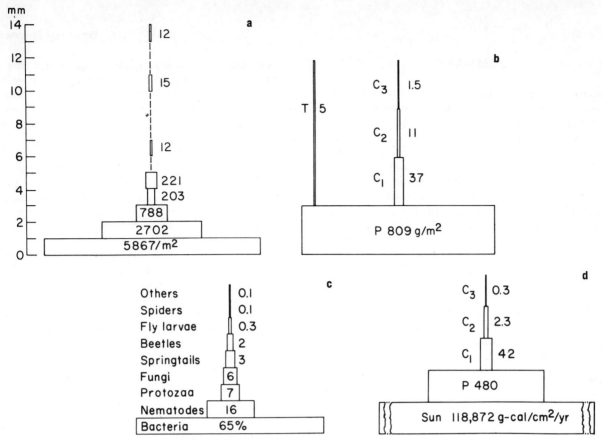

Fig. 10-8 Ecological pyramids of (a) numbers by size classes (soil animals of tropical broad-leaved forest, Williams 1941); (b) dry biomass, Silver Springs stream (Odum 1957a); (c) per cent total metabolism by different taxa of a hypothetical soil population in a meadow (Macfadyen 1957); (d) gross productivity, Lake Mendota (Lindeman 1942).

among the tertiary consumers (C_3), because of the great difference in size of individual animals at these two levels (Odum 1959). In Michigan, the ratio between numbers of rodents and their hawk and owl predators was found to be approximately 1000:1 (Craighead and Craighead 1956). A density of 100 deer is required to support one wolf (Pimlott 1967).

In evaluating the importance of trophic levels in the community, biomass is a more important index than number of individuals since it integrates size with numbers. Of 13 ratios of fresh weights between different trophic levels (Lindeman 1941, Juday 1942, Turček 1952, Birch and Clark 1953, Odum and Odum 1955), the majority fall between 2:1 and 12:1. These early studies furnish no certain evidence that the ratio varies between different trophic levels; this needs further investigation. Although these are not given as established limits, biomass ratios falling far outside this range will represent inadequate or incomplete sampling of populations, populations out of balance, or quite special situations. Thus the 23:1 ratio between P and C_1 trophic levels in Cedar Lake Bog in Minnesota (Lindeman 1941) is correlated with large annual accumulations of vegetation in a very old pond where winter stagnation under the ice kills off much of the animal life that might otherwise consume the plant material. The ratio of 22:1 in dry weight between P and C levels in Fig. 10-8b is due, in part, to a significant fraction of the produce level being exported downstream. In balanced communities, not only must the total biomass of plants ordinarily be larger than the biomass of herbivorous animals that feed upon them, but, in turn, the biomass of each consumer level must ordinarily be greater than the biomass of the succeeding level that feeds on it.

Metabolism is even more accurate than biomass, for it represents the rate at which energy is being utilized and work performed. Figure 10-8c evaluates a hypothetical soil population on this basis; we need actual measurements of this factor in every community.

Finally, a pyramid may be drawn to show the rate at which new organic matter is produced through reproduction and growth at each trophic level (Fig. 10-8d). The basic principle in this pyramid, fundamental in all types of pyramids, is that

the rate of production cannot be less and will almost certainly be greater than the rate of primary consumption, which in turn cannot be less and will almost certainly be greater than the rate of secondary consumption, etc. [Lindeman 1942: p. 408].

SUMMARY

In respect to feeding behavior, animals are herbivores, fungivores, carnivores (predators), omnivores, or saprovores, and have special adaptations for securing particular kinds of food. Chemical adaptation (unpalatability of a plant or prey species, adaptation to that unpalatability on the part of the consumer) is widespread. Various methods have been developed for analyzing the kinds of food that animals consume. Animals discriminate in their choice of food, depending on its nutritional values, digestibility, size, abundance and availability, and the protective devices that the food possesses. Relatively few species, however, are restricted to feeding upon a single species of plant or animal prey.

Concealing coloration is of various types: protective, disruptive, obliterative, and aggressive resemblance. Bright coloration may be warning of some disagreeable feature that the animal possesses, be mimicry of another species that possesses such features, or be deflective or directive.

Food chains commonly contain three to five links: a plant, an herbivore, and one or more carnivores. In each community the large number of food chains interconnect in various ways to form a food web. The relative number of organisms in the different links tends to be constant to give a balance of nature.

All organisms making up a link in the food web may be considered together as constituting a trophic level, of which there may be five: producer; and primary, secondary, tertiary, and quaternary consumers. Decomposers work over the excreta and dead organisms from all levels. In lower trophic levels, there is generally a greater variety of species, larger number of individuals, higher rates of reproduction, and individual animals are usually but not always smaller in size. A plot of the number of individuals in different size classes or the number of individuals, biomass, or productivity in different trophic levels takes the form of a pyramid. The relationships shown in these pyramids are fundamental to an understanding of community structure and dynamics.

Chapter 11

ENERGY EXCHANGES, PRODUCTIVITY, AND YIELD FOR MAN

The food coactions of organisms are important since animals can obtain energy only from consumed food. All activities of organisms constitute work and require energy. The amount of work performed depends on the amount of energy the organism can mobilize. Hence the flow of energy through the ecosystem and the manner and efficiency with which it is used is an ecological process of utmost importance. All of biology is dependent on the earth taking in energy as sunlight, using it, and then re-radiating it back to outer space (Goldman 1965, Petrusewicz 1967, Odum 1971).

ENERGY FLOW THROUGH THE ECOSYSTEM

Unlike nutrients, energy does not circulate indefinitely through the ecosystem. Energy is continuously and rapidly lost, although a certain amount of energy may pass through the ecosystem more than once before it is entirely dissipated. Hence energy must continuously enter into the ecosystem from the outside.

Acquisition

The basic source of energy for all trophic levels is solar radiation (Fig. 11-1). A surface exposed at right angles to the sun's rays, and outside the earth's atmosphere, would receive energy at the rate of 1.94 g-cal/cm^2/min. This is the solar constant. On the earth's surface, solar energy is effective only during the daylight hours and, because of absorption in passing through the atmosphere, scattering by smoke, dust particles, and cloudiness, only about 46 per cent of daylight radiation reaches the earth's surface (Fritz 1957), although this varies with latitude, season, and locality. At Columbus, Ohio, located at latitude 40°N, the average amount of solar energy is estimated as follows (Shaw 1953):

	kcal/m^2/day
Summer	6263
Spring and Autumn	3628
Winter	1604
Average	3781

The basic means by which solar energy is trapped is the formation of sugar by plants that contain chlorophyll (Rabinowitch 1945–46). In terms of moles (Brody 1945),
Photosynthesis:

$$6CO_2 + 6H_2O + 709 \text{ kcal} \longrightarrow 6O_2 + C_6H_{12}O_6$$

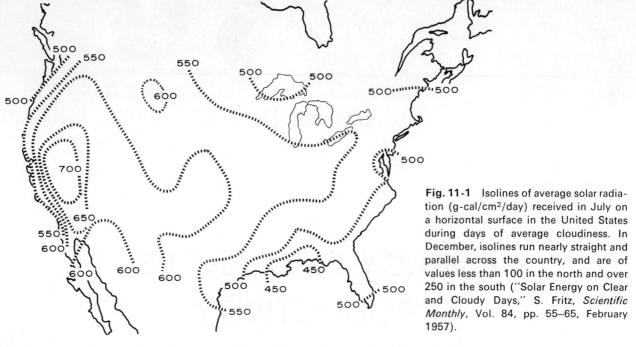

Fig. 11-1 Isolines of average solar radiation (g-cal/cm²/day) received in July on a horizontal surface in the United States during days of average cloudiness. In December, isolines run nearly straight and parallel across the country, and are of values less than 100 in the north and over 250 in the south ("Solar Energy on Clear and Cloudy Days," S. Fritz, *Scientific Monthly*, Vol. 84, pp. 55–65, February 1957).

Respiration:

$$C_6H_{12}O_6 + 6O_2 \longrightarrow 6CO_2 + 6H_2O$$
$$+ \; 674 \text{ kcal (free energy)} + 35 \text{ kcal (entropy)}$$

Respiration is going on at all times to furnish energy for the plant's activities, and this energy is derived from oxidation of the sugars formed in photosynthesis. Under ordinary daytime light intensities, the amount of sugar formed by photosynthesis greatly exceeds the amount oxidized in respiration. Photosynthesis ceases during darkness, but loss of sugar at night, because of respiration, continues. Sugars may be converted to starch or fat, or the carbon, hydrogen, oxygen in combination with nitrogen, sulfur, and phosphorus may form into proteins. The amount of sugar or new organic matter resulting from photosynthesis represents the *primary production* of the ecosystem. Accurate measurement of the rate of primary production is one of the most important problems of trophic ecology, for the activities of all plant and animal organisms in the community depend on the energy thus supplied. It is necessary to distinguish between *gross primary production,* the total amount of energy captured, and *net primary production,* the amount that remains after that used for respiration. Primary production is commonly expressed in terms of glucose or carbon, indirectly in the amount of oxygen released or carbon dioxide absorbed, or directly as calories of energy. *Production* during a period of time is *productivity,* and both production and productivity are commonly related to the *standing crop* or *biomass* (B) at a particular trophic level (n).

Use

Net energy at the producer level becomes available for use of animals when it is transferred to the higher trophic levels through *predation* (I_{n+1}), here considered also to include consumption of plants (Fig. 11-2) (Lindeman 1942, Macfadyen 1957). When predators consume their prey completely, as do fish feeding on plankton, there is no wastage. With many predators, however, for example wolves feeding on deer, the prey is eaten piecemeal, and much is not used. Predatory kill must therefore be separated into the *gross energy intake* (I) and the *energy wasted* (W).

Certain of the food taken in *secondary productivity* is indigestible or may simply be undigested, or if digested and absorbed is not completely metabolized in the tissues, so that it is eliminated in feces, urine, and other excreta. This is designated *excretory energy* (E). *Assimilated energy,* sometimes called *metabolized energy,* is the energy of the food actually absorbed and utilized ($I - E$).

A good portion of the energy ingested is used for existence, that is, for basal metabolism, temperature regulation, procurement and digestion of food, and other normal activities. There is almost continuous loss of heat energy from the body, and in homoiotherms this must be compensated for by increased heat production. Energy is used for the production of eggs and sperm, reproductive behavior, and other activities. Even the process of converting raw food into protoplasm is work and requires energy. In transfer of energy from one form to another or into work, there is always loss of free energy. No transfer is 100 per cent efficient except as it is degraded into heat. This is the second law of thermodynamics. The total energy that is utilized to perform work and to produce heat is called *respiratory energy* (R). Energy is continuously being lost from the ecosystem in this manner, hence needs to be replaced if the ecosystem is to continue to function.

Net production per unit time at the consumer levels is gross energy intake minus the losses of excreta and respiration ($I - E - R$). It is the actual production of animal tissue and involves growth.

Aside from being killed by predators, organisms die from a multitude of other factors, such as disease, extreme weather, starvation, combat, old age, and accident. In order for the populations of the different trophic levels to be maintained at a more or less constant level, organisms that die *non-predatory deaths* (*D*) must be replaced by the reproduction and growth of new individuals.

Energy lost from a trophic level through excreta, non-predatory deaths, and wastage from kills is used by saprovores or *decomposers* (*Dec*). This allows for the conversion of nitrogen and other compounds into forms suitable for reabsorption by plants, as shown by the nutrient cycles (pp. 144–145). Mixed in this level in some ecosystems are also the photosynthetic autotrophic bacteria, which use solar energy to manufacture a small amount of carbohydrates. Energy from the decomposer level recirculates into higher trophic levels when detritus mixed with the decomposing organisms are consumed. Note that the energy-flow diagram (Fig. 11-2) thus has two routes to the primary consumers, the relative importance of each varying between ecosystems.

Ecosystems are seldom entirely isolated one from another. Energy entering into an ecosystem by means other than photosynthesis is an *import* (*im*); energy

leaving an ecosystem other than by respiration is an *export* (*ex*).

When populations of different trophic levels are in balance, the total net production of each trophic level, after losses from non-predatory deaths and wastage from kills have been subtracted, is consumed by predators of the higher trophic levels. In unbalanced populations, predatory consumption may not equal the available net production, so that the population of that trophic level increases. If the predatory kill exceeds the available net production, the population decreases. A *change in the biomass* (*b*) of a population may, therefore, be either plus or minus.

Increase in biomass comes with the growth of individuals. When an individual organism grows, it increases in size and weight by adding organic matter. When reproduction takes place there is an increase in number of individuals, but not necessarily an increase in biomass, which takes place only if the offspring increase in size. Since individuals of most species have limits of growth increase, reproduction increases the potential productivity of a community by adding to the number of individuals capable of growth.

MEASUREMENT OF PRODUCTIVITY

Productivity may be measured during any reasonable period of time. Because of essential metabolic differences between day and nighttime, however, the 24-hour day is the smallest practicable unit. Similarly, because of seasonal changes in the environment and in community populations, the measurement of annual production is probably most useful. Since primary production is basic and concerns the capture of energy by plants, it will be considered first.

Primary Production

Various methods are employed for measuring primary productivity, each procedure having certain advantages and disadvantages (Ryther 1956, Mann 1969). Further work in evaluating and improving these methods or developing new ones is desirable.

A common procedure for analyzing aquatic habitats, dating back to Gaarder and Gran (1927), is to supend during daylight hours equal samples of green phytoplankton, ordinarily inseparably mixed with bacteria and zooplankton, in both transparent and blackened bottles at the same depth at which obtained (Fig. 11-3). Photosynthesis of course does not occur in the blackened bottle, and there is a *loss* of oxygen, resulting from respiration, *R*, and decomposition, $E + D + W$. In the transparent bottle, photosynthesis occurs in addition to respiration and decomposition, bringing a production of carbohydrates. There will either be an *increase* in oxygen concentration, or the loss of oxygen will not be

Fig. 11-2 Energy flow through an ecosystem the trophic levels of which are in balance with each other; symbols are explained in text. The widths of the arrows are intended to suggest the relative proportions of energy flow in the various directions, but this is schematic because the proportions vary widely in different ecosystems.

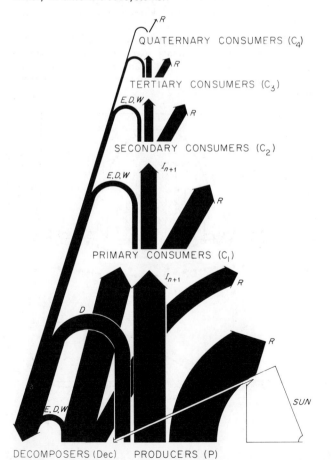

QUATERNARY CONSUMERS (C$_4$)

TERTIARY CONSUMERS (C$_3$)

SECONDARY CONSUMERS (C$_2$)

PRIMARY CONSUMERS (C$_1$)

DECOMPOSERS (Dec)   PRODUCERS (P)

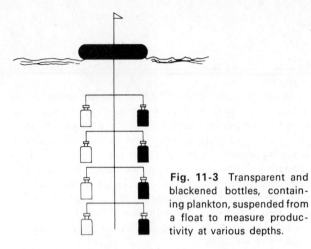

Fig. 11-3 Transparent and blackened bottles, containing plankton, suspended from a float to measure productivity at various depths.

so great as in the blackened bottle. The difference in the final oxygen content of the two bottles will be a measure of *gross production: I_n*.

If the oxygen content of the water is measured at the beginning of the experiment, the loss of oxygen in the blackened bottled subtracted from the difference in oxygen content of the two bottles at the end of the experiment will represent the net productivity. This net productivity may also be determined from the difference in the oxygen content of the transparent bottle between the beginning and the end. To obtain net production for an entire daily cycle, the consumption of oxygen for respiration and decomposition over 24 hours must be subtracted from the gross photosynthetic output during daylight hours.

One may use the amount of carbon dioxide absorbed during a period of time as a measure of photosynthesis, if correction is made for the carbon dioxide given off in respiration and decomposition. Changes in the amount of CO_2 in the water may be calculated from the differences in pH, the hydrogen-ion concentration (Beyers 1963). As an illustration of the use of radioisotopes, net production during daylight hours may be measured by introducing a known amount of $^{14}CO_2$ into a volume of water where the amount of carbon dioxide already present is known. The amount of ^{14}C absorbed by the phytoplankton can be accurately determined by use of counters applied to phytoplankton collected and dried at the end of the period. Then the proportion of the radioactive carbon absorbed to the amount introduced can be applied to the total CO_2 initially present to get the total amount absorbed (Nielsen 1964).

In fertile eutrophic lakes there is a continual sinking of dead organic material, derived chiefly from the plankton, from the epilimnion into the hypolimnion. The decomposition of this material absorbs oxygen from the hypolimnion and liberates CO_2 to produce a stagnation period during the summer months. The amount of oxygen deficit, or carbon dioxide increment, and the rate at which it forms can be measured to furnish a rough index of the lake's net productivity during the period between spring and autumn overturns (Hutchinson 1938, Ruttner 1953). Such estimates are in error by the import or export of organic material by streams, and they will vary in comparative useful-

ness depending on the volume ratio of hypolimnion to epilimnion (Hutchinson 1957).

Since nitrogen and phosphorus are metabolized more rapidly by plants in the manufacture of food during the growing season than they are regenerated from decomposing material, the rate and extent of the depletion of nitrates and phosphates in freely circulating bodies of water or in the epilimnion of stratified lakes serve as an index of the amount of organic matter produced (Hutchinson 1957). The rate of accumulation and regeneration of these substances in the hypolimnion from the dead organisms that sink into it may also be used to get an approximation of primary production (Waldichuk 1956). These measurements are not exact since they do not account for the repeated regeneration and reutilization of the substances in the photic zone during the season or their transference to and storage in the bodies of animals.

The rate of photosynthesis varies in relation to the amount of chlorophyll present and to light intensity. The amount of chlorophyll in the standing crop phytoplankton may be determined photometrically for all depths and calculated in terms of unit area of surface (Ryther and Yentsch 1957, Flemer 1969).

Determination of the rates of photosynthesis and oxygen use in streams and ponds may be made by direct measurement of changes in the concentration of oxygen and carbon dioxide in the water as between day and night (Fig. 11-4). Because of photosynthesis, there is a net increase in oxygen concentration during the daytime. At night the oxygen loss gives a measure of the rate of respiration and decomposition, and this

Fig. 11-4 Photosynthesis, community respiration, and diffusion in an English chalk stream: (a) curves; (b) cumulative values (Edwards and Owens 1962; Crown copyrighted, reproduced by permission of the Controller of H. M. Stationery Office).

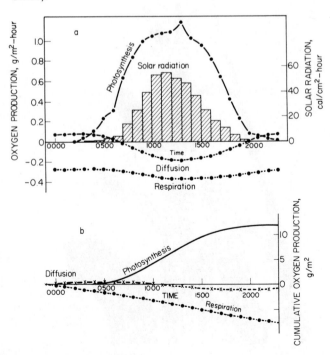

presumably remains the same throughout the 24-hour daily cycle. Adding the average hourly night loss to the average hourly gain during the day and multiplying by the hours of daylight gives the total gross production for the 24-hour day. To obtain the net production for the entire day, the hourly loss at night must be multiplied by 24 and subtracted from the total gross production. Corrections need to be made, however, for possible diffusion of oxygen from the air into the water, particularly at night when oxygen concentration in the water is lowered, and diffusion out of the water during the day, if supersaturation occurs. Additional corrections will also be necessary for import of oxygen from groundwater and surface drainage and export of oxygen and carbon dioxide downstream by swift currents (Odum 1956, 1957, Beyers 1963).

On land, the annual net production of herbaceous plants that grow from seed or underground parts is approximately equivalent to the biomass of the vegetation at its maximum stage of growth, provided no appreciable amount has been lost or consumed by animals (Fig. 11-5). Exclosures may be erected to prevent grazing of above-ground vegetation by larger animals. The measurement of total production is inexact when the maximum standing crop occurs before the terminus of the growing season. The annual crop divided by the length of the growing season gives net productivity in terms of average daily increment. Daily productivity varies, however, with the stage of growth. At one site, the cumulative productivity of common cattail, from March 29 to July 2, when the largest crop was reached, averaged 8.17 g C/m²/day, but for a short period of maximum growth, May 4 to 28, was 23.48 (Penfound 1956). Seasonal biomass production of grasses is increased by a moderate amount of grazing, so in measuring primary productivity under natural conditions, the stimulating or inhibiting effect of animal consumption should be given proper evaluation.

The annual woody increment of trees and shrubs is proportional to the increase in diameter or width of growth rings. Mature trees may be felled, unit samples of branches, trunk, and roots dried and weighed, and the annual woody production since germination calculated (Ovington 1957). Logarithmic regressions may be calculated for felled trees, relating their biomass and production to diameter at breast height or other variables. These regressions can then be applied to the trees in a study plot of forest to estimate its biomass and net primary productivity (Woodwell and Whittaker 1968, Whittaker and Woodwell 1969). Standing live trees may also be climbed and appropriate measurements taken to calculate annual productivity (Reiners 1972). Attempts have been made to measure respiratory losses (Moller *et al.* 1954), the annual production of foliage, seeds, acorns, and nuts (Downs and McQuilken 1944), and the total uptake of carbon dioxide (Botkin *et al.* 1970).

Secondary Production

When an animal species is represented by a low overwintering population, or an immature stage, the maximum biomass obtained in each generation is the approximate net production for that generation (Fig. 11-6). However, this does not account for continued reproduction and growth of individuals after the maximum biomass of the population is attained, nor does it account for excreta, natural deaths, or the kill of predators. If the population of the species is maintained at a more or less uniform level throughout the year, the mean biomass times the number of generations gives the net production, again with the exception of the factors mentioned above. Lindeman (1941) considered the phytoplankton turnover, or the production of a new generation, to occur every week from May through September and every 2 weeks through the rest of the year, the zooplankton to replace itself bi-weekly through the year, *Chaoborus* to have three generations per year; midge flies, two; and various aquatic beetles and bugs, one generation per year. Juday (1940) estimated that the mean standing crop of both phytoplankton and zooplankton replaced itself every 2 weeks throughout the year. To obtain gross productivity, the respiratory rates of these animals must also be measured.

Although measurement of energy flow through the total phytoplankton, zooplankton, and smaller soil organisms is often practical and sufficient, measurements are required on individual species of larger size in the higher trophic levels before the total utilization of energy by the trophic level can be determined (Southwood 1966). These measurements should be

Fig. 11-5 Hypothetical growth and metabolism curves for an annual plant (Westlake 1965; originally published by the University of California Press, reprinted by permission of the Regents of the University of California).

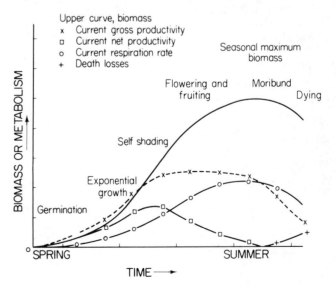

based on careful field observations and experiments of food requirements, reproductive rates, growth rates, mortality, and so on. The energy intake and requirements of individual species can often be measured under experimental conditions by presenting a known amount of food to one or several individuals and determining the amount consumed during a period of time (Fig. 11-7). This is preferable to measuring the oxygen intake of resting animals. The influence of various environmental factors, size and age of the animals, density of populations, and so on, on the food consumption can be ascertained and often the proportion utilized for existence, growth, and other activities (ground animals, Bornebusch 1930; *Tubifex*, Ivlev 1939; grasshoppers, Smalley 1960; rotifers, Edmondson 1946; *Daphnia*, Richman 1958; insects and spiders, Van Hook *et al.* 1970; fish, Gerking 1954, Mann 1965; birds, Kendeigh 1970 (Fig. 11-7); mammals, Grodzinski and Gorecki 1967, Trojan 1970).

Mortality from non-predatory causes (D) has rarely been measured. Combined rates of mortality from predatory and non-predatory causes may be computed from life-table data (p. 204) or from recaptures of marked animals. If one knows the population of predators in an area and their average food requirements, non-predatory deaths may be calculated as the difference.

Radionuclides (radioisotopes) of common elements, ^{45}Ca, ^{14}C, ^{60}Co, ^{64}Cu, ^{131}I, ^{32}P, ^{3}H, and several others, are useful as a tool for tracing mineral and energy transfer between populations or trophic levels or their use by individuals. A known amount is introduced into the ecosystem and the rate at which it passes through the food chain is measured by different kinds of detectors or counters—Geiger tubes, gamma crystals, beta crystals. Or a known amount may be provided an individual and the speed at which it is excreted used as a measure of the rate of metabolism. These procedures

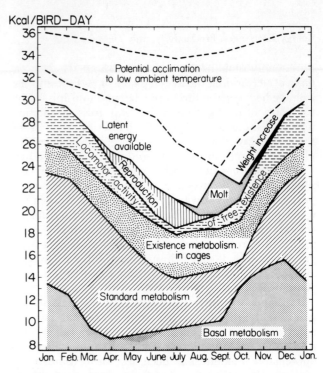

Fig. 11-7 Monthly energy budget of the house sparrow throughout the year. The birds feed principally on mixed grain (modified from Kendeigh 1973).

are potentially of great importance in ecosystem research but must be used with caution because of technical difficulties and possible hazards (Odum and Golley 1963, Reichle 1963, Odum 1971).

The International Biological Program sponsored a series of handbooks describing methods for measuring both primary and secondary productivity of various ecosystems (F. A. Davis, Philadelphia). See also Petrusewicz (1967).

Energy-flow Equation

These many factors may be brought together in the following equation to show the flow of energy through any trophic level of consumers:

$$I_n + im - ex = E + R + D + I_{n+1} + W + b$$

One can make a number of derivatives from this equation. Net production (P), for instance, includes more than just an increase in biomass:

$$P = b + D + I_{n+1} + W$$

The equation for predation, if we disregard imports and exports, is

$$I_{n+1} = I_n - E - R - D - W - b$$

To measure each factor in these equations, it is desirable but not always practical first to determine the energy requirements of the individual animal, then to multiply these requirements by the population of

Fig. 11-6 Cumulative energy budget (without egestion) of a beetle, *Tribolium castaneum*, maintained at 29°C, 70 per cent relative humidity, and consuming flour (Klekowski *et al.* 1967).

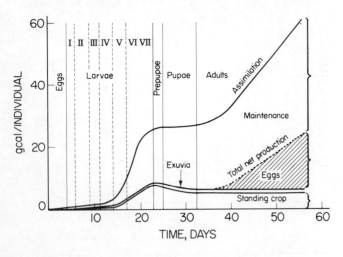

individuals making up the species, and finally to combine the quantities for all the species. When in steady state ($b = 0$), the energy flow through the ecosystem as a whole may be measured either by the amount of food photosynthesized by the producers or by the total respiration of all trophic levels. The exchange of oxygen and carbon dioxide is in balance. This can be demonstrated in balanced aquaria or experimental microcosms (Beyers 1964). If the ecosystem is not in steady state, corrections must be made for the change in biomass.

ENERGY-FLOW VALUES

The *photosynthetic efficiency* of plants in the use of solar radiation under natural conditions is very low. Net primary productivity in terms of visible radiation is generally around 0.4 per cent on land and 0.2 per cent in the ocean (Vallentyne 1965). Even in agriculture, for instance with rice production in Japan and wheat in Denmark, efficiency is only 2.0 to 2.5 per cent. Photosynthesis increases with light intensity but soon reaches a maximum because of the low carbon dioxide content of the air, lack of nutrients or water in the substratum, or unfavorable temperature. An efficiency of 4 per cent has been obtained in culture with an artificially increased concentration of CO_2 (Bonner 1962). Only certain wavelengths of solar radiation are effective in photosynthesis. Likewise, the growing season of plants is limited to only a portion of the year. Much radiation is reflected back from ground and water surfaces or is absorbed by the ground or water or other non-living material and later radiated back into the atmosphere (Gates 1962).

The utilization of solar radiation received at Lake Mendota, Wisconsin, has been apportioned in per cent as follows (Juday 1940), last two figures corrected:

Melting of winter ice	2.9
Annual heat budget of water	20.4
Annual heat budget of bottom	1.7
Evaporation	24.7
Reflection	24.0
Conduction, convection, radiation	25.9
Gross biological intake	0.4

In both fresh water (Verduin 1957) and the sea (Shimada 1958), maximum rates of photosynthesis occur in the early morning (Fig. 11-8). Monthly variations show a relationship between photosynthesis and solar radiation, although it is observed that maximum photosynthesis occurs in June rather than in July, for reasons that are not clear (Fig. 11-9).

The extent to which solar radiation is captured in *primary productivity* depends considerably on the luxuriance of the vegetation (Table 11-1). According to Whittaker (1970), desert scrub and tundra have net

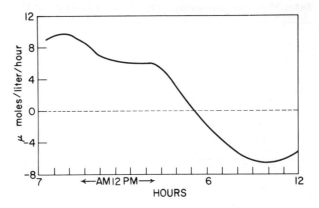

Fig. 11-8 Rate of carbon dioxide removal, photosynthesis, at different times of day in western Lake Erie. Negative values at night mean that carbon dioxide is being added rather than removed (after Verduin 1957).

Fig. 11-9 Rate of carbon dioxide removal, photosynthesis, in different months of the year in western Lake Erie correlated with intensity of solar radiation (after Verduin 1956a).

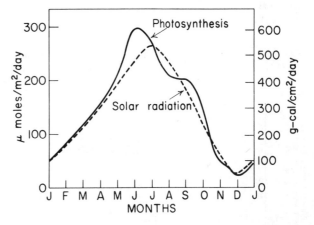

dry weight productivity of less than 150 g/m²-yr; temperate grassland, 500; woodland, shrubs, savannah, and agricultural land, 600–700; boreal forest, 800; temperate deciduous forest, 1300; and tropical forest, 2000. In aquatic habitats, net productivity is for the open ocean, 125 g/m²-yr; the continental shelf, 350; lakes and streams, 500; swamps, marshes, attached algae, and estuaries, 2000.

Annual production in marshes and grassland commonly equals the above ground biomass of vegetation, since this growth is renewed each year (Fig. 11-10). Annual productivity of woody perennial vegetation, however, may be only a small percentage of the standing biomass. Instead, of all the annual production becoming litter at the end of the year, as in marshes and grassland, only about 50 to 60 per cent enters the detritus pathway in forest ecosystems (Reiners 1972).

Some intensive studies of productivity have been carried out with *phytoplankton* in marine waters (Riley *et al.* 1949, Riley 1952, Deevey 1952). It is calculated that the total annual fixation of carbon by photosynthe-

Table 11-1   Net primary productivity for the world about 1950 (from Lieth 1972).

Vegetation unit		Size of unit (10⁶ km²)	Annual energy fixation (10⁶ g-cal/m²)	Annual energy fixation (10¹⁸ g-cal/area)
Forest		50	5.5	277.0
Woodland		7	2.8	19.6
Dwarf and open scrub		26	0.4	10.2
Grassland		24	2.5	60.0
Extreme desert		24	$+$	0.1
Cultivated land		14	2.7	37.8
Swamp and marsh		2	8.4	16.8
Lake and stream		2	2.3	4.6
	Total continental	149	2.9	426.1
Reefs and estuaries		2	9.0	18.0
Continental shelf		27	1.6	42.6
Open ocean		332	0.6	199.2
Upwelling zones		0.4	2.5	1.0
	Total oceanic	361	0.7	261
	Total earth	510	1.3	687

Fig. 11-10   Semidiagrammatic profile of natural ecosystems in Minnesota, together with estimates of their above-ground biomass and annual net production (Reiners 1972).

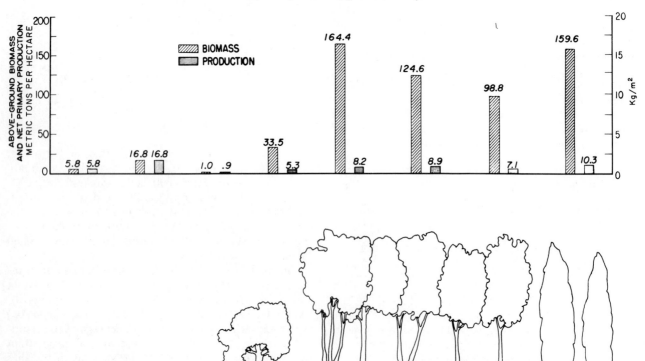

sis in Long Island Sound is about 470 g/m². Over half of this amount (265 mg) is used in the *respiration* (R) of phytoplankton (56 per cent). Of the 205 g/m² net production, 26 per cent appears to be used by the macrozooplankton, 43 per cent by the microzooplankton and bacteria, and 31 per cent by the benthic fauna and flora (Riley 1956). In the sea off Plymouth, England, the zooplankton is required to assimilate daily an equivalent of 4 per cent of its dry weight in vegetable matter just to meet respiratory needs, and 7-10 per cent for growth and to offset the amount consumed by other animals. Thus, of the energy intake at this level, approximately 30 per cent was used for respiration and 70 per cent for growth. On the other hand, pelagic fish, at a higher trophic level, used 90 per cent of assimilated food for respiration and converted only 10 per cent into body tissue (Harvey 1950).

In a Montana reservoir, of the total gross production of the phytoplankton during the summer months, 17 per cent was dissipated in respiration, 4.5 per cent was converted into increase of phytoplankton, 71 per cent was consumed by macrozooplankton, and 7.6 per cent utilized by bacteria, microzooplankton, and bottom fauna. The energy intake of the macrozooplankton was divided 90 per cent for respiration and 10 per cent for population increase (Wright 1958).

From measurements made at Lake Beloe, in the Soviet Union, *non-predatory deaths* (D) of *Tendipes plumosus* and other *bottom fauna* is reported to be twice predation (Ricker 1946). In a Massachusetts spring-pool, however, non-predatory deaths of another species of midge and of planarians amounted to one-fourth predation (Teal 1957). In a small Indiana lake, non-predatory deaths of bluegills amounted to 64 per cent of the average protein content of the standing crop during the year, but only 29 per cent of the total turnover of protein (Gerking 1954).

One does not ordinarily expect much if any *wastage* (W) with fish feeding on plankton or in consumption of plants by herbivores generally. Herbivores ordinarily consume only parts of the plant system, the remainder of which proceeds to regenerate. This differs from most predator-prey coactions, where the whole animal is eaten, or at least killed by the predator. However, in Poland, it was estimated that destruction of short-leaved grasses by grasshoppers amounted to 6-10 times more than they consumed, and destruction of taller sedges was 15-25 times greater (Andrzejewska and Wojcik 1970). Likewise, lemmings in the Arctic tundra consume only a portion of the vegetation that they destroy. Wastage in predatory kills at the consumer trophic levels are often considerable. Adult lions in Africa kill, on an average, 20 kg (44 lb) of food per day; they require only about half of this amount for existence. The unused portion of the kill supports a large population of bird, mammal, and invertebrate scavengers (Wright 1960). Harvestmen are predaceous on insects. Males waste nearly three-fourths of their prey, females about one-third. Of the food ingested, about 46 per cent of the energy is assimilated; the remainder is eliminated in the feces (Phillipson 1960).

Predation (I_{n+1}) by bluegills on the bottom fauna, chiefly midge fly larvae, was found in the same Indiana lake mentioned above to be more than represented by the standing crop or which emerged from larvae into adults. This implies that the rate of production was substantial. On a yearly basis, the production of fish food was at least 10 times the net production of the bluegills. Growth of the fish, however, was limited to the 5 summer months and that of the bottom fauna and entomostraca, on which the bluegills also fed, extended throughout the year (Gerking 1962). In an artificial outdoor pond, using a radiophosphorus tracer for measuring the energy flow, the ratio of biomass from the producer through two secondary consumer levels was 24:1.7:1. This corresponds to a ratio of productivity between the three levels of 193:23:1 (Whittaker 1961).

Table 11-2 Per cent of available net production of the producer taken by herbivores in the primary consumer level.

Ecosystem	Consumer	Per cent consumed	Authority
Salt marsh	Invertebrates	7	Teal 1962
Marsh grass only	Grasshoppers	<1	Smalley 1960
Sedge meadow	Grasshoppers	8	Andrzejewska and Wojcik 1970
Wet meadow	Grasshoppers	10	Andrzejewska and Wojcik 1970
Field	Insects	<13	Odum *et al.* 1962, Golley and Gentry 1964
Field	Insects	10	Van Hook 1971
Field	Voles	0.5-1.3	Trojan 1970
Desert scrub	Rodents, lagomorphs	<2	Chew and Chew 1970
Pine forest	Rodents	0.6-19	Ryszkowski 1970
Oak forest	Insects	11	Bray 1964
Maple-beech forest	Insects	6	Bray 1964
Beech forest	Rodents	4	Grodzinski *et al.* 1970

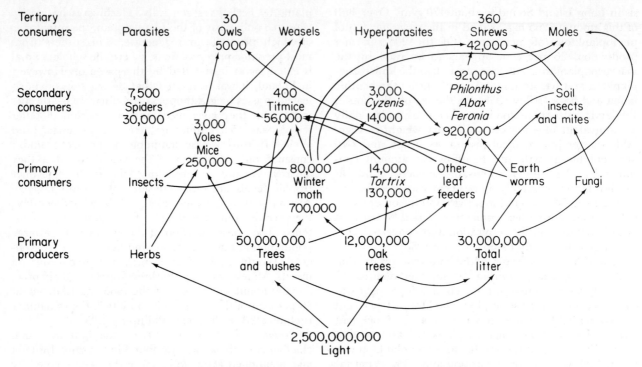

Fig. 11-11 Simplified energy-flow diagram for Wytham deciduous forest, near Oxford, England. Estimates of consumption and production in kcal/hectare-yr are inserted above and below each link for which figures are available (Varley 1970).

In the aquatic ecosystems described above, the standing crop biomass of the producers is relatively low, and high percentages of the annual net production are consumed by the primary consumers. However, in many semiaquatic and terrestrial ecosystems where the standing crop biomass is relatively high, the consumption by primary consumers takes a relatively low percentage of the annual net production (Table 11-2, Fig. 11-11). The percentage that is not consumed accumulates as litter. This is evident in the extensive leaf fall from deciduous trees and the stands of dead grass and forbs in the prairie and marsh in the autumn. This litter then becomes food for the decomposers which are abundant in and on the ground. The amount of litter production, mostly leaves, in proportion to production of woody tissue in the trunks and large roots of trees and in the rhizomes of perennial herbs and grasses, tends to increase toward the tropics, where there is greater light energy available during the growing season, and toward hot deserts, where lack of moisture becomes a problem (Jordan 1971). Because of more rapid decomposition with higher temperature, however, accumulated litter mass tends to decrease from cold-temperate forests to the tropics.

Three explanations have been suggested for the lack of complete utilization of the net production of the producers by the primary consumer trophic level: (a) the abundance of herbivores is not regulated by the abundance of food but is held at relatively low densities by predation from the higher consumer levels (Table 11-3) (Hairston *et al.* 1960, Slobodkin *et al.* 1967), (b) lack of ability of the herbivores to increase the size of their population during the year commensurate with the increase in producer biomass, and (c) progressive change in the plant tissue as the season advances so that it becomes less edible and attractive to the consumers (Odum 1964). The first two points

Table 11-3 Relation of spider predators (*Araneaus quadratus*) to consumption by grasshoppers of the primary net production of a grassy meadow. A uniform number of grasshoppers was confined in different cages in each of which a different number of spiders was introduced (Andrzejewska *et al.* 1967).

Number of spiders	Number of grasshoppers remaining	Consumption of grasses (g dry weight)
0	75	73
6	55	56
8	40	43
11	34	37
13	22	24
14	24	27
17	24	27

will be considered in Chapters 14 and 15; the third point warrants further attention here.

Plants have evolved various defenses against consumption by herbivores, such as spines and thorns, tissue toughness, nutritional deficiencies, biochemical inhibitors (Beck 1965). For instance, adult fecundity and larval survival of the moth, *Hyphantria cunea*, feeding on apple tree leaves decreases as the season progresses, owing to increasing nutritional stress (Morris 1967). Other species of moth caterpillars concentrate their feeding on oak leaves to the spring when leaf proteins are highest and before the leaf tannins render the tissues tough and less palatable (Feeny 1970). As a result of genetic changes and selective pressure from herbivores, plant species may develop resistant strains to insect attacks, for instance wheat to the Hessian fly, alfalfa to the pea aphid, and sorghum to the chinch bug (Pimentel 1968).

The assimilation ratio of small soil or ground animals feeding on detritus is often quite low: about 15 per cent in the millipede, *Narceus americanus* (O'Neill 1968), 33 per cent in the woodlouse, *Tracheoniscus rathkei* (White 1968), 20 per cent in oribatid mites (Engelmann 1961), and 10.6 per cent in the aquatic stonefly naiad, *Pteronarcys scotti* (McDiffett 1970). These ratios are low because of the large amount of cellulose in the diet, requiring special enzymes to digest. Some wood-eating

arthropods, such as termites and wood roaches, possess flagellate protozoans in their guts that change the cellulose into sugar, which is then used both by the host and the symbiont (p. 192).

The nourishment that some animals obtain from the litter is largely from the fungi, actinomycetes, and the bacteria that the litter contains rather than the litter itself. After the maceration that the litter attains in passing through the digestive tract of the animals, its decomposition by the soil bacteria and fungi is rendered more effective. Perhaps the most important function of these small soil animals in the dynamics of the ecosystem is this production of excreta and acceleration of decomposition rather than providing biomass for the feeding of higher trophic levels (Edwards *et al.* 1970).

In some ecosystems, detritus is the chief source of energy for the higher energy levels (Fig. 11-12). Mudflats on the California coast contain at least 39 g/m^3 dry weight of bacteria. Assuming that this biomass increases only 10 times per day, there would be 390 g/m^3 produced per day available for nourishment of the animal population (ZoBell and Feltman 1942). Ingestion rates by detritus feeders are often very high compared with ingestion rates of herbivores (Table 11-4).

If the annual accumulation of detritus is not completely decomposed, consumed, and recirculated

Fig. 11-12 Litter consumption by arthropods in a tuliptree deciduous forest community in Tennessee, estimated from radiocesium turnover rates. The average concentration of ^{137}Cs in the litter is 25 pCi/mg; consumption is given as pCi/day and mg/day (= I). Equilibrium refers to the amount of ^{137}Cs maintained in the body; T_b (days) is the biological half-life or number of days to eliminate 50 per cent of a unit input of ^{137}Cs from the body, and k (day^{-1}) is the rate of outgo per day in the excreta (= E) (from Reichle and Crossley 1965).

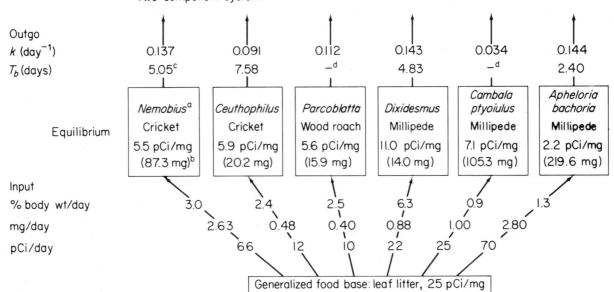

$^a T_b$ Approximated using *Acheta domesticus*

^{b} Mean dry weight per individual

^{c} Corrected to an environmental temperature of 16°C

^{d} Two-component system

through the ecosystem, there is an accumulation in the soil. The rate at which this unused net production is added to the substratum may be higher in favorable than in unfavorable habitats. In an Arctic lake, subject to freezing for a large part of the year, the accumulation of organic bottom deposits has averaged less than 20 g/m²-yr during the 6000 years since recession of the Pleistocene glacier, compared with 80 g/m²-yr in a Connecticut lake during a comparable period of time (Livingston *et al.* 1958).

Succession

It is characteristic for the total annual net production of seral ecosystems to be greater than can be utilized. Fixation of nitrogen commonly exceeds denitrification, there is absorption of minerals from underlying rocks, and in ponds and lakes there is import of nutrients from the surrounding drainage basin. With increase in carrying capacity of the habitat, there is a progressive increase in the standing crop biomass. Succession occurs when species formerly unable to become established because of low energy or mineral resources or other intolerable situations now find conditions favorable. Their invasion forces those resident species that cannot withstand competition to disappear. Oligotrophic lakes become eutrophic and then marsh. Bog mats change to bog forests and eventually to the climax (Lindeman 1942). Sand and rock habitats become covered with dense forests.

In succession, the ratio of primary production to biomass is at first high but decreases as the climax approaches (Fig. 11-13). The ratio of biomass accumulation to annual net productivity increases from

close to 1:1 in an annual herb stage to 2–5:1 in perennial herb and low shrub stages, to 5–10:1 in high shrub and young forest stages, to 30–50:1 in very mature forests. It is not net productivity but the biomass standing crop in relation to productivity, that reaches a maximum in the climax of most terrestrial successions. Plankton communities in stabilized aquatic ecosystems in contrast may have biomass productivity ratios as low as 0.2–0.5:1 (Whittaker and Woodwell 1972).

In early stages, food chains are based principally on the photosynthesizing plants, but in later stages food chains based on the detritus become more important (Margalef 1968, Odum 1969). In the climax, all trophic levels tend to become balanced with one another; all the net production of one level is consumed by other levels, and total annual use equals total annual production. Such a stable relationship occurred some years ago in western Lake Erie, when measurements of CO_2 exchange between plants and animals indicated that photosynthesis of the producers and respiration of the total aquatic community were approximately equal (Verduin 1956). Likewise, in a coral reef community in the Pacific Ocean, a balance in the ecosystem was demonstrated with a level of energy exchange between producers and consumers of approximately 96 kcal/m²-day (Odum and Odum 1955).

EFFICIENCY RATIOS

Studies are accumulating where measurements have been simultaneously obtained for several factors in the energy-flow equation. To obtain an evaluation of the comparative efficiency of energy flow through populations, trophic levels, and ecosystems, the ratio of one parameter to another is useful. Ratios commonly used are the following (Table 11-4), which are usually expressed as percentages (Wiegert 1964, Kozlovsky 1969, Van Hook 1971):

GROSS INTAKE ("ecological efficiency"): I_{n+1}/I_n = percentage of the energy intake at one trophic level (I_n) passed on to the next (I_{n+1}). The percentage is lower in endotherms, because of their high respiratory rate, than in ectotherms (Turner 1970).

ASSIMILATION: $(I - E)/I$ = percentage of ingested energy (I) that is assimilated ($I - E$) in the same trophic level. These percentages are generally higher for carnivores than herbivores.

RESPIRATION: $R/(I - E)$ = percentage of energy lost in all activities (R), as measured by total oxygen absorbed, carbon dioxide eliminated, or food metabolized to energy assimilated ($I - E$). These percentages are higher in endotherms than in ectotherms.

NET PRODUCTION: $(I - E - R)/I$ = percentage of new tissue formed ($I - E - R$) to total food ingested (I). This varies widely between species. The ratio of net

Fig. 11-13 Plant production and biomass in forest succession following fire on Long Island, New York. Production (A) increased rapidly through herb and shrub stages to oak-pine forest, where it became stabilized. Biomass (B) increased steadily throughout the period but appeared to be approaching stable level near 40 kg/m² after 200 years (Whittaker 1970).

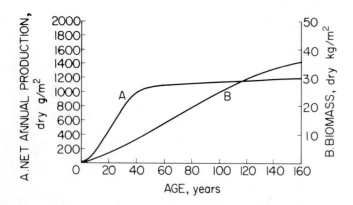

Table 11-4 Efficiency percentages for various populations.

Species	Habitat	Intake (kcal/m²·yr)	Ecological, I_{n+1}/I_n	Assimilation, $(I-E)/I$	Respiration, $R/(I-E)$	Net production, $(I-E-R)/I$	Authority
Detritus-feeders (including contained micro-organisms):							
Annelids, *Tubifex*, *Limnodrilus*	Lake sublittoral and profundal zones	236	—	53	40	31	Kitazawa 1959
Bivalve, *Scrobicularia plana*	Tidal mud-flat	554	2.4	61	79	13	Hughes 1970
Snail, *Littorina irrorata*	Salt marsh	645	—	45	86	6.3	Odum and Smalley 1959
Soil mites, *Oribatei*	Grassy field	10	4.2	20	83[a]	4.2	Engelmann 1961
Detritus- and plant-feeder:							
Cricket, *Pteronemobius fasciatus*	Grassy field	22	8.8	41	68	13	Van Hook 1971
Herbivores:							
Grasshopper, *Orchelimium fidicinium*	Salt marsh	84	—	33	63	12	Odum and Smalley 1959
Grasshoppers, *Melanoplus*, *Conocephalus*	Grassy field	113	4.3	54	52	26	Van Hook 1971
Butterfly larva, *Pieris rapae*	Gardens	83	—	63	69	16	Kitazawa 1959
Spittlebug, *Philaenus spumarius*	Grassy field	1.6	0.8–17	58	92	4.8	Wiegert 1964
Mirid bug, *Leptopterna dolabrata*	Grassy field	0.6	—	31	43	17	McNeill 1971
Meadow vole, *Microtus pennsylvanicus*	Grassy field	25	—	70	97	2.4	Golley 1960
Bank vole, *Clethrionomys glareolus*	Beech forest	4.1	—	83	97	2.2	Grodzinski *et al.* 1970
Yellow-necked field mouse, *Apodemus flavicollis*	Beech forest	3.2	—	86	97	2.3	Grodzinski *et al.* 1970
Insectivore:							
Spider, *Lycosa*	Grassy field	6.8	—	91	63	34	Van Hook 1971
Carnivore (feeding on *Microtus*):							
Weasel, *Mustela rixosa*	Grassy field	0.5	—	96	98	2.2	Golley 1960

[a]Engelmann gives this percentage as 96 but indicates a possible error of 13 per cent.

Table 11-5 Comparison of energy-flow values (kcal/m²-yr) in five aquatic ecosystems (modified from Kozlovsky 1968 and based on original data from Juday 1940, Lindeman 1942, Teal 1957, 1962, and H. T. Odum 1957).

Parameter	Lake Mendota	Cedar Bog Lake	Silver Springs	Root Spring	Salt Marsh
Producers					
Insolation	1,188,720	1,188,720	1,700,000	1,095,000	600,000
Assimilation, A	5,017	1,200	20,810	710	36,380
Respiration, R	1,253	300	11,977	55	28,175
Net production, $A - R$	3,764	900	8,833	655	8,205
Predation, I_{n+1}	554	196	—	—	—
Non-predatory deaths, D	3,210	704	—	—	—
Change in biomass, b	—	—	0	0	—
Export, ex	—	—	—	—	3,548
Primary consumers					
Gross intake, I	554	196	—	—	—
Import, im	0	0	486	2,368	—
Assimilation, $I - E$	443	168	3,368	2,318	767
Respiration, R	169	64	1,890	1,746	596
Net production, $I - E - R$	274	104	1,478	572	171
Predation, I_{n+1}	34	34	—	—	—
Non-predatory deaths, D	240	70	—	—	—
Change in biomass, b	—	—	0	−4	—
Export, ex	—	—	—	31	112
Secondary consumers					
Gross intake, I	34	34	—	—	—
Assimilation, $I - E$	31	31	383	208	59
Respiration, R	18	18	316	89	48
Net production, $I - E - R$	13	13	67	119	11
Predation, I_{n+1}	3	0	—	0	—
Non-predatory deaths, D	10	13	—	121	—
Change in biomass, b	—	—	0	−4	—
Export, ex	—	—	—	2	11
Tertiary consumers					
Gross intake, I	3		—		
Assimilation, $I - E$	3		21		
Respiration, R	2		13		
Net production, $I - E - R$	1		8		
Predation, I_{n+1}	0		0		
Non-predatory deaths, D	1		8		
Change in biomass, b	0		0		

production to assimilated energy $(I - E - R)/(I - E)$ varies inversely with the respiration ratio.

The ultimate goal is determination of the rate and efficiency at which energy is used in each trophic level, and how it flows from one trophic level to another through the ecosystem. Since each trophic level consists of many species populations, calculation of the total flow is difficult. Although some of the early attempts to do so have been criticized (Slobodkin 1963), they are, at least, indicative of the general magnitude and relationships (Tables 11-5 and 11-6). Ecological efficiency shows no consistent trend between trophic levels. The higher assimilation ratios found in the higher trophic levels are offset by higher respiration ratios and lower ratios of net production.

YIELD

The proportion of a population or trophic level that is removed by another population or trophic level is its *yield*. It is the amount taken through predation. In such applied fields of ecology as agriculture, forestry, and wildlife management, man attempts to harvest the available net production of food, timber, or game animals for human benefit rather than to let it be used by other organisms or to accumulate in the habitat.

The potential yield decreases quantitatively at progressively higher trophic levels, since excretory and respiratory losses and non-predatory deaths bring an accumulative dissipation of energy. The ratio of net productivity of one level compared to the next,

Table 11-6 Efficiency percentages for two ecosystems (calculated from Table 11-4).

Trophic level	Lake Mendota	Cedar Bog Lake
Gross intake (ecological efficiency), I_{n+1}/I_n		
Producers (I_{n+1}/A)	11.0	16
Primary consumers	6.1	17
Secondary consumers	8.8	—
Assimilation, $(I - E)/I$		
Producers (A/insolation)	0.4	0.1
Primary consumers	80	86
Secondary consumers	91	91
Tertiary consumers	100 (—)	—
Respiration, $R/(I - E)$		
Producers (R/A)	25	25
Primary consumers	38	38
Secondary consumers	58	58
Tertiary consumers	68	—
Net production, $(I - E - R)/I$		
Producers ($A - R)/A$	75	75
Primary consumers	49	53
Secondary consumers	38	38
Tertiary consumers	33	—

$(I - E - R)_{n+1}/(I - E - R)_n$, varies widely but averages about 10 per cent between all trophic levels (Kozlovsky 1968). Thus 100 calories of net production available at the producer level becomes 10 calories at the C_1 level, 1 calorie at the C_2 level, 0.1 calorie at the C_3 level, and only 0.01 calorie at the C_4 level. This is why there are generally not more than five trophic levels or links in a food chain, often fewer. The animals in the higher trophic levels are usually larger, have higher rates of energy utilization, and there is an effective limit to the size of the area over which they can hunt for prey.

Yield should never exceed net production, lest with reduction of the standing crop the productive potential becomes exhausted. On the other hand, the demand for energy by man or predator populations is often limited by food or other resources, in which case there is continuous pressure to harvest the maximum yield that prey populations can provide. The determination of *maximum sustained yield* or *optimum yield* without jeopardizing continued production year after year is one of the most important and complicated problems in both theoretical and applied ecology (Russell 1931, Watt 1968).

There is a point in the growth curve of all populations at which the species is making the greatest use of the energy resources of the ecosystem and growing most rapidly without the depressing effects of intra- and inter-specific strife, predation, and disease becoming excessive. This is the point of inflection between the accelerating and inhibiting phases of the growth curve, the point at which the increment curve reaches

its highest level (p. 211). In most species investigated, this occurs at a population size approximately half that of the asymptote. Theoretically if the yield were so great that the population is kept below this point, total production would be reduced because of the small parent breeding stock that is left. If the population were allowed to go beyond this point of inflection, there would be wastage. Fewer offspring would be brought to maturity because of increased competition and other factors. It appears that the optimum yield should be such as to maintain the population continuously at this level, and thus balance maximum annual production (Hjort *et al.* 1933, Ketchum *et al.* 1949, Scott 1954).

It may well be that balanced ecosystems have evolved under natural conditions so that predation is of such intensity as to maintain populations of prey at this level of maximum productivity. In a small pond in southern Michigan, in which no predatory fish were present, the benthos biomass increased two- or three-fold during the season to an upper asymptote, after which there was no net productivity. In another similar pond with fish present, the benthos biomass eaten by fish never reached this asymptote, and productivity was maintained continuously at such a high rate that the production during the growing season amounted to 17 times the standing crop (Hayne and Ball 1956).

Experimental work has not so far demonstrated a relation between optimum sustained yield and the point of inflection in the population growth curve. In laboratory cultures of flour beetles, productivity increased progressively with rates of exploitation that brought the surviving population far below the point

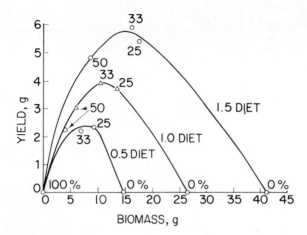

Fig. 11-14 Biomass and yield of guppy populations at three different diet levels. The percentages of exploitation are shown near the points on each curve (Silliman 1968).

of inflection (Watt 1955). With *Daphnia pulicaria*, maximum sustained yield occurred over a period of time when 90 per cent of newborn animals were removed at regular intervals (Slobodkin and Richman 1956). In experimental populations of guppy fish, the standing crop was reduced but the yield was greatest when the tri-weekly exploitation removed 30–40 per cent of the individuals, and the population mass was at one-third its asymptotic level (Fig. 11-14). An exploitation rate of 75 per cent brought extinction of the population (Silliman and Gutsell 1958). In an 8-year study of fish production in a small lake in Pennsylvania, the optimum sustained yield of 14.56 g/m²-yr averaged 43 per cent of the standing crop before exploitation (Cooper *et al.* 1971). As many as 78 woodchucks can be harvested annually from a total population that numbers 100 at the start of the breeding season (Davis *et al.* 1963).

The lack of agreement between experimental results and theory may be due in part to the fact that the age distribution of the population after such harvests is not the same as under conditions of normal population growth. For instance, in the above guppy experiment, with a yield of only 10 per cent, there was an increase in the number of adults followed by a decrease in the number of juveniles. With yields between 25 and 50 per cent, there was an initial decrease in the number of adults, but this was followed by an increase in the number of juveniles. With a yield of 75 per cent, there was a rapid decline of both adults and juveniles.

Certainly much more study is required to determine practical means of estimating optimum yield, to understand the factors involved, and to put proper harvesting procedures into operation if man is to make maximum use of natural resources (Beverton and Holt 1957, Ricker 1958, Gross 1969, Schoener 1969). In North America, several species have been exterminated through overuse; the passenger pigeon, for instance. On the other hand, there is evidence that in some localities the yield annually taken of fish, muskrats,

and deer is not as great as populations of these species could support.

With organisms that have no specific adult size but continue growth throughout life, such as fish, yield should be calculated in terms of biomass. With these animals there is the additional problem of determining the minimum size limit of individuals which would provide the greatest sustained yield in weight for the population as a whole (Saila 1956, Ricker 1958).

ENERGY FOR MAN

Man's traditional sources of energy for food and industry are the current yields from natural and agricultural ecosystems founded on photosynthesis by green plants; on fossil fuels in the form of coal, petroleum, and gas; and from harnessing physical forces in the environment, such as water power (Howard Odum 1971). There is considerable concern that sources of energy will soon become inadequate in at least those areas where there will be continuous heavy demands.

One main purpose in the extensive damming of our river systems at the present time is to derive power for industry. There is a limit, however, to this source of energy, and much of the potential new sources of hydroelectric power lies in relatively undeveloped countries or is not readily accessible. Some development is possible for deriving energy from the tides, wind, and geothermal steam, but this also is limited (Hubbert 1969).

Fossil fuels represent solar energy captured by photosynthesis and deposited during past geological periods when net production by green plants exceeded its utilization by animals and decomposers. This supply may be exhausted in the next 100 to 300 years. During the last 30 years, 90 per cent of the increase in power consumption has been due to per capita utilization, only 10 per cent to increase in population size (Ayres and Kneese 1971).

Lack of energy may not, however, be the critical factor eventually limiting the population growth of man if two new sources of energy can be fully exploited. Solar radiation supplies the United States with 2000 times as much energy as is now being used (Daniels 1949). Because of its diffuse nature, there are technical difficulties involved, however, that render it unpromising as a direct source of power or heat to industry (Dresser 1956, Hubbert 1969). There is much research underway to unravel the chemical processes of photosynthesis. Once this is discovered, the artificial manufacture of food may supply more than enough for as large a population as the world can accommodate. In addition to energy, photosynthesis, of course, also requires the essential elements carbon, hydrogen, oxygen, nitrogen, and phosphorus, among others.

If atomic energy can be obtained on a fully efficient basis, this would be sufficient, it is estimated, for over

20 billion people (Weinberg and Hammond 1970). This figure should not be seriously considered as a potential world population, however, as other factors, such as pollution and psychological effects, will limit human populations far below this level (pp. 238–240). Power from the *fission* reaction of uranium-235 is already an accomplished fact, but if all the present-day needs for power were derived solely from this source, the world supply of uranium-235 would be exhausted within a few years. If, however, the *breeder-reactor* can be perfected whereby non-fissile isotopes are converted into fissile ones, very much larger supplies of source material will be available.

Enormous amounts of energy are potentially available in the *fusion* of hydrogen isotopes into helium. This has already been achieved in an uncontrolled explosive manner in the hydrogen or thermonuclear bomb. Research is underway to obtain a controlled fusion reaction, but it is not now possible to estimate when, if ever, this can be accomplished. Inherent difficulties with both the fission and fusion processes for mobilizing atomic energy is the dissipation of the heat that is liberated and the disposal of the radioactive wastes. Unless these problems can be solved, they may set limits on the amount of energy that can be used from the vast resource (Hubbert 1969). The electric-generating capacity of modern industry is now increasing 7 per cent per year. It is estimated that at this rate, the mean global temperature would rise 1°C in 90 years and 3°C in 108 years. Unless offset by compensating factors (p. 154), the rise would be sufficient to melt the ice caps and drastically alter the world's geography (Cole 1969).

Until man can mobilize other sources of energy, feeding the increasing world population becomes critical. The problem is partly one of transportation and economics for supplying people in poor agricultural countries with surpluses from the highly productive ones. At present there are about 450 million well-fed people living in comparative luxury and 2,400 million who are under- or malnourished. About one-fourth of the world's 32 billion acres (128 billion hectares) are potentially suitable for cultivated crops, another one-fourth is suitable for grazing, and the remaining half is in part forested or of very limited use for food production. About one-half, the better half, of the potentially arable and grazing land is now being used. The value of grazing lands is that harvesting is done by herbivores which are able to use plants that are otherwise wasted to man (Hendricks 1969).

Increased food productivity depends on (a) expanding the amount of land under cultivation and (b) increasing the yield per unit area. Although there are extensive areas in Africa and South America not now under cultivation, their utilization for crop production will depend on considerable improvement in our ability to manage tropical soils and to maintain their fertility (p. 382). Once desalinization of seawater

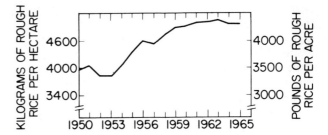

Fig. 11-15 Rice yields in Japan appear to be reaching an upper limit; plotted as a 3-year sliding average (Brown 1967; copyright 1967 by the American Association for the Advancement of Science).

becomes practical and extensive, there could be large-scale irrigation of the extensive deserts of the world. Yields from agricultural lands have been greatly improved in recent years by heavy use of chemical fertilizers, pest control, and new varieties of crops (Tables 11-7 and 11-8). There are limits, however, to how much productivity may be raised by these means (Fig. 11-15), and the "green revolution" has been attained at the expense of pollution and deterioration of environmental quality (p. 153) (Brown 1967, Hendricks 1969, Borman 1970, Boerma 1970, Smith 1972). Farming the sea has only limited potentials for food (pp. 407–408).

Agriculture maintains predominantly monospecies seral communities of high energy and mineral produc-

Table 11-7 Indexes of fertilizer use in the United States[a] over 3-year periods referred to 1950–52 (index = 1.00) (Hendricks 1969).

Years	Nitrogen	Phosphate	Potash
1950–52	1.00	1.00	1.00
1953–55	1.33	1.13	1.19
1956–58	1.66	1.09	1.30
1959–61	2.15	1.23	1.42
1962–64	3.02	1.52	1.72

[a]Hawaii, Alaska, and Puerto Rico included.
Source: U.S. Department of Agriculture, 1966.

Table 11-8 Indexes of yield of several crops in the United States over 3-year periods referred to 1950–52 (index = 1.00) (Hendricks 1969).

Years	Wheat	Rice	Corn	Cotton	Soybeans
1950–52	1.00	1.00	1.00	1.00	1.00
1953–55	1.08	1.13	1.04	1.32	0.93
1956–58	1.36	1.34	1.27	1.54	1.10
1959–61	1.41	1.44	1.44	1.64	1.15
1962–64	1.50	1.66	1.66	1.83	1.15
Maximum attained	12.2	4.00	7.8	7.3	4.5

Source: U.S. Department of Agriculture, 1966. Maximum attained calculated from diverse sources.

tivity. They are, however, ecosystems, although relatively simple ones. A comprehensive mathematical model worked out for, say, a cornfield would be useful in simulating energy and material exchanges and for testing the influence of various factors. The same model would also be adaptable to the more complex ecosystems of nature. Such an expanded model would be a powerful tool to help man order his priorities in regard to plant or plant-animal communities to determine how far they can serve his needs, be it food production, water conservation, climate modification, or aesthetic enjoyment (Lemon *et al.* 1971, Van Dyne 1969).

Supplying man with food is not only a problem in productivity but may also involve modifications in diet. Man is an omnivore and enjoys a high proportion of meat, milk, and eggs in his diet. His per capita consumption of beef is said to have increased 75 per cent in the last 20 years. These animal products come from the consumer levels in the food chain. Man could harvest more energy from natural and agricultural ecosystems if he restricted himself to yield from the producer level. If he were to change to a diet wholly of rice, wheat, corn, potatoes, and other plant products, it is estimated that the amount of land now in cultivation could support a population six times the present size (Borman 1970), assuming there were no other limiting factors on population growth. The fewer the links in the food chain from which man or any other predator takes his yield, the greater is the harvest that can be obtained. A farm pond, for instance, will yield a considerably larger crop of herbivorous carp than of fish-eating bass, more muskrats than mink, and so on. Nutritional problems for man on a pure plant diet, however, involve the lack of certain essential amino acids, vitamins, as well as calcium, that are deficient or absent in most plant food but common in animal food. Soybean meal contains some of the missing amino acids, or if the essential amino acids and vitamins were factory produced and supplied as supplements to the plant diet, health could be maintained (Brody 1952). Although providing all of mankind with adequate energy and nutrients in the future poses serious problems, the solution of these problems is still within the realm of possibility—if the human population explosion can otherwise be contained.

SUMMARY

Energy, unlike nutrients, does not circulate indefinitely through the ecosystem. It is continuously dissipated to perform work and produce heat, and hence must be continuously replaced. The chief source of energy is solar radiation. This energy is captured by green plants in photosynthesis. Some of this energy is transferred to higher trophic levels through predation, but the amount that is transferred decreases at each higher trophic level until it becomes so small that the food chain terminates. A variety of methods has been developed to measure quantitatively the flow of energy through the individual animal, the population, the trophic level, and the ecosystem.

Gross productivity is the total energy intake per unit area and unit time at any trophic level. It is called primary productivity at the producer level and secondary productivity at the consumer levels. Gross energy intake minus excretory and respiratory losses is the net production.

Net production not lost in predatory and nonpredatory deaths produces growth and increase of biomass. Energy lost in excreta, non-predatory deaths, and wastage of predatory kills is utilized by the decomposers to reactivate the nutrient cycles, or it may recirculate again through the ecosystem. In some ecosystems, especially in those with large standing crops of producers, much of the flow of energy is from producer to detritus to primary consumer rather than from producer directly to consumer.

In seral stages, annual net production exceeds total utilization, so that energy and nutrients accumulate. This increases the fertility of the substratum. In climax communities total utilization may balance total production so that the ecosystem is at an equilibrium or steady state.

The percentage of gross energy intake at one trophic level passed on to the next higher trophic level does not vary consistently between trophic levels but is lower with endotherm than with ectotherm populations. Higher percentages of gross energy intake are utilized for respiration at the higher trophic levels and in endotherms, but this may be compensated for by higher rates of assimilation of the gross energy intake. The ratio of net production to assimilated energy varies widely between species populations but tends to be lower at the higher trophic levels.

Productivity of populations is sustained at a faster rate over a longer time if the surplus production above a certain level is removed by predators, or man. Theoretically, the population level giving greatest absolute productivity should come at the point of inflection in the growth curve of the population, but disturbance of age ratios or other conditions may place the level of optimum sustained yield at some other point.

Man may be faced with a food and energy shortage if his population continues its rapid rise, at least until new sources of energy can be mobilized. Agriculture is applied ecology and all the fundamental processes at work in natural ecosystems also apply to man-made ecosystems. Increased agricultural and industrial productivity may be limited by the increased environmental deterioration that it causes. As food becomes scarce, man may be forced to a wholly plant diet.

Part Four

POPULATION ECOLOGY

Chapter 12

INTRA- AND INTERSPECIFIC COACTIONS BETWEEN INDIVIDUALS

Population ecology is concerned with interrelations or coactions between individuals within and between species. Coactions may either be beneficial to the participants, *cooperation* (Allee *et al.* 1949, Allee 1949, 1951), or harmful, *disoperation*. Interspecific cooperation includes mutualism, commensalism, and many of the interrelations that establish the community as a dynamic unit. Coactions that are harmful, to at least one of the participants, include parasitism, predation, and competition. Many of these coactions act to regulate density or, along with climate, cause fluctuations or even more drastic changes in number of individuals from time to time. Established behavioral interrelations between individuals, characteristic of a species, are its *ethics* and as such are found in animals as well as in man.

INTRASPECIFIC COOPERATION

An early manifestation of cooperation in the evolution of animals is the grouping of free-living protozoans to form colonies, and the further development of such colonies into multi-cellular metazoans that thereafter behave and respond as unit organisms. Whether the first gathering of protozoan cells to form colonies developed for better protection from some predator or environmental condition, improved utilization of food supplies, or more efficient reproduction, it is impossible to say. The colonial form, however, must have had survival value to persist.

Colonization quickly led to *division of labor* between somatic and reproductive cells, as occurs in *Volvox*, and later to division of labor between somatic cells themselves, so that different cells or organs became specialized to serve the particular functions of digestion, respiration, circulation, and so on. Cooperation among cells, tissues, and organs gave greater metabolic efficiency to the whole individual and resulted in evolution to the highest types of animals. Similarly, the aggregation of individuals must have survival value, because it persists. Hundreds, sometimes thousands, of spotted lady-beetles hibernate under leaves at the forest-edge. Mayflies, midges, and mosquitoes swarm for mating purposes. Millions of bats roost together in large caves, notably in the Carlsbad Caverns in New Mexico. The migratory locust moves from one locality to another in immense hordes, and birds usually migrate in flocks. Highly organized societies are found in such insect groups as termites, ants, bees and wasps, as well as in some breeding colonies of birds and mammals. Even the biotic community exhibits division of labor in that dominant species control the microclimate and conditions in which other species live; producer organisms capture energy and manufacture food, which is then divided between all members of the community; predators help to control population densities of their prey; and so on.

Aggregations

All organisms react on their habitat in one way or another, and when they occur in numbers these reactions produce a conspicuous effect. *Water conditioning* occurs when physical or chemical changes occur as the result of organisms living in it. Compared with unconditioned water, these changes may have either a harmful or a beneficial effect on organisms introduced into the water after the original organisms have been removed. Water is said to be *homotypically* conditioned when the changes were previously produced by individuals of the same species as being studied and *heterotypically* conditioned when the changes were produced by a different species.

Experimental studies have demonstrated that goldfish grow faster in water that has been homotypically conditioned for 24 hours than in unconditioned water. Both fish and amphibian larvae also do better in water conditioned by the presence of mollusks than in unconditioned water (Shaw 1932). The marine flatworm *Procerodes wheatlandi* will survive much longer when transferred to fresh water conditioned by the presence of either live or dead individuals of the same species or by fresh-water species of flatworms than they do in unconditioned fresh water. The longer survival in toxic solutions, faster growth, and greater reproduction of protozoans, snails, flatworms, cladocerans, amphibian larvae, and fish occurring in aggregations rather than as isolated individuals is attributable to water conditioning.

Colloidal silver is toxic to fish. Ten goldfish were simultaneously exposed to 1 liter of water dosed with colloidal silver. They lived an average of 507 minutes each. Fish individually exposed to a similar concentration of silver in the same volume of water lived an average of 182 minutes. The slime from the grouped fish was sufficient to precipitate much of the colloidal silver and render the solution less toxic (Allee and Bowen 1932). Photosensitive animals survive longer when exposed to excessive illumination in groups than singly because of partial shading of one by another, but fresh-water planaria exposed to ultraviolet live longer in groups, even when no shading is involved (Allee and Wilder 1939). Marine flatworms *Procerodes* survive longer in fresh water in groups than singly because the first worms that die from the group release calcium into the water, conditioning it and giving protection to the animals that remain (Oesting and Allee 1935).

Various factors are involved in producing favorable conditioning: minute organic particles in suspension resulting from excreta, regurgitated food, or disintegration of dead animals previously present may become concentrated in the alimentary tract of the animals and serve as an unsuspected food resource (Allee and Frank 1949); mucus or slime secreted by organisms may coagulate, precipitate, or reduce the potency of toxic substances; salts liberated from the body may change the osmotic properties of the culture medium; or there may be liberation of growth-promoting substances from one animal that affects other animals (Allee *et al.* 1949). Many of these effects, not all of which may be favorable (pp. 197–198), are doubtlessly at work in natural habitats and should be carefully studied as part of the internal dynamics of the biotic community. It may well be, for instance, that during the course of evolution organisms have become adapted to tolerate or take advantage of these external metabolites given off by their neighbors, with the result that the metabolites have become an important part of their environment (Lucas 1947).

Benefits derived from aggregating are shown in other ways by terrestrial animals. In honeybees, when hive temperatures drop below 14°C (57°F) during the winter, they form clusters and maintain a mass temperature several degrees above outside temperatures. This is brought about by increased metabolic oxidation of honey in their bodies and by increased muscular activity. Furthermore, the compact cluster presents a surface area for heat loss that is less than the total surface area of the individuals separately (Milum 1928). When there is danger of overheating, the bees in the hive spread out on the combs and fan with their wings to create a circulation of air. They will also carry water into the hive and place small quantities both outside and inside the comb cells. The forced air circulation evaporates the water and cools the hive. Bees also cool themselves by constantly moving their tongues in and out of their mouths, exposing to evaporation the moisture that is present on them as a thin film (Lindauer 1955). Temperature regulation is less well developed in other social Hymenoptera (Himmer 1932).

Coveys of bobwhite quail roost in close circles, at night. Perhaps this enables detection of predators approaching from any direction, but by that behavior the birds can tolerate lower air temperatures and for a longer time than isolated birds can (Gerstell 1939). Similarly, mice huddle in low air temperatures, a behavior that reduces heat radiation and consequent need for frequent feeding (Prychodko 1958).

A single muskox or bison may succumb to a pack of wolves. When in a group, the males form a circle facing outward with the females and young inside, whereby they are usually able to ward off the attack. By the same token, a single wolf has difficulty killing a deer; a single coyote, killing a pronghorn antelope. But in packs the wolves can overpower a deer, and by individually taking turns in relay fashion, a pack of coyotes can chase a pronghorn to exhaustion.

Whether an animal occurs singly or in groups may affect its learning rate and behavior. The common cockroach and the shell parakeet learn simple mazes less rapidly when other individuals are around than when alone, but goldfishes, minnows, and green sunfishes learn mazes faster in groups, phenomena spoken of respectively as negative and positive *social facilitation*. Many animals are more active and alert in groups than alone; in groups, individual imitations of others' behavior are common. Cormorants and pelicans fish more proficiently in groups than alone because group behavior is organized and each individual plays a certain role (Allee 1951).

The beneficial effects of aggregation are lost if the aggregation is either too small or too large. For instance, the longevity of *Drosophila* is greatest with a population density of 35 to 55 flies per 1-ounce (28-g) culture bottle (Pearl, Miner, and Parker 1927). Smaller densities are unable to control the growth of the yeasts on which they feed; greater densities exhaust the food supply and excessive amounts of

excreta accumulate. Likewise an initial population of 4 *Tribolium* beetles per 32 g of flour reproduces more rapidly during the 25 days following than smaller or larger initial populations (Park 1932). For all kinds of animals, competition for food and other resources of the habitat becomes more and more intense as populations increase in size above an optimum. The benefits resulting from an increase in the size of aggregations up to the optimum represents cooperation; the harmful effects resulting from aggregations that are too large is disoperation.

Social Organization

The simplest animal aggregations exhibit little social organization, for the individual organisms are brought together more or less ephemerally by chance, by sexual attraction, for reproduction, or because of a similar response to environmental factors. An evolution of organization may, however, be traced through intermediate stages to the complex division of labor found in some insect societies (Fig. 12-1). Specializa-

Fig. 12-1 Model of a royal cell of the termite, showing different castes. The queen has an enlarged abdomen; her head is turned to the right. The king is in the left center. Two soldiers with pointed heads are in the upper right. Most of the rest are workers (courtesy Buffalo Society of Natural Sciences).

tion occurs both in morphology and behavior. The three primary castes of termites and ants are the winged reproductive males and females, the wingless sterile soldiers that possess large mandibles and irritating glandular secretions, and the smaller, wingless, often sterile workers. The soldiers defend the colony against predaceous enemies; this function is assumed by workers in bees and wasps, among which a distinct soldier caste is lacking. In termites, the soldiers may be either males or females; in ants, they are females. The worker caste in ants is usually female, but in higher termites it may consist of either males or females. In primitive termites the nymphs of other castes substitute for the workers. The workers collect food, cultivate gardens of fungi, take care of domesticated aphids or coccids, feed the other castes, and build shelters. The earliest organized social life of primitive man was perhaps neither so highly organized nor so far advanced in an evolutionary sense as these complex societies of insects, even though it was from the greater psychological potentialities of primitive man that modern civilization arose (Allee *et al.* 1949, Allee 1951).

In these social relations, indeed in all sorts of symbiotic relations between individuals, one or both partners must have specialized behavior to effect and maintain the relationship. Chemical stimuli are important in this respect and have received much study to date, but physical stimuli, such as color, shape, texture, temperature, and so on, may also have primary integrative importance as releasers for specific behavior responses, the products of long evolution (Davenport 1955).

MUTUALISM

Mutualism is an association between two or more species in which all derive benefit in feeding or in some other way. The term symbiosis has often been applied to this relationship, but *symbiosis* properly refers to the intimate association of two or more dissimilar organisms, regardless of benefits or the lack of them, and hence includes mutualism, commensalism, and parasitism.

Mutualism, as is true also with commensalism and parasitism, may be *facultative*, when the species involved are capable of existence independent of one another, or *obligate*, when the relationship is imperative to the existence of one or both species. Considerable study and experimentation is sometimes required to decide whether a particular relationship is facultative or obligate, or even whether it is truly mutualistic. Many examples of ecological interest of mutualism, commensalism, and parasitism are cited by Pearse (1939) and Allee *et al.* (1949); only a few will be given here.

Mutualism in plants is demonstrated in the associations of fungi and algae to form lichens, of nitrogen-fixing bacteria with the roots of legumes, and of fungal mycorrhizae with the roots of many flowering plants (Went and Stark 1968).

There are many intimate relations between plants and animals (Buchner 1953). Mutualism is suspected in the presence of photosynthetic algal cells in the protective ectoderm of green hydra, and those associated with turbellarians, mollusks, annelids, bryozoans, rotifers, protozoans, and the egg capsules of salamanders. The algae give off oxygen, benefiting the animals, which in turn supply carbon dioxide and nitrogen to the plants. The thick growth of algae often found on the carapace of the aquatic turtles is important mostly as camouflage for the turtles (Neill and Allen 1954). Certain beetles, ants (Bailey 1920, Weber 1957), and termites cultivate fungi for food. Bacteria in the caeca and intestine of herbivorous birds and mammals aid in the digestion of cellulose. The cross-pollination of flowers by the agency of insects and birds seeking nectar and pollen is of such great importance that many structural adaptations in both plants and animals fit the one to the other to ensure the success of the function (Robertson 1927, Dorst 1946).

Animals, especially birds and mammals, are of great importance as agents of plant distribution (McAtee 1947). Seeds, fruits, even entire plants become attached to feathers or fur, or ingested seeds are eaten and eliminated unharmed with the feces. When bare seeds are eaten they are usually macerated, digested, and entirely destroyed unless they have very hard coats. But fruits are fed upon primarily for pulp, and most of the seeds pass through the alimentary tract unharmed. Animal transportation of ingested seeds is perhaps the most important means by which fruit species are dispersed (Taylor 1936). Furthermore, germination of the seeds is frequently improved by mechanical abrasion in the stomach and thinning of the seed coat by digestive juices, making them more permeable to water and oxygen (Krefting and Roe 1949). Germination of acorns and nuts is improved if they are buried in the ground rather than left lying on the ground surface. Squirrels, chipmunks, wood rats, and some birds, particularly jays (Chettleburgh 1952) and woodpeckers, cache acorns and nuts as a winter food supply, hiding them in cavities and nooks or burying them in the soil. Perhaps most are recovered and eaten; Cahalane (1942) found that 99 per cent of the acorns buried by the fox squirrel in a locality where the animals were numerous were recovered by the animals, largely through the sense of smell. One per cent of the thousands of nuts produced by it during the lifetime of a tree that are buried but not recovered would be adequate to ensure the continuance of the forest. Invasion of oak and hickory trees into sandy areas is greatly accelerated by, and is sometimes dependent on, this coaction of squirrels (Olmsted 1937), and the dispersal of forests up the slopes of mountains against gravity may also depend in large part on transportation of the heavy seed by animals (Grinnell 1936). The interesting concept involved here is that plants have evolved fruits and nuts that are highly attractive to animals as food substances. However, the production of prodigious numbers of fruits and nuts during their lifetimes ensures that at least some will escape consumption and will be more widely and effectively dispersed.

Large populations of such herbivores as rabbits and deer sometimes do considerable damage to new propagation of herbs and trees, but the effects of over-browsing cannot be dismissed as all bad (Webb 1957). Removal of the lower branches of established trees by deer may not seriously affect the vigor of the trees; indeed, such pruning may actually increase their value as lumber. Deer pawing the leaf-litter may thereby plant, so to speak, some seeds that would not otherwise become established. Thinning dense stands of young trees may allow residuals to grow more rapidly; much of new growth is doomed anyway because of root competition, and shading cast by established trees. Some species of shrubs and trees actually produce more annual growth under heavy than light browsing. Other species, however, may be killed when small by heavy browsing, although they tolerate considerable browsing when mature. Detrimental effects of both browsing and grazing become evident in an area in the form of excessive invasion of new species which are little used as food, and disappearance or stunting of the food species that are desirable (Graham 1954).

Some tropical acacias have evolved foliar nectaries or other food bodies as well as enlarged hollow stipules, spines, or other structures to attract stinging ants. In return, the plants obtain protection from herbivorous mammalian and insect enemies (Brown 1960).

Interspecific mutualism is nicely demonstrated by the flagellate *Trichonympha*, an obligate anaerobe in the gut of several species of wood-eating termites (but not in the family Termitidae) where it digests cellulose (Cleveland 1924). *Trichonympha* and related species also occur in the alimentary tract of the wood-eating roach *Cryptocercus* (Cleveland 1934). The termite and roach reduce the wood to small fragments, passing them through the alimentary canal to the hindgut, where the protozoans digest the cellulose, changing it into sugar. The host benefits the protozoa by removing harmful metabolic waste products and maintaining anaerobic conditions in the intestine (Hungate 1939).

The ruminant stomach and the horse caecum contain enormous numbers of ciliates and bacteria, some of which digest cellulose. The microorganisms reproduce the equivalent of their biomass each day. Digestion of these organisms provides the host with about 20 per cent of its nitrogen requirement (Hungate 1960).

COMMENSALISM

Commensalism defines the coaction in which two or more species are mutually associated in activities centering on food and one species, at least, derives benefit from the association while the other associates are neither benefited nor harmed. It is often difficult to establish definitely the nature of the relations between species; and phenomena considered at one time to be commensalism have been later found to be parasitism or mutualism. The concept of commensalism has been broadened, in recent years, to apply to coactions other than those centering on food; cover, support, protection, and locomotion are now frequently included (Baer 1951).

The remora fish are remarkable for having the spinous dorsal fin modified to form a sucking disk on top of the head by means of which they become attached to the body of the shark, swordfish, tunny, barracuda, or sea turtle. They are of small size and are not burdensome to the host. The host benefits the remora, however, for when the host feeds, the scraps of food floating back are swept up by the remora. Many small animals become attached to the outside of larger ones, such as the protozoans *Trichodina* and *Kerona* on *Hydra*, vorticellids on various other aquatic organisms, branchiobdellid annelids on crayfish, and so on. Commensals may also be internal; consider, for instance, the harmless protozoans that occur in the intestinal tract of mammals, including man.

The yellow-bellied sapsucker drills holes into the phloem of trees from which oozes a flow of sap. The woodpecker uses the sap for soaking or mixing with the insects that it captures for food or may drink the sap separately. The sap and the insects that it attracts are also fed on extensively by other birds, especially hummingbirds, and by other insects and a few mammals (Foster and Tate 1966).

The pitcher of the pitcher plant found in bogs furnishes a breeding site or home for certain species of midge flies, mosquitoes, and tree toads. Many kinds of microorganisms, both plant and animal, live in the canal system of sponges.

The nest of one species often furnishes shelter and protection for other species as well. Ant nests may contain guest species of various other insects. Large hawk nests sometimes have nests of smaller species tucked in their sides (Durango 1949); some birds place their nests close to wasps, bees, or ants for the protection offered by these insects (Hindwood 1955). Woodchuck burrows are used also by rabbits, skunks, and raccoons, especially in the winter. During dry periods the water in crayfish burrows, a meter below the ground surface, often teem with entomostraca (Creaser 1931).

COMMUNITY ORGANIZATION

The final stage in the evolution of cooperation is the biotic community. Analogous to a multicellular individual, the community is composed of organic units, in this case organisms and species rather than cells and tissues. It has a definite anatomy in its stratification, niches, and food chains. The community, too, is a thing born, and it exhibits the same characteristics of growth and maturity as do individuals. There is succession of stages to the climax community like the series of instars in the life cycle of an insect. If the community is injured, it heals the wounds in its structure through secondary succession. The community is self-sustaining in that it absorbs energy from the sun and metabolizes it at various trophic levels in order to do work. There is division of labor, analogous to the functions of the various organs in the body of a single individual; plant species manufacture the food that animals need, and dominant species create environment conditions within the community suitable for the existence of other species. There is transmission of stimuli, intercommunication, between individuals and species by voice, odor, sight, and contact. There is control over the numbers of individuals of each species in the balance of nature. The result is that the biotic community is a highly integrated recognizable unit in which species exhibit various degrees of interdependency. The existence of each component depends to a certain extent on cooperation between them all, so that the community functions as if it were an organic entity. These complicated interrelationships have come about through evolution because of their survival value for the component species involved.

PARASITISM

Parasitism is the relation between two individuals wherein the *parasite* receives benefit at the expense of the *host*; parasitism is therefore a form of disoperation. Parasitism is mainly a food coaction, but the parasite derives shelter and protection from the host, as well. A parasite does not ordinarily kill its host, at least not until the parasite has completed its reproductive cycle. Were the parasite to kill its host immediately on infecting it, the parasite would be unable to reproduce and would quickly become extinct. The balance between parasite and host is upset if the host produces antibodies or other substances which hamper normal development of the parasite. In general the parasite derives benefit from the relation while the host suffers harm, but tolerable harm.

Classification

Parasites are commonly classified as *ectoparasites*, those which live on the outside of the host, and *endoparasites*, those which live in the alimentary tract, body cavities, various organs, or blood or other tissues of the host (Baer 1951). Ectoparasites may be parasitic only in the immature stages—the hairworm larvae, parasitic in aquatic insects; only the adults parasitic—fleas, on birds and mammals; or both larvae and adults may be parasitic—the blood-sucking lice and flies, biting

lice, mites, and ticks that occur on birds, mammals, and sometimes reptiles, and the monogenetic trematodes on fish. Similar relations obtain among endoparasites, although it is more common to have all stages parasitic: entozoic amoebae, trichomonad flagellates, opalinid ciliates, sporozoans, pentastomids, nematodes, digenetic trematodes, acanthocephalons, cestodes, and some copepods.

Animals may also be parasitic on plants. Nematodes infest the roots of plants. Galls are formed by wasps or gnats especially on oaks, hickories, willows, roses, goldenrods, and asters. Mites stimulate formation of witches' brooms in hackberry. A variety of insects the larvae of which are leaf miners, wood borers, cambium feeders, and fruit eaters, should be included here. Plants themselves may be parasites either on other plants or on animals. Bacteria and fungi are among the most important disease-producing organisms in animals.

Social parasitism describes the exploitation of one species by another, for various advantages. Old World cuckoos and the brown-headed cowbird of North America do not build nests of their own; rather, they deposit their eggs in the nests of other species, abandoning eggs and young to the care of foster parents (Weller 1959). The bald eagle sometimes robs the osprey of fish that it has just caught. One species of ant waylays foraging workers of another species and snatches away the food they are transporting, or the robber species may deliberately rob another nest of food. Some species of ants make slaves of the workers of other species. Various other types of dependency of one species on another have evolved, not only between ants, but also in other social insects, such as termites, wasps, and bees. Social insects are apparently the only animals other than man to have succeeded in domesticating other species, and of cultivating plants, particularly fungi, for food (Wheeler 1923).

Evolution and Adaptations

The ancestors of ectoparasites were clearly free-living forms. It is not difficult to imagine how a small organism living freely in water or vegetation could accidentally have settled on the outside of a larger species and found conditions favorable for survival. There would even be selective advantage in such a niche if the organism found a rich source of food. The biting lice probably evolved from psocid insects that live beneath the bark of trees. They may have transferred from this niche to bird nests and then to the birds themselves. Most ectoparasitic insects probably are derivatives of carnivores, saprovores, or suckers of plant juices.

Endoparasites may in some cases have evolved from ectoparasites; more likely, they came directly from free-living ancestors or from commensals. For example, free-living nematodes and scavenger beetles both feed

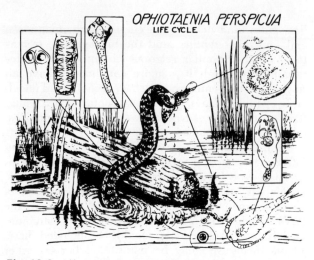

OPHIOTAENIA PERSPICUA
LIFE CYCLE

Fig. 12-2 Life cycle of a snake tapeworm. The eggs are voided into the water with the feces of the snake, where they are ingested by the copepod *Cyclops* (lower right). A procercoid (middle right) develops in the copepod, from the egg. If the copepod is eaten by a fish, the procercoid changes into a plerocercoid (upper right) and becomes encysted in the liver or mesenteries. When the fish is eaten by a water snake, the mature tapeworm develops (upper left). Other intermediary hosts are tadpoles and frogs (Thomas 1944).

upon decaying organic material, and it is easy to visualize how the beetles could have accidentally consumed one or more nematodes. Many kinds which have since become parasites, such as protozoans and flatworms, could have had their first entrance into the alimentary tracts of prospective hosts via drinking water, and subsequently invaded other organs in the body. The invaders would have found their hosts abundant food sources, but would have needed some preadaptation to live at the low oxygen concentrations characteristic of digestive tract, to resist being consumed by the digestive juices of the host, and to keep from being carried out with the feces. As succeeding generations of parasites became increasingly adapted to live either on or in their hosts, many kinds lost the capacity for a free-living existence. Specialization to internal parasitism has cost the loss of locomotor, sensory, and digestive organs, none of which is needed, and led to the development of organs of attachment, increased reproductive capacity, and, in several forms, to polyembryony, intermediate hosts, and a complicated life-cycle (Lapage 1951).

Some parasitic species are more highly evolved than others. Many parasites, for instance, pass their entire existence in a single host; others require one, two, even three intermediate hosts (Fig. 12-2). It is of ecological significance that both primary and intermediate hosts of a parasite occur in the same habitat or community. Even then the hazards to successful passage from one host to another are so great and mortality so high that prodigious quantities of offspring are produced to ensure that at least a few individuals will complete the cycle.

Parasites are transferred from one host to another by active locomotion of the parasite itself; by ingestion, as one animal sucks the blood of or eats another; by ingestion, as an animal takes in eggs, spores, or encysted stages of the parasite along with its food or drinking water; as a result of bodily contact between hosts; or by transportation from host to host by way of vectors. As an illustration of vectors, the bacteria that cause tularemia in man are carried from rabbit to rabbit by ticks. Man contracts the disease when he handles infected rabbits, but the incidence of infection is greatly reduced in the autumn when cold weather forces the ticks to leave the rabbits and go into hibernation (Yeatter and Thompson 1952).

Host Specificity

Copepods are of all animal parasites the most ubiquitous in their host relationships, being reported from various invertebrate groups and from fish. Most parasitic genera, however, are adapted to hosts of one phylum only. The acanthocephalans *Gracilisentis* and *Tanarhamphus* are yet more specific, normally found only in the gizzards of shad fish; *Octospinifer* is found only in catostomids; *Eocollis*, only in centrarchids. Each order of birds possesses its own particular species of tapeworms; this is true even when several orders of birds live in the same habitats, as do, for instance, grebes, loons, herons, ducks, waders, flamingoes, and cormorants (Baer 1951). Species of flagellate protozoans that occur in termite alimentary tracts are largely host-specific (Kirby 1937). However, considerable caution needs to be exercised in assigning host specificity to protozoans. Many species have invaded more than one taxonomic host group; and often several species of a single genus of Protozoa frequent the same host species.

Some species of gall wasps attack only one species of oak. Where a single species parasitizes two or more host species, the shape and structure of the gall formed around the egg and larva on both hosts is essentially similar. When several insects are found on the same oak, each kind of parasite produces its own characteristic gall form. Apparently the characteristics of the gall that develops depend more on the kind of enzyme secreted by the parasite than on differences of host tissues (Kinsey 1930).

Segregation of parasites to special niches is demonstrated by species of biting lice restricted to the head or body regions of birds. Some nematode species are found throughout the body in connective tissue, but not in the gut; some occur only in the digestive tract and associated organs; certain species occur in the glandular crop of birds, but others only in the caecum; many species occur exclusively in the lungs or in the frontal sinuses. Such fine restriction of parasites to particular hosts or organs is a consequence of precise physiological and morphological adaptations that permit the parasite to survive and complete the life cycle only under very special conditions. Likewise, the life cycle of the parasite is often closely synchronized in time with that of its host (Foster 1969).

Host-specificity can make the taxonomies of many parasites useful for corroborating phylogenetic relationships of their hosts (Kellogg 1913). The South American bird *Cariama cristata* has been shifted from one order to another, and was at one time even put into a special order. A study of its helminth parasites disclosed two species of nematodes and two genera of cestodes present which occur together elsewhere only in Eurasian bustards. The occurrence of these forms in groups so far removed geographically from one another could be coincidental or the result of parallel evolution, but for a number of reasons it seems more likely that *Cariama* and the bustards are derived from a common ancestor which became infected with these parasites, the parasites persisting in spite of evolutionary divergence and geographic separation of hosts. It is interesting to note that this relationship of the hosts is sustained by recent taxonomic study of them by ornithologists (Baer 1951).

Effect on Host: Disease

By *disease* we mean a condition which so affects the body or a part of it as to impair normal functioning. Parasites may not cause immediate mortality, but they cause damage to body structures which, should it become excessive, may cause death. We may perhaps better visualize the role parasites play in producing disease by listing some of the more common agents of mortality in organisms, in addition to predators and parasitoids, which will be described beyond.

1. Worm parasites, such as tapeworms, nematodes, and acanthocephalans may wander through the host's body doing mechanical injury as well as destroying and consuming tissues. The host may respond by forming a fibrous capsule or cyst around an embedded parasite.
2. Protozoan parasites are especially important in the alimentary tract and in the blood. A sporozoan species of *Eimeria* damages the walls of the intestine in upland game birds, producing coccidiosis; *Toxoplasma* becomes encysted in the brain of rodents; *Leucocytozoon* is a blood parasite common among waterfowl and game birds.
3. Bacteria cause a variety of diseases, notably tularemia, paratyphoid, and tuberculosis among birds and mammals, as well as other diseases in lower types of organisms.
4. Viruses are so submicroscopic in size that many kinds pass through the finest filters. They are the potent agents of hoof and mouth disease in deer, spotted fever in rodents, encephalitis and distemper in foxes and dogs.
5. Fungus spores of *Aspergillus* that occur in moldy pine litter may be drawn into the lungs of ground-feeding birds, where they germinate and grow, causing aspergillosis. Fungus may also develop on the external surface of animals.
6. External parasites such as ticks, fleas, lice, mites, and flies do not commonly produce serious mortality by themselves, but they are often vectors transmitting protozoa, bacteria, and viruses from one animal to another. Heavy infestations of external parasites may, however, lower the vitality or vigor of an animal and cause diseases of fur (mange) or feathers.
7. Nutritional deficiencies in vitamins or minerals, or improper balance among carbohydrates, proteins, and fats may produce malformations, lack of vigor, even death. Variations in amount, composition, and intensity of solar radiation may affect the vitamin content of the food an animal consumes. Long restriction to emergency foods of low energy content and outright starvation often cause considerable loss of life during periods of climatic stress.
8. Food poisoning, *botulism*, occurs when certain foods become contaminated with the toxins released by the bacterium *Clostridium botulinum*. Many waterfowl are stricken in some localities. Waterfowl also often pick up and swallow gun-shot from marshes in which there has been much hunting, and get lead poisoning.
9. Physiological stress (Selye 1955) is a term that has come to be applied to changes produced in the body non-specifically by many different agencies which may accompany any disease. Effects of stress include loss of appetite and vigor, aches and pains, and loss of weight. Internally, the stress syndrome is characterized by acute involution of the lymphatic organs, diminution of the blood eosinophiles, enlargement and increased secretory activity of the adrenal cortex, and a variety of changes in the chemical constitution of the blood and tissues.

   Stress gives rise to abnormal conditions, but it simultaneously elicits from the body defense mechanisms against those abnormal conditions. It is presently believed that the anterior pituitary gland and the adrenal cortex are chiefly responsible for integrating the defense mechanisms. Three stages are involved: the *alarm reaction*, in which adaptation has not yet been acquired; the *stage of resistance*, in which the body's adaptation is optimum; and the *stage of exhaustion*, in which the acquired adaptation is lost. Characteristic of the exhaustion phase are, among others, hypoglycemia, adrenal cortical hypertrophy, decreased liver glycogen, and negative nitrogen balance.
10. Accidents, ageing, starvation, and so on, must also be included as important causes of mortality.

Organisms that produce disease generally fall into one or two categories. They are either present in the body at all times but not normally virulent, or they are normally absent but are virulent from the moment the host is infected by them. Even the healthiest animals chronically entrain many parasites and noxious organisms in the body, but these organisms wreak overt harm only when they become unusually abundant, when virulent mutant strains develop, or if, for one or another reason, the host's vitality and resistance decline to the point where the host is no longer able to withstand the effects of their presence. Any animal suffering an unusually heavy infestation of parasites will show the tax thus put upon its vitality as a loss of vigor and weight, decreased growth rate, and low resistance to vicissitudes of its natural environment. Normally, a more or less mutual tolerance exists between host and parasite such that the demands of the parasite are in equilibrium with the host's capacity to meet them. Host-tolerant parasites have been naturally selected for; mutant strains that are exceptionally virulent quickly die out because they kill the host, without which they cannot survive.

A single attack, even a mild one, of some diseases often confers a partial or complete *immunity* from further attacks of the same disease, even though the agent of the disease may still be carried in the body of the recovered victim. Immunity is an acquired physiological adaptation by which the immune is able to withstand the presence of an otherwise noxious organism, suffering little or no deleterious consequence of that presence. The fact of immunity is demonstrated when parasites not conspicuously harmful to their normal hosts are introduced into a species to which they are normally exotic. The novel host has had no prior occasion or opportunity to adapt immunitively to the alien parasite, and may sicken, even die, of

the effects of the parasite's presence, the same effects in kind and intensity which the normal host easily takes in stride. For instance, a trypanosome that is a natural parasite in many of the larger wild mammals of Africa evokes no spectacular effects in its usual hosts. But when the parasite is vectored by the tsetse fly to man, it causes sleeping sickness; to cattle, nagana. *Zoonosis* is the study of diseases common to animals and man. Animals often serve as reservoirs for disease organisms that may under proper circumstances be conveyed to man. The threat of plague is ever present in some parts of the world, where the organism prevails among the resident rodents (Kalabukhov 1965). A good discussion of diseases in wild animals is a symposium edited by McDiarmid (1969).

PARASITOIDISM AND PREDATION

Some Diptera and Hymenoptera deposit their eggs in the immature stages of other insects; the larvae on hatching feed on the host until they are fully grown (Fig. 12-3). The relation of the larva to its host is frequently described as one of parasitism. But it is fundamentally different from parasitism in that the host generally dies of the larval depredations before the larva emerges, but the larva generally lives in spite of the host's death. The relationship resembles that of predator to prey, except that, unlike the true predator, the larva lives within the body of its prey and kills it slowly as it feeds, not suddenly before it feeds. Such larvae are, for these reasons, best thought of as *parasitoids*.

Parasitoids may in turn be infested with *hyperparasitoids*. In the Chicago, Illinois, region *Samia cecropia*, a saturniid moth, suffers the destruction of nearly 23 per cent of its cocoons by an ichneumonid parasitoid, *Spilocryptus extrematis*, which deposits an average of 33 eggs on the inside of each cocoon or on the surface of the larva. The host larva dies in a few hours after the parasitoid hatches, and the ichneumonid larva moves about freely, feeding on the cuticle or burrowing into the tissues to drink the body fluids. Another ichneumonid, *Aenoplex smithii*, was found as a secondary parasitoid, feeding on the larvae of *S. extrematis* in about 13 per cent of the cecropia cocoons infested by the latter species. A chalcidid, *Dibrachys boucheanus*, fed both upon *S. extrematis* and, as a tertiary parasitoid, upon *A. smithii*. Another chalcidid, *Pleurotropis tarsalis*, infected cocoons containing *D. boucheanus* and eventually killed the larva as a quaternary parasitoid (Marsh 1937). To have five links in an inverted parasitoid food chain is perhaps unusual, but hyperparasitoidism is common and of importance in controlling the size and interrelations of animal populations.

Predation is a form of disoperation, at least in point of immediate effects, since one animal kills another for food. Predation is important in community

Fig. 12-3 Development of a parasitoid black digger wasp: (a) eggs in position on the host larva; (b) the developing larva; (c) the fully grown larva devouring the remainder of the host (courtesy Illinois Natural History Survey).

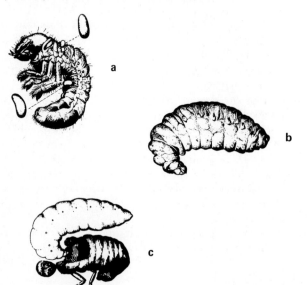

dynamics in so many ways that we will divide discussion of it to food coactions (Chapter 10), productivity (Chapter 11), and regulation of population size (Chapters 14 and 15).

COMPETITION

Competition is the more or less active demand in excess of the immediate supply of material or condition exerted by two or more organisms (Clements and Shelford 1939: p. 159). The materials and conditions sought by animals include food, space, cover, and mates. When these materials are in more than adequate supply for the demands of those organisms seeking them, competition does not occur; when they are inadequate to satisfy the needs of all the organisms seeking them, the weakest, least adapted, or least aggressive individuals are forced to do without, or go elsewhere. Competition may result in death for some competitors, but this is from fighting or being deprived of food or space rather than being killed for food as in predation, or by disease as in extreme parasitism.

Competition may be either direct or indirect. It is *direct* where there is active antagonism, struggle, or combat between individuals; *indirect*, when one individual or species monopolizes a resource or renders a habitat unfavorable to the establishment of other organisms having similar requirements. Direct competition, or *interference*, is evident in the fighting of bull seals for larger harems and of grouse for a better position in the social hierarchy; in chasing and color displays (a sort of saber-rattling) by fish and birds for defense of territories; in the singing and calling of birds, some mammals, and frogs as bids for mates; and in the excretion of chemicals that affect the behavior or health of other organisms (allelochemistry).

Indirect competition, or *exploitation* (Brian 1956), is common among plants when certain species monopolize the water and nutrient resources of the soil or available light so that competing species cannot maintain themselves (Clements, Weaver, and Hanson 1929). Once an area is well saturated with established individuals, it is often more economical of energy for new individuals to seek homes elsewhere, even in less favorable situations, than to intrude. To be successful by indirect competition, a species needs to get established in an area first, or if the invasion of various species is nearly simultaneous, then to have a more rapid rate of reproduction and growth, or a greater longevity, so as to utilize the resources of the habitat to the fullest possible extent (Crombie 1947). It is desirable, but not always possible, to distinguish between and quantify the relative roles of interference and exploitation when species compete (Culver 1970).

Competition is usually keenest between individuals of the same species, *intraspecific competition*, because they have identical requirements for food, mates, and so on,

and because they are more nearly equal in their structural, functional, and behavioral adaptations. *Interspecific competition* occurs where different species require in common at least some materials or conditions. The severity of competition depends on the extent of similarity or overlap in the requirements of different individuals and the shortage of the supply in the habitat. It is generally the case that the more unlike the kinds of competing organisms, the less intense the competition. Yet birds compete with squirrels for acorns, nuts, and seeds; insects and ungulates compete for food in grassland; the bladderwort plant competes with small fish for entomostraca and other plankton.

The immediate test of success in competition is survival; the ultimate test is leaving the largest number of established offspring. Aside from allelochemistry, which is only beginning to be understood, competition has five important effects in the animal community:

1. Establishment of social hierarchies
2. Establishment of territories
3. Regulation of population size
4. Segregation of species into different niches
5. Speciation

The first two effects are chiefly intraspecific and, along with allelochemistry, will be considered in this chapter. Regulation of population size involves both intra- and interspecific competition, and many other types of coaction as well, and will be considered in Chapters 13, 14, and 15. The last two effects are interspecific; they will be discussed in Chapters 16 and 17. It is important to realize that, when these effects are fully manifested, there is a decrease in tension and intensity of competition as each individual takes its place in the structural and functional organization of the community.

Allelochemistry

Included here are the coactions whereby chemicals secreted by one organism affects the growth, health, or behavior of other organisms. *Allelopathy* is produced in plants when toxins are liberated that inhibit seedling growth in the vicinity. This may affect succession of plant species, especially important in the early stages (Muller 1966). Some pioneer species in the abandoned field sere produce substances inhibitory to nitrogen-fixing and nitrifying bacteria. This retards invasion of other species that require higher nitrogen concentration in the soil (Rice 1964). Volatile inhibitors are generally more prevalent than water-soluble ones, and relatively more prevalent in arid than humid climates (Moral and Gates 1971). Antibiotics produced by bacteria, fungi, actinomycetes, and lichens are widespread in nature and may be one of the reasons why bacteria pathogenic to man cannot multiply well in

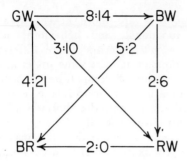

Fig. 12-4 Peck-dominance between the lowest four birds in a flock of seven common pigeons. All four birds were dominated by the three other birds of the flock. The ratios show the proportion of times each bird was successful in its encounters with other individuals (Masure and Allee 1934).

soils. A number of antibiotics, such as penicillin, have been used extensively in human medicine (Burkholder 1952). The use of allelochemic coactions in agriculture has possibilities but has yet to be exploited.

Among animals, overcrowding of tadpoles in culture dishes is associated with the occurrence of peculiar round vacuolated cells in the intestinal tract and feces that appear responsible for curtailment of further growth (Rose 1960). Under laboratory conditions, killer stocks of *Paramecium aurelia* produce a toxin, paramecin, at the rate of one unit-particle per animal per 5 hours. One unit-particle is enough to kill one individual of sensitive stock of the same species as well as being lethal to other species of *Paramecium* (Austin 1948). Conditioning that becomes unfavorable homotypically may sometimes be favorable, or at least tolerable, heterotypically. Thus in protozoan infusions there is a microsere of one species succeeding another.

Allelochemic effects are of great variety in both plants and animals: repellants, escape substances, suppressants, venoms, inductants, counteractants, attractants, signals, stimulants, autotoxins, antoinhibitors, and so on. *Pheromones* are chemical messages between members of a species especially important in reproductive behavior, social regulation and recognition, alarm and defense, territory and trail marking, food location, and so on. Many of these effects are beneficial to the individual; others serve for competitive purposes (Whittaker and Feeny 1971).

Social Hierarchies

When groups of individuals of certain animal species are confined to limited areas, frequent fights or pecking of one another occur. By way of these encounters, the more aggressive and successful individuals establish a hegemony to which the more submissive individuals acquiesce. A *social hierarchy* is thus established; the phenomenon was first clearly described for the domestic fowl (Schjelderup-Ebbe 1922).

The peck-order in the domestic fowl is a linear one.

Close observation of marked individuals showed that, in a flock of 13 birds, one bird became the supreme despot of the whole flock; another bird was submissive to the first but despotic over the remaining 11; and so it went on down to the last bird, which had the right to peck none but was pecked by all. This type of social aggressiveness or *despotism* is called *peck-right*. In practice, certain individuals establish the right to peck others and not get pecked back. In the middle of a series, the order is sometimes less fixed, and reversals or triangles occasionally occur. Although most easily demonstrated in the crowded conditions of captivity, peck-right has also been observed to obtain under free natural conditions. The peck-right type of social hierarchy has been found to occur in various degrees of perfection in several other species of birds, in several species of mammals and fish, and in a few lizards, frogs and toads, crayfish, and insects.

Possession of the following characteristics usually gives an individual at least some advantage in gaining a high position in the despotic order: strength, good health, maturity, relatively large size, hegemony over own territory, responsibility of acting to protect young, accompaniment by members of his own group when meeting a stranger, male over female during the non-breeding season, female mated with a strong male, the hormone testosterone, and innate aggressiveness (Allee *et al.* 1949: pp. 413–414). A high position in the social order is advantageous to the individual, as it gives him priority over food, mates, territory, and other resources of the habitat (Collias 1944) and is sometimes, but not always, correlated with leadership in the group.

In some species the social hierarchy is not as overt as that we have described. In *peck dominance* the individual that is usually subordinate is successful in a certain number of conflicts (Fig. 12-4). Position in the despotic order is a function of ratios of success in continuing conflicts rather than of the results of the initial contact an individual has with each member of the group. Figure 12-4 shows that bird BR successfully subdued GW in 21 encounters but was subdued by GW in 4 encounters. Thus, an individual can occupy any position in the hierarchy as long as he is able to maintain that position against all challengers. The individual who has successfully challenged a higher position from a lower moves up to the higher. Plainly the positions in the hierarchy are fixed in order, but occupancy of those positions is fluid. A still more fluid form of social aggressiveness is *supersedence*, in which a successfully challenging individual usurps the position of another individual, momentarily possessing special advantages in the presence of food or some other thing. This type of relation has been described for the golden-crowned sparrow (Tompkins 1933) and may likely be found in many other species.

There have been few studies of social despotism as an interspecific phenomenon (Fig. 12-5) (Neuman 1956). The range of aggressiveness between individuals

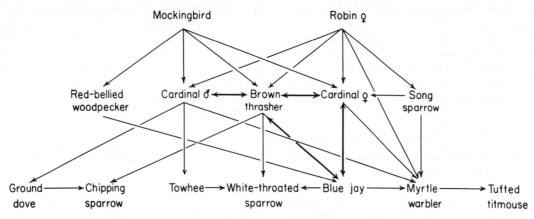

Fig. 12-5 Social hierarchy among species of birds visiting a feeding station during the winter (Dennis 1950).

within any species is so wide that strong individuals of one species may be despotic over weak individuals of another even though the majority of individuals in the first species are submissive. However, the sharp-tailed grouse is usually dominant over the ring-necked pheasant, and the latter is usually dominant over the prairie chicken (Sharp 1957). The manner in which different species fit into a social hierarchy may be a key to structure and organization of communities.

Territory and Home Range

The establishment of territories, especially during the breeding season, is another expression of despotism, but a special one in that it determines the spatial relations between motile animals (Brown and Orians 1970). A *territory* is any area defended against intruders. It may be the entire home range over which the animal is active, or only a small portion around the nest (Fig. 12-6). Although many animals tend to be gregarious during the non-breeding seasons, they frequently take up isolated positions and become intolerant of the close presence of others when undertaking reproduction. A *home range* is that area regularly traversed by an individual in search of food and mates, and caring for young but is not defended (Fig. 12-7).

The establishment of territories is best developed in birds (Hinde 1956), but also occurs in some other vertebrates (Carpenter 1958), including man (Ardrey 1967) and certain invertebrates (marine limpets, Stimson 1970; wood ant, Elton 1932; dragonflies, Jacobs 1955; killer wasps, Lin 1963; field crickets, Alexander, 1961; and others, Brown and Orians 1970). There is increasing evidence that most adult animals, except for small aquatic species, establish home ranges, if not territories, at least during the breeding season (snails, Edelstam and Palmer 1950, Cook *et al.* 1969; crayfishes, Black 1963; fish, Lewis and Flickinger 1967;

toads, Bogert 1947; frogs, Bellis 1965, Dole 1968; salamanders, Madison and Shoop 1970; uta lizards, Tinkle *et al.* 1962; turtles, Cagle 1940; mammals, Seton 1909, 1925–28). Immature animals, species in migration, or shifting populations during the non-breeding seasons commonly do not have definite areas to which they confine their activities. An area should not be called a territory unless one can ascertain that it is defended against intruders of the same species or at least exihibits exclusive possession in the presence of others. A territory is usually defended by individuals, but it may be defended by a closely integrated group of individuals, even over a period of years, as in lemurs (Klopfer and Jolly 1970). The size of home ranges and territories varies proportionally to weight (W) in lizards ($W^{0.95}$), birds ($W^{1.16}$), and mammals ($W^{0.69}$). Larger animals have higher metabolic requirements

Fig. 12-6 Theoretical relation between home ranges (area enclosed within solid lines) and territories (area enclosed within broken lines). The black dots represent nesting sites (Burt 1943).

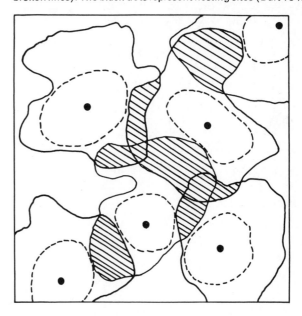

and therefore need larger areas over which to hunt for food (Turner *et al.* 1969).

Territoriality has become so ingrained in the behavior of some types of animal that simple advertisement of possession constitutes adequate defense. Such advertisement takes the form of song or other vocal expression in birds, some mammals, and some frogs, or the deposition of scent or chemical cues (pheromones), as in many mammals (Mykytowycz 1968) and insects. If an intruder persists in invading a territory, however, the owner will variously display bright threatening coloration, scold or growl, give chase, or actually engage in physical combat.

Maintenance of a definite territory has several benefits: a definite breeding location in which the nest can be confidently established and protected is afforded; it aids the acquisition of mates; it ensures an area of sufficient size to provide food both for the adults and, later, for the young; and it frees the possessor of the onus of despotic interference by other individuals. The extent to which these advantages are attained varies with the species (Nice 1941). Although competition for territory is most keen between individuals of the same species, it also occurs between

different species with similar requirements for food and reproduction (pp. 250–254; Orians and Willson 1964, Low 1971). A home range, on the other hand, only provides a breeding location. Possession of territory lessens the pressure of competition during the reproductive period, particularly for the female, when the entire energy and attention of animals needs to be devoted to the production of offspring.

SUMMARY

Beneficial cooperation is evident in division of labor between cells, tissues, and organs within the individual, between individuals in societies, and between species living together in communities. Benefits derived from cooperation are physiological and behavioral and may affect survival, reproductive success, and more efficient use of natural resources. Cooperation between species that is intimate and beneficial to both participants is called mutualism; where only one participant benefits, commensalism. These relations may be either facultative or obligate. Where one or more of the participants is harmed there is disoperation, of which parasitism, parasitoidism, competition, and predation are the examples. Distinction is made between true parasites, social parasites, and parasitoids. True parasites and their hosts have evolved adaptive interrelations so that coexistence occurs for varying lengths of time. The host is generally weakened, however, and virulent strains of the parasite may cause high mortalities. Causes of mortality or disease among organisms are predators, parasitoids, worm parasites, protozoan parasites, bacteria, viruses, fungi, external parasites, nutritional deficiencies, toxication, physiological stress, and accidents.

Competition may be exerted directly through interference in the activities of one organism by another, or indirectly in the form of excessive exploitation of natural resources. It may be either intraspecific or interspecific. Allelochemic secretions from one organism that affects the growth, behavior, or health of other organisms are a subtle means of competition, cooperation, or intercommunication. Competition may result in establishment of social hierarchies, establishment of territories, regulation of population size, segregation of species into different niches, or speciation. The overall effect of competition is to relegate the individual and species to an orderly place in the structure and organization of the community with the result that there is decrease in tension and disturbance.

Fig. 12-7 Travels of a box turtle over its home range during a week, July 7–14 (Stickel 1950).

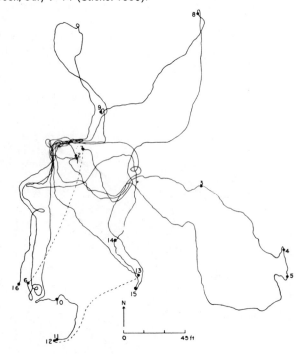

Chapter 13

REPRODUCTIVITY
AND POPULATION STRUCTURE

The rate at which a species reproduces, or its *reproductivity*, and the frequency of its population *turnover* can affect the speed with which it occupies new areas, becomes adapted to new niches, or evolves into new races. In order to analyze the population dynamics of a species, it is necessary to know its life history. This involves the stages in its life cycle, mortality rates of each stage, longevity, sex and age ratios, age at which individuals become sexually mature, fecundity, factors causing mortality, and so forth (Cole 1954). The proportion of different ages and sexes gives the population a definite structure. All these essential data may be conveniently summarized in the form of life tables.

FECUNDITY

Species vary greatly in the characteristic number of generations, broods, or litters produced per year, and in the sizes of them. Protozoans often divide so rapidly that they produce a new generation every few hours. Plankton organisms, less fecund, may produce a new generation every few days. Many vertebrates breed but once a year, some large animals only once every two or three years. Several species of small birds and mammals have two or more broods per year. The female woodland white-footed mouse in Michigan may produce three litters between early April and early June, and two more between middle August and early October (Burt 1940). Under favorable environmental conditions, rodents may continue to breed throughout the winter, so that their reproductive potential is enormous (Kalabukhov 1935).

Innate Capacity

The maximum size of a litter is determined by the physiological and morphological characteristics of the species. With mammals, which produce viviparous young, the size of the uterus and body cavity as well as the number of mammary glands for suckling the young after birth are limiting factors. With birds there is a limit on the number of eggs that one individual can cover and successfully incubate. In species that do not take care of their eggs after laying, the number produced may be limited only by the energy resources of the parent. This is indicated in part by the inverse relation between number of eggs produced and their average size (Lack 1954).

Table 13-1 Reproductivity in the starling in relation to brood size (from Lack 1948).

Brood size	Number banded	Per cent recovered after 3 months	Brood size × per cent recovered
1	65	0	0
2	328	1.83	3.7
3	1278	2.03	6.1
4	3956	2.07	8.3
5	6175	2.07	10.4
6	3156	1.68	10.1
7	651	1.54 ⎱	10.2
8	120	0.83 ⎰	
9	18	0	0
10	10	0	0

Parental Care

The number of eggs or young produced per litter is correlated inversely with the amount of attention that they require. When parental care is altogether lacking, invertebrates may lay 1,000 to 500,000,000 eggs at one maturation; where there is some protection afforded by brood pouches, 100 to 1,000 eggs may be laid; with a high degree of brood protection, 1 to 10 or more eggs may be laid. Mammals seldom have more than a dozen young in a single litter and, in larger species, usually only one. Characteristic clutch size among birds varies from 1 to 15; rarely, 20.

There is a limit on the size of the brood or litter that adult warm-blooded animals can successfully feed and raise to maturity. There is no advantage, for instance, for starlings to have broods larger than five (Table 13-1). In larger broods, each individual receives less food, and hence has less vigor and weight on leaving the nest. Mortality increases either before fledging or in immediately subsequent months (Perrins 1965). In those species that feed their young in the nest, the clutch size has evolved through natural selection to the greatest number that can be hatched and raised successfully through efforts of the adults (Lack 1967). In those species whose young leave the nest and feed themselves at hatching, the clutch size depends in large part on the capability of the female to mobilize energy in her body to produce eggs of a particular size (Kendeigh 1941, Ryder 1970). The variability in clutch and litter size for most species allows them to take advantage of temporarily improved conditions.

Weather

Clutches laid by birds during periods of hot weather are usually smaller than those laid when temperature is moderate (Kendeigh 1941). Clutches laid by related species in temperate latitudes tend to be larger than those laid in the tropics (Moreau 1944, Lack 1947–48). The fecundity of white-tailed deer is higher with good forage than with poor forage (Cheatum and Severinghaus 1950). Reproduction is generally more successful after periods of high mortality than during years of abundance.

The mobilization of energy, usually within a definite period of time, is a limiting factor in warm-blooded organisms. The house wren, for example, lays 5, 6, or 7 eggs per clutch, the total weight of which is 7.0, 8.4, or 9.8 g respectively; yet the adult female herself weighs only 11.5 g. It is estimated that under average conditions about one-third of the daily energy intake of the bird, above its needs for existence, is deposited in the eggs being produced. Any appreciable change in temperature or rate of feeding thus affects the size of the egg, the number laid, or whether laying is undertaken at all (Kendeigh 1941).

Among invertebrates, clutch size also varies under different conditions and in different localities. The copepod *Eudiaptomus gracilis* commonly carries 11 eggs in April, 3 in early August, 9 in early November, and 5 or 6 over the winter. There is a decrease between

Table 13-2 Relation of reproductive to mortality rates per year (Lack 1954).

Local differences in same or related species

Species and locality	Young produced per pair	Adult mortality rate (%)
Starling		
Switzerland	5.8	63
England	4.7	52
Blue tit		
Britain	11.6	73
Spain and Portugal	6	41
Canary Isles	4.3	36
Wall lizard		
Italian mainland	24	40
Italian islands	11	20
California fence lizard		
Plains	8.5	80
Mountains	3.3	30

Species differences in same locality

Species and locality	Young produced per pair	Average further life of half-grown young (days)
White-footed mice in California		
Peromyscus californicus	6.2	275
Peromyscus truei	11.7	190
Peromyscus maniculatus	20.0	152

spring and summer in the number of eggs carried in its brood pouch by the cladoceran *Daphnia*. *Diaptomus siciloides* carries but 4 eggs in mountain lakes of California, as many as 18 in the Illinois River These variations appear to be correlated with differences in temperature and food supply (Hutchinson 1951).

Death Rate

Death rates vary among species and are correlated with rates of reproduction (Table 13-2). The death rate of a species is influenced by a number of factors, but of fundamental importance is the number of young that are born in relation to the carrying capacity of the habitat. When more young are born than the habitat can support, the surplus must either die or leave the area. When populations are stabilized at a constant level, the death rate must fluctuate with the birth rate. Evolutionary adaptation tends to lower the frequency at which the population replaces itself and to raise reproduction to the highest rate compatible with the energy resources both of the species (Table 13-1) and of the habitat (Lack 1954).

SURVIVAL OF YOUNG

Success in raising young depends not only on the ability of the adults to care for the young, but also on the vitality of the embryo and on the chance destruction of nests, eggs, or young by storms, wind, floods, predators, accidents, and desertion of the parents. Considerable data are available in this connection with birds (Fig. 13-1).

Nest failures in birds are most frequent early in the nesting cycle and decrease progressively as nesting proceeds: 2.4 per cent per day during nest-building, 2.2 per cent per day during egg-laying, 1.2 per cent per day during incubation, and 0.5 per cent per day while the young are in the nest (Kendeigh 1942). Location of the nest is a factor in the successful raising of young (Table 13-3).

Table 13-3 Correlation between type of nest or nest location in birds and percentage of fledglings raised from eggs laid (Kalmbach 1939, Nice in Spector 1956: pp. 93–94).

Category	Number of studies	Per cent successful
Precocial gallinaceous species nesting on the ground	17	44
Open nests of altricial species	27	46
Waterfowl in aquatic habitats	22	60
Hole-nesting altricial species	32	66

Fig. 13-1 Average life equation of a stabilized ruffed grouse population in New York State (Bump *et al.* 1947).

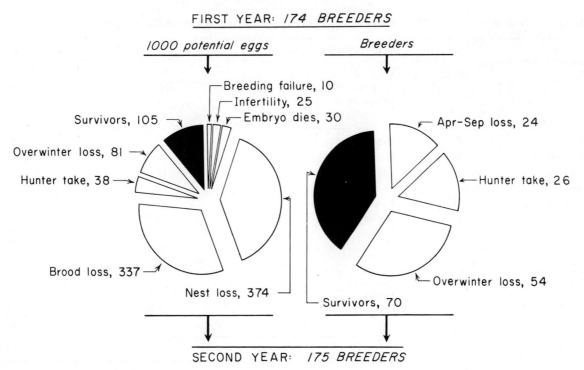

FIRST YEAR: *174 BREEDERS*

1000 potential eggs   *Breeders*

Breeding failure, 10
Infertility, 25
Embryo dies, 30
Survivors, 105
Overwinter loss, 81
Hunter take, 38
Apr-Sep loss, 24
Hunter take, 26
Brood loss, 337
Nest loss, 374
Overwinter loss, 54
Survivors, 70

SECOND YEAR: *175 BREEDERS*

The relatively low percentage of nests that produce young successfully in many species is not a true index of annual reproductivity, since birds commonly make a second or even a third attempt if earlier nestings were failures. The ring-necked pheasant in Iowa has maximum nesting success averages of only 41 per cent; yet, before the season is over, by making repeated efforts, between 70 and 80 per cent of the hens are successful in raising broods. Full reproductive success is not assured, however, in raising the young to the stage of leaving the nest. In the study of the ring-necked pheasant, the average number of young hatched in successful nests was 8.7; after 1 to 3 weeks the average size of the brood was reduced to 6.7; after 4 to 5 weeks to 5.9; after 6 to 7 weeks to 5.3; and after 8 to 10 weeks to only 4.9 (Errington and Hamerstrom 1937).

LIFE TABLES

Species differ widely in the number of young produced each year, in the average age to which they live, and in their average rate of mortality. When sufficient facts about a species are known, a *life table* that tabulates the vital statistics of mortality and life expectancy for each age group in the population may be formulated (Table 13-4, Pearl 1923). Age is usually represented by the subscript index x and is some convenient fraction of a species' mean life span, such as a year or stage of development. The life table is set up on the basis of an initial cohort of 100, 1000, or 100,000 individuals; and the number living to the beginning of each successive age interval is symbolized as l_x. Plotting these data gives a *survivorship curve* for the species, and a number of

survivorship curves have been worked out for different species (Bourliere 1959, Ito 1959). The number dying within each age interval is designated as d_x and gives a *mortality curve*. The rate of mortality during each age interval is commonly expressed as the percentage of the number at the beginning of the interval $100(d_x/l_x)$ and is indicated as q_x. Survival rate is the difference between the mortality rate and 100 per cent $(100 - q_x)$ and is expressed as s_x (Hickey 1952). Life expectancy (e_x) is the mean time that elapses between any specified age and the time of death of all animals in the age group.

Life tables are also useful for computing the average longevity of a population, for showing the age composition of a population, for indicating critical stages in the life cycle at which mortality is high, for showing differences between species, for showing the success of the same species in different biotopes, for furnishing information of value in game and fish exploitation (yield), and in control of pests (Quick, in Mosby 1960).

Information for constructing life tables may be obtained from a knowledge of age at death of a random but adequate sample of the population; information on the age ratios of the living population, provided it is stabilized; or from data on a single cohort, adequately identified, followed throughout its life span (Southwood 1966).

Curves of survival plotted from life tables may be of three types (Fig. 13-2; Pearl and Miner 1935, Deevey 1947). In type 1 a cohort finds environmental conditions ideal; all members, born at the same time, live out the full physiological life span characteristic of the species, and all die at about the same time. The survivorship curve of modern industrialized man approaches type 1 (Fig. 13-3; Pearl 1922). In type 2, the

Table 13-4 Life table for the 1952–53 generation of the spruce budworm in New Brunswick, Canada (after Morris and Miller 1954).

x	l_x	Factor responsible for d_x		d_x	q_x
Eggs	1000	Parasites		17	2(−)
		Predators		86	9(−)
		Others		6	1(−)
			Totals	109	11
Instar I	891	Dispersal, etc.		428	48
Hibernacula	463	Winter		79	17
Instar II	384	Dispersal		242	63
Instars III–VI	142	Parasites		51	36
		Disease		3	2
		Birds		20	14
		Others		61	43
			Totals	135	95
Pupae	7	Parasites		0.6	8
		Predators		0.7	10
		Others		1.3	18
			Totals	2.6	36
Moths	4.3		Totals for generation	995.7	99.5

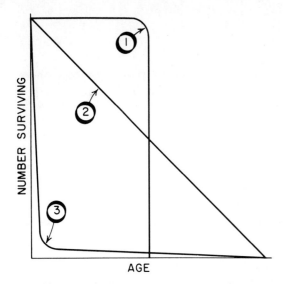

Fig. 13-2 Schematic representation of different types of survivorship curves. The vertical scale may be graduated, arithmetically or logarithmically. If graduated logarithmically, the slope of the line will show the rate of change; a straight diagonal line is indicative of a mortality rate equal at all ages (Deevey 1947).

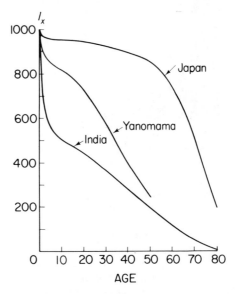

Fig. 13-3 Survivorship curves for three types of human populations: a highly urbanized and industrialized country (Japan in 1960); a densely populated, primarily agricultural country (India in 1901); and an unacculturated tribe of South American Indians, Yanomama (Neel 1970; copyright 1970 by the American Association for the Advancement of Science).

rate of mortality is constant at all age groups, so that an individual's chance of living another year is just as good at one age as another. This is approximately true for some adult birds (Farner 1945a). Type 3 shows extremely heavy mortality early in life, as with such forms as the oyster, where most of the millions of eggs spawned never hatch, but those few individuals that survive have a high life expectation thereafter. Most survivorship curves observed under natural conditions are intermediate between these three types (deer, Taber and Dasmann 1957; annelid worms, Lakhani and Satchell 1970; insects, Ito 1959; isopods, Paris and Pitelka 1962).

SEX RATIO AND MATING BEHAVIOR

The *primary* sex ratio at the time the eggs are fertilized should be approximately 50 ♂♂ : 50 ♀♀ in most species, although it has seldom been measured. This ratio may be displaced in one direction or the other by differential mortality of the two sexes during the period of growth, become manifest in the *secondary* sex ratio at the time of hatching or birth, and even more pronounced in the *tertiary* sex ratio of the adults (Mayr 1939). Recognition of sex and age in living animals is often difficult, although criteria have been worked out for many species (Taber, in Mosby 1960).

The sex ratio of the adults is especially important in understanding mating relations and reproductive potentials. For instance, the adult sex ratio in ducks is often in the neighborhood of 60 ♂♂ : 40 ♀♀ (Johnsgard and Buss 1956). Since these birds are largely *monogamous* (♂♀) under natural conditions, a population of 100,000 birds does not furnish 50,000 breeding pairs but only 40,000. In *polygynous* (♂♀♀) species, such as pheasants, some grouse, turkeys, deer, and fur seals, breeding potential is probably not diminished

under natural conditions if there are two, three, or even ten times as many females as males. On the other hand, *polyandry* (♂♂♀) has become characteristic of some tinamous and bustards, and this is correlated with a preponderance of males (Kendeigh 1952).

The tertiary sex ratio is not a constant factor. In the California quail it was found to vary monthly from 51 ♂♂ : 49 ♀♀ in early autumn to 53 ♂♂ : 47 ♀♀ during winter to 56 ♂♂ : 44 ♀♀ in June (Emlen 1940). Game birds that are monogamous or only slightly polygynous in the wild may become highly polygynous after hunting seasons or in captivity, situations in which there is a preponderance of females over males. Yearly variations in the ratio of males to females in the house wren are correlated inversely with tendencies toward polygyny, and positively with tendencies toward polyandry (Kendeigh and Baldwin 1937). Mating behavior is therefore to some extent adaptable in order to compensate for lop-sided sex ratios and to maintain high reproductive capacity.

BREEDING AGE

The age at which young animals first attain the ability for reproduction affects the reproductive capacity and rate of growth of populations. Planktonic entomostracans are sexually mature in a few days; insects, often in a few weeks. Among birds, a tropical sparrow is known to reach full reproductive level in 6 to 8 months (Miller 1959). Small non-tropical song birds commonly nest during spring and summer of the year following

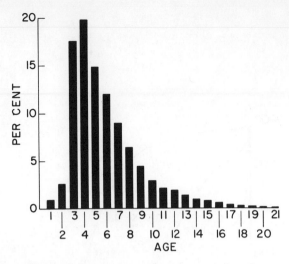

Fig. 13-4 Age composition of a breeding population of common terns (Austin and Austin 1956).

that in which they hatched, but banding of nestling house wrens indicated that 12 to 18 per cent failed to do so until the second or third year (Kendeigh and Baldwin 1937). Upland game birds probably nest as yearlings; geese and wild turkeys do not nest until they are 2 years old; common terns, commonly only after 3 years (Fig. 13-4). Some lizards and snakes require 2 to 3 years to reach sexual maturity; turtles much longer.

Females of the European voles may mate at 13 days, even before they are weaned, and give birth to their first litter when only 33 days old (Frank 1957). Woodland white-footed mice born in spring may produce young late in summer; but most small and medium-sized mammals do not breed until 1 year old (Fig. 13-5). Beaver, wolf, lion, and whale breed when 2 years old. Big game mammals, such as deer, bison, and bear, reach maturity only after 3 years. The elephant is said to require 8–16 years, and the rhinoceros 20 years (Spector 1956: pp. 115, 119).

NON-BREEDING POPULATIONS

Although the breeding population of a particular mammal, bird, or other animal is the only fraction of the total population of a species concerned with its reproductivity, there is often present in an area a substantial, though inconspicuous, non-breeding population that must be considered in any understanding of community dynamics (Zimmerman 1932, Watson and Jenkins 1968). In a 16-hectare (40-acre) tract of spruce-fir forest in northern Maine, there were, in 1950, 308 individuals (154 pairs) of nesting birds present during the first half of June. By the use of fire-arms the population was reduced to 21 per cent by June 21, and held at this level until July 11. This involved a removal of not only 228 breeding birds plus 49 of uncertain status, but also of 250 new birds appearing to take over the territories and places of the nesting birds that were removed (Hensley and Cope 1951). This surprisingly high non-breeding reserve may not be typical of all species of birds (Bendell 1955). Other studies have shown that the non-breeding population, especially of

birds, consists principally of young that have been slow to reach sexual maturity, of surplus individuals of either sex in monogamous species, and of adults which, for one reason or another, have lacked reproductive vigor or have been unsuccessful in establishing proper breeding conditions.

LONGEVITY AND MORTALITY RATE

When protected in captivity, animals are capable of living surprisingly long periods (Spector 1956: p. 182). Definite physiological limits of life are characteristic of each species and are occasionally realized under natural conditions (Cooke 1942), but invariably the potential longevity of a species is many times greater than the mean longevity actually attained by wild populations (Bourliere 1946). Protozoa and some coelenterates and flatworms are potentially immortal when they reproduce by fission.

Finding the mean length of life for wild populations requires the working out of life tables, accomplished for only a few species. In birds older than the juvenile stage, it commonly varies from 1 to 5 years (Farner 1955), although in some large species it is considerably longer. In rodents, usually not more than 6 per cent of the population reaches 1 year of age (Blair 1953). Rats of different species or races live on the average from 3 to 9 months (Bourliere 1959). The larger Dall mountain sheep has a mean length of life of 7.09 years. Adult barnacles have a mean life of 12.1 months (Deevey 1947), and different species of rotifers variously from

Fig. 13-5 Monthly changes in the density and age structure of a population of prairie deer mice in Michigan. As one population dies out a new one takes its place, and there is a population turnover (Howard 1949).

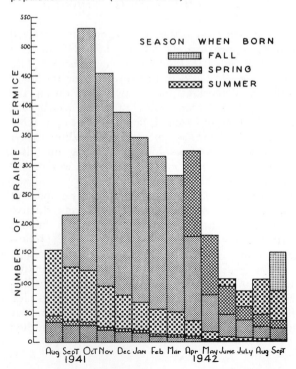

3 to 35 days (Edmondson 1946). Longevity may often differ between the sexes. Thus in the male flour beetle, *Tribolium madens*, it is 199 days, in the female, 242 days; in the male *T. confusum*, 178 days and in the female, 196 days (Park 1945).

The rate of mortality in many species varies from one age level to another; thus, a mean death rate has only general significance. In birds, however, the death rate is nearly constant once they become adult, and it is then apparent that it varies inversely with adult longevity (Table 13-5). In adult penguins, pelicans, shorebirds, gulls, and swifts, the annual mortality rate is commonly between 12 and 30 per cent; in herons, hawks, and owls it is about 30 per cent; in ducks, doves, and song birds it is between 40 and 68 per cent, while in gallinaceous birds it is the highest, 60 to 80 per cent. These rates for game species include mortalities from hunting (Farner 1955).

AGE RATIOS

The life table gives the number or percentage of individuals in a brood or litter surviving to the next age level. From such data, as well as by occasional direct observation, it is possible to determine the age structure of a population at any one time. Table 13-6 gives the percentage of each age class in populations of adults having three different mean survival rates in all age classes. It is at once apparent that the number of age classes in a population is greater when survival rates are high than when they are low. It is also evident that there is less difference between number of individuals in succeeding age classes when survival rate is high than when it is low.

The exact age of the sexually mature adult is usually at best difficult to determine unless one can band or mark the young when they first appear, or unless there are growth rings, such as in the scales and otoliths of fish and in the shells of clams, or other criteria of age that can be used. Immature animals are often distinguishable from adults (Thompson 1958), so that adult-young ratios are usually obtainable and often yield important information. The proportion of immatures to adults is highest at the end of the breeding season, and then usually declines until the beginning of the next period of reproduction, because of the higher mortality rate of the young compared with that of adults. In the California quail, the ratio of immatures to adults in October was 70:30. During the following months the ratio progressively decreased as follows: November and December, 62:38; January, 58:42; February, 56:44; March, 54:46; and the breeding season, April to June, 50:50 (Emlen 1940).

Age ratio is of practical value in wildlife management (Alexander 1958). A low ratio of immatures to adults indicates a poor reproductive season and should caution against excessive take or yield, as the population is declining. The continued public concern for the

Table 13-5 Relation between annual mortality and longevity in birds (after Lack 1951).

Bird species	Average longevity (years)	Average annual mortality (%)
Starling	1.1	63
California quail	1.5	50
Song sparrow	1.7	45
Lapwing	2.0	40
Barn swallow	2.8	30
European swift	5.1	18

Table 13-6 Theoretical age composition of stabilized populations with three different survival rates, assuming that the rate of mortality is the same for each age group. The figures are the percentage or number of animals (l_x) in each age class in a population of 100 (from Nice 1937).

Age (x) in time intervals	Survival rate		
	75	50	25
1	25	50	75
2	19	25	19
3	14	13	5
4	11	6	1
5	8	3	0
6	6	2	0
7	5	1	0
8	4	0	0
9	3	0	0
10	2	0	0
11	1	0	0
12	1	0	0
13	1	0	0
Totals	100	100	100
Average life span	3.8	2.0	1.3

survival of the whooping crane is justified, for instance. The entire wild population of the species winters in the Aransas Wildlife Refuge in Texas. From 1949 to 1953 there were only 3 to 4 young birds each year compared to 21 to 34 adults. Even though the species population has increased, in 1972 there still were only 5 young and 54 adults. Low ratios of young to adults also occur with overpopulation, but overpopulation is usually easily detected. Bag limits may ordinarily be increased if the ratio of young remains consistently high. Here again, however, high ratios of immatures to adults are characteristic when populations are recovering from catastrophes. When a population of rusty lizards in Texas was reduced by drought in 1954, the percentage of 1-year-olds changed from 63 in the relatively stabilized population to 85 in the subsequent expanding population (Blair 1957). When a population is stabilized, the young and middle-aged classes are more nearly equal in numbers, the decline in size occurring progressively throughout life.

ADAPTATION TO BIOTOPE

In the stabilized population of any species, whatever the number of eggs or young produced per pair of adults, the average number of offspring reaching reproductive status can never be greater than two in sexual forms, which is the number required to replace the parents on their death. With each new generation there is, therefore, a *population turnover*, with newly born individuals replacing the adults that die. In a stabilized population, the rate of increase of a population through the course of several reproductive cycles must equal the death rate, so that the value of one factor is also a measure of the other. Either factor is indicative both of the rate of population turnover and of the intensity of environmental resistance.

The intrinsic growth rates for populations (p. 211) of several species under optimum conditions is given in Table 13-7 by the factor r_m, which represents the potential rate of increase per individual per day. There is a general inverse relation between growth rate and longevity, T. However, if growth rate were dependent only on the longevity of the species, then $r_m T$ would be a constant. Obviously this is not true. It appears that different growth rates may correlate with various intensities of environmental resistance in the different habitats occupied by different species. If there were a habitat offering no environmental resistance, and all offspring therefore survived, then a female would need to produce only one female offspring to replace herself when she died. This would be the net production rate of Table 13-7. Actually, the number of female offspring that must be produced to offset mortality caused by the environment is always more than one, attesting the rigor of the natural environment in spite of the species adaptations for life in it. The method used in calculating the net production rate, $\Sigma l_x \cdot m_x$, the sum of number alive at age x, l, times rate of reproduction at age x, m, for all age groups, is explained by Evans and Smith (1952).

It is interesting that as a result of long evolutionary processes, the low net reproduction rates for the herbivorous vole, the omnivorous rat, and the parasitic human louse indicate that they are in much better balance with what to them are optimum environments than are the graminivorous flour beetle and rice weevil.

The vole and rat are viviparous; the other three, oviparous. It would be very interesting to have similar data on other species to show the degree to which adjustments to particular environments have become perfected and the reproductive strain imposed upon related species for occupying different habitats.

SUMMARY

Reproductivity is the rate at which a species reproduces. The number of offspring raised to maturity per unit of time is generally characteristic of a species, and varies with fecundity and survival of the young. Fecundity depends upon the morphological and physiological capacities of the species, the amount of parental care that the offspring receives, and weather conditions. Death rates correlate directly with the number of young produced. Reproduction cannot be considered successful unless the young reach sexual maturity.

Life tables tabulate, in condensed form, the vital statistics of survival and mortality by time intervals. They provide essential data for calculating longevity and age composition of populations. Survivorship curves show three characteristic survival patterns, but most populations exhibit a relatively high death rate early in life and a lower, more constant death rate thereafter.

Sex ratios are often correlated with mating behavior. The age at which full reproductive maturity is attained varies widely between species. Young birds, surplus adults of either sex, and birds unsuccessful in establishing breeding relations sometimes constitute a relatively large non-breeding population in addition to the more conspicuous breeding one. Perhaps this is true also for other animals, but evidence is scanty.

Ratios of young animals to adults often indicate whether a population is expanding, contracting, or is stabilized. In stabilized populations the number of offspring reaching reproductive maturity can never be greater or less than the number of adults themselves. The number of young that must be produced to permit such a population turnover gives a measure of the rigor of the environment, and how well adapted a species is to its niche.

Table 13-7 Comparison of intrinsic growth rates and other data on the populations of different species (compiled from various sources by Evans and Smith 1952).

Species	Intrinsic growth rate per day (r_m)	Average longevity (days), T	$r_m \cdot T$	Net production rate
Short-tailed vole	0.0125	141.75	1.772	5.90
Norway rat	0.0147	217.57	3.198	25.66
Flour beetle	0.101	55.6	5.616	275.00
Rice weevil	0.109	43.4	4.731	113.56
Human louse	0.111	30.92	3.432	30.93

Chapter 14

REGULATION OF POPULATION SIZE AND PEST CONTROL

It is seldom possible to measure the total worldwide population of any species, unless it is one of restricted distribution and is readily accessible to censusing. For most species, population size can only be expressed in terms of number per unit area (population density). The abundance of a species in a geographic region is termed its *average* or *regional density*. A region, however, usually includes unfavorable environments from which the species is absent, as well as suitable biotopes in which it is populous. The abundance of a species within its particular biotope is called its *economic* or *biotope density*, never less and almost always higher than its regional density.

The regional density of a species depends on the prevalence of its favored biotope in the area and the density which the species maintains within its biotope. Muskrats may be very numerous in a marsh, but if there are few marshes in the region, their average density will be low. We will here be primarily concerned with why an animal species attains a particular level of abundance within its biotope.

Species obviously vary in the level of abundance that they attain. Springtails may occur in hundreds per square meter, large mature snails as one per m², and white-footed mice as only one individual per 400 m². Reasons for these differences in population levels are the size classes of the animals, their position in food chains, pyramids, and trophic levels, and their behavior.

The population of any one species may be said to be *stabilized*, or *regulated*, when it fluctuates in an irregular but restricted manner from the mean (Fig. 14-1). If environmental conditions temporarily become unusually favorable or unfavorable, population size may fluctuate accordingly, but with stabilization there is always the tendency to revert again to the average level, when the unusual conditions have disappeared. The dynamic resiliency of populations is evident in the high rates of increase that occur with the beginning of recovery after a population has been depleted, and the progressive diminution in the rate of increase as the population approaches its characteristic level (Figs. 14-2,3; Tanner 1966). Fundamental for the analysis of population dynamics is an understanding of the growth curve. At the beginning of the population growth curve, increasing numbers may sometimes bring a cooperative effect evident in increased rates of growth and reproduction (Odum and Allee 1954). However, as the population continues to increase, cooperation is soon replaced by disoperation in that the reproductive rate then varies inversely and the mortality rate varies directly with the density of the

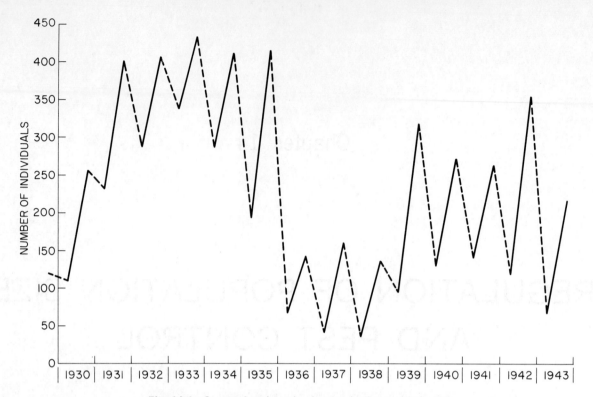

Fig. 14-1 Seasonal and yearly changes in the population of bobwhite on 1800 hectares (4500 acres) in Wisconsin. The solid line shows the net reproductive increase each year from spring to autumn; the dashed line, the mortality over winter (from Errington 1945).

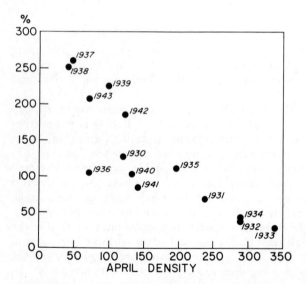

Fig. 14-2 Per cent yearly increase in population size of bobwhite in relation to April densities (Errington 1945).

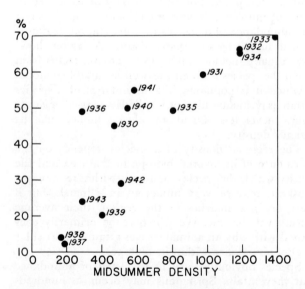

Fig. 14-3 Per cent loss rates of bobwhite from midsummer to early winter in relation to midsummer densities (Errington 1945).

population in terms of mature offspring raised. As a necessary corollary, the mean longevity of a population also varies inversely with its density (Davis 1945).

GROWTH CURVE

The characteristic sigmoid shape of the cumulative growth curve of organisms (Fig. 14-4) was demonstrated by Verhulst in 1839, and the most widely used description of the curve is the logistic equation (Pearl and Reed 1920). This curve has been used to describe the population growth of such diverse organisms as yeast, *Paramecium*, *Drosophila*, *Tribolium*, and man (Pearl 1927, Park 1939), and even the growth of communities in particular habitats as represented by the increase in number of species. Under natural conditions, however, growth of animal populations is subject to so many variable factors, including the change from one morphological stage to another in the life cycle of many species, and the change in the physical environment both daily and seasonally, that the curve is often not fully expressed, even though its trend is present inherently (Slobodkin 1961).

The sigmoid curve shows that a finite population grows slowly at first, then at an *accelerating rate* which is at maximum at the *point of inflection*, after which the population continues to increase but at a *decelerating rate*, finally becoming stabilized at the upper *asymptote*. Logistic growth curves are symmetrical, and the point of inflection is one-half the value of the asymptote. The lower concave part of the curve is called the *accelerating phase of growth* and the upper convex part of the curve, the *inhibiting phase of growth*.

If the number of new individuals added during each unit of time, *absolute growth rate*, is plotted against time midway in each period, a bell-shaped curve is obtained, the peak of this curve coinciding with the point of inflection on the sigmoid curve. However, the number of individuals involved in the absolute growth rate varies with the length of the time unit used, and the time unit of greatest significance varies from one species to another. Comparisons of growth rate of different populations are difficult unless instantaneous growth rates are obtained.

The *instantaneous growth rate* (r) is the rate of growth at a point on a time scale and is usually expressed in terms of increase per individual or unit biomass per unit of time. It cannot be measured, but it can be calculated by the differential equation (Park 1939, Andrewartha and Birch 1954)

$$r = \frac{dN}{dt} = r_m N \cdot \frac{K - N}{K}$$

where N is the size of the population at any time t; dN/dt stands for the instantaneous rate of change in the size of the population (dN) during an interval in time (dt) and hence may represent the growth rate at any desired time on the growth curve; r_m is maximum instantaneous growth rate, the biotic potential, innate capacity, or the *intrinsic rate of increase* per individual per unit of time in an environment where there are no limiting factors; and K is the maximum size of the population reached at the asymptote. This equation means that the rate of growth equals the potential rate of increase in the size of the population ($r_m N$), multiplied by the fraction of the maximum population size (carrying capacity) still remaining to be filled ($K - N$)/K. In its integrated form,

$$N = \frac{K}{1 + e^{a - r_m t}}$$

where the constant a is the natural logarithm of ($K - N$)/N when t is zero.

In order to solve the equation for the logistic growth curve, it is necessary to determine the intrinsic growth rate, r_m. In an environment without limiting factors, population growth is logarithmic. The factor r_m, the value of which varies with species, is the exponent that indicates this potential growth rate. The elephant, for instance, has a very slow growth rate. It has been estimated, however, that if all offspring survived and in turn reproduced, a single pair could give rise to 19,000,000 elephants in 750 years. On the other hand, a single stem mother of the common cabbage aphid gives rise to an average of 41 young, and there may be

Fig. 14-4 Annual ecesis of the invertebrate community in the herb and shrub strata of a deciduous forest (Trelease Woods, University of Illinois).

$$K = \frac{223 + 266 + 218 + 273 + 272}{5} = 250$$

$$a = \log e \frac{250 - 3.5}{3.5} = 4.25; \quad r = \frac{\log e(20/3.5)}{28} = 0.062$$

$$N = \frac{250}{1 + e^{(4.25 - 0.062t)}}$$

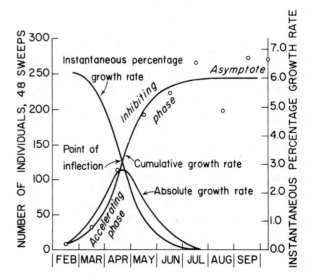

12 generations per year between March 31 and August 15. If they all lived, the progeny resulting would number 564,087,257,509,154,652 individuals in only 4.5 months (Herrick 1926). It is of considerable ecological value to determine both the maximum potential rate at which a species could increase under ideal conditions and the factors that prevent this increase from being realized.

The intrinsic rate of increase, r_m has been defined as *the maximal rate of increase attained at any particular combination of temperature, moisture, quality of food, and so on, when the quantity of food, space, and other animals of the same kind are kept at an optimum and other organisms of different kinds are excluded from the experiment* (Andrewartha and Birch 1954 p. 33). The age distribution of the population needs to be stabilized. Such an ideal environment may be set up under controlled experimental conditians. Actually, it is sometimes approximated under natural conditions during the very early stages of the accelerating phase of growth. Under such conditions the value of r_m may be approximated from the equation

$$r = \frac{\log_e (N_{t_2}/N_{t_1})}{t_2 - t_1}$$

Thus if a population is doubled in a period of 3 weeks, or if the mean length of life of a generation is 3 weeks, then $r_m = \log_e 2/3 = 0.2310$ per individual per week. A few values of r_m are given in Table 13-7. A number of factors affect the intrinsic growth of a species: number of young at each reproduction, the number of reproductions in a given period of time, the sex ratio of the species, the age distribution of the population, their age at reaching sexual maturity, and so forth (Birch 1948). There are other statistical procedures for calculating equations both for the logistic curve and for other variations of the growth curve (Ricklefs 1967, Caughley and Birch 1971).

In Fig. 14-4 the logistic curve has been fitted to the annual ecesis of the invertebrate community of the herb and shrub strata of a deciduous forest. The upper asymptote, K, was obtained by averaging the five randomly fluctuating monthly density values for June through October. The value of r_m was derived from the increase in community size for 28 days in February–March, but different values of r_m can be substituted in the integrated equation above until a curve is obtained that best fits empirical data.

Although r_m is an important constant, it occurs only under special conditions in nature. The equation for the actual or observed instantaneous growth rate (r) was solved for different parts of the curve to give the following values: March, 1.08 individuals per day; April, 3.55; May, 2.65; June, 0.59. By plotting intermediate times it appears that the highest growth rate, 3.87, comes about April 24, at the point of inflection of the growth curve, and also at the time of greatest

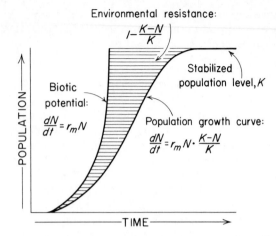

Fig. 14-5 Relation between biotic potential and environmental resistance in determining the population level attained by a species.

absolute growth. Since the instantaneous growth rate varies at different points along the growth curve, it is necessary to take equivalent stages in the growth curves for making rate comparisons between different populations. The point of inflection is of considerable significance in this regard; it represents the same equivalent age of populations whether they respectively attain to the asymptote in a matter of hours, days, months, or years.

The *instantaneous percentage growth rate*

$$100 \cdot \frac{dN/dt}{N}$$

declines progressively with time.

In a study of the process of population growth there are many advantages to working in the laboratory with populations of a single species held under experimental conditions where environmental factors can be closely controlled and varied at will (Park 1941). In such studies, the rate of growth of the flour beetle in experimental cultures has been found to vary with temperature, humidity, and light; according as whether fresh flour is added each day; whether competing forms or predators are introduced; and so on. The final population density turns out to be the same regardless of the number of beetles originally introduced into the flour, but it varies with the volume of the medium and other factors. The accelerating phase of growth is probably induced by frequent and successful mating contacts between individuals as populations increase in size. The inhibiting phase of growth is a result of decreasing food supply, accumulation of excreta which correlates with reduced fecundity of the adults; lowered rate of metamorphosis of the immature and increased mortality of the larvae, and cannibalism of the adults on the eggs.

Under natural conditions growth curves for populations of single species are clearly evident when a species invades a new area that is favorable; when a species is recovering from a catastrophe or cyclic depression; and as a species builds up its population in the spring after the termination of a winter dormancy or migration. Many factors in natural environments modify rates of population growth and determine the levels at which populations reach the asymptote. At the asymptote, the environmental resistance equals the biotic or reproductive potential and the population is stabilized (Fig. 14-5).

DENSITY-DEPENDENT FACTORS

The processes affecting levels attained by populations may be conveniently divided into two groups; those that are density-dependent and those that are density-independent. The first group of factors are biotic in that they depend on coactions between individuals within the same population or between populations of different species. Density-dependent factors may stabilize populations at an asymptote, the level of which is determined by the carrying capacity of the environment. Density-independent factors determine the potential level which populations may attain, and are basically physical in nature. They do not normally produce stabilization, since they vary in intensity largely independently of the size of population or community. All factors taken together are commonly considered to constitute the environmental resistance, a convenient if not entirely accurate term.

Density-dependent factors (Howard and Fiske 1911: p. 107, Nicholson 1933, 1954, Smith 1935, Solomon 1949, Richer 1954) are those that vary in the intensity of their action with the size or density of the population, but not all density-dependent factors are density-stabilizing. Only if the *percentage* of a prey species destroyed by predators, for instance, increases with the size of the population and decreases as the population declines, is natural control preventing indefinite population expansion, yet preventing extinction, too. This action then tends to *stabilize* the population size (Table 14-1). If, however, the percentage of prey taken remains approximately the same at all population levels the effect is *proportional*. If the percentage of prey or host affected actually decreases as the population increases, the effect is *inverse* (Tothill 1922). Obviously the proportional or inverse effects of a factor, whether predation or something else, cannot inhibit the continuous expansion of a population. Complete quantitative data are required in order to classify and evaluate the effect of any factor. The density-dependent factors that will be considered in respect to their stabilizing effect on population size are competition, reproductivity, predation, emigration, and disease. The effects of density-dependent factors have been much studied from a mathematical viewpoint, but the present approach will be largely introductory and

Table 14-1 Different effects of density-dependent factors on the size of animal populations, assuming number surviving is composed equally of males and females, and each pair gives birth to 6 young. The size of the prey population each generation includes both the young and adults. Thus, of 3 individuals surviving, the number of young produced (1.5 × 6 = 9) plus the 3 adults totals 12.

Factor	Generation						
	1	2	3	4	5	6	7
Stabilizing (assuming an increase of 10% mortality each generation until the population becomes stabilized)							
Size of prey population	4	12	32	72	128	180	180
Number destroyed	1	4	14	40	83	135	135
Per cent mortality	25	35	45	55	65	75	75
Number surviving	3	8	18	32	45	45	45
Proportional (assuming a constant rate of mortality)							
Size of prey population	4	12	36	108	324	972	2,916
Number destroyed	1	3	9	27	81	243	729
Per cent mortality	25	25	25	25	25	25	25
Number surviving	3	9	27	81	243	729	2,187
Inverse (assuming a decrease of 2.5% mortality each generation)							
Size of prey population	4	12	36	116	384	1,304	4,564
Number destroyed	1	3	7	20	58	163	456
Per cent mortality	25	22.5	20	17.5	15	12.5	10
Number surviving	3	9	29	96	326	1,141	4,108

non-mathematical (Andrewartha 1961, Boughey 1968, McArthur and Connell 1966, Southwood 1966, Watt 1968, Williams 1964).

Competition

The definition and basic principles of competition have already been considered (Chapter 12). We are here concerned with how competition helps to stabilize a population at a particular level. In this respect competition is primarily for space, cover, and food.

Every terrestrial green plant requires a volume of soil for its root system and a volume of air in which it can display its foliage to receive solar radiation. In a dense forest the individual tree grows tall because of competition with its neighbors. Trees unable to keep up with this competition become overtopped by other trees and, lacking sunlight, die. Sessile marine animals, such as corals, mussels, and barnacles, may crowd into close physical contact, even growing on top of one another, but there is undoubtedly a limit to the number that can survive and carry on normal activities in an area of restricted size (Fig. 14-6).

Competition for space is well demonstrated in those species that defend territories (Kendeigh 1947, Watson and Moss 1970). With increase in number of birds in an area, for instance, there is, at first, some accommodation as the size of territories varies inversely with the size of the population and amount of competition involved (Fig. 14-7; Kendeigh 1947). With decrease in size of territories, however, comes intensification of competitive singing, scolding, chasing, and fighting. On a 6-hectare (15-acre) area there were no instances of destruction of nests, eggs, or young in the 6 years through which the population of male house wrens did not exceed 11, but during the 13 years when such acts of destruction did occur, the male population had ranged from 11 to 16 (Kendeigh 1941b). A pair of birds requires a specific minimum territory for successful nesting. When an area becomes saturated with territories compressed to this limited size, disturbances occur in nesting and other individuals attempting to invade the area are forced to go elsewhere. Thus the population density becomes limited by the space available. Even with species possessing undefended

Fig. 14-6 Competition for space by barnacles; a median longitudinal section through a hummock (after Barnes and Powell 1950).

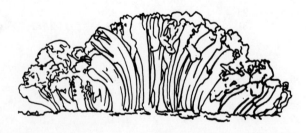

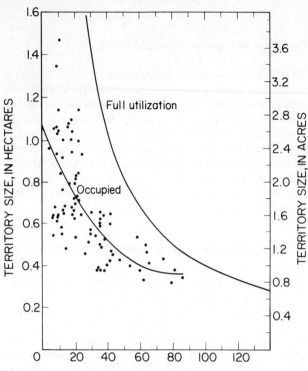

Fig. 14-7 Territory size in dickcissels as a function of the density of males. Note that the males' territory is compressible only to about 0.36 hectare (0.9 acre) and that at high populations the total space occupied approaches more closely the total utilizable (Zimmerman 1971).

and overlapping home ranges, aggression may be directed against strangers attempting to enter the social unit (Bustard 1970, Healey 1967).

Related to the competition for space is the competition for the most favorable portions of the biotope, those offering maximum food and protection. In Holland, three species of tits (*Parus major*, *P. coeruleus*, *P. ater*) prefer mixed woods to pine woods. In years when they are scarce, the species are mostly confined to the mixed woods; when populations increase in size, they do so first in the mixed woods, until the birds become intolerant of further crowding. Then they spill over into pine woods to nest but never become as abundant (Kluijver and Tinbergen 1953). High-quality environment for red squirrels is in cone-producing evergreens, where they must defend territories both for breeding and survival over winter. Surplus individuals are forced into poorer-quality deciduous forest types, where winter survival is greatly reduced (Kemp and Keith 1970).

For fish, the number of individuals may continue to increase in the presence of a limited food supply, but each individual becomes stunted in size. There is a tendency for the biomass of a species to be regulated by the food supply, with size or weight of individuals varying inversely with number. Thinning the population artificially usually results in increased growth of remaining individuals (Parker 1958). Body size in *Daphnia* (Frank *et al.* 1957) and in several species of mammals also appears to be to a certain extent density-dependent, in that smaller-size individuals are characteristic of larger populations (Scheffer 1955). On the other hand, with overcrowding in the medaka fish, a social hierarchy becomes established and only the dominant individuals maintain free access to the food (Magnuson 1962).

Competition is one of the most important factors in regulating population size, and its role is to ensure an adequate supply of food, space, and other resources necessary for the existence and reproduction of the individual. In those species in which direct competition (interference) results in the establishment of effective territories or social hierarchies, the population becomes stabilized at a level below that at which the food resources become exhausted and at which an adequate rate of reproduction is ensured. Individuals unsuccessful in obtaining territories or a high rank in a social hierarchy are forced to go elsewhere and may starve. At best, they survive but do not reproduce. Conventionalized displays or social behavior of other sorts, such as the dancing aerial swarms of gnats and midges, milling of whirligig beetles on water surfaces, and the vocalization of vertebrates and insects (Wynne-Edwards 1962), may also serve as feedback mechanisms signaling the level of population density and thereby preventing overpopulation.

When territorial spacing or social hierarchies are absent or inadequate for fully controlling population size, competition exerts effects indirectly, in reducing reproductivity, exposing individuals to higher rates of predation, and bringing emigration of surplus individuals.

Reproductivity

A study of the reproductive rate of a species in relation to population density requires separate studies of the number of eggs or young produced, called *fecundity* or *natality*, and the number reaching sexual maturity or *survival*.

In cultures of *Paramecium*, a decrease in the volume of culture fluid for the same initial number of individuals, or an increase in the initial number of individuals for a given volume of fluid, decreases the rate at which cell fission occurs (Myers 1927). Birth rate and growth rate in cultures of *Daphnia magna* vary inversely with the density of population even when a surplus of food is present (Pratt 1943, Frank *et al.* 1957).

There is an inverse relation between density of adults and fecundity in the pond snail *Lymnaea elodes*. However, when additions are made to the food supply, both the number of eggs laid and the number of young produced greatly increases (Eisenberg 1966).

Female *Drosophila* fruit flies and other insects (Fig. 14-8) crowded into small bottles do not lay as many eggs as they do when not crowded. This has been attributed to the competition of females for space, and to frequent disturbing contacts with other flies, so that

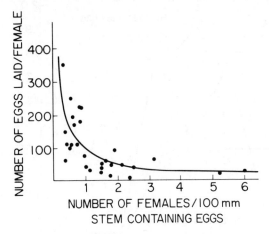

Fig. 14-8 Effect of crowding on the number of eggs laid by a jumping plant-louse on the stem of a plant (Watmough 1968).

Fig. 14-9 Effect of crowding on the mortality of jumping plant-louse nymphs (Watmough 1968).

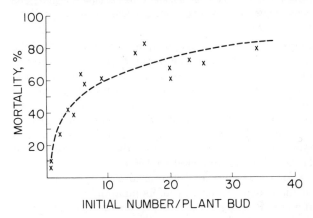

they do not feed adequately. It could also be attributed to their energy being dissipated and to their ovipositing being too often interrupted (Pearl 1932, Bodenheimer 1938, Chiang and Hodson 1950). On the other hand, the reduction in fecundity may not be so much a result of disturbance as one of reduction both in the quantity and quality of food that is available per individual (Robertson and Sang 1945).

Fertility of eggs appears to be high as they are laid under natural conditions. Egg viability, the capacity to hatch, has been shown in *Drosophila* cultures, however, to be modified by the same factors that affect fecundity, particularly the amount of food available to the adult (Robertson and Sang 1945).

The survival of young is greatly affected by the number of animals present (Fig. 14-9). When larvae of *Drosophila* are reared at different densities in containers of equal size and with equal amounts of food, the percentage that succeeds in pupating drops in an almost straight line with increase in density of the larvae (Fig. 14-10). However, because of the larger initial numbers of larvae present, the actual number pupating

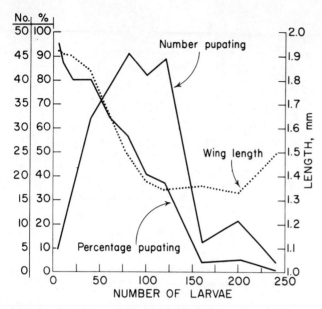

Fig. 14-10 Effect of density of larval populations of *Drosophila melanogaster* on number pupating, percentage pupating, and wing length of resulting female adults (after Chiang and Hodson 1950).

is greatest at intermediate densities. The size of the adults emerging from the pupae decreases abruptly as density increases. It has been shown experimentally that effects produced were due to the exhaustion of food at progressively earlier growth stages as the population densities increased. The continued growth of some individuals even after the original food was gone was apparently because they devoured dead larvae (Chiang and Hodson 1950).

For the grain weevil there is also an optimum intermediate density for rate of population increase even though the progeny raised per female decreases progressively as the population increases (Table 14-2). There is a similar relationship among fish (Herrington 1947).

Experimental studies of populations of the flour beetle *Tribolium confusum* show that, as they increase in size and modify the flour in which they live, there is a decrease in the number of eggs deposited, an increase in the length of the larval period before pupation, an increase in larval and pupal mortality, and a decrease in the weight of both the pupae and adults. Apparently these effects are produced partly by decreased fecundity of the individual females and partly by cannibalism of larvae and adults upon the eggs and pupae,

presumably induced by accumulation of excreta and deterioration of the food supply. When the modified flour is replaced by fresh flour at 48-hour intervals, the rate of reproduction rises, even when the beetle populations become very large (Park 1934, 1938, Park and Woollcott 1937, Hammond 1938–39, Rich 1956). Similar effects of crowding have been demonstrated on several other species of insects (Miller 1964, Kiritani *et al.* 1967, Watmough 1968).

Overcrowding of pink salmon in small impoundments causes retention of many eggs within the female at spawning, and perhaps also mechanical injury to the eggs already deposited from excessive stirring of the gravel (Hanavan and Skud 1954).

The average growth rate of tadpoles in a limited volume of water is inversely proportional to the number of individuals. However, some individuals grow at normal rates at all densities; the decreasing average growth rate at higher densities is due to the larger number of individuals that become stunted (Rose 1960). This effect of overcrowding is related to the presence of algal-like cells eliminated with the feces. This water conditioning effect on growth is interspecific (Licht 1967).

A study of a European bird, the great tit, carried on for 5 years in 16 different areas, revealed a striking inverse relation between density and fecundity (Kluijver 1951, Lack 1960, Krebs 1970). The average number of eggs laid per pair of birds per season varied from 13 to 20 at population densities of 8 to 12 pairs per 40 hectares (100 acres), to only 7 or 8 at population densities of 9 to 19 pairs per 40 hectares. The percentage of pairs having second clutches during a year varied between 40 and 100 at population densities of less than 16 pairs, but decreased to less than 10 at higher densities. Similar results have been obtained on the North American house wren (Kendeigh and Baldwin 1937). Apparently the lowered fecundity at high population densities in these cases is in part the result of frequent disturbance and conflicts resulting from the crowding of territories and in part to less food available per pair on the smaller-sized territories. The nonbreeding population of birds is doubtless high only when the breeding population is sufficiently dense that it occupies all of the most favorable territories or, as with the tawny owl, when there is a dearth of food (Southern 1959).

In the vicinity of Ithaca, N.Y., during 3 years of population increase, the average number of embryos

Table 14-2 Reproductivity of the grain weevil at different densities (Maclagan 1932).

Grain weevil	Density							
Weevils per gram of grain	0.25	0.50	1	2	4	8	16	32
Number of grains per weevil	100	50	25	12.5	6.25	3.12	1.56	0.78
Population size after 64 days	69	95	138	167	192	77	51	29
Progeny per weevil	17.2	11.8	8.6	5.2	3.0	1.2	0.4	0.1

Table 14-3 Relation of reproduction to density of laboratory mice during a 4-month period (after Retzlaff 1939).

Groupings	Number of groupings	Litters per female	Offspring per litter	Offspring per female
1♂, 1♀	12	4.7	7.8	35.2
2♂, 2♀	6	4.3	7.2	31.1
4♂, 4♀	3	4.0	6.6	26.2
8♂, 8♀	2	3.6	7.4	26.8
12♂, 12♀	1	3.4	6.4	21.9

per pregnant female in the meadow vole was 6 to 6.2, but in the year of decline following the peak, only 4.5 to 5.5 (Hamilton 1937a). Similar changes in size and frequency of litters have been found with the montane vole in California (Hoffman 1958) and with the snowshoe rabbit (MacLulich 1937, Rowan and Keith 1956, Green and Evans 1940).

Experimental studies of populations of the laboratory mouse lend support to high population as a curtailing influence on reproduction (Table 14-3). With an increase in the number of mice crowded into cages of uniform size there was a decrease in the number of litters produced, in the size of each litter, and in the total number of young. At the higher densities there was considerable fighting, resulting in serious wounds and even death for some individuals. A social hierarchy was established, and it appeared that only the despots at the top of the bite order were able to reproduce at a normal rate. Those at the bottom of the order produced few young or none at all (Crew and Mirskaia 1931, Retzlaff 1939, Crowcroft and Rowe 1957). It has also been shown in the coccid insect *Lepidosaphes ulmi* that a decrease in fecundity at high population levels was not a result of a decrease in number of eggs laid by fertile females but of a decrease in the percentage of females that were fertile (Smirnov and Polejaeff 1934).

An experimental population of house mice was established in a large enclosure with cover and water supplied in excess but with food allotments held to a constant daily amount. The population increased in size until the per capita consumption of food was cut by one-fourth as a result of the increased number of animals, and then reproduction stopped altogether. It appears that in times of stress the limited energy resources of animals are diverted from reproduction to individual survival (Strecker and Emlen 1953). In another experiment, food was supplied in excess but space and cover were restricted. Population increase was finally limited by litter mortality from cannibalism and desertion. In some of the populations there was a decline in fecundity. This appeared to be the result of a social hierarchy becoming established so that subordinated individuals failed to get adequate amounts of food, even though a surplus of food was available,

and were prevented from completing their mating behavior (Southwick 1955). These types of responses to overcrowding have been found in several other species of small mammals (Petrusewicz 1963, Lidicker 1965).

The amount of disturbance of females with suckling young in crowded experimental populations of the house mouse decreased the number of litters successfully weaned. Females abandoned or devoured their young, and when the disturbance factor became sufficiently severe, all successful weaning of litters ceased (Brown 1953).

It is desirable, when considering reproductivity as a density-stabilizing factor, to distinguish clearly between fecundity and success in raising young to maturity. There is increasing evidence that changes in fecundity are density-dependent in many species, but nevertheless are not usually sufficiently great to be of major importance in stabilizing populations at any definite level (Lack 1954a). However, mortality of the immature stages induced by intraspecific competition, predation, and disease becomes extensive with overcrowding and is usually much more important.

Predation

It is well known that variations in the population level of predators often coincide or closely follow variations in the population of prey species (Fig. 14-11), but it is not always certain whether the number of predators depends simply on the abundance of prey serving as food, or whether the predators by their feeding regulate the number of prey animals. Experimental studies amply demonstrate that under certain conditions, at least, both true predators and parasitoids greatly affect the numbers of the species on which they feed, and hence similar relationships may be looked for under

Fig. 14-11 Relation of changes in the populations of long-tailed mealybugs and their predators, both living on citrus trees in California during 1946 (DeBach 1949).

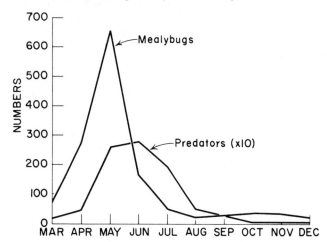

natural conditions. Basic mathematical treatment and equations to show variations and fluctuations in the number of individuals and in animal species living together as result of predation, competition, and parasitism goes back to the pioneer work of Lotka (1925) and Volterra (1931). For a more recent general treatment, see Slobodkin (1961), Wilson and Bossert (1971).

Predators may respond to an increase of prey "functionally," by an increase in number of prey consumed per individual predator, "numerically," by an increase in number of predators, or by a combination of the two (Solomon 1949, Holling 1959). The number of prey taken depends on the characteristics of both prey and predator and on the density and quality of alternative foods available to the predator. Each of these components may be further subdivided until a complex but realistic and precise computer model of the predator-prey system is formulated (Holling 1959, 1966).

Quantitative determination of the significance of predation in controlling populations under natural conditions is difficult to make, since it requires accurate measurement of the number of prey per unit area, the number of predators in the same area, and the number of prey taken by the predators. Some excellent studies of predator-prey coactions between vertebrates have been concerned with the relations between a single predator and its various prey species (Tinbergen 1933, Errington et al. 1940, Errington 1943, Murie 1944, Tinbergen 1946, Fitch 1947, Dunnet 1955, Craighead and Craighead 1956), or between a particular prey species and all its predators (Errington 1945, 1946, Bump et al. 1947, Koford 1958), but a complete understanding of the role of predation in regulating population levels requires a knowledge of coaction between all prey species and all predator species within community limits, since interrelations between any two species are affected by the interrelations of each species with others in the community.

Table 14-4 Percentage of different prey species taken by wood owls, during a winter (1930–31) in which populations of European meadow voles were high, compared with the following winter (1931–32) when vole populations were reduced (from Tinbergen 1933).

Species	Winter 1930–31	Winter 1931–32
Rodents		
European meadow-mouse	88.0	52.0
House mouse, European woodland mouse	5.8	14.2
European red-backed mouse	1.1	2.1
Norway rat	0.5	1.9
Birds	4.1	27.1
Miscellaneous	0.5	2.7

When meadow voles are abundant, an owl will feed largely on that single species, but when populations of meadow voles become reduced, the owl will prey to a greater extent on other species if they are available (Table 14-4). The plaice fish crops the siphons of a bivalve clam, when the clam is numerous, and thereby limits its population increase. If its actions reduces the prey population below a certain threshold, the fish switches its feeding mainly to polychaete worms (Steel et al. 1970). Other species that take the brunt of predation when a preferred species becomes reduced in availability are called *buffer species* (Bump et al. 1947).

Not only is there variation in the food of predators dependent on the availability of several potential prey species, but with any one prey species there is variation in the number and kinds of predators affecting it, dependent on its density and vulnerability. During outbreaks of insect or mouse plagues, predator species of many kinds converge on the easily obtainable food supply (Piper 1928, McAtee 1922, Kendeigh 1947). Predatory pressure is therefore very flexible, shifting its major impact from species to species and from one locality to another.

When only a small number of species are involved in the food web, as in arctic communities or with insect pests infecting cultivated crops, stability of population levels is difficult to attain. An increase or decrease in the abundance of any one species produces changes in all other species. On the other hand, when a large number of species are involved, as in tropical communities or in complex stands of temperate zone vegetation, each predator has so much choice of prey and each prey species is subjected to attacks from such a variety of predators that a sudden change in the population level of any one species is absorbed without greatly affecting the stability of the community as a whole (Voûte 1946, Craighead and Craighead 1956, MacArthur 1955). Cycles of population are much more prevalent, therefore, in far northern communities where the variety of species is scanty than in the highly complex communities of southern latitudes.

The importance that predation may have to maintenance of health and vigor in prey populations, aside from regulating their numbers, is of significance. In careful observations of 688 attacks by hawks on other birds, only 7.6 per cent were successful in the capture of the prey, but of these successful captures, over 19 per cent of the victims had previously shown injuries, abnormalities, or unusual behavior (Rudebeck 1950–51). In another similar study, the abnormal individuals among the victims varied from 14 to 33 per cent (Burckhardt 1953). Water boatmen with one or more legs artificially amputated were destroyed by fish in an experimental setup at a faster rate than were normal individuals (Popham 1942). Apparently predation exerts a selective force and less-fit individuals are eliminated at higher rates than are the fit. Fifty per cent of the kills that wolves make of caribou and mountain lions of wapiti and mule deer are of crippled or sick individuals, although the incidence of such individuals among caribou is less than 2 per cent. Very young and very old individuals are also more vulnerable than others (Crisler 1956, Hornocker 1970).

Vertebrate predators often also play a role in the control of invertebrate prey. This role is usually more effective when the prey population is at a low to medium density (Fig. 14-12). The rate of reproduction of insects and other invertebrates is so much higher than that of vertebrates that once the prey population surpasses a certain abundance level, the percentage taken by the vertebrate predator becomes less and less (Buckner 1966).

Herbivores often consume small amounts of plant growth and usually have no evident control over plant growth (Chapter 11). Marine fishes, however, may exert control over the algal growth along the shore (Randall 1961) and, similarly, littorine snails and limpets will keep the growth of diatoms in check (Castenholz 1961). When terrestrial herbivores are abundant, so that high grazing pressure is exerted on certain preferred species, the composition of the vegetation may be drastically altered (Harper 1969).

The ultimate result of parasitoids is the death of their hosts. The relations between host and parasitoid are just as varied and complex as between a true predator and its prey (Nicholson 1933).

The normal relation between host and *parasitoid* is one of equilibrium, where neither becomes overly abundant or overly scarce. This means that the abundance of parasitoids must also be controlled by density-stabilizing factors. Parasitoids may be infected with *hyperparasitoids*, and the relations between the two are similar to those between the pasasitoid and the original host (Nicholson and Bailey 1935).

At low host densities, the reproductivity of the parasitoid tends to vary proportionally with the density of the host (Varley 1947). However, it has been shown experimentally that with increase in population density of the parasitoid, there is decreased reproductivity for it, increased difficulty in finding individual hosts not already infected (DeBach and Smith 1941, Walker 1967), and increased competition between duplicate infestations in the same host individual (*superparasitism*), so that neither parasitoid survives (Fiske 1910): Of interest in this regard is that in one experiment 50 parasitoids during a limited period of time were able to find and eventually kill 80 per cent of the hosts, but that it required 100 parasitoids to find 95 per cent of the hosts and 200 parasitoids to find them all. This phenomenon is comparable to the law of diminishing returns and is doubtless one reason why a parasitoid rarely exterminates a host (DeBach and Smith 1947). In order for a particular parasitoid to regulate the numbers of a particular host species, it ordinarily needs to have a high intrinsic rate of increase, at least equal to its host (Muir 1914), and to have high searching

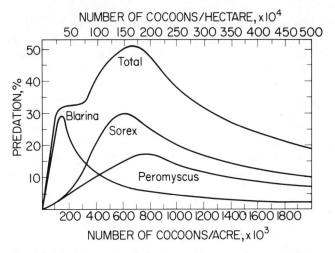

Fig. 14-12 Per cent predation by small mammals on cocoons of the European pine sawfly. The relation is density-stabilizing up to a certain population level, then density-inverse (Holling 1959).

ability for locating host individuals (Andrewartha and Birch 1954).

Even though the density of host or prey population greatly affects the success with which a parasitoid or predator finds its victim, searching is not random as far as the individuals are concerned. Predators in general have evolved many adaptations in sense organs, methods of attack, and special behavior patterns designed to facilitate the finding of specific prey, avoid unsuitable objects, save time, and increase efficiency (Thompson 1939). However, searching is largely at random as far as the area covered by the entire population is concerned since the individuals mostly hunt independently of each other and may cover the same or different areas indiscriminately. Many parasitoids avoid placing their eggs inside the bodies of prey that are already infected, but this behavior tends to break down when the ratio of number of parasitoids to number of uninfected prey is high.

The coaction between host and parasitoid may be complicated by differential effect of environmental conditions, such as temperature, on the two species. Experimental studies have shown that the greenhouse whitefly, a homopteran, at temperatures below 24°C lives longer, lays more eggs, has a higher rate of oviposition, and consequently increases more rapidly in abundance than does its chalcidid parasitoid. At 24°C, the rate of population increase is about the same in the two species, but above 24°C, the parasitoid population increases more rapidly than does the host species. The result is that the percentage of hosts infected increases markedly with rise in temperature (Burnett 1949). Similar relations have been demonstrated for other species of hosts and parasitoids (Payne 1934).

Buffer species may be as important with parasitoids as with true predators. Prior to 1925 in the Fiji Islands, the zygaenid moth *Levuana iridescens* was a serious pest of coconuts, defoliating trees over extensive areas.

Outbreaks terminated only when its supply of food became exhausted. In 1925, the tachinid fly *Ptychomyia remota* was introduced from Malaya and within a year reduced *Levuana* to a rare species, a status which it has had ever since. However, *Ptychomyia* requires alternative hosts to maintain its existence when *Levuana* becomes reduced in numbers. Such alternative hosts, or buffer species, occur on most of the Fiji Islands, but on one island from which they are absent the death rate among the predators became so high that *Leuvana* has been able again to increase in numbers (Andrewartha and Birch 1954).

Emigration

The pressure of overpopulations can be relieved by emigrations of individuals from particular localities as well as by their death. This has been shown experimentally with a house mouse population (Fig. 14-13). It is of interest that those individuals which remained continued their normal rates of reproduction. This contrasts with the drastic reduction, even cessation, of reproduction in other colonies from which emigrations were prevented (Strecker 1954).

Two species of aphids placed in their optimum niches, one at the top, the other at the bottom of a single barley plant, multiplied to saturation and dispersed downward and upward on the plant until both species came to exist side by side. Continued reproduction and overcrowding forced surplus individuals to emigrate to surrounding plants over 7.5 cm away, leaving the two populations in equilibrium on the original plant. In another experiment where plants were within 3.0 cm of each other, the aphids spread to the preferred sites on the second plant rather than to less favorable spots on the first plant (Ito 1954).

Emigrations under natural conditions occur when there is overcrowding in the migratory locust, lemming, grouse (Fig. 14-14), snowy owl (Gross 1947), snowshoe rabbit (Cox 1936), Arctic fox (Braestrup

Fig. 14-14 The 1932 emigration of sharp-tailed grouse from northern Ontario and Quebec (Snyder 1935).

1941), gray squirrel, and occasionally in other species (Heape 1931, Dymond 1947). The emigrations of the European lemming in the Scandinavian countries are spectacular (Elton 1942) and often lead to occupation of new areas elsewhere (Kalela and Kopenen 1971). Emigrations on a reduced scale are known to occur also with lemmings in North America (Thompson 1955).

Emigration has survival value for the species or otherwise it would have disappeared in the course of evolution. Emigration into less favorable environments may permit individuals to reproduce which would have failed to do so had they remained in the preferred habitat. The maximum productivity of tits, mentioned in the above section on *Competition*, occurs in a region when 65 per cent of breeding pairs occur in the preferred mixed woods and 35 per cent in the pine forest. Even though the rate of reproduction in the pine forest is lower than in the mixed forest, it more than offsets the depression which high populations have on fecundity of the birds in the preferred habitat (Brown 1969). Likewise by dispersing into new localities, there is opportunity gained for interbreeding with other populations leading to more genetic heterozygosity and adaptability. Populations of animals inhabiting marginal habitats often also avoid population crashes that frequently occur in congested areas (Evans 1942, Lidicker 1962).

It is, of course, population pressure that is responsible in large part for the dispersal of young and extension of ranges into new areas. Under normal conditions adult animals, especially among the higher vertebrates, are well established on their territories and the young

Fig. 14-13 Relation between population size of house mice, as expressed by average amount of food consumed per day, and emigration of mice away from the colony. The colony was started with five pairs of mice in January (Strecker 1954).

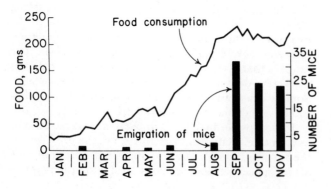

are forced to seek homes elsewhere. Among insects, there is a relation between emigration and inherited behavior tendencies. Individual tent caterpillars, both larvae and adults, differ innately in the extent to which they show activity even within the same colony. In the development of populations of excessive size, spread of infestations of the insect into new regions is largely by the more active individuals. The outbreak finally terminates when the proportion of sluggish individuals comes to predominate in the population (Wellington 1966). Genetic differences between dispersing and resident voles have also been demonstrated (Myers and Krebs 1971).

Disease

Although *infectious disease* in some form is a common cause of mortality, it is less important as a stabilizing factor than the others already considered, because it reduces the population size in an important manner only when epidemics, or more accurately *epizootics*, occur. The mortality may then be extreme so that the population falls way below the level of stabilization, and a period of recovery follows.

Whether or not epizootics occur depends on the virulence of the disease-producing organism, the rapidity with which it is transmitted from individual to individual, and the resistance of the hosts (Sladen and Bang 1969). Worm and protozoan parasites, bacteria, and viruses may be transmitted through body contact of host individuals, by the host ingesting contaminated food or water, or by vectors which are commonly external parasites themselves. It is obvious that ease and rapidity of transmission increase with the size of host populations. Overcrowding, scarcity of food, and inclement weather often also lower the vigor of the hosts so that they become more susceptible (Herman 1971).

In the course of time, natural selection tends to evolve tolerable relations between hosts and the disease organisms that they harbor. Mutations of disease organisms to greater virulence result in more rapid or extensive die-offs of the host, with the consequence that the mutant strains disappear. During upswings in host populations, extra virulent mutations may persist for a time, because of the abundance of host individuals to which they can spread, but when the host population declines, only those host individuals will survive that are not infected with the virulent strain or that develop immunity to it.

Epizootics among overcrowded wild animals are often severe. Among mammals, epizootics have been observed in voles, lemmings, mice, rats, beavers, squirrels, rabbits, moles, foxes, deer (Elton 1942), birds, fishes, and reptiles. Incidence of them is often sporadic in that they do not always appear with high densities of host populations or with declines in cyclic species so that their importance as a regulating factor on animal populations has been difficult to evaluate (Chitty 1954).

Physiological stress is not infectious but becomes pathological when extreme and may bring considerable mortality in the population. States of stress have been produced experimentally by allowing confined populations of albino and house mice and meadow voles to increase to high levels of abundance, even with a surplus of food and water present. Evidence that a state of stress exists is demonstrable in experimental populations by increase in the weight of the adrenals, by decrease in the weight of the body, testes, thymus, preputial glands, and seminal vesicles; by decrease in the number of circulating eosinophils in the blood; and by aberrant maternal behavior (Clarke 1953, Christian 1956, Louch 1956, 1958). An increase in adrenal weights, especially the adrenocortical tissue, as population rises, has also been shown in wild populations of the Norway rat (Christian and Davis 1956), and the behavior symptoms usually associated with high physiological stress have been observed in wild populations of European meadow voles (Frank 1953). Increased adrenocortical secretion leads to increased mortality, indirectly by lowering resistance to disease, parasitism, and adverse environmental conditions, or, directly, through shock disease. Stimulation of pituitary-adrenocortical activity is usually accompanied by inhibition of reproductive function and occurs regularly with the increased competition between individuals concomitant with large populations. This stress factor has been studied most intensively in mammals (Christian and Davis 1964). *Shock disease* appears to be a manifestation of the stress syndrome and occurs in snowshoe rabbits when the liver degenerates, leaving inadequate reserves of glycogen available for emergencies. Under these conditions, any undue exertion or excitement may cause animals to go suddenly into convulsions, sink into a coma, and die (Green and Larsen 1938).

Relaxation of Density-dependent Effects

With so many decimating factors acting on populations, one wonders why species do not become extinct more often than they do. The explanation is that there is relaxation in the intensity of action of the factors as the affected population becomes reduced in size. This is an aspect of density dependence, although of concern is control of the floor instead of the ceiling of population size. Hunting pressure, even by man, becomes greatly reduced as the time and effort required to bag prey exceeds the rewards obtained (Jordan 1971). Extinction is prevented by the heterogeneity of the environment so that at least some individuals escape the full force of the factor; hyperparasitoidism or overcrowding reducing the population of parasitoids and predators themselves; shift of predatory attacks to buffer species;

development of immunity to or tolerance of the factors involved; change in behavior so that the decimating factor is avoided; survival of dormant eggs, pupae, or encysted stages after the active stages in the life cycle perish (Solomon 1949).

DENSITY-INDEPENDENT FACTORS

Variations in *space or cover*, favorable weather, and food occur independently of population densities and may cause drastic changes in the abundance of animals. Heavy silting of estuaries along the coast from erosion of the surrounding upland may smother oyster spat and reduce the amount of hard surface available for setting quite independently of the number of oysters already there. The amount of solid surface available also determines the population density reached by sessile rotifers (Edmondson 1946). Variations in water level of a stream affects the availability of suitable spawning areas for fish, and consequently their abundance (Starrett 1951). A drought may dry up a marsh, making it unsuitable for muskrats and waterfowl. Space is suitable for many animals only if it provides adequate cover, usually vegetation, as shelter from adverse weather and refuge from enemies. The same amount of space but without cover will induce much fighting and mortality among meadow voles, which otherwise would occupy it peacefully (Warnock 1965).

Differences among species in their relative demands for space, food, and shelter affect the population levels that they attain. Species of small body size require less space than those of large size. In similar fashion, species that get along in small territories will be more numerous than those requiring large territories.

The physiological adjustments of animals to *weather* was described in Chapter 2. There are limits of tolerance to various climatic factors beyond which organisms cannot survive, and these limits vary with the species. A severe winter freeze may kill all but a few hardy individuals, regardless of the size of the original population (Kiritani 1964). Likewise, reproduction and growth take place more effectively under some climatic conditions than others and affect the population level attained (Table 14-5). Populations subjected to intolerable weather conditions commonly fluctuate violently and erratically. Weather can never bring stabilization of level in a population for any appreciable length of time (Klomp 1962).

Evidence is extensive that available *food* sets an upper limit to the size of populations when other conditions are favorable (*Paramecium*, Hairston 1967; rotifers, *Euchlanis dilatata*, King 1967; snails, *Lymnea elodes*, Eisenberg 1970; cladocerans, *Daphnia obtusa*, Slobodkin 1954; insects, Moore 1967; Ayala 1968; guppy fish, Silliman 1968; birds, Lack 1966). This principle was known to Thomas R. Malthus back in 1798 and influenced Charles Darwin in his develop-

Table 14-5 Effect of different combinations of temperature and humidity on the levels attained by populations of flour beetles in experimental cultures (Park 1954).

Temperature (°C)	Relative humidity (%)	Mean ± S.E.
34	70	38.25 ± 1.53
34	30	9.61 ± 1.07
29	70	50.11 ± 3.40
29	30	18.79 ± 1.32
24	70	45.15 ± 2.77
24	30	2.63 ± 0.35

ment of the theory of evolution. The supply of food varies from place to place because of substratum, type of vegetation, fertility of soil, climate, and other factors. Likewise, failure of a food crop any year or season because of weather or other cause will deplete populations that depend on it for subsistence. On the other hand, the deliberate increase in food, as in agricultural crops, has allowed some species to become abundant that once were scarce, for instance, many species of crop pests. In the food chain, herbivores are more abundant than consumers and consumers in the lower trophic levels more abundant than those in the higher levels, because of the greater biomasses of food available in the lower trophic levels.

Outbreaks of spruce budworm do not occur in the coniferous forests of Canada until the succession to white spruce and especially balsam fir develops a sexually mature evergreen canopy overtopping the aspen and birch. Insect larvae newly emerged from hibernation feed on the flowers, especially the male flowers, before the leaf buds open; then they move down to consume the current foliage, eventually defoliating and killing the trees. Millions of dollars worth of timber is destroyed. Native parasitoids are incapable of preventing outbreaks. The outbreak dies out in a few years, by which time all the mature trees are destroyed and the insects' food supply is exhausted. Another outbreak will not occur in the locality until another generation of spruce and balsam matures in the area (Prebble 1954).

Although there is no doubt that lack of space, unfavorable weather, and starvation may often limit increase in numbers or deplete populations immediately and directly, many, perhaps most species have evolved control mechanisms that stabilize populations at a level below that at which these factors would produce catastrophic effects. The dunlin sandpiper defends a territory in the high Arctic tundra, where food is relatively scarce, 5 times larger in size than it does in the subarctics, where food is more abundant (Holmes 1970). If it is not competition that stabilizes the population, it may be one of the other density-dependent factors. The *density-independent or environmental factors set the ultimate limit or carrying capacity* which cannot be permanently exceeded; the *density-dependent or biotic factors generally stabilize populations at or somewhat below that ultimate level.*

Although space, weather, and food are largely determined by factors extraneous to the population, their importance or intensity of effect may often be density-responsive in an indirect or passive way. Severe weather may decimate small populations relatively less than large ones because only with large populations are individuals forced into cover of inferior quality or become fully exposed. This relation also holds with some forms of predation. In stored grain, the heat produced by the infesting insects may raise the temperature beyond their limit of tolerance and thereby limit population growth (Solomon 1953). Insects and rodents at high populations tend to have reduced vigor and health and to be affected by weather conditions which at low population levels are easily tolerated (Chitty 1960, Wellington 1960). When populations of brown lemmings are low, they utilize less than 1 per cent of the annual production of the grasses and sedges which are their favorite food. At peak populations, however, they use nearly 100 per cent of the growth, thereby destroying their cover and greatly reducing the potential for food (Thompson 1955a).

INTERCOMPENSATIONS

The difficulty of evaluating, under natural conditions, the role of any factor in regulating the size of animal populations is in large part a result of the fact that it seldom acts alone. The *time* in the life cycle of an organism at which a factor takes effect influences the importance of it. Normally, the earlier in the life of the organism at which a factor is effective, the more nearly its apparent controlling role is a real one. Thus 60 per cent of the mature larvae of an insect may be fatally infested with parasitoids, but if 82.9 per cent of the original output of eggs have already failed to reach this stage for other reasons, the influence of these parasitoids must be evaluated at only 10.2 per cent, $0.60 \times (100 - 82.9)$, instead of the full 60 per cent (Table 14-6). On the other hand, a 10 per cent apparent mortality resulting from egg parasites which comes

at an early stage in the life cycle may produce a nearly equal real mortality (9.5); but a 10 per cent parasitization of pupae late in the cycle may cause only 0.47 per cent real mortality.

The *variability* of a factor also affects its importance. If a factor consistently produces, say, a 60 per cent mortality year after year, it will influence the size of the population any particular year less than will another factor that varies from, say, 20 to 30 per cent. However, a variation of 10 per cent in a factor that averages 60 per cent mortality is more important that the same variation in a factor that usually produces only 20 per cent mortality. An increase in mortality from 60 to 70 per cent reduces the surviving population 25 per cent $(40 - 30)/40$, but an increase from 20 to 30 per cent reduces the surviving population only 12.5 per cent $(80 - 70)/80$. Factors that cause a variable, though perhaps smaller, mortality are largely responsible for the observed change in population levels and are called *key factors* (Morris 1959).

Aside from time and variability, the influence of any factor is dependent on the level of population size at which it first comes to exert an effective or critical role. This level represents a *threshold of vulnerability* of the population for that particular factor. The threshold of vulnerability varies between species and within the species, depending on the amount of protective cover that is present, the movements and activities of the species, its protective coloration, and the aggressiveness and capabilities of the predators themselves.

The bobwhite is relatively safe from its important predator, the great horned owl, when it can find plenty of cover with food close at hand. However, when the prey population attains densities above the carrying capacity, so that surplus individuals are forced to use inferior cover or go greater distances in search of food, predation becomes intensified (Errington 1946). Parasitoids respond quickly to minor fluctuations in abundance of the spruce sawfly, but if the sawfly population escapes their control it soon reaches the threshold at which a virus disease become effective. There have been no outbreaks of the sawfly, since these parasitoids and the virus were introduced into the country (Neilson and Morris 1964). A population threshold of at least two aphids, *Drepanosyphum pla-*

Table 14-6 Evaluation of mortality factors effective at different stages in the life cycle of an insect (Thompson 1928).

Stage	Factor	Apparent mortality (%)	Real mortality (%)
Eggs at deposition	Sterility	5.0	5.0
Eggs after deposition	Egg parasites	10.0	9.5
Young larvae	Intrinsic factors	80.0	68.4
Mature larvae	Larval parasites	60.0	10.2
Mature larvae	Agricultural factors	30.0	2.05
Pupae	Pupal parasites	10.0	0.47
Adults	Meteorological factors	54.86	2.30
		Total	97.92

Table 14-7 Evaluation of horned owl winter predation on bobwhite (from Errington 1937).

Number of locality-winter records	Density of prey (% of carrying capacity)	Intensity of predation (% of owl pellets containing quail remains)
9	36–100	1.9 (0.0–6.3)
10	106–123	6.7 (0.0–16.0)
3	133–150	14.6 (10.4–19.0)
4	155–197	11.0 (0.0–20.0)

tanoides, per leaf is required on sycamore to maintain a beetle predator, *Adalia bipunctata* (Dixon 1970).

There is also an upper *limit of vulnerability* or *escape phase* (Voûte 1946) above which a factor no longer exerts effective restraint on population increase. This is well shown in the predation of small mammals on cocooons of the European pine sawfly (Fig. 14-12). With increase in the number of cocoons beyond the peak capacity of small mammals to attack them, the percentage of predation decreases and the sawfly escapes any significant controlling influence from this factor (Holling 1959, Beaver 1967). Bird predation on insects after a peak effectiveness is reached may even become a hindrance in that it destroys parasitoids that then become the most effective controlling factor (Thompson 1929, Betts 1955).

Competition between individuals appears at rather low population levels among vertebrates. With crowding, a social hierarchy may become established, or territories and home ranges become compressed in size. When fighting becomes intense, reproductive activities are disturbed and surplus individuals are forced to emigrate. Surplus individuals forced into inferior cover become especially vulnerable to predation (Table 14-7). In other species, such as insects, competition may be less effective, and the population quickly reaches a level at which predation or parasitism becomes significant. Emigration and epizootics among insects ordinarily do not occur unless competition, predation, and parasitism fail to control the increase and very high population thresholds are reached (Severtzoff 1934, MacKenzie 1951). The rise of populations beyond the predominant controlling influence of a factor does not mean that the factor no longer exerts any effect. Actually, population control is accumulative (Fig. 14-16), as with rise in population levels more and more of the density dependent factors come into play (Milne 1957).

For some organisms, no density-dependent force is effective; the population is never stabilized, and it continues to increase until there is exhaustion of space or food or curtailment by bad weather. The rose thrip, for instance, an insect that inhabits rose blossoms, multiplies rapidly during spring and early summer, as does the number of rose blossoms available to it. The favorable weather period normally ends long before the thrips have time to saturate the niche. Summer drought brings high mortality and a decline to the low densities of the species characteristic of late summer and winter. The rise of the thrip population is a race against time, the increase in density greatest in those years when the favorable period lasts longest; but it never reaches the point where competition becomes important. Annual variations in maximum densities in this species are almost entirely the result of density-independent climatic factors (Davidson and Andrewartha 1948). The fact, nevertheless, that the population increase over the course of years follows a sigmoid curve (Fig. 14-15) suggests that biotic factors are involved to some extent (Smith 1963). Climatic factors may be more important with many insect species than with other groups of animals (Uvarov 1931, Bodenheimer 1938, Thompson 1939, Andrewartha and Birch 1954).

The various stabilizing and limiting factors act in an *intercompensatory manner* (Fig. 14-16). All stabilizing factors, for instance, are in temporary abeyance following catastrophes of weather, drought, floods, or other factors until there is recovery of normal population levels again (Nicholson 1954a). One of the most thorough studies in this connection has been made on the muskrat (Errington 1946, 1951). This species is subject to such density-dependent mortality factors as intraspecies competition or fighting; predation, especially by mink and foxes; emigrations from overcrowded habitats; and epizootics. Overpopulations may be reduced by one of these factors singly, or by two or more working simultaneously. If fighting or predation keeps the population at a low level, disease is unimportant; but if fighting or predation is negligible some one particular year, then disease may

Fig. 14-15 Control of population size of thrips inhabiting rose blossoms by density-independent factors. The total population tends to increase each year, as indicated by the sigmoid curve, but never reaches saturation of the available niches because of the onset of summer drought. The dotted lines indicate decline from the maximum population size for each of the years 1932–37 (Davidson and Andrewartha 1948).

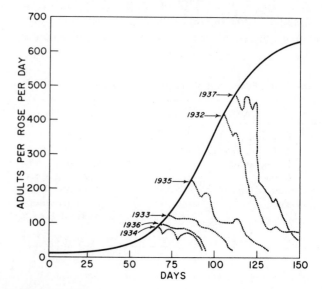

reduce the numbers of animals. Emigration to other areas occurs when a marsh becomes overcrowded or drought reduces the carrying capacity. If freezes, violent storms or floods, drought, or trapping reduces the population excessively, there is compensation by increased breeding activity, and for a time all other regulating factors are held in abeyance. The fur yield of a muskrat marsh cannot, therefore, be increased simply by destroying the predators, for other controlling factors become proportionately more effective. Trapping for fur, if not excessive, is economically profitable and can be carried on year after year, if the animals eliminated through trapping are restricted to the numbers that would be destroyed anyway by natural factors. The general trend is to maintain the population within the carrying capacity of the habitat. Improvement of yield is brought about only by increase in the carrying capacity in respect to food, cover, and space. These concepts are fundamental not only to an understanding of population dynamics, but also to wildlife management.

RELATION TO DISTRIBUTION

Variations in abundance of a species and the manner in which population levels are controlled are closely related to distribution. Three zones of abundance may be recognized. There is an inner *zone of normal abundance*, where climatic and other conditions are ordinarily favorable and high populations of the species are characteristic. Surrounding this inner area is a *zone of occasional abundance*, where climatic or other conditions are usually severe enough to hold populations at a low level, but where occasional years occur in which high populations may be reached. On the outside is a *zone of possible abundance*, where the normal environment is such that the species cannot maintain a permanent population but where the species may occur during favorable years by emigration from the inner zones (Cook 1929). Populations can become stabilized only in the innermost zone. Where climate, suitable space or cover, and food continually vary from year to year, as in the middle and outer zones, stabilization is never attained for any appreciable length of time (Swenk 1929). Ordinarily, therefore, density-dependent factors are most effective and become fully expressed only in the center or optimum habitat of the range of a species. Populations decline and fluctuate to an increasing extent toward the periphery of the species' range, where density-independent factors exert a relatively greater and more direct influence.

PEST CONTROL

"Pests are species whose existence conflicts with people's profit, convenience, or welfare" (Clark *et al.* 1967).

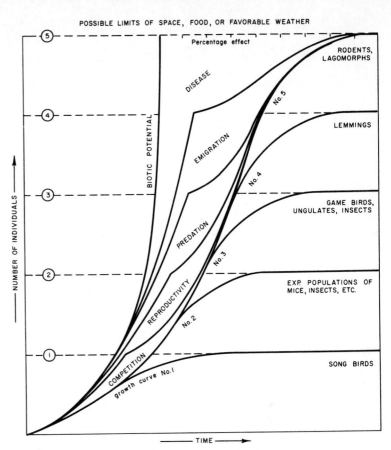

Fig. 14-16 Interrelations of various density-dependent and density-independent factors ("environmental resistance") in the regulation of population size at various possible levels. Considerable variation occurs between species in the relative importance and position of the different factors. Note that the limits of population growth set by space, food, or favorable weather may occur for different species at 1, 2, 3, 4, or 5, and that the influence of density-dependent factors is cumulative.

Any species may become a pest when it surpasses a certain level or threshold of abundance so that it exerts a depredating influence on man's interests. Pest control consists in keeping populations below this threshold (Fig. 14-17), and a variety of procedures have been and are being used (Kilgore and Doutt 1967). Most of the following discussion is concerned with insect pests, but basic principles of control apply to all forms of life, including vertebrates (McCabe 1966).

Species seldom reach pest proportions under natural conditions, because of the action of density-dependent factors and the great diversity of plant and animal species coacting in various ways. Species become pests when the long-established patterns of intra- and interspecific relationships become upset, as when man introduces exotic food plants in extensive monocultures, or when his pest-control measures exclude the possibility of natural biological controls becoming established.

Most pest species are phytophagous. Monocultures of plants invite rapid multiplication of those species that find the plants especially suitable as food. Return to *crop rotation*, as in the past, would reduce the cumulative buildup of these insect populations over the years.

Biological agents capable of maintaining insect

225

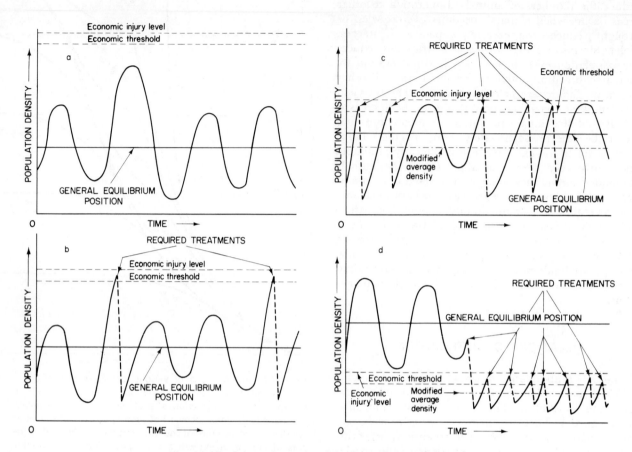

Fig. 14-17 Schematic graphs of the fluctuations of theoretical arthropod populations in relation to their general equilibrium position, economic thresholds, and economic-injury levels. (a) Noneconomic population whose general equilibrium position and highest fluctuations are below the economic threshold, e.g., *Aphis medicaginis* Koch on alfalfa in California. (b) Occasional pest whose general equilibrium position is below the economic threshold but whose highest population fluctuations exceed the economic threshold, e.g., *Grapholitha molesta* Busck on peaches in California. (c) Perennial pest whose general equilibrium position is below the economic threshold but whose population fluctuations frequently exceed the economic threshold, e.g., *Lygus* spp. on alfalfa seed in the western United States. (d) Severe pest whose general equilibrium position is above the economic threshold and for which frequent and often widespread use of insecticides or other control measures are required to prevent economic damage, e.g., *Musca domestica* in Grade A milking sheds (Stern *et al.* 1959).

populations at a lower average density level than would occur in their absence are commonly parasitoids, predators, and pathogens (Franz 1961). Hedge rows or weedy fence rows would provide diversified vegetation that could maintain populations of these biological agents ready to attack crop-destroying species when they first appear. The continuance of *clean culture* should be reevaluated. Clean culture and artificial pest control, along with heavy chemical fertilizers, give higher crop yields but at the expense of deteriorating environ-

ments. The problem requires economic reorientation and limiting the human population to a size that can be supported (Pimental 1961).

Resistant crop variaties can sometimes be developed for protection against the pest if the proper genes from mutant strains or related wild forms are bred into the cultivated ones. This often requires a worldwide search for the desirable hereditary factors.

Animal and plant pests are often exotic species introduced intentionally or accidentally. Finding an abundance of food, favorable climate, and in the absence of their traditional natural controls, these species quickly become destructive. More rigid *quarantine* against foreign species would reduce the number of new pests invading the country.

Attempts are often made with exotic pests to search out and introduce their natural *predators, parasitoids, or viruses*, but this program is fraught with difficulties. Over the last 80 years, about 520 such control agents have been imported into the United States. Of these some 115 have become established, but only about 20 have provided significant control of the pest. A difficulty in this procedure is the mass production required of the agent to get it established and functional. For effective control over the cotton bollworm in one operation, the release was made of 500,000 aphid lions per hectare (200,000 per acre) over a sustained period (Irving 1970). In the Hawaiian Islands, the sugar cane leafhopper is, however, controlled by the imported capsid bug *Cyrtorhinus mundulus*, and the sugar cane borer by the tachinid fly *Ceromasia sphenophori*. The black scale in California is successfully controlled by the chalcidid *Scutillista cyanea* imported from South Africa (Fleschner 1959, Varley 1959). The silk industry of Italy was apparently saved by importations of *Praspaltella berlesi*, which controlled the scale that was destroying mulberry trees, on the leaves of which the silkworm feeds (Thompson 1928).

These parasites and predators are usually specific in action, and so far there appears to have been no known damaging effects on non-pest species. Extreme care must be exercised, however. This requires not only detailed knowledge of the controlling agent but also of the dynamics and interactions within the community where it is to be introduced. Pests most adequately controlled are usually scale-insects, mealybugs, aphids, and leaf-hoppers of the order Homoptera which are sedentary, gregarious, and limited in the number of host species that they attack. Beneficial parasitoids and predators belong chiefly to the Hemiptera, Diptera, Coleoptera, and Hymenoptera (Sweetman 1952).

The relations between some insect herbivores and plants are somewhat similar to those between parasitoids and their hosts. *Cactoblastis* and *Dactylopius*, introduced from California, are credited with destruction of large areas of tree cactus in the Hawaiian Islands (Huffaker 1957). The prolific introduced aquatic alligator weed of southeastern United States, introduced from South America, is being brought under control by a flea beetle, also introduced, that came from the native home of the weed (Maddox *et al.* 1971).

A disease, myxomatosis, has been used for the suppression of the rabbit population in Australia. Caused by a filterable virus, it is highly contagious among the introduced European rabbits, but apparently it is not transmissible to man or other animals. The virus is carried between rabbits principally by mosquitoes.

Death occurs about 15 days after exposure. In 1950, extensive field trials with the myxoma virus were undertaken in eastern Australia, and by the end of the year mortalities locally as high as 99.8 per cent occurred. Epizootics continued in later years but with somewhat lower virulence. The prognosis of the disease is uncertain. In some regions of Australia, rabbits have recovered from less virulent strains of the virus, or there has been selection of genetically more resistant individuals, so it is possible that some degree of immunization may arise. The disease, however, may be successful in keeping the population at a low level. The virus was introduced into France in 1952 and by 1961 had spread to almost every country in Europe (Fenner and Ratcliffe 1965).

The controlling of pest populations through release of *sterilized males* has shown promise. The male screw worm fly, for instance, which infests the wounds of domestic animals, mates repeatedly but the female only once. If the female mates with a sterile male, infertile eggs are laid. The fly can be sterilized in the pupal stage by irradiation with x-rays or gamma rays, and large numbers have been reared in cultures, sterilized, and distributed by airplanes. This procedure has eradicated the pest from several regions where it was formerly prevalent (Baumhover *et al.* 1955). Another procedure involves the application of chemosterilants to native populations at a central source and then allowing the treated insects to disperse. This use of sterilized males is more effective when the population is still low and to prevent pest levels from being attained than in reducing populations once they have reached pest levels. The natural population needs to be overflooded by sterilized individuals by a factor of 25 or more for the procedure to be effective (Knipling 1963, Irving 1970).

The use of *attractants*, including chemical phenomones, light, sound, and so on, to lure animals to their doom is being investigated. In one experiment, the synthetic lure methyl eugenol was used to eradicate the oriental fruit fly from a small island within 6 months. Small squares of fiberboard containing the attractant and fortified with an insecticide were distributed by aircraft over the infested area at intervals of 2 weeks (Irving 1970).

There is a possibility, also, of using *hormones* to prevent normal growth or metamorphosis. A hormone-like fatty acid derivative from balsam fir has been found effective with the European bug *Pyrrhocoris apterus*. This chemical is specific for the insect family Pyrrhocoridae, which contains many harmful species. Why it is synthesized by the tree is not known, but possibly it serves a protective function (Williams 1967).

Chemical insecticides (Chapter 9) have been very important in the control of pests detrimental to agriculture and to human health and comfort. Their continued use would be desirable if their potential for harm to the environment were reduced. Effort should be made to render them more quickly degradable, more efficient in affecting only specific pests, less accumulative as they pass through food chains, and less toxic to man.

Of promise for the future are *integrated control programs* in which a variety of procedures are used, each in moderation and under conditions where it is most effective. Natural biological controls will be used as much as possible to restrain populations from rising above the threshold of economic harm (Fig. 14-17). Chemicals will be used only when the use of biological control fails (Stern *et al.* 1959, Geier and Clark 1960, Irving 1970).

SUMMARY

The regional density of a species depends on the prevalence of its favored biotope and its density within this biotope. The growth of populations forms a sigmoid curve and may become stabilized at the asymptote by density-dependent factors. The logistic equation provides a convenient description of many such growth curves. Density-dependent factors increase in intensity as the population level rises and decrease as the population level declines. The most important density-dependent factors are competition, fecundity and survival of young, predation, emigration, disease, and physiological stress.

The level at which populations become stabilized is determined by such limiting factors as space or cover, prevailing weather, and food supply. These factors are largely density-independent and determined by the physical conditions of the environment. At times their action is also responsive to the size of the population, as the amount of space, protection from weather, and food available per individual decreases as the population increases.

The influence of any factor upon a population is determined by the time in the life cycle of the organism at which it is effective, its variability, and its threshold and upper limit of vulnerability for the population. Intercompensations occur so that when one factor becomes ineffective in controlling the density of a population, another factor becomes more effective.

A species normally attains a stabilized level of density only in the center of its range, where physical conditions are optimum. Toward the periphery of its range, its population density becomes increasingly unstable and fluctuating.

Species become pests when they affect man's interests and need to be controlled. Maintenance of potential pest species below levels that are economically harmful may be aided by changing agriculture practice from continuous clean monocultures back to crop rotation and diversification of surrounding vegetation. More rigid guarantees against introduction of foreign species decrease the hazard of new exotic pests. Resistant varieties of crop species can sometimes be found or developed. Introduced predators, parasitoids, and viruses are sometimes successful in preventing or reducing outbreaks. Dispersal of sterilized males lower the chance of outbreaks occurring. Attractants may lure animals to their death. If biotic procedures fail, chemical pesticides are available. Intensive research is required for well-integrated control programs.

Chapter 15

IRRUPTIONS, CATASTROPHES, CYCLES, AND THE POPULATION EXPLOSION OF MAN

Abundance may change continuously and progressively in one direction over a long period of time, or variations in abundance may take the form of irruptions, catastrophes, or cycles. An understanding of how and why such changes in abundance occur is of considerable academic interest, and is of the utmost importance for the economic management of fish and game, preservation of wildlife, and in animal husbandry, agriculture, and forestry.

Minor fluctuations of less than 2:1 or 3:1 are often the result of sampling errors in estimating the true size of the population. When the ratio of population sizes from one period to another is greater than can be explained by errors of sampling, the fluctuations have meaning for which we should know the causes. Population ratios from one year to another are commonly of the order of 10:1, 100:1, or in insects, up to 10,000:1 or more (Solomon 1949).

PROGRESSIVE CHANGE

Populations that continue to increase or decline over a period of years are said to change progressively. The phenomenon is demonstrated as a species invades a new habitat or region or is becoming extinct. Progres-

sive change in numbers also occurs with seasonal growth of populations. Long-time climatic change may produce gradual changes in abundance and distribution. Thus the amelioration of winter temperatures in northern Europe since the mid-nineteenth century correlates with the northward dispersal and increase in abundance of several species of birds and mammals (Kalela 1949).

IRRUPTIONS, OUTBREAKS, PLAGUES

The phenomenon of a population suddenly exploding to supersaturate an area is called an *irruption*, *outbreak*, or *plague*. These terms are considered here to be synonymous and to represent the time when an animal is abundant or injurious enough over an appreciable area to be noticed and recorded by untrained observers (Carpenter 1940b). The number of rodents may be in the hundreds or thousands per hectare, of insects in the millions. Outbreaks are known to have occurred since the beginning of recorded history in Europe, Africa, and North America, especially in insects and rodents. Plagues of European meadow voles were recorded 18 times in France between 1792 and 1931 (Elton 1942). There were 16 invasions of the common crossbill into southwestern Europe between 1900 and 1965, the birds coming all the way from the boreal forest of Norway and northern Russia (Newton 1970).

The cause and control of plagues or outbreaks have concerned man since civilization began (Fig. 15-1). Outbreaks are commonly associated either with favorable weather and food conditions so that survival and reproduction is increased, or with a disturbance in predator-prey relations. Rapid increase in population size may thus occur with almost any kind of animal in any habitat. Irruptions of the bean clam occurred several times between 1894 and 1955 in the intertidal zone at La Jolla, California. The abrupt decline of the last outbreak in 1951–52 was the result of an epizootic associated with a minute unicellular organism of uncertain identity, found in the tissues of the clam (Coe 1955).

CATASTROPHES

Catastrophes occur at more or less widely spaced intervals and bring marked depressions in the population level of a species. Figure 15-2 shows annual populations of the house wren over 53 years, first in Ohio, then in Illinois. Decidedly low points in this curve occurred in 1918, 1926, 1940, 1958, and since 1961. Information from observers indicated that low populations were widespread in eastern North America both in this species and in many other song and game birds during most of these years. There is considerable evidence that these conspicuous variations in abundance, as well as

some less pronounced, were the result of severely low winter temperatures. It is of interest that, in England, severe winters, causing high mortality among such song birds as thrushes, blackbirds, and tits, were recorded in 1111, 1115, 1124, 1335, 1407, 1462, 1609, 1708, 1716, 1879, and 1917 (Elton 1927). To birds that feed on the ground, the depth of snow is as critical a factor as low temperature in determining the number that survive (Fig. 15-3).

Among mammals, fluctuations in the population of the common hare in Denmark have been correlated with the varying effects of summer rainfall, spring temperatures, and the number of days of frost during the winter (Andersen 1957).

Catastrophes may occur with practically any type of animal life. The severe winter of 1917–18, for instance, produced a marked reduction in the numbers of many species of marine invertebrates in the region of Woods Hole, Mass. (Allee 1919). In seeking correlations between catastrophes and weather or other environmental conditions, one needs to determine the critical period in its life cycle during which the species is most vulnerable, and the weather conditions coincident with that period that exert greatest effect.

CYCLES

Populations are said to be cyclic when they alternatively irrupt and subside in a more or less uniform manner between high and low levels of density.

Fig. 15-1 Onset and subsidence of an outbreak of chinch bugs in Illinois during the 1930s. Areas supporting the densest populations are indicated by the darker pattern (Shelford and Flint 1943).

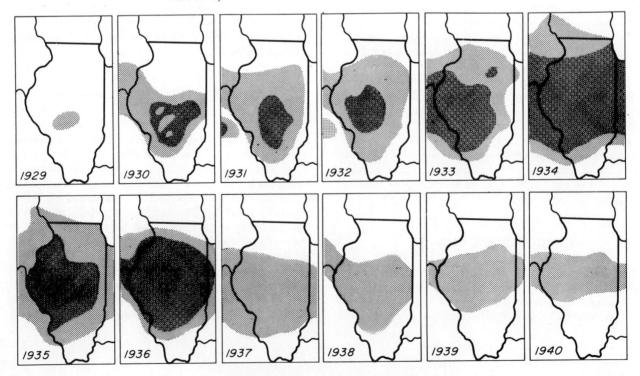

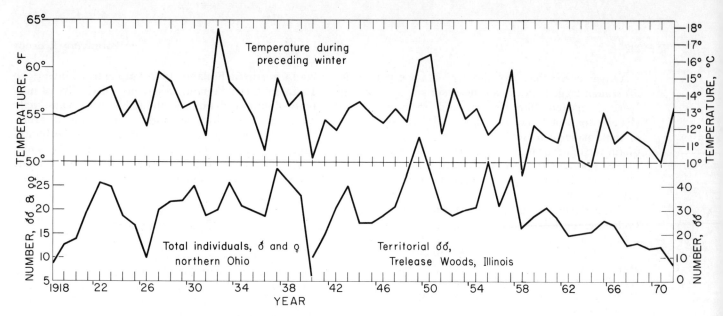

Fig. 15-2 Yearly variations in populations of the house wren. The data for northern Ohio are for total males and females on a 15-acre (6 hectare) estate, most of the males being banded and captured at their nest-boxes. The data for Trelease Woods, in central Illinois, are for territorial males censused on a 55-acre (22 hectare) tract by the spot-map method. Mean temperatures are for the wintering range only, based on monthly weather data for Tampa and Jacksonville, Florida; Savannah, Georgia; Montgomery and Mobile, Alabama (Kendeigh and Baldwin 1937, Kendeigh 1944).

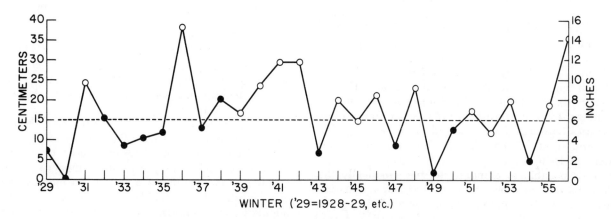

Fig. 15-3 Fluctuation in the average snow depth in the winter in southwestern Finland and population fluctuations of the partridge during the autumns following. The dotted line shows the critical snow depth, below which increases (solid circles) and above which decreases (open circle) in population usually occur (Siivonen 1956).

Types

Although cycles of different duration have been postulated for many species at different times, the best established cycles are those of periodicities of 3–4 years and 9–10 years (Speirs 1939, Elton 1942, Dymond 1947, G. R. Williams 1954, Siivonen 1957, Wing 1961, Keith 1963, Dewey 1965, Pearson 1966, Migula *et al.* 1970).

The best known 3–4 year cycles are demonstrated in the following species:

Birds
Snowy owl
Willow ptarmigan
  (northern Europe)
Capercaillie
Blackgame
Hazel grouse

Mammals
European lemming
Siberian lemming
Brown lemming
Collared lemming
Meadow vole (several
  species)
Arctic fox

Other species that may also vary in a 3–4 year cycle are rough-legged hawk, northern shrike, red fox (far North), marten (far North), and sockeye salmon (Pacific coast of North America). All these are species of the northern hemisphere. However, the California quail, introduced into New Zealand, showed a clearly regular and synchronized periodicity of 4 years between 1948 and 1961 in these southern hemisphere islands (Williams 1963).

Species well recognized as showing the 9–10 year cycle are the following:

Birds	*Mammals*
Ruffed grouse	Snowshoe rabbit
Sharp-tailed grouse	Muskrat
Willow ptarmigan	Canada lynx
(North America)	

In addition, a number of other species may have a 9–10 year cycle: rock ptarmigan, goshawk, great horned owl, red fox (South), marten (South), fisher, mink, and Atlantic salmon.

The 9–10 year cycle is not common in Europe, but fluctuations of approximately this periodicity occur in Greenland and Iceland. Fluctuations in populations, commonly of 5–6 years (Mackenzie 1952), occur in the British Isles in red grouse, rock ptarmigan, black game, and capercaillie, but there is question as to whether they are regular and definite enough to be truly cyclical. Among invertebrates, insect pests of coniferous forests in Germany fluctuate in periods variously from 6 through 18 years (Eidmann 1931); grasshoppers in Manitoba, 7 through 16 years (Criddle 1932); and chinch bugs in Illinois, from about 3 through 16 years (Shelford and Flint 1943). Subjective estimates of damage by the starfish *Asterias forbesi* on mollusk fisheries between New York and Cape Cod suggest a periodicity of 14 years for this marine species (Burkenroad 1946).

It is possible to demonstrate mathematically that an apparent cycle or a series of irregular fluctuations may actually be compounded of several distinct periodicities, each of different duration (Wing 1935). Cycles can thus be postulated in the population fluctuations of many species, in the migration of birds, and in human economics (Wing 1935, Huntington 1945), but the biological significance of these hidden periodicities remains to be demonstrated.

The 3–4 year cycle is better expressed in the far North and the 9–10 year cycle in more southerly latitudes. The 3–4 year cycle may change to a 9–10 year cycle, correlated with latitude, even in the same species (red fox, marten). South of latitudes 45°–50°N in North America and about 60°N in Europe, variations in population size appear progressively less extreme and cyclic, more irregular or random in character. Thus the numbers of four species of gallinaceous birds during peak years divided by their counterpart numbers during low years changes from 3.8 in Lapland, to 2.4 in northern Finland, to 2.0 in central Finland, to 1.7 in southern Finland (Siivonen 1954). Cycles may be distinct and definite in the far North because only a few species are involved and the environment is relatively monotonous and severe; in more southerly latitudes, population fluctuation becomes more irregular and uncertain because of the interaction of many species and a more moderate environment.

Reality of Cycles

To designate fluctuations as cyclic implies considerable regularity for them. What ecologists call cycles are really *oscillations* because both phase and amplitude are inconstant. The variability that is evident, however, especially in phase, is less than is to be expected by chance, and reasonably accurate predictions can be made of the course of future variations in population size (Davis 1957, MacLulich 1957).

The short-term cycle is commonly 3, 4, or 5 years long, although it may be as short as 2 years, or as long as 6 years (Elton 1942). The snowshoe rabbit cycle varies between 8 and 11 years; the lynx cycle, between 8 and 12 years (MacLulich 1937). The coefficient of variation, standard deviation divided by the mean, for different species having the short cycle varies from 30 to 50 per cent, and is of the same order of magnitude for the longer cycle (Cole 1951). It is of interest that by drawing numbered cards from a well-shuffled deck or rolling dice (Palmgren 1949, Hutchinson and Deevey 1949) or plotting random numbers (Cole 1951) short and long cycles may be obtained of about the same relative lengths and variation coefficients as animal population cycles.

There has been considerable controversy about the true nature of cycles (Cole 1954, Hickey 1954, Pitelka 1957), that is, whether they are caused by a master factor or factors or by the random impact of a large number of factors acting more or less independantly of each other. Before oscillations in the size of natural populations are considered cycles, they should be tested statistically for randomness (Keith 1963). Even then, the cause of cyclic populations can be expected always to contain some random components inextricably mixed with any systematic ones (Cole 1956). The reality of cycles may also be tested by the amount of synchrony that they exhibit in different localities and on different species. The presence of such synchrony would indicate the presence of broad coordinating influences or mechanisms rather than haphazard effects of many random factors.

Synchrony

With the 9–10 year cycle, local areas may show peak populations that are out of phase with other local areas

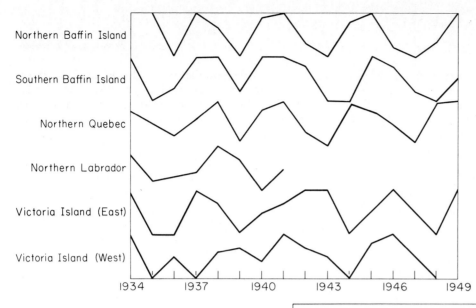

Northern Baffin Island

Southern Baffin Island

Northern Quebec

Northern Labrador

Victoria Island (East)

Victoria Island (West)

1934   1937   1940   1943   1946   1949

Fig. 15-4 Relation between peaks and lows of the lemming cycle at different localities in eastern and northern Canada (Chitty 1950).

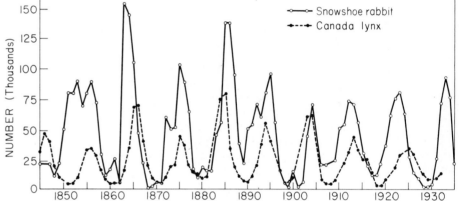

Fig. 15-5 Population cycles of the snowshoe rabbit and one of its chief predators, the Canada lynx, in northern Canada (data adjusted for years 1912 to 1920), based on number of pelts handled by the Hudson Bay Company (from MacLulich 1937). More recent peaks of the snowshoe rabbit have occurred in 1942–43, 1953, 1961, and 1971 or 1972, and of the Canada lynx about 1935, 1944, and 1954 (Keith 1963).

by 1, 2, or 3 years. But when large regions are considered, the peak is manifested over 3 or 4 years in the course of which most local areas reach maximum populations while the following trough in the regional cycle may spread over 5 or 6 years when very few, if any, local areas have large populations (Butler 1953). A similar relation probably holds between local and regional fluctuations with the 3–4 year cycle (Fig. 15-4; Hayne and Thompson 1965). Synchrony is sometimes evident in local populations that are isolated by a hundred or more miles (161 km) from other populations of the species (Brooks 1955).

Some variation in cyclic tempo occurs in different parts of the world, although generally they are close to being in phase. Local areas out of phase with the main cycle commonly come back into phase by the time the next peak is reached. The main cycle of grouse and ptarmigan over most of Canada and in northern United States has shown peaks in 1896, 1905, 1914–16, 1923, 1932–33, 1941–42, and 1950–53. In the maritime provinces of Canada—Newfoundland, New Brunswick, Nova Scotia, and Prince Edward Island—the cycle is advanced 3 years ahead of the main cycle; in Alaska, the cycle lags by 3 years. The isolated population of rock ptarmigan in Iceland has peaked in 1871, 1880, 1891, 1902–19 (two cycles merged?), 1924–27, 1935–36, 1945, and 1954 (Gudmundsson 1960). In North America, the grouse cycle is nearly synchronous with

the cycle of the snowshoe rabbit (Fig. 15-5), but in Scandinavia it coincides with the cycle of the lemmings (G. R. Williams 1959).

The lemming cycle in Canada is similar to that in Norway (Elton 1927). Peaks in the 3–4 year cycle for small rodents in Finland, Norway, and eastern North America have occurred in 1923, 1926, 1930, 1934, 1937–38, 1941–42, 1945–46, 1948–49, and 1953, with deviations of one or more years for particular areas (Siivonen 1954). Peaks occurred in northern Alaska in 1946, 1949, 1953, 1956, 1960, 1965 (Schultz 1969), and either 1971 or 1972. Rodent peaks of other species at southern latitudes are often out of phase with the lemming peaks by 1 or 2 years (Pearson 1966, Migula et al. 1970).

The predator cycle is dependent on the cycle of the herbivorous mammal or bird prey species (Fig. 15-5). The correspondence in the cycles of predator and prey is usually close, although that of the predator often lags 1 or 2 years behind that of its prey (Chitty 1950, Butler 1953). The snowy owl emigrates in large numbers from Canada into the United States within a year after the decline in the lemming population (Shelford 1945). In those parts of Greenland where the fox population lives largely on lemmings, the 3–4 year fox cycle is very pronounced, but this is not true for the coastal areas, where the fox depends on a variety of food other than lemmings (Braestrup 1941).

233

Intrinsic Causes

Instrinsic factors affecting animal cycles are those inherent in the populations themselves and in the interrelations of different species. They are primarily biotic in nature (Hutchinson and Deevey 1949). A number of general theories of possible intrinsic mechanisms have been found inadequate, and still other theories have limited application (Slobodkin 1954, Dymond 1947, Thompson 1941). However, disease, predation, food, and genetical or physiological explanations merit consideration.

Disease An older but common theory of the general cause of cycles hypothesizes that animal populations build up to a peak, at which time an epidemic disease occurs that reduces numbers until the disease can no longer spread. The cycle then starts over again. Among epizootics observed in cyclic species are those caused by the blood-sucking stomach worm *Obeliscoides cuniculi* in the snowshoe rabbit; the blood protozoan *Leucocytozoon bonazae* in the ruffed grouse, and protozoan infection of the brain caused by *Toxoplasma* in rodents. However, these diseases have been encountered in some cyclic declines but not in others and offer no explanation of the regular recurrence of the cycles. *Toxoplasma*, for instance, was reported in three early population declines of rodents, but was not demonstrably present in later declines (Elton 1942).

Predation According to early mathematical theories of Lotka and Volterra (D'Ancona 1954) and of Nicholson and Bailey (1935), a population consisting of a single prey species and a single predator or parasitoid species occurring together in a limited area, with all external factors constant, automatically displays periodic oscillations or cycles in the numbers of both species. As the predator population increases, it will consume a progressively larger number of prey until the prey population begins to decrease. As the number of prey diminishes, there will be less food for the predator, and they will thus decline. After a time the number of predators will be so reduced that the high reproductive rate of the remaining prey will more than compensate for the loss from predation, and the numbers of the prey species will again increase. This will be followed shortly by an increase in numbers of the predator. The cycle would thus continue indefinitely. According to the differential equations, the predator will never be able completely to destroy the prey, nor will the predator species ever completely disappear by reason of starvation.

There is considerable controversy concerning this theory (Andrewartha and Birch 1954). Gause (1934, 1935) conducted a test in a classic series of experiments with protozoan cultures. An experimental food chain was established: boiled oatmeal $\longrightarrow$ bacteria $\longrightarrow$ *Paramecium caudatum* $\longrightarrow$ *Didinium nasutum*. When five *Paramecium* were introduced 1 day and three *Didinium* 2

days later, the population of *Paramecium* was exterminated by the predator. The predator, left without food, disappeared soon after. In another experiment, cover sediment was introduced into the microcosm, in which the *Paramecium* could hide, thus to escape the attacks of *Didinium*. The same number of each species was introduced at the same time. The number of predators increased, and they devoured many of the prey. However, the remaining prey escaped into the cover, and the predators died of starvation. When this happened, the prey, now unchecked, increased in an unlimited manner. When, on the other hand, a microcosm was prepared in which there was no refuge, and one *Paramecium* and one *Didinium* were introduced on every third day, a series of oscillations resulted. It appeared that continual cycling of prey and predator populations could be maintained only with immigration of individuals from the outside. In other experiments, interrelated cycles of *Paramecium* and the yeast on which it fed were established (Gause 1935). Another predator on *Paramecium*, *Woodruffia*, encysts when the prey population is reduced below a certain level, thus allowing its recovery (Salt 1967).

In experimental greenhouse plots of strawberries, populations of an herbivorous mite, *Tarsonemus pallidus*, and its predator, another mite, *Typhlodromus*, fluctuated regularly in relation to each other. At low populations, the prey species was relatively secure in the cover offered by hairs, spines, and leaf crevices, thus avoiding annihilation. The predator species survived because it utilized honeydew and other nourishment as substitute food until the prey species again increased in numbers (Huffaker and Kennett 1956).

Reciprocal fluctuations in the density of the azuki bean weevil and its larval parasitoid, a braconid wasp, were sustained experimentally under constant conditions for 112 successive generations. Apparently the prey was able to survive in the low of the cycle because of the difficulty the parasitoid experienced in finding the surviving individuals; the parasitoid, however, never became extinct (Utida 1957).

These examples indicate that oscillations in the populations of predators and prey can be sustained for relatively long periods of time only if such factors as cover for the prey, buffer food, temporary abatement of predation, or immigration are introduced into the experiment, which exempts the prey population from attack at the low point of the cycle. This background of experimental studies is useful in the analysis of oscillations of animal numbers that are observed under natural conditions.

In an area near Point Barrow in northern Alaska, Siberian lemmings were scarce from 1949 to 1951, increased in 1952, and were near or at a peak in 1953. Associated with this cyclic rise in the lemming population was a marked increase in the number of predators. There was no breeding in 1951 of pomarine jaegers, snowy owls, and short-eared owls; very few were even seen. In 1953, however, breeding pairs were recorded in densities respectively of about 18, 0.3, and 3–4 per 250 hectares (per square mile). Least weasels and Arctic and red foxes increased from scarce or no record to common. Correlated with this heavy predation, the lemming population became reduced by mid-July of 1953 to one-tenth or less of what it had been when the snow cover melted in early June (Pitelka *et al.* 1955). Cyclic changes between 1929 and 1940 in the collared lemming at Churchill, Manitoba, were accompanied by marked fluctuations in breeding populations of snowy and short-eared owls and of the rough-legged hawk (Shelford 1943). This rapid build-up of predator populations must be attributed to their ability to shift from one region to another according to availability of local prey. The lemming becomes more vulnerable to predation when large populations consume the vegetative cover. Influxes of predators sufficient to exert a role in controlling outbreaks of mice (Banfield 1947), ruffed grouse, and snowshoe rabbit (Morse 1939), and bobwhite (Jackson 1947) have been reported for regions as far south as Toronto, Minnesota, and northwest Texas.

Cyclic fluctuations of the California vole and western harvest mouse seem to be caused by the activities of predators; island populations are not known to cycle in the absence of carnivores (Pearson 1966, 1967). However, after studying three lemming peaks in Alaska,

Pitelka (1964) concluded that while predators are the chief source of mortality in lemming peaks, these high populations are doomed anyway, and the effect of predation is superimposed on a more basic cycle of interaction between lemmings and the vegetation (see below).

Food   Animal populations are well known to decrease either through starvation or emigration when their food supply fails. The cyclic fluctuations of predators may well depend more on the cyclic fluctuations of the herbivore populations upon which they feed than on their producing the fluctuations (Fig. 15-6). If this is so, the basic problem becomes an explanation of the regularity in the fluctuations of the herbivore populations. However, the coaction between herbivore and vegetation is essentially similar to that of a predator-prey interaction (Lack 1954). These interrelations have been more intensively studied for the 3–4 year cycle than for the 9–10 year cycle.

In the *year 1* of a lemming cycle, reproduction is maximum and the population reaches a peak. The vegetation has just attained its maximum productivity and its nutrient content is high, especially of calcium and phosphorus. Soon, however, overgrazing reduces the amount of vegetation, the insulating ground litter becomes minimum, and the ground becomes thawed down to the mineral soil. This is aided by the burrow-

Fig. 15-6   Carnivore predation on *Microtus* on 14 hectares (35 acres) during 3 years. Divide left-hand vertical scale by 4 to obtain number of *Microtus* eaten per month per carnivore (Pearson 1966).

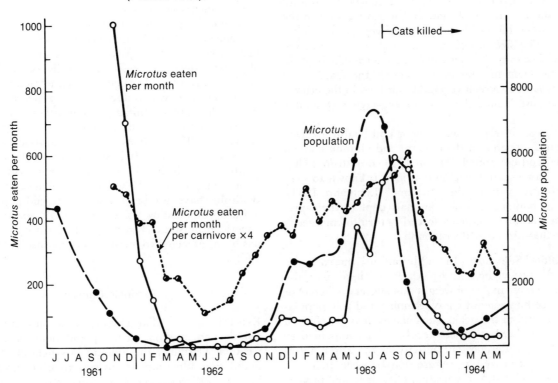

ing and nest-building of the animals. By the end of the summer, the population crash occurs, so that by *year 2* the lemming population is minimum. With the ground unfrozen, the nutrients resulting from what decomposition there is sink below the root zone of the plants. Plant growth is therefore poor and lemming reproduction minimum. By *year 3*, however, the vegetation is producing more litter, the ground does not thaw so deeply, and the new vegetation becomes richer in nutrients. With more and better-quality plant food, the lemming reproduction increases. During *year 4* these processes are accelerated, the thicker litter permits only shallow thawing of the ground, the greater amount of decomposition concentrates nutrients where the vegetation can easily absorb them, and all this leads toward the maximum reproduction of the lemmings attained early the following year, year 1 of the next cycle (Schultz 1969). Northern plants are often unable, in the short growing season, to make a luxuriant vegetative growth and produce seed every year. A period of 2 or more years may be required following a population irruption for full recovery of the vegetation both quantitatively and qualitatively (Thompson 1955a, Watson 1956, Lauckhart 1957, Klein 1970).

Important in the increase of populations to outbreak levels, both for lemmings in the north and for voles in the south, are sufficient reproductive vigor and proper environmental conditions for winter breeding. Snow is of considerable importance in this respect to protect the animals from winter predation and to maintain subnivean temperatures not much below 0°C (Shelford 1943, Kulicke 1960, Pearson 1963, Krebs 1964, Mullen 1968, Keller 1970). Important also is an early disappearance of snow and warm weather in the spring to permit quick growth of the vegetation (Fuller 1967). The size of egg clutches and the vigor of the hatched young in grouse and other tetraonid birds in Finland seem to depend on whether the females are able to get new green vegetation for food in the critical period just before the start of egg-laying (Siivonen 1957).

Changes in reproductive vigor and health are well known to depend on the mineral and vitamin content of the food consumed (Mason 1939, Braestrup 1940). The vitamin content of animal food is known to vary quantitatively from time to time (Lehmann 1953). During population peaks animals deplete their preferred food plants and are forced to turn to emergency foods that are not self-sustaining.

Natural Selection In addition to increases in reproductive vigor and growth rates, there may also be an increase in size or weight and aggressive behavior as succeeding generations of lemming and vole populations ascend to the peak. The changes may be affected by the weather but are not necessarily accompanied by observable differences in activity of the adrenopituitary endocrine system. Aggression between males is, however, related to maturation of their sexual organs,

which varies in rate with the stage of the cycle (Christian 1971). With increasing competition as the population nears the peak, opportunity arises for selection of the more vigorous and aggressive individuals for survival. This aggressive behavior leads to excessive interference between individuals, which results in lowered success of reproduction, emigration, or death, and the population declines. With reduced populations, there is no longer natural selection for aggressiveness and weight, and vigor declines. The qualitative differences between individuals from populations in the ascending and declining phases of the cycle are not certainly genetic, but they may be, as the changes in physiological vigor and behavioral aggressiveness exhibit trends that are continued over a number of generations, after the initiating conditions have disappeared (Chitty 1960, Krebs 1964, 1964a, 1966, 1970). Change in reproductive vigor between the ascending and descending limbs of the peaks in some insect populations can also be traced through two or three generations (Sander 1962, Shorrocks 1970).

Physiological Stress It is obvious that at times of high population densities animals are subject to increased stresses of various sorts in their search for food and cover and escaping predators. They may have to go longer distances to find the essentials for existence and to fight with other animals for possession of them. All of this puts an extra drain on their energy resources at the same time that they may be compelled to subsist on inferior food or tolerate nutritional deficiencies of one sort or another. One theory claims that the body adapts physiologically to these stresses under the stimulus of increased hormone secretion from the pituitary and adrenal glands, but when the stresses for existence and reproduction become too great, there is first a depression of reproductive functions, inhibition of growth, decreased resistance to disease, and finally death may result. It has been postulated that the die-off at the end of a cycle is due directly to such an exhaustion of the adrenopituitary system rather than to such external factors as lack of food, disease, or predators (Christian 1950). "Shock disease" is a manifestation of this stress syndrome. It has been repeatedly observed in Minnesota during the decline of the snowshoe rabbit cycle. Symptoms of the exhaustion phase of the stress syndrome have also been observed in wild populations of European meadow voles (Frank 1953) and the North American muskrat (Fig. 15-7). This theory, however, has been subjected to important criticisms (Negus *et al.* 1961).

Extrinsic Causes

Extrinsic factors that may affect cycles are largely abiotic and external to the populations, such as the weather. Certainly the amount of synchrony between population fluctuations over extensive areas and between different species or events that are otherwise

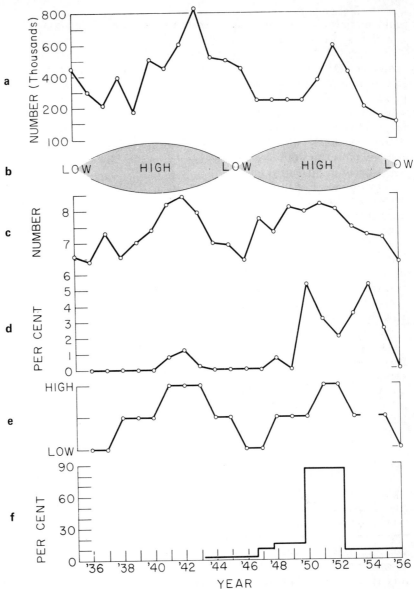

Fig. 15-7 Changes in population and reproductive and physiological vigor of muskrats in Iowa, correlated with the 9–10 year grouse-rabbit cycle (peak years 1941–42, 1950–51) elsewhere over North America: (a) autumn populations; (b) grouse-rabbit cycle; (c) size of litters; (d) young breeding in year of their birth; (e) tolerance of crowding; (f) resistance to disease (Errington 1957).

clearly unrelated (Dewey 1960) is greater than can be explained by chance alone and suggests that some climatic or extraordinary factor may be effective, either directly or in controlling the time schedule at which the intrinsic factors function. For example, widespread synchrony in red squirrel populations in Alberta, Canada, is correlated with production of white spruce cone crops, which in turn is correlated with low rainfall during the summer of the preceding year (Kemp and Keith 1970). We showed above that catastrophes may be caused by failure of food crops, extreme weather conditions, or a combination of the two. There can be no question that weather is important, either directly or indirectly, in producing fluctuations in animal and plant populations. However, for weather to exert a role in inducing cyclic fluctuations in population densities would presuppose that weather must also be cyclic.

Weather is affected by variations in solar radiation (Stetson 1947, Fainbridge 1961). Sunspots, chromospheric eruptions, solar coronal disturbances, and ionospheric and geomagnetic disturbances are indices of solar activity, but not direct measurements of it. The agent presumably effecting changes in the atmosphere and in living organisms may be short-wavelength ultraviolet radiations or emitted charged particles. Emissions from the sun are constantly undergoing great fluctuations, with maximum intensities for short periods being 100 or 1000 times the minimum intensities.

Sunspots are the only expression of solar activity that has been measured over a long period of time (Bray and Loughhead 1965). The intervals between peaks in the mean daily number of sunspots per year fluctuate from 7 to 17 years, average 11.2 years. A correlation between number of sunspots and temperature, rainfall, and cloudiness is sometimes indicated but needs more complete substantiation before it can be fully accepted (Thomson 1936). For instance, an analysis of records spanning 109 years for the period May through October in southern Wisconsin indicates lower temperatures, greater precipitation, and less sunshine in years when sunspots were increasing or at a maximum than when they were decreasing or at a minimum (Morris 1947). A similar correlation between summer rainfall and the sunspot cycle has been demonstrated for the Toronto, Canada, area (Clayton 1943).

The sunspot cycle is as variable in length as are the cycles of grouse and snowshoe rabbit, and there have been repeated unsuccessful attempts to correlate these cycles. Since 1750, the sunspot cycle has coincided with the snowshoe rabbit and lynx cycles part of the time, but then has gone out of phase until one became the inverse of the other (MacLulich 1937). If solar radiation is responsible for population cycles, it is clear that the number of sunspots is not a reliable index for judging the intensity of radiation (Fig. 15-5).

Since the growth of trees and the width of the annual rings they form is largely dependent on rainfall, one may conjecture the weather record back 3000 years by measuring the width of annual rings in the giant sequoias. This analysis of tree rings indicates the possibility of a variety of weather cycles, some important ones being in the neighborhood of 9–10 years (Douglass 1928). Weather cycles of 3–4 years are difficult to demonstrate, but an analysis of changes in barometric pressure and other characteristics of the annual atmospheric circulation over the British Isles indicates that they may exist (Goldie 1938). Cycles or outbreaks resulting from climatic factors are not necessarily absolutely synchronous over large areas. There is a limit to the size of the area over which a change in weather produces a single common effect. Outbreaks of spruce budworm progressed eastward in Canada between 1945 and 1949. These outbreaks were probably less a result of spontaneous dispersal of the moth, although this was a contributing factor, than of the progressive eastward circulation of favorable polar air masses (Greenbank 1957).

Solar radiation may also affect animals and plants in other ways than through the weather. A little ozone in the atmosphere at the earth's surface is stimulating to animals, but a high amount is harmful. There is some experimental evidence that ionization of the air, that is, the conversion of neutral gas molecules into electrically charged ions, may affect the health and the vigor of animals. The height of the ionosphere and ozone layers above the earth is controlled by the intensity of solar radiation, and it is possible that cyclic variations in the height of these layers may affect organisms in ways that are little understood at the present time (Reiser 1937, Huntington 1945).

Fluctuations in solar ultraviolet radiation vary the extent of ionization of air and the rate of ozone for-

mation. The ozone layer serves as a protective blanket which prevents ultraviolet rays from destroying all life on the earth. In small doses, ultraviolet is anti-rachitic, germicidal, and erythemal; in some species ultraviolet also affects skin pigmentation (Luckiesh 1946). Animals obtain vitamin D either as a product of direct radiation of the skin or in the food that they consume. A few studies indicate that medium intensities of ultraviolet radiation combined with optimum conditions of rainfall or weather correlate with highest populations of chinch bugs, bobwhite quail, prairie chickens, pheasants, cottontail rabbits, pronghorns, and the amount of butterfat in cow's milk (Shelford 1951a, 1952, 1954a).

Solar radition is not received in equal intensities in all parts of the world. Its intensity varies because of inclination of the earth's axis relative to the sun, differences in terrain, amount of cloudiness, and so forth. Solar radiation likewise does not have an identical effect on all species because of differences in their sensitivity to it and because critical periods in their life cycles come at different times of the year. These theories are conjectural at the present time, but the influence of extramundane forces on animal and plant life should not be dismissed until their potential effects are thoroughly explored.

POPULATION EXPLOSION OF MAN

The world population of man increased from only 2–3 million at the time of Christ to approximately 1.5 billion in 1900. It then began to explode, reaching 3.6 billion in 1970 and may exceed 6 billion by the year 2000. This irruption is due to reduction in mortality rates without compensating decrease in birth rates.

In his early history, man had to depend on solar energy for the production of both food and heat. With the discovery of fire, he used the burning of wood to furnish heat when solar energy was deficient, to cook his food, and to expand his capacity to use machines for locomotion, farming, and industry. This allowed his dispersal into parts of the world hitherto closed to him and the production of more extensive food supplies. The next step came with the mining of coal about eight centuries ago and production of petroleum just over one century ago. World consumption of energy from fossil fuels has increased about 4 per cent per year during the last century. The increase in world population is doubtless a response in large part to this increase in energy resources, and if atomic energy becomes generally available, there will be no restraint on continued population expansion on this account (Hubbert 1969).

Early man through development of superior intelligence and the use of tools soon eliminated predation as a significant mortality factor. With improved medical practice and expertise, many common diseases have been eliminated or at least greatly reduced in incidence or severity. Emigration has extended to nearly all parts of the world, although less extensively into deserts, tundra, and thick tropical forests. The oceans have no resident human population, although they could conceivably be occupied if there were sufficient demand and engineering ingenuity. By mechanically controlling the microclimate to which he is exposed, he has become in large part independent of, although still responsive to, the macroclimate. He is now contemplating procedures for controlling the macroclimate. By the "green revolution" man has greatly increased production of food, although there is considerable question whether increasing food productivity can keep up with the accelerating number of mouths to feed.

Although man can make more efficient use of the space he has available, the world is still a finite body. Sears (1958) states:

> Man's physical body occupies space, somewhere between two and four cubic feet . . . At his present rate of increase . . . it would be less than 700 years—say 22 generations—until there is standing room only, with each space of 3 by 2 feet or 6 square feet occupied. On this basis there is room for exactly 4,646,000 people in each square mile. I have perhaps been over-generous in estimating the per capita area, but I did wish to leave space enough to permit each individual to reach in his pocket for the rent money when it becomes due. A little after this the hypothetical human population would weigh more than the planet.

There have been many advances in medicine and in the prevention of disease, yet the possibility of new mutant strains is ever present. With the density of man steadily increasing and with his great mobility, the potential for a worldwide pandemic has never been greater. For instance, if an especially virulent strain of influenza should appear, it is doubtful that the United States and other developed countries could produce enough vaccine fast enough to save their own populations, let alone those in underdeveloped countries. The panic that would arise can hardly be imagined (Ehrlich and Ehrlich 1970). As man evolved through the hunting and gathering stages in the development of civilization, there was little opportunity for infectious diseases to occur, but as he increased in numbers and began to live in villages, such diseases became more common. The evolution of genetic immunity against infectious diseases or development of measures for their prevention and control are essential for survival (Armelagos and Dewey 1970).

Man in his various cultures has neither eliminated nor significantly decreased the role of competition, either between individuals striving for higher positions in the social hierarchy, between nations for space and recources, or between different ideologies in politics, economics, and religion. There is every likelihood that as populations expand, these conflicts will become even more drastic and devastating. A basic characteristic of man, as with all organisms, is desire for space, food, reproduction, and comfort. Without such drives, existence in the face of scarcity is impossible—it is the driving force of evolution.

Reproductivity, as expressed both in fecundity and success of young in attaining maturity, generally decreases in lower animals as populations expand, but in many species the decrease, especially in fecundity, is insufficient to prevent the populations from reaching thresholds where starvation, disease, or competitive struggling becomes disastrous. Such thresholds will be reached also by man unless a determined effort is made to reduce fecundity to a point where the population can be stabilized. Man is unique among animals in that his intelligence has evolved to an extent that he can anticipate the future and plan for his welfare, although it remains to be demonstrated whether he has the will and the discipline to do so. It is encouraging that European man's birthrate in 1800 was about 35 per thousand adults per year, while by 1970 it had dropped to about 17 per thousand. This, however, is not enough, and annual birthrates in the developing countries remain high, around 45 per thousand.

The average number of offspring per family in 10 European countries averages 2.7; in the United States it has been 3.3. A stable population could be established if the number were reduced to 2.3 per family. However, even when this number is reached, the population will continue to rise for a while. A high proportion of the present-day world population is under 15 years of age. As these young people mature, the size of the childbearing fraction of the population will increase and more than offset the decrease in birth rate for at least one generation (Ehrlich and Ehrlich 1970).

A crucial problem is the size of the world population that should be considered optimum or desirable. An estimate has been made that about 1 billion people is the maximum supportable at U.S. levels of affluence by present agricultural and industrial resources (Hulett 1970, Eugene P. Odum 1970, Taylor 1970). The much larger populations now occurring are possible only because over 80 per cent of the world population is living well below these levels. These underprivileged people seek a larger share of the world's resources and a rise in their standard of living. The option is either ruthless suppression or accommodation—the first is unthinkable, the second is possible but only with effort and concessions from all (Fig. 15-8).

The first concession is recognition that parenthood is a privilege, rather than a right (Hardin 1970). If voluntary family planning is inadequate, then stringent government regulation may be required. Continued population growth will lead only to disaster. Involved necessarily with deliberate control over population numbers is control over its genetic makeup. Possibilities exist not only to eliminate factors from the germplasm that induce physical and mental defects but also to breed individuals or races with special characteris-

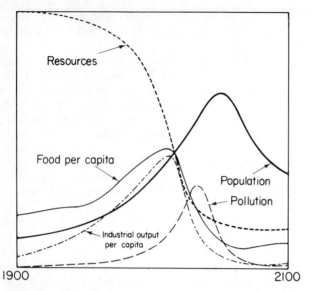

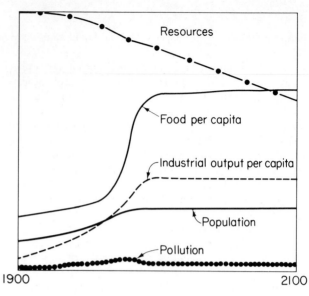

Fig. 15-8 This "standard" world model run assumes no major change in the physical, economic, or social relationships that have historically governed the development of the world system. All variables plotted here follow historical values from 1900 to 1970. Food, industrial output, and population grow exponentially until the rapidly diminishing resource base forces a slowdown in industrial growth. Because of natural delays in the system, both population and pollution continue to increase for some time after the peak of industrialization. Population growth is finally halted by a rise in the death rate due to decreased food and medical services (Meadows *et al.* 1972).

Fig. 15-9 Stabilized world model possible only with the human population held constant after 1975, industrial capital investment equal to depreciation after 1990, resources recycled, pollution controlled, and improved methods of restoring eroded and infertile soil. Even then, natural resources are gradually depleted but at a rate so slow that there may be time for technology and industry to adjust (Meadows *et al.* 1972).

tics that seem desirable. Man has control over his future fate in more ways than one (Bajema 1971).

Human society is faced with the problem of setting new values for personal conduct (Cailliet *et al.* 1971), establishing new modes of thought, and devising new forms of collective self-control (Caldwell 1970). This may require much loss of privacy and personal freedom. All evidence indicates that the maintenance of a high quality of living will require that our materialistic and economic growth, as well as our population, become stabilized at an asymptote consistent with the resources available and the protection of environmental quality (Fig. 15-9; Meadows *et al.* 1972).

Environmental deterioration varies inversely with economic and technological growth, which in turn varies directly with population growth. The "key factor" for the future survival of mankind on earth is curtailment and stabilization of the world population at an appropriate level (Sears 1958, Kirk 1967, Hardin 1968, Howard 1969, Bormann 1970, Ehrlich and Ehrlich 1970, Hertz 1970, Hulett 1970, Wagar 1970, Ayres and Kneese 1971).

SUMMARY

Populations may increase or decrease in size progressively through a period of time; they may suddenly irrupt when conditions become favorable, or decline precipitately with unfavorable weather or failure of food supply, or they may vary cyclically. The most apparent cyclic phenomena are the 3–4 year rodent cycle and the 9–10 year grouse-rabbit cycle. Populations of several predators display correlated patterns. Cycles are most clearly developed in the far North.

Many theories have been advanced as to the immediate cause of population cycles; disease, coactions between prey and predator, depletion of food supplies in quantity or quality, physiological stress, and natural selection, but a complete and satisfactory explanation is not yet at hand. There is enough synchrony in the timing of cycles over the world as to suggest that variations in weather or solar radiation may be involved.

The population explosion of man is due to elimination of predation, emigration into new habitats, increased survival of young, reduction of disease, increased productivity of food—all related to man's increasing knowledge and technical skill in the exploitation of natural resources. Mankind is headed toward catastrophe caused by starvation, pandemics, or competitive struggling unless there is a determined reduction in fecundity, a reorientation of social and individual objectives, and maintenance of environmental quality.

240

Part Five

EVOLUTIONARY

ECOLOGY

Chapter 16

NICHE SEGREGATION, SPECIES DIVERSITY, AND ADAPTATION

Up to now we have been mostly concerned with *proximate* factors or how organisms are adapted morphologically, physiologically, and behaviorally to their environment. We will continue to be concerned with proximate factors but we need also to consider *ultimate* factors or how these adaptations have evolved because of their survival value to the species (Allee *et al.* 1949, Orians 1962, Lack 1965). It is consideration of these ultimate factors that constitutes the essence of *evolutionary ecology*.

The evolution of distributional patterns and communities will be considered in Chapters 18 and 19. Time and space do not permit adequate attention to the evolution of territorial and colonial nesting, time of breeding, sex ratios, mating systems, and any number of other behavioral and social phenomena. See, for example, the symposium on *behavioral evolution* as it relates to bird nests, spider webs, caddisworm cases, termite shelters, and bee nest-cells (Collias 1964), *co-evolution* of adaptations in insects and squirrels and in the seeds on which they feed (Janzen 1969, Smith 1970), the interrelation between mortality and natality (Willson 1971), and the evolution of parental care (Kendeigh 1952). *Natural selection affects behavior and social patterns as it does structure and function*, to render the organisms more fit and able to survive in specific

environments and to live together in particular relations one with another. In this chapter and the next we will consider ecological niches and taxonomic species, how they have evolved together to give structure to the community, and how behavioral interrelations between individuals have evolved social patterns, ethics, and culture, culminating with man.

The *ecological niche* is a particular combination of physical factors (microhabitat) and biotic relations (role) required by a species for its life activities and continued existence in a community.

Although Steere (1894) of the University of Michigan had an appreciation of niches in his explanation of the way bird species within a genus were segregated on different islands in the Philippines, the term and concept were really first developed by Joseph Grinnell (1917, 1924, 1928) of the University of California. He considered the ecological niche the ultimate distributional or spatial unit occupied by just one species, or subspecies, to which that species is held by structural and instinctive limitations such as climatic factors, kind and amount of food, suitable nesting-sites, and cover. He wrote of animals generally and birds in particular as having *preference* for a particular niche, *choosing* surroundings consistent with their needs. At about the same time Charles Elton (1927), in England, independently defined the niche in more functional terms as an animal's *place in the biotic environment, its relation to food and enemies.* The present-day concept of the niche is an elaboration of these basic ideas, with emphasis on all the relations of the organism to both the physical and biotic factors of its environment. These parameters on relations are subject to objective and concrete definition and measurement.

Hutchinson (1957) suggested treating the environmental variables, both physical and biotic, affecting a species population as a set of *n* coordinates. The ranges of these variables within which the species can survive and reproduce then define an *n*-dimensional abstract volume or hypervolume, which he terms the *fundamental niche.* If two species in a biotope overlap in their fundamental niches, only one species may survive in stable interaction in the volume of overlap where they are in direct competition. If the two fundamental niches differ in part, the two species can survive together and coexist in the biotope. The *realized niche* of each species is, therefore, restricted to that portion of its fundamental niche to which it is better adapted. A community can thus be conceived as an assemblage of species in a biotope that have divided up the "niche space" (the full range of environmental variables affecting those species in the community) so that they occupy niche volumes that differ, though they may overlap.

The restriction of a species to a particular niche depends on its structural adaptations, physiological adjustments and tolerances, and behavior patterns. At the same time, living in a particular niche brings increasing evolutionary perfections of adaptations for living there rather than somewhere else. The variety of species living together in different niches gives that ecosystem *species diversity.* The concepts of species and niche can hardly be considered one without the other, and both concepts need to be approached from the point of view of *evolution* and *adaptation.*

CHARACTERISTICS OF THE NICHE

Microhabitat

The physical features of the microhabitat—substratum, space, and microclimate—are basic determinants of whether a particular niche can be occupied by a given species. The basic differences in marine, fresh-water, and terrestrial habitats restricting the distribution of communities is immediately obvious. The features differentiating microhabitats are less apparent, but assume major significance when particular occupants of it, rather than the community as such, are under investigation (Fig. 16-1) (Prosser 1955).

In earlier chapters we have described how species segregate in aquatic habitats according to whether the bottom is rock, sand, or mud and how the swift current limits the inhabitants of streams to species possessing clinging structures and proper orienting behavior. We also described how the characteristic of the soil, which the females first test with their ovipositors, determine where grasshoppers and tiger beetles lay their eggs. In Scotland, four species of ants occur in the same pine forest, but they show preferences in where they are found to different positions on decaying stumps, to hard or soft wood, to wood with different moisture content, to ground away from stumps, and were distinguished by whether they form galleries or mounds, the size of the galleries, and their aggressiveness in their defense (Brian 1952). Millipedes in deciduous forests also show fine distinctions in their preferences for substratum (Table 16-1).

Of 18 species of rodents studied in Utah, 4 were found only in rocky situations, 2 only in gravelly soils, and 2 only in sandy soils. The niches of the other 10 species were not defined by the type of soil (Hardy 1945). The soil preference of all four species of pocket gophers in the western United States was deep light soil, but the species with the strictest requirements was also the most aggressive and displaced the less aggres-

Table 16-1   Percentages of each millipede species found in different microhabitats (O'Neill 1967).

Habitats	*Eurparus erythropygus*	*Pseudopolydesmus serratus*	*Narceus americanus*	*Scytonotus granulatus*	*Fontaria virginiensis*	*Cleidogonia caesioannularis*	*Abacion lacterium*
Heartwood at center of logs	94	0	0	0	0	0	0
Superficial wood of logs	0	67	4	7	0	14	0
Outer surface of logs beneath bark	0	21	71	0	0	0	0
Under log, but on log surface	3	8	7	60	0	0	16
Under log, but on ground surface	3	4	13	0	97	14	37
Within leaves of litter	0	0	0	27	0	43	0
Beneath litter on ground surface	0	0	5	7	3	29	47

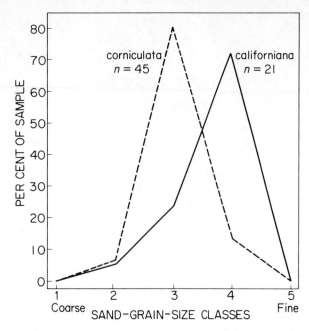

Fig. 16-1 Differential selection by two species of beach-hoppers of different-sized sand particles in an experimental gradient. The preference of *Orchestoidea californiana* for finer sand that packs more firmly and holds more moisture agrees with its natural habitat of exposed flat beaches. In contrast, *O. corniculata* is found on short steep beaches made up of poorly sorted, often coarse, sands (Bowers 1964).

sive species to less preferred but tolerable soil habitats (Fig. 16-2).

Congeneric species of ectoparasitic mites and fleas on small mammals commonly are distributed among several host species. When they occur on the same host, they are segregated by species on different parts of the body, a given species is present only when the host occurs in a particular type of vegetation, or they occur at different seasons (Jameson and Brennan 1957). This is also true with water mites parasitizing Coleoptera and Hemiptera in shallow ponds (Lanciani 1970). Segregation of parasitoid insect species on the cocoon of the jack pine sawfly is effected, in part, in another way. One species predominates at high host densities, another at low host densities, and others at medium host densities (Price 1970).

The importance of microclimate in niche segregation of species is shown by a study made in Danish bogs (Nørgaard 1951). The low humidities and high temperatures obtaining at mid-day on the surface of the peat mat restrict one species of spider to the sub-surface stalk region of the sphagnum. Another species of spider tolerates the surface conditions, so the two species divide the habitat between them.

Microclimate is often a major factor in determining whether a species can maintain itself against competition in a particular microhabitat. This has been shown experimentally. When equal numbers of two related species of beetles are introduced into the same flour container and placed at a particular combination of temperature and relative humidity, one species becomes established, the other is eliminated. The particular species favored in the various microclimates are as follows (Park 1954):

34°C–70 % R.H.	*Tribolium castaneum*
34°C–30 % R.H.	*Tribolium confusum*
29°C–70 % R.H.	*Tribolium castaneum*
29°C–30 % R.H.	*Tribolium confusum*
24°C–70 % R.H.	*Tribolium confusum*
24°C–30 % R.H.	*Tribolium confusum*

Similar reversal of dominance has been found to take place at high and low temperatures with different species of grain beetles (Birch 1953), *Drosophila* flies both in Europe (Timoféeff-Ressovsky 1933) and in North America (Moore 1952), two insect parasitoids using the same host (DeBach and Sisojevic 1960), and turbellarian flatworms (Beauchamp and Ullyott 1932). Usually the species favored by a given microclimate has a higher rate of population growth at that particular temperature or humidity.

There is a positive correlation between high oxygen tensions required by trout for saturating their blood hemoglobin and the oxygen-rich waters that they select. The restriction of these fish to cold waters is correlated with the fact that a rise in temperature decreases the oxygen-loading capacity of the hemoglobin. Bullheads and carp, common to warm waters of low oxygen and high carbon dioxide content, have hemoglobin that loads and unloads at low oxygen tensions and is less sensitive to changes in carbon dioxide tension and temperature (Haws and Goodnight 1962).

Arctic mammal species differ in thickness and density of fur, which insulate against loss of body heat, and this determines whether they can sleep above ground and be active during the winter or whether they must confine themselves to nests and runways below the snow level (Scholander *et al.* 1950).

In central Illinois, the short-tailed shrew is largely subterranean in habit and occurs in moist habitats; the woodland white-footed mouse is nocturnal and inhabits the forest floor; the prairie vole is restricted to grassland and the most arid of the habitats of

Fig. 16-2 Soil preferences and tolerances of the pocket gophers *Geomys bursarius* (GB), *Cratogeomys castanops* (CC), *Thomomys bottae* (TB), and *T. talpoides* (TT). The species are arranged in decreasing competitive ability, which is correlated with their increasing tolerances of less preferred soil types (Miller 1964)

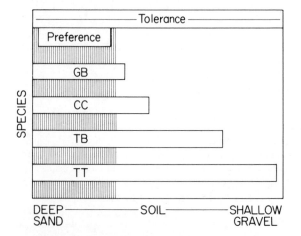

the three species. There is a connection between the amount of water available in the habitats and the level of water exchanges in the animal. At 19°C, for instance, the rate of water absorption and loss in the shrew is twice that of the mouse. The rate of total water turnover in the vole is about the same as in the mouse, but water loss through the lungs and skin is much lower, indicating acclimatization to a drier habitat (Chew 1951).

Diurnation

When two or more species are competing for the same resources of a single habitat, this competition is reduced or eliminated if one species makes use of these resources at a different time of day, or in a different season, than the other. Time then becomes a dimension in which species occurring in the same community may differ in niche.

The white crappie and black crappie fish are very similar in habits, food requirements, and local distribution, but the white crappie feeds considerably more during the daylight period than does its more aggressive black cousin, and this slight difference in timing may be sufficient to permit it to occur in the same areas as the black crappie (Childers and Shoemaker 1953).

Six species of the cladoceran, *Daphnia*, occur in one lake in Maine, but they differ in the time of day and season of year for their major activity as well as by depth, food, and temperature adaptations (Tappa 1965).

The females of the butterfly *Colias eurytheme* may be either orange or white; all males are orange. There is partial separation of the two color phases in that the white females are relatively more active in the early morning and the yellow females later in the day (Hovanitz 1948).

During a winter in England, when birds were coming to banding traps for food, it was noted that the European robin did so most frequently just after sunrise and just before sunset, the European blackbird just before and after midday, while the blue tit had peaks of feeding between the feeding times of the other two species (Lees 1948).

Of four rodents occurring together in California, one is a carnivore. Of the three herbivores, one is diurnal and the other two are nocturnal, but one of these hibernates during the autumn and winter (Chew and Butterworth 1964).

Aspection

The two grasshoppers *Arphia sulphurea* and *A. xanthoptera* occupy similar niches except that *A. sulphurea* overwinters in the nymph stage, reaching maturity from April to late July, while *A. xanthoptera* overwinters in the egg stage and hence requires a longer time to mature in the spring. The adults of the latter species occur from late July to early November (Blatchley 1920). Several species of butterflies that feed on the nectar of flowers exhibit a more complicated pattern (Fig. 16-3).

Predatory arthropods segregate out seasonally in Polish forests and fields in that large species of spiders predominate from March to May, medium-sized spiders during much of May and June, and carabid beetles from July to September (Breymeyer 1966).

Tropical species of dragonflies and damselflies which have recently extended their range northward confine their main periods of flight to the warmest months and to the middle of the day. Native species that are better adapted to colder climates are active in early spring

Fig. 16-3 Flight periods of nine hespirine butterflies for feeding on flower nectar. Note the temporal replacement of the members within each set and the coincidence of peaks of one set with the intervals of other sets, especially the relation between sets 1 and 2. Species within each set have flight and feeding habits more nearly similar than species in different sets (from Clench 1967).

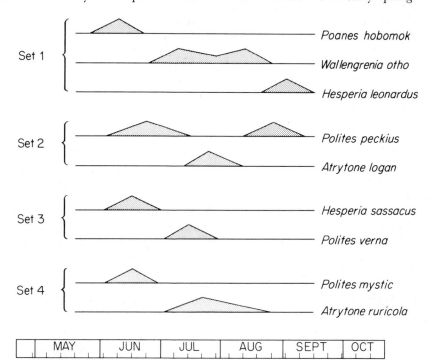

and autumn and in the twilight hours of the day (Kennedy 1927). Times of emergence of stream insects are also correlated with their ancestral places of origin and the extent of their acclimatization to temperature.

Three kinds of sockeye salmon occur in Cultus Lake, British Columbia (Ricker 1938): (1) the normal anadromous stock, whose offspring may either migrate out to sea or remain as (2) residuals, and (3) the land-locked salmon, which remain continuously in the lake. The anadromous and residual populations differ in breeding coloration, but both spawn from October to December. The land-locked forms, which closely resemble anadromous sockeye in breeding coloration, spawn only in August and September.

The tern *Sterna virgata* nests on the Kerguelen Islands in October and November; *S. vittata* uses the same nesting area in January and February. The niche of each species differs somewhat further in that *S. virgata* feeds to some extent in inland waters, but *S. vittata* is strictly marine (Murphy 1938).

Shelter and Vegetation

Animals require shelter or *cover* as a protection against unfavorable weather and enemies. Caves, overhanging ledges, deep valleys or canyons, or burrows in the ground may serve as shelter for terrestrial animals. The darkness of night is a protection against diurnal predators; daylight is a protection against nocturnal ones.

Vegetation is an important source of shelter for animals. Some animals cannot tolerate too much solar insolation, hence seek shade. In arid habitats, jack rabbits shift the location of their forms on the ground at different times of the day to stay in the shade of bushes, and it is a common sight in prairie or desert regions to see horned larks, meadowlarks, or other birds lined up in the narrow shadow cast by telephone poles or fence posts. Burrows of all sorts, whether in the ground or in trees, give the animal good insulation against both winter cold and summer heat. The foliage of trees, shrubs, and even grasses and vines reduces the amount of heat radiated from the bodies of animals, especially warm-blooded ones, on cool clear nights, and vegetation in general serves as a windbreak. Birds keep to the lee side of exposed patches of woods during cold windy weather.

By staying under cover, prey animals may escape notice of passing predators, or if detected, may more easily avoid capture. Dense vegetation, thorny thickets, burrows, and other situations impenetrable to predators are sometimes called *escape cover*. We also speak of *nesting cover*, *winter cover*, and *roosting cover*, depending on the particular purpose which the cover serves. Animals are often protectively colored to conceal themselves better from enemies in particular kinds of cover.

When the beetles *Tribolium confusum* and *Oryzaephilus surinamensis* are introduced experimentally into a flour medium, *Tribolium* is ordinarily successful, *Oryzaephilus* is eliminated. But when the flour medium contains pieces of glass tubing of such bore as will exclude *Tribolium* but let the larvae of *Oryzaephilus* enter and pupate, both species survive (Crombie 1946).

Most vertebrates and some invertebrates, especially insects and spiders, build nests, usually of plant material. The type of nest it builds is grossly characteristic of a species and dependent on inherited behavior patterns, yet individual nests are uniquely modified to fit into particular situations. Nests protect eggs and young against weather, and are usually well concealed from enemies.

Bird species are commonly found at different heights or in particular strata of the vegetation associated with the characteristic location of their nests (Beecher 1942), where they seek refuge from enemies (Dunlavy 1935), where they do their feeding (Hartley 1953), or the location of their song-posts (Kendeigh 1947). Birds nesting in the tree-tops often feed outside the forest (Colquhoun and Morley 1943). It is often obvious, from the arrangement of toes, length of the legs, and other characteristics, whether a bird scratches the ground for its food, gets its food in the air, wades in marsh, is a swimmer and diver, a percher, or a tree-trunk climber. However, the minor adaptations of toes, legs, and wings in closely related species that enable them to occupy different niches within the same general type of vegetation are more difficult to detect (Dilger 1956, Osterhaus 1962). Segregation to a niche may involve, in addition to obvious external characters, many adaptations throughout the body in skeleton, musculature, and other organs (Burt 1930, Richardson 1942). Many types of animals other than birds, for instance mosquitoes (Snow 1955) and lizards, are also segregated by strata to where they most commonly occur (Snow 1955).

Warblers are numerous in the evergreen-deciduous forest ecotone of eastern North America because they nest and feed in so many diverse niches (Kendeigh 1945, MacArthur 1958):

Blackburnian warbler: Top level of evergreen trees
Black-throated green warbler: Middle level of evergreen trees
Magnolia warbler: Low level of evergreen trees
Redstart: Secondary deciduous growth
Black and white warbler: Tree trunks
Black-throated blue warbler: Shaded shrubs
Chestnut-sided warbler: Sunlit shrubs
Canada warbler: Wet shaded ground
Yellowthroat: Wet sunlit ground
Ovenbird: Dry shaded ground
Nashville warbler: Dry sunlit ground
Louisiana waterthrush: Stream margin
Northern waterthrush: Bog forest

Mammalian adaptations to different strata have already been discussed. When given a choice between a grassy habitat and a tree-trunk habitat, the short-tailed forms of *Peromyscus* mice selected the grassy habitat; the long-tailed forms, the tree-trunk habitat (Harris 1952). It has been demonstrated that the long tails of some species and subspecies give them a greater proficiency in climbing than their shorter-tailed relatives exhibit (Horner 1954). Similarly, the chipmunk *Eutamias umbrinus* occurs only in areas when the trees are sufficiently close together that they can escape the attacks of the more aggressive *E. dorsalis* (Brown 1971). In the arid country of southern California the giant kangaroo rat is predominant in flat country covered with brush; on brushy slopes and rolling hill-tops the Fresno kangaroo rat replaces the giant kangaroo rat; the Heermann's kangaroo rat is forced to live on the open plains since it cannot compete successfully with the other two species on brush-covered land (Hawbecker 1951).

Considerable evidence was presented in Chapters 6, 7, and 8 to show how animal distribution correlates *locally* with types of vegetation, and more will be presented later with respect to *geographic* distribution. Except for a few herbivorous and parasitic species, animals do not respond to the taxonomic composition of vegetation when they seek cover or food, but rather to life-form of plants; or they respond to the microclimatic conditions established by the vegetation.

In northern Europe, the kinglet *Regulus regulus* occurs with the chickadee *Parus atricapillus* in spruce and pine forests, but is mostly absent from the birch forests which the chickadee frequents. The kinglet is unable to feed extensively at tips of the pendulous birch twigs because, unlike the chickadee, it is less able to hold itself in an inverted position, because of poor development of certain muscles in the leg (Palmgren 1932).

The ovenbird is absent from coniferous forests unless a few deciduous trees are also present, since the bird requires broad leaves for construction of its oven-shaped ground nest. The red-eyed vireo feeds on insects taken from the leaves and the smaller stiff twigs of deciduous trees. It is mostly absent from coniferous forests, where the needle-shaped leaves are attached on all sides of flexible twigs and the bird finds difficulty in obtaining a footing (Kendeigh 1945). When birds of different species were given a choice between the branches of coniferous trees and those of deciduous trees, there was evident a direct correlation between length of foot-span, that is, the distance from the tip of the middle front toe to the tip of the hind toe, and the frequency of perching on the evergreen branches. Birds with small foot-spans greatly preferred the branches of deciduous trees (Palmgren 1936).

Three species of garter snakes are found together in Michigan, but *Thamnophis butleri* is restricted to grasses and sedges near water, *T. sauritus* prefers bushy areas near water and is a frequent climber, while *T.*

sirtalis occupies a variety of habitats regardless of proximity to water (Carpenter 1952).

The evidence indicates that if the type of cover required by a species is missing, that species will not occur even if all other conditions are favorable. This is of particular concern to the wildlife manager. He must learn to control succession, either by accelerating or retarding it, to give species of game the cover that they need (Leopold 1933, Elton 1939).

Food and Predators

Since most organisms select their food from that most easily available to them, it is usually more important in characterizing their niches to indicate the type or characteristics of food consumed and the stratum or exact microhabitat from which it is obtained than merely to give a list of species that are taken. Thus fresh-water fishes are best classified as mud-eaters, plant-eaters, plankton-eaters, mollusk-eaters, insect-eaters, fish-eaters, detritus-eaters, or omnivora (Forbes 1914). In a similar manner birds have been categorized into aerial-soaring, or perching insect-eaters; those which feed on foliage insects, seeds, or nectar; timber-searchers or drillers; feeders on ground insects or seeds; and predators (Salt 1957). Sympatric bats differentiate into fruit-, nectar-, insect-, and fish-eaters (McNab 1971).

The accurate description of feeding niches requires careful attention to details. Two or more species may feed together in the same community but be segregated from each other because they search for their food from different plant species, from different parts of the same plant, or they take different foods from the same parts of the same plants. Large-billed birds, for instance, husk large seeds more quickly and with less effort than do small-billed birds. Differences in seed or food selec-

Fig. 16-4 Separation by height of foraging male and female red-eyed vireos (Williamson 1971).

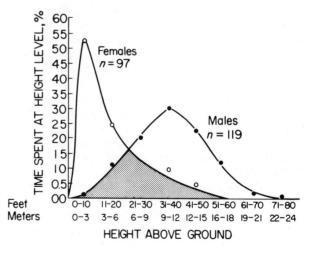

tion are related to the shape and dimensions of the bill (Kear 1962, Newton 1967). Evolution of food niches dependent on the size of the food item appears feasible only when the number of potentially competing species is low. When the number is higher, the tendency instead is to evolve differences in feeding habitat or foraging behavior (Schoener 1965). Species may overlap broadly in their feeding habits during most of the year but be clearly segregated during periods when food is scarce (Gibb 1954, Betts 1955, Keast 1970).

The two sexes of birds often differ in size of body, bill length, and other characteristics, correlated with different subniches that they occupy in the vegetation and in the food that they consume (Fig. 16-4). This is especially well marked in woodpeckers (Selander 1966). In grassland where the vegetation and food resources are more uniform, the sexes may forage in different directions and at different distances from the nest (Robins 1971). Such cooperation between the sexes has an obvious advantage where the food supply is limited.

The difference in type of vegetation inhabited by each of the three species of Michigan garter snakes mentioned above correlates with differences in kinds of food each consumes: *Thamnophis butleri* feeds almost entirely on earthworms and leeches; *T. sauritus*, on amphibians, fish, and caterpillars; and *T. sirtalis*, on both earthworms and amphibians, as well as a few mammals, birds, fish, caterpillars, and leeches (Carpenter 1952). Several species of *Drosophila* flies may occur in a single region, especially in the tropics, but each species feeds preferentially on a different species of yeast (Dobzhansky *et al.* 1956, Cunha *et al.* 1957). The spider-wasps *Anoplius semirufus* and *A. apiculatus* have many very similar behavior patterns, but they avoid competition at a critical point. *A. apiculatus* hunts for its food in the sandy areas where both species nest, while *A. semirufus* moves to woods or shrubby areas to feed and feeds on a different group of organisms at that (Evans 1953). Of four species of triclad flatworms found together in a lake in England, two feed mainly on oligochaete worms, one on snails, and one on isopods (Reynoldson and Davis 1970). Two species of fish having very similar requirements occur in the same type of habitat in British Columbia, but *Cottus rhotheus* has a larger mouth than *C. asper*, and feeds on larger kinds of food (Northcote 1954). Size differences even occur commonly between related species of copepods that live together in the same body of water (Hutchinson 1951). Selective dependence of fish on planktonic species of different sizes helps to determine the relative abundance of the planktonic species and even their presence in the community (Brooks and Dodson 1965).

For describing the position a species occupies in the food chains of a community, it is necessary to indicate not only the kinds of food eaten but also what species prey on it and the manner in which they do so. Some parasitoids are quite specific as to the kinds of animals in which they deposit their eggs. If a parasitoid that is specific to a prey species is present in a niche, the simple presence of that parasitoid may determine the success with which the prey species will compete for and fully occupy the niche.

BEHAVIOR ADJUSTMENTS

Somehow an animal must get into a favorable niche out of the multitude of niches available to it. Doubtless, some animals accomplish this passively by the random dispersal of spores, eggs, or larval stages, some of which by chance reach favorable locations, there to mature and survive. When immature stages of two or more species find a particular location favorable, that species which settles there first, in the greatest numbers, or grows faster is likely to acquire dominance over the area and force the other species to take a marginal position or disappear (Connell 1961).

Higher animals with their more complex nervous systems, greater intelligence, and sense organs more highly evolved, are equipped to search actively for and recognize niches suitable to them either by sight, smell, contact, or other means. For instance, the intricate migratory behavior of birds and other animals is such that they seek out nearly constant climatic environments throughout the year. Migratory behavior may have evolved as the result of groups being unable to solve the problem of niche isolation locally (Cox 1968).

The apparent ease and speed with which a new generation of individuals discriminates a hospitable niche means that they must in some way recognize the merits of that niche by definite characteristics of it that may be in the nature of sign stimuli (Table 16-2). Such characteristics are usually prominent, though they need not necessarily be the most essential features of the niche (Lack 1937).

Probably most animals exercise a deliberate, although not necessarily conscious, evaluation process in choosing one niche from those available. This has been tested experimentally by exposing the animals to gradients of environmental factors, either in the laboratory or field (Harris 1952); a variety of apparatus and procedures is available for such tests (Shelford 1929). Usually there is a coincidence of the species' experimentally ascertained preferendum and its natural preferendum. For instance *Elipsocus melachlani* and *E. westwoodi*, both psocid insects, occur abundantly on larch trees, but *E. melachlani* frequents those dead branches heavily encrusted with lichens, and *E. westwoodi* frequents living branches covered with the alga *Pleurococcus*. Laboratory experiments showed clearly that when each species was given a choice, each selected its customary habitat. Furthermore, the feeding of *E.*

Table 16-2 Percentage evaluation of sign stimuli for recognition of niches in different species of pipits (*Anthus*) (Svärdson 1949).

Species: Habitat:	*A. trivialis* Forest-edge	*A. pratensis* Dry meadows	*A. campestris* Sandy areas	*A. spinoletta* Mountains, rocky shores
Light and open country	30	40	40	35
High outlooks	37	14	20	26
Green color	20	20	5	5
No vegetation on ground	5	5	30	15
Water nearby	2	15	1	15
Conspecific males nearby	4	4	2	2
Other external stimuli	2	2	2	2
Totals	100	100	100	100

westwoodi was restricted almost entirely to the alga, although *E. melachlani* would feed on both the alga and on lichens (Broadhead and Thornton 1955).

For certain species, niche preference can be attributed to appropriate behavioral patterns alone. Isopod species occur in water and on land but only in places where the humidity is high. What success the group has achieved on land appears to be the result of their avoidance of the rigors of ordinary terrestrial conditions by means of behavior mechanisms that retain them in these moist cryptozoic niches, rather than to the development of any special morphological or physiological adaptations (Edney 1954).

A stereotyped behavior pattern appears to make the magnolia warbler build a nest supported in the interlocking leaves or twigs of a conifer rather than in the vertical fork of a tree or shrub, as the redstart regularly does. The black-throated green warbler originally had a nest-building behavior similar to that of the magnolia warbler, but in some regions it has taken to building in forks, a behavior which has expanded its range into both deciduous and coniferous forests. Why is the American robin restricted to localities where it can get mud to put into its nest? Others members of the family Turdidae do not use mud in their nests. There is, on the other hand, an advantage for barn and cliff swallows to use mud in constructing their nests because it enables them to use locations on the vertical sides of cliffs or buildings free of competition from other species. There appears to be no physical reason why a barn or cliff swallow could not build its nest in crevices or holes like other swallows, why a bank or rough-winged swallow could not build a mud nest like the barn or cliff swallow, or why a robin could not build like other thrushes. Such niche segregations are apparently consequences of restrictions imposed by behavior patterns alone, although one can never be sure but that each species has some hidden adaptation that keeps its characteristic kind of nest the best nest for it, and its preferred niche the best niche for it.

It is possible that certain species of birds are confined to coniferous forests because they are of northern origin and coniferous forest was the original community available to them; similarly, the broad-leaved deciduous forest is conjectured to have been the original community inhabited by species of southern ancestry. Presumably each group evolved heritable, instinctive behavior patterns which continue to drive it back to the ancestral community in which it belongs, so to speak, even though other types of communities have become available (Lack and Venables 1939).

Where two species with similar niche requirements come into competition, one species must possess better adaptation to it if it gains full possession of the niche to the exclusion of the other species (Lack 1944). Preadaptation or possession of suitable adaptations also appears a necessary prerequisite for a species to invade a new niche or habitat and successfully displace another species already occupying it, as frequently occurs (Simpson 1953a). The search for such microadaptations as might give one species an advantage over another in a competition for a particular niche is a real challenge to ecological research.

INHERITANCE

The fact that all individuals of a species behave almost identically in many of their activities indicates that these behavior patterns or instincts are in some form passed on or inherited from one generation to the next. Behavior that is learned after birth is much more variable between individuals. Behavior patterns are rooted in the structural arrangements of neurons and synapses and in the functions of hormones and enzymes. Once a stimulus is received a definite action results. Predisposition for a species' behavior patterns could well be inherited genetically through chromosomes and genes like any other structural or functional characteristic. These *inherited behavior patterns* are doubtless subject to evolutionary development as much as are structural and functional adaptations; indeed, the one may have developed synchronously with the others (Kendeigh 1952, Spieth 1952).

The often large and elaborate nests built by termites

are really manifestations of behavior patterns inherent in the species. The nests of higher termites are built specifically by the sterile workers; plainly, whatever is involved in the capacity the workers have to build a certain type of nest must be transmitted genetically, through the sexual forms only. Adaptive modifications in nest structure occur and phylogenetic sequences in nest structure that correlate with phylogenetic sequences in morphological characters can be demonstrated (Emerson 1938).

The two toads *Bufo americanus* and *B. woodhousei* differ in rate of embryonic development and embryonic temperature tolerance but interbreed freely. Embryos of hybrids show an intermediate rate of development (Volpe 1952). The call of a naturally occurring hybrid toad (*Bufo americanus* x *B. woodhousei*) was found to be intermediate between the calls of the two parent species (Blair 1956).

Three sibling species of the cricket *Nemobius fasciatus* differ slightly in structure and color but are scarcely distinguishable except by the songs of the males. Although they do not interbreed under natural conditions, they were induced to do so under experimental conditions. The song of the F_1 males was intermediate between those of its parents, indicating genetic influence (Fulton 1933).

The spinning behavior of the flour moth has been shown to be inherited on a Mendelian genetic basis; it is also dependent on light and food factors. At least two genes are involved; in the F_1 generation the nonspinning behavior is almost, but not quite, dominant; in the F_2 generation there is segregation of spinning individuals (Caspari 1951).

The segregation of the prairie deer mouse and woodland white-footed mouse into different niches is very definite, as is also the segregation of related species occurring in chaparral (McCabe and Blanchard 1950). The same segregation to different habitats holds even between prairie (*bairdii*) and woodland (*gracilis*) subspecies of the same species (*Peromyscus maniculatus*). It is of interest that laboratory-reared individuals not previously conditioned to their natural habitats were given a choice under experimental conditions; *bairdii* selected grass habitat, *gracilis*, a tree-trunk habitat. This suggests genetic inheritance of habitat preference (Wecker 1963). In hybrids between the two subspecies, choice of the grass habitat was dominant over choice of the tree-trunk habitat (Harris 1952).

There may be transmission of behavior patterns to succeeding generations by *tradition* rather than by genetic mechanisms, that is, training of young, young imitating parents, conditioning or imprinting. It has been demonstrated that some parasitoid wasps lay their eggs only in the same kind of larvae as they themselves became conditioned to during their early growth and feeding. Young birds and other animals become imprinted to their own parents, to their own species, and perhaps to their proper niche, at critical stages in their development (Baldwin 1896, Cushing 1941, 1944, Thorpe 1945, Klopfer 1964).

It is very probable that the manner in which niche segregation is passed on to succeeding generations is not the same in all species. We may believe, however, that most behavior of animals has a genetic basis, but may become highly modified through practice, imitation, and experience.

INTERSPECIFIC COMPETITION

Segregating Effect

Charles Darwin stated in *The Origin of Species* the case for interspecific competition as an instrument segregating species into different niches as follows (Crombie 1947):

1. The reproductive capacity of organisms is greater than the carrying capacity of the environment.
2. The range of an organism's tolerance of physical conditions and choice of food is limited.
3. The failure of an organism to survive, or be born at all, may be a result of the direct action of unfavourable habitat, predators, parasites, or competitors.
4. When competition occurs, it is severest between organisms with the most similar requirements.
5. In general, the closer the taxonomic relationship between them, the more similarity there is in needs and habits of species.
6. When new forms appear in a given locality, either by evolvement there or by invasion after evolutionary divergence elsewhere, they will either eliminate or be eliminated by their nearest relatives if they compete with them, unless.
7. Each form becomes adapted to a different niche, in which case competition between them will cease, and they may occur in proximity.

Evidence that interspecific competition is the most critical factor confining a species to this or that niche (Fig. 16-5) is available with the expansion of the species beyond the usual limits of its niche when this competition is removed. This expansion is often evident in geographic differences in the niche characteristics of a species (Table 16-3). In Scotland, the mountain hare occurs at high elevations, the common hare at lower ones. In Ireland, which was isolated as an island before the common hare could reach it, the mountain hare occurs at both high and low elevations and is differentiated into distinct subspecies (Huxley 1943). Success in competition between species of turbellarian flatworms depends on temperature and water current, but when the competing species is absent, the remaining species disperses far into the microhabitat usually occupied by its competitor (Beauchamp and Ullyott 1932).

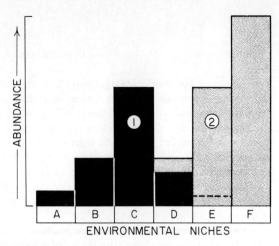

Fig. 16-5 Species 1 finds optimal conditions in niche C, and reaches greatest abundance in that niche. It is also able to utilize niches B and D, but with less efficiency, and niches A and E only very poorly. Species 2 cannot utilize niches A, B, and C at all, finds D only partially suitable, but E and especially F very favorable. Species 1 and 2 overlap in niche D, but species 2 prevents any occupancy of niche E by species 1. The absence of competition in niche A makes it open to a species evolving adaptations to it (from Mayr 1949).

Restriction of a species to a particular niche may result from either competition or lack of suitable conditions elsewhere (MacArthur 1965). In interspecific competition, the innate aggressiveness of one or the other species helps to determine the outcome. There are four species of chipmunks, each in somewhat different vegetation and at different elevations, on the east slope of the Sierra Nevada in California. *Eutamias alpinus* occurs at the higher elevations and is aggressively dominant over *E. speciosus* adjacent below. The food of *E. alpinus* is limited in amount but supplies of it can be effectively defended, and aggressiveness in the species has been selected for in evolution. The food of *E. speciosus*, on the other hand, is abundant, populations are limited by predation, and aggressiveness is not an important characteristic of the species. At the edge of the desert, *E. amoenus* is again more aggressive than *E. minimus* and prevents its occupancy of the lower

mountain slopes. *E. minimus* is largely limited to the hot desert, where aggressive activity is not metabolically feasible. *E. minimus* is better adjusted to the desert than the other three species because it makes use of body hyperthermia and retreats quickly, when disturbed, to its cooler burrow (Heller 1971, Heller and Gates 1971).

The general effect of interspecific competition is restriction of a population more closely within its optimum niche. Intraspecific competition exerts pressure impelling individuals to disperse into less favorable situations. The relative pressure exerted by these two forces determines whether at any particular time the species is contracting or expanding its habitat range (Svärdson 1949). Because of these interacting forces, the total population of all species in a region is constant, conditioned by the food and other resources, although the population of individual species may depend on the number of competing species composing the fauna (Crowell 1962).

Interspecific competition is reduced or eliminated altogether when the combined requirements of all species are less than the supply of materials available. Species of land snails, insects, aquatic clams, and copepods sometimes occur together in considerable profusion but with little evidence of competition between them (Boycott 1934, Fryer 1957, Ross 1957). Voles, during upswings in population, may become superabundant for periods of 2 or 3 years. During such periods several species of hawks and owls may feed in the same field without competing because there are more than enough voles to supply all. During downswings in the rodent population, however, competition does occur, and some or all the predator species are forced to turn to other prey (Lack 1946).

Parasites or predators may keep the populations of competing species below the level which available food resources of the habitat can sustain so that competition is reduced or disappears. For instance, when two species of weevil are placed together in a limited amount of food, one species is eliminated by the other

Table 16-3 Two species of robins in Victoria, Australia, divide the habitat in their feeding, but in Tasmania, where only *Petroica* has dispersed, it has added a significant ground-feeding component to its ancestral feeding niche and acquired longer legs in the process (Keast 1970).

	Eopisaltria australis, Victoria	*Petroica rosea,* Victoria	*P. rodinogaster,* Tasmania
Areas of feeding activity (%)			
Aerial hawking	—	15	8
Canopy	—	30	12
Twigs	—	20	5
Branches, outer	—	25	23
Branches, inner	10	10	20
Trunks, upper	10	—	—
Trunks, lower	5	—	—
Ground	75	—	32
Tarsus length (mm)	19.6–21.9 (20.8)	13.8–14.8 (14.4)	17.4–18.8 (18.2)

in about five generations. However, when a wasp equally parasitic on the larvae of both species is introduced into the mixed culture, the populations of both species decline markedly below their normal levels in single pure cultures, and they both continue to exist together indefinitely (Utida 1953, Pimentel 1961, Slobodkin 1964).

Several species may occur together in an intermediate environment where conditions are continually fluctuating so that first one species, then another, is favored, and no one species can establish consistent dominance. This occurs frequently where food is abundant but weather conditions vary from month to month, from year to year, or from one locality to another (Ross 1957, Istock 1965, Ayala 1969).

A microhabitat may also be occupied simultaneously by two or more species if their combined populations do not exceed the carrying capacity of the microhabitat and if the populations of the species involved are limited by conditions in some other part of their niches (Lowe-McConnell 1969). Thus the niches of house wren, bluebird, black-capped chickadee, tufted titmouse, white-breasted nuthatch, and tree swallow overlap because they all nest in small cavities in trees or boxes. Usually the number of nesting sites is ample to the demand, so that there is no competition for them. The population of each species is restricted by factors other than available nest-sites, perhaps by food supply, unfavorable climate, parasites, or predators. The wren feeds mostly on the ground under bushes, the bluebird in open fields, the chickadee on the smaller tree branches, the tufted titmouse on the larger branches and on the ground, the white-breasted nuthatch on the trunk of trees, and the tree swallow in the air. Competition occurs between them only when one or more species temporarily increases in abundance so that there are not enough nest-sites to go around.

When competitive species occupy the same microhabitat, they sometimes set up mutually exclusive territorial relations, based on responses to similarity of body form and behavior (Simmons 1951) or call notes (Dilger 1956, Lanyon 1957). This divides the space and reduces competition. The suggestion has been made that interspecific territoriality occurs when insufficient time has elapsed for the species to evolve ecological divergencies that would allow for territories to overlap, or where the environment in some manner prevents the evolution of fine niche distinctions (Orians and Willson 1964). An intermediate step leading toward niche segregation is where one of the two species is usually successful in establishing its territory in the optimum biotope and the other species is forced into a less desirable one to which perhaps it eventually becomes adapted and to prefer (Murray 1971).

Because of parasites, unfavorable weather, predators, or other causes, species do not continuously saturate their habitats. It is only when they do that competition becomes clearly evident. Competition in the house wren takes various forms, including destruction of eggs and young and even killing of adults. In a 19-year study of a 6-hectare (15-acre) plot, such drastic competitive acts, both intra- and interspecific, occurred only when the number of male birds was more than 10. When the number of breeding house wrens was reduced to 10 or less, in consequence of heavy over-winter mortalities, no such competition occurred (Kendeigh 1941b).

The niche relationships and the amount of competition between species sometimes varies geographically. The vertical range of the salamander *Plethodon glutinosus* in the southern Appalachians is sharply defined from those of three different subspecies (*jordani, shermani, metcalfi*) of *Plethodon jordani*, but completely overlaps the ranges of two other subspecies (*clemsonae, melaventris*) (Hairston 1951).

During a spruce budworm insect outbreak in the coniferous forests of northern Ontario, three species of warblers—bay-breasted, Tennessee, and Cape May—greatly increased in numbers, partly because of their aggressiveness and partly because they were more accustomed to feeding on this type of food. Other warblers were held in check and one species, the magnolia warbler, actually decreased in numbers because of competition with the three favored species. Competition between birds for song posts and territorial space was considerable with nearly a third of all conflicts observed occurring between individuals of different species (Kendeigh 1947). It is probable that niche segregation of species becomes established at times of stress or crises like this, and after the behavior pattern once becomes fixed in the species, only sporadic attacks of one species on another are thereafter sufficient to check random variations away from the standard pattern. At times other than those of stress, direct conflict between species is not often observed, so that its importance in segregating species to particular niches is occasionally not fully appreciated (Lack 1944, Udvardy 1951, Andrewartha and Birch 1954).

Competitive Exclusion Principle

The evidence presented in this chapter demonstrates the important concept (Gause's Rule) that has come to be known as the "competitive exclusion principle" (Hardin 1960): *an ecological niche cannot be simultaneously and completely occupied by stabilized populations of more than one species.* In other words, two or more species with closely similar niche requirements cannot exist indefinitely in the same area. Two species with expanding populations attempting to occupy the same niche will sooner or later come into competition for possession of it. Rarely, if ever, will they be equally adapted, and ordinarily the one with the better adaptations or greater aggressiveness will win out and occupy the

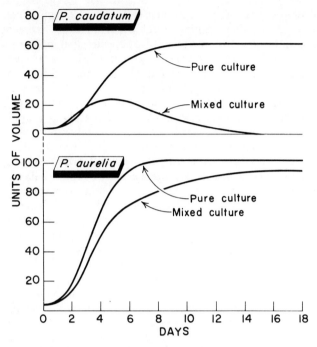

Fig. 16-6 Growth of populations of *Paramecium caudatum* and *P. aurelia*, cultivated separately and in mixed populations (from Gause 1934).

niche to its full carrying capacity. Species thus become segregated in different ranges, habitats, or mode of feeding (Lack 1971). The basic idea of this rule was understood from observations of natural distribution long before it was verified experimentally (Steere 1894, Grinnell 1904, Jordan 1905, Volterra 1931, Gause 1934). A corollary is the *coexistence principle*: *different species which coexist indefinitely in the same ecosystem must have different ecological niches.*

In experimental cultures, *Paramecium caudatum* and *P. aurelia* maintain separate populations at a high level, but when the two species are mixed, they quickly come into competition (Fig. 16-6). As long as the food supply is ample, both species increase in biomass, but as the food supply approaches exhaustion, *P. aurelia* persists and *P. caudatum* declines until it finally disappears. An analysis of the relative adaptation of the two species shows that *P. aurelia* is capable of faster population growth and is more resistant to the accumulating waste products than is *P. caudatum* (Gause 1934). Similarly, *Daphnia pulicaria* in mixed cultures causes the extinction of *D. magna* when oxygen and food become limited (Frank 1957), and *Tetrahymena pyriformis* persists while *Chilomonas paramecium* disappears in mixed cultures of these protozoans (Mucibabic 1957).

When two species of flour beetles with similar requirements for food, space, and other conditions are cultured together in the same volume of flour, one species always becomes extinct and the surviving species then establishes a stabilized population. In cultures

free of parasites, *Tribolium castaneum* is the successful species, but in cultures containing the sporozoan parasite *Adelina tribolii*, *T. castaneum* becomes extinct and *T. confusum* persists, since it is less susceptible to the parasite (Park 1948). Parasites or predators may then influence the success of competition between species by affecting one more than the other (Crombie 1947).

Unusual direct observation of competitive exclusion activity was recorded with two species of blackbirds in California (Fig. 16-7). Redwings had fully occupied a marsh with established nesting territories during February and early March. Then suddenly a colony of tricolored blackbirds invaded, leaving only a few intact redwing territories around the margin. As a rule, competitive exclusion is achieved more frequently, quickly, and efficiently through interference, such as this, than through exploitation. There is an evolutionary trend among birds, for instance, to establish formalized mechanisms of interference, such as song, display, and chasing behavior in the defense of territories (Miller 1968).

When two species with similar niche requirements meet in competition under natural conditions, one of three things will generally happen (Lack 1944):

1. One of the two species will be so much better adapted that it will spread rapidly through the range of the other and exterminate it.
2. One species will be better adapted to a portion of the range, in which it will eliminate the other species, but the other species will be better adapted to the remainder of the range and will occupy it exclusively. Thus the two species will occupy adjacent geographic regions with perhaps a zone of overlapping occupancy.

Fig. 16-7 Displacement of redwinged blackbirds by tricolored blackbirds in a marsh in California (Orians and Collier 1963).

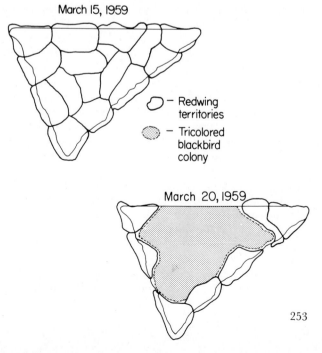

3. Each species will be better adapted to a different portion of the niche, to which it will become restricted; with this separation, each species will then spread through the range of the other.

Advantages of Niche Segregation

Probably the major advantage animals gain by occupying different niches is escape from continuous intense competition. It is also true that the niche occupied is favorable to the species physically in furnishing suitable substratum and microclimate, although many species have the ability to live elsewhere were competition not involved. Automatic segregation of a species into its niche through inherited behavior patterns or imprinting avoids the great expenditure of energy and loss of time that would be required if this segregation had to be worked out anew each year or each generation. Segregation into niches avoids confusion of activities between organisms in the community and permits a more orderly and efficient life cycle on the part of each species. Furthermore, the segregation of each species into different niches permits the occupancy of the area by a larger number of species, since they will better divide the available resources between them. Similarly, the more distinct the niche of a species is, the more it can avoid conflict with its neighbors and lead a life that is orderly, productive, and efficient. Competition is thus a potent factor in giving ecological structure to the community.

Community Segregation of Related Species

Closely related species usually have somewhat similar adaptations and niche requirements; this affects their distribution in communities in relation to each other.

In a study of 55 animal and 27 plant communities from a wide variety of habitats (Elton 1946), it was found that 86 per cent of the genera of insects and other animals and 84 per cent of the genera of plants were represented by a single species. The average number of species per genus of animal and plant in each community was 1.38 and 1.22 respectively, compared with an average of 4.23 species per genus for 11 large insect groups that occur in the British Isles as a whole. This means that while closely related species may occur in the same regional fauna, related species are more apt to be segregated into adjacent communities or habitats than into the same one. This conclusion has been disputed on a mathematical basis because of the small size of the communities and the small number of species involved (Williams 1947) but is supported by other studies (Lack 1944, Bagenal 1951, Hairston 1964).

It is true that in larger habitats or more diversified communities the larger number of niches ordinarily present should permit related species to occur. Yet, where related species could occur in large tropical communities of Africa, bird species belonging to the same genus occurred together in the same community in only 26 per cent of the communities, and if the weaver finches Ploceidae are excluded, only 16 per cent of possible overlap in habitat distribution is realized. This segregation of related species into different communities applied to an equal extent with species belonging to the same family (Moreau 1948). There are some genera of birds and other animals, however, in which a large variety of related species may be found together; notably, certain wood warblers *Dendroica*, buntings *Emberiza*, white-eyes *Zosterops*, whistlers *Pachycephala*, weavers *Ploceus*, hawks *Accipiter*, and the insects *Drosophila*, *Anopheles*, *Aedes* (Mayr 1947). Certain habitats, such as Lake Baikal in Africa, are unusual in possessing a great variety of closely related aquatic species (Brooks 1951).

The tendency for related species to be segregated into different communities or regions is an old concept and may be stated: *given any species in any region, the nearest related species is not likely to be found in the same region nor in a remote region, but in a neighboring district separated from the first by a barrier of some sort* (Jordan 1905). The principles and reasons underlying this law require an understanding of speciation, which will be taken up in the next chapter.

Ecological Equivalents

Communities in general are of fundamentally similar organization, and the facts we have already considered about food chains, trophic levels, and pyramids of numbers and biomass attest this. Consequently, a coniferous forest has many niches similar to those available in a deciduous forest, a grassland has some niches that are much the same as forest niches, a pond has some of the same niches as do lakes. The niches in a forest, a grassland, or a pond on one continent are very much the same as those to be found in similar communities on other continents. Similar niches in different communities or in different regions are commonly occupied by species possessing similar but not necessarily identical habits, adaptations, and adjustments (Table 23-1). Such species are called *ecological equivalents* (Friedmann 1946, Dirks-Edmunds 1947). Equivalent species are not necessarily closely related taxonomically.

SPECIES DIVERSITY

Communities could well be analyzed, compared, and evaluated in respect to the niches of which they are composed rather than the species that they contain. Each taxonomic species, however, generally occupies

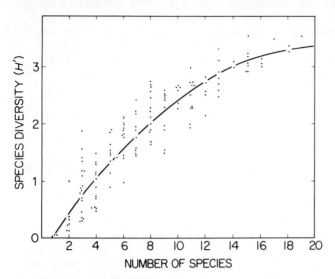

Fig. 16-8 Relation of the species diversity index to the number of tree species (from Monk 1967).

a distinctive ecological niche, and the population of individuals furnishes an index of the prevalence of a particular niche in an ecosystem. Therefore, much can be learned about community structure and composition from an analysis of species diversity.

The Shannon-Weaver species diversity index has two components: the number of species present and their "equitability" or the relative distribution of individuals between the species (Chapter 3). The index becomes higher with increase in number of species or more equal number of individuals in each species or both. The relation between the species diversity index and the number of species in natural communities is curvilinear (Fig. 16-8). After a certain number of new species are added, the index tends to level off, as new species added are generally rare and represented by only a few individuals each (Monk 1967).

Predominance

It is characteristic of the taxonomic structure of communities that a few species furnish the greatest bulk of the population entirely out of proportion to the rest of the species. Thus in stream riffles, two species make up 85 per cent of the total riffles populations and another two species constitute a similar percentage of the mud-bottom-pool populations (Table 4-1). In the littoral zone of Lake Erie, one species furnishes 85 per cent of the population on cobble and gravel bottoms, and another species makes up 68 per cent of the populations on mud bottoms (Table 5-2; Sanders 1960). In populations of 79 species of birds nesting on a large tract, including advanced stages of succession in New York State, 5 per cent of the species (4 species) included 37 per cent of the individuals, 10 per cent (8 species) included 56, and 50 per cent (40 species) included

96 per cent (Evans 1950). Expressed in a different manner, 2 of 41 invertebrate species found in decaying oak logs in England contributed 50 per cent of the total individuals; 3 species, 25 per cent; 24 species, 24 per cent; and 12 species, only 1 per cent (Fager 1968). The abundant species are ordinarily of small size, herbivorous in their food habits, and at the bottom of the food chains and pyramids. Predominance is better expressed in communities occurring in "harsh" environments, that is, in environments where relatively few species find conditions favorable, than in those occurring in mesic ones that are favorable for many species (Fig. 16-9; Whittaker 1965, McNaughton and Wolf 1970).

Succession

Succession is the process whereby a series of communities and habitats replace one another, beginning in an extreme habitat and ending in a mature climax ecosystem. Several examples of seres have been analyzed in Chapters 4–8. With succession on land there is generally a progressive increase in height of the dominants, stratal differentiation of the vegetation, elaboration of internal microclimates, diversity of plant life-forms, biomass, and development of soil structure and fertility (Whittaker and Woodwell 1972). Since the original habitat is bare of all life, at least in a primary succession, there is necessarily a progressive invasion of new species as each community develops in the sere. Species diversity tends to be greater in each successive stage, but the highest diver-

Fig. 16-9 Relation between predominance of plant species in a community and the position of that community in a habitat gradient from dry prairie (1) through mesic meadow to wet marsh (10). Predominance is stronger in the more extreme environments. Predominance is calculated as the percentage of abundance of the two top species to the total number of individuals (from McNaughton and Wolf 1970; copyright 1970 by the American Association for the Advancement of Science).

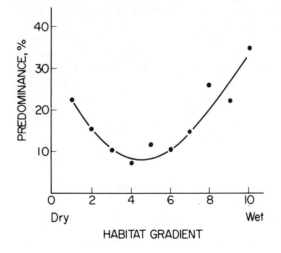

Table 16-4　Changes in avian and habitat diversity with succession on abandoned strip-mine areas (Karr 1968, Karr and Roth 1971).

Stage	Bird species diversity, H_i'	Standing crop biomass diversity, H_b'	Existence energy diversity, H_p'	Plant species diversity, H_i'	Foliage height diversity, H'	Per cent vegetation cover
Bare ground	1.537	1.406	1.467	2.180	0.000	8
Early shrub	2.742	2.408	2.646	4.596	0.807	114
Late shrub	3.182	2.979	3.152	6.220	0.920	145
Young bottomland forest	3.315	2.582	3.127	4.137	1.070	225
Older bottomland forest	2.917	—	—	—	—	—
Climax forest	2.554	—	—	—	—	—

sity index both for richness of species and equitability may sometimes occur in an intermediate or subclimax rather than the final climax stage (Table 16-4, Fig. 16-10; Auclair and Goff 1971). Successional trends may likewise vary between strata within the community (Whittaker 1969).

In the sere following the retreat of a glacier in Alaska, richness of plant species increased rapidly during the first hundred years as succession proceeded through the first six stages to the spruce forest, then more gradually through the hemlock forest to a maximum in the muskeg steady state. Equitability was erratic but also reached a maximum in the final stage (Reiners *et al.* 1970). Tree species diversity has also been shown, both in Wisconsin (Loucks 1970) and Florida (Monk 1967), to increase from pioneer to subclimax or climax communities.

By combining studies done in abandoned fields in Georgia and Illinois (Fig. 16-10), it is apparent that bird species diversity increases as time proceeds and the prevailing vegetation changes from grass to shrubs to mature climax forest. There is a suggestion here that the highest index occurs preceding the climax, where there is still a remnant of seral species mingling with a full complement of species belonging to the climax forest. Karr (1968) likewise found a higher bird species diversity in the strip-mine sere in the early bottomland forest than in the climax (Table 16-4). In both studies species diversity varied consistently with richness of species, while equitability showed no definite or reliable trend. The fairly constant and high equitability index (mostly greater than 0.80) found generally in bird communities may be a reflection of the even dispersion and resistance to crowding of nesting birds as the result of their intraspecific territorial behavior (Tramer 1969).

Latitudinal Gradients

Spatial gradients of habitat are often joined with gradients in species diversity. For instance, the species diversity of fish increases downstream correlated with greater depths of water (Sheldon 1968), species diversity of benthic macroinvertebrates in streams increases from polluted to clean areas (Wilhm and Dorris 1968), species diversity of benthic Foraminifera, Mollusca, Arthropoda, and Echinodermata increases at greater depths (Buzas and Gibson 1969), and diversity of ants, greatest in the tropics, decreases with latitude, altitude, and aridity (Kusnezov 1957).

Of considerable interest is the progressive latitudinal increase in number of species per unit area from the arctics to the tropics found in most groups of both plants and animals (Fig. 16-11) on land and in the sea, and in geological as well as present time (Fischer 1960, Preston 1960, Simpson 1964, Cook 1969, Low-McConnell 1969, Stehli *et al.* 1969). A number of hypotheses, some of them overlapping, have been suggested to

Fig. 16-10　Regression of bird species diversity on chronological age (succession) in oldfields, combining data from the Georgia Piedmont region (Table 7-5) for the early stages (+) and from central Illinois for the later stages (o) (modified from Roth 1967).

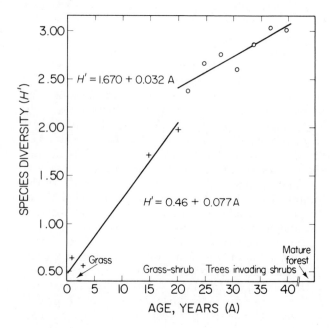

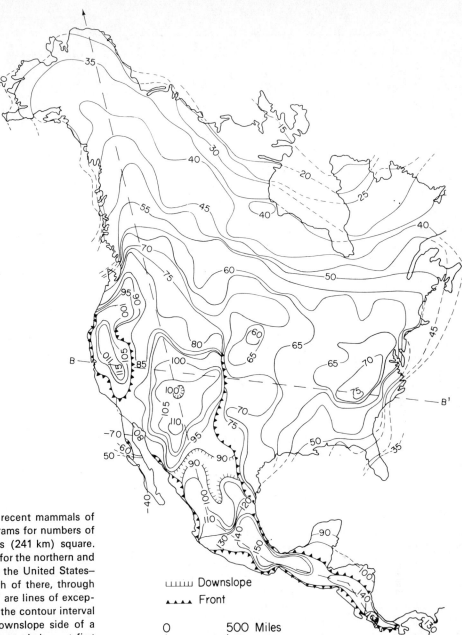

Fig. 16-11 Species density contours for recent mammals of North America. The contour lines are isograms for numbers of continental species in quadrats 150 miles (241 km) square. The interval between isograms is 5 species for the northern and central parts of the map, approximately to the United States—Mexican border—and 10 species for south of there, through Mexico and Central America. The "fronts" are lines of exceptionally rapid change that are multiples of the contour interval for the given region. Indication of the downslope side of a contour is given in two areas where it is not obvious at first sight (Simpson 1964).

⊔⊔⊔⊔⊔ Downslope
▲▲▲▲ Front

0   500 Miles
⊢────────⊣
         805 Kilometers

explain this greater species diversity found in the tropics and controversy exists as to which are more likely to be correct (Pianka 1966):

Time Tropics have existed, undisturbed, for a very long geological time, thus allowing for more complete evolution of species and occupancy of available niches (Valentine 1968, Whittaker and Woodwell 1972).

Spatial Heterogeneity There is a greater complexity and diversification of topography and microhabitats toward the equator (Janzen 1967, Mayr 1969), which leads to a greater diversity in the vegetation and in turn provides a wider variety of niches for animals (Pianka 1967). Spatial heterogeneity, leading to speciation, was particularly pronounced during Pleistocene and post-Pleistocene periods when the tropical forest was broken up into a number of scattered refugia (Haffer 1969).

Competition In temperate and arctic regions, evolution of species is limited by fluctuations in climate to which they must adapt. In the tropics, physical conditions are more favorable, and evolution is more subject to interspecific competition, and this brings a finer distinction between niches (Dobzhansky 1950).

Predation More predators and parasites in the tropics hold prey populations to lower levels, hence allowing additional species to evolve and divide up the resources (Paine 1966). Higher predation of herbivores on seeds may lead to greater distance between parent plants. With increasing distance from the parent, the density of seeds decreases, but the probability of seeds not being found by an herbivore increases (Janzen 1971).

Climatic Stability Stable climates and constancy of resources permit survival of species having narrow

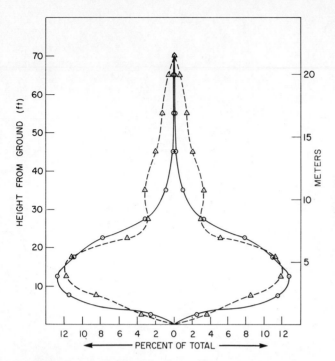

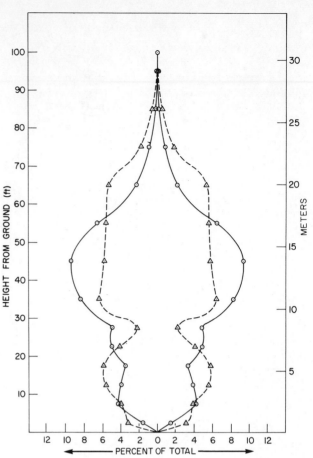

Fig. 16-12 Use by all bird species of tree foliage in the oak-juniper woodland (left) and in the ponderosa pine montane forest (right). The solid line represents the per cent of foliage volume present at various heights above the ground. The broken line is the per cent of total observations of birds at each height (Balda 1969).

ranges of habitat tolerances. This leads to evolution of finer specializations and adaptations (Janzen 1967, Sanders 1968, Ashton 1969).

Productivity Higher productivity provides resources for more species, a sedentary mode of life is encouraged, interspecific coactions increase, and species tend to become more specialized. Maintenance of high productivity throughout the year also permits utilization of resources at different times by different species (Connell and Orias 1964, MacArthur 1965). Some studies, however, have found an inverse correlation between productivity and diversity, as in early successional stages, and there may be no significant relation over-all between the two (Whittaker 1965).

Community Stability

Experiments show that monocultures, or plantings of single species, are subject to rapid and extreme fluctuations in the animal populations that feed upon them. The predominance index is high among consumers. Food chains are short, and there is high productivity per unit biomass of living organisms. There is little resiliency against disturbances in the physical environment, invasion of new species, predation, or disease (Piementel 1961). Such open communities, although ordinarily including more than one producer species, are typical of harsh habitats: pioneer stages in succession, arctic tundra, desert.

On the other hand, mature or stable communities

are more complex, composed of many species, closed or resistant to the invasion of new species, have long and anastomosing food chains, possess a low ratio of productivity to biomass, and maintain relatively constant-sized populations (Margalef 1963, Woodwell and Smith 1968). With the acquisition of stability, communities change from a composition and structure based predominantly on adaptations to the physical habitat to one that is biologically accommodated (Sanders 1968). Community stability increases toward the climax in succession, and tropical or mesic communities are thought to be more stable than arctic or desert ones (MacArthur 1955, Hutchinson 1959).

However, the relationship of community stability to trophic-web structure is not a simple one. High species diversity and complexity of niche interrelations in one level of the trophic web may, indeed, produce stability at that level but instability at another level (Watt 1968: pp. 39–50, Hurd *et al.* 1971). For instance, if one or more species of herbivores suddenly builds up in numbers because of a favorable period of weather or increase in food supply, the carnivore level that ordinarily keeps it under control may have too much stability to allow compensatory build-up in its numbers, and the herbivore "escapes" from control. Too much interspecies competition between carnivores may depress the effectiveness of all attacking species.

Progression toward stability within a trophic level or whole community depends not only on increase in richness of species but also on higher equitability of species populations. Greater richness of species depends in part on the greater productivity or carrying capacity of the ecosystem and in part on the way the resources are divided as the result of competition (McNaughton and Wolf 1970). In stable ecosystems, both primary and secondary consumers are more specialized in their food habits, but at the higher consumer levels, flexibility in food habits still remains (Paine 1963). Greater species diversity means either that the niche breadth, especially in respect to habitat, is narrower, that there is more extensive overlapping of the niches, such as in food taken (Klopfer and MacArthur 1961), or that there are more different kinds of niches dependent on differences in feeding habits and reproductive requirements (Schoener 1965, Orians 1969, Karr 1971). Greater species and niche diversity may thus be expressed either "within habitat" or "between habitat" (MacArthur 1965), that is, by dividing the natural resources of a particular habitat or community between a larger number of species or by fitting a larger number of species into an environmental gradient or geographic area (Whittaker and Woodwell 1972). The narrowing of niches and increased specialization in the use of particular resources implies that these resources are more continually available and that species have attained greater efficiency in their utilization.

In relation to use by man, a greater yield can usually be obtained from fertile ecosystems having only short or simple food chains than from stable ecosystems possessing complex food chains. This greater yield, however, is apt to be less constant from year to year than are the smaller yields obtainable from stable ecosystems (Watt 1968).

Vegetative Structure

Because of their diurnal conspicuousness and ease of observation, birds have been extensively studied for correlation of species diversity, niche size, and vegetative structure. Type of vegetation and life-form of predominant plants are important in segregating species into different communities, as shown elsewhere in this book, but the main consideration here is on vertical stratification. Different species of birds have their activities of nesting, feeding, and singing at different levels in the vegetation, and they vary in their use of trees of different foliage types. The total activities of all species of birds tend, however, to conform approximately to the vertical profile of the combined tree species (Fig. 16-12).

An index of foliage height diversity may be calculated, using the same Shannon-Weaver equation as for species diversity,

$$FHD = -\sum_{i}^{s} p_i \log p_i$$

(p. 23), but in this case p_i is the proportion of the total foliage which lies in each of several strata in the

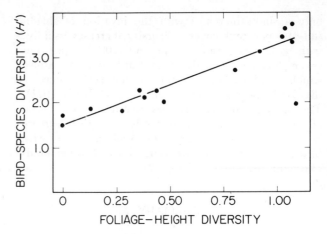

Fig. 16-13 Regression of bird-species diversity on foliage-height diversity (Karr and Roth 1971).

vegetation. The foliage coverage at each stratum may be estimated by vertical sightings, by counting the leaves touched by a vertical pole, or in other ways, a number of such estimates being made at intervals along transects through the vegetation. The number and ranges in height of strata recognized will depend on the type of vegetation but is commonly one to four. A direct linear regression or correlation occurs between bird species diversity and foliage height diversity calculated in this manner (Fig. 16-13; MacArthur and MacArthur 1961). Likewise, a direct linear relation occurs between number of lizard species in flatland desert regions and plant volume diversity, calculated in a slightly different manner (Pianka 1966). With small desert rodents, presence and density are related not only with plant growth form and foliage density but also with the soil's resistance to sheer stress, important for ease in burrowing (Rosenzweig and Winakur 1969).

If instead of calculating foliage height diversity, the accumulative percentage of the ground covered by the various strata is considered, an even more useful correlation with bird species diversity may be obtained (Fig. 16-14). On bare ground, bird species diversity is low. With the addition of grasses, there is

Fig. 16-14 Relation of bird species diversity to per cent vegetative cover (Karr and Roth 1971).

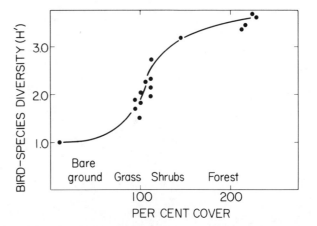

only a slight increase; with the invasion of shrubs (80–100 per cent cover), diversity increases rapidly; with both shrubs and trees present (100–150 per cent cover), diversity quickly reaches an asymptote above which addition of more strata does not further increase the bird species diversity (Karr and Roth 1971). The increase in bird species density as succession advances (Fig. 16-10) is dependent on this increase in vegetative cover. Relations, such as these with birds, can be expected to occur also with other kinds of organisms.

SUMMARY

The ecological niche is a particular combination of microhabitat and biotic relations required for the activities and existence of a species in a given ecosystem. The niche of a species is defined by the features of the substratum and microclimate to which that species is peculiarly adapted, the time of day and season of year when it is mainly active, the type of shelter or cover it requires, the manner in which it uses vegetation in its reproductive performance, the type of food it consumes and where it gets it, and the predators that prey on it.

Animals, particularly higher types, have more or less stereotyped behavior patterns associated with their restriction to particular niches. These behavior patterns may be genetically inherited and subject to evolutionary development, or they may be transmitted to succeeding generations through training or conditioning (imprinting) of the young. Niche requirements consequently have a major role in determining the kinds of environments (and communities) in which species occur.

Segregation of species into different niches is doubtless the result of interspecific competition. According to the competitive exclusion principle or Gause's rule, an ecological niche cannot be simultaneously and completely occupied by stabilized populations of more than one species. Occupancy of different niches reduces interspecific competition, furnishes the species with microhabitats to which they are especially adjusted, reduces confusion and disturbance, and permits a greater variety of species to occur in a given community and in the same region.

Since communities in various parts of the world are fundamentally similar in organization and structure, species occupying similar niches in them make similar adjustments. Although not necessarily related taxonomically, these species are ecological equivalents. Communities commonly have a predominance of a few species composing the bulk of the populations. Predominance tends to be stronger, and equitability and diversity lower, in communities of extreme environments. Closely related species tend to be segregated into different but adjacent regions separated by barriers.

Species diversity generally increases with succession toward late seral or climax stages, from arctic to tropical regions, and from harsh to mesic habitats, correlated with increasing diversity in the vegetative structure. Community stability is established only with high species diversity accompanied by necessity with complex and intercompensating relations between the species.

Niche segregation, species diversity, and vegetative structure, as these change in relation to one another and different environments, are major aspects of community organization that have come about through the long process of evolution.

Chapter 17

SPECIATION
AND CULTURAL EVOLUTION

As shown in the preceding chapter, when a species fits into a niche it evolves interrelations with other species. Individuals within the species establish a pattern of behavior to which all adhere, so that the species functions in an efficient organized manner. Much of this behavior is learned in each generation and thus constitutes a mores or ethics, the prototype of culture. The evolution of culture in human society is an outgrowth of these intra- and interspecific interrelations in the behavior of lower animals and developed concurrently with speciation of man.

Speciation is the process of evolutionary differentiation between populations, which may result in one older species becoming split into two or more new ones. Speciation also entails divergence of the new species into different niches or geographic areas. An understanding of the basic principles of speciation is therefore a prerequisite to an understanding of both local and geographic distribution of organisms (Tax 1960).

TAXONOMY

As the term is used by taxonomists, a *population* is a local aggregation of individuals that differs slightly, but characteristically, from other local aggregations of the same species. Geneticists define a population as a reproductive community sharing a common, characteristic gene pool. Every local population is different from every other one; but they are not easily distinguishable from one another, and therefore are given no formal nomenclature. What a population includes may vary from the entirety of a species to but a few individuals, according as the rate and extent to which individuals interdisperse between localities to make a common genepool.

Among bisexual forms, a *species* is a group of populations capable of successfully interbreeding and is reproductively isolated under natural conditions from other such populations. These criteria are often difficult to apply, however, so reliance is commonly based on superficial or morphological characters (Sokal and Crovello 1970). Species are usually morphologically distinct, but the distinguishing characteristics are sometimes barely discernible. Hybridization may occasionally occur between clearly defined species under captive or unnatural conditions, but does not occur with any significant frequency under natural conditions.

Fossils found separated in different geological strata, or living populations separated in space, are considered capable of successful interbreeding, and therefore to be of the same species, if essentially similar structures, functions, and behaviors can be adduced for them. It is unfortunate that species cannot presently be recognized by entirely objective means, but even if this were readily possible there would be difficulties, inasmuch as populations may be at various stages in the differentiation of complete reproductive isolation. Species are distinguished by the familiar bionomial nomenclature standardized by international rules (Mayr *et al.* 1953).

A *subspecies* is a geographically defined group of populations which differs in color, size, or some other taxonomic characteristic from other populations within the same species but nonetheless interbreeds with them freely, regularly, and successfully where their ranges come into contact. Subspecies are commonly distinguished by a trinomial nomenclature, although the desirability of this has been questioned (Wilson and Brown 1953, Owen 1963). The taxonomic differences between subspecies are usually less pronounced than those between even closely related species, but they are genetically fostered differences nonetheless (Sumner 1924, Huxley 1943). Race and variety are terms sometimes used as synonyms for subspecies, or for populations even less well defined than a subspecies. It is difficult to determine whether populations that differ phenotypically are subspecies or species when their ranges do not verge, for in the absence of opportunity for them to commingle, it remains uncertain whether interbreeding would or could occur. The assignment of taxonomic rank under these circumstances is dependent upon considerable subjective inference. Boundaries between adjacent subspecies are frequently arbitrary in that they fit differences between the populations for some characters but not others. Differences between subspecies or populations are often correlated with differences in topography, soil, climate, or vegetation, and, at least in some instances, appear to be adaptive to these differences in the environment.

If a subspecies becomes isolated by a barrier so that it is prevented from interbreeding with the rest of the species, variations in taxonomic characters may accumulate and the population evolve so distinctively as to pass beyond the rank of subspecies into that of species. However, not all or even most subspecies change into species, a process which depends on effective reproductive separation and on the forces of natural selection. Speciation is the process of differentiation between populations of the same species in consequence of reproductive isolation (Simpson 1953).

Populations that do not differ by clearly defined or conspicuous taxonomic characteristics, but nevertheless do not interbreed because of physiological or behavioral differences, are described as *sibling species*. Sibling species have been noted especially among Diptera (*Drosophila, Anopheles*), Hymenoptera (ants), Lepidoptera (especially moths), and Protozoa (*Paramecium*) (Mayr *et al.* 1953). It is apparent that the evolution of physiological and behavioral differences often precedes the differentiation of recognizable taxonomic distinctions (Krumbiegel 1932, Thorpe 1930).

Two species the distributional ranges of which do not overlap (that is, there is geographic isolation between them) are said to be *allopatric*. Two species are said to be *sympatric* when their ranges overlap, even though they may locally be ecologically segregated into different habitats; for example, the situation in which one species of snail is limited to floodplain and another species to upland habitats, or one species of rodent confined to the foothills and a closely related species to a higher zone on mountains only a short distance away. Populations should be considered sympatric if they occur within the common dispersal ranges of their young, so that there is at least possible a continuous and appreciable interbreeding and flow of genes among them.

ISOLATION OF POPULATIONS

Sympatric species do not interbreed because one or more isolating mechanisms keep them separated. We will proceed to examine what isolating mechanisms are, and how geographic isolation permits them to arise.

Isolating Mechanisms

Isolating mechanisms that prevent sympatric species from freely interbreeding are largely or entirely biotic factors (Mayr 1970, Huxley 1943, Allee, Emerson *et al.* 1949, Dobzhansky 1951, Patterson and Stone 1952). They are the following types:

ECOLOGICAL: Segregation of species into different habitats, communities, or niches by reason of structural adaptations, physiological adjustments, or behavior responses.

ETHOLOGICAL: Difference in sign stimuli or behavior patterns required for successful species and sex recognition and for mating.

MECHANICAL: Lack of physical conformity of sexual organs, chemical incompatibility of sperm and egg.

GENETIC: Hybrid sterility or decrease of fertility.

Ecological isolation was described and illustrated in the discussion of niches (Chapter 16).

Animals may fail to find breeding partners; that is, they may remain *ethologically* isolated. Careful studies in all vertebrate groups and in such invertebrates as insects, spiders, crustaceans, and snails indicate that animals identify the sex of individuals of their kind only by actively recognizing special clues or sign stimuli (Tinbergen 1951). These clues may be special color markings, shape or outline, scent (Moore 1965), song or call-notes (Stein 1963, Waldron 1964), touch, behavior patterns, or some combination of these. Courtship leading to copulation is often complex and involves a number of steps, each step in the behavior serving as a releaser for the next (Fig. 2-7). If a step is not performed properly and in its sequence, the courtship

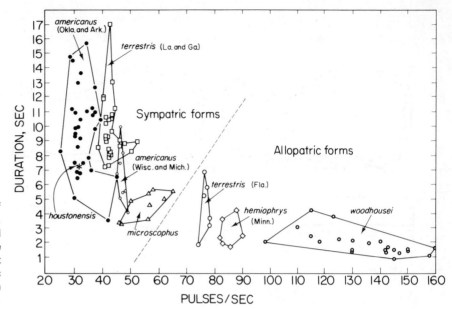

Fig. 17-1 Duration and pulse rate of calls of the toad, *Bufo americanus* group, as plotted from tape recordings and sound spectrograms. Note greater divergence from *woodhousei* of sympatric forms that overlap it in distribution than of allopatric forms that do not occur in the same region (Blair 1962).

performance ceases forthwith, and there is no sexual consummation. Different species may be effectively isolated from interbreeding simply because the sexes of one possess different sign stimuli than the sexes of the other, and the sexes of one kind characteristically pursue patterns of pairing behavior that are not stimulating to the sexes of the other. Mating calls or song are important means for isolating the breeding of frogs and toads (Fig. 17-1), birds (Dilger 1956), crickets (Alexander 1962), and other forms.

Ethological isolating mechanisms (Spieth 1958), effective under natural conditions, often lose efficacy when those conditions are disturbed; by so much they are difficult to work with experimentally. Where their home ranges overlap, two species of mice, *Peromyscus leucopus* and *P. gossypinus*, frequently occur in the same habitat. Yet, very few hybrids have ever been found. When brought into the laboratory, however, the two species not only hybridize freely, but produce fertile offspring. It appears that ethological factors keep them separated under natural conditions (Dice 1940).

Two closely related budworm species, *Choristoneura fumiferana* found on balsam fir, and *C. pinus* found on jack pine, are isolated ecologically on different host trees, and because the first species completes its mating season before adults of the second species appear on the wing. Ethological isolation occurs when occasionally their mating periods overlap. Females will mate only with males of their own species, even though males of both species attempt to mate indiscriminately (Smith 1954).

Species related but of different sizes may be unable to fit their copulatory organs together, because of structural incompatibilities. Such hindrance to interbreeding is a *mechanical* isolation. Failure of male toads to clasp females of larger or smaller species results in reproductive isolation between species of *Microhyla* (Blair 1955). Polygyrid snails of the genus *Stenotrema*

have definite behavior patterns prerequisite to copulation. A careful study of several species in this genus (Webb 1947) showed that differences in these behavior patterns are sufficient to keep some species separated, but that in other instances it is differences in the structure of the copulatory organs that apparently prevent interbreeding.

Genetic isolation occurs when there is inability to produce offspring because of incompatibility of spermatozoa and eggs, abnormalities of growth, or the offspring are sterile. Sperm of the sea urchin *Strongylocentrotus franciscanus* sufficient to give 73 to 100 per cent fertilization of eggs of the same species produced only 0 to 1.5 per cent fertilization of *S. purpuratus* eggs. Eggs of one species may sometimes be successfully fertilized by sperm of another, but all sorts of disturbances may occur in the zygote, such as chromosome elimination during cleavage, arrest of gastrulation or organ formation, and death of embryos in advanced stages. A well-known example of a usually sterile hybrid is the mule, the result of a cross between a male ass and a female horse.

Genetic and mechanical isolation usually furnish more certain reproductive separation between species than do ecological or ethological isolation. The latter two forms probably represent early steps in the process of speciation; the former two, the culmination of speciation.

Geographic Isolation

Physiographic barriers such as land masses, mountain ranges, and bodies of water can effect complete or nearly complete isolation of populations. Population segments of a species may become geographically isolated when reproductive individuals cross a barrier by chance. For instance, individuals may be blown by

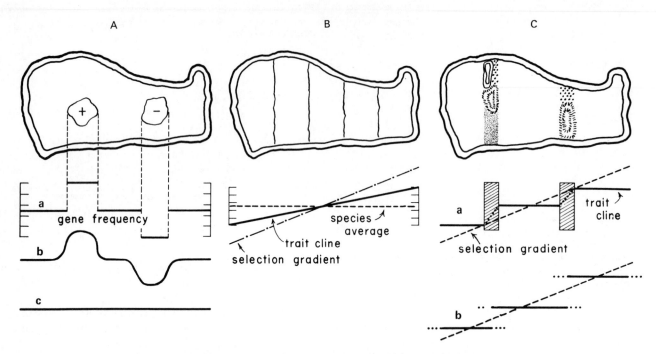

Fig. 17-2 (A) Local plus and minus variations in any characteristic of a species occupying a uniform habitat eventually disappear because of gene flow throughout the population (panmixia). (B) When the habitat characteristics of a geographic range occupied by a species gradually change from one extreme to another, a selection gradient is established that partially counteracts the tendency toward panmixia, producing a *cline* in the characteristics of the species. (C) When partial barriers to otherwise free gene flow occur in an environment or selection gradient, a stepped cline (a) produces distinct races or subspecies. If the barriers become complete, the isolation of population may result in divergent evolution and complete speciation, so that later, after the barriers disappear (b), biological isolating mechanisms prevent the populations from interbreeding (after Womble 1951).

storms or carried by rafts to outlying islands. A barrier may arise subsequent to the dispersal of a species, such as when species disperse into a new area by way of land bridges which later disappear (Fig. 17-2).

When barriers are only partially effective or distances so great that gene flow between adjacent populations is hindered but not stopped, and natural selection goes on independently in each area, some differentiation of the populations may occur but not above the level of subspecies (Fig. 17-3). The rate of gene flow through the species population is more or less proportional to the rate at which individuals disperse from birthplaces. Less than 5 per cent of young pied flycatchers surviving to 1 year of age return to their places of birth in order to nest; the other 95 per cent disperse widely. Only 4 subspecies have differentiated in this species, since there is widespread promiscuous interbreeding between individuals. On the other hand, 63 per cent of young song sparrows surviving to sexual maturity return to nest in the vicinity of their

birthplaces; only 37 per cent disperse elsewhere. This results in a limited flow of genes from one locality to another and is correlated with the development of 28 subspecies (Haartman 1949). In the range from Maine to Florida, populations of leopard frogs readily interbreed with adjacent populations. But when individuals from Maine are brought together with individuals from Florida, the two cannot interbreed successfully (Moore 1946). There is a progressive change from north to south in several characteristics of the leopard frog, and it is of interest that if populations intermediate in the range should be eliminated, the populations at each of the two extremities of the range would be considered separate species, which at present they are not.

Only through geographic isolation do populations ordinarily differentiate into distinct species. Any change in habitat, behavior, or genetics acting singly or in combination, in the absence of geographic isolation, is insufficient to prevent at least some significant gene exchange with the rest of the species, with resultant preservation of the species. One apparent exception to this principle is where there occurs simultaneous development of polyploidy in certain individuals, which renders them sterile with normal members of the species, although not with each other. Polyploidy rather commonly gives complete genetic isolation in plants, but is rare in animals, occurring in some parthenogenetic forms.

In the presence of geographic isolation, genetic variations and natural selection may bring the affected population to a different course of evolution than in the parental species, especially if adaptation to a new environment is also involved. Biotic isolating mechanisms may develop in the process. If the geographic barrier formerly separating the population should disappear, and the hitherto isolated population again comes into contact with the rest of the species, interbreeding will not then occur. This is the process of speciation, the details of which will now be examined more carefully.

VARIATION IN POPULATION CHARACTERISTICS

Observable differences in structure, function, and behavior between individuals belonging to the same species are common. Actually, no two individuals, except perhaps identical twins, have exactly similar characteristics. It is the gradual accumulation of many small variations over many generations which eventually gives a population reproductive isolation and, consequently, species identity.

Non-heritable Variations

Not all variations of organisms are of direct significance in speciation. Non-heritable changes in body structures, functions, and behavior are common. If muscles are used continually and intensively, they become thicker and stronger; if one kidney is removed, the other becomes hypertrophied; skin subjected to frequent rubbing or pressure thickens and becomes horny; and so on through a lengthy catalog. Animals progressively exposed to ever more severe temperatures or lower oxygen concentrations will tolerate extreme conditions, which, had they been suddenly presented, would have been fatal. Insect larvae transferred to a new type of food often become so conditioned to it that they produce a strain that prefers that food to other more usual food of the species. Chimney swifts in wilderness North America nested in hollow trees, but with settlement of the country and the construction of chimnied buildings during the last two centuries, the species has changed its behavior almost completely, accepting chimneys as a satisfactory nest-site location.

Many phenotypic adaptations persist generation to

generation, either as similar responses made by each generation to constant environmental conditions; as the result of imitation of parents, conditioning of young or imprinting (Cushing 1941, 1944, Thorpe 1945); or because the particular genes responsible for these characteristics have been sorted out (canalized) from the general gene pool of the species (Waddington 1957). One would expect a change in behavior or function, arisen in consequence of exposure to the new conditions, usually to presage the evolutionary development of a new structure, for natural selection cannot bring about the structural adaptation or perfection of an organ unless the organ is being used for the new purpose (Prosser 1957, Ford 1964).

Polymorphism

When individuals of a population occur in more or less discrete groupings of color phases, body sizes, or other character variations that are not due to sex, age, caste, or continuous mutation, the population is said to display *polymorphism*. Polymorphism of a character arises

Fig. 17-3 Speciation through distance in the salamander species *Ensatina eschscholtzii* in California. The coastal subspecies *picta* may represent the ancestral type and demarcate the center of dispersal from which clinal lines became dispersed southward in the coastal and interior mountains separated by a barrier. Recently, the subspecies *xanthoptica* crossed this barrier. Interbreeding occurs between adjacent subspecies, but partial reproductive isolation obtains between *xanthoptica* and *platensis*, and complete isolation similar to that of species obtains between *eschscholtzii* and *klauberi* (Stebbins 1949).

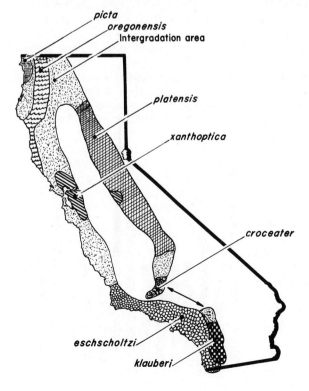

in a species when heterozygotes persist in an environment in the face of natural selection and homozygotes are reduced or eliminated. For instance, individuals with certain characteristics may be better adapted to environmental conditions during the spring, while individuals with other characteristics are superior in summer or autumn. This results in a mixture of types in the population more or less segregated by seasons (Fig. 17-4). Likewise populations may vary in characteristics as adaptations to local habitat conditions (Fig. 17-5). An increase or decrease in the frequency of a given characteristic appears to be a result of variations in the selective pressure of the environment, permitting individuals with certain characteristics or gene combinations to survive at one time or place; other individuals, at other times and in other places. In an environment that presents the same set of selective pressures year after year, there commonly occurs a stability in the ratios of the different forms in which a character is manifested (Ford 1964, Dobzhansky 1951, 1956, 1958, Sheppard 1958, Kennedy 1961).

Most species differ not by single genes but by hundreds certainly, possibly by thousands, of genes. When *panmixia* (free interbreeding) obtains in a species, these genes may be arranged in all sorts of combinations to form an almost infinite mixture of character modifications (Caspari 1951). Heterozygotes are, therefore, much more flexible in adaptively responding to the environment than are homozygotes. The more characters for which an individual is heterozygous, the more adaptable its offspring are likely to be. Adaptive polymorphic populations are more efficient in exploiting the environment than are genetically uniform ones. Conversely, species that are widespread geographically through many habitats are genetically more diversified than are those restricted to few or specialized habitats. There are limits, however, beyond

which a character cannot change. The continual tendency for characters to fluctuate around a mean or intermediate condition gives a population *genetic homeostasis* (Lerner 1954).

Genetic Drift

Although certain characters may result from the action of a single gene or pair of genes, many, perhaps most, characters within a species are polygenic; that is, they are affected by a multiplicity of genes. The exact form in which a character is expressed depends on the particular combination of genes which the individual or population possesses (Waddington 1957).

If a fertilized female, a single pair of animals, or at most a few hundred individuals become separated from the rest of the species, there will be represented in them a considerable decrease in the number of genes available to the main body of the species, since no individual or small group of individuals can possess all the genes that occur within the species' pool. Inbreeding within small isolated populations may thus bring into prominence traits which are expressed only irregularly and inconspicuously within the species as a whole. Establishment thus of restricted genotypes in small populations by loss of genes or accidental changes in frequencies at which certain genes occur is called *genetic drift*, or the *Sewall Wright effect* (Wright 1931). Character variations formed in this manner are usually nonadaptive. It is often difficult to be sure whether local character differences are due to genetic drift or to subtle differential selection by predators or habitat and hence are adaptive, indicating a polymorphic species (Goodhart 1962).

Similar fixation of special characteristics may occur in species subject to catastrophic or cyclic reductions in abundance. The genotypes of the few survivors will determine the genetic makeup of the entire new population that develops in the area (Elton 1930, Timoféeff-Ressovsky 1940). For example, the arctic fox is a cyclic northern species possessing a white and a "blue" color phase. Over most of the fox's range, the blue phase is much less common than the white, but on certain islands only the blue phase occurs. It is possible that at some time in the past, at the bottom of a population cycle, only homozygous blue foxes survived; reproduction of these animals and their offspring rendered the entire new population blue. Likewise, island populations, differing somewhat from related forms on the adjacent mainland or on other islands, may be the result of peculiar gene combinations of the first colonizers. This is the *founder principle* (Mayr 1970).

Mutations

Mutations may be the result of chemical changes in the individual gene or of chromosomal aberrations,

Fig. 17-4 Seasonal changes in relative frequency of the third chromosome with the standard (black), arrowhead (stippled), and Chiricahua (cross-hatched) gene arrangements in a natural population of *Drosophila pseudoobscura* (Dobzhansky 1951).

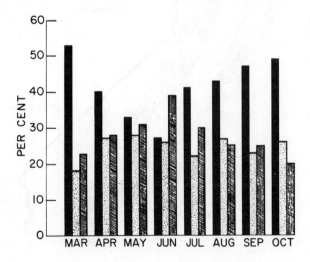

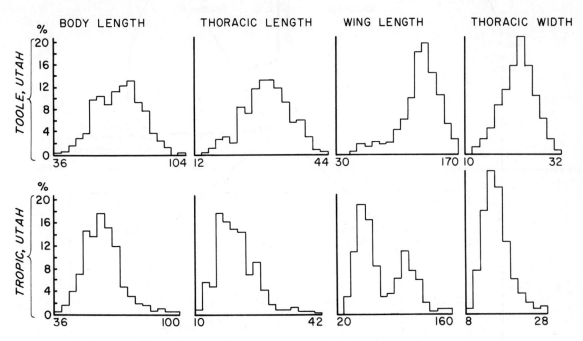

Fig. 17-5 Variations in body and wing measurements between two local populations of the gall wasp. Abscissas are in micrometer scale units; ordinates, in percentages of the population in each measurement class (after Kinsey 1942).

the latter in the form of changes in the number of chromosomes (haploidy, polyploidy) or of arrangement of genes on the chromosomes (deletion, duplication, translocation, inversion) (Dobzhansky 1951). The rate at which any one gene mutates varies greatly from one kind of gene to another, but the average rate is of the order once in every 100,000 or 1,000,000 individuals. Between 0.4 and 10 per cent of the individuals in each generation may possess mutated genes (Schmalhausen 1949). Other estimates of frequency of gene mutation are even higher. Natural populations may therefore be well supplied with small mutations of differing potential values to the organism. Mutations of different sorts apparently occur haphazardly and are not influenced by environmental conditions; only accidentally do they give special advantages to an organism. Adaptation of a species to a particular habitat or niche is effected through natural selection of the favorable mutations out of the many that occur.

The size of the population and the rate at which a particular gene mutates affect the odds that a mutation of that gene will become established in a population. In a stabilized population, two offspring must survive and mature to replace the parents on their death. If in one parent gene A mutates to the non-lethal gene A', the odds are 1: 1 that genotype AA' will appear in one of their two offspring and that the mutant gene will be transmitted to the following generation. If both offspring are heterozygous, the odds for continuation of gene A' into the next generation are increased to 3: 1. On the other hand, if the parent carrying the mutant gene fails to reproduce or if all the offspring die, the

mutation is lost. The odds that a single mutation will persist through 127 generations is estimated to be only 1: 67 (Fisher 1930).

If the mutant gene A' is a dominant, the character is immediately expressed in the phenotype; if it is recessive, the character will not appear in the phenotype until two heterozygous individuals mate to give rise to the homozygous recessive $A'A'$. In a small population, inbreeding between heterozygous individuals will quickly produce both homozygous dominant and homozygous recessive genotypes, as well as the heterozygous line, and provide a variety of phenotypes upon which natural selection may work. In a large population, mating between heterozygous individuals will be less frequent because these individuals will constitute a lower proportion of the total population. However, if the gene mutates repeatedly in different individuals, the high mutation rate will greatly increase the number of individuals carrying the gene and increase the chances that the mutant character will become expressed in the population.

Hybridization

The critical test of whether or not speciation has occurred comes when a barrier between two geographically isolated populations breaks down, so that the formerly isolated populations again come into contact (Fig. 17-6). If speciation is complete, they will not interbreed; if it is not, hybridization will occur. The detection of hybridization between two popula-

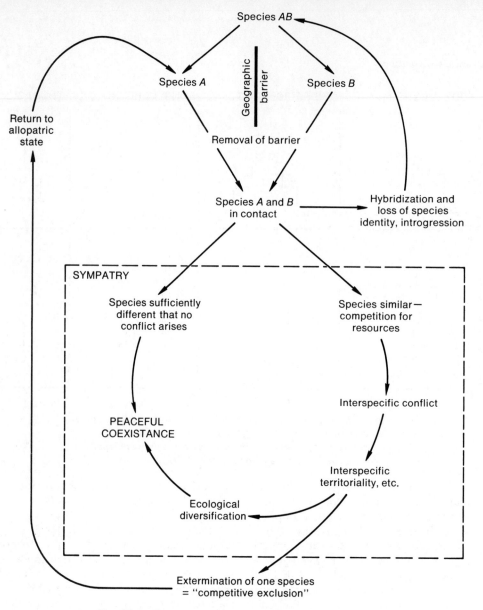

Fig. 17-6 Sequence of events that may occur when two, previously allopatric, congeneric species meet (from Rowley 1967).

tions may be difficult, for *introgression* commonly occurs; that is, the hybrids backcross with either or both parental populations, and the backcrosses resemble the parental populations very closely. The result of introgression is the gradual intrusion or transfer of the characters of each population into the other, so that all distinction between them disappears (Anderson 1949).

If two populations differing in habitat requirements interbreed, the first-generation hybrids will often show best adjustment to an intermediate environment. A second generation, if it occurs, will then consist of individuals of both ancestral and hybrid types, each of which requires its own peculiar habitat for optimum development. If there is a paucity or absence of intermediate habitats, the intermediate forms will be selected *against*, and die out. If the hybrids are sterile or have a lower reproductive capacity than the parental populations, they also will be selected against. Any ecological, ethological, or genetic divergence between the two

populations that reduces the gamete wastage in hybrids will be selected for, with consequent *reinforcement* of the isolating mechanism. It happens often, therefore, that niche segregation or differences in behavior between closely related species will become decidedly more pronounced in the overlap zone of their ranges than elsewhere. The result, of course, is continued divergence of the two populations until interbreeding between them ceases, and they form distinct species (Brown and Wilson 1956).

On the other hand, there are circumstances in which fertile hybrids are not selected against. If the area in which the hybrids are formed offers a variety of habitats, some of which are different than the habitats occupied by the parent population, the hybrids may find themselves fully as well adapted to them, and by so much be fully as well adapted to that area as are their parents. Under these circumstances the hybrids will survive, introgression will occur, and the formerly isolated parental populations will fuse into one. Introgression between populations has been observed and recorded in areas where man by removing forests or producing other disturbances in the environment has destroyed barriers that maintained a sharp ecological isolation of populations; the phenomenon doubtless occurred repeatedly in the geological past with changes in physiography, climate, and vegetation (Anderson 1948, Blair 1951, Sibley 1954, Hubbs and Strawn 1956). In extreme cases, the hybrids may obtain some advantage or show better adaptations than their parents. They will then be selected *for* and may eventually replace both parental populations. This appears to be taking place with *Colias* butterflies in northern Canada at the present time (Hovanitz 1949).

Still another alternative occasionally occurs where two species come together. The species retain their reproductive isolation but converge in color, form, or behavior that facilitates mixed flocking, more effective aggressiveness or protection against the same predators, and other advantages (Moynihan 1968).

Asexual and Self-fertilizing Forms

Asexual and self-fertilizing organisms include half or more of the Protozoa and many invertebrates and plants. These organisms offer problems of species recognition and evolution that are in many respects different from those in bisexual forms. Each individual is reproductively isolated from every other individual, giving rise to offspring that are genetically alike by fission, budding, sporulation, or self-fertilization. Mutations, including polyploidy, however, occur, and if favorable, may transform a strain or clone as the result of natural selection. Local clones may differ therefore in phenotypic and genotypic characters in the same manner as bisexual populations, even though there is no opportunity for variation to occur through assortment and recombination of the genes. Clones that are genetically distinct are in the nature of sibling species. Such clones, however, are for convenience considered as belonging to one and the same taxonomic species, if they show similar morphological characters that are not readily distinguishable (Meglitsch 1954, Boyden 1954, Sonneborn 1957).

NATURAL SELECTION

Natural selection is a continual force exerted on each successive generation. Before natural selection can take place, however, there must be phenotypic *variations* between individuals, from which selection can be made. In order for these selected variations to have significance in speciation, they must be genetically fixed and *heritable*.

In most species, many more offspring are produced than can possibly survive. Because of this *overproduction*, there is competition between the offspring for the necessities of life, which, together with the strife between predator and prey and between organisms and their physical environment, creates a *struggle for existence*.

There is differential survival in this struggle for existence because some individuals have structural, functional, or behavioral variations that give them advantages over individuals lacking those variations. The superior genotypes will make a relatively larger contribution to the gene pool of the next generation. The result of *differential survival* and *differential reproduction* is popularly known as the *survival of the fittest*.

The accumulation of favorable variations in a population brings the species generally to a better *adaptation* to the physical conditions of the habitat, avoidance of predation, more efficient physiological functioning, and new behavior patterns. *Natural selection* favors those variations that are adaptive, and thereby fosters the continued existence and improved reproduction of the species. These adaptations may never give a perfect adjustment of the organism to its environment, since the environment seldom remains constant for any length of time. If the population undergoing these changes is geographically isolated so that the favored changes do not spread throughout the species, then differentiation of characters leading to speciation occurs.

When the ratio of one character to another changes from $1.00 : 1.00$ to $1.01 : 1.00$ in each generation, the character is being selected for; more individuals with the character are surviving than are individuals without it. With a selective advantage of 1 in 100, a dominant character will become established in 99 per cent of a population in about 1200 generations (Huxley 1943). This rate is considered rapid evolution. A selection pressure of even 1 in 1000 represents fairly rapid change, but when selection is decreased to 1 in 1,000,000 or more, evolution is relatively slow. A good mathematical analysis of selection pressures is given by Li (1955).

The action of natural selection is evident in the way conspicuous odd-colored house mouse mutants are eliminated from a mixed population when subjected to cat predation (Brown 1965); in the matching of the color of local natural populations of small mammals and snails with the color of the soil or vegetation (Dice and Blossom 1937, Blair 1951a, Sheppard 1954); in the correlation of the pigmentation of butterflies with local differences in temperature, moisture, and solar radiation (Hovanitz 1941); in bringing parallel variants in many kinds of fish when in the same kind of environment (Hubbs 1940); and generally in establishing the many other adjustments and adaptations of organisms to their particular niches.

Natural selection is relatively less effective in small populations than in large ones. Small populations may be locally restricted, come into conflict with few competing species, and experience only a favorable habitat. Because of the low selection pressure, chance combination of genes (genetic drift) may produce characters of little or no adaptive value that yet have a good chance to persist while really adaptive characters may be lost. In the Hawaiian and Society Islands, there is a great variation in the characteristics of snails that variously occur on the different islands and locally in different isolated valleys or regions on the same island (Clarke and Murray 1969). This is apparently a result of the fixation of random variations in small populations not subjected to any considerable selective pressure. In populations that are increasing rapidly in size, in the upswing of cycles, say, there is little selection, and non-adaptive variations may persist as well as adaptive ones. As populations come to saturate habited niches and disperse into new or less favorable habitats, competition, predation, and parasitism increases, and individuals become exposed to ever more severe physical and climatic conditions. Natural selection then functions, and characteristics that are adaptively advantageous will tend to persist while those less favorable or even harmful are eliminated (Carson 1968).

Mutations upon which natural selection works are often recessive. They do not become fully expressed in the phenotype except when the individual is homozygous for the character. Nevertheless they are important, and in stabilized populations tend to persist indefinitely in constant proportion to the dominant alleles (Hardy-Weinberg law). A recessive character will become more prominent in a population if that particular gene continues to mutate toward the recessive, or if the homozygous recessive phenotype has adaptations that give it selective advantages. In this latter case, natural selection may ultimately result in complete suppression of the dominant allele so that the hitherto recessive allele becomes permanently fixed in the population as the only gene for the character.

Since emergent species usually entrain adaptations to new environments, it would appear that those characteristics by which we distinguish species and higher taxonomic categories generally are such as serve some useful purpose to the organisms either structurally, functionally, or in point of behavior. It is often very difficult to determine a useful function for all distinctive characteristics of a species, yet one can seldom be positive that a seemingly minor character does not serve, say, as a releaser for some critically necessary behavior or is not vitally important in other unsuspected ways. However, not all characters that distinguish species or higher taxonomic categories are necessarily adaptive (Robson and Richards 1936, Simpson 1953). Some characters originally adaptive may have lost their usefulness, although they persist in the organism. With natural selection no longer acting on them, they usually retrogress and may eventually disappear, as have skin pigmentation and eyes in many cave animals, for instance. Other characters may even have a slightly unfavorable value, but be closely linked genetically with selectively preferred favorable characters and thus continue in the organism.

Before a population can occupy a new region or even expand its niche it must show at least some *preadaptation* for it (Allee *et al.* 1949, Simpson 1953). Preadaptation may take the form of a wide range of tolerance that can encompass the conditions of the new habitat as well as the old; or it may take the form of a new use for a structure, different from its original function (Bock 1959). The European rabbit certainly showed a good deal of preadaptation to the Australian environment; it became a local pest within 3 years of the introduction of 24 individuals in 1859. We have earlier described why preadaptation must have been necessary for the origin of parasitism. Preadaptation permits individuals to exist in new habitats or to perform new functions, but subsequent perfection of an adaptive trait depends on the accumulation and selection of additional favorable genetic variations over many subsequent generations.

Adaptive Radiation

When a species bypasses or surmounts a dispersal barrier, it may penetrate an area having a variety of niches novel to the species. A plastic species may quickly differentiate adaptively into a number of new species, each becoming established in an unoccupied niche or, if sufficiently aggressive, displacing an original but less adaptive occupant. Such *adaptive radiation* is known to have occurred in the case of ancient marsupials, which crossed from Asia to Australia and differentiated into the variety of species now found there.

The invasion and occupancy of the Hawaiian Islands by snails, insects, and birds is of special interest. There are some 3722 insect species endemic to the islands. All these species appear to be derived from some 250 ancestral forms that arrived in 14 separate invasions since Pliocene time (Zimmerman 1948).

The ancestral prototype of the honeycreeper birds reached the Hawaiian Islands sometime within the last 5 million years (Baldwin 1953). Different populations became isolated on different islands, as a result of which there arose the red and black nectar-eating species that are grouped in the subfamily Drepaniinae. As the nectar-feeding niches became fully occupied, a population diverged in its behavior, feeding more heavily on insects than on nectar. The new niche allowed redispersion of the population through the various islands, and there ensued a second burst of speciation yielding the green insect-eating forms belonging to the subfamily Psittirostrinae. Additional speciation produced short-

and long-billed species of insect-eaters. Somewhere in the lineage of the latter group, the birds acquired seed- and fruit-eating habits, and the long bill also became a thick bill. Rapid evolution in this family still appears to be in process.

Rate of Evolution

There is evidence that, under natural conditions, small-scale variations in local populations may sometimes be manifested within a surprisingly few generations (Huxley 1943). Melanistic forms of butterflies and moths now occur in industrial areas in both Europe and North America, where vegetation has become coated with dark-colored debris, although 100 years ago such melanistic forms and the industrial soot as well were virtually absent (Owen 1961). Conspicuous adaptive differentiation in color and size has occurred in the house sparrow since its introduction into North America and the Hawaiian Islands during the latter half of the nineteenth century (Johnston and Sealander 1964). Full speciation in some small passerine species in the tropics is believed to have occurred in 20,000–30,000 years, but for larger species, a much longer time is required (Haffer 1969). Two species of harvest mice in the San Francisco Bay area appear to have differentiated in the last 25,000 years (Fisler 1965).

The geological record indicates that the rate of change in skeletal characteristics of several lineages of mammals since Pleistocene time is of the order of 0.2 per cent per 1000 years (Kurten 1958). Subspecies more often require 10,000 years to become well defined and may continue to evolve for 500,000 years before rising to the species level. The evolution of a fully defined species most commonly requires at least 50,000 years and frequently a very much longer time (Fig. 17-7). Some living mammal species are 1,000,000 years old, and some lower vertebrate and invertebrate species have persisted relatively unchanged for 30 million years (Simpson 1949, 1953).

The rate of change, divergence, or evolution of populations into new species depends partly on the rate at which new mutations are occurring in the species' gene pool, and partly on the rate and extent to which the environment is changing. In a long-continued uniform environment, a species becomes stabilized in a favorable relationship with the habitat and community. The various ecological niches are effectively occupied, and little evolution occurs. New mutations can add little to perfected adaptation. If the habitat changes, however, established adaptations may no longer be appropriate, and variations hitherto rejected may now prove beneficial. A mutation selected for in one or more species may initiate a chain of events that alters the internal balance of the whole community, with resultant rapid evolution (Olson 1952). For instance, there

has been considerable differentiation of animals into subspecies during and since the Pleistocene glaciation, but probably most of our present-day species originated in pre-Pleistocene time.

CULTURAL EVOLUTION

Webster's dictionary defines culture as the complex of distinctive attainments, beliefs and traditions that constitute the background of racial, religious, or social groups. The *culture* of a species includes its social or ethical as well as its intellectual and aesthetic attainments, especially as they are symbolized or expressed in many different ways (White 1959). By *ethics*, reference is to definite behavior patterns concerned with interrelations between individuals that have become established for the welfare of a species. *Morals* refer to the ethical relations and conduct of individuals. Animals have ethics and culture insofar as they have evolved characteristic behavior and social interrelations, but culture is more complex in the higher vertebrates, particularly in man, correlated with the development of a higher degree of intelligence.

Different levels of social organization may be recognized: (1) those where individuals simply tolerate each other's presence when crowded into a restricted space, (2) those that show some integration so that the group responds more or less as a unit, and (3) those that have established division of labor and definite social organization (Allee 1931). Division of labor started evolving very early, first with differentiation of somatic and reproductive cells in colonial protozoans, then differentiation of tissues in coelenterates, differentiation of

Fig. 17-7 Age and time of origin of present-day species of reef corals and mollusks as shown by their percentage of occurrence in fossil faunas of the East Indies (after Umbgrove 1946).

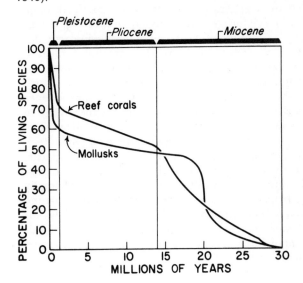

organs in vermes, separation of the sexes, specialization of castes in social insects, and finally segregation of species into niches in the community. Among the social insects, for instance, ants and termites have for millions of years been building elaborate dwelling places, domesticating other species, raising vegetables, storing food, practicing sanitation, and living under elaborate social hierarchies—analogous to what man was to evolve independently and in a different manner later on. With division of labor, each individual or species contributes to the whole, and the welfare of the social unit or community depends on this cooperation. These cooperating units develop high efficiency in using natural resources, with the result that the constituent individuals survive and reproduce more successfully and the species is more likely to continue to exist.

In lower animals, social systems are poorly developed if one can say they occur at all. Species multiply to the limit of their food or other resources except as held in check by predators, weather, disease, or decreased fecundity. With development of competition between individuals in higher species there becomes established degrees of dominance and subordination expressed in social hierarchies, territorial systems, and niche segregation. These form a basis for more effective community cooperation, so that food resources become divided and utilized on a continuous self-sustaining basis.

Dominance is also expressed in the social and political hierarchies of man and probably evolved for the same reason. Leadership is essential to social stability, more efficient exploitation of natural resources, and advances of one social system in competition with others (Tiger 1970). Likewise the imperative of man for possession and defense of property and a place in the social organization of the species is similar in purpose to territorialism among lower animals (Ardrey 1967, Tinbergen 1968).

Behavior patterns, characteristic of species, are passed from generation to generation partly through genetic mechanisms and partly by imprinting, conditioning, or training of the young to behave in particular ways. For instance, in birds, the capacity for fear is inherited, but the young need to learn in most cases what to fear; feeding is instinctive, but the young must learn through imitation of their parents and experience what is the proper food for the species; the ability to sing is inherited, but the young must hear the song of their species to perform properly; and so on (Nice 1943, 1963). In the lower invertebrates behavior appears relatively stereotyped and automatic, but in higher vertebrates, such as fish, birds, mammals, and finally man, behavior becomes more complex with learning relatively more important. This complexity allows for more variability between the behavior of individuals and innovations to develop faster in the mores of the species.

There can be no doubt that evolution of social systems has taken place (Allee 1940). Probably this evolution has not been by group selection of whole systems as such, as has been suggested (Wynne-Edwards 1962), but more likely by selection of those individuals that show capacity for particular kinds of interrelations and cooperative use of resources which result in their greater survival and reproductive success (Crook 1965, Lack 1966, Wiens 1966). This is *cultural evolution*, often contrasted to biological evolution concerned chiefly with structure and function rather than behavior. Behavioral patterns, social systems, and community interrelations can be traced phylogenetically through animal groups in the same manner as can structural characteristics. By the same process, human ethics, culture, and religious beliefs can be traced from earlier beginnings, going all the way back to lower animals. In certain respects, evolution of social systems or mores among animals is comparable to evolution of vegetative systems or physiognomies among plants (Whittaker and Woodwell 1972).

We have shown how competition is of prime importance in regulation of population size, segregation of species into different niches, and speciation. The competitive drive for advantages is also evident in modern civilization in the development of new techniques in industry, procedures in economics, and policies in government. Nations and ideologies compete in the same manner and for the same purpose. There is selection of those techniques, procedures, policies, and ideologies that are most fit and rewarding to the individual. It is difficult to imagine how competition can ever be eliminated without the advancement of civilization coming to a standstill. Man, unlike animals, may through his intelligence deliberately change the course and rate of cultural evolution and through ill-advised action reduce fitness as often as he increases it. Anthropological and historical record shows that civilizations have risen and fallen depending on the vigor and wisdom of their activities. Over the course of time the healthier, more ingenious, and more intelligent civilizations have persisted longer, with successive steps leading to the complex highly developed culture of the present day.

However, cooperation is also rewarding to the individual and the species (Allee 1938), as is shown in the division of labor that has evolved in the functioning of various organs of the body for the betterment of the individual, in the caste system of social insects, in the evolution of parental care in many kinds of animals, in acceptance by individuals of their position in social hierarchies, by segregation of species into different niches for the better functioning of the ecosystems, and in many other ways. Similarly, man becomes specialized by his skills and knowledge, each to perform a particular function better than others, so that with exchange of products or services exploitation of resources takes place more fully and efficiently for the benefit of all.

Cooperation is profitable not only economically but also in man's more personal interrelations. In order to live together in limited areas, both lower animals and the human animal must have rules of conduct, recognize the rights of others, and balance competitive ambitions against cooperative objectives. Such codes of ethics are products of evolution that have met the test of time. Although religion, as practiced in various cultures, involves a superstructure of theology, its basic function is the promotion of codes of ethics to control competitive frictions within a framework of cooperation so that individuals may live together with some degree of peace and harmony. These codes of ethics, as they apply both to man and lower animals, are continuing to evolve to meet new situations and to provide more adaptive interrelations between individuals and between individuals and their changing environment. Important to the evolution of a sophisticated code of ethics and a high-level intellectual and aesthetic attainment is improved means of communication.

Language probably developed as man began to live together in social groups and evolved through natural selection as it improved efficiency in hunting, defense against enemies, and coordination of other activities. Written language came much later than speech. Cro-Magnon man probably had speech 30,000 years ago, and it is quite possible that even very early man had some crude language. Not much is known about how speech originated. Many kinds of animals have means of intercommunication, and vocalization is common in birds, mammals, and primates. Communication by means of cries, grunts, call-notes, and song mostly indicate *signs* of emotional states and differs from true language, which uses *symbols*. A young ape, for instance, could utter a cry of hunger which would be responded to, but he could not say "banana" for what he wanted to eat. The evolution of speech must have been coordinated with the development of conceptual thinking and of abstraction. Cro-Magnon man demonstrated symbolic and conceptual thinking in his cave paintings (Critchley 1960).

Aside from family units, which actually may have come later, the most primitive social organization of man was the *band*. Commonly bands consisted of 30 to 100 interrelated people, involved in a hunting and gathering economy at a subsistence level, and with only a beginning of social integration. This stage of social development persisted a very long time and all later stages have come only during the last few thousands of years. With the beginning of agricultural practice, bands fused into *tribes* varying from 500 to 2000 individuals and then into *chiefdoms* up to 10,000 persons. The chiefdom consisted of a social hierarchy centering around a chief who was believed to be a direct descendent of the original ancestor of the society.

Subordinates had social status according to their relationship to the chief. *States* evolved as populations expanded and productivity of food increased to provide surpluses above the needs of those participating. With these surpluses came the differentiation of urban and rural economics. The cities were composed primarily of nonfood producers who specialized in crafts, economics, government, and religion. They were dependent on the surplus food production of the surrounding hinterland, which in turn obtained goods and services from the cities.

Religious beliefs and practices had a primary function in teaching ethical standards and inducing individuals to adhere to them. Government established rules and a body of police to enforce these rules. Ample resources, a large labor force, and a prosperous economy stimulated construction of elaborate buildings, monuments, and temples, often inspired by religious fanaticism. Close contact between diversified groups encouraged communication through language, art, and music. Civilization thus became established (Sanders and Price 1968).

The point made here is that the modern culture of man, "civilization," is an evolutionary outgrowth of social behavior in lower animals, is continually evolving as the result of interplay of competition and cooperation with natural selection of what best serves the welfare of the individual, is a natural consequence of population expansion, and is unique only in that the high intelligence of man, also a product of evolution, allows expression of aesthetic, economic, communicative, and inventive capacities that are latent or poorly developed in lower forms (Tiger and Fox 1971).

SUMMARY

Speciation is the process of evolutionary differentiation often leading to species formation, a process usually also involving separation or divergence of populations into different ecological niches. Sympatric species do not interbreed because of ecological, ethological, mechanical, or genetic isolating mechanisms.

Geographic isolation of two populations of the same parental species appears prerequisite to complete speciation. As long as a significant amount of gene flow occurs between populations, they diverge no further than subspecies.

Populations may show different characteristics as the result of non-heritable variations, heritable polymorphism, genetic drift, mutations, and hybridization.

With natural selection, those individuals possessing adaptive variations obtain a greater chance for survival and reproduction, and, if the variations are heritable, contribute more to the gene pool of the population. This leads to a change in the characteristics of the population and possible speciation. Exposure to new habitats or niches may bring adaptive radiation into a variety of new species.

Evolutionary ecology is concerned not only with speciation and its accompanying niche segregation but also with the manner in which all ecological structure and function has evolved. Part of this evolution is *biological* in that species characteristics are transmitted genetically by means of chromosomes and genes, and part of it is *cultural* in that behavior and social systems or ethics are transmitted through the training and learning experiences of the young. Human culture, including ethics and religion, has its evolutionary roots in the behavior and social systems of lower animals.

Part Six

GEOGRAPHICAL
ECOLOGY

Chapter 18

DISPERSAL DYNAMICS
AND
DISTRIBUTIONAL PATTERNS

Through the study of *geographical ecology* we attempt to understand how organisms are distributed over the world and what forces brought about this distribution. The explanation of why a particular species has come to occupy a particular niche in a particular part of the world requires a knowledge of where it originated, how it dispersed, and how it evolved its present adjustments and characteristics. When populations invade new regions and environments and become geographically isolated, they commonly differentiate into new species. Each species thus becomes the product or visible expression of a particular combination of environmental factors, interactions, and locality, and as such the species is the practical taxonomic unit with which the ecologist must deal.

David Starr Jordan (1928) once stated that the general laws governing the distribution of animals can be reduced to three simple propositions: a species of animal will be found in any part of the earth having conditions suitable for its maintenance unless (1) its individuals have been unable to reach this region because of barriers; or (2), having reached it, the species has been unable to maintain itself because of inability to adapt to the region or to compete with other forms; or (3), having arrived and survived, it has subsequently so evolved in the process of adaptation

as to have become a species distinct from the original type.

In this chapter we will be concerned both with dispersal dynamics and with the distributional patterns of floral and faunal units and ecological communities that result.

DISPERSAL DYNAMICS

Dispersal is the spread of individuals away from their homesites. Dispersal movements are usually slow, and cover relatively short distances in the life time of an individual. The cumulative result of short dispersals by successive generations, however, may become conspicuous in the course of years, decades, or centuries, when it constitutes *range expansion* of the species into a new habitat or area. Some remarkable instances of range expansion have resulted from the purposeful introduction of the house sparrow into North America in 1852–53 and the European starling in 1890–91 (Fig. 18-1); and the accidental introductions of the black rat, Norway rat, and numerous insect pests, at various times. In similar fashion the gray squirrel and muskrat have been introduced into Europe (Elton 1958). Once man had helped them to overcome the ocean barrier

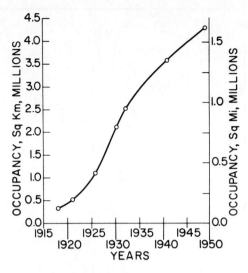

Fig. 18-1 Dispersal of the European starling in North America between 1918 and 1949 (compiled from Kessel 1953).

that previously held them in check, these species spread unusually rapidly because of the optimum environment on the continent. There is reason to believe, however, that under entirely natural conditions the rates of dispersal of all species, once a barrier is passed, would be similar, varying only with respective reproductive potentials, characteristic speeds of locomotion, and relative abilities to find unoccupied niches, overcome competition, acclimatize to new climatic conditions, and acquire new behavior patterns.

Manner and Means of Dispersal

Animals find suitable habitats and niches in various ways (Udvardy 1969). In a uniform environment, dispersal movements radiate in all directions from the home area (Fig. 18-2). The greater the density of individuals in the home area, the more quickly distant areas are invaded, and the farther away do individuals move (Fig. 18-3).

A common method of achieving dispersal is the *broadcasting* of enormous numbers of eggs, spores, encysted stages, or young so that they scatter into a wide variety of places in a more or less random manner. Those that come by chance into suitable environments persist and become established; those that enter into unfavorable locations are destroyed or never develop. Broadcasting is a wasteful procedure; it is especially characteristic of aquatic species. The fresh-water clam annually produces hundreds, perhaps thousands, of eggs. Only two fertilized eggs need mature that the two parents be numerically replaced when they die, and thus the population of the species be maintained at a constant level. Contrastingly, in those forms, such as birds and mammals, that have developed a high degree of parental care, the number of eggs or young annually produced is commonly a half-dozen or less, and the offspring exercise considerable discrimination in their choice of suitable habitat.

Although the dispersal of broadcast offspring is not under the control of the parents or the young, it is not

often truly random. Water and wind currents and other agents of dispersal may channel eggs and spores in restricted directions. Such dispersal is described as *passive conveyance.* In streams, all agents of passive conveyance, except some other animals, direct dispersal downstream. Upstream dispersal must be the result of active locomotion.

Eggs of insects, snails, fish, or other aquatic organisms will sink unless they are buoyed up by currents, possess flotation mechanisms, or are attached to some floating object. Logs, masses of vegetation, and other debris are sometimes torn loose from the banks of rivers and float out to sea carrying the smaller animals attached to or trapped on them. It is estimated that

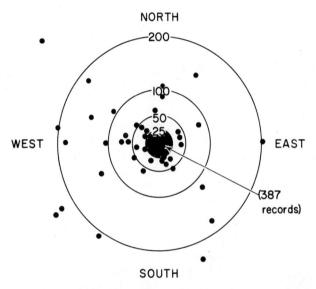

Fig. 18-2 Relation of breeding localities to birthplaces among robins. Numerals on concentric circles are distances from origin in miles (1 mile = 1.6 km). Data points beyond the 200-mile radius are not placed to scale (Farner 1945).

Fig. 18-3 Effect of increase in population at source (200, 500, 1000 eggs deposited) on the dispersal of European corn borer larvae (Wolfenbarger 1946, after Neiswander and Savage).

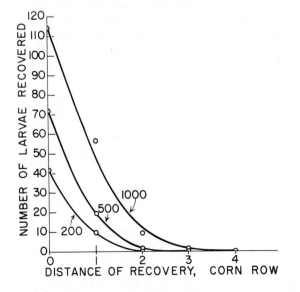

over 300 debris rafts of significant size are formed each century and float out to sea from the mouths of the larger tropical rivers (Matthew 1915). Of 59 pieces of flotsam picked up in the Caribbean Sea, 0.5–16 km from shore, 25 per cent contained at least one live terrestrial animal and 6 per cent had three or more species (Heatwole and Levine 1972). The passengers on such rafts and bits of flotsam may come to be colonists of remote islands or even other continents.

The dispersal of fresh-water animals from one river to another is sometimes effected by erosion, when the process permits one river system to rob a branch of another river system (Crosby 1937). Following the recession of the continental glacier in North America, the proglacial Great Lakes had outlet down the Illinois and Mississippi Rivers. Later, a new outlet was established over Niagara Falls, and the Lakes became a part of the St. Lawrence River system.

Wind or air currents act in a manner somewhat similar to water currents. Not only disseminules but also adult forms may be transported. Some terrestrial spiders have evolved a special mechanism allowing them to use mild air currents for dispersal. The young climb up on a clod of earth or other object and spin out long threads or flocculent masses from glands in their abdomen. This continues until enough buoyancy is created to lift the spider and carry it away, sometimes distances of hundreds of kilometers. Strong winds will often carry insects and even birds great distances away from their usual courses. Crop pests may be blown north during the summer and cause damage, but may never become permanently established because they are killed by the northern winter cold. Hurricanes are an important means of colonizing islands far at sea with mainland species (Elton 1925, Darlington 1938). There are authentic records of rains of fishes and other aquatic species that were sucked up and transported appreciable distances by tornadoes (Gislen 1948).

In studies done in England (Freeman 1946), it was estimated that the number of insects drifting through a rectangle 91 m (300 ft) high and 1610 m (1 mi) long amounted to 12,500,000 per hour. The number was highest during May, June, and September, at temperatures above 18°C. The aerial population over the forests and swamps of Louisiana has been measured (Glick 1939), by means of traps placed on the wings of airplanes, and found to average the following number of individuals per 1000 m³ of air:

Altitude	Daytime	Night
6 m (20 ft)	10.3	—
61 m (200 ft)	5.2	—
152 m (500 ft)	—	6.2
305 m (1000 ft)	2.2	2.6
610 m (2000 ft)	1.1	1.1
914 m (3000 ft)	0.6	0.5
1524 m (5000 ft)	0.3	0.4

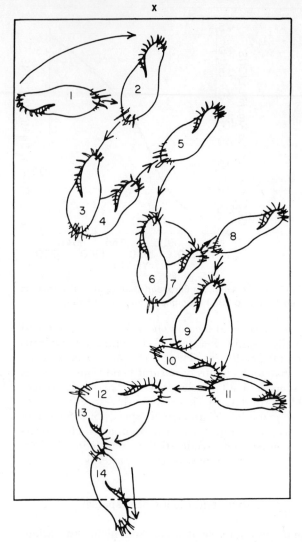

Fig. 18-4 Trial and error response of the hypotrichian *Oxytricha fallax* to heat. The slide is heated at X. An *Oxytricha* at station 1 changes position as indicated by the arrows, repeatedly moving backward, turning to the right, then moving forward. Finally, at stations 13–14, it directs away from the heat, and moves in a straight path toward a cooler region (from Jennings 1906).

Diptera were most numerous, followed during the daytime by Coleoptera, Homoptera, Hymenoptera, Araneida, Hemiptera, and others. Spiders and wingless insects were greatly reduced in numbers at night because of the lack of vertical convection currents. Generally those species tend to be most numerous that have the greatest wing area per unit weight, hence the greatest buoyancy. Species of insects in the upper air are not distinct from those flying in the lower layers, but their densities decrease (Johnson 1957, Taylor 1960). Improved methods for obtaining aerial densities employ a suction pump to strain known volumes of air per unit of time (Johnson 1951).

Other animals serve as vehicles for passive transportation. Bits of vegetation, small animal life, and the eggs or adults of worms, entomostracans, rotifers, insects, snails, and fish may cling to the feet or feathers of water-birds such as ducks, rails, and herons and be carried many kilometers when the birds migrate or they may become attached to mammals, resident birds, or insects for more local dispersal (Maguire 1963). A species of *Succinea* snail, native only to St. Croix and Puerto Rico, was found alive in the feathers of a bobo-link shot in Cuba. *Ferrissia* snails have been found attached to the wing covers of aquatic beetles and sphaeriid clams clamped on their legs. These insects occasionally fly from one body of water to another. Crustacean eggs have also been found viable after passing through the digestive tract of different kinds of birds (Proctor and Malone 1965). Plant seeds may be dispersed thousands of kilometers in the digestive tracts of birds (Proctor 1968). Fruits become conspicuous, abundant, and nutritious, thus attracting birds at the proper time to aid in their seed dispersal (Snow 1971).

Some dispersal movements are determined by the manner in which the animals respond to environmental factors. Such *directed movements*, or taxes (p. 14), may induce the dispersal of animals upstream positive to current rather than downstream.

Trial and error is involved to a considerable extent in the dispersal movements of animals (Fig. 18-4). If, initially, an individual progresses into an unfavorable habitat, it may be able to withdraw and proceed in a new direction. This may continue many times until by chance it discovers an area that is favorable. Trial and error movements are manifested in all groups of animals. In higher types, and even in many lower types, the individual may learn by experience and reduce the number of false trials that it makes. Thus initially random movements may eventually evolve into directed behavior patterns.

Barriers

Dispersal continues until a barrier that makes further movement difficult or prevents successful colonization is encountered (Udvardy 1969). Barriers are of different sorts and are classified as *physiographic, climatic,* and *biotic.* The difficulty of surmounting or bypassing any type of barrier varies greatly among species. River valleys, for instance, may be barriers to animals frequenting mountains: but to lowland species, river valleys are important dispersal highways.

To fresh-water organisms in river and lake habitats, intervening land masses are usually effective barriers. The headwaters of different river systems may lie only a few kilometers apart, yet contain quite different species because the continuous water route down to the mouth of one river and back up to the other may be a distance of several hundreds or thousands of kilometers. A waterfall may be a barrier for all non-flying aquatic species, and even riffles or swift water may prevent upstream dispersal of pond or lake species. The salt water intervening adjacent rivers flowing into the sea is a barrier to most fresh-water forms that might otherwise invade one river from the other.

Terrestrial organisms are hemmed in by a great variety of barriers. The oceans constitute the major physiographic barriers since they separate the faunas of the several continents and isolate islands from each other. Lakes are not effective barriers because they can be readily skirted, but wide rivers coursing long distances between banks of dense vegetation, as does the Amazon River through the tropical rain forests of South America, may limit the range of forest mammals, butterflies, flightless beetles, land snails, and even birds (Mayr 1942: pp. 228–229). The Grand Canyon of the Colorado River separates the ranges of the Kaibab and Abert squirrels and several other species, even in country that is semi-arid and open (Goldman 1937). Mountains are sometimes considered barriers to lowland species, and valleys barriers to mountain forms. This is only true if the change in climate and vegetation that such barriers produce are unfavorable to the species.

Deserts are important climatic barriers since they are hot and dry. Temperature affects animals directly, since species have definite limits of tolerance, comfort, and efficiency. Precipitation is important because it controls the type of vegetation that occurs in a region, and the animals adapted to that type of vegetation. A low relative humidity may directly limit the dispersal of moist-skinned species. Excessive solar radiation limits some species to forest habitats, excluding them from open country; without recourse to shade, such animals would experience overheating and critical loss of water from the body. Short photoperiods may limit distribution northward during the winter, especially if low temperature obtains, and may influence southward distribution of northern animals in the summer. The length of the season between spring and autumn killing frosts, or between dates when its limits of temperature tolerance are reached in the spring and autumn, may determine whether or not a species can complete its life cycle at a given latitude. A corollary consideration is whether the accumulation of heat is sufficient to furnish cold-blooded animals and plants sufficient energy for growth and reproduction.

Biotic barriers consist of changes in vegetation, food, competitors, and predators. The adaptations and behavior patterns of many animals fit them to niches in specific types of vegetation; should the previously amenable vegetation change, the animals may have great difficulty in adapting to it. Tree squirrels, for instance, are replaced by ground squirrels in prairies and deserts. Most animals have a great adaptibility to food, so food is not so limiting a factor. Some insects, however, such as certain aphids, are narrowly limited to particular species of plants as a source of food. Where their food plant is not present, they cannot exist.

Competition between species is also a potent force in controlling distribution. The boundary between the ranges of the house wren and Bewick's wren in eastern North America is not sharply defined, varying as a function of competition between the two species. Either species can live in the range of the other, but in the North, the house wren usually wins in competition for territory and nest-sites. In the South, the Bewick's wren is more successful (Kendeigh 1934). Predators, such as the great horned owl and the swifter hawks, tend to confine the bobwhite to a forest-edge habitat, where it is less vulnerable to attack. Trypanosome parasites carried by the tsetse fly are effective barriers against successful introduction of domestic ungulates into certain parts of Africa, and the rabbits introduced into Australia have limited the range and greatly reduced the abundance of several species of native marsupial.

Dispersal of Young

The dispersal of a species is primarily accomplished in the immature stages. This is obviously true of eggs and spores, but banding and marking studies have shown that among the higher animals—birds and mammals— it is also the young which disperse the species. Once a bird has reached sexual maturity and nested, it has strong tendencies to return to the same area in following years. The distribution of young birds is not random, however, as they tend to return to the general vicinity of their birthplaces rather than uniformly over the range of the species. Thus only 0.5 per cent of 557 adult house wrens recovered a year after banding (Kendeigh 1941b) nested farther than 3.3 km (2 miles) from the site where they had nested the year of banding, but 15 per cent of the 182 birds banded as nestlings were recovered at greater distances, the longest of which were 32 km (20 miles), 56 km (35 miles), 80 km (50 miles) and 1120 km (700 miles). Dispersal distances for young of other species are proportionally comparable (Haartman 1949, Berndt and Sternberg 1968). Apparently the young are at a disadvantage in the competition for nesting territories with adults that have previously nested in the area, hence are the ones forced to go elsewhere.

Of small mammals, it is characteristic that once an individual has selected a homesite, it rarely leaves it for another (Burt 1940, Blair 1953). It was observed that in the months following the time at which they had been captured, marked, and released, 95 per cent of 133 adult woodland white-footed mice resumed habitation within 183 m (200 yd) of the site of capture.

Rate of Dispersal

If dispersal from birthplace were typically limited to one direction, then a simple mean of the distances to which the young disperse before they breed would give the dispersal rate per generation. It is the case, however, that dispersal proceeds peripherally in all available

directions and may extend to surprising distances (Bateman 1950).

The area exposed to invasion and the average time required to saturate that area increase proportionally as the square of the linear distance (d^2) from the center, since the total area within which the individual could settle is πd^2. A number of equations have been suggested for computing mean dispersal distance (Haldane 1948, Haartman 1949, Dice and Howard 1951, Ito and Miyashito 1965). Consider the data on house wren nestling recovery distances, presented above. Excluding the truly extraordinary distance of 1120 km, and observing that only about 93 per cent of young female wrens nest when they are 1 year of age, we compute an annual dispersal rate of approximately 8 km (5 miles) for this species. The mean dispersal distance of one group of 154 young woodland white-footed mice (Burt 1940) is about 176 m (192 yd). However, mice born in the spring mature sexually very quickly and breed in late summer or autumn, so the annual dispersal rate must be somewhat greater than this figure indicates. The annual dispersal of marked snails resulting from random movements along a linear bank of uniform habitat averaged only 5.5 m (6 yd) (Goodhart 1962).

Our dispersal rate data so far have described the outward diffusion of a local population through an area already occupied by the species. Once it has surmounted a barrier the dispersal rate of a species into a suitable area previously unoccupied by it should be faster. The European starling was introduced into North America about 1890. From a central locus around New York City it spread at an accelerating pace until, in 1940, it had become established over 6,500,000 km^2 (2,500,000 mi^2) (Wing 1943), a mean rate of about 130,000 km^2 (50,000 mi^2) per year. With amelioration of the climate in Finland during recent years, the lapwing spread northward between 1899 and 1954 at a mean annual rate of 7 km (4.3 miles) (as computed from Fig. 9 in Kalela 1955); the roe deer, between 1850 and 1945, at a mean annual rate of about 9.5 km (5.8 miles) (computed from Fig. 1 in Kalela 1948). The Norway rat invaded southwestern Georgia and virtually displaced the previously established black rat at a rate of about 430 km^2 (167 mi^2) per year (Ecke 1954). By way of contrast, it has taken the fresh-water amphipod *Gammarus pulex* the last 6000 years to disperse across 12 river systems from southern England into Scotland (Hynes 1954).

Causes of Dispersal

The reproductive rate of any species is so great that if all offspring survived the world would be overrun with that species within relatively few generations. Because species produce a surplus of young in most years, there is continuous pressure on individuals to move into all suitable niches, and to seek out new areas in which to settle. The impact of large numbers of individuals struggling for survival is described as *population pressure*, and is doubtless the most potent force inducing dispersal. It should be recognized, however, that population pressure is not uniformly constant year after year. When because of poor breeding conditions or catastrophe there is a reduction in the over-all population of a species, that decimated species may withdraw into its optimum habitat and be less put upon to exploit new or less desirable areas. In years favorable to the production of large surpluses of young, a species will often be found in less favorable habitats, even regions it would not otherwise occupy at all (Kluyver and Tinbergen 1953). The broadcasting of eggs or offspring, or the passive conveyance of them to other regions, varies directly with the size of the population producing them, and is hence as much an expression of population pressure as the active search for new areas engaged in by individuals under their own locomotion. In addition to relieving population pressure, dispersal is advantageous to the species for faster spread of mutations through the entire population and for occupation of new or depopulated areas. The tendency to disperse may even have been selected for in evolution so that it has become innate in young animals (Howard 1960).

Animals cannot disperse successfully, if at all, into new areas to the characteristics of which they are not structurally, functionally, and behaviorally adapted. If an area, which has excluded a species, changes so that the species is then adaptive to it, that species can successfully invade. The American robin, song sparrow, chestnut-sided warbler, house wren, and prairie horned lark have invaded Georgia only in recent years as the logging of forests, initiation of early seral, grassy, and shrubby stages, and extensive general cultivation of the land have produced habitats meeting the requirements of these birds (Odum and Burleigh 1946). Likewise, if an area remains unchanged but the species acquires new structural, functional, or behavioral traits by which it can adapt to that area, the species can invade the area. If the food supply fails, homesites or vegetation are destroyed, or a pernicious change in climate occurs there, animals may be forced to leave an area to which they were well adapted and disperse, more or less temporarily, into an area to which they are less well adapted. The snowy owl, for instance, depends heavily on lemmings and mice for food in its usual range, the Arctic tundra. In correlation with the cyclic decline of the lemming population there, the owls invade our northern states.

DISPERSAL PATHWAYS

An understanding of how animal and plant groups have dispersed over the face of the earth involves the controversy as to whether all the continental land masses were at one time intimately connected and, if so, how long ago they drifted apart to their present locations. Separated continents need at least temporary land bridges if there is to be dispersal and interchange of fauna between them.

Continental Drift Theory

This theory postulates that throughout the Paleozoic and much of the Mesozoic the presently distributed continents were grouped into two great land masses: a northern Laurasia was separated from a southern Gondwana by the vast sea of Tethys, although narrow connections between the two might have occurred for short periods. In the Jurassic or Lower Cretaceous the land masses fragmented, and the fragments subsequently drifted apart (Fig. 18-5). Laurasia is supposed to have split into North America, Greenland, Europe, and most of Asia; Gondwana, into South America, Africa, Arabia, Madagascar, India, Australia, and Antarctica (Wegener 1924, Du Toit 1937). This theory is suggested by the shapes of the continents that could conceivably be fitted together; by the characteristics of the Atlantic Ocean basin that make it appear to have been formed by a rifting apart of land masses; by new paleomagnetic studies; by the similar geological stratigraphy shown by characteristic assemblages of rocks and similarity in their ages, and by invertebrate fossils in Africa and South America; by some similarities of present-day fauna and flora at the same latitudes on different continents; and by the difficulty tropical species have of dispersing over arctic land bridges.

Evidence in favor of the continental drift theory has been increasing in recent years—in fact, drift is apparently still going on. Deep sea drilling shows that the sea floor has been spreading outward from the mid-Atlantic ridge at a uniform rate of 2 cm/yr (Maxwell *et al.* 1969). However, the continents have certainly been separated throughout most of Tertiary time and perhaps also the latter part of the Mesozoic (Table 19-1), so that dispersal of modern organisms over the world must have taken place in other ways (Mayr 1952, Runcorn 1959, Maxwell 1968, Hurley 1968, Allard and Hurst 1969).

Land Bridges

An alternative theory is that the continents have been situated in their present positions throughout geological time. They have changed their boundaries with rise and fall of ocean levels and with upheaval and subsidence of land masses, but there has been no significant lateral drifting (Fig. 18-6). This theory was generally held to be correct by most geologists and biogeographers up until the last few years. Regardless of their ancient history, the continents have been largely separated during the last 60 to 100 million years, and

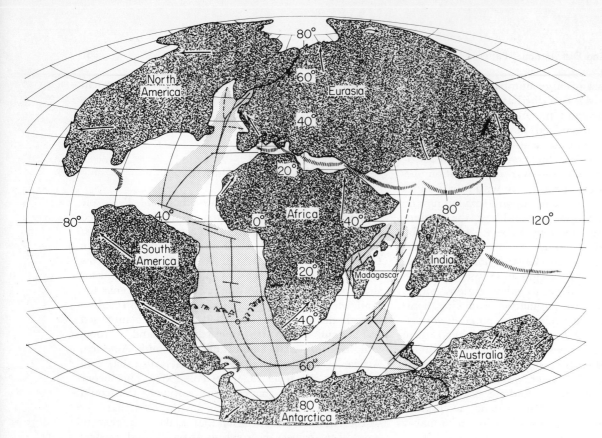

Fig. 18-5 Continental drift at the beginning of the Tertiary, 65,000,000 years ago. The South Atlantic has widened into a major ocean, a new rift has carved Madagascar away from Africa, the rift in the North Atlantic has switched from the west to the east side of Greenland, the Mediterranean Sea is clearly recognizable, and Australia still remains attached to Antarctica. The shaded areas on the ocean floor are new areas created by sea-floor spreading; the black lines represent zones of slippage; the black arrows indicate vector movements of continents since the drift began, and the hatched lines in black are zones of crustal uptake (Dietz and Holden 1970; from *Scientific American*).

Fig. 18-6 Paleogeography of the world during the Upper Cretaceous, assuming permanency of continents rather than continental drift, showing land bridges and epeiric seas (Ross 1951).

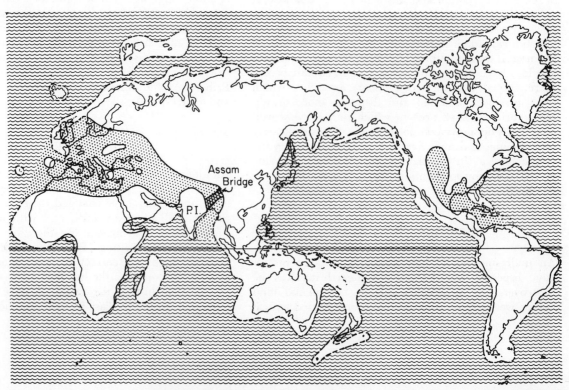

movements of animal and plant groups during this latter period have been most important in determining their present distribution.

The separated continents, however, have been connected by at least temporary land bridges at various times during the period. A land bridge now connects North and South America at the isthmus of Panama, and Eurasia and Africa are connected by the isthmus of Suez. One of the best known and generally accepted land bridges of the past, but not now in existence, connected Asia and North America across the Bering Strait and Bering Sea (Hopkins 1967). Possibly bridges or extensive archipelagos occurred in the Atlantic Ocean area in the past that permitted dispersal of organisms not possible at the present time, but concepts of this sort are highly hypothetical at the present time (Fig. 18-7).

Land bridges may exist for only a short time in the geological sense, but they serve as important dispersal routes for those land animals and plants able to cross them. When a land bridge allows free passage of most animals and plants in either or both directions, it is called a *corridor* (Simpson 1940). If the land bridge is narrow, has an unfavorable climate, a lack of suitable niches, or too many competitors, it is called a *filter*—only a few specially adapted species are able to pass over it. A *sweepstakes route* of dispersal is one over which only a few species pass, more or less by chance. Because of its general unfavorableness, it is generally only one-way. The island-hopping pan-tropical dispersal of organisms from southeastern Asia through the south Pacific has been accomplished by relatively few species; the more distant the island, the fewer the species that have reached it.

A glance at a globe quickly shows that the continental land masses are concentrated mostly in the northern hemisphere. There are three broad land extensions southward below the equator: the Malay Peninsula, East Indies, and Australia; Africa; and South America. The southern hemisphere otherwise consists largely of vast expanses of oceans. On a land route, North America intervenes South America and Eurasia. Eurasia is the largest continent, is central to all the others, and always has had a great diversity of climate and terrain. The size, arrangement, and positions of the continents are of importance to interpretation of the past evolution and dispersal of animals.

Considerable evidence (Matthew 1915) indicates that some large groups of animals, notably mammals, first evolved in Eurasia and then spread to other parts of the world. North America was a less important center of origin and dispersal. Warm, moist, uniform climates prevailed in the early Carboniferous, Jurassic, mid-Cretaceous, and Eocene (Table 19-1). During periods of continental emergence, climates on the northern continents changed from moist and warm to more arid and cold. Periods of aridity and glaciation are known to have occurred in the northern continents

during the Permian, at the end of the Triassic, at the beginning and end of the Cretaceous, and during the Pleistocene. Animals adaptively limited to moist, warm environments were restricted to tropical regions or dispersed outward into the southern land extensions. New, more advanced animal types adapted to the new conditions in the north appeared. Monotremes entered Australia at an early date. Marsupials probably originated in Eurasia and dispersed into Australia and South America during late Mesozoic or early Tertiary, although there is no fossil evidence for these suppositions (Darlington 1957). An oppossum still occurs in North America, but no marsupial is now present in Eurasia. Marsupials did not get into Africa, but are present in South America and have adaptively radiated into a variety of forms in Australia. The predominance of marsupials in Australia is probably due to their coming over at an early date on a land bridge or their chance success in surmounting the sweepstakes route

Fig. 18-7 Crests of major ridges less than 1220 m (4000 ft) deep in the Atlantic Ocean (wavy lines) represent possible sites of exposed lands in the Mesozoic and early Tertiary, at least as extensive archipelagos (Axelrod 1960; from Sol Tax, *Evolution after Darwin*, University of Chicago Press, copyright 1960 by the University of Chicago).

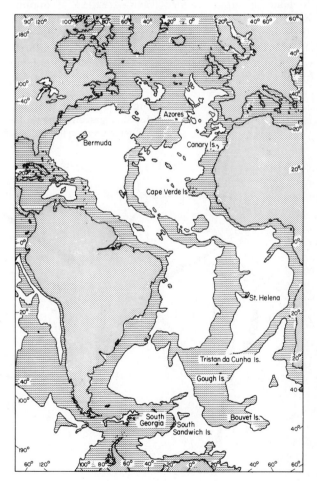

of islands from the Asian mainland. Bats and members of the rat family arrived much later. There have been successive waves of dispersal of higher mammalian forms into Africa and South America, but many recently evolved mammalian types are largely confined to the northern hemisphere, where most of the primitive forms there have long since become extinct. The origin and dispersal of birds may have followed the same pattern as that of mammals, but their geological history is less easily traced because birds are not easily fossilized.

A modification of Matthew's theory appears necessary, at least for the cold-blooded vertebrates (Fig. 18-8). Freshwater fishes, amphibians, and reptiles are most richly developed in tropical regions: all are definitely handicapped in dispersing into temperate or cold regions by the climatic barriers. Fresh-water fishes have evolved a richer north temperate and arctic fauna than amphibians and reptiles, since their aquatic habitat protects them better against extreme cold than does the terrestrial or semi-terrestrial habitat of the other two groups. Amphibians extend farther north than do reptiles. Evaluating all types of evidence, Darlington (1957) concludes that these three groups of cold-blooded vertebrates, and probably also the warm-blooded groups, reached their greatest taxonomic diversification during the Cretaceous and early Tertiary, and not in temperate Eurasia but in the Old World tropics, especially in the Orient. Uniform, warm, humid climates and a great variety of available niches, along with many possibilities for geographic

isolation, appear to have induced evolution in these groups, stimuli quite different from those suggested by Matthew for the warm-blooded mammals. From the Old World tropics, dispersal proceeded into Africa, into Eurasia, across the Bering land bridge into North America, and finally across the Panama land bridge into South America. Subsequent evolutionary radiation of new forms occurred in each continent.

A northern route between Asia and North America by way of the Bering land bridge as a route of dispersal for tropical species presents problems in respect to climate. We know, however, that during the Cretaceous and early Tertiary the climate in these northern regions was much warmer than it is now. We may suppose that the bridge shut off the cold Arctic Sea from the Pacific Ocean, and that the southern shores of the land bridge were washed by the warm Japanese current. This would have made it possible for warm temperate species to use the bridge, but probably not tropical species, unless we suppose that the tolerance to cold of the ancestral stock of our present tropical fauna and flora was greater than it is at the present time.

Another difficulty that southern species would encounter on the bridge would be the very long days of the summer and the very short ones of the winter. Tropical species are adjusted to fairly equal photoperiods at all seasons of the year. Seasonal differences in photoperiod are due to the inclination of the earth's axis. It has been suggested that this inclination has changed during geological time (Schwarzbach 1963). Although it appears very likely that the Bering land

Fig. 18-8 Dispersal of toads (*Bufo*) over the world—except for Madagascar, some of the East Indies, and Australia—from a center of origin in the Old World Tropics (after Darlington 1957).

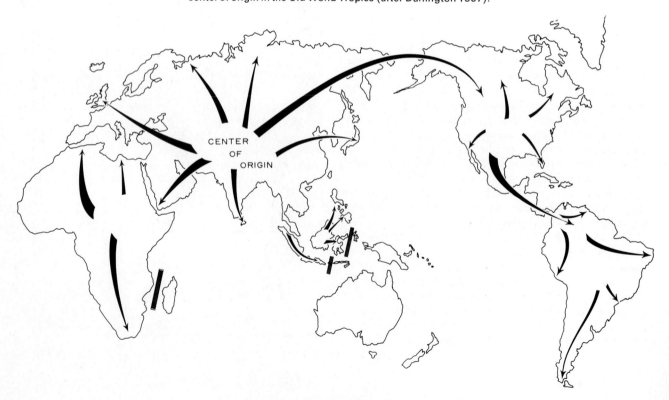

bridge was an important route of dispersal between Asia and North America, considerably more study is required before we will satisfactorily understand how this was accomplished by various kinds of animals.

Centers of Origin

The tracing of dispersal routes presupposes a starting point where the taxonomic group, whatever its size, first evolved. These starting points are called *centers of origin*. Various criteria for determining centers of origin have been suggested (Savage 1958), but caution must be exercised in applying them (Cain 1944). Of the many criteria proposed, the following two are especially important, although neither one is infallible:

1. *Location of greatest differentiation of the type or the greatest variety of endemic races, species, and genera, including primitive forms or fossils.* The older a group is and the longer it has occurred at a particular location, the more chance it has had to radiate into different habitats, become isolated, and evolve into new varieties. However, a shift of climate or a drastic change in physiography may render an original locality unhabitable and the group moves elsewhere. Also, a group losing its vitality may contract its range into some area other than the one in which it originated.
2. *Continuity and convergence of lines of dispersal.* Dispersal from a center radiates in all directions in which conditions are favorable and until insurmountable barriers are encountered. Lines of dispersal may be readily distinguished where one or more taxonomic characters can be traced from primitive or generalized types through more and more specialized types the greater the distance involved. However, once a species filters through a break in a barrier and invades an extensive, new, and favorable habitat, there may be increased evolution of new types, and a secondary dispersal center formed. This has happened repeatedly in the geological past, so there is often difficulty in distinguishing which center is the original one for a group.

Island Dispersal

Dispersal of animals from continents to islands and from one island to another presents special problems. Many of the world's prominent islands occur on the continental shelf and are separated from the mainland only by shallow seas. At times of land emergence, as when glaciers lock up quantities of water as snow and ice enough to lower the level of the seas, these islands become connected by land bridges to the mainland, and dispersal of forms occurs. The British Isles have been thus connected to Europe; Japan to Korea and

Siberia; Sumatra, Java, and Borneo to Malaya; New Guinea and Tasmania to Australia; and Newfoundland to Labrador. On the other hand, the Bermudas, Azores, Hawaiian and other small Pacific islands, and possibly New Zealand, the West Indies, and Madagascar, could not have had mainland connections and thus must have received their present faunas by some means other than overland dispersal (Chapter 20).

Islands adjacent to continents, unless very small or long separated, generally have faunas similar to that on the nearby mainland. Oceanic islands are more difficult to colonize, however, and often have unique unbalanced faunas or chance assemblages of species. Mammals, amphibians, and fresh-water forms are often absent or scarce. Flying birds, bats, and insects blown in by strong winds, and small invertebrate forms readily transported on rafts are better represented.

Since islands are colonized with difficulty, the number of species on them is generally small compared with the mainland, and interspecific competition is reduced. A single established genus or family on islands free of close competitors may adaptively radiate into new niches to form a variety of species or races, as did the group of finches in the Galapagos Islands observed by Charles Darwin and later studied by Lack (1945) and Bowman (1961), and insects, honey creepers (birds), and other forms in the Hawaiian Islands (Zimmerman 1948). Animals on islands are often smaller in size than their relatives on the mainland, for instance, ponies on the Shetland Islands, Key deer on the Florida Keys, and the gray fox on Catalina Island. Islands are generally too small to support adequate populations of large animals requiring extensive home ranges. At the same time, there may be selective advantages operating in evolution for becoming smaller. Because of lack of competitors and predators, primitive animals isolated on islands may survive long after their relatives on the mainland have perished, as has, for instance, the reptilian *Sphenodon* on New Zealand. Confinement of a species to a limited range permits extensive inbreeding so that the population becomes more homozygous in its various genetic traits, hence much less adaptable to new situations than are larger heterozygous populations. Traits that would be eliminated by predation pressure on mainlands sometimes become established in populations on islands. All these conditions render oceanic island species liable to extinction by invasion of mainland forms, and renders them impotent to reinvade the mainland.

There is continual invasion of individuals of the same or new species onto islands; at the same time there is continual extinction of individuals and species. A biotic equilibrium is reached only when the two rates are equal. The relation between rates of invasion and extinction varies with the distance from the mainland. Successful colonization after invasion is more likely the higher the ratio of birth rate to mortality rate. The number of established species commonly varies

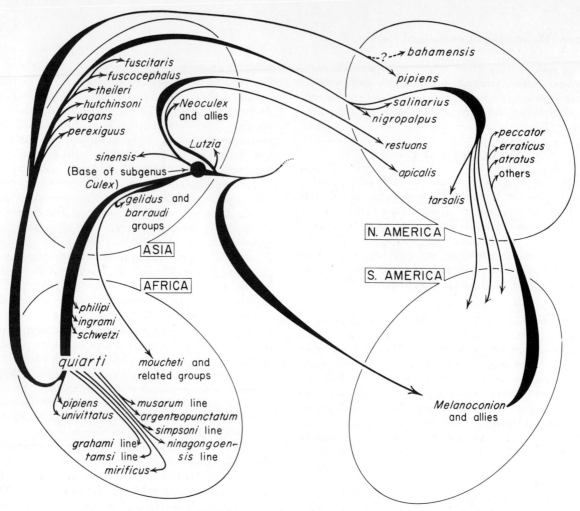

Fig. 18-9 Phylogeny and dispersal of the mosquito genus *Culex* (Ross 1953).

with the size of the island and diversity of biotopes. On small islands with restricted number of biotopes, the ratio of genera to species is high. All these rates, relationships, and processes can be quantified mathematically, and they are of significance not only to island biogeography but also to dispersal of organisms generally (MacArthur and Wilson 1967).

Origin of North American Fauna

A dry land bridge connected North America to Asia across the Bering Sea during most of early and middle Tertiary and during glacial periods in the Pleistocene (Hopkins 1967). During early Tertiary, Alaska had a temperate climate, but as the climate became progressively colder in late Tertiary, there was increased filtering of animal groups having access to the bridge. At the present time all land connection has, of course, disappeared, although a few forms, particularly birds, are able to hop the narrow straits. Across this bridge came a heavy traffic of Asiatic mammals (Simpson 1947, Savage 1958), birds (Mayr 1946), reptiles, amphibians, and fish (Darlington 1957), modern insects (Fig. 18-9) (Ross 1951, 1953), and other groups. There

was also some dispersal from North America into Asia, but this reverse movement was much less strong than the one from Asia into North America.

During the early part of the Tertiary, all major groups of mammals appear to have moved freely across the land corridor so that there was considerable similarity between the two continental faunas. However, from early Eocene to early Oligocene, new orders, families, and subfamilies arose that were distinctive to each continent. Later in the Tertiary, there were fewer groups dispersing across the bridge, and these were largely confined to genera within the higher taxa already established on each continent.

There is no evidence for a land bridge between North America and Europe via Greenland and Iceland since early Tertiary, although it is possible that one existed earlier. A few animal forms may have been able to hop from island to island across the North Atlantic, but the faunas of North America and Europe are not sufficiently similar to suggest any recent close connection of the continents. Some pan-tropical forms may have crossed the Pacific Ocean from island to island to reach the western hemisphere, but former land bridges across either the Pacific or South Atlantic oceans are highly unlikely.

A continuous land connection occurred between North and South American in late Cretaceous time. Ancient types of mammals, birds, reptiles, amphibians, insects, and other groups got into South America over this land bridge from the north and differentiated into distinct families and other taxonomic categories (Dunn 1931). During most of the Tertiary, no land bridge existed between the two continents, although there were scattered islands separated by relatively narrow water gaps which some groups, particularly birds, may have been able to use (Fig. 18-10). The land connection now in existence was apparently formed in the late Pliocene or early Pleistocene.

Previous to late Pliocene there were about 29 families of land mammals confined to South America and 27 confined to North America; the two continents had not more than one or two families in common. During the Pleistocene, after the land bridge had been in existence for some time, 22 families were represented on both continents; 7 had dispersed from South to North America, 14 from North to South America, one is of uncertain origin. Some families have become extinct. The present faunas of the two continents contain 14 families in common, 15 families found only in South America. and 9 families found only in North America. Thus there has been considerable dispersal over the land bridge in both directions, but not a complete exchange or unification of faunas (Simpson 1940).

The modern fauna of North America is thus derived principally from (1) Eurasia, (2) South America, and (3) by autochthonous development. By *autochthonous* we refer to species evolved from very old indigenous types that may or may not be represented by related forms on other continents. The proportion of any local fauna that is derived from one or the other of these sources varies with each geographic area and with each group of animals (Mayr 1964). Northward on the continent and in the western mountains, the Eurasian bird element is strongly represented. The South American element becomes greater southward, especially in the lowlands of California, Mexico, and Central America. Many Eurasian forms have dispersed through North America into South America, but no modern forms, at least, of South American origin have been able to invade Asia through North America.

MIGRATION

Migration, like dispersal, involves movements and the invasion of new areas. Migration, as here defined, differs from dispersal in that it is a *periodic* movement back and forth between two areas (but see Urquhart 1958). In contrast, dispersal is a one-way outward movement. Migratory invasions of areas are temporary and repetitive, but invasions resulting from dispersal may be permanent.

Migration is best known in birds, as an invasion of

breeding area alternating with an invasion of wintering area, annually. Representatives of other groups of animals also migrate, particularly mammals, fish, and insects (Heape 1931). Migration may be classified as annual, diurnal, or metamorphic (Clements and Shelford 1939). Annual and diurnal cycles are correlated with the two most pronounced time cycles in the physical environment. Metamorphic migrations are movements from one habitat to another in different stages of an animal's life cycle.

Annual Migrations

Annual migrations may involve a change of latitude, or altitude, or be more local in extent.

Latitudinal migrations may traverse only a few kilometers or may extend almost from pole to pole (Fig. 18-11). In terms of their occurrence in an area bird species are described as *permanent residents*, species represented in an area throughout the year even though some individuals migrate; *summer residents*, species present only during the warmer part of the year, which includes a breeding season that may extend from early spring to late autumn; *winter visitors*, species present only during the winter or non-breeding period; *transients*, species present only during migration periods, neither breeding nor wintering in the area; and *accidentals*, species that are rare or irregular in occurrence. The bird population in most communities reaches peaks during the vernal and autumnal aspects, as transients arrive and temporarily swell populations. The autumnal peak is usually the greater because adult populations are incremented by the large number of young birds hatched during the breeding season. Latitudinal bird migrations also occur in the southern hemisphere but are less conspicuous because of the small continental land masses there and small populations of birds involved.

There are many causes of bird migration, varying

Fig. 18-10 Southern Central America and northwestern South America in middle Miocene time (Whitmore and Stewart 1965; copyright 1965 by the American Association for the Advancement of Science).

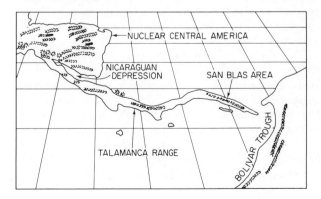

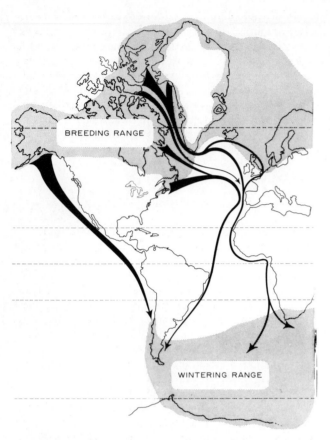

BREEDING RANGE

WINTERING RANGE

Fig. 18-11 Migration routes, the longest known, of the arctic tern (Lincoln 1950).

in relative importance with different species. Aquatic species must leave northern areas before their food supply is cut off by the freezing of the lakes, ponds, and rivers. Insectivorous species unable to change to other types of food must migrate before insects go into hibernation or disappear. The metabolism and food requirements of many song birds are so high that even when food is abundant they cannot eat enough during the short winter day to give them sufficient energy to survive the long, cold winter night. Migration northward in the spring escapes the high summer temperatures of the south and gets the migration into latitudes where the days are long (Kendeigh 1934). Whatever its immediate causes, migration presumably evolved because survival was more successful among those individuals that departed than among those that remained (Lack 1954a).

The timing of migration is not usually regulated, however, by the factors just listed as causes, for most birds migrate days, weeks, even months before the beginning of intolerable conditions in the autumn and after they disappear in the spring. The annual stimulus to migration is complex and involves changes in physiological state, energy balance, and hormones (Kendeigh *et al.* 1960). The chief environmental stimuli are changes in length of day and night and changes

in temperature with the progress of the seasons. The regularity of migration by which species arrive at a given point about the same date year after year is probably a response to the regularity of change in day length. The fact that species may arrive a few days early or late of the usual arrival date is a result of the superimposed effect of variations in temperature, to which the birds are also responding.

The mechanics of migration, fly-ways, flocking behavior, migration routes, orientation, and so forth, are too intricate for detailed analysis here (Lincoln 1950). Much research is now in progress analyzing the factors involved; it is sufficient merely to add that migratory behavior is organized as an instinctive behavior pattern in the bird's nervous system. If a stimulus is not presented, the behavior will not be expressed. The stimulus arises when the interaction between internal physiological rhythms and environmental cycles reaches a critical stage.

Bison regularly migrated from northern parts of the Great Plains to pass the winter in southerly reaches, traversing a distance of 300–600 km (200–400 miles). Some bats, particularly the hoary and red bats, regularly migrate between Canada and the northern states. The fur seal breeds in the Pribilof Islands in the Bering Sea during the summer, and migrates southward as much as 4800 km (3000 miles) for the winter.

The monarch butterfly breeds in the northern states and migrates several thousands of kilometers to winter as far south as the Central American tropics. A small proportion of individuals successfully make the return migration in the spring. Evidence is accumulating to indicate two-way migratory behavior in other species of butterflies and insects (Uvarov 1928, Fraenkel 1932, Williams 1958).

Migratory locusts or grasshoppers occur both in the eastern and western hemispheres. *Schistocerca gregaria* inhabits the arid grassland and semi-deserts of Africa and southern and western Asia (Uvarov 1928); *Melanoplus mexicanus* occurs in the northern Great Plains of North America. In both species, solitary and migratory phases occur which differ in size, wing length, and coloration. The migratory phase apparently develops under conditions of higher temperatures and good breeding conditions so that overpopulations occur. When migration begins, immense swarms of adult flying individuals move great distances. Migration in the nymphal hopper stage is more limited. Vegetation is ravenously devoured wherever the swarm stops. Such migrations were extensive in North America between 1876 and 1879 when populations moved from the northern Rocky Mountain area into the states immediately west of the Mississippi River. Eggs were deposited at the terminus of the migration flights and at least some of the succeeding generation exhibited return flights in following years.

Altitudinal migrations are movements of no more than a few kilometers up and down the slopes of mountains.

By descending to lower altitudes in the autumn, an organism obtains some of the same benefits secured by those species that undertake latitudinal migrations— less snow, higher temperatures, and more food. Birds restricted to alpine habitats in the summer are common winter residents of lowland areas. Some of the larger mammals, such as the mule deer (Russell 1932) and the American wapiti (Altmann 1952), have very regular migration habits in respect to herding, timing of movements, and migration routes. They move to the high alpine meadows soon after the vegetation renews its growth in the spring, and come back down to the valleys in time to avoid the deep winter snows of the higher slopes.

Local migrations do not necessarily involve a change of latitude or altitude and are often quite limited in distance covered. However, in tropical grasslands and savannas, where wet and dry seasons bring great changes in available water, vegetation, and food, there is a great exodus of both mammals and birds during the dry season and an influx during the rainy season.

The Atlantic salmon, after reaching sexual maturity, may ascend fresh-water streams in subsequent years to spawn and return each time to the sea. Many deep-water fishes spawn annually in shallow waters and then return to deep water again. Turtles come onto the land to lay their eggs; snakes disperse from their winter dens with the advent of warm weather in the spring; tree frogs go to small pools to mate and spawn; and resident birds move onto their breeding territories.

Insects perform regular migrations both into hibernation and out of hibernation. In the autumn, forest species migrate downward and may be found in peak numbers first in the shrubs, then in the herbs, then in their hibernacula in the ground. Many insects of the forest-edge, meadows, and agricultural crops also hibernate in the forest, usually a few meters in from the south-exposed edge, where they derive some heat from the winter sun. These insects usually migrate into the forest in the same stratum, herb or shrub, in which they occur during the summer, then downward into the soil. These flights into the forest occur with declining temperatures and sometimes include spectacularly large numbers of individuals. As they come out of hibernation in the spring the direction of movement the insects take is just the reverse that taken in the autumn, that is, upward into their proper stratum, then horizontally outward into open country (Weese 1924).

Daily Migrations

Ascent of plankton toward the surface at night and descent to greater depths during the day occur both in the sea and in lakes. The lake-dwelling culicid larva *Chaoborus* lies on the bottom during the day but becomes pelagic at night. Snails, slugs, and millipedes in the deciduous forest lie quiescent under logs or litter during the day, but come out at night to crawl around on the forest floor or even climb up on the vegetation to a height of perhaps a few meters. Although these excursions are restricted in range, they are more or less regular and periodic, and may be thought of as migrations.

Metamorphic Migrations

Aquatic larvae and naiads of several orders of insects eventually change into adults that leave the aquatic habitat and become aerial. The adult stage is often short-lived. The eggs are deposited in water, or the immature stages return again to water to begin the cycle over again. The length of the cycle may be part or the whole of a year, or a longer period. The seventeen-year cicada is a well-known insect whose nymphs spend 17 years underground feeding on juices from the roots of trees. The adults appear above ground in large numbers in late May, mate, lay their eggs on twigs of various trees, and then disappear, all in a few weeks. The eggs hatch in about 6 weeks, and the nymphs drop to the ground and bury themselves for another long period of years.

The anadromous Pacific salmon ascends fresh-water streams but once to spawn. The breeders die; it is their offspring that return to the ocean to develop for a period of years before they make the migration. Many adults return to spawn in the same stream or even in the same headwaters where they were hatched. The Pacific salmon apparently uses the sun as a compass mechanism to find its way from the ocean to the stream, then swims upstream against the current, rejecting all odors until it arrives at the home tributary to which it was initially conditioned (Hasler 1960). The two species of catadromous eels, of the western hemisphere and of Europe, migrate to the open sea in the region northeast of the West Indies in order to reproduce. The immature stage of the eel not the adults, returns to the two continents.

ECESIS

Dispersal or migration of individuals into new areas is without great ecological significance unless those individuals become established and can build up significant populations. The process by which organisms become established in new areas is called *ecesis*. Ecesis will occur if a species disperses into a habitat favorable to it and can secure its proper niche or become adjusted to a new niche, new competitors, predators, parasites, and disease organisms (Maguire 1963). Ecesis, to go to completion, requires first the establishment of individuals in an area, then the growth of the population that they form, and finally or simultaneously, the maturing of community structure with the invasion of many other species (Ross 1962, Olson 1966).

Temporary ecesis is the rule with migrant species, as the periodic change of location is normal in their life behavior. Ecesis as range expansion is not surely permanent until the species demonstrates that it can survive critical years in weather cycles. For instance, northward dispersal of tropical forests is thwarted by frosts that occur only at rare intervals. Insects may continuously expand their ranges for a period of years, only to be forced back hundreds of kilometers by a severe drought or cold spell.

Under natural conditions, growth curves for populations of single species are clearly evident when a species invades a new area that is favorable; when a species is recovering from a catastrophe or cyclic depression; and as a species builds up its population in the spring after the termination of a winter dormancy or migration (Fig. 18-12). The many factors in natural environments that modify rates of population growth and determine the levels at which populations become

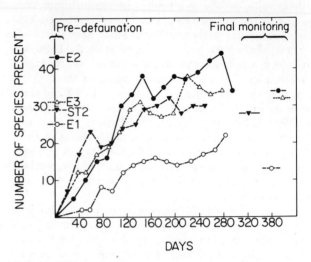

Fig. 18-14 Colonization curves of insects arriving on four mangrove islands after removal of the original fauna by fumigation. The size of the original populations is shown at the left (after Simberloff and Wilson 1969).

stabilized at the asymptote were considered in Chapter 14.

The total community is an aggregation of many species. With seasonal onset of favorable weather or when a new habitat becomes available with progressive development of substratum, microclimate, and vegetation, the coordinated invasion and ecesis of these many species also follows the sigmoid or logistic curve (Fig. 18-13). When, however, the new habitat is fully developed and ready for colonization by a group of species, as for instance islands (MacArthur and Wilson 1967), the lower self-accelerating portion of the sigmoid curve may be eliminated and only the upper or self-inhibiting portion expressed. Furthermore, colonization curves for insects arriving on small mangrove islands in the Florida Keys, after removal of the original fauna by fumigation, show that a larger number of species invade than can become established (Fig. 18-14). Extinction of less aggressive or more poorly adapted species results from various causes, including competition and predation (Yount 1956, Simberloff and Wilson 1969). The asymptote of the curve is reached when the species present establish interactions between them so as to form a closed or stabilized community that discourages further invasion. Thus the process of growth or ecesis is much the same whether it is at the level of the individual cell, organism, species population, or community.

With succession, a series of communities become established one after another. This is best represented by a sequence of sigmoid curves, each indicating a particular stage in the sere. As a new community invades, the previously established one disappears, probably at a rate indicated by a reverse sigmoid curve. Ordinarily a larger number of distinct plant communities can be recognized in a particular sere than animal

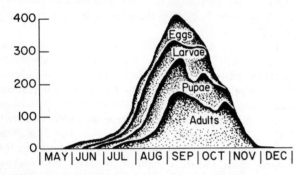

Fig. 18-12 Seasonal growth and decline of a nest colony of common hornets, *Vespa crabo*, in France (Bodenheimer in *Biol. Rev.*, 1937, after Janet).

Fig. 18-13 Ecesis of nesting bird species in central Illinois during the pre-vernal and vernal aspects (compiled from Smith 1930).

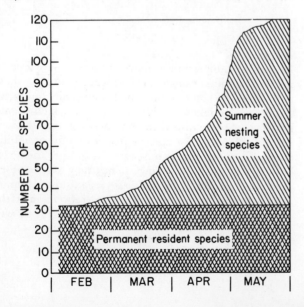

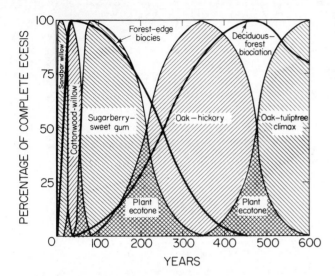

Fig. 18-15 Generalized model of ecesis of five successive plant communities (cross-hatched) and two animal communities (heavy lines) in a floodplain sere. Plant communities and time scale are based on Shelford (1963).

communities. The ecesis of a single animal community may overlap several plant stages. Thus the deciduous forest biociation begins its development on the Mississippi River floodplain with the establishment of the first trees in the cottonwood-willow associes (Fig. 18-15), continues to develop through the sugarberry-sweet gum, oak-hickory associes, and into the climax oak-tuliptree association. However as the climax reaches its full dominance and all remnants of earlier plant stages are eliminated, the decrease in diversity of plants is often correlated with a decrease in species diversity of animals. The ecesis of distinct animal communities does not correspond, therefore, with ecesis of distinct plant communities, but the ecesis of animal communities is correlated only with the development of distinct types of vegetation. A review of Chapters 4–8 will show that this model holds also for all kinds of seres. A background of understanding of the processes of dispersal and ecesis and the problems encountered by species in their invasion and establishment in different parts of the world is desirable for proper evaluation of present-day distribution of organisms and the distributional units that are recognized.

DISTRIBUTIONAL UNITS

Only a few groups of animals are cosmopolitan or nearly so, in having a worldwide distribution, notably cyprinid fish, frogs of the genus *Rana*, colubrid snakes, passerine birds, rodents, and man. Most organisms are restricted in distribution because of barriers of one sort or another or because of past history of origin and dispersal. An understanding of their present-day dis-

tribution takes us into biogeography, climatology, and paleoecology. Major units of distribution are the zoological regions, areas defined largely by the past and present relations of the continents to each other. Each region may then be subdivided into faunal or ecological units, depending on the criteria used.

Zoological Regions

Zoological regions have long been separated one from another by water, mountain, or desert barriers (Fig. 18-16) so that they have evolved distinctive and characteristic orders and families of animals. These major units were first recognized by Sclater (1858), modified by Huxley (1868), extended by Wallace (1876), and best described in a modern way by Newbigin (1950), Beaufort (1951), and Darlington (1957).

The Australo-Papuan region has been isolated the longest and is marked by the absence of most groups of placental mammals, although some have been introduced. The original mammalian fauna, derived from Eurasia when Australia and New Guinea (Papua) connected to that continent by a land bridge, consisted of monotremes and marsupials. Murid rats and mice and bats invaded later and in a different manner. It is of interest that woodpeckers among birds, true frogs among amphibians, and fresh-water fish are absent or scarce.

The *Neotropical region* contains the largest number of unique endemic forms of birds and mammals (Keast 1969), perhaps because it is the farthest by a land route from the oriental tropics where the major taxonomic groups probably originated and because the region is almost completely surrounded by water. The occurrence of many related forms in South America, Africa, and Australia, and the absence or poor representation of these forms, even as fossils, in North America and Eurasia may be explained either by the continental drift theory or by connections in the geological past with a much warmer Antarctica continent (Glenny 1954). For instance, the ratite bird, rhea, and the lungfish, *Lepidosiren*, occur only in South America; the ostrich and the lungfish *Protopterus* only in Africa; and the emu and cassowary birds and the lungfish *Epiceratodus* only in Australia and adjacent islands.

The *Ethiopean region* is noted especially for giraffes, antelopes, zebras, elephants, and other ungulates that wander around in large herds; by the large hippopotamus and rhinoceros; by a number of cats and dogs; and by the primate lemurs, baboons, macaques, chimpanzees, and gorillas. There are several endemic families of birds and other forms. Absent are deer and bears among mammals and salamanders and tree frogs among amphibians.

The fauna of the *Oriental region* shows relationship to that of the Ethiopean region in several respects. India was connected with Africa at various times. During

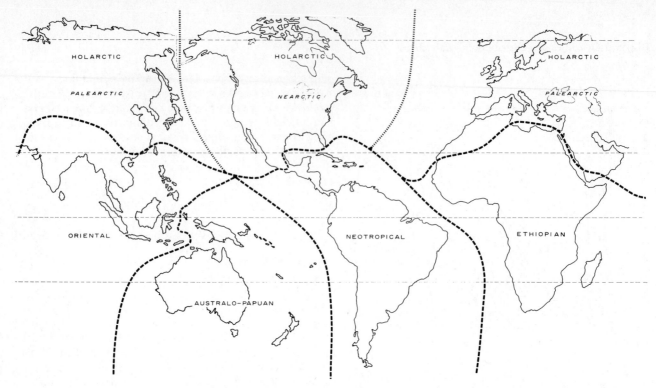

Fig. 18-16 Zoological regions of the world (adapted from Map No. 201 HA, Goode Base Map Series; copyright by the University of Chicago).

the Miocene and Pliocene (Table 19-1), a uniform moist climate prevailed, and forest and savanna vegetation extended through northern Africa, Arabia, and India allowing the fauna to move back and forth. The ocean barrier developed later as well as the present desert regions. The Oriental region has a tiger instead of the lion of Africa, its own elephant, the gibbon, orangutan, and tarsier among the primates, two rhinoceroses, several species of deer and antelopes, and, like Africa, lacks tree frogs and salamanders.

The broad Bering land bridge connected North America and Asia through much of the Tertiary, so that many genera and even species of animals occur on both continents, especially in their northern portions. Salamanders are largely limited to this *Holarctic region*, but primates, other than man, are largely missing. Farther south on the two continents the fauna becomes progressively more different (Udvardy 1958). Antelopes, sheep, goats, and certain other groups are in greater variety of species in the *Palaearctic subregion*; water moles of the subfamily Desmaninae are unique to the subregion. On the other hand, the *Nearctic subregion* has several families of birds, such as the vultures, turkeys, mockingbirds, vireos, and wood warblers that have not spread into Asia and Europe. Rattlesnakes, salamanders (Ambystomidae), suckers, and catfishes common to the Nearctic are either absent in the Palaearctic or poorly represented.

Faunal Areas

When populations become isolated geographically, they tend first to differentiate into species, then as time goes on, into higher taxonomic units. Each zoological region is divisible into smaller units on the basis of only the lower taxonomic units: genera, species, or even subspecies.

The first attempt to subdivide North America into geographic units of biological significance was made by Schouw in 1823, for plants (Kendeigh 1954). His work stimulated zoologists to undertake a number of similar efforts, an especially appropriate one being that of J. A. Allen (1892). The *life-zone* system was developed between 1890 and 1910 by C. Hart Merriam and was used extensively for several decades.

The concept of biotic provinces first began to attract attention about 1911 and became defined and mapped by Dice in 1943 and again in 1952a. A *biotic province* is a continuous geographic area that possesses a fauna distinguishable, at the species and subspecies levels, from the fauna of adjacent areas, at least to a certain degree. Both faunal areas and provinces are commonly delineated by mapping the distributions of species and drawing boundaries where distributional boundaries are most prevalent (Udvardy 1963, Hagmeier 1966). The boundaries of biotic provinces are more likely to coincide with physiographic barriers than with types of vegetation. Unlike faunal areas, life zones, and biomes, biotic provinces never occur in discontinuous geographic fragments since they are intended to show regional areas of evolutionary differentiation. A biotic province which includes a mountain may have several types of vegetation or life belts, each serving as a center of differentiation for its characteristic fauna when compared with similar life belts in other provinces. The biotic province system is being used at the present time by some mammalogists, ornithologists,

292

and herpetologists in the study of particular taxonomic groups, but there has been no general synthesis of these studies and of plant groups to render the provinces truly "biotic" in nature.

Faunal Groups or Elements

In all the systems above described, faunal distribution has been analyzed in terms of geographic areas, and the chief criterion for determining the limits of an area has been that of distinctiveness between the fauna of different areas. A different approach is that stimulated by the herpetofaunal studies by Dunn (1931); avifaunal studies by Stegmann (1938), Mayr (1946), and Miller (1951); Simpson's studies of mammalian fauna (1947); and the plant studies of Wulff (1943) and Cain (1947). The fauna or flora is analyzed in terms of groups of species with similar centers of origin, dispersal routes, geological histories, and habitat preferences. The objectives are to learn the place where and time at which these groups of species originated, how the groups came to occupy a particular part of the world, became adapted to live in new environments, and how they evolved into the present-day living species. A type of vegetation, life zone, or biotic province may contain species from several different faunal elements that have come to live together. Thus in the bird composition of California one can recognize boreal, Great Basin, Sonoran, and authochthonous elements (Miller 1951) and in the bird fauna of Europe as many as 24 elements (Voous 1963) (see also Table 21-2).

Biomes

A still different approach for the study of distribution is the ecological one, involving the recognition and classification of natural communities. A number of different ecological systems have been suggested (Whittaker 1962, Olson 1970), but we will here be concerned only with the biome concept (Clements and Shelford 1939). The systems described earlier in this section are zoogeographical, in their attention to centers of origin, dispersal routes, and evolution. The biome system emphasizes community dynamics and environmental relations. This does not mean that zoogeographers do not consider the relations of climate and terrain in controlling dispersal and evolution, nor that ecologists are not concerned with the geological history of the forms with which they deal, but the viewpoints and objectives are different.

Basic to the understanding of the biome system is the recognition of communities composed of characteristic combinations of animal and plant species, of successional relations between communities, and that succession in all local habitats eventually converge into a climax community pattern, the most important ecological characteristic of a geographic area (Cain 1939, Whittaker 1953). The development of an understanding of these ideas has been a major concern of this book, especially Chapters 4–8.

A *biome* is a biotic community, characterized by distinctiveness in life-forms of the important climax species. On land, the most important climax species are usually plant dominants that occur in distinctive vegetation and landscape types; in the ocean, the important organisms that define biomes are usually the predominant animals, which are sometimes also dominants.

Seral communities are developmental stages and are consequently considered part of the biome. The animal and plant constituents of seral stages are, however, more widely distributed than are species belonging to climax communities, since the habitats in which they belong are more nearly alike in different parts of the world than are the habitats that contain the climax. Seral species are not generally useful, therefore, to defining the limits of biomes. The majority of animal constituents of the climax community, however, are characteristic only of the climax vegetation or habitat and therefore of restricted distribution.

The principal biomes of the world, or *biome-types* of some investigators are the following:

Terrestrial	Marine
Temperate deciduous forest	Oceanic plankton and nekton
Coniferous forest	Balanoid-gastropod— thallophyte
Woodland	
Temperate shrubland	Pelecypod-annelid
Tundra	Coral reef
Grassland	
Desert	
Tropical savanna	
Tropical forest	

The vegetational portion of the biome is sometimes called the *plant formation* (Weaver and Clements 1938). A *plant association* is a subdivision of a biome or formation, distinguished by uniformity in the species composition of the climax plant dominants. The *associes* is the equivalent seral plant community, regardless of whether it belongs to the same or a different type of vegetation than the climax. The important point to remember here is that the biome is distinguished by the life-forms of the climax dominants, but subdivisions of the biome are recognized principally by taxonomic composition.

The type of climax in a terrestrial area is determined mainly by the conditions of climate (Fig. 18-17), although secondary correlation also occurs with major soil groups. Compare Figs. 18-18 and 18-19 with the biome maps inside the front and back covers. Were climate the only factor involved and the terrain uniform, the climax community would be monotonous in its composition and structure, except as one community graded into another with change in climate. Where the composition and character of the prevailing vegetation varies more or less permanently with changes in physiography, soil, or fire factors, we may speak of physiographic, edaphic, or fire *subclimaxes* or faciations.

This is in agreement with the *monoclimax* viewpoint; that is, that the most representative climax in an area is determined by the prevailing climate. An opposing concept is that of *polyclimaxes*. Proponents of the latter viewpoint give nearly equal significance to climate, soil, topography, and other factors, and believe that each major variation in composition or structure of the mature vegetation should be considered as equally important. Hence, several climaxes, or at least a complex community pattern that varies in structure and species composition from one site to another, may be claimed for an area (Whittaker 1957). The controversy is largely one of emphasis and semantics.

A *biociation* is a subdivision of a biome distinguished by uniformity and distinctiveness in the species composition of the climax community, particularly of the animal predominants. The *biocies* is the equivalent seral community. With the biome, we recognize the primary importance of the life-form of the principal climax organisms for establishing the major units into which the geographical distribution of organisms are naturally divided. The biociation concept does not regard differences in the species composition of plant dominants, such as are recognized in the plant associations, as establishing the fundamental subdivisions of the biome for animals. So long as they are of the same life-form, variations in the plant species composition

produce only minor differences in dominance, insufficient, for the most part, to induce striking changes in the species composition of the rest of the community. The relation between biocies, associes, and changes in vegetation that occur with succession (Fig. 7-4) obtain also between biociations, associations, and changes in vegetation that occur geographically. Actually, the animal ecologist has no absolute need to identify the species composition of the plant dominants if he can describe the vegetation accurately in other ways (Dansereau 1951). If one wishes not to define plant communities on the basis of the taxonomic composition of the dominants alone, the concept of biociations may be equally useful for the analysis of distribution of plant species. The "natural" areas of Cain (1947) are a step in this direction.

A biociation may originally derive its species components from several faunal elements (Table 21-2), but once the community constitutes a unit, it may thereafter serve as an element itself, and a geographic area may be described in terms of the biociations and biocies it contains. Biociations differ from biotic provinces in that the latter are geographic areas, rather than communities, and in mountainous areas may contain several life belts or different types of communities. Furthermore, biotic provinces may be characterized in part by the presence of particular subspecies. Subspecies are not used in defining biociations.

It is clear, then, that two sets of factors basically control the present local and geographic distribution of organisms. The first set is ecological, including the physical factors of the substratum and climate for terrestrial organisms and the composition of the water for aquatic forms; the biotic factors, especially of food, cover, reproductive requirements, competition, and predators; and the psychological factors of behavior

Fig. 18-17 Climographs of principal biomes of the world (Hammond 1972). (Copyright 1972 by the American Association for the Advancement of Science).

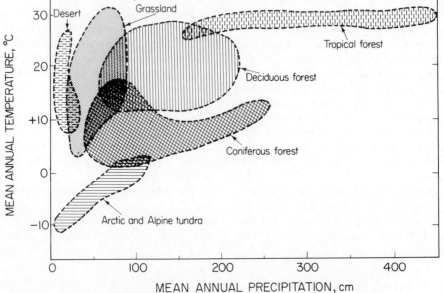

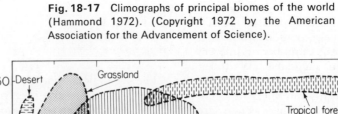

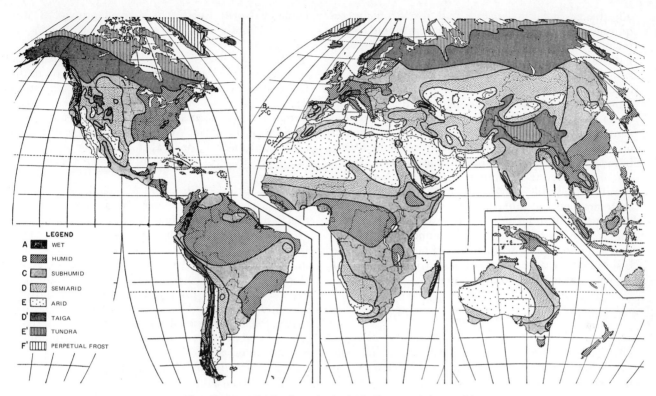

Fig. 18-18 Distribution of principal climates of the world
(Blumenstock and Thornthwaite 1941).

Fig. 18-19 Distribution of the principal soil groups of the
world (Simonson 1957).

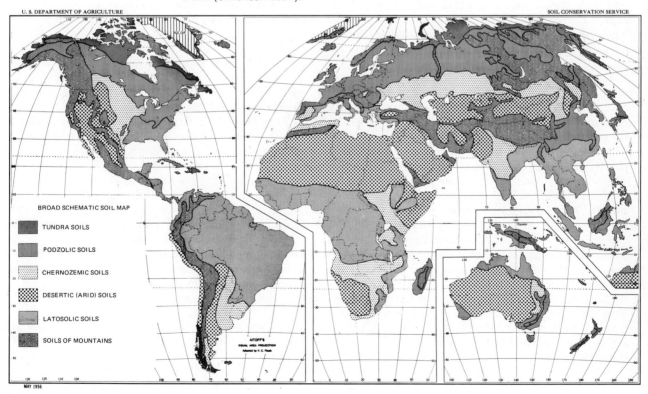

adjustments, inherited mores, and specific niche requirements. On the other hand, zoogeographic factors include the considerations of place of origin; dispersal pathways, rates, and barriers; and evolutional acquisition of new structural, functional, or behavioral adjustments that permit invasion of new areas, surmounting of old barriers, or incorporation into new communities. Actually a third set of factors is also concerned in the present-day distribution of animals and plants. These are the disruptive or modifying influences of man in the destruction of natural ecosystems, deliberate cultivation of modified ones for his own benefit, in the transportation and mixing of fauna and flora between different parts of the world, and in other ways. In order properly to evaluate the changes that man has wrought in his environment, it is important to understand what the environment was like before man became dominant, and it is with this that we will deal in the following pages.

SUMMARY

Dispersal of animals into new areas may be a range expansion of a species if individuals find unoccupied niches, are able to overcome competition, or are able to acclimatize and adapt to the conditions. Dispersal is chiefly by the immature stages that are broadcast randomly in all directions, conveyed passively by wind, current, or other animals, respond to environmental factors by directed movements or taxes, or find their way by trial and error. Dispersal continues at measurable rates until a physiographic, climatic, or biotic barrier is reached. Factors that induce dispersal include population pressure, failure of food supplies or loss of favorable homesites, opening up of new areas elsewhere, and pre-adaptation for new and different niches.

The origin of taxonomic groups of organisms may be traced to various centers, usually distinguished by the occurrence of the greatest amount of differentiation within the group and the convergence of lines of dispersal. According to Matthew, the major vertebrate groups first evolved in temperate Asia; according to Darlington, that event took place in the Oriental tropics. The present distribution of primitive forms over the face of the world has been influenced by continental drift, but the distribution of modern forms is largely accounted for by dispersal over land bridges that have periodically come and gone during more recent geological time.

The fauna of continental islands is derived mainly from the adjacent mainland over such bridges in the past, but the faunas of the more distant oceanic islands are often unique and unbalanced, and dependent on the accidental dispersal of miscellaneous species. The fauna of North America is derived principally from Eurasia, South America, and by autochthonous development.

Migration, like dispersal, involves movements and invasion of new areas, but differs from dispersal in that the movements are periodic back-and-forth movements between two areas. Migrations may be annual or daily, or may be in the form of changes of habitat at different stages in a life cycle. Annual migrations may be latitudinal, altitudinal, or local. Annual migration is best developed in birds but also occurs in mammals, fish, and insects.

Ecesis is the establishment of organisms in an area into which animals have come by dispersal or migration. This involves the establishment of individuals, the growth of populations, the invasion of more and more species, and finally the development of mature communities. The growth of species populations and of complex communities commonly follows a sigmoid or logistic curve. The disappearance of populations and communities follows an inverse or negative growth curve. Succession may be represented by a series of positive and negative growth curves as one community replaces another.

There are five zoological regions into which the world is divided. Free dispersal between regions is prevented by the major barriers of oceans, mountain ranges, and deserts; each region is thus characterized by distinct orders and families of organisms. Within each region secondary barriers separate divisions of lesser rank characterized by genera, species, and subspecies. Of the Nearctic region of North America, such differentiation has prompted the recognition of faunal areas, life zones, and biotic provinces.

In contrast to the analysis of fauna in terms of geographic units is analysis in terms of the elements that it possesses. A faunal element is a group of species coming from the same center of origin and having similar geological histories.

Zoogeography is the study of animal distribution in terms of centers of origin, dispersal routes, and evolution. The ecological system emphasizes environmental relations and community dynamics.

The principal geographic unit in the ecological system is the biome. A biome is a major biotic community characterized by distinctiveness in the life-forms of the important climax species. Seral communities are developmental stages of the biome. The biome is divided into plant associations, distinguished by uniformity in the species composition of the climax plant dominants, and into biociations, identified by uniformity and distinctiveness in the species composition of the climax community, particularly of the animal predominants. Some nine terrestrial biomes and four marine biomes are recognized.

Chapter 19

PALEOECOLOGY
AND THE EMERGENCE OF MAN

With the biome system as the point of departure for analyzing distributions, we are in a position to study the geological history of these community units: how they were first formed, when they first became well defined, how they dispersed over the world, and why they came to occupy their present locations. When we know the origin and geological history of vegetational communities on land, we will be better able to understand the origin, differentiation, and present-day distribution of the species of these communities, including man (Epling 1944, Rhodes 1962, Ross 1962). A review of geological succession will be helpful (Table 19-1), although we will be mostly concerned with tracing the history of the biomes during the Tertiary and Quaternary eras. Primary attention will be given to North America.

PHYSIOGRAPHIC CHANGES

At the beginning of the Tertiary, some 60–70 million years ago, the interior of the North American continent was still widely inundated by the epicontinental seas of the Cretaceous period. As these seas gradually receded, the continent acquired its modern topographic appearance. The Mississippi embayment area is an extension of the coastal plain that continues around the Gulf of Mexico and northward along the Atlantic coast. This coastal plain emerged progressively throughout the Tertiary era. Its general character is much the same now as it has always been —tidal salt marshes and estuaries intermingled with shallow lagoons bounded by off-shore bars.

The Appalachian Mountain system first appeared near the end of the Paleozoic era and had become eroded to a peneplain by the beginning of the Tertiary. A new uplift then occurred, and erosion again followed. In the Miocene, only the Schooley peneplain, a nearly level surface only slightly above sea level, remained. However, monadnocks, hills of resistant rock rising some hundreds of meters, were left projecting out of the Schooley peneplain. Mount Monadnock, White Mountains, Great Smokies, Cumberland Mountains, among others, are such formations. The Schooley peneplain subsequently underwent a series of archings and uplifts until it reached some 1200 m (4000 ft) above sea level along the central axis to give the region its present-day character. In New England, Pleistocene glaciation covered these mountains, rounded them off, scraped away the old soil, and left a poorer soil full of boulders.

The Ozark and Ouachita Mountains were also formed at the close of the Palaeozoic, underwent subsequent erosion, and experienced minor uplifts.

Table 19-1 Major steps in the geosere of the earth (geological time table), especially as it applies to North America (modified from Dunbar 1949). Pre-Cambrian time began with the origin of the earth as a planet, 3–5 billion years ago; later came the formation of oceanic depressions and continental platforms, water, a thin atmosphere, and the beginning of life. The first evidence of plant life consisted of bacterial and algal-like marine forms and of animal life—sponges and segmented marine worms. The sequence of abundant fossils permits dividing the last half-billion or more years into eras, periods, and epochs.

Paleozoic era: 600+ million years ago

Cambrian period: marine invertebrates only.

Ordovician period: marine invetebrates continue predominance; rise of armored fishes.

Silurian period: first invasion of land by plants; rise of air-breathing scorpions and millipedes and of fresh-water fishes.

Devonian period: first forests and extensive land floras; diversification of fresh-water fishes, rise of labyrinthodont amphibians, and increase of land fauna, especially spiders, mites, and wingless insects.

Mississippian period: increase of amphibians.

Pennsylvanian period: luxuriant swamp floras, mostly of spore-bearing types; fresh-water clams and amphibians abundant and on land, giant insects, spiders, centipedes, snails, and first reptiles.

Permian period: (elevation of Appalachian and Ouachita Mountains); decline of ancient flora and rise of conifers and modern fern families; modern insects and advanced types of amphibians and reptiles appear.

Mesozic era: 220 million years ago

Triassic period: plants, mostly ferns, cycads, conifers; stegocephalian amphibians and dinosaurs numerous; archaic mammals appear.

Jurassic period: "age of cycads"; reptiles evolve higher and more diversified forms, first toothed birds and frogs appear.

Lower Cretaceous period: woody angiosperms spreading over world.

Upper Cretaceous period: (great inland seas, warm climate worldwide); modern genera of deciduous hardwood trees predominant, sedges and grasses appearing; clams and snails common in fresh water; culmination of dinosaurs, toothed birds, archaic mammals (elevation of Rocky Mountains).

Cenozoic era: 65 million years ago

Tertiary period

Paleocene and Eocene epochs: [inland seas recede, Appalachian region peneplained (Schooley) but later again uplifted, climate warm and humid]; hardwood forests predominant, palms abundant; modern mammals and birds replace archaic forms.

Oligocene epoch: 36 million years ago (continent peneplained); turtles, alligators, crocodiles at maximum.

Miocene epoch: 25 million years ago (western mountains becoming elevated, climate turning drier and colder); grasses disperse over open plains; insects reach full development and mammal fauna expands.

Pliocene epoch: 13 million years ago (continued elevation of western mountains, especially Sierra Nevada; lower Great Basin becomes arid); grasslands become extensive and semi-desert vegetation develops in Southwest; mammals at maximum and man-ape changing into man.

Quaternary period

Pleistocene epoch: 2+ million years ago (continental glaciation); emergence of man, great mammals disappear.

Recent epoch: 20+ thousand years ago (glaciers recede); man becomes dominant, rise of civilization.

Elsewhere in the central interior between the Appalachian Mountains and the Great Plains, peneplanation was the dominant force throughout the Mesozoic and Tertiary. Low coastal marshes extended around the Mississippi embayment, and marshes and swamps were frequent elsewhere. During the Pleistocene, glaciers moved tremendous quantities of soil and rock from Canada southward and from mountain ridges into valleys. The retreat of the ice front proceeded haltingly with alternating retreats and advances. When it was stationary but melting, the glacier formed concentric terminal moraines; when in active retreat, the glacier left a more uniform layer of till in its wake.

The Laramide orogeny, which occurred at the end of the Cretaceous period and continued into the Eocene, formed a series of mountain ranges in the Rocky Mountain system, including the Big Horns, Wind River, Black Hills, Uintas, and the series of more or less parallel ridges in the Great Basin. Rapid erosion filled the deep basins between the mountains with debris. By the Oligocene, peneplanation was nearly complete, although the surface was several hundred meters above sea level and monadnocks remained.

Beginning in the Miocene, increasing in intensity through the Pliocene into early Pleistocene, but decreasing since, mountain formation was widespread not only in the Rocky Mountains but also in the Cascades, Sierra Nevada, and Coast Ranges. Volcanic action was extensive in the West, especially in Nevada, Idaho, Oregon, and Washington. Highly fluid basalt welled out of long fissures in the earth's crust, filled valleys, altered drainage systems, accumulated sheets up to 1500 m (5000 ft) thick over 80,000 km² (200,000 mi²), and formed the Columbia River basalt plateau. Much younger vulcanism along the Cascade axis formed Mount Rainier, Mount Hood, and Lassen Peak. During early Tertiary the Sierra Nevada was only a low range which was peneplaned by the Miocene, but then the Sierra was uplifted by faulting and tilting so that the eastern edge of the block was 4000 m (13,000 ft) above sea level.

During early Tertiary, the area of the present Coast Ranges was in part an archipelago, separated from the coast by a deep sea that is now the interior of California. Folding and faulting in the Coast Ranges began in the Miocene and were most active in the Pleistocene. Such activity is still going on, as evidenced by the recent earthquakes.

CLIMATIC CHANGES

The climate of the early Tertiary can be deduced from types of plant and animal fossils found in the proper rock strata. After peneplanation of the western mountains there was little to obstruct the warm, moisture-laden, westerly Pacific winds sweeping across the continent. Rains were heavy and fell frequently throughout the year. The Mississippi embayment helped to maintain a uniform oceanic climate. Sub-tropical conditions extended as far north as the Dakotas, and temperate conditions obtained far to the north. With the elevation of the western mountains in the Miocene and Pliocene, especially the Sierra Nevada and Cascades, the westerly winds were forced to high elevations, cooled, and lost much of their moisture as precipitation on the windward western slopes during the winter; a dry season began to develop during the summer. On the lee eastern mountain slopes, arid conditions developed because the winds, warmed as they descended the mountains, retained what moisture remained in them, thus producing a *rain-shadow*. Dry plains and desert then developed in the Great Basin. More moisture precipitated as the winds crossed the Rocky Mountains. Mixing of

the westerly winds with winds from the North and South, however, produced more moisture east of the Rockies than in the Great Basin, and east of the Great Plains the western mountain rain-shadows had little effect.

Concurrent with increasing aridity over the continent was cooling of the air. This began in middle or late Oligocene and brought a gradual southward shift of climatic belts, which culminated in the very severe glaciation of the Pleistocene. The actual cause of the glaciation is obscure, but there is no doubt that the glaciation was accompanied by a drop in average temperature and an increase in annual precipitation.

EARLY TERTIARY GEOFLORAS

The geological record of Tertiary plants is good, particularly in the western United States. Volcanic ash, lake deposits, coastal swamps, and river basins preserved fossils more or less *in situ*. Fossils and many clear identifiable impressions of different kinds of leaves left in the rock indicate that there were three principal floras, geofloras, or groups of plants in North and Central America that maintained identity over wide ranges of space and time (Figs. 19-1 and 19-2; Chaney 1947). It was from these floras and the faunas that they contained that the modern vegetation types, biomes, plant associations, and biociations differentiated. Doubtless there was some latitudinal and altitudinal zonation in the early floras, but the development of present-day community units is the result of later, rigorous climatic zonation, and a more extensive physiographic diversification over the con-

Fig. 19-1 Distribution of geofloras in the Early Tertiary. Diagonal lines show areas where semiarid geofloras were commencing to appear (Axelrod 1960; from Sol Tax, *Evolution after Darwin*, University of Chicago Press, copyright 1960 by the University of Chicago).

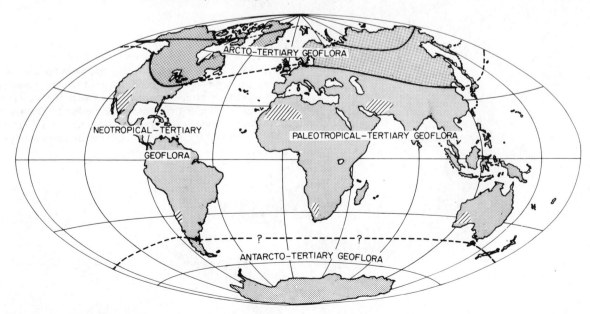

tinent than prevailed in the early Tertiary. Instead of evolving new tolerance limits, species, with some exceptions, dispersed into those regions that continued favorable to them and became extinct in regions that became unfavorable. Species with similar ranges of tolerance thus came into association and, as interdependent coactions became established, into closely knit communities.

Neotropical-Tertiary Geoflora

This flora is known from several Paleocene, Eocene, and Oligocene deposits (Berry 1937). It was composed of tropical and subtropical plants now limited largely to southern Florida, Mexico, and the tropics. Trees characteristically had broad, thick, evergreen leaves. The laurel family, Lauraceae, was particularly well developed in North America, and some modern descendants are found in the temperate flora. The Neotropical-Tertiary Geoflora extended from the

tropics as far North as at least 49° latitude in the West and 37° latitude in the East. With the drying and chilling of the continent that began in late Oligocene and Miocene, this flora retreated southward and eastward to the localities in which it is found today.

Its counterpart, the *Paleotropical-Tertiary Geoflora* of the Old World, contains many of the same families and genera. This suggests an interchange of taxa across the tropical oceans in the past, probably in the Mesozoic era (Axelrod 1970).

Arcto-Tertiary Geoflora

The Arcto-Tertiary Geoflora completely encircled north temperate regions with little variation in composition or character. Plants migrated freely across the Bering land bridge between North America, Asia, and Europe. There was a land bridge between North American and Europe across the North Atlantic during the Mesozoic, but it is uncertain whether this bridge

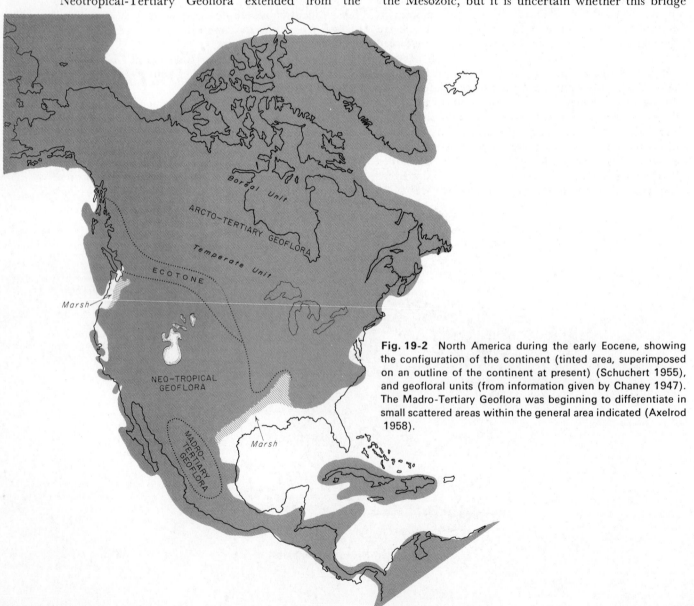

Fig. 19-2 North America during the early Eocene, showing the configuration of the continent (tinted area, superimposed on an outline of the continent at present) (Schuchert 1955), and geofloral units (from information given by Chaney 1947). The Madro-Tertiary Geoflora was beginning to differentiate in small scattered areas within the general area indicated (Axelrod 1958).

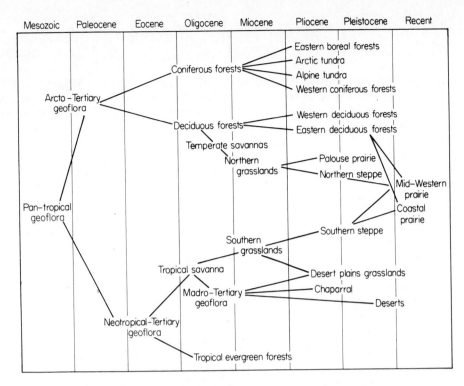

Fig. 19-3 Approximate time relationships among the evolving vegetational units (exclusive of deserts) in North America (from Dix 1964).

persisted appreciably into Tertiary time (Lindroth 1957, Laughton 1971). On the Arctic islands and in northern Siberia, the flora reached North to within 8° latitude of the North Pole. How these species tolerated the long seasons of darkness is a problem; daily photoperiods, as well as temperature, could have been on a different pattern then than now if the North Pole had occupied a different location (Schwarzbach 1963). An ecotone with the Neotropical-Tertiary Geoflora began at 57° latitude on the Pacific coast and 51° in southern British Columbia, Alberta, and Saskatchewan. In Asia, an ecotone began at 42° latitude in Manchuria; in central Europe, at 48°–50°.

Much of the present similarity between eastern North American and Eurasian floras and faunas may be traced to the continuous and extensive distribution of the Arcto-Tertiary Geoflora during the early Tertiary (Wang 1961). Types that occur prominently in eastern North America and eastern Asia, for instance, are, among the plants, tuliptrees, magnolias, sweetgums, sassafrases, witch-hazels, and partridge-berries; among the animals, paddlefish, alligators, fresh-water turtles, lizards (*Leiolopisma*, *Eumeces*), snakes (*Natrix, Opheodrys, Elaphe, Agkistrodon*), and hellbenders (Schmidt 1946); as well as various birds and mammals.

The chief differentiation of the Arcto-Tertiary Geoflora was latitudinal into *boreal* and *temperate units*. The boreal unit contained such trees and shrubs as the *Metasequoia*, baldcypress, pines, spruces, willows, birches, and hazels. The temperate unit included maples, alders, birches, hornbeams, chestnuts, dogwoods, hawthorns, beeches, ashes, walnuts, pines, sycamores, poplars, oaks, willows, baldcypress, bass-

woods, elms, and *Metasequoia*. Although there was some mixture, deciduous species predominated in the temperate unit; the boreal unit contained more coniferous species. These same latitudinal relations obtain at the present time (Fig. 19-3). The climate of the temperate unit was probably humid, with summer rainfall and moderate temperatures (Chaney 1948).

With the elevation of the western mountains in the Miocene and Pliocene and the drying and cooling of the climate, the Arcto-Tertiary Geoflora retreated from the far north and from the interior, and the American portion lost its contact with Asia and Europe. In North America, the main movement of the temperate unit was to the south and east. Beech, basswood, elm, and hornbeam disappeared from the west, probably because of the lack of summer rain there, but were prominent in the movement to the southeast. *Metasequoia, Cercidiphyllum, Ailanthus,* and *Ginkgo* became extinct in North America but are still to be found in central China. The boreal unit followed closely behind the temperate unit; and in the higher elevations of the mountains, it extended as far southward as did the temperate. There were changes in the taxonomic composition of the Tertiary forest during this long period. Some genera were lost entirely, some new ones were added, others were changed by evolution; but all these changes were conservative. The present-day mixed mesophytic forests of the Cumberland Mountains of eastern Tennessee most nearly resemble the Arcto-Tertiary forests of the past. The original Geoflora, however, included some broad-leaved evergreen tree species that are largely eliminated from the present mixed mesophytic forest of North America, although only partially so from the Asian forest (Wang 1961).

The *Antarctic-Tertiary Geoflora*, derived from the Cretaceous flora in the southern hemisphere, evolved independently of the Arcto-Tertiary Geoflora. It was made up of southern conifers and evergreen dicots like those now in southern Chile, New Zealand, and southeastern Australia-Tasmania.

Madro-Tertiary Geoflora

The Madro-Tertiary Geoflora in North America was not as well defined as were the other geofloras. Its center of origin was on the Mexican plateau, perhaps beginning in the late Cretaceous-Eocene, in scattered dry sites on the lee sides or rain-shadows of high ridges and mountains, but its history previous to the Miocene is obscure. The flora contained a variety of small trees, shrubs, and probably some grasses, although the fossil record of grasses is poor. These species seem to have been derived principally from the Neotropical-Tertiary Geoflora in response to increasingly arid environments (Axelrod 1958).

Minor elements of this flora extended into the Great Basin area, but its main movement northward occurred in the Miocene and Pliocene. During the latter epochs, the flora came to occupy large areas in southern California, the Great Basin, and the western Great Plains, areas which were being vacated by the other two Tertiary Geofloras because of the increasing aridity. Derived in large part from this flora are the present-day communities of Mexican pine forest, woodland, chaparral, sagebrush, subtropical scrub (thorn forest), desert, and arid grassland. These communities are relatively young, as distinct entities; the North

Fig. 19-4 Glacial and interglacial stages of the Quaternary. The synchrony of the stages in North America and Europe is suggestive only. (Temperature fluctuations and time scale according to Ericson and Wallin 1968; copyright 1968 by the American Association for the Advancement of Science.)

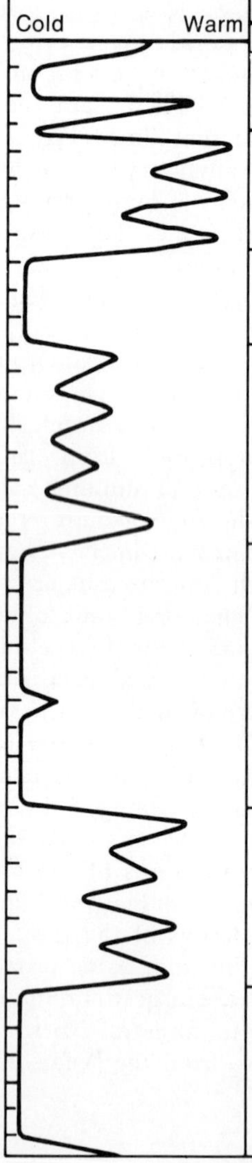

Cold Warm	North America	Years Ago	Alps	Northern Germany, Holland
	Recent	0	Holocene	Postglacial
	Wisconsin glacial	100,000	Würm	Weichsel
	Sangamon interglacial	200,000	Riss-Würm	Eem
		300,000		
	Illinoian glacial	400,000	Riss	Saal
		500,000		
	Yarmouth interglacial	600,000	Mindel-Riss	Holstein
		700,000		
		800,000		
		900,000		
	Kansan glacial	1,000,000	Mindel	Elster
		1,100,000		
		1,200,000		
		1,300,000		
	Aftonian interglacial	1,400,000	Günz-Mindel	Cromer
		1,500,000		
		1,600,000		
		1,700,000		
	Nebraskan glacial	1,800,000	Günz	Weybourne
		1,900,000		
		2,000,000		

American desert biome, for instance, is probably not older than Pleistocene (Axelrod 1950). Semiarid Tertiary geofloras also appeared at low to middle latitudes in the western parts of other continents, later to develop into desert-type vegetation.

PLEISTOCENE

Physical Conditions

The Pleistocene was marked by a series of great climatic fluctuations throughout the world (Flint 1971). Thirty-two per cent of the land area of the world became buried under glacial ice; 10 per cent still remains ice-covered. In places, this ice reached a thickness of at least 1500 m (5000 ft), roughly the thickness of the ice sheets now on Greenland and Antarctica. Outwash from the glaciers carried sediment hundreds of kilometers beyond the farthest ice boundaries. Sea level fell 100 m (328 ft) or more below the present level, as water became bound in glacial ice. This resulted in presently submerged coastal plains becoming exposed all around the world.

This glaciation must have been initiated by an increase in precipitation. The Arctic Ocean, according to one explanation, was open water, and winds picked up moisture and deposited great masses of snow over the northern part of the continent which when compressed changed to ice and flowed southward into the lower latitudes. Glaciation terminated only after the Arctic Ocean again froze over and snowfall decreased (Donn and Ewing 1966). Poleward movement of land masses as the result of sea-floor spreading apparently occurred late in the Tertiary period. This reorientation of land and sea distribution induced changes in air and sea currents, provided colder climates, and prime conditions for glaciers to form (Crowell and Frakes 1970).

There were a number of successive advances and retreats of glacial ice in North America, in northern Europe, in the Alps, and elsewhere (Fig. 19-4). Glaciation was less extensive in Asia probably because of its continental climate and lower precipitation. The terminology of glacial and interglacial ages varies with the region and the authority, and there is further confusion since the different glacial advances in different areas may not have been entirely synchronous. Dating of geological periods can only be approximated, and different methods give widely varying results (Broecker 1965). The Pleistocene period probably began at least 2 million years ago (Emiliani 1966b, Ericson and Wollin 1968).

During the warm interglacial ages the biota reoccupied the newly uncovered areas as the glacier melted back, only to be driven out as the glacier again advanced. At the present time we are in an interglacial period. According to one prediction, a new glacial period will begin in a few thousand years and reach its peak about 15,000 years from now (Emiliani 1966a). This prediction does not include consideration of accelerated cooling of the atmosphere due to emission of dust and vapor from industry, discussed earlier.

We are chiefly concerned with the last glaciation, which began perhaps 100,000 years ago. In North America, the Wisconsin glaciation is divided into several substages, representing secondary advances of the glaciers, preceded and followed by warm periods. Generally, the glacial retreat during a warm period exceeded its temporary advance during the next cold period. The last of the glaciers still persists in northern Canada and Greenland.

North America was connected by a land bridge to Siberia during much of the Wisconsin period as a result of the drop in sea level, exposing the Bering-Chukchi continental shelf. The presence of the bridge permitted invasion of Asiatic species of animals, including man, into North America. It is probable that during one or more of the warm substages within the Wisconsin period, there occurred a narrow ice-free corridor, produced by the melting back and separation of the Laurentide ice sheet and the Cordilleran glacial system, that permitted these invading species to proceed on to more southern latitudes (Hopkins 1967). Of the rest of the world, the British Isles were connected at this time to continental Europe (Antevs 1929); the Indo-Malayan Archipelago extended as dry land to include Sumatra, Borneo, and Java; and New Guinea and Tasmania were connected to Australia (Mayr 1944).

It is estimated that, in northern Ohio, the ice advanced during certain stages of the Wisconsin glaciation at the rate of 100 m (350 ft) per year, and in southern Ohio at from 12 to 33 m (38–108 ft) per year (Goldthwait 1959). There is evidence that the advancing ice lowered the temperature sufficiently ahead of it—for a distance of 800 m (0.5 mile)—to decrease the annual growth of spruce and other coniferous trees but not to kill them until the glacier actually overrode and destroyed the forest (Fig. 19-5; Burns 1958).

The drop in mean annual temperature over temperate North America is estimated to have been 5° to 10°C, but was doubtlessly greater at the edge of the glacier (Dillon 1956). Superficial oceanic water layers in the tropics dropped approximately 6°C (Emiliani 1955). The gradient from low to higher temperatures at the glacial front was probably steep. Storm tracks in North America during maximum glaciation extended from the west and southwest to the east and northeast; thus, warm winds were brought against the front of the glacier. There is controversy as to how far south of the glacier the high-pressure anticyclonic conditions developed by the great ice mass were felt (Hobbs 1926).

Precipitation appears to have been comparatively heavy during the glacial stages over much of the world. Even in areas where continental glaciation did

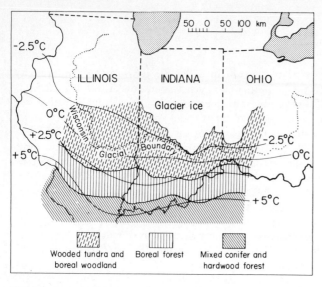

Fig. 19-5 Mean annual isotherms and vegetational zones at the time of a major ice advance about 1000 years after the Wisconsin glacial maximum (Wayne 1967).

Fig. 19-6 Conjectural map of vegetation at the time of Wisconsin glaciation. An ice-free corridor east of the Canadian Rockies may have existed during warm substages. The four coniferous forest refugia are numbered: 1, Appalachian; 2, Rocky Mountain; 3, Pacific; and 4, Alaskan (map compiled from information given by Meinzer 1922, Hansen 1947, Braun 1950, Hobbs 1950, Flint 1952, 1957, Martin and Mehringer 1965, Whitehead 1965, Wells 1966, 1970, Wright 1970).

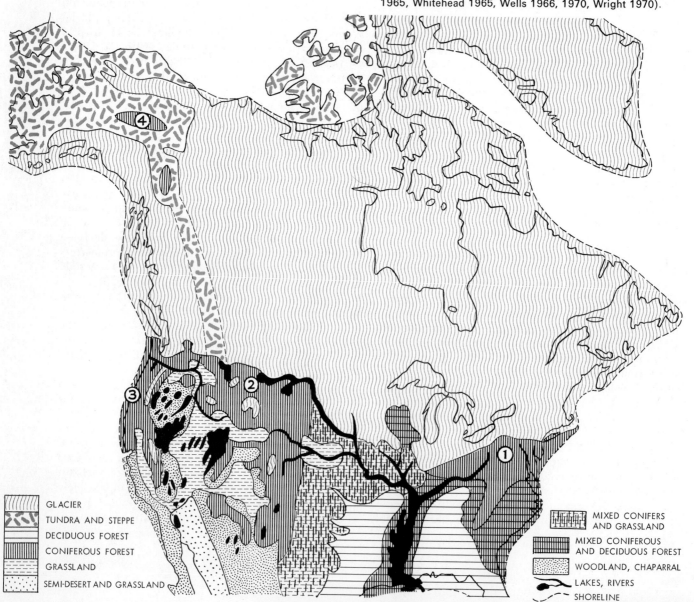

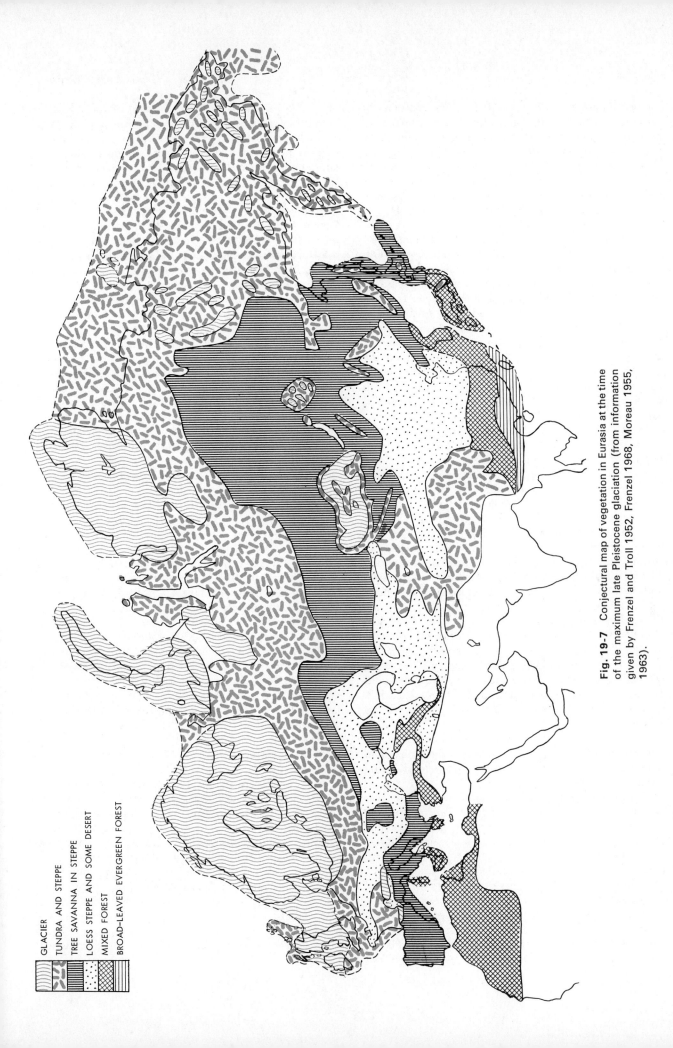

Fig. 19-7 Conjectural map of vegetation in Eurasia at the time of the maximum late Pleistocene glaciation (from information given by Frenzel and Troll 1952, Frenzel 1968, Moreau 1955, 1963).

GLACIER

TUNDRA AND STEPPE

TREE SAVANNA IN STEPPE

LOESS STEPPE AND SOME DESERT

MIXED FOREST

BROAD-LEAVED EVERGREEN FOREST

not occur (tropical America, Africa, and Australia, for instance), variations between pluvial and interpluvial periods probably coincided with the glacial and interglacial periods and produced far-ranging effects on the geographic distribution of organisms (Moreau 1963). Lakes Bonneville and Lahontan, as well as many smaller ones, were formed in the Great Basin of North America during these pluvial periods (Meinzer 1922, Hubbs and Miller 1948).

There was considerable alteration of drainage patterns over the northern part of the continent. Old river valleys were filled or dammed by ice or moraines. New outlets were formed. Rivers previously separated became connected. The retreat of the glacier left vast level areas without drainage so that many lakes, swamps, and bogs remain in northern glaciated areas. In other places the large amounts of melt water cut new channels or widened old valleys, through which the surplus water was transported to the sea. Silt, sand, and gravel were spread out in outwash plains, from which winds picked up the finer material and deposited it elsewhere as loess in layers up to 2.5 or 3 m (100 in.) thick over hundreds of square kilometers. The treatise of Thienemann (1950) is an extensive account of what happened to the fresh-water fauna.

Virtually all the fossil plants and mollusks during the Pleistocene are represented by living relatives (Baker 1920). Changes of ecological significance are, for the most part, in point of geographic distribution rather than organic evolution. But insects, especially beetles, which are well represented in the fossil record, evolved rapidly into new forms. Many large herbivorous mammals, their ecologically dependent carnivores, and their scavengers became extinct all over the world. Large mammals present during early stages of the Pleistocene, but no longer occurring in North America, include camels, horses (one species later re-introduced), ground sloths, two genera of muskoxen, peccaries, a giant bison, a giant beaver-like animal, a stag-moose, several kinds of cats, mammoths, and the mastodon. It is an interesting hypothesis that this dramatic extinction was not due to an unfavorable climate or an epizootic disease, but to the appearance of hunting man. The timing of major extinctions in North America and elsewhere appears to correspond to the times when man entered areas previously free from his hunting (Martin and Wright 1967).

By the beginning of the Pleistocene, there was doubtless a broad zone of coniferous forest across the northern part of the continent and perhaps some tundra. Deciduous forest covered the eastern states; grassland occurred in the central part of the country; and coniferous forest was extensive in the western mountains and on the Pacific slope, much as at present.

Glaciation displaced the coniferous forest over vast areas in the north, but there is considerable controversy as to the area south of the glacial boundary in which the deciduous forest and grassland were affected

(Figs. 19-6 and 19-7). Interpretation of the probable climate, of past and present distribution of plants and animals, and of the pollen record in bogs indicates that, in North America, the deciduous forest and grassland may not have been entirely displaced, but that they became extensively mixed with coniferous species to varying distances south of the margin of the glacier. The ranges of many animal species certainly extended farther south during the height of glaciation than they do at the present time. However, the occurrence of snails in Pleistocene deposits in Kansas indicates that during Wisconsin time, deciduous woodland extended along streams much as it does at the present time (Fry and Leonard 1952). With the greater precipitation that was generally prevalent, much of what is now desert in the Great Basin and the Southwest was probably grassland then (Epling 1944, Hansen 1947, Braun 1950, Thomas 1951, Frey 1955, Graham and Heimsch 1960, Martin and Mehringer 1965, Whitehead 1965, Wells 1966, 1970, Wright 1970).

Enormous amounts of cold water, melted from the glacier during the summer months and, perhaps, carrying chunks of ice, drained down the Delaware and Susquehanna Rivers in the East, the Ohio, Wabash, Illinois, Missouri, Platte, and Mississippi Rivers in the central part of the continent (Hobbs 1950), and the Snake and Columbia Rivers in the Northwest. Water filled these wide river valleys from the present bluffs on one side to the bluffs on the other, and doubtless extended the boreal microclimate for many kilometers (Wolfe 1951), perhaps permitting the establishment of coniferous trees and other northern species on their banks. Cold glacial waters draining southward between the coast and the Gulf Stream helped to create a microclimate permitting northern species, including conifers, to invade as far south as Florida.

A tree line existed at 1200–1500 m (4000–5000 ft) on the higher Appalachian peaks. Coniferous forests that are now limited to the higher elevations of the Appalachian Mountains descended the mountain slopes perhaps as much as 600 m (2000 ft) and covered large areas. In the mountains all over the world, the snow-line (Klute 1928) and biotic zones (Murray 1957) were at least 500 m (1600 ft) and in some humid localities possibly as much as 1500 m (5000 ft) lower than they presently are.

Evidence is scanty for the existence of tundra in North America south along the ice front during its advances, although tundra and steppe occurred in Alaska during the last Wisconsin stage. Tundra mammals such as the muskox and woolly mammoth are well represented as fossils along the old glacial margins, and a few fossils of these species have been found as far south as Texas, Mississippi, and Florida (Potzger 1951). Tundra, however, developed extensively along the front of the glacier as it retreated northward.

In North America the coniferous forest survived

glaciation in four separated refugia (Adams 1905, Halliday and Brown 1943). The centers of these refugia were in the middle Appalachians, the Rockies of the United States, the Pacific slope of the Cascades, and Alaska (Hulton 1937). The *Appalachian refugium* during the Wisconsin epoch extended westward south of the Great Lakes. How well it was separated from the Rocky Mountain refugium is uncertain. Coniferous forest occurred widely across the northern and western portions of what is now extensive interior grassland, but whether during the Pleistocene it occurred as a continuous forest or only as scattered groves around bodies of water and along streams is controversial. The *Rocky Mountain refugium* was separated from the Pacific refugium by the Great Basin, the high peaks of the Cascades and Sierra, and by seasonal differences in precipitation. Probably this separation was only partially effective as a belt of coniferous forest extended around the north border of the Great Basin. Coniferous forest in the *Pacific refugium* extended some hundreds of kilometers farther south than it does at the present time. The *Alaskan refugium* was small and surrounded by tundra and steppe. Several of the islands in the Arctic Archipelago may also have been free of ice. Fossil remains found in frozen muck and silt indicate the occurrence in Alaska during interglacial, and perhaps glacial periods, of woolly mammoth, muskox, reindeer, and many other forms (Flint 1952).

Our description of how biotic communities were affected by the Pleistocene glaciation is tentative. According to Deevey (1949) and Dorf (1960), the climate everywhere south of the glacial front was considerably cooled, deciduous forest was driven into refugia in Florida and Mexico, and tundra and coniferous forest prevailed everywhere in between. Griscom (1950) believes that continental refrigeration extended well into Mexico, causing extensive southward dispersal of northern birds and the elimination of the South American element previously present. Speciation supposedly occurred in populations of certain birds (Huntington 1952) and amphibians, reptiles, and mammals (Blair 1958) that became fragmented and isolated from each other in the southeastern and southwestern refugia. There is more likelihood, however, that these animals were segregated into different populations, not by closed coniferous forests, but by grassland or a tenuous savanna type of habitat in Texas (Martin and Harrell 1957) and by the broad cold waters of the Mississippi River when melting of the glaciers was at its height.

In Europe and Asia, westerly winds were diverted south into the Mediterranean region, and high anticyclonic barometric pressures developing over the glacier brought dry, cold, northeasterly and easterly winds. Loess was deposited in a broad belt from western France east into the Balkans and northeast into Siberia and China. Unlike North America, true tundra graded into loess steppe over extensive areas, and coniferous and deciduous forests became mixed together. The Mediterranean climate was cooler and moister than it is at the present time (Zeuner 1945).

POST-PLEISTOCENE

Retreat of the Glacier

The melting of the ice was quite rapid in North America (Fig. 19-8), perhaps 134 m (440 ft) per year in the Great Lakes region (Flint 1971). Melt water filled depressions to form vast pro-glacial lakes. Sea-level rose over the submerged Atlantic coastal plains rapidly at first and then more slowly (Milliman and Emory 1968). The Great Lakes, in their early stages, had outlets down the Hudson and Mississippi Rivers and had different interconnections from those at present (Fig. 19-9). Still later, Lakes Agassiz, Ojibway-Barlow, and others were formed in the north (Fig. 19-10). A knowledge of these lakes and the history of past drainage systems is prerequisite to interpretations of present-day distributions of aquatic organisms.

With the melting back of the ice, local glaciers were left in the Catskill Mountains of New York, on Mount Katahdin in Maine, in the Shickshock Mountains on the Gaspé Peninsula, in Newfoundland, and in Labrador. The glacier receded faster in the western interior of Canada between the Rocky Mountains and Hudson Bay than it did to the East. There is evidence that the glacier disappeared from the Hudson Bay region while still persisting over the highlands of Quebec and the Labrador Mountains. The last remnants of the ice sheet still remains on the mountains and plateaus of Baffin, Devon, Ellesmere, and Axel Heiberg Islands, and in Greenland (Flint 1947).

The Clisere

As the glacier retreated, vast areas were freed for reinvasion by plants and animals (Adams 1905, Gleason 1922). The land must have been a barren, sterile expanse of raw parent soil material, deficient in nitrogen. The first plants to invade were probably vascular species the root nodules of which bore bacteria, fungi, or actinomycetes capable of fixing nitrogen from the air, thereby enriching the soil (Lawrence 1958).

Identification of kinds of pollen and comparative counts of pollen grains from various depths in relict peat bogs gives us a picture of the predominant vegetation in the surrounding upland at various times in the past (Sears 1942, Deevey 1949). To make such studies, a core of peat is obtained from the deepest part of the bog by means of a special hand auger (Fig. 19-11). The lowest portion of the core is the oldest; the most recently formed is at the top. Samples of the core at

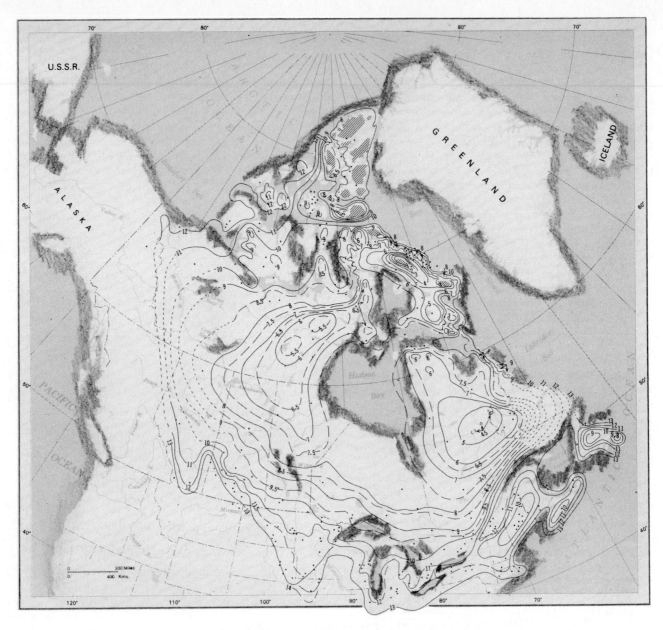

Fig. 19-8 Radiocarbon isochrones of the retreat of the Laurentide ice. Isochrone location based on ^{14}C dates, coastline location, moraine orientation, and other field evidence. Numbers on lines are thousands of years before present (Bryson *et al.* 1970).

Fig. 19-9 Great Lakes at the time of the Port Huron substage glaciation (Hough 1963).

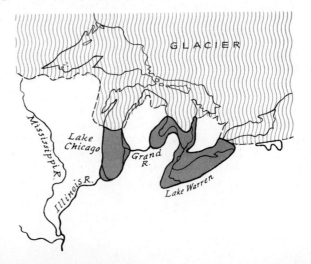

various depths are suitably prepared, examined under a microscope, and the pollen grains identified and counted. The predominant kinds of pollen at any level of the core represent the prevailing species of plants in the vicinity at the corresponding period of time, although there is no exact relation between kinds of pollen and abundance of the various species (Davis and Goodlett 1960). Additional evidence of past conditions can be obtained from lake sediments, stratification of soil, and macrofossils in the form of seeds, diatoms, mollusks, bones, and so on.

The climate is conjectured from the relation of similar flora to climate at the present time. The complete sequence of stages, or clisere, occurs only in regions near the southern limit of the reach of the glacier. The later stages have not developed in more northern localities where the glacier has been gone for a shorter time and where the climate has not warmed up sufficiently.

A chronology for post-Pleistocene time is given for North America and Europe in Table 19-2. The time scale is determined, in part, by measuring the radioactivity of carbon, C^{14}, obtained from samples taken at various depths in glacial deposits. Radioactive carbon disintegrates in non-living matter at a progressive rate; its half-life is 5568 years. The age of any

Fig. 19-10 Glacial Lakes Agassiz and Ojibway-Barlow and the outlets of each (after Flint 1947).

Fig. 19-11 Types of trees which have lived and died during the past few thousand years in Quebec. Instrument taking borings of lake bottom for pollen samples is operated from a boat. Symbols representing different pollen grains and the percentage which each species constitutes in the total are shown at the left. The type of climate indicated by the prevailing vegetation at the time is shown at the right (Wilson 1952).

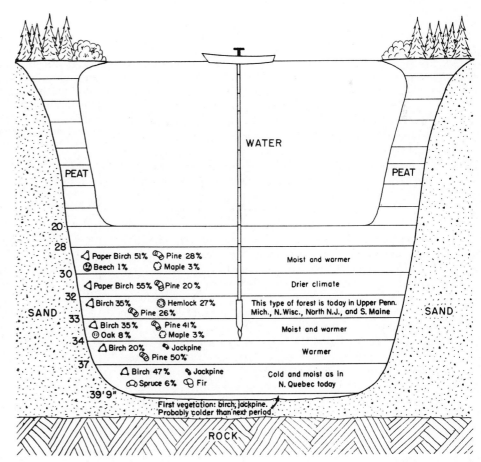

Table 19-2 Late glacial and post-glacial chronology. There is considerable variation in estimated time of beginning and ending of each stage between localities and authorities (Sirkin 1967). The table is based on data from Deevey 1949, Deevey and Flint 1957, Smith 1965, Cushing 1965, Davis 1957, 1959, Geis and Boggess 1968, Flint 1971.

Time	Stages and Climate	Eastern North America	North Germany	Baltic Basin
1,000 A.D. 0	*Sub-Atlantic* Cooler, moister	Oak, chestnut, beech, hemlock	Beech, oak	Mya Baltic Sea Limnaea
1,000 B.C. 2,000 B.C. 3,000 B.C.	*Sub-boreal* Warm, dry (Xerothermic period)	Oak, hickory, elm Prairie peninsula	Oak, beech	Littorina Sea
4,000 B.C. 5,000 B.C.	*Atlantic* Warm, moist (Climatic optimum)	Oak, beech, hemlock	Oak, elm, linden	
6,000 B.C. 7,000 B.C.	*Boreal* Warmer, drier	Pine, oak	Pine, hazel	Ancylus Lake
8,000 B.C.			Birch, pine Park-tundra	Yoldia Sea
9,000 B.C. 10,000 B.C.	*Pre-boreal* Cool moist	Spruce, fir	Pine, birch	Baltic ice lake
11,000 B.C.	*Sub-arctic*	Tundra	Tundra	
12,000 B.C.	Cold	Deglaciation	Park-tundra	
13,000 B.C.			Tundra	
14,000 B.C.				

(Vertical label spanning Sub-boreal through Boreal: Hypsithermal (thermal maximum))

sample, up to 50,000–60,000 years, can be determined on the basis of the extent to which it has degenerated (Libby 1961).

The recession rate of the glacier was ordinarily faster than the advance of vegetation and animal life. The belt of tundra that developed may have been 160 km (100 mi) wide as the ice retreated through New England but was probably much narrower west of the Appalachians. Coniferous forest, consisting of spruce, fir, and later pine, advanced and replaced the tundra along its southern margin but at a rate slower than the tundra expanded northward. Deciduous forest, requiring a better soil, amelioration of climate, and competitive displacement of the already established coniferous forests, moved northward even more slowly. There is evidence that this northward movement of the biota is still in progress, and that the great belts of vegetation are not yet stabilized in respect to each other and to the climate. However, much of this instability is due to fluctuations in the climate itself. The occurrence of pine pollen between that of spruce and fir and deciduous hardwoods in peat cores from eastern North America indicates a swing toward a drier as well as warmer climate, but it was soon followed by a warm, moist climate.

During the warm, moist *climatic optimum*, when conditions for forest growth were most favorable, eastern hemlock spread from the northern Appala-

Fig. 19-12 The largest black area shows the present main range of the eastern four-toed salamander. The smaller spots in the southern states represent boreal relicts of a wide southern distribution during the height of the glacial advance. The isolated group in Nova Scotia may also represent a relict from a more northern dispersal of the species during the post-glacial climatic optimum (Smith 1957).

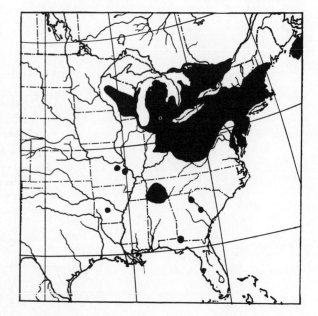

chians and became firmly established in New Jersey, New York, New England, and, to a lesser extent, in Ohio. Beech appeared early in New Jersey and spread through New York and Ohio. Animal species extended their ranges northward and withdrew from

the South. When later forced to withdraw from over-extended ranges to the North, some species left relict populations, which persist to the present time (Fig. 19-12).

Following the climatic optimum came a warm, dry climate, termed the *xerothermic period*. The forest vegetation prevailing from the Midwest into New England consisted of drought-tolerant oaks and hickories. Most of the beech withdrew from Ohio, but remained established in the East, where hemlock declined for want of moisture. A *prairie peninsula* penetrated at least as far as Ohio, and scattered patches of prairie occurred beyond (Fig. 19-13). Grassland animals penetrated far to the East (Schmidt 1938, Smith 1957). Boreal forest retreated northward; sugar maple-basswood forests extended far into Manitoba (Jenkins 1950). In Saskatchewan and Alberta, the northward withdrawal of coniferous forest left groves of aspen trees in the moister and more sheltered locations, while grassland invaded the drier areas (Moss 1944). The numerous lakes in the Great Basin shrank in size or entirely dried up, and desert biota spread both far to the North and high up onto the mountains. Northern species were eliminated from the tops of many southern mountains (Whittaker 1956).

With the coming of cooler, moister climate in the subsequent sub-Atlantic period, the prairie peninsula receded, leaving behind relict populations of biota still present today (Fig. 19-14). Beech once again spread westward, followed by hemlock; hemlock

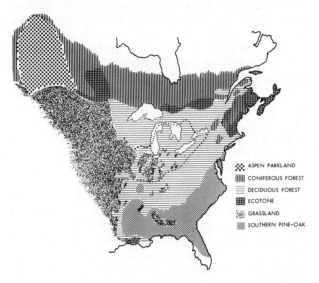

Fig. 19-13 Conjectural map of vegetation during the xerothermic period in eastern North America (from information given by Transeau 1935, Clements 1942, Jenkins 1950, Halliday 1937). The broken line shows the probable extent to which the prairie massasauga dispersed eastward during this period (Schmidt 1938).

Legend:
- ASPEN PARKLAND
- CONIFEROUS FOREST
- DECIDUOUS FOREST
- ECOTONE
- GRASSLAND
- SOUTHERN PINE-OAK

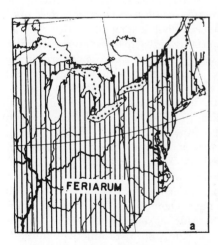
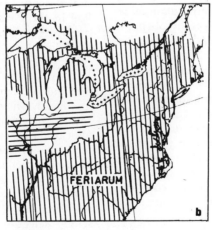
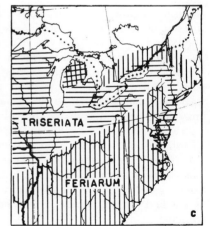
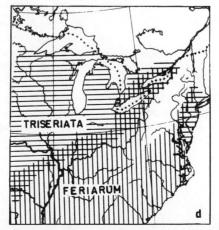
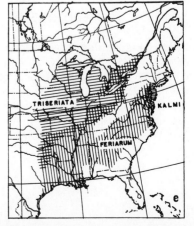

Fig. 19-14 Postulated post-glacial dispersal movements of two subspecies of the chorus frog, *Pseudacris triseriata*, in (a) the climatic optimum, (b) early xerothermic, (c) xerothermic maximum, (d) post-xerothermic, (e) the present. The population of *triseriata* left behind in New Jersey has recently been recognized as the subspecies *kalmi* (Smith 1957).

re-established its dominance in the Northeast. In the northern states there is some evidence that spruce again spread southward. Many of the lakes of the Great Basin refilled with water.

Within historic time, smaller fluctuations in climate are known to have occurred. These have been determined from advance and retreat of mountain glaciers, growth rings of the giant sequoia trees, lake levels, records of past civilizations such as that of the Maya of Yucatan, as well as inferences from historical documents. In western United States fluctuations in moisture have been dated as follows (Brooks 1949):

Wet	500–250 B.C.	Dry	A.D. 1100–1300
Dry	250–100 B.C.	Wet	A.D. 1300–1400
Wet	100 B.C.–A.D. 200	Dry	A.D. 1450–1550
Dry	A.D. 300–800	Wet	A.D. 1550–
Wet	A.D. 900–1100		

Mountain glaciation, especially in the Alps and Iceland, was extensive between 1600 and 1850, but glaciers all over the world have been shrinking since then at a very rapid rate. During the last hundred years, mean annual temperatures have increased 0.5° to 2.2°C, and the sea level has risen about 6 cm (2.5 in.) (Flint 1971, Baum and Havens 1956). This amelioration of climate has permitted the northward dispersal of birds and other animals in recent years into Ontario (Urquhart 1957), Iceland (Gudmundsson 1951), northern Europe (Kalela 1949, Haftorn 1958), and in the sea (Taylor et al. 1957). Other species will doubtless follow in the future; northern communities are not presently saturated with the variety of species they could support. This is true of aquatic communities as well as terrestrial ones. For instance, the fresh-water fish fauna of North America is most highly developed in the Mississippi River system. The impoverished variety of the fish fauna northward and northeastward may be caused in part by the cold waters and low nutrients there but is also due in large part to the failure of fish species to bypass land barriers and to disperse into these regions since the retreat of the glacier. A northward movement of fauna may be expected to continue until the carrying capacity of the ecosystems is reached, or until there is another reversal in the climate.

EMERGENCE OF MAN

From the australopithecine ape-men of the beginning of the Pleistocene evolved early man, *Homo erectus*, who persisted throughout most of the epoch. During the late Eem and early Weichsel or Würm periods (or 150,000 years ago), *H. erectus* was replaced by *H. sapiens neanderthalensis*, who in turn during middle or late Würm, the time of the last major glaciation, was replaced by *H. sapiens sapiens*. This evolution in physical structure was accompanied by evolution in his culture (Chapter 17). Rough or chipped stone implements gave way to polished stone tools, pottery, bow and arrows, and to domestication of plants and animals and cultivation of the land. Increasing populations led to development of villages, urbanization, and finally the complex social systems of modern times.

Early evolution of the hominids took place in equatorial regions, especially in Africa, and somewhat later in southern Asia. Important in this early evolution was the acquisition of adaptations in structure, function, and behavior to live in a savanna-type ecosystem, as increasing aridity converted the closed tropical forest into more extensive areas of mixed trees and grassland. Walking erect may have evolved for traveling long distances in semi-open country. In the forest it had been sufficient to cover short distances rapidly and to climb. With the acquisition of greater mobility came dispersal to practically all parts of the tropical Old World. With geographic isolation, evolution brought the differentiation of subspecies as we see them in living man: australoids (aborigines of Australia, New Guinea, and adjacent islands), mongoloids (Asiatics, American Indians), caucasoids (Europeans), congoids (Negroes and Pygmies), and capoids (Bushmen and relatives) (Coon 1962, Goldsby 1971).

Early man found an optimum habitat in the tropical savannas where edible roots, plants, and game were plentiful, and the climate was favorable. Actually he did not come to settle in the tropical rain forest until late in the Pleistocene. For a long time man was an omnivore or *unspecialized food collector*, living on fish, shell food, nut grass, berries, nuts, coconuts, turtle meat, and the flesh of any other small animals he could kill. Agriculture came later, probably 9000 years ago and independently in Southwest Asia and in Central and South America, as he learned to cultivate the wild species of plants, especially cereals (Harris 1967). Wheat may have been the first plant cultivated for food, possibly even 50,000 years ago. The early origins of present-day agricultural crops can only be surmised for the most part. Certainly most of our modern cultivated plants have a complex genetic history resulting from much hybridizing and deliberate or unconscious selection of the more productive mutants (Anderson 1960). The history of animal domestication and husbandry is equally complex and interesting (Epstein 1955). Dogs, cattle, sheep, goats, and pigs may have been domesticated 9000 to 10,000 years ago (Protsch and Berger 1973).

Early hominids were covered with dense hair and were deeply pigmented. Hairlessness developed, perhaps as an adaptation to prevent overheating while active over long periods in the hunting of game during the warm part of the day (Morris 1967).

Skin color may be related to the need for synthesis of vitamin D through the action of sunlight. Manufacture of this vitamin takes place in the inner layer of the skin, and the amount of short-wave solar radiation at this depth is determined by the outer layer. As pigmented man dispersed out of the tropics, from where solar radiation is always adequate, to northern latitudes, where solar radiation is reduced during the winter months, he became subject to rickets. White skin, which is mostly depigmented and dekeratinized, evolved at around 40°N latitude to maximize ultraviolet penetration and vitamin formation. Yellow skin, which is mainly keratinized, is intermediate; black skin, which is mostly pigmented, may have protected man from penetrating short-wave radiation in the tropics but presented a problem of overheating from the long-wave-lengths (Hamilton 1973). It is of interest that Eskimos in the far north obtain their vitamin D from the fish oil that they consume in abundance and are not dependent on solar radiation (Loomis 1967).

Peking man in Asia apparently used fire for cooking his meat and vegetables during the Elster period, perhaps 400,000 years ago. This was of benefit especially to the young and to older individuals whose teeth had begun to decay. Fire was probably captured when started by lightning strokes. The ability to make fire came much later, perhaps not long before the beginning of agriculture. Evidence for the use of fire in Africa dates back only 50,000 years. Man is the only animal that has mastered the use of fire. This was essential for progress, not only in agriculture for rendering food more palatable, but also for dispersal into cold climates, for industry, and in other ways. The use of animal skins for keeping warm in cold seasons and for dispersal into cooler climates led to the development of clothes (Stewart 1956, Leopold and Ardrey 1972).

Correlated with the relative freedom of the forelimbs, even the apes had some capacity for *using* tools. The ability to *fabricate* tools represents an advance that developed rapidly as man and his intellect evolved. The earliest tools were made from stone and bones. Later they were mounted on wooden shafts for more effective use as weapons. Following the Stone Age came "Ages" of copper, bronze, iron, steel and steel alloys, and now nuclear power. Man's ability to invent tools to aid him in all his activities has been fundamental to the mastery of his environment, for exploiting its resources, and for the development of civilization itself.

During the last interglacial (Eem) and early glacial (Würm) periods, tool-craftsmanship and population density reached levels in Europe well above those prevalent in Africa. Through improvement in clothing, use of fire, and shelter in caves, man became able to keep himself comfortable even in subarctic climates (Fig. 19-15). By the middle Würm he began to make open-air shelters of crude tents or huts. During this period came the first movement of populations across the Bering land bridge into Alaska, with later dispersal southward over the American continent.

The focus of cultural evolution was primarily in the temperate mixed savanna and tundra-steppe rather than in the warmer woodlands of the Mediterranean region. The former ecosystem, like the tropical savannas of Africa, was richly supplied with game, and man developed more of a carnivorous appetite. He became a *specialized hunter-gatherer* and in both the Old and the New World, wreaking considerable havoc among the large gregarious herbivores. Much of the food economy of man in the Old World centered around the reindeer, and in the New World around the mammoth and other large grazing animals.

By the end of the Pleistocene and the beginning of the post-glacial Holocene, the reindeer became relatively less plentiful, there were food shortages, and man turned back to a more extensive use of plants for food. A low subsistence type of *agriculture* developed and animal domestication expanded, particularly in the Near East, in Southeast Asia, and in Meso-America and the Andean highlands. In the southwestern desert region of North America, this stage may not have begun until about 2000 B.C. (Woodbury 1958–59).

During the warm, moist Atlantic climate prevailing from 6000 to 4000 B.C., agricultural practices improved, new tools were invented, and villages became established. Food became more plentiful, populations increased, and more extensive dispersal took place over Europe (Fig. 19-16), Asia, and middle America. Early agriculture was largely "slash and burn," clearing of the land of its original vegetation, cultivating it until the soil lost its fertility, then moving to a new area, leaving the old soil to lie fallow for several years until it regained its fertility. Out of this, however, came the more sophisticated practices of the present time.

The deterioration of the natural environment probably began in a limited way with man's use of fire but became serious with the development of agriculture and has continued at an accelerating pace through the *industrial revolution* of the last two centuries. Native flora and fauna were decimated and replaced with non-indigenous domesticated plants and animals. Accelerated erosion, damming of rivers, drainage of lakes and ponds, and other activities changed the physical environment. A new cultural landscape was created (Thomas 1956).

Agricultural productivity was especially high in the floodplains of the major rivers, the Tigris and Euphrates of the Near East, the Nile in Egypt, and the Indus in western India, and it is in these regions that urban development first took place (Mumford 1956). With surplus food available, commercial contacts were made with more distant areas for exchange

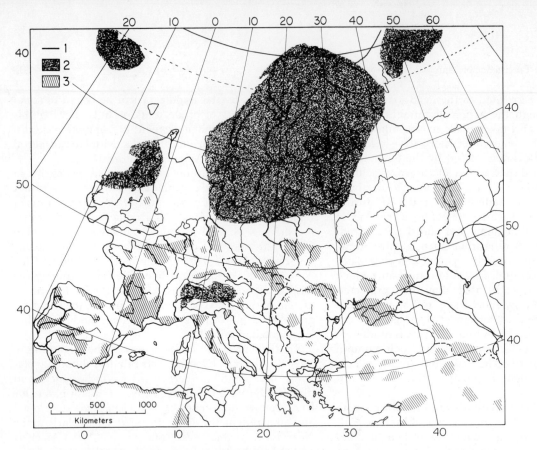

Fig. 19-15 European settlement during the early Würm. (1) Approximate position of coastlines, (2) glaciers, (3) distribution of Mousterian and related industries of early man (reprinted from Karl W. Butzer, *Environment and Archeology*. Chicago: Aldine-Atherton, Inc., 1971; copyright © 1971 by Karl W. Butzer, reprinted by permission of the author and Aldine-Atherton, Inc.).

Fig. 19-16 Early village farming cultures in Europe and adjacent areas *ca.* 4000 B.C. (reprinted from Karl W. Butzer, *Environment and Archeology*, Chicago: Aldine-Atherton, Inc., 1971; copyright © 1971 by Karl W. Butzer, reprinted by permission of the author and Aldine-Atherton, Inc.).

314

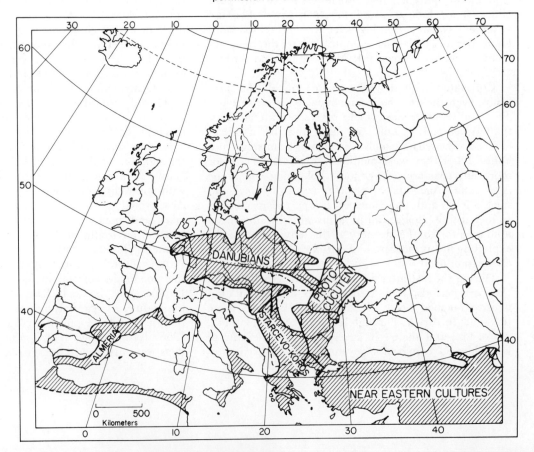

of commodities; food and pottery for flint, metal, and stone. Craftsmen, tradesmen, and clerks necessarily evolved as specialized occupations within the villages and cities, and along with these technological and economic changes came all the later complexity of society, politics, military and police forces, art, music, and religion that we recognize as civilization (Butzer 1964, Boughey 1971).

SUMMARY

At the beginning of the Tertiary Era, 65 million years ago, the North American continent was partly covered with epicontinental seas, marshes, and lakes. Scattered mountain ranges occurred in the Rocky Mountain region, but these had been greatly eroded by Oligocene time. Rainfall was heavy and temperatures mild. Semi-tropical conditions extended across the continent to 49° North latitude in the West, and 37° North latitude in the East; temperate climates obtained far to the North.

Three principal floras occurred during early Tertiary time. The Neotropical-Tertiary Geoflora was co-extensive with the tropical climate. The Arcto-Tertiary Geoflora consisted of a temperate unit, largely deciduous forest, and a boreal unit, preponderantly coniferous species; this geoflora extended to within eight degrees latitude of the North Pole and across the Bering land bridge into Eurasia. The Madro-Tertiary Geoflora first appeared during the Eocene in scattered dry sites on the lee sides of high ridges in northern Mexico and southwestern United States, but did not become well developed until the Miocene.

Beginning in the Miocene and increasing in intensity through the Pliocene into early Pleistocene, mountain-building was extensive in the Rocky Mountains, Appalachians, Ozarks and Ouachitas, Cascades, Sierra Nevada, and Coast Ranges. The epicontinental seas receded. The climate in the rain-shadows of the mountain systems became increasingly arid, particularly in the Southwest, Great Basin, and on the Great Plains. Concurrently, the climate became progressively cooler, a trend culminating in the severe glaciation of the Pleistocene.

As a result of these changes in physiography and climate, the Neotropical-Tertiary Geoflora retreated to the present tropics to constitute the tropical forest and tropical savanna biomes of today. The Arcto-Tertiary forest withdrew southward and eastward to form the temperate deciduous forest, coniferous forest, and tundra biomes. Into the areas vacated by these two floras the Madro-Tertiary Geoflora expanded to form the woodland, chaparral, grassland, and desert biomes.

At maximum glaciation during the Pleistocene, the tundra biome was greatly restricted in North America and the coniferous forest biome was destroyed, except for refugia in Alaska, the middle Appalachians and less extensively westward into the grassland, the Rocky Mountains of the United States, and on the Pacific coast west of the Cascade Mountains. Everywhere it extended to lower elevations in the southerly mountain areas. The deciduous forest and grassland were extensively modified in the north by intrusion of coniferous forest species. Because of heavy precipitation, grassland and woodland were more widely distributed through the Great Basin and the Southwest.

With the retreat of the glacier in post-Pleistocene times, the tundra and coniferous forest biomes reoccupied most of northern North America. In the northern states from Minnesota and Illinois eastward, pollen data indicate changes of climate from cool-moist to warm-moist to warm-dry, then back to the cooler, moister conditions of the present time. Accompanying these climatic changes was a succession of vegetation from spruce-fir to pine and oak to oak-hemlock-beech to oak-hickory and the prairie peninsula, then back to oak-beech-chestnut-hemlock.

Comparable changes in climate and vegetation occurred in Europe and Asia throughout the Tertiary and Quaternary eras. These changes in climate and vegetation had a profound effect in both the New and Old Worlds on the evolution and dispersal of animals; and on the development of present-day biotic communities.

Primitive man, first restricted to the highly productive and undemanding tropical savannas of Africa dispersed during early and middle Pleistocene into Europe and Asia as he acquired clothes, the use of fire, and improved tools. With further development of his culture from an unspecialized to a specialized hunter-gatherer economy and the use or construction of shelters he came by late Pleistocene to occupy the sub-arctics and spread over the Bering land bridge into North, Central, and South America. It is likely that as he spread, man hunted many of the large Pleistocene mammals to extinction. With decrease in big game, he developed agricultural skills, food surpluses, exchanges of commodities, larger populations, urbanization, and modern civilization.

Chapter 20

TEMPERATE
DECIDUOUS FOREST BIOME

In this and the following chapters we will try to gain an understanding of the geographic distribution of plants and animals as they occurred in primeval time before the colonization of the continents by white man, using biomes as our units of analysis. Each biome will be considered in respect to its distribution, vegetation, and plant associations; the constituents of its various biociations; the relative abundances of the principal animal species, especially mammals and birds; the adaptations and adjustments, especially behavioral, to the biome as demonstrated by the predominant animals; and human uses made of it. We will devote most of our study to biomes of North America, but the rest of the world will not be neglected. A general reference which the reader will find valuable is *Ecology of North America* (Shelford 1963).

The temperate deciduous forests of North America, western Europe, eastern China, and Japan are related as developments of the Arcto-Tertiary Geoflora, which at one time was practically continuous around the world in north temperate climates. In North America, the deciduous forest is best developed in the eastern United States, although elements of it are mixed with conifers in the North, West, and through the mountains of Mexico into Guatemala (Sharp 1953).

Mean annual precipitation for the biome in North America varies from 75 to 125 cm (30–50 in.); mean annual precipitation for the Gulf states is occasionally as high as 150 cm (60 in.). For the most part, rain falls periodically throughout the year; in many places, precipitation also falls as snow in wintertime. Mean monthly temperatures from north to south vary from January minima of −12° to 15°C (10°–60°F) to July maxima of 21°–27°C (70°–80°F). Average mid-day relative humidities during July range from 75 per cent in the East to 50 per cent where the biome contacts prairie in the West. The annual frost-free period varies from about 150 days in the North to as much as 300 days in the South (Kincer 1941).

The climax of the deciduous forest biome is a community dominated by broad-leaved trees that are leafless during the winter over most of the area. In the South, the dominant trees are mostly evergreen. The trees usually form relatively dense forests with a closed canopy, but where the biome verges on prairie, the forest gives way to savannas containing scattered groves. The shrub stratum is often poorly developed within the forest because of the deep shade there, but is well formed at the forest-edge. The herb stratum has a rich variety of flowering plants, which are especially conspicuous in the spring. All seasonal aspects are well defined. The leaves of the trees and shrubs, as well as those of most herbs, are intolerant of freezing temperatures over-winter and hence are shed in the North during the autumnal aspect. Consequently, there is considerable seasonal change in forest microclimates, to which animal life must respond. The growing season is sufficiently long to permit full development of new foliage and maturation of seed each year, although the size of the seed crop, upon which many animals depend, varies greatly from year to year.

PLANT ASSOCIATIONS IN NORTH AMERICA

The principal plant communities (Fig. 20-1) are the following (Braun 1950, Shelford 1963):

1. *Liriodendron-Quercus* association: Tulip-oak or *mesophytic forest.* Centrally located on unglaciated Appalachian Plateau. Contains the richest mixture of tree species. White basswood and yellow buckeye are good indicators of the association. This association best represents the original characteristics of the temperate division of the ancient Arcto-Tertiary Geoflora from which most of the other associations were derived in modified form.
2. Western mesophytic forest: Ecotone area containing a mosaic of climaxes similar to adjacent associations.
3. *Quercus-Carya* association: Oak-hickory forest. Center of distribution in Ozark and Ouachita Mountains but radiating far into the prairie along river valleys and into Gulf and South Atlantic states.
4. *Quercus (Castanea)* association: Formerly called oak-chestnut forest, but chestnut now largely destroyed by blight and its place in canopy taken by oaks and other species, best developed in Appalachian Mountains (Woods and Shanks 1959).
5. *Pinus-Quercus* associes: Southeastern pine forest. Southern species of pines, often mixed with oak. Form extensive subclimax stands in south Atlantic and Gulf states (Fig. 20-2). When fire is prevented,

Fig. 20-2 Frequent ground fires prevent the southeastern pine forests from succeeding into a deciduous forest climax (courtesy U.S. Forest Service).

pines disappear and the forest becomes one of mixed hardwoods (Quarterman and Keever 1962).

6. *Magnolia-Quercus* association: Magnolia-oak forest. Found in southern portions of Gulf states and most of Florida. Dominant trees are coriaceous, broad-leaved, and evergreen; forests often dense, with deep shade, with Spanish moss and other epiphytes hanging from branches; grading southward into tropical forest (with royalpalm) in Everglades and Florida Keys. Early seral stages include fresh-water marshes and cypress swamps, pine flatlands, scrub oak, patches of prairie, coastal dunes, and salt marshes (Davis 1940).
7. *Fagus-Acer* association: Beech-maple forest. Mostly northern in distribution; two principal climax dominants only.
8. *Acer-Tilia* association: Maple-basswood forest. Occurs mainly in Wisconsin and Minnesota and southward to northern Missouri.
9. *Tsuga-Pinus*-northern hardwoods ecotone: Mixture in southern Canada and in the Appalachian Mountains of beech, sugar maple, and basswood with eastern hemlock, various northern species of pine, and yellow birch.

ZONATION

Climate varies with altitude as it does with latitude; most notably, air temperature varies inversely with altitude. Because of this, there are corollary changes

Fig. 20-1 Distribution of plant associations (Braun 1950), numbered as in text, showing tree species diversity (Monk 1967).

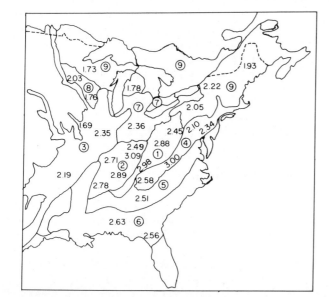

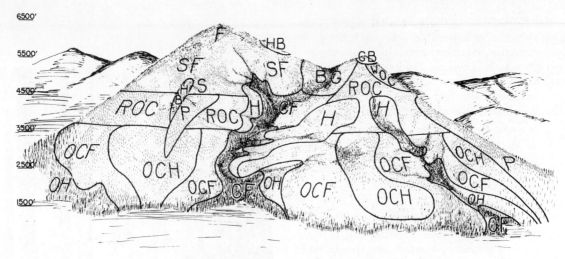

Fig. 20-3 View of an idealized mountain and valley of the Great Smoky Mountains, looking east (Whittaker 1956).

BG, beech gap forest	OCH, oak (chestnut) heath
CF, cove forest	OH, oak-hickory forest
F, Fraser fir forest	P, pine forest and pine heath
GB, grassy bald	ROC, red oak (chestnut) forest
H, hemlock forest	S, spruce forest
HB, heath bald	SF, spruce-fir forest
OCF, oak (chestnut) forest	WOC, white oak (chestnut) forest

in vegetation such that conspicuous *zonation* is apparent. Zonation of vegetation and differences in climate profoundly affect animal distributions.

In the Great Smoky Mountains of eastern Tennessee, there are two zones, differentiated essentially by temperature (Fig. 20-3). Each is characterized by a pattern of different but intergrading communities, distinct from the pattern of the other. On north slopes of the mountains the demarcation between them is approximately the 1400 m (4500 ft) elevation. The lower zone is mostly deciduous forest, grading laterally from moist mixed mesophytic or cove forest on the north slopes through oak-hickory and oak (chestnut) to southern pine forest and grassy balds on the warm, dry, south slopes (Fig. 20-3). The vegetation of the upper zone also changes, north to south, as moisture conditions change: gray beech forest on the north and in the moist mountain gaps gives way to spruce-fir forest, which in turn changes into heath balds on the exposed southern slopes (Whittaker 1952, 1956).

Contrastingly, each of the several zones of New York's Catskill Mountains is characterized by a vegetation pattern less heterogeneous than those in the Smokies and different from that of the other zones. Below 230 m (750 ft) oak-chestnut prevails; between 230 m and 610 m (2000 ft) there is an ecotone of beech-maple-hemlock; then comes a zone where hemlock drops out and the forest is principally gray beech, sugar maple, and yellow birch. Above 980 m and extending to 1280 m (3200–4200 ft), the deciduous forest is replaced by spruce-fir coniferous forest (Kendeigh 1946).

ANIMAL COMMUNITIES

North American Deciduous Forest Biociation

This biociation occurs in the climax and late seral stages throughout the deciduous forest proper. It extends into the pine-hemlock-hardwoods ecotone, although locally within the ecotone there is rather sharp segregation of many animal species between deciduous and coniferous forest (Kendeigh 1946, 1948). The community is represented as a biocies in the aspen-birch seral stage of the boreal forest. The biociation penetrates well into the magnolia-oak association in the Gulf states, but becomes progressively more diluted southward with species from the southeastern mixed biocies. To the west, the community occurs in the wider strips of forest along the streams, but as the forest diminishes in density westward, the forest-edge biociation replaces the forest biociation.

Mammal species that occur or formerly occurred through the deciduous forest biociation include:

Opossum	White-footed mouse
Short-tailed shrew	Gray fox
Eastern mole	Black bear
Eastern chipmunk	Raccoon
Gray squirrel	Mountain lion
Southern flying squirrel	Bobcat

The mountain lion, bobcat, and black bear are also common in other biomes, but the other species listed are characteristic inhabitants of the deciduous forest. Seton (1909) estimated original populations of mountain lions and bears at one per 26 km² (1 per 10 mi²), and gray foxes at one per 10 km² (1 per 4 mi²). Gray squirrel populations vary greatly by time and place, but when common may average 2.5+ per hectare (1+ per acre). Chipmunks vary in numbers from year to year, depending on the abundance of nuts and seeds that they can store in their underground burrows to supply them over winter. In beech-maple forests of northern Ohio they average 25 or more per hectare (10 per acre) during the autumnal aspect of good years (Williams 1936). The combined autumn populations of mice and shrews vary from less than 25 per hectare (10 per acre) in poorer forests having little ground humus to ten times as many during good years in a good habitat. A gradient of increasing populations, from west to east, depending largely on moisture availability as well as abundance of humus, is marked in shrews (Wetzel 1949).

Birds prominent in the deciduous forest biociation include:

Broad-winged hawk (formerly)	Black-capped chickadee
Ruffed grouse	Tufted titmouse
Wild turkey (Fig. 20-4)	White-breasted nuthatch
Great horned owl	Wood thrush
Barred owl	Yellow-throated vireo
Whip-poor-will	Red-eyed vireo
Pileated woodpecker	Cerulean warbler
Red-bellied woodpecker	Worm-eating warbler
Hairy woodpecker	Hooded warbler
Downy woodpecker	Ovenbird
Great crested flycatcher	Louisiana waterthrush
Acadian flycatcher	Redstart
Eastern wood pewee	Scarlet tanager

The ovenbird and red-eyed vireo are usually the two most abundant species in deciduous forest stands.

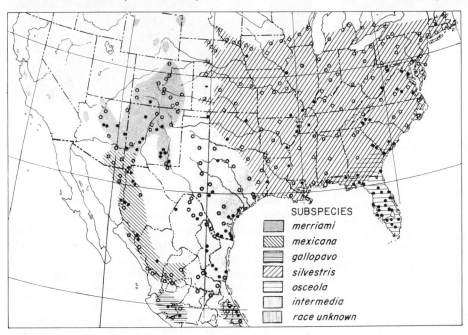

SUBSPECIES
- merriami
- mexicana
- gallopavo
- silvestris
- osceola
- intermedia
- race unknown

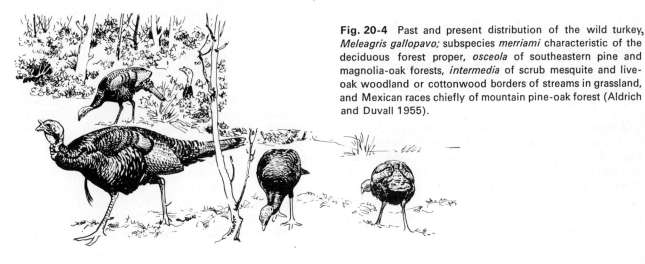

Fig. 20-4 Past and present distribution of the wild turkey, *Meleagris gallopavo;* subspecies *merriami* characteristic of the deciduous forest proper, *osceola* of southeastern pine and magnolia-oak forests, *intermedia* of scrub mesquite and live-oak woodland or cottonwood borders of streams in grassland, and Mexican races chiefly of mountain pine-oak forest (Aldrich and Duvall 1955).

Fig. 20-5 Two predators of the deciduous forest biociation: the timber rattlesnake and the great horned owl (courtesy U.S. Forest Service).

An average population of each is 35 to 40 pairs per 40 hectares (100 acres). A 40-hectare plot of average deciduous forest supports approximately 200 pairs of birds, representing all species, as an average. The breeding ranges of most of the species listed above coincide rather closely with the deciduous forest, although some species, such as the downy woodpecker, are distributed more widely and are represented by different subspecies in other biomes (Pitelka 1941, Udvardy 1963). The avifauna is nearly equally divided in origin between Eurasian, North American, and South American elements, but the North American fauna has the largest breeding population (Table 21-2).

Reptiles and amphibians are represented by:

Timber rattlesnake (Fig. 20-5)	Marbled salamander
Copperhead	Slimy salamander
Black rat snake	Red-backed salamander
Red-bellied snake	Common newt
Five-lined skink	Wood frog
Box turtle	Tree frogs

Invertebrates are too numerous and varied to be mentioned specifically (see Chapter 8). Snails and slugs are especially abundant in the moist mixed mesophytic forests of the southern Appalachians, but decrease in abundance and variety as the forest becomes drier and approaches the prairie (Shimek 1930). Millipedes are numerous in the rich humus of the forest floor. Insects and spiders are represented by a multitude of species in all strata.

North American Deciduous Forest-edge Biociation

Eastern North America, prior to white colonization, had thousands of kilometers of contact between deciduous forest and prairie, with tongues of forest extending far into the prairie along the river valleys (Fig. 20-6). Deciduous forest even bordered the prairie on the north where the aspen grove ecotone intruded in front of the boreal forest. A characteristic forest-edge type of vegetation and distinct animal community occurs along these contacts and where the forest confronts ocean or large lakes. The forest-edge community also permeates the deciduous forest in the role of a seral community or biocies on rock, sand, abandoned fields, and around water (Chapters 7 and 8).

A different faciation of the forest-edge biociation occurs west of the Great Plains. As the interior of the continent grew arid in the Miocene and Pliocene, many species of deciduous trees together with their associated animals were able to persist in local habitats throughout the western part of the country. A distinct plant community—riparian woodland—of willows, cottonwoods, sycamores, aspens, alders, and other broad-leaved deciduous trees presently occurs along streams, bodies of water, and elsewhere. It appears to be seral to coniferous forest or woodland over most of the West, but reaches out into grassland and desert in a manner similar to the tongues of forest in the East, thus greatly extending the linear distance of the forest-edge.

Fig. 20-6 Forest-edge between deciduous forest and prairie (Shelford 1963).

The animal species composition reflects the relationship obtaining between the riparian woodland in the West and the forest-edge community in the East (Fig. 20-7). Nearly half of the species listed below pervade both faciations, albeit represented by different subspecies. Several species are confined to one or the other faciation, as indicated. Speciation among forest-edge forms was doubtless encouraged by the virtual isolation of both faciations when the grassland biome evolved. Common species (Ingles 1950, Miller 1951, Walchek 1970) are:

Mammals

Eastern mole (East)	Long-tailed weasel
Eastern cottontail (East)	Striped skunk
Woodchuck (East)	Wapiti
Fox squirrel (East)	Mule deer (West)
Gray wolf	White-tailed deer (East)
Red fox	(Fig. 20-8)

Birds

Turkey vulture	Red-shouldered hawk
Sharp-shinned hawk	(East)
Cooper's hawk	Sparrow hawk
Red-tailed hawk	Bobwhite (East)
Swainson's hawk (West)	Mourning dove

Yellow-billed cuckoo
Black-bellied cuckoo (East)
Screech owl
Common nighthawk
Chimney swift (East)
Ruby-throated humming-
bird (East)
Hummingbirds (several
spp., West)
Red-headed woodpecker
(East)
Yellow-shafted flicker
(East) (Fig. 20-7)
Red-shafted flicker (West)
Eastern kingbird (East)
Western kingbird (West)
Cassin's kingbird (West)
Barn swallow
Violet-green swallow
(West)
Common crow
Blue jay (East)
Black-billed magpie
(West)
House wren
Catbird (East)
Brown thrasher (East)

Eastern bluebird (East)
Robin
Chestnut-backed
chickadee (West)
Cedar waxwing
Loggerhead shrike
Starling
Warbling vireo
Bell's vireo (West)
Blue-winged warbler
Yellow warbler
Yellowthroat
Yellow-breasted chat
Brown-headed cowbird
Baltimore oriole (East)
Bullock's oriole (West)
Common grackle (East)
Brewer's blackbird (West)
Indigo bunting (East)
Lazuli bunting (West)
Rufous-sided towhee
American goldfinch
Black-headed grosbeak
(West)
Chipping sparrow
Field sparrow (East)
Song sparrow

Reptiles

Blue racer	Brown snake
Smooth green snake	Garter snake
Milk snake	Ribbon snake

Fig. 20-7 Flicker bird populations isolated during late Pleistocene glaciations in eastern and western forest-edge refugia developed into a *yellow-shafted* and a *red-shafted* "subspecies." Following Wisconsin glaciation, the ranges of the two forms dispersed and met along the north-south line shown. Hybridization and introgression is now occurring (Short 1965).

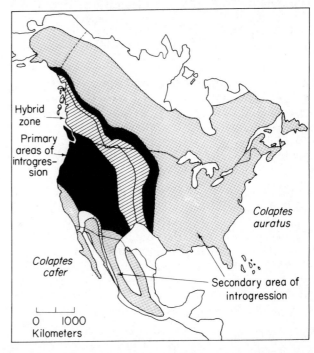

Southeastern Mixed Biocies

A number of animal species have their centers of distribution in the south Atlantic and Gulf states and are associated with the southeastern pine forest, the magnolia-oak forest, or with seral stages. There is, doubtless, more than one community involved, but until more detailed analysis can be made the species may conveniently be listed together. Common terrestrial vertebrates of climax and seral communities are:

Mammals

Southeastern shrew	Cotton mouse
Florida least shrew	Golden mouse
Marsh rabbit	Florida mouse
Swamp rabbit	Hispid cotton rat
Southeastern pocket gopher	Eastern wood rat
	Pine vole
Marsh rice rat	Round-tailed muskrat
Eastern harvest mouse	Eastern spotted skunk
Oldfield mouse	Florida skunk

Birds

Black vulture	Mockingbird
Swallow-tailed kite (formerly)	Blue-gray gnatcatcher
	White-eyed vireo
Mississippi kite (formerly)	Prothonotary warbler
	Swainson's warbler
Turkey	Bachman's warbler
Carolina parakeet (formerly)	Parula warbler
	Yellow-throated warbler
	Pine warbler
Chuck-will's widow	Prairie warbler
Red-cockaded woodpecker	Kentucky warbler
	Orchard oriole
Ivory-billed woodpecker	Boat-tailed grackle
Scrub jay	Summer tanager
Fish crow	Cardinal
Carolina chickadee	Painted bunting
Brown-headed nuthatch	Seaside sparrow
Carolina wren	Bachman's sparrow

Reptiles and Amphibians

Rough green snake	Eastern fence lizard
Chicken snake	Six-lined racerunner
Corn snake	Brown skink
Kingsnake	Chameleon
Southern hog-nosed snake	Spadefoot toad

In addition to these species, many of those listed for the deciduous forest and forest-edge are also common, but frequently represented here by different subspecies from those that occur in the North. The seral relations of many of the mammals (J. C. Moore 1946, Pournelle 1950), birds (Nelson 1952), as well as certain insects (Rogers 1933, Friauf 1953) have been worked out for various areas of northern Florida.

Several of these species of mammals, birds, reptiles,

Fig. 20-8 White-tailed deer in forest-edge habitat (courtesy U.S. Forest Service).

and amphibians have dispersed from the Southeast far into the deciduous forest and forest-edge communities. Their distributional ranges, in many cases, extend westward into Texas and southward into Mexico. The closest related forms of some of the more restricted species also lie to the west and south, for instance the scrub jay. The evidence is suggestive that this biocies and the corresponding plant communities did not originally belong to the deciduous forest biome. It seems more likely, rather, that they are derived from the sclerophyllous woodland and pine forests of the Madro-Tertiary and from the broad-leaved forests of the Neotropical-Tertiary Geofloras. During the Pliocene or earlier, the Madro-Tertiary biota was apparently continuous around the north side of the Gulf of Mexico (Pitelka 1951a) but later separated into eastern and western portions with the development of grassland through Texas to the Gulf of Mexico.

The terrestrial fauna indigenous to the southern tip of Florida is predominantly deciduous forest-edge species, species of the southeastern mixed biocies, and invasion of tropical species from the West Indies. Among birds, the white-crowned pigeon, zenaida dove, smooth-billed ani, gray kingbird, black-whiskered vireo, as well as races of nighthawk and yellow warbler, are recent newcomers (Robertson 1955). There is also a rich and varied aquatic avifauna that is for the most part tropical in origin. The manatee and the alligator formerly extended from Florida around the north side of the Gulf of Mexico; the crocodile was limited to southern Florida.

European Deciduous Forest Biociation

Dominants of the plant associations in Europe are different species of the same genera that occur in North America, particularly beeches, maples, oaks, hornbeams, and basswoods. Many mammals and birds of the European deciduous forest and seral stages also belong to the same genera as North American species. The similarity in genera may be traced back to the continuity of the Arcto-Tertiary forest between the two continents during the Tertiary; the dissimilarity of species to divergent evolution since the two communities became separated.

Mammals common to the European forest and forest-edge include both the common and white-toothed shrews, European mole, common hare and European rabbit, several mice, wolf (same species as in North America), red fox (perhaps the same species as in North America), weasels, wildcat, wild boar, two deer, and European bison.

The bird fauna (European fauna of Stegmann, 1938) includes some falcons, kites, and eagles, a pigeon and a cuckoo, owls, several woodpeckers, a jay, crows, several tits, a nuthatch, a creeper, a wren, several thrushes, a rich variety of Old World warblers only poorly represented in North America, an Old World flycatcher not found in the New World, an oriole, and various finches. Absent are the tyrant flycatchers, vireos, and wood warblers that are so prominent in the North American deciduous forest (Lack and Venables 1939, Turček 1951, 1952, 1955). Lists of invertebrates, especially of ground animals, are given by Kühnelt (1944). It is possible that Pleistocene glaciation disturbed this biociation much more than that in North America (Moreau 1963).

Asiatic Deciduous Forest Biociation

The broad-leaved deciduous forest of eastern China, Formosa, Korea, and Japan contains many species of plants and animals belonging to the same genera as occur in Europe or North America. During the early Tertiary, this forest was in direct contact, via the Bering land bridge, with that in North America, and deciduous trees still maintain a narrow and tenuous contact along the southern edge and in seral stages of the coniferous forest with the deciduous forest of Europe. In addition, there are some endemic genera of plants and animals confined to the area. A number of Indo-Malayan species penetrate into the biociation as far as northern China and Japan. Stegmann (1938) gives a list of bird species occurring in this area that belong to what he calls the Chinese fauna, but he does not distinguish among those characteristic of forest, forest-edge, and seral communities. Uramoto (1961), in Japan, has pioneered in analyzing the species composition, biomass, and caloric food consumption of the bird population in a deciduous forest community.

ANIMAL ADJUSTMENTS

Animals are adapted structurally, functionally, and behaviorally to live in or under trees. They may use the trees directly as lookouts, singing posts, nest-sites, for cover and protection, and as a source of food; or they may simply take advantage of the rich humus created by the annual fall of leaves, or the shade, greater humidity, and equable temperatures of the forest habitat. Some animals, for instance the eastern chipmunk, die within a few minutes if exposed directly to the sun. Snails and slugs are most active and carry on their reproductive activities during the moist vernal aspect, but may be conspicuous throughout the summer when they are able to maintain the necessary water balance.

Special adaptations for arboreal habits and for climbing are the sucking discs on the toes of tree frogs, the sharp claws and opposable toes in woodpeckers and squirrels, the prehensile tails of opossums and white-footed mice, the parachutes and bushy tails of squirrels as well as the movable scales of some of the snakes, the many legs of the millipedes, and the slimy feet of slugs and snails.

Hearing and voice are well developed in many forest animals, although vision is less perfected since visibility is limited anyway. The rich and almost constant singing of forest birds throughout the breeding season is well known, but the voice, or songs, of squirrels, chipmunks, and wolves are also well developed for mammals. The loud singing of tree frogs is noteworthy, and the nightly chorus of insect voices, especially those of orthopterans, is remarkable. Most of these sounds serve to attract mates or advertise territories.

The regular and pronounced changes in photoperiod and temperature bring full development of the breeding season of most animals to its peak during the spring and early summer. Deer, bats, and a few others, however, characteristically mate during the autumn, and some of the squirrels and owls during the winter.

All species must meet the severe winter conditions of short days, low temperatures, and scarcity of food in one way or another. In those forms of mammals and birds that remain active over winter and in those insects that hibernate in exposed situations there is considerable increase in resistance to cold by internal physiological adjustments, and they live either on kinds of food that are not usually concealed by snow or on food cached when it was plentiful. Mammals den up in hollow logs or trees during short severe cold periods, coming out again when the weather is mild. Flocking is common in most birds during the winter season in contrast to their isolation in territories during the breeding season. Flocks commonly seek shelter on the lee side of forest areas or in river valleys to get protection from cold winds. Populations and variety of birds are supplemented during the winter as northern species come south.

Those species of birds, mammals, reptiles, amphibians, insects, and snails that cannot maintain activity in winter conditions either migrate or hibernate.

Migration among birds commonly reduces the population to less than one-third the number of individuals present during the early summer, but during the spring and autumn migratory periods, populations are temporarily greatly increased. Various insect species, including the monarch butterfly, migrate many kilometers southward. Other species move much shorter distances from open country into the forest-edge preparatory to hibernation (Weese 1924).

Woodchucks, bats, and possibly chipmunks hibernate in the true sense; the black bear enters a pseudo-hibernation state, remaining quiescent over winter but maintaining temperature and other body functions at near normal. Reptiles and amphibians bury themselves in decaying stumps or logs, in the ground below the frost line, or in the mud bottom of ponds. Nearly all insects and other invertebrates migrate out of the trees, shrubs, and herbs to the forest floor, where they hibernate. Some species move up and down in the soil to keep below the frost line. Other species overwinter only in the egg or some other immature stage. Farther south, especially in the magnolia-oak forest where there is less need, hibernation and migration of populations that breed in the region are much less pronounced.

HUMAN RELATIONS

The original condition of the forest and its wildlife has, of course, been greatly modified by man. The American Indian should probably be considered a native inhabitant of the deciduous forest, and the modifications he produced (Day 1953) as a normal influence comparable to that of other large mammals. The white man, however, is equipped with a large variety of tools that renders his influence extreme. As a consequence, some forest and forest-edge species, such as the mountain lion, gray wolf, eastern bison, wapiti, passenger pigeon, Carolina parakeet, probably the ivory-billed woodpecker, and others have become extinct. With agriculture and lumbering, seral stages have become more prevalent, so that there has been considerable shift in the relative abundance and importance of species from what occurred originally (Bennett and Nagel 1937, Allen 1938).

White man finds in the climate of the deciduous forest biome conditions favorable for the highest efficiency of his various activities, for his greatest health and energy, for maintenance of high population densities, and for high development of modern civilization (Huntington 1924). The chief and most profitable occupations of man in the deciduous forest biome are agriculture, mining, and industry. In eastern Asia, the broad-leaved deciduous and evergreen forests are occupied by Mongolians, and like the white man this yellow race early developed a high degree of civilization and large populations.

Forests early became essential to white man as a source of lumber, fuel, and raw materials of industry. Trees furnish him shade from the hot summer sun and protection from the cold winter winds. In the early settlement of North America, forests were cleared for farming purposes with difficulty, but forest land was considered more fertile than grassland because it grew trees instead of grass. As man dispersed westward across North America into the grassland biome, he first built his home in the fringes of forest along the streams or in outlying groves (Hewes 1950). As settlement increased, however, surplus people were crowded onto the prairie as they were crowded also into other biomes. It is of interest that in his invasion of grassland man planted trees around his home and thus tried to bring the forest environment with him.

SUMMARY

The temperate deciduous forest biome is derived from the Arcto-Tertiary forest and is best developed in eastern North America, western Europe, and eastern Asia. In those places, precipitation is moderate and temperatures mild during the summer growing season, but the winter season is generally unfavorable for the activity of most organisms.

Animal communities of major significance are the North American deciduous forest biociation, North American deciduous forest-edge biociation (often a biocies), southeastern mixed biocies, European deciduous forest biociation, and Asiatic deciduous forest biociation.

Animals are adapted and adjusted in various ways to live in and under trees. Reproduction takes place principally in the spring and early summer. The severe winter season is adjusted to by increase in physiological hardiness, hibernation, or migration. The variable climate contributes to good health and high energy for man, and some of the maximum developments of civilization have occurred in deciduous forest areas.

Chapter 21

CONIFEROUS FOREST, WOODLAND, AND TEMPERATE SHRUBLAND BIOMES

CONIFEROUS FOREST BIOME

The coniferous forest is a continuous, often dense, forest of needle- or scale-leaved evergreen trees. The sclerophyllous leaves prevent excessive evaporation of water during winter and dry periods, and are adapted to withstand freezing. The evergreen leaves take full advantage for photosynthesis of short summer growing seasons, intermittent warm periods of autumn and spring, and the warm winter rains of the Pacific coast. The flexible branches bear snow-loads without breaking; snow-loads tumble easily off the cone-shaped tree. The dead, dry needles which cling to the trees feed devastating crown fires, much more common in coniferous than deciduous forest.

Distribution and Origin

Coniferous forests are largely confined to the northern hemisphere. They are transcontinental in Canada (Halliday 1937) and in higher elevations on the mountains through Mexico and Guatemala, into Honduras and Salvador. In Eurasia there is also a northern transcontinental coniferous belt with disjunct patches of coniferous forests on all higher mountains southward. The main mass of coniferous forest species is derived from the boreal element of the Arcto-Tertiary Geoflora, and is much older geologically than is the deciduous forest. There is some evidence, however, that the eastern hemlock is a segregate from the temperate rather than the boreal unit of the Arcto-Tertiary Geoflora (Braun 1950, Oosting and Bourdeau 1955, Whittaker 1956), and that the western arid-tolerant ponderosa pine and Mexican pines come from the Madro-Tertiary Geoflora.

Climate

In the transcontinental forest of North America, precipitation varies between 38 and 100 cm (15–40 in.) and is mostly summer rain. Mean monthly temperatures vary from a winter low of about −30°C to a summer high of 20°C (−20° to +70°F). The summer period between killing frosts varies from 60 to 150 days. The northern boundary of this forest coincides with the summer position of the Arctic frontal zone, the southern boundary with its winter position (Bryson 1966). On the Pacific slope of the high western mountains, because of the westerly winds coming from the warm Japanese current, precipitation is higher (125 to 225+cm, 50 to 90+in.); most of it falls as winter rain. Mean monthly temperatures are more uniform (2° to

Fig. 21-1 Above, montane forest in Oregon—a virgin stand of ponderosa pine (courtesy U.S. Forest Service). Right, forest-tundra in northern Manitoba, composed of spruce and tamarack with the ground covered with a thick layer of moss and lichens (courtesy W. P. Gillespie).

18°C, 35° to 65°F) and the frostless season is 120 to 300 days long. Humidity is high, and fogs are frequent in this region. In the northern Rockies, Cascades, and Sierra Nevada, heavy winter precipitation falls as snow that accumulates to several meters in depth; winter temperatures are considerably lower. Snowfall is not as heavy in the central Rockies, and declines steadily, southward.

Plant Associations of North America

Pinus-Tsuga association (pine-hemlock forest): Eastern hemlock is the climax, but eastern white, red, and jack pines are of wider distribution; northern white-cedar and yellow birch are prominent. The forest has been badly disturbed by logging and fire, factors which, with climatic succession, have permitted a wide penetration of hardwoods to form an ecotone between deciduous forest and boreal forest. The association extends from Minnesota to New England, and south into the Appalachian Mountains.

Picea-Abies association (boreal forest): White spruce and balsam fir are most important (related species in Appalachians), but black spruce and tamarack also prominent; extends across southern Canada to the northern Rocky Mountains, north into Alaska, and south in Appalachian Mountains; alder thickets common in wet areas and heath shrubs in forest openings; quaking aspen and paper birch occur extensively as seral stages (La Roi 1967). Very little of the western part of the boreal forest is climatic climax, since most of the area is swept by intermittent fire, which is then followed by secondary succession (Dix and Swan 1971). *Aspen groves*, or *parklands*, form a broad ecotone between forest and grassland from Minnesota to the Rocky Mountains (Bird 1930). The forest decreases in height and density northward, its floor carpet of lichens and mosses increases in depth and extent, and it becomes interspersed with numerous bogs or muskegs as it approaches tree-line. Lichen woodland is especially well developed east (Harper 1964). and muskegs west of Hudson Bay (Fig. 21-1). The location of tree-line is unstable. In some localities at the present time, as in Alaska, it appears to be advancing into tundra (Griggs 1936), elsewhere it may be retreating southward (Raup 1941). West of Hudson Bay, radiocarbon dating indicates that the tree-line may have been 280 km (170 miles) north of its present location 3500 years ago during the Climatic Optimum. Lack of tree regeneration after fire caused a retreat of tree-line to its present position only a few hundred years ago (Bryson *et al.* 1965). This whole area is *forest-tundra*, as distinguished from the denser, taller boreal forest; it is equivalent to the Hudsonian zone of Merriam *et al.* (1910).

Picea-Pinus association (petran subalpine forest): Extends southward at higher elevations in Rocky Mountains to Arizona, New Mexico, and higher peaks of Mexico; contains Engelmann and blue spruces, subalpine fir, and several species of pine.

Tsuga-Pinus association (Sierran subalpine forest): Occurs chiefly in Cascade Mountains and Sierra Nevada; mountain hemlock as well as various pines, subalpine larch, and red fir prominent; trees tall and narrowly cylindrical at lower elevations but dwarfed, gnarled, and misshapened at tree-line; aspen and lodgepole pine extensive as seral stages after fire in both Sierran and petran subalpine forests.

Pinus-Pseudotsuga association (petran montane forest): At lower elevations in the Rocky Mountains; ponderosa pine, Douglas-fir, blue spruce, and white fir most important; ponderosa pine most aridity-tolerant; trees often widely spaced with grass stratum underneath, sometimes forming savannas (Fig. 21-1).

Pinus-Abies association (Sierran montane forest): Contains species listed for petran montane forest and also sugar pine, incense-cedar, and giant sequoia (central Sierras); chaparral develops after fire.

Pinus-Pinus association (Mexican pine forest): An extension of montane forest, chiefly pines, at higher elevations in Mexico.

Thuja-Tsuga association (coast forest): A luxuriant humid forest on the Pacific slope of mountains from northern California to Alaska (Day 1957, Fonda and Bliss 1969); western hemlock, western redcedar, Alaska-cedar, Douglas-fir, Sitka spruce, and redwood most characteristic; trees sometimes 90 m (300 ft) high and to 6 m (20 ft) diameter; deep shade in climax forest but in openings there may be dense tangles of shrubs, lianas, tall ferns; moss often thick over ground and fallen logs; forest in North extends to west slopes of Rocky Mountains in Idaho, Montana, and British Columbia to form a *Coast forest ecotone* with petran montane and subalpine forests, in which grand fir, western white pine, and western larch are prominent.

Animal Communities

There are three principal biociations in this biome, two in North America and one in Eurasia. There is overlap in their species compositions. Species occurring in seral or climax stages of both North American biociations, although less common in the Mexican pine forests, include (Fig. 21-2):

Mammals

Water shrew	Deer mouse
Snowshoe rabbit	Porcupine
Red squirrel	Gray wolf
Northern flying squirrel	Black bear

Birds

Goshawk	Red-breasted nuthatch
Pigeon hawk	Brown creeper
Ruffed grouse	Winter wren
Great horned owl	Hermit thrush
Saw-whet owl	Swainson's thrush
Yellow-bellied sapsucker	Golden-crowned kinglet
Hairy woodpecker	Ruby-crowned kinglet
Black-backed three-toed woodpecker	Solitary vireo
Northern three-toed woodpecker	Orange-crowned warbler
	Wilson's warbler
Traill's flycatcher	Purple finch
Olive-sided flycatcher	Pine grosbeak
Gray jay	Pine siskin
Common raven	Red crossbill
	Lincoln's sparrow

North American Boreal Forest Biociation

This biociation extends from the Atlantic Ocean to the Rocky Mountains in Canada and south on the Appalachian Mountains to northern Georgia (Shelford and Olson 1935, Kendeigh 1947, 1948, Munroe 1956). There is a broad overlap or fusion between the boreal and montane forest biociations in the northern Rockies, where species of one biociation penetrate into the other (Rand 1945, Drury 1953).

Fig. 21-2 Species common in the coniferous forest biome: (left to right, above) porcupine; gray jay; (below) moose, boreal forest; wapiti, montane forest (courtesy U.S. Forest Service).

Characteristic mammals that occur generally through the boreal and pine-hemlock forests, in addition to those listed in the above section, are:

Arctic shrew	Rock vole
Masked shrew	Meadow jumping mouse
Smoky shrew	Woodland jumping mouse
Pigmy shrew	American marten
Star-nosed mole	Fisher
Hoary bat	Ermine
Least chipmunk	Least weasel
Northern bog lemming	Wolverine
Gapper's red-backed	Lynx
mouse	Moose
Ungava phenacomys	Woodland caribou

Bird species found in this biociation are listed in Table 21-1. This biociation is especially notable for the large representation of wood warblers in the avifauna, each with its own specialized niche (MacArthur 1958). In northern Ontario, warblers constitute 69 per cent of the breeding bird population in the spruce-fir forest; in northern Maine, 63 per cent.

As one proceeds south from Ontario and Maine into Minnesota, Michigan, New York, and along the Appalachian Mountains to Tennessee, species both of mammals and birds drop out, apparently as they reach limits of tolerance to climatic factors. Perhaps the elimination of these competing species, or possibly the change in climatic conditions, allows other species to become more abundant. This is especially noticeable among birds—the red-breasted nuthatch, brown creeper, winter wren, golden-crowned kinglet, solitary vireo, black-throated green warbler, blackburnian warbler, Canada warbler, and slate-colored junco attaining much larger populations in the Smoky Mountains of Tennessee than in northern Ontario. In addition the veery, black-throated blue warbler, and often black and white warblers become numerous. This constitutes a variation in the boreal forest biociation (Stewart and Aldrich 1952) which may be designated the *Appalachian faciation*.

When hemlock and spruce-fir forests occur in the same region, as in eastern Tennessee, some bird species adaptable to both show a definite preference for one over the other (Table 21-1). A similar differentiation of

Table 21-1 Comparison of boreal forest avifaunas and population densities (number per 40 hectares, or 100 acres) of breeding birds in the Black Sturgeon Lake area of northern Ontario (Kendeigh 1947), in Aroostook County, northern Maine (Stewart and Aldrich 1952), and in the Great Smoky Mountains of eastern Tennessee (Fawver 1950).

Bird species	Northern Ontario Spruce-fir	Northern Maine Spruce-fir	Eastern Tennessee Spruce-fir	Eastern Tennessee Eastern hemlock
Goshawk	+	0	0	0
Broad-winged hawk	+	+	0	0
Pigeon hawk	+	0	0	0
Spruce grouse (Fig. 21-3)	1	+	0	0
Ruffed grouse	2	+	+	+
Yellow-shafted flicker	4	2	0	1
Pileated woodpecker	+	+	0	+
Yellow-bellied sapsucker	4	+	0	0
Hairy woodpecker	2	2	3	3
Downy woodpecker	2	+	0	0
Northern three-toed woodpecker	2	+	0	0
Black-backed three-toed woodpecker	+	0	0	0
Yellow-bellied flycatcher	2	5	0	0
Acadian flycatcher	0	0	0	7
Least flycatcher	6	+	0	0
Gray jay	4	+	0	0
Blue jay	1	3	0	0
Common crow	0	+	0	0
Black-capped chickadee	2	4	2	7
Boreal chickadee	2	8	0	0
Red-breasted nuthatch	3	8	20	10
Brown creeper	9	+	38	3
Winter wren	5	4	34	7
Robin	0	6	3	4
Wood thrush	0	0	0	22
Hermit thrush	1	4	0	0
Swainson's thrush	4	21	0	0
Veery	0	+	18	7
Golden-crowned kinglet	8	12	38	2
Ruby-crowned kinglet	2	8	0	0
Solitary vireo	2	9	24	39
Red-eyed vireo	7	6	0	1
Black and white warbler	0	+	0	0
Tennessee warbler	59	9	0	0
Nashville warbler	8	8	0	0
Parula warbler	0	5	0	0
Magnolia warbler	6	55	0	0
Cape May warbler	28	28	0	0
Black-throated blue warbler	0	2	6	64
Myrtle warbler	3	20	0	0
Black-throated green warbler	6	+	0	55
Blackburnian warbler	6	22	0	44
Chestnut-sided warbler	0	+	2	0
Bay-breasted warbler	92	69	0	0
Ovenbird	10	2	0	48
Mourning warbler	2	+	0	0
Canada warbler	0	1	0	28
Scarlet tanager	0	+	0	6
Rose-breasted grosbeak	0	+	0	2
Evening grosbeak	+	0	0	0
Purple finch	2	4	0	0
Pine grosbeak	0	2	0	0
Pine siskin	1	2	0	0
Slate-colored junco	3	10	125	29
White-throated sparrow	18	8	0	0
Totals	319+	349+	310+	389+

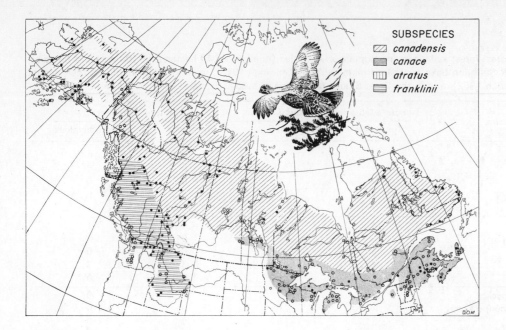

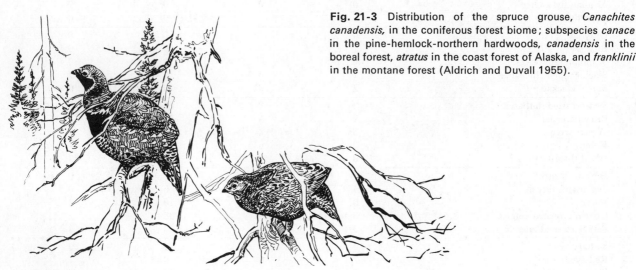

Fig. 21-3 Distribution of the spruce grouse, *Canachites canadensis,* in the coniferous forest biome; subspecies *canace* in the pine-hemlock-northern hardwoods, *canadensis* in the boreal forest, *atratus* in the coast forest of Alaska, and *franklinii* in the montane forest (Aldrich and Duvall 1955).

bird populations in these two forests is evident in Algonquin Park in southern Ontario (Martin 1960). The long association of hemlock with deciduous forest may also have permitted the invasion into the former of the Acadian flycatcher, wood thrush, overbird, and scarlet tanager.

In the Catskill Mountains of New York, but not in the Cheat Mountains of Virginia (Brooks 1943), a zone of deciduous forest intervenes between the lower hemlock and the higher spruce-fir forests. Several warblers and other species, normally characteristic of the hemlock forest, occupy niches in the deciduous forest as well and attain high populations therein (Saunders 1936, Brooks 1940, Kendeigh 1945). Such intermingling of species in an ecotone is to be expected and is especially characteristic of the Appalachian faciation. This intermingling of species between coniferous and deciduous forests is also shown with insects in the Smoky Mountains (Whittaker 1952).

Some of the bird species listed above reach larger populations in a seral shrub or forest-edge biocies (Adams 1909). The Philadelphia vireo, palm warbler, Wilson's warbler, rusty blackbird, and Lincoln's

sparrow are largely limited to shrubs or second growth; the northern waterthrush occurs in bogs; the savannah sparrow, in marshes and grassy areas; and the white-winged crossbill, irregularly through the climax. These species extend to the northern tree-line. Seral aquatic stages in the boreal forest contain beaver, muskrat, and nesting horned grebe, black duck, common goldeneye, Canada goose, and the common and hooded mergansers (Hanson *et al.* 1949). Actually, the coniferous forest does not develop a recognizable forest-edge along its southern border because these borders grade by steps into deciduous forest, aspen parkland, woodland, and chaparral. The closest resemblance to an edge are shrubby openings within the forest or the subseres that develop in bogs, burns, and logged areas. The aspen parkland contains a fauna in which grassland, boreal forest, deciduous forest, and deciduous forest-edge species are all represented (Bird 1930) and is essentially an ecotone. Invertebrate composition of the seral stages bears a strong resemblance to that occurring in seral stages of the deciduous forest.

Along its northern border, the coniferous forest comes in direct contact with open tundra to form a broad ecotone. Boreal forest biociation species reach their northern limits of distribution and tundra species begin to appear. Total breeding bird populations decrease to less than 100 pairs per 40 hectares. There are no distinctive mammals, but several birds are characteristic of this subarctic, lichen woodland and muskeg, Hudsonian, or (most apt) *forest-tundra faciation* (Manning 1952, Harper 1953, 1956, 1964, Preble 1908, Gillespie 1960):

Solitary sandpiper	Blackpoll warbler
Lesser yellowlegs	Pine grosbeak
Rough-legged hawk	Hoary redpoll
Boreal owl	Common redpoll
Hawk-owl	Tree sparrow
Great gray owl	Harris' sparrow
Northern shrike	White-crowned sparrow
Gray-cheeked thrush	Fox sparrow
Bohemian waxwing (west)	Smith's longspur

The pine grosbeak and white-crowned and fox sparrows extend southward at tree-line on the western mountains, and the gray-cheeked thrush and blackpoll warbler extend southward at high elevations in the northern Appalachians.

North American Montane Forest Biociation

The main body of the North American montane forest biociation, considering the forest-interior and forest-edge together, occurs principally in the coast forest (Storer *et al.* 1944, Miller 1951, Macnab 1958, Gashwiler 1970) and is less developed in the montane and subalpine forests of the Rocky Mountains, Cascades, and Sierra Nevada (Rasmussen 1941, Hayward 1945, Munroe 1956, Snyder 1950, McKeever 1961). Because

of the mountainous terrain and the many possibilities for populations to become partially or wholly isolated from each other, there are many local subspecies and species of mammals and birds (Findley and Anderson 1956). The following lists include only common species of wide distribution through the biociation:

Mammals

Shrews	Long-tailed vole
Mountain beaver	Western jumping mice
Western chipmunks	Grizzly bear
Yellow-bellied marmot	Western marten
Golden-mantled ground squirrel	Mountain weasel
Douglas' squirrel	Wolverine
Bushy-tailed wood rat	Mountain lion
Red-backed mice	Bobcat
Heather vole	Mule deer
	Wapiti

Birds *(Fig. 21-4)*

Golden eagle	Pigmy nuthatch
Blue grouse (Fig. 21-5)	Dipper
Flammulated owl	Varied thrush
Pygmy owl	Mountain bluebird
Calliope hummingbird	Townsend's solitaire
Williamson's sapsucker	Virginia's warbler
White-headed woodpecker	Audubon's warbler
Dusky flycatcher	Townsend's warbler
Hammond's flycatcher	Hermit warbler
Western flycatcher	Western tanager
Western wood pewee	Evening grosbeak
Steller's jay	Cassin's finch
Clark's nutcracker	Oregon junco
Mountain chickadee	Gray-headed junco

Fig. 21-4 Foraging niches of birds in the montane forest biociation of the central Rocky Mountains (Salt 1957).

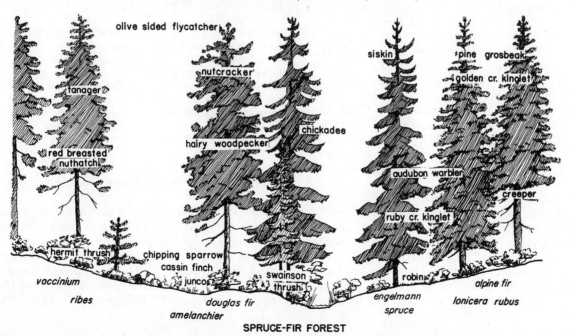

SPRUCE-FIR FOREST

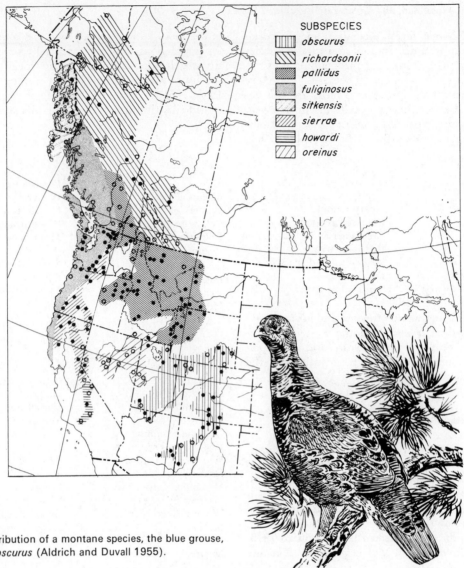

SUBSPECIES
- ⊞ *obscurus*
- ⊠ *richardsonii*
- ◩ *pallidus*
- ▨ *fuliginosus*
- ▨ *sitkensis*
- ▨ *sierrae*
- ⊟ *howardi*
- ▨ *oreinus*

Fig. 21-5 Distribution of a montane species, the blue grouse, *Dendragapus obscurus* (Aldrich and Duvall 1955).

In Colorado, the population of breeding birds is less than one-sixth what it is in the boreal forest (Snyder 1950). In the northern and central Rockies there is considerable mixture with species from the boreal forest, both in birds and mammals but these boreal species drop out progressively southward and very few of them cross into the Coast forest. The woodland caribou, for instance, ranges only to northeastern British Columbia and the moose to central British Columbia, eastern Idaho, and western Wyoming. The western facies of the deciduous forest-edge biociation penetrates widely as a seral stage through the western forest biociation, and certain of its species may persist into the climax (p. 321).

Some initial studies, especially of birds, indicate that a *Mexican faciation*, or possibly a biociation, is distinguishable. This faciation contains many species of southern origin. The faciation includes the Mexican pine forest and much of the pine-oak woodland, both of which are derived from the Madro-Tertiary Geoflora rather than the Arcto-Tertiary Geoflora as is the rest of the biome. Although several species from the

above lists occur, the more characteristic species include (Marshall 1957, Balda 1967):

Whiskered owl	Beardless flycatcher
Broad-tailed hummingbird	Mexican jay
Rivoli hummingbird	Mexican chickadee
Blue-throated hummingbird	Olive warbler
Coppery-tailed trogon	Grace's warbler
Arizona woodpecker	Painted redstart
Sulphur-bellied flycatcher	Red-faced warbler
Olivaceous flycatcher	Hepatic tanager
Coues' flycatcher	Mexican junco

Eurasian Boreal Forest Biociation

The tree dominants of the Eurasian plant associations are different species but the same genera of pines, firs, larches, spruces, poplars, and birches that occur in North America. This biociation is best developed in Asia, from whence the biota is dispersed across the northern part of the continent into Europe (Berg 1950, Jahn 1942, Kalela 1938, Palmgren 1930, Pleske 1928,

Stegmann 1932, 1938, Haviland 1926, Schäfer 1938, Soveri 1940, Turček 1956).

The mammal fauna contains shrews, a varying hare, flying and red squirrels, a chipmunk, red-backed mice, the wolf and red fox, a brown bear, martens, weasels, wolverine, lynx, a moose, and a deer. Several of these species (wolf, red fox, wolverine, lynx) are considered by some taxonomists to be conspecific with North American forms (Rausch 1953).

This biociation is equivalent to the Siberian bird fauna of Stegmann (1938) and includes several species of grouse, owls, woodpeckers, crows and jays, and tits, a creeper, several thrushes, several Old World warblers, kinglets, a wagtail, waxwings, and several finches or sparrows. The wood warblers, abundant in the boreal forest of North America, are absent. There is also a forest-tundra faciation of birds in Eurasia as occurs in North America (Johansen 1963).

Paleoecology

In the early Tertiary we may suppose that the boreal unit of the Arcto-Tertiary forest had a fairly uniform animal composition from eastern Canada into Asia and Europe. As the forest progressed southward during the middle and later Tertiary, a large segment became separated in consequence of the submergence of the Bering land bridge, becoming the Eurasian biociation. Forms now peculiar to the Eurasian and to the North American biociations must have evolved after this separation took place (Udvardy 1958).

In North America, as the Arcto-Tertiary biota retreated southward with the progressive chilling of the continent, it was separated into eastern and western portions by the northward invasion of grassland over the Great Plains, except as it had contact through the boreal forest across Canada in the north. During the Pleistocene even this northern contact was reduced with each major advance of the glacier. Although coniferous forest extended at these times well into the grassland biome, perhaps as a continuous unbroken forest it was limited to a narrow belt in the north. Southward it may have taken the form of savanna or woodland. Its continuity across the continent was further interrupted by the barren outwash plains from

the glacier into the Mississippi valley and by the grass-covered Dakota badlands and Nebraska sandhills (Wright 1970). Furthermore, the western part of the continent was thrown up into mountains beginning in the Miocene, and the climate there became more diversified and rigorous. Plant and animal species tended to segregate into either the western or the eastern section of the forest, depending on where habitat conditions and community coactions were more favorable to them, and isolation encouraged divergent speciation. The western section continued to have sporadic contact with the Eurasian biociation, especially during glacial periods of the Pleistocene, but the eastern section was too far away. Hence came the differentiation of the boreal forest biociation in the eastern lowlands of the continent and the montane forest biociation in the western mountains and on the Pacific coast.

In this connection it is of interest to compare the origin of the breeding bird faunas in different coniferous forest communities and in the deciduous forest (Table 21-2). Southward in the western mountains there is a decrease in importance of species of Eurasian ancestry and an increase of species of North and South American origin. The North American element is of relatively greatest importance and the South American element of least importance in the boreal forest.

Pleistocene glaciation enhanced the differentiation of boreal and western forest biociations since it allowed independent subspeciation and even speciation in the four refugia (Fig. 19-6). The boreal forest became compressed with each glaciation into the Appalachian refugium, but the western forest was segregated three ways into the Rocky Mountain, Pacific, and Alaskan refugia. At these times the Alaskan refugium was probably connected by the Bering land bridge to Asia.

The present-day distribution of the four subspecies of moose suggests that they were isolated during at least Wiscosin glaciation in the Appalachian, Rocky Mountain, and Alaskan refugia, and in the unglaciated area of Wisconsin, Minnesota, and Illinois (Fig. 21-6). This area probably served also as the refugium for the western subspecies of the woodland caribou, while the eastern subspecies was isolated in the Appalachian refugium (Vos and Peterson 1951). Different subspecies of arctic shrews had refugia to the south and east of the glacier and in Alaska. The American marten and the

Table 21-2 Origin of avifauna in coniferous forest communities of the western hemisphere. Percentages of species and breeding pairs of different ancestry (Mayr 1946, Snyder 1950, Balda 1967).

Forest	Locality	Eurasia		North America		South America		Unanalyzed	
		Species	Pairs	Species	Pairs	Species	Pairs	Species	Pairs
Spruce-Douglas-fir	Colorado	69	85	15	8	0	0	16	8
Spruce-Douglas-fir	Southern Arizona	50	57	31	25	8	11	12	7
Ponderosa pine	Colorado	67	55	13	38	7	4	13	1
Ponderosa pine	Southern Arizona	35	40	26	36	23	14	16	10
Boreal forest	Ontario, Maine	52	20	30	79	3	+	15	1
Deciduous forest	Ohio	28	23	32	52	28	23	12	2

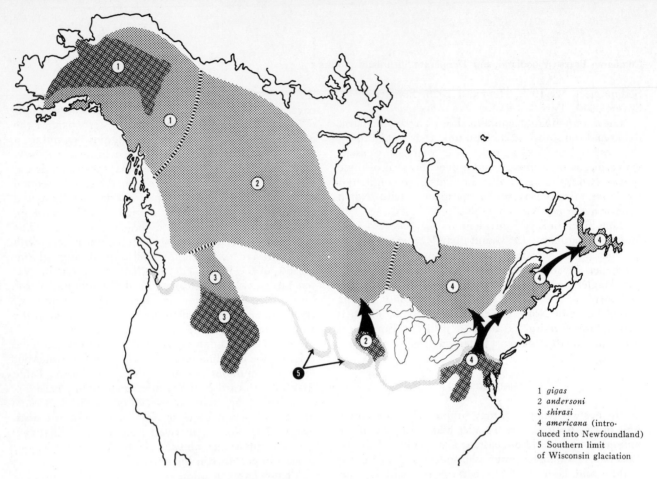

1 *gigas*
2 *andersoni*
3 *shirasi*
4 *americana* (intro-
duced into Newfoundland)
5 Southern limit
of Wisconsin glaciation

Fig. 21-6 Present distribution of four subspecies of moose, *Alces americana,* in North America. Post-Pleistocene dispersal routes from distributions at the time of Wisconsin glaciation are shown by arrows (from Peterson 1955).

red squirrel apparently survived in the Appalachian refugium; the western marten and Douglas' squirrel, in the Pacific refugium. Of the red-backed mice, *Clethrionomys gapperi* has apparently dispersed from the Appalachian refugium, *C. dawsoni* from the Alaskan refugium, and *C. wrangeli* from islands off the coast of British Columbia (Rand 1954).

Various subspecies or closely related species of birds apparently differentiated as populations were isolated in one or more of the four refugia. This seems to have occurred with the spruce grouse, sapsucker, gray jay, boreal chickadee, certain warblers (Fig. 21-7), slate-colored and Oregon juncos, and white-crowned sparrow (Rand 1948, Drury 1953, Hubbard 1969).

In each interglacial period, the coniferous forest fauna previously isolated in the Appalachian, Rocky Mountain, Pacific, and Alaskan refugia doubtless dispersed centrifugally from each center until they came into contact with each other. Such segregation and dispersal of the biota must have occurred four times during the Pleistocene; the dispersal from refugia after the Wisconsin glaciation is still going on. The four subspecies of moose have come into contact with each other (Peterson 1955), the least chipmunk has entered Ontario and Quebec (Peterson 1953), and the evening grosbeak has spread across Canada from the western mountains only within the past hundred or so years. Some of these changes in range may have been

hastened as a result of human interference. The woodland caribou was formerly the principal large ungulate present in the boreal forest (Vos and Peterson 1951), but as logging and fires opened up the forest, the caribou has greatly declined in numbers and the moose has become more abundant. The white-tailed deer has also spread from the deciduous forest well into the boreal forest in recent years, and other species of mammals and birds appear in the process of doing so.

ANIMAL ADJUSTMENTS

Animal adaptations for life in coniferous forest are similar in many ways to those for life in deciduous forest. Ecological niches in these two forests are similar, although the species that occupy them are different. Important differences are the stiff needle-shaped character of conifer leaves and their arrangement around all sides of the twigs, which hinder the movements and feeding of some birds, and the poor decomposition of the needle leaves that accumulate on the ground, not favorable to high populations of many species of invertebrates. In contrast to the aspects of the deciduous forest, the vernal and autumnal aspects are less well developed since most of the trees retain their foliage throughout the year.

The woodland caribou is largely restricted to the

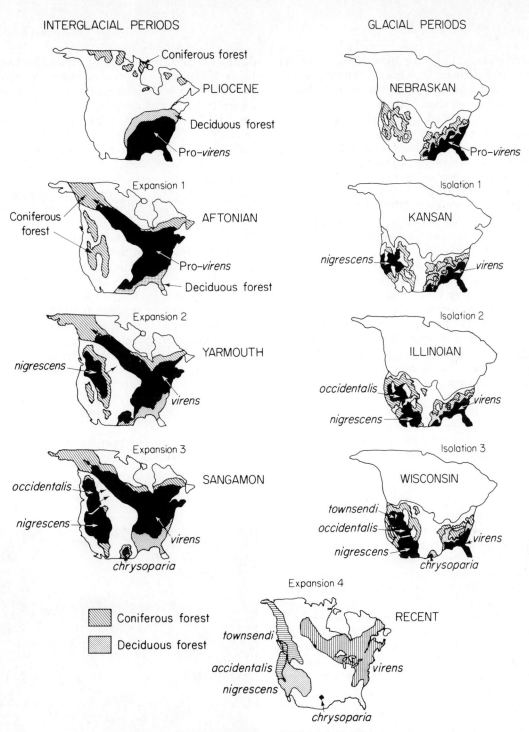

Fig. 21-7 Evolution of the black-throated green warbler group of closely related species during the Pleistocene. The ancestral "pro-*virens*" was probably an inhabitant of the deciduous forest. Populations ebbed and flowed with glacial thrusts and when restricted in range and isolated from one another differentiated both in species and in preferred vegetative niche. The five species involved are *Dendroica virens,* black-throated green warbler, boreal forest; *D. nigrescens*, black-throated gray warbler, woodland; *D. occidentalis*, hermit warbler, Sierran montane forest; *D. townsendi*, Townsend's warbler, coastal forest; *D. chrysoparia*, golden-cheeked warbler, mixed deciduous and cedar forest on Edwards Plateau, Texas (modified from Mengel 1964). This same model of evolution and dispersal is also evident with some other groups of wood warblers, flycatchers, and juncos (Mengel 1970).

climax forest, where it feeds on reindeer moss, a ground lichen, and on tree lichens. Moose are found throughout seral stages as well as in the climax. During the summer they commonly feed on water lilies, pondweeds, sedges, and grasses; during the winter, on the tips of birch, aspen, cedar, balsam fir, and various other shrubs and small deciduous trees (Shelford and Olson 1935). Small mammals are abundant. In northern Michigan, the populations of two species each of mice, chipmunks, and shrews varied from 6.2 individuals per hectare (2.5 per acre) in jack pine to 12.5 (5.0 per acre) in black spruce, 16.0 (6.4 per acre) in hemlock, 19.5 (7.8 per acre) in a white-cedar swamp, and 28.2 (11.3 per acre) in white birch (Manville 1949).

Perennial animals that remain active over winter have a high tolerance of low temperatures and use food not readily obscured by snow (Snow 1952). The large mammals become browsers in the winter (Fig. 21-8). Wapiti chew patches of bark off aspen trees when other forage is difficult to find. Scars thus formed are ideal sites for the development of fungus disease (Packard 1942). Birds feed on seeds extracted from the cones of the coniferous trees, on buds, and on bark insects. When the seed crop fails, large numbers of pine siskins, pine and evening grosbeaks, red crossbills, and white-winged crossbills emigrate southward into the United States. Small ground and subterranean animals are well insulated under the snow, where temperatures even in the far north may drop only a few degrees below freezing (Pruitt 1957). Some birds, such as the grouse, roost at night in cavities formed in snowbanks.

Less than half of the nesting bird population of the montane forest biociation migrates for the winter, and then only to lower altitudes on the mountains. In contrast, the birds of the boreal forest are acclimatized to warm climate, and over 80 per cent migrate hundreds of kilometers to the south. A few mammals also migrate, such as the hoary bat in the East and the wapiti and mule deer down the mountain slopes in the West.

Insects virtually dominate the forest, at times. Vast numbers of mosquitoes and flies force moose to spend much of the summer submerged in water, and are generally annoying to other animals and man. The larch sawfly has spread across Canada and the northern states during the last 85 years and caused considerable

Fig. 21-8 Adjustments of mammals and birds for feeding in deep snow (Siivonen 1962).

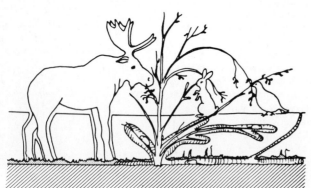

defoliation and destruction of tamarack (Coppel and Leius 1955). The spruce budworm (a lepidopteran larva) feeding on the leaves has killed balsam fir and spruce trees on vast areas at repeated intervals in the past: 1807–18, 1870–80, 1904–14 (Swaine and Craighead 1924), and again in the 1940s. Several kinds of bark beetles, wood borers, and long-horned beetles are also destructive forest insects (Figs. 21-9, 10).

Beetles, ants, aphids, jumping plant-lice, leaf-hoppers, spiders, and other invertebrates—notably snails, annelids, and millipedes—are not numerous over most of the biome (Rasmussen 1941, Hayward 1945, Blake 1945). Most ground invertebrates have higher population densities in the seral aspen and birch stages than in the coniferous climax (Hoff 1957). Reptiles are few; only the garter snake extends very far north. The northern wood frog, leopard frog, and mink frog are widely dispersed in suitable habitats throughout the boreal biociation. Because of its greater humidity and more equable temperatures, invertebrates and cold-blooded vertebrates are generally more numerous in the Coast forest than elsewhere through the biome.

Human Relations

Only the lower, warmer portions of the coniferous forest biome are permanently inhabited in large numbers by

Fig. 21-9 Gallery pattern of a bark beetle in lodgepole pine: (a) nuptial chamber; (b) egg gallery; (c) egg niche (Reid 1955).

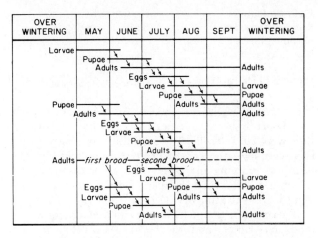

Fig. 21-10 Seasonal history of a bark beetle (Reid 1955).

white man throughout the year. Logging for pulpwood and lumber is an important occupation. Over the more rugged northern portions of the coniferous forest, the population is scattered and, in North America, there are more Indians than whites, at the present time. The Indians engage in hunting and fur-trapping for support. Larger settlements of white men occur where minerals may be mined or oil obtained. These regions, as well as the higher conifer-clad mountains, are resorted to for fishing and other recreational activities during the warm summer months.

WOODLAND BIOME

In contrast to forest, woodland is an open stand of trees with the ground nearly continuously covered with grasses. The trees are usually short, 6 to 15 m high (20 to 50 ft), but may have a dense crown. In favorable local habitats, the trees form a closed canopy, but in arid situations they are scattered. The trees vary widely in leaf structure, but nearly all species are evergreen and tolerant of low moisture. In North America, woodland of different types extend from Washington and Wyoming well down through California, Arizona, and Texas into Mexico. A similar type of broad-leaved, evergreen sclerophyllous woodland, together with chaparral, occurs around the Mediterranean Sea in Eurasia and Africa.

Climate

In Utah, precipitation in this biome ranges from 4 to 6 cm (10 to 15 in.) per year, and mean monthly temperatures from —5° to 21°C (23° to 69°F) (Woodbury 1947). Precipitation is often higher in Mexico, but this is offset by a higher rate of evaporation. West of the Sierra Nevada precipitation comes principally during the winter months, as it also does in the Mediterranean region.

Plant Associations in North America

Quercus-Quercus association (oak woodland): Broad-leaved mostly evergreen oaks with Digger pine in certain habitats (Miller 1951); mostly west of the Sierra Nevada but extending north into Oregon and Washington.

Pinus-Juniperus association (piñon-juniper woodland): Pines and junipers of several species from eastern slopes of Sierras and Cascades across Great Basin to Montana and New Mexico (Woodbury 1947, Woodin and Lindsey 1954).

Pinus-Quercus association (pine-oak woodland): On mountain slopes of central and northern Mexico into southern California, Arizona, New Mexico, and Texas; oak scrub, mostly evergreen, at lower elevations grades into pine-oak woodland with some juniper at higher elevations and then into the Mexican pine forest; contains a rich variety of species (Gentry 1942, Leopold 1950, Marshall 1957).

Paleoecology

The various types of woodlands in North America were derived from the mixed pines, junipers, and oaks of the Madro-Tertiary Geoflora. The piñon-juniper woodland is a segregation that became adapted to the cold winter climates of the Rocky Mountains and Great Basin. The oak woodland during lower Pliocene was widely distributed over the central and southern portions of the Great Basin but with the trend toward colder winters and decreased rainfall came to be restricted to the moister mountain habitats within the desert and to Pacific coast regions with winter rain and mild temperatures. Woodlands were more widely dispersed northward and at lower altitudes during the pluvial Wisconsin periods and again during the post-glacial altithermal period than they are at present (Axelrod 1950, 1957, Martin 1963). There is some evidence that woodland vegetation including oaks once extended around the north side of the Gulf of Mexico as far as Florida (Pitelka 1951).

Woodland Biociation

The animal life of the oak and piñon-juniper woodland communities in western North America (Fig. 21-11) is not highly distinctive. The trees, being broad-leaved or needle-leaved, attract species from both the riparian woodland and montane forest biociations. Since the trees are scattered and interspersed with grass or shrubs, chaparral, grassland, and desert species also penetrate well into the community. In respect to species composition, therefore, these woodlands in North America to a large extent are an ecotone. The animal constituents

Fig. 21-11 Piñon-juniper woodland in Utah.

of the pine-oak woodland are similar to those of the Mexican forest, as indicated above.

Mammals

Of larger mammals, the mule deer, mountain lion, and coyote commonly occur during the winter months in the piñon-juniper woodland of Utah and Arizona, although most of these species spend the summer high in the mountains. The bobcat also occurs and the grizzly bear was formerly not uncommon. The rock squirrel, cliff and panamint chipmunks, desert and dusky-footed wood rats, and piñon mouse are found in both the piñon-juniper and the petran bush but show preference for broken country, rocky hillsides, and cliffs (Woodbury 1933, Rasmussen 1941, Deacon *et al.* 1964). In southern New Mexico, four species of mice—deer, brush, rock, and piñon—occur more or less together (Dice 1942). The open floor of the oak woodland in California is relatively devoid of mammal life, with only the California and brush mice common in the vicinity of brushy growth. The western gray squirrel is probably most common in this community (Vaughan 1954).

Birds

Certain bird species appear to be more characteristic of woodland than are mammals. Species occurring rather widely in Arizona (Rasmussen 1941), Utah (Hardy 1945), California (Miller 1951), and Mexico (Marshall 1957) are:

Band-tailed pigeon	Nuttall's woodpecker
Black-chinned hummingbird	Ash-throated flycatcher
Acorn woodpecker	Gray flycatcher
Lewis' woodpecker	Coues' flycatcher
Ladder-backed woodpecker	Yellow-billed magpie
	Scrub jay
	Piñon jay

Plain titmouse	Hutton's vireo
Brindled titmouse	Black-throated gray warbler
Common bushtit	
Blue-gray gnatcatcher	Grace's warbler
Scott's oriole	House finch
Western bluebird	Lawrence's goldfinch

Rattlesnakes, lizards, and horned toads invade from the desert but are not particularly characteristic of the woodland itself. Invertebrate populations are relatively low, and consist principally of spiders, ants, termites, jumping plant-lice, and a sprinking of ichneumonids, flies, leafhoppers, beetles, and banded-wing locusts (Rasmussen 1941).

There is a distinct Mediterranean avifauna of woodland, chaparral, and desert grassland that is best developed in northern Africa but extends north through Spain, Italy, the Balkans, Turkey, and southwestern Asia (Stegmann 1938).

TEMPERATE SHRUBLAND

Two kinds of shrubland occur, evergreen and deciduous. Chaparral, in the strict sense, consists of xeric broad-leaved evergreen bushes, shrubs, or dwarf trees, usually not more than 2.5 m (8 ft) high, and occurring in dense more or less continuous stands. Leaves may be thick and heavily cutinized, odorous, glaucous, or hairy. Beneath the bushes and shrubs there may be abundant ground litter. Chaparral is less dense where there are rock outcroppings and grass. The occurrence of fire every few years aids in the perpetuation and dispersal of chaparral (Hanes 1971). Most species readily produce sprouts after their tops are destroyed by fire, provided fire does not occur too frequently; germinatian of some seeds is hastened by the heat of the fire. Chaparral tends to spread as a seral stage into areas of montane forest and woodland when the latter is destroyed by fire. Although chaparral is doubtless seral over much of its range, it appears to be climax over fairly large areas in southern California and northern Baja California, and a narrow belt on the slopes of the Sierra Nevada and southern Rockies. This is a region with winter rains and dry summers—the so-called Mediterranean climate. As one progresses inland out of the winter-rain region, the shrubs become deciduous. Broad-leaved evergreen chaparral also occurs, with woodland, around the Mediterranean and elsewhere on other continents.

Plant Associations in North America

Coastal chaparral occurs from southern Oregon to northern Baja California and eastward into Nevada and Arizona (Weaver and Clements 1938).

Petran bush (or interior shrubland) occurs as a lower zone on the mountains from South Dakota to Texas and westward into Nevada and Arizona (Fig. 21-12). Although the shrubland in Arizona is partly evergreen, that in most other areas is deciduous.

Both associations are derived from the Madro-Tertiary Geoflora (Davis 1951) and have a phylogenetic history similar to the oak woodland and piñon-juniper woodland respectively, with which they are closely associated.

Chaparral Biociation

There are no mammals peculiar to the chaparral in North America, although in Utah the bobcat, rock squirrel, and cliff chipmunk reach relatively large populations in the petran bush (Hayward 1948). In California, the brush rabbit and the dusky-footed wood rat are numerous in heavy brush, along with other mammals also found in woodland (Vaughan 1954). There may be 6 to 12 occupied houses of the white-throated wood rat per hectare (2–5 per acre) in southern Arizona (Hanson 1957). The mule deer becomes common (about 10 per km² or 25 per mi²) during the winter when it migrates down from the higher elevations in the mountains. The chaparral fauna, like that of the woodland, is largely ecotonal between montane forest, woodland, grassland, and desert scrub (Fig. 21-13). Mammals are generally able to avoid direct mortality from fires, but with destruction of the vegetation their numbers, especially of the smaller species, become reduced for several years thereafter (Lawrence 1960).

The coastal chaparral is extensive enough, however, so that these birds show preference for it (Miller 1951):

Mountain quail	*Gray vireo
California quail	*MacGillivray's warbler
Anna's hummingbird	*Lazuli bunting
Allen's hummingbird	*Rufous-sided towhee
Wrentit	Brown towhee
Bewick's wren	Rufous-crowned sparrow
California thrasher	Black-chinned sparrow

Fig. 21-12 Petran bush in Utah.

Those species marked with an asterisk are found also in the petran bush of Utah and, in addition, the broad-tailed hummingbird, Virginia's warbler, and green-tailed towhee are characteristic there (Hayward 1948). The petran bush, even more than the coastal chaparral,

Fig. 21-13 Distribution of important mammals on the Kaibab Plateau in Arizona. Shaded areas represent relative distribution in different communities and at different elevations (Rasmussen 1941).

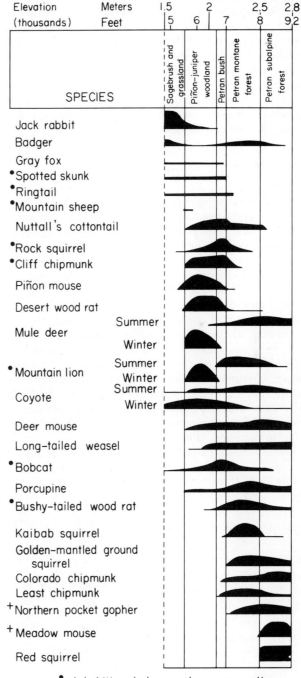

contains many species from the deciduous forest-edge and montane forest biociations.

No snake or lizard is particularly characteristic of chaparral, although at times they may be numerous. For the most part, the reptiles occurring in chaparral belong more properly to the desert or grassland and reach their upper altitudinal limits in this biome. Among invertebrates in the petran bush, mites, ants, leafhoppers, locusts and grasshoppers, beetles, aphids, and flies are conspicuous, and large numbers of parasitic and gall-forming hymenopterans depend on the oaks for completion of their life cycles. A few millipedes and centipedes are to be found under rocks, but snails are scarce (Hayward 1948). In the coastal chaparral of southern California, the period of greatest activity for most invertebrate species comes in March and April, toward the end of the winter rains, and is correlated with the flowering season of plants and the period of greatest soil moisture. During the hot dry summer many invertebrates aestivate (Ingles 1929).

ZONATION

With increase in elevation there is a decrease in temperature and increase in wind velocity, depth of snowfall, and total precipitation (Table 21-3). Mean annual temperature tends to drop about 0.6°C (1.0°F) for each rise of 100 m (325 ft); hence, by going up a mountain a few hundred meters one encounters similar, but not identical, climates and biota as occur at lower elevations many kilometers northward on the continent. In deep valleys and canyons, on the other hand, there is often cold air drainage at night, so that colder types of vegetation and fauna occur than on the nearby ridges. Because mean temperature decreases northward on the continent, the vertical zonation of communities on the mountains occurs at lower elevations northward. Tundra, for instance, that is found only at the tops of the higher mountain peaks in the south occurs at sea level in the north (Fig. 21-14).

Alpine tundra and coniferous forest at the higher

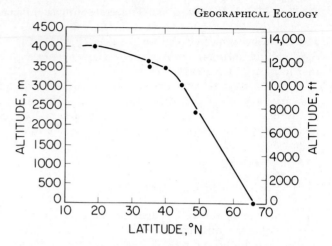

Fig. 21-14 Curvilinear relationship between altitude of the tree-line and latitude from the American subtropics in Mexico, along the Rocky Mountains, to the arctic tree-line near the mouth of the Mackenzie River in Canada. The straight portion of the curve has a slope of 144 m per degree of latitude (Bamberg *et al,* 1968).

elevations represent southward dispersal of the Arcto-Tertiary Geoflora, while woodland, chaparral, grassland, and desert at lower elevations represent northward dispersal from the Madro-Tertiary Geoflora. These dispersals doubtless began as these western mountains became elevated in mid-Tertiary time. During the Pleistocene period, glaciation occurred extensively in the higher mountains and forced all communities to lower elevations (Fig. 21-15). Pluvial climates generally accompanied glaciation so that grassland and desert in the lowlands were succeeded by coniferous forest. More continuous zones of forest in foothills and through valleys encouraged wide latitudinal dispersal of animals. With recession of the glaciers and rewarming of the climate, the forest and alpine tundra again withdrew to higher elevations, and mountain ranges assumed the isolation from each other that we see at the present time. This isolation has induced considerable speciation, but the similarity or relationship between animal life in different mountain regions is explicable from the paleoecological history of the area.

Table 21-3 Climate in different zones on the west slope of the Wasatch Mountains, Utah (Price and Evans 1937).

Community	Elevation (m)	Depth of snow on March 1 (cm)	Total precipitation (cm)	Precipitation as snow (%)	Frost-free period (days)	Mean temperature (°C)		
						May-Oct.	Oct.-May	Annual
Piñon-juniper woodland	1700	3.8	29.7	45	—	17.0	2.0	8.3
Petran bush	2333	63.2	44.4	—	90	14.4	−0.2	5.9
Petran montane forest	2698	120.1	74.9	70	87	11.7	−2.6	3.3
Petran subalpine forest	3078	134.4	71.1	80	80	8.7	−5.7	0.3

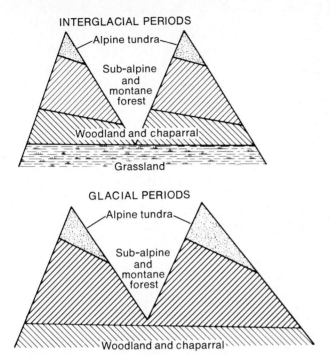

Fig. 21-15 Changes in zonation and continuity of communities between interglacial and glacial periods. Biota isolated on different mountain ranges during interglacial periods may disperse widely during glacial periods.

SUMMARY

Coniferous forests are largely confined to the Northern Hemisphere. Over most of the area, the summer growing season is short and the winter long and cold. On the Pacific coast of North America, however, precipitation occurs mostly during the winters, which are mild.

Some mammal and several bird species occur widely through the biome, but the following animal communities are recognized: North American boreal forest biociation, North American montane forest biociation, and Eurasian boreal forest biociation. Appalachian and forest-tundra faciations of the North American boreal forest biociation are well-marked for birds as is also a Mexican faciation of the montane forest biociation.

In early Tertiary, the boreal unit of the Arcto-Tertiary forest doubless had a fairly uniform fauna from eastern Canada across the Bering land bridge to western Europe. This fauna became differentiated into the three biociations with the disappearance of the Bering land connection between Alaska and Asia, and the separation of the North American fauna into western and eastern sections by the northward penetration of grassland and the southward movements of the Pleistocene glaciers. Differentiation at the subspecies and species levels was further encouraged by the segregation of the North American fauna during Pleistocene glaciation in the Appalachian, Rocky Mountain, Pacific, and Alaskan refugia.

Animal adaptations are similar to those in the deciduous forest, but often more extreme. Boreal species, for instance, must have greater physiological tolerance to cold to remain active over winter. Migration is also more extensive, although in mountainous areas migration is largely alittudinal rather than longitudinal. Browsing is more common, for the ground during winter months is generally covered with snow. Legs are adapted as poles for tramping through snow, as snowshoes for treading on top of the snow, or as diggers for burrowing through the snow. Insects often cause considerable damage to forest trees, and mosquitoes and flies are annoying to both animals and man.

White man is relatively less numerous in this biome in North America and Indians more numerous. Chief occupations are logging, hunting and trapping, and mining.

Woodland and chaparral biociations have some distinctive animal species but in many areas tend to be ecotonal between coniferous forest and grassland or desert.

The zonation of communities on mountains depends principally on decrease in temperature and increase in precipitation with increase in elevation. In western North America, desert or grassland, woodland or chaparral, coniferous forest, and alpine tundra occur at successively higher elevations.

Chapter 22

TUNDRA BIOME

Tundra comprises the communities that extend north of the tree line, or above the tree line on mountains, to the zone of perpetual snow and ice. Tundra communities are essentially similar in North America and Eurasia (Berg 1950), but different and more varied when isolated on temperate and tropic mountains. A little tundra occurs on Antarctica (Castri *et al.* 1966, Holdgate 1970).

CLIMATE, SOIL, AND TOPOGRAPHY

Arctic regions in North America have an annual precipitation less than 4 cm (10 in.), although east of Hudson Bay it occasionally reaches 8 cm (20 in.). Most of it comes as rain during the summer and early autumn; snowfall is generally light (Koeppe 1931). Humidity is high and evaporation low during the summer.

Mean monthly temperatures vary between extremes of −35°C and +13°C (−30° and 55°F). The July isotherm of 5°C (41°F) is sometimes used to separate the so-called *high arctic* and *low arctic*, and the isotherm of 10°C (50°F), which corresponds closely to tree-line, to separate the low arctic from the *subarctic*. Frost may occur at any time in the North, but there is usually a frost-free period of about 60 days in the South.

During the summer, the surface of the ground commonly thaws to a depth of only a few centimeters; permanently frozen soil, *permafrost*, underlies (Ray 1951). The soil becomes wet and soggy, and the accumulation of water in depressions forms numerous shallow ponds. Freezing and thawing are potent forces in arctic regions, since they may occur daily for long periods of time. This action fragments large boulders into small rocks; forms polygon shapes on level ground surfaces varying in diameter from a few centimeters to several meters; develops large ground ice or peat mounds or smaller hummocks (frost heaving); causes *solifluction*, a downward slumping of soil on slopes to form terraces, or, a gradual creep of rocks and soil downslope with the consequent rounding off of ridges and other irregularities in the topography. The general molding of the landscape by frost action is called *cryoplanation* and is of ecological importance because it makes the soil unstable and limits the kind of vegetation that can develop on it.

During the winter, the soil freezes down to the permafrost, except under streams, on stream banks and narrow floodplains, and in sandy areas. Lakes are frozen for 9 months of the year with ice a meter or more thick. Under the ice, oxygen almost disappears, so that conditions are usually critical for animal survival (Andersen 1946). Ponds less than 1 or 2 m deep freeze to the bottom.

In the spring, absorption of solar radiation causes

the mean temperature of the surface of the soil to rise above freezing 3 or 4 weeks before the mean temperature of the air, and it is at this time that plants and ground animals renew their growth and activity (Sorensen 1941). Below the surface, ground temperature rise lags behind rising air temperature. Temperature just below the surface is higher than at greater depths in the summer, but lower in the winter, the turnover coming in April or May in the spring and September or October in the autumn (Beckel 1957).

Photoperiods north of the Arctic Circle vary from zero hours during mid-winter to 24 hours during mid-summer. Elsewhere, the length of periods depends on latitude. Even in the summer, however, light intensity is low compared with tropical latitudes.

In contrast to the arctic plains, alpine habitats on the higher mountain slopes are on rugged, often precipitous, terrain. Each mountain top is an island isolated from other mountain tops by intervening forested lowlands. Dispersal and ecesis of organisms between mountain tops follow much the same rules as between oceanic islands (Vuilleumier 1970). High plateaus may have more or less level surfaces, but the size of such tundra-covered areas, except for the Tibetan plateau of Asia, is generally very limited. These alpine habitats also lack a permafrost in the subsoil, except in the far north, and the extreme change in photoperiod during the year that arctic regions experience. Soils are thin and unstable except in small pockets on the slopes, in valleys, or on protected flat surfaces. Average temperature is low, but the range between daily maximum and minimum during the summer is sometimes as great as 32°C (58°F). North slopes are colder than south slopes. The length of the growing season between killing frosts is similar to what it is in the arctic. Precipitation and humidity are commonly high, and the mountain tops are frequently shrouded in fog. Snowfall in some areas, as in the Cascades, may reach 18 m (60 ft) and is generally much greater than in the arctic. Water runoff is rapid because of the severe topographic relief. Winds are strong. On clear days, light intensity, notably ultraviolet, and evaporation may be high because of the thin air. A unique characteristic of the alpine habitat not shared by the arctic is low barometric pressure and oxygen concentration, which probably does not affect plants as much as it does some animals. Altogether, the alpine environment imposes greater severities on plant development than does the arctic environment (Bliss 1956), and this is doubtless true for animal activities also.

VEGETATION

Tundra has the appearance of short-grass plains, but differs in that the vegetation consists of sedges, rushes, lichens, mosses, ericaceous or decumbent shrubs, and flowering herbs as well as grasses. The plants are generally of small size, stunted growth, and compact structure adopted to resist desiccation and mechanical abrasion from wind, snow, and sand. Germination of seeds is poor, and most species require several years to produce the first flowering. Most tundra plants are therefore perennials, and vegetative reproduction is important. Seral vegetation varies in composition depending on whether it develops around ponds, in low wet places, or on clay, sand, gravel, or rock. Flowering herbs are often abundant with different species coming into bloom progressively during the year (Sørensen 1941, McClure 1943). Although primary productivity calculated on a year-round basis is very low, on a daily basis during the short growing season (50–70 days) it is comparable to that in many temperate communities (Bliss 1962).

Succession has been studied in only a few areas, and the true nature of the climax is unknown over much of the arctic. There is doubt as to whether or not a stable climax, as understood for southern latitudes, actually develops (Raup 1951, Sigafoos 1951, Britton 1957). This is due to the instability of the soil, varying depths to which the soil thaws out in the summer, depth and duration of the snow cover, exposure to wind, and grazing and trampling by animals. Although tundra associations are not recognized in this book, we give special consideration to the arctic tundra and alpine tundra of the northern hemisphere. On mountains in the tropics and the southern hemisphere, alpine communities of paramo, puna, and tussock grasslands also occur.

Viewed broadly, the tundra is the terminus of latitudinal trends wherein (a) species diversity both in floras and faunas decreases from the tropics northward, (b) individual species are more strongly predominant, (c) niche breadth increases over a wider range of microhabitats, (d) successions are shorter and less important (Shelford 1963), and (e) community biomass and productivity decrease (Whittaker 1962). Much of the same changes occur with increasing altitudes in mountains at lower latitudes.

Arctic Tundra

The "barren" grounds of the far north are divisible into four significant types in respect to animal distribution. *Bush* or *mat tundra* contains dwarf trees, decumbent shrubs, or heath, usually mixed with mosses and lichens. Near Churchill in northern Manitoba, much of the area is muskeg; but climax vegetation is interpreted as a mixture of low ericaceous heath and *Cladonia* lichen growing in a mat of sphagnum and other organic material 7 to 10 cm (3–4 in.) thick. This climax develops on wet ground, clay, sand, and on gravel and rock ridges (Shelford and Twomey 1941). A variety of dwarf shrub-lichen-grass-sedge types have been described in Alaska (Hanson 1953, Churchill 1955) and in the eastern arctic (Polunin 1934–35, 1948, Holttum 1922). *Grass tundra* is largely limited to deeper mineral and organic soils. The soils are more fertile, and in places thaw out in the summer to a depth of 1 m. Different species of grasses and sedges are dominant in recognizable seral and climax stages (Hanson 1951). *Lichen-moss barrens* (Tanner 1944, Hanson 1953) have been called desert tundra or rock desert by various investigators. The soil is thin, and there is much exposed rock. Vegetation is scant and consists of crustose and foliose lichens, mosses, and scattered short herbs or very small shrubs. In the low arctic, this may be a seral stage, but in the high arctic it is often the only vegetation able to tolerate the severe climate. In the extreme north there is *perpetual snow and ice*, a polar desert; vegetation is practically absent, and animal life is restricted to marine forms along the ocean coast. The slight development of tundra in the antarctic is of the lichen-moss barrens; most of the antarctic continent is covered with ice (Lindsey 1940, Holm-Hansen 1940).

Alpine Tundra

Tundra extends into the tropics on the high western mountains and into New England on a few of the higher Appalachian peaks. This alpine tundra consists chiefly of grasses and sedges without conspicuous development of Ericaceae or the great masses of foliose lichens and mosses found in parts of the arctic (Cox 1933, Daubenmire 1943, Hayward 1945). About 37 per cent of 170 vascular species collected in the alpine tundra of the Colorado Rockies also occur in the arctic, and about half of these are circumpolar in distribution. Most of the remaining species are endemic to North America, and many species are uniquely endemic to the Rockies (Holm 1927). The alpine vegetation of the high mountains in New England is more closely related floristically to the Arctic tundra than to the alpine vegetation of the Rocky Mountains (Bliss 1963). The taxonomic composition of alpine vegetation varies

Fig. 22-1 Tree-line in Rocky Mountain National Park, Colorado. Sub-alpine forest below, alpine tundra above.

greatly from place to place, but most of the plants are perennials. The dwarfness of the shoots in proportion to the flowers and fruits that they bear is very striking. As one ascends the mountain slopes, the grass tundra gives way to lichen-moss barrens, and then to perpetual snow and ice. On the downslope side, there is often bush tundra and at tree-line the trees are dwarfed and misshapen into a dense shrub cover (*krumholz*) by the wind and cold. Flowering herbs are often abundant and conspicuous.

The occurrence of krumholz is evidence that trees have extended up the slopes of mountains as far as they are able under present climatic conditions (Griggs 1946). The alpine tree-line is usually very irregular (Fig. 22-1). Outlying trees may occur at some distance in advance of the forest proper if they can secure the protection of an embankment, or find other suitable microhabitats. In some mountain areas trees advance to higher altitudes on ridges than in valleys because snow accumulates to great depths in the valleys and takes longer to melt.

ARCTIC TUNDRA BIOCIATION

Composition

There is enough uniformity in the animal life of the arctic tundra in North America and Eurasia so that only one biociation is presently recognized. Circumpolar distribution is characteristic of both vertebrates (Udvardy 1958) and invertebrates (Netolitzky 1932). Common animals of the tundra are those listed below, which are conspecific or represented by equivalent species on the two continents. Marine or strictly coastal species and those of more limited distribution are omitted (Bailey 1948, Banfield 1951, Bee 1958, Harper 1953, 1956, Manniche 1910, Manning 1946, 1948, Porsild 1943, Preble 1908, Rausch 1953, Salomonsen 1950–51, Soper 1944, 1946, Stegmann 1938, Taverner 1934, Taverner and Sutton 1932):

Mammals

Masked shrew	Grizzly bear
Arctic shrew	Polar bear (limited to
Arctic hare	coast)
Arctic ground squirrel	
(suslik)	Ermine
Tundra vole	Wolverine (glutton)
Brown, European,	Barren ground caribou
Siberian lemmings	(reindeer)
Collared lemming	Peary's caribou (limited
Gray wolf	to North)
Arctic fox	Muskox

The muskox, formerly of wide distribution, is now restricted to North America and Greenland (Fig. 22-2). The tundra in North America is richer in species, both of mammals and birds, west of Hudson Bay than it is eastward, and richest in Alaska.

The most abundant mammals on the tundra are lemmings, and in peak years their numbers are enormous. Among the larger animals the caribou form large herds and are important in the food and economics of Eskimos and Indians. Seton (1912) estimates their original number at 30 million, but they are much reduced at the present time. There are also fewer muskox now (C. H. D. Clarke 1940).

Fig. 22-2 Muskox herd in a defensive ring against wolves (Shelford 1963).

Birds

Yellow-billed loon	Sanderling
Arctic loon	Baird's sandpiper
Red-throated loon	Pectoral sandpiper
White-fronted goose	Purple sandpiper
Oldsquaw	Dunlin
Common scoter	Red phalarope
Rough-legged hawk	Northern phalarope
Gyrfalcon	Pomarine jaeger
Peregrine falcon	Parasitic jaeger
Willow ptarmigan	Long-tailed jaeger
Rock ptarmigan	Herring gull
Sandhill crane	Glaucous gull
Semipalmated plover	Arctic tern
Black-bellied plover	Snowy owl
American golden plover	Horned lark
Long-billed dowitcher	Common raven
Whimbrel	Water pipit
Ruddy turnstone	Lapland longspur
Knot	Snow bunting

In addition to these species, the *North American faciation* contains the whistling swan and snow goose in the north; Canada goose and semipalmated, least, and white-rumped sandpipers rather generally distributed; and in the west, the Eskimo curlew (now probably extinct), Hudsonian godwit, stilt and buff-breasted sandpipers, and Smith's longspur. The *Eurasian faciation* also contains some species limited to it: two species each of swans and geese, several plovers and sandpipers, and another pipit and bunting.

Less than one-third of the above species of birds are entirely terrestrial in their life requirements. Many species get their food from the fresh-water ponds and lakes or on the margins of these bodies of water, so characteristic of at least the low arctic tundra. The population of small birds on the upland tundra is low compared with forest populations (Manning *et al.* 1956, Watson 1963).

Reptiles and amphibians are poorly represented where not absent, and the invertebrate fauna is comparatively restricted in variety. In the ponds on the west side of Hudson Bay occur a stickleback fish, a flatworm, a leech, an annelid, a few snails, a couple of phyllopods, a few species each of Cladocera, Copepoda, Ostracoda, and Amphipoda; a good representation of dytiscid and hydrophilid beetles, and an abundance of midge fly larvae. Since the lakes and

rivers thaw out for only a few weeks, annual productivity is low (Frey and Stahl 1958). Fish are more numerous in rivers, and are largely migratory salmonids. Pond life in the Alaskan tundra is essentially similar to that near Hudson Bay (Johansen 1922).

On land, the snails *Succinea* and *Vertigo* are found in wet tundra on the west side of Hudson Bay. Spiders and mites are well represented. Springtails and flies are especially numerous among the insects, and there are a few species of Lepidoptera, Coleoptera, and Hymenoptera, but species of Hemiptera, Homoptera, Orthoptera, Odonata, and Neuroptera are scarce or absent. Ants are scarce on the tundra but bumblebees are conspicuous. Especially noteworthy are the vast devastating hordes of mosquitoes, black flies, and deer flies that reach a peak of numbers in mid-July (Seton 1912, Shelford and Twomey 1941, McClure 1943). The invertebrate life of western Greenland is essentially similar (Longstaff 1932). Quantitative studies of the soil fauna in eastern Greenland showed that springtails and mites, especially Oribatidae, reached populations of 780,000 per m^2 in bush tundra, but only 3000 per m^2 in the lichen-moss barrens (Hammer 1937). Seven different societies of invertebrate fauna have been differentiated here (Macfadyen 1954). Both lumbricid and enchytraeid annelid worms are reported in tundra soils of the Old World (Cherkov 1961).

In comparison to the arctic, the antarctic supports a limited fauna. Among the invertebrates are found several peculiar species of Protozoa, 16 species each of rotifers and tardigrades, two fresh-water crustaceans, mites, and at least 18 species of insects. Mites and springtails are predominant. Vertebrates are primarily marine, although several species, especially birds, nest on land (Lindsey 1940, Holm-Hansen 1963, Castri *et al.* 1966).

Food is more abundant along the shores of northern oceans; the association of sea and land provides niches for various species not found abundantly inland (Freuchen and Salomonsen 1958). During the winter the sea is covered with ice, there is little or no light, and phytoplankton is scarce or absent except for reproductive spores and eggs. However, nutritive salts, such as nitrates and phosphates, accumulate in large supply, so that in May when the ice disappears and light returns there is an almost explosive development of phytoplankton followed by microcrustaceans and other zooplankton. This is the key to the teeming abundance of fish, sea birds, and marine mammals that occur at this time. Large colonies of fulmars, cormorants, auks, murres, guillemots, gulls, and others nest on ledges of precipitous cliffs or in some cases on islands or shores down close to the water. Eider ducks and other waterfowl are frequently numerous. Vegetation is best developed in and around these colonies because of the rich nutrient added to the' soil from the excreta of the birds. One of the most common seals is the ringed seal, which remains over winter, even in the high arctic, by keeping blow-holes open through the ice. The harbor seal and harp seal overwinter in the more open waters of the low arctic. Other species of seals, whales, walruses, and polar bears occur during the summer throughout most of the maritime areas of the arctic.

Animal Adjustments

The arctic environment is exceedingly rigorous and oscillating in respect to nearly all factors, particularly temperature, moisture, and light. These factors vary to extremes in winter found nowhere else in the world and in summer can be tolerated only by those species that have evolved appropriate adaptations. The Arctic fauna contains relatively few species, but they exhibit high fecundity, large body size, slow growth, and the resulting ecosystems are simple and unstable—just opposite to what one finds in the tropics (Dunbar 1968).

White coloration is common, especially over winter, in several mammal species (artic hare, collared lemming, gray wolf, arctic fox, polar bear, ermine, Peary's caribou) and in a few birds (willow and rock ptarmigans, snowy owl). When the ground is covered with snow, white coloration, of course, conceals both the prey and predators. Many of these species acquire darker coloration during the months between May and September. Melanism may absorb more heat at the body surface and at the same time protect underlying tissues from excessive ultraviolet. The latter would be especially important at high altitudes. The white winter color apparently does not give special protection against heat loss from the body, as has sometimes been thought (Hammel 1956).

A major habitat problem that tundra animals must solve is tolerating or avoiding the long severe cold of the winter season (Morrison 1966). Cold-blooded animals are generally acclimatized so that they remain active at temperatures down to freezing much better than their relatives in temperate and tropical zones. This is particularly true for aquatic species (Scholander et al. 1953, Bullock 1955). Invertebrates commonly pass the winter in the larval or pupal stage that is especially resistant to freezing, although beetles,

spiders, and some other forms may over-winter as adults. Rotifers, tardigrades, midge fly larvae, and dytiscid and hydrophilid beetles may be frozen in the ice for months or even years, yet resume activity immediately on thawing (Lindsey 1940, McClure 1943, Andersen 1946). Because of the short growing season and slow development, many tundra insects require two or more summers to complete their development.

The larger mammals and over-wintering birds have good insulation in long, dense pelage or plumage, and in fat. Heat production in their bodies is not greatly increased until very low air temperatures are reached (Scholander et al. 1950). The tarsi and legs of ptarmigan and snowy owls become well feathered in the winter.

Voles, lemmings, and ermines escape the winter cold by staying in their runways and nests under the snow. The ptarmigan also digs tunnels into snowbanks, where it roosts protected from the cold (Wetmore 1945), sometimes for days at a time. Only the arctic ground squirrel truly hibernates. This it does by excavating burrows into sandbanks or hills which, because of exceptional drainage, possess an area that remains unfrozen between the deep permafrost and the winter frost at the surface (Mayer 1953). The bears den up during the cold weather but remain active to the extent of giving birth to their young in the middle of the winter.

Those species that cannot tolerate or escape the winter cold and lack of food migrate. The barrenground caribou on the mainland migrates in long strung-out armies to the southern portions of the tundra, even well beyond the tree-line into the forest-

Fig. 22-3 Summer and winter distribution of the barrenground caribou in North America (Scotter 1967).

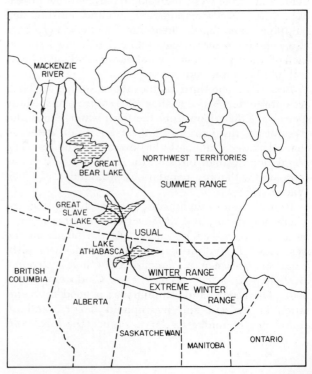

tundra, to pass the winter (Fig. 22-3), and their trails remain conspicuous throughout the year (C. H. D. Clarke 1940, Harper 1955). The caribou on Greenland, Spitzbergen, and the northern islands of the Canadian Archipelago are necessarily resident throughout the year. Migration of the bird fauna in this biociation is nearly complete; during the winter only an occasional hawk, ptarmigan, raven, or owl will be encountered over the land, although marine birds occur wherever there is open sea. When the birds return in the spring they are quick to get nesting started. Often they are already mated, carry through their nesting cycle quickly, and then leave promptly again for southern latitudes.

In general, the melting of the snow and the break-up of ice is the signal of transition from winter to summer. Although it is possible to recognize all of the four aspects (Sørensen 1941), the change from winter to summer and back again to winter is so rapid that all aspects are abbreviated except the hiemal and aestival. May is the usual month of parturition among the larger mammals, and the peak of bird nesting comes in late June and early July (Fig. 22-4).

Owls, hawks, water birds, and some passerine species do not breed in those years in which scarcity of food or delayed freeing of nesting grounds of snow and ice are detrimental to survival. Failure of hawks and owls to breed occurs especially in years when the lemming population is low (Marshall 1952).

Ptarmigan, hare, voles and lemmings, and their predators, particularly the fox and snowy owl, are subject to oscillations in abundance that come at intervals of either 3–4 or 9–10 years. These oscillations are more pronounced in the arctic tundra than in any other biome.

The food coactions of the herbivorous animals are of interest. Caribou feed on lichens, including reindeer moss, especially in the winter, which they uncover by pawing through the snow. During the summer they also consume shrubby growth and sedges (Harper 1955). Grass is the principal food of the muskox, but it also eats willow brouse and, less frequently, lichens and mosses (Jackson 1956). Ptarmigan feed on plant material left behind by these large animals. The chief food of ptarmigan, however, is the buds, leaves, and tender branches of willow and other shrubby plants not easily obscured by snow. Their winter food appears to be richer in fats and contains less protein than the food they consume during the summer (Gelting 1937). Gulls are known to eat warble fly larvae rising from the skin of caribou (Scalon 1937) but depend mainly on dead fish that they find in open water, or on carrion. Most of the small passerine birds are seed-eaters or mixed seed- and insect-eaters. Berries become abundant in August, and are much sought after by birds. Exclusively insectivorous land birds, such as warblers, would find great difficulty in surviving and reproducing on the open tundra. Shorebirds are largely insectivorous

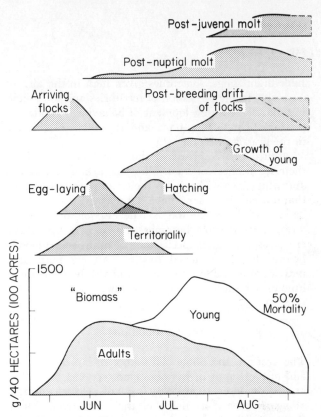

Fig. 22-4 Variation in total biomass and activities of the dunlin sandpiper during the summer season at Barrow, Alaska (Holmes 1966).

but depend mainly on insect larval forms found along the margins of the numerous ponds (Holmes 1970).

Carnivores, particularly the arctic fox, exert a potent influence on the distribution of birds, especially in regions or at times when lemmings, the fox's preferred food, is absent or scarce. Arctic birds are adapted in behavior to reduce predation by commonly nesting on islands, cliffs, holes in the ground, between rocks, and in dense vegetation. Some species sit closely on their nests ("freeze") to avoid detection on the approach of predators; others practice "injury feigning" to distract predators away from their nests (Larson 1960).

Although the continuous arctic summer light permits activity throughout the 24-hour day, most mammals and birds need periods of rest. Small rodents have a short feeding and resting cycle lasting 2 or 3 hours. Birds appear to rest most frequently in the hours around midnight, but the periods of rest are much shorter than in southern latitudes (Cullen 1954, Peiponen 1970). In one study conducted above the Arctic Circle, adult robins fed their young for 21 hours per day and the young birds grew so rapidly that they left the nest in 8.8 days instead of the 13 days usual in more southerly latitudes (Karplus 1949). During the winter, there is at least a faint glow of light for an hour or two at noon. It is during this period that ptarmigan and probably other birds feed most heavily (Gelting 1937). As a rule, northern birds also tend to lay larger clutches of eggs than their closest relatives in southern latitudes (Lack 1947–48). Related to their nesting is the tendency for many species to have flight songs, sometimes given high in the air, for announcing the possession of territories and for soliciting mates. This development of behavior is related to the lack of high song posts and is similarly exhibited in the grassland biome.

A final characteristic of these northern animals is their fearlessness of man. The few Eskimos, Indians, and white men who inhabit the tundra are so scattered that animals in general have not learned to fear them. To a certain extent this is true also of some boreal forest species; snow buntings, snowy owls, grosbeaks, crossbills, and other species may be approached closely before they are moved to flee. Among mammals, individual arctic foxes not infrequently linger close to human habitation for days at a time.

Human Relations

The arctic is the home of Eskimos in North America and of the Lapps in Eurasia (Hadlow 1953, Freuchen and Salomonsen 1958). The Eskimos are concentrated along the coast, as much of their food comes from the sea: fish, walrus, seal, and polar bear. During the summer, caribou flesh, bird eggs, and berries are eaten. Caribou fat and seal oil are burned in the Eskimo igloos to furnish light and heat, and pelts from these animals are made into clothing and blankets. Meat is eaten either cooked, dried, or raw, and some of it is frozen and buried in the ground for the winter days of scarcity. The Eskimo gets his transportation during the summer in light boats made of sealskin stretched over frameworks of driftwood or bone, and during the winter in sleds drawn by dogs. Fur trapping is the chief source of income. Many of these primitive habits are now becoming highly modified as the Eskimo adopts the ways of the white man.

The Lapp lives much like the Eskimo, although he more commonly lives in a tent made of reindeer skin than in an igloo made of snow. The reindeer is used by the Lapp for pulling his sled, for meat, and for milk. The Lapp may also keep goats. Lapps move up and down the mountains in summer and winter in search of pasture for their animals, northward and southward with the seasons for fishing; below tree-line they occasionally grow meager crops.

ALPINE TUNDRA BIOCIATION

Tibetan Faciation

The largest alpine areas of the world lie on the Tibet Plateau and in the adjacent Himalayan Mountains of Asia (Schäfer 1938, Mani 1968). Occurring here are related forms of pikas, pipits, rosy finches, and horned larks also found in mountain areas of the western hemisphere, while the marmot and sheep may be conspecific with North American forms (Rausch 1953).

Stegmann (1938) gives a long list of birds especially characteristic of the Tibetan fauna. The Tibet Plateau may represent an important center of origin of alpine species, some of which then became dispersed into the higher mountains of Europe and North America. The Tibetan fauna evolved independently from that of the Arctic tundra and there are few or no bird species common to the two. There is some overlap of species, however, with the Asiatic grassland (Mongolian fauna), Asiatic deciduous forest (Chinese fauna), and Ethiopian desert (Mediterranean fauna) biociations, which suggests their possible remote derivation.

North American Faciation

Because of its small total area, rugged terrain, and discontinuity between mountain peaks, there are only a few species characteristic of the alpine tundra in North America.

Mammals and birds are conveniently divided into *northern* and *southern faciations*. Mammals occurring in the high tundra from Alaska to British Columbia are the collared pika, hoary marmot, singing vole, barren-ground caribou, mountain goat, and Dall's sheep. Species limited to the southern mountains are common pika, yellow-bellied marmot, and mountain sheep. The common pika is differentiated into over 30 subspecies in the various mountain areas. Shrews, bears, coyotes, weasels, badgers, mice, wapiti, and mule deer of the western montane biociation range up into the alpine tundra during the summer. Ground squirrels (Fig. 22-5) and pocket gophers reach this community south of Canada by extending their ranges from the low elevation grasslands through seral stages in the intervening coniferous forests. Subspecies of mountain sheep formerly occurred regularly on the Great Plains and in desert regions at low altitudes (Buechner 1960).

Of birds, the white-tailed ptarmigan (Fig. 22-6), water pipit, and gray-crowned rosy finch are characteristic and widely dispersed. In the central and south-

Fig. 22-5 Columbia ground squirrel in alpine tundra, Glacier National Park (courtesy R. L. Day).

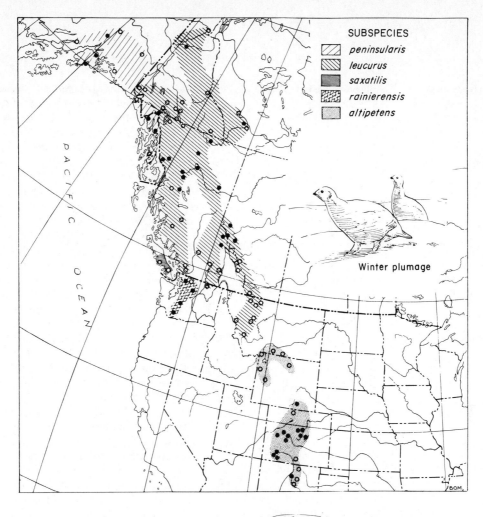

SUBSPECIES

⟋⟋ *peninsularis*
⟍⟍ *leucurus*
▓ *saxatilis*
▦ *rainierensis*
⣿ *altipetens*

Winter plumage

Fig. 22-6 Distribution of the white-tailed ptarmigan, *Lagopus leucurus*, in the alpine tundra (Aldrich and Duvall 1955).

Summer plumage

ern Rocky Mountains the brown-capped and black rosy finches replace the gray-crowned rosy finch. Rock wrens and horned larks are occasional summer visitors. In the alpine meadows of the far North, savannah sparrows and upland plovers are common, although they belong principally to the prairie biociation and grassland seral stages of the boreal coniferous forest. Also in the north, around ponds, occur lesser yellowlegs, herring gulls, short-billed gulls, and Bonaparte's gulls (Drury 1953).

Except for the caribou and pipit, the alpine and arctic tundra biociations have no important species in common. There appears less taxonomic interrelationship between animals in alpine and arctic tundras than there is with plants. Probably most of the species listed above are of northern origin and, entering North America over the Bering land bridge, dispersed southward on the mountains as they became elevated and the alpine tundra differentiated.

Reptiles and amphibians are uncommon. In contrast to the arctic tundra, flies are comparatively few except in the vicinity of ponds, but there is an abundance of springtails, ground-dwelling beetles, leaf-

hoppers, grasshoppers, true bugs, butterflies, ants, bumblebees, mites, and spiders (Hayward 1945, 1952).

Alpine ponds have an impoverished fauna. In a small pond at 3507 m (11,500 ft) in the Colorado Rockies of maximum depth 1 m and which freezes solid in the winter, plankton was scant during the summer after the ice melted, but midge fly larvae reached populations of over 1900 per m², *Pisidium* fingernail clams, 1470 per m², and tubificid worms, 168 per m². A fairy shrimp was the most characteristic metazoan in the open water, although a small number of aquatic insects occurred along the shoreline (Neldner and Pennak 1955). Phyto- and zooplankton tend generally to be represented by fewer species and a smaller number of individuals than in temperate or tropical lakes (Thomasson 1956).

Animal Adjustments

As in the arctic tundra, most of the residents in the alpine tundra that remain active over winter are white in color: mountain goat, mountain sheep, Dall's sheep, and white-tailed ptarmigan.

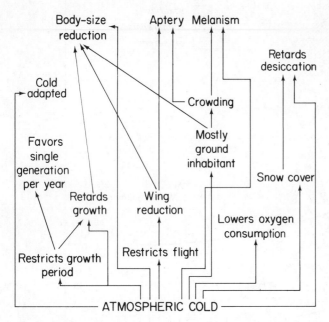

Fig. 22-7 Effect of atmospheric cold on insect life at high elevations (from Mani 1968).

The pikas inhabit masses of loose rock rather than the climax meadow itself and also occur well below tree-line. During the summer they gather stacks of tundra vegetation and during the winter subsist on this hay. Pocket gophers spend most of their existence in underground burrows feeding on roots and bulbs and hence are well protected from winter cold. The hoary and golden-mantled marmots and ground squirrels hibernate. Wapiti and deer migrate to the lower mountain slopes and valleys for the winter as do most of the birds. There is often some downslope movement of mountain sheep and goats for the winter, but it is not so extensive as with the wapiti and deer (Hayward 1952). The behavior and interrelations of wolves, Dall's sheep, caribou, and other species in the high tundra of Alaska are described in detail by Murie (1944).

Low temperature slows up the development of insects and other invertebrates and reduces the number of generations possible during the year (Fig. 22-7). Animals, however, are generally acclimatized to be active at low temperatures. For example, springtails are sometimes abundant on the snow, where they freeze at night and thaw out during the day. An interesting food chain occurs with their feeding on conifer pollen falling on the snow and being fed upon by mites.

Because of the usually strong winds, insects and even birds stay close to the ground and fly as little as possible. Many insects are wingless. Birds commonly feed and build their nests on the sheltered side of obstacles, or they crawl into holes and crevices. When in the open, they persistently face the wind (Hingston 1925).

In the Colorado Rockies, grasshoppers are among the most numerous species. However, of 28 species recorded, only 11 are truly alpine species, the rest being found only as adults which have flown or been blown up from lower altitudes and do not reproduce successfully there (Alexander 1951). There is a progressive decline in number of orthopteran insect species from low to high elevations comparable to the decline from southern to northern latitudes. Equivalent stages in their yearly life cycles occur progressively later at higher elevations (Fig. 22-8).

In the Himalayan Mountains, plant cover extends to 5150 m (17,000 ft), with the highest known plant at 6100 m (20,130 ft). At these elevations, salticid spiders prey on anthomyiid flies and collembola, which in turn feed on fungus and decaying plants. Mites, centipedes, and machilid fleas also occur. The nest and eggs of a partridge bird was found at 6000 m (19,000 ft). Life is possible at these extreme heights when wind clears the ground of snow, temperatures warm up on sunny days, melting snow supplies moisture, and the organisms can tolerate freezing temperatures at other times (Snow 1961).

During the late spring, south slopes on the mountains become free of snow before the north slopes because they get direct radiation of the sun. Consequently, plant and animal life become active on south slopes before they do on north slopes. Individuals of the same species will also begin growth and reproduction on the lower slopes before they do on the higher slopes. Birds appear to construct more compact nests, as insulation against cold, at high than at low altitudes, and they build these nests on the exposed south sides of thickets or trees, where they benefit from the heat of the sun (Heilfurth 1936).

The low oxygen pressure at high altitudes appears to be more critical for the warm-blooded mammals than it is for birds, which are adapted to fly at high elevations anyway, or for invertebrates and plants that have much lower rates of metabolism and oxygen requirements (Hall 1937, Kalabuchov 1937). Mammals moving up to high altitudes may become temporarily acclimated through increases in rate of respiration, in rate of heart beat, in number of red blood cells, and in hemoglobin, but these adjustments are seldom as effective as in those species which are permanent residents at high altitudes. Species acclimatized to low oxygen pressures are affected in a reverse manner when they move to low altitudes. Some mammals that have a wide altitudinal range, particularly the pocket gopher *Thomomys*, are, like plants, smaller in size at high than at low elevations with a continuous gradation between the extremes (Davis 1938).

Human Relations

There are few humans living permanently in the high mountainous regions of North America, but white man goes up there, taking sheep and goats to summer pasture, and for recreational or sight-seeing purposes. Man does occupy the Tibetan plateau of Asia. There, he depends on the yak to plough his fields and to furnish meat, butter, and milk. Although much of the plateau is too rugged and cold for crops, millet, corn, and wheat are grown in sheltered valleys.

PALEOECOLOGY

The origin of the tundra *flora* is uncertain (Raup 1941), but the species involved may be segregates from seral stages, especially bogs, of the Arcto-Tertiary Geoflora that were tolerant of arctic and alpine environments. With the cooling of the continent and the coming of the glaciers we may suppose that these species were left behind when the rest of the flora retreated southward. During Pleistocene glaciation, these species survived in the Alaska refugium (Hultén 1937), in possibly unglaciated islands of the Arctic Archipelago, and along the southern margin of the glacier. Pollen profiles indicate that climate and vegetation in the Alaska refugium were not greatly different then from now (Livingstone 1955). During the warm interglacial periods, tundra was probably limited to far northern regions and high mountains, with forests covering much of what is now the low arctic.

When the glaciers retreated in post-Pleistocene time, there appears to have been a period when many tundra species were continuous in distribution from the arctic plains onto the mountain slopes in the northern Rockies. Probably at this time also arctic species were able to disperse farthest southward, as alpine tundra occurred more extensively at lower elevations, and intervening forests were less extensive (Daubenmire 1943). As the forests dispersed northward through the valleys and lowlands and then gradually up onto the mountains, alpine vegetation retreated to the higher elevations and became separated from the arctic tundra proper. This northward dispersal of coniferous forest was especially rapid during the thermal maximum period, but in some localities, as in Alaska, it is not yet complete (Griggs 1936). On the other hand, the tree-line may be retreating southward at the present time in other areas (Raup 1941). The coming of the forests interrupted the complete colonization of alpine slopes by arctic species, but the forests brought a new element into the mountain flora derived from refugia south of the glacier (Raup 1947). Since the coniferous forest contained seral grassy stages with species intruding from the grassland biome, some of these species also penetrated the alpine vegetation and became part of it. Furthermore, tundra and grassland probably came into direct contact during the glaciation periods, so there is intermingling of tundra and grassland species in arctic as well as alpine regions (Hayward 1945).

The tundra *fauna* probably began to evolve late in the Tertiary, along with the tundra flora, as the continent cooled and the Arcto-Tertiary Geoflora retreated southward. The true arctic fauna is apparently derived from previously wide-ranging forms able to tolerate cold climates (Johansen 1956–58) and from coniferous forest and grassland, fresh-water marshes, the seacoast (Stegmann 1938), and mountainous or upland regions (Larson 1957). Adaptations in the avifauna proceeded along lines either of toleration of the many hours of darkness and severe cold of the winter climate, or to the development of extensive migrations, sometimes far into the southern hemisphere as in some shorebirds and terns.

During glacial periods of the Pleistocene, the tundra of the Alaska refugium was isolated from the tundra south of the glacier but in direct contact over the Bering land bridge with the unglaciated tundra of Asia. During these times there was an exchange of fauna with emphasis on invasion of Eurasian forms into North America. With each recession of the glacier and disappearance of the land bridge, the Alaska tundra again became isolated from Asia and connected to the North American tundra. This has allowed Asiatic species to disperse over North America to varying degrees (fresh-water triclads: Kenk 1953, birds: Cade 1955). Isolation of caribou in refugia in Alaska, on Arctic islands, and south of the glacier permitted evolution of various subspecies (Banfield 1963), and isolation in the Alaskan and Appalachian refugia correlates with distribution of the lemming *Dicrostonyx*

Fig. 22-8 Stages in the life cycle of the grasshopper *Melanoplus dodgei* in relation to altitude (Alexander and Hilliard 1969).

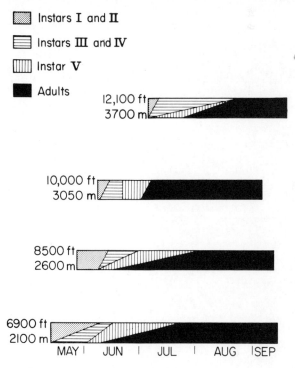

groenlandicus west of Hudson Bay and *D. hudsonius* east of Hudson Bay (Guilday 1963). Some species, such as the grizzly bear, doubtless invaded the tundra from the western coniferous forests since Pleistocene time. Other species now on the tundra are probably derived from refugia in the Arctic islands and from south of the glacier (Rand 1954, Johansen 1956–58).

SUMMARY

The tundra biome extends beyond the tree line in the far north and on high mountains. It has low precipitation, low temperatures, a short growing season, and, in the arctic, extreme seasonal changes in length of day and night and a permafrost in the ground. Arctic vegetation is bush or mat tundra, grass tundra, or lichen-moss tundra. Perpetual snow and ice occur in extreme areas. The tundra flora is probably derived from seral stages in the Arcto-Tertiary Geoflora which became segregated as the continent cooled in the Pliocene and Pleistocene.

The arctic tundra biociation is fairly uniform faunistically in North America and Eurasia, although two faciations are distinguishable. Mammals, birds, mosquitoes, and flies are the most conspicuous animals, with springtails and mites predominant in the soil. White color is common among mammals and birds, especially during the winter. Acclimatization to cold is highly developed in many resident forms. Some small mammals remain active under the insulating cover of snow during the winter months. Most birds migrate. Cycles of abundance are pronounced in several species. The Eskimo and Lapp mainly hunt and fish for a living.

The alpine tundra biociation is best developed in the Tibetan plateau in central Asia. Only a few species of mammals and birds are peculiar to the biociation. Mammals are physiologically adjusted to the low oxygen pressure. Man finds habitation in this area difficult except during the summer months, when he brings his sheep and goats for pasture or comes for recreation.

Chapter 23

GRASSLAND BIOME

Grassland presently occurs on all the continents, and at one time covered 42 per cent of the earth's surface. Grasslands everywhere possess marked similarities in points of climate, physiognomy, and animal mores. In Russia this community is termed the *steppe*, in Hungary, the *puszta*, in South Africa, the *veld*, and in South America, the *pampas* (Carpenter 1940). In North America the tall, dense grasslands with their rich fertile soils in the eastern portion of the biome are called the *prairie*; in the West, the short grasses and shallow soil characterize the *plains*.

In North America, grasslands extend from northern Saskatchewan and Alberta and central British Columbia to central Mexico, and from Indiana to California. The eastern portion is a huge expanse, continuous except for forest strips in the river valleys, but the continuity of the western portion is broken by the many mountain ranges. In general, the terrain is flat or rolling, green in summer and brown in autumn and winter (Weaver and Clements 1938, Shelford 1963).

CLIMATE

Precipitation in North America may be as high as 100 cm (40 in.) per year adjacent to deciduous forest, but trees cannot spread into the grassland because of high rates of evaporation, late summer and autumn droughts that are particularly severe and prolonged during some years, and intermittent fires that kill seedling trees but from which grasses quickly recover. Fire sweeping over the flat upland terrain burned extensive areas at frequent intervals and was a contributing factor in maintaining grassland (Wells 1965). Rainfall decreases and becomes more irregular and evaporation increases in a gradient from east to west or southwest, because of the general pattern of air circulation over the continent (Kincer 1923, Borchert 1950). An isohyet of 2–3 cm (5–8 in.) precipitation separates grassland from desert.

Few species of grasses can tolerate the entire range of precipitation, and differences in the moisture requirements of species are the main reason for the subdivision of the biome into its various plant associations. Winds are normally heavy, and there occasionally are severe winter blizzards.

Temperature is not as critical a factor as moisture, as is evident in the great north-south distribution of grasslands. Temperature, however, helps to separate temperate grasslands from tropical grasslands. In the North, mean monthly winter temperatures drop to —15°C(5°F), while summer temperatures in the South may exceed 32°C (90°F). There are great seasonal and daily ranges of temperature. The frostless period in the North may be only 100 days long, but in the South frost rarely or never occurs. In the California and bunch grass prairies, most of the rainfall comes in the winter months, and in southern California some grasses start their growth in late autumn and come into bloom in December, although others wait until spring.

VEGETATION

Grassland owes its characteristics to the perennial grasses that constitute the dominant climax vegetation. Annual grasses are largely confined to seral stages. The climax grasses may be *tall* (1.5–3 m), *mid* (0.5–1.5 m), or *short* (less than 0.5 m) and grow in *bunches* or as *sod*. Forbs occur mixed with the grasses, and variation in the time of blooming of these broad-leaved and mostly perennial herbs as well as of the grasses gives the grassland a variety of aspects like the forest (Weaver 1954, Weaver and Albertson 1956). The prairie has beauty, character, and a history all its own (Craig 1908a, Weaver 1944, Costello 1969).

Grasses grow quickly after the onset of warm and rainy weather and are adapted for long quiescent periods of dryness and cold. The leaves or tops of the grasses die down during unfavorable seasons, but underground buds regenerate new growth during the next favorable period, even if this be delayed for some years. The grasses and forbs are deeply and extensively rooted, except in arid climates, where a hardpan occurs near the surface (Fig. 9-7). Competition is primarily for the limited water supply and only secondarily for light. Since grasses grow from the base of the leaf, they can tolerate considerable grazing by large herbivorous animals, and this is an important factor in their dominating the prairie.

Several of these, the most important genera of grasses in the North American grasslands, occur also in other parts of the world (Clements and Shelford 1939):

Mostly tall and mid grasses	Mostly short grasses
Andropogon—blue stem	*Aristida*—triple-awned grass
Agropyron—wheat grass	
Elymus—wild rye	*Bouteloua*—grama grass
Festuca—fescue	*Buchloë*—buffalo grass
Koeleria—June grass	
Panicum—panic grass	
Poa—blue grass	
Sporobolus—drop-seed	
Stipa—needle grass	

PLANT ASSOCIATIONS OF NORTH AMERICA

Stipa-Sporobolus association (true prairie): Mostly tall and mid grasses in a long strip extending north and south in the eastern more humid part of the biome next to deciduous forest. Much of this prairie was marshy and poorly drained before white man came (Hewes 1951). Oak-hickory forests occur as scattered groves in better drained areas on hills, in sandy areas, and along streams making a savanna. The coastal prairies of Texas constitute a faciation of this community.

Stipa-Bouteloua association (mixed prairie): Mid grasses confined to the moister low areas; short grasses, to drier hillslopes.

Bouteloua-Buchloë association (short grass plains): On Great Plains east of Rocky Mountains. The climate here is so dry that mid grasses are inconspicuous except during wet years. Pronghorn and bison in former days consumed mid grasses as fast as they appeared, in preference to the short grasses, and hence should be considered co-dominants along with the grasses (Larson 1940).

Agropyron-Festuca association (bunch grass prairie): In northern half of the Great Basin and into British Columbia, mostly isolated from rest of biome by mountains and desert. Precipitation comes chiefly as snow and rain during winter months. Dominant species mostly mid grasses which grow as bunch grasses. Overgrazing by domestic animals has permitted the less palatable sagebrush and related species to spread widely and give character to the landscape.

Stipa-Poa association (California prairie): Located in central valley of California almost completely isolated from rest of grassland, dominants are mostly mid and bunch grasses. This is a region of winter rains. Much of area is now cultivated or overgrazed and contains many weedy annuals and exotic species.

Aristida-Bouteloua association (desert plains): Most arid of grasslands, composed mostly of short and bunch grasses. It occurs from southeastern Texas to southern Arizona and extends well down into Mexico. Because of overgrazing and control of fire, desert and tropical shrubs, such as mesquite, creosote bush, *Opuntia* cactus, are conspicuous throughout the association. The sugary pods of mesquite are eagerly eaten by cattle although the bony seeds resist digestion and are dispersed widely.

PALEOECOLOGY

Grasses did not evolve until the Upper Cretaceous period of the Mesozoic, and did not become important in North America until the elevation of the Western mountains in mid-Tertiary produced a semi-arid climate in the middle of the continent. Several of the mid grasses are circumpolar in distribution and may have constituted seral stages in the Arcto-Tertiary Geoflora. They probably segregated out to form the true, mixed, and bunch grass prairies when the forests belonging to this flora retreated southward and eastward. The close relation of grass species on the prairies of North America and northern Eurasia is doubtless due to their similar derivation from the Arcto-Tertiary Geoflora. The tall grasses of the Andropogoneae may be of tropical origin. The short grasses were probably derived from the Madro-Tertiary Geoflora to form the short grass and desert plains. Although one cannot be sure because of the paucity of grasses in the fossil record, it appears that the grassland is of mixed and relatively recent origin (Axelrod 1950, 1952).

During portions of the Pliocene and Pleistocene rainfall was heavy; grassland probably extended through the Great Basin and into the Mohave desert (Fig. 19-6). There is little information available to indicate whether the California grassland was ever in broad contact with the rest of the biome, but there was probably a narrow and irregular contact through mountain valleys at either or both the southern and northern extremities.

As the glaciers advanced during the Pleistocene, vegetation was at first affected only a short distance beyond its front, and the glaciers probably advanced out onto the grasslands in the north. With the persistent cooling of the climate, however, coniferous forest became established in scattered favorable situations over much of the grassland area. Relict stands of conifers still persist (p. 306).

During the post-glacial xerothermic period, grassland doubtlessly retreated in the southwestern states and in Mexico as the desert biome became extended, but it is difficult to draw boundary lines for the extremes of these advances and retreats. In the eastern part of the continent during the xerothermic period, prairie advanced as a peninsula far into the deciduous forest (Fig. 19-13; Wright 1968). Relict patches, including the Illinois prairie, still remain after later cool and more humid climate permitted the forest to recover much of the area it had previously lost. Doubtless the prairie also advanced northward into the area now dominated by the boreal forest of central Canada during the xerothermic period, where relict patches of grassland may still be found (Love 1959). Reinvasion of prairie areas was slow at first because of frequent fires caused by lightning and Indians and to poor drainage, but since settlement of the area by white man and artificial lowering of the water table it has become rapid (Gleason 1922, Box *et al.* 1967).

In spite of its vast extent, only a single grassland biociation can be recognized at present in North America, although it varies in composition between different regions.

NORTH AMERICAN GRASSLAND BIOCIATION

The North American grassland biociation extends in reduced form as a biocies or seral stage into the deciduous and coniferous forest biomes. Common species are the following:

Mammals

Eastern cottontail (East)
Nuttall's cottontail (South)
Desert cottontail (West)
White-tailed jack rabbit (North)
Black-tailed jack rabbit (South)
Thirteen-lined ground squirrel
Franklin's ground squirrel
Black-tailed prairie dog
Northern pocket gopher
Plains pocket gopher
Olive-backed pocket mouse
Plains pocket mouse
Silky pocket mouse
Hispid pocket mouse
Banner-tailed kangaroo rat (South)

Plains harvest mouse
Western harvest mouse
Deer mouse
Northern grasshopper mouse
White-throated wood rat (South)
Meadow vole
Prairie vole (East)
Meadow jumping mouse
Coyote (Fig. 23-6)
Swift fox
Long-tailed weasel
Black-footed ferret
Badger (Fig. 23-6)
Prairie spotted skunk
Pronghorn (Fig. 23-5)
Bison (Fig. 23-4)

Birds

Ferruginous hawk
Prairie falcon
Greater prairie chicken (Fig. 23-1)
Lesser prairie chicken
Mountain plover
Long-billed curlew
Upland plover
Marbled godwit
Wilson's phalarope
Franklin's gull
Burrowing owl
Short-eared owl
Horned lark

Sprague's pipit (North)
Western meadowlark
Dickcissel (East)
Lark bunting
Savannah sparrow
Grasshopper sparrow
Bairds's sparrow (North)
Vesper sparrow
Lark sparrow
Cassin's sparrow
McCown's longspar (North)
Chestnut-collared longspur (North)

Only a few mammal and bird species found in the grassland are endemic within the geographic area of the climax (Figs. 23-1, 23-2, 23-3). Many occur in the grassy seral stages of surrounding biomes (Hoffman and Jones 1970, Mengel 1970). This is also the case with many insect species (Ross 1970). Likewise many desert species extend their ranges into the grassland biome, where sagebrush, mesquite, and other shrubs come into overgrazed areas. Because of these various influences, the long north-south range through various temperatures, and the isolated nature of some portions of the grassland, several faciations may be recognized (Carpenter 1940). They have not yet been clearly defined but may correspond to the biotic provinces described by Dice (1943) and Blair (1954): Illinoisan in the East; Saskatchewan in the northern Great Plains; Kansan in the central Great Plains; Texan and Comanchian in the South; Navahonian in contact

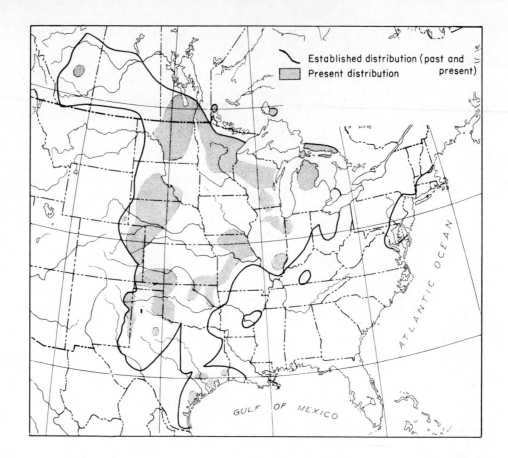

Established distribution (past and present)
Present distribution

Fig. 23-1 Present and past distribution of the greater prairie chicken, *Tympanuchus cupido* (Aldrich and Duvall 1955).

Fig. 23-2 Past and present distribution of the bison. The areas of present herds are considerably exaggerated in size (Petrides 1961, after Hall and Burt).

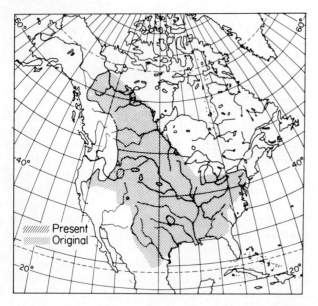

Present
Original

with the Southwestern desert; Palusian corresponding with the bunch grass prairie; and Californian, including the California prairie. The student should re-read Chapter 8, which gives additional data on the grassland fauna, especially in regard to reptiles and invertebrates.

OTHER BIOCIATIONS

Grasslands in other parts of the world (Brehm 1896, Haviland 1926, Stegmann 1938) are occupied by species ecologically equivalent to those that inhabit the grasslands of North America. The parallel evolution of adjustments and behavior in animals that occupy similar habitats, although often quite unrelated taxonomically, is of particular interest and may best be shown in table form (Table 23-1).

ANIMAL ADJUSTMENTS

It is in the grasslands throughout the world that the large herbivorous ungulates reach their largest popula-

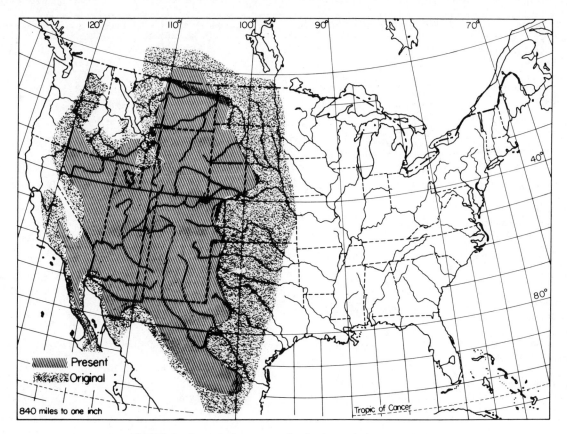

Fig. 23-3 Past and present distribution of the pronghorn (Petrides 1961, after Einarson).

tions. Their adaptations for feeding on grasses and the high productivity of grasses, which in fact is stimulated by moderate grazing, gives an efficient food coaction of high energy utilization. A forest cannot support such large populations of grazing animals since the herb layer is less luxuriant, composed more of the broad-leaved herbs, and shrub and tree foliage cannot tolerate continued browsing. The evolutionary development of these large ungulate populations had to await the evolution of these extensive grasslands in mid-Tertiary time (Stirton 1947). Actually, the bison (Fig. 23-4) did not appear on the grassland until relatively late in the Pleistocene. The plains bison, related to the wisent of Eurasia, is probably a mixture of a form that crossed the Bering land bridge during the early Illinoian glacial period and a form that came over during the Wisconsin glacial period (Flerov and Zablotski 1961).

It is common for large ungulates to feed in large herds. The primitive population of bison in North America is estimated at 50–60 million animals, perhaps an average of 6 per km² (15 per mi²). Few occurred west of the Rockies even in colonial times, but to the east and north the species penetrated far into forested areas. Herds containing 20,000 individuals were common and an occasional herd reached a population as high as 4 million animals (Seton 1909). When attacked

by wolves or other predators the bulls formed a circle facing outward with the cows and calves inside. The animals shed their fur in the summer and were greatly annoyed by flies, mosquitoes, and the penetrating seeds of the needle grass. They relieved their miseries by rolling in wallows, covering themselves with mud. Some calves were killed by wolves, coyotes, bears, and mountain lions. Adults died in consequence of bogging down in sloughs or swamps, breaking through thin ice when crossing rivers in winter, and of old age. Disease apparently was never common. The animals regularly grazed until mid-morning, traveled sometimes 16 km (10 miles) to a water hole, rested and chewed their cud during mid-day, and grazed again in the evening. A herd would graze an area intensively for several days or weeks, then move to some other area. In these movements and migrations, the animals commonly traveled in single file. There is question, however, whether north-south migrations were very regular and extensive (Roe 1951).

The pronghorn antelope (Fig. 23-5) occurs in the drier grasslands, including California, and their primitive population is estimated at 4 per km² (10 per mi²). They traveled in herds of 100–200, sometimes 2000. When they were scattered for feeding, the approach of a predator was quickly signaled from one to the other

Table 23-1 Equivalent species in grassland communities around the world (modified from Allee *et al.* 1949: pp. 470–471).

Ecological niche	North America	South America	Asia	Africa	Australia
1. Saltatorial (leaping) herbivores					
a. grasshoppers	*Melanoplus spretus* *Melanoplus differentialis* *Melanoplus maculipennis* *Melanoplus ponderosa*	*Schistocerca paranensis*	*Locusta migratoria*	*Locusta migratorioides* *Stauronotus macroccanus* *Schistocerca peregrina*	
b. mammals	Jack rabbits		Jerboa	Springhaas	Red kangaroo
2. Fossorial (burrowing) herbivorous mammals					
a. feeding in herb stratum	Ground squirrels Prairie dogs	Viscacha Pampas cavy	Souslik Hamster Bobak	African ground squirrels	Wombat European rabbit (introduced)
b. feeding in subterranean stratum	Pocket gophers	Tucotucos	Mole rat	Golden moles	Marsupial mole
3. Cursorial (running) herbivorous birds		Rhea		Ostrich	Emu
4. Cursorial gregarious herbivorous mammals	Bison Pronghorn	Pampas deer Guanaco	Saiga Goitered gazelle and other gazelles Maral Wild ass Wild horse	Quagga Burchell's zebra Eland Springbok Gazelles (several species) Black wildebeest Blue wildebeest Bubal Bontebok 30 other genera of antelopes	Pig-footed bandicoot
5. Cursorial predators					
a. snakes	Blue racer Bull snake Prairie rattler	*Cyclagras gigas* *Rhadinea merremii*	Common cobra Asian moccasin Elaphe	Puff adder Black-necked cobra Rock python	Death adder Tiger snake
b. mammals	Swift fox Coyote (Wolf) (Cougar)	Pampas cat Red wolf	Pallas cat Corsac fox Cheetah Jackal	Lion Serval Caracal Cheetah Cape hunting dog	Tasmanian wolf (marsupial)

by raising the hair in the white rump patch so that it flashed like a tin pan reflecting the sun.

Safety for these animals depended largely upon fleetness of foot, and some of the fastest-running animals in the world occur in this biome (Craig 1908a, Visher 1916). Joined with this ability was long-range vision to discern the approach of danger from a distance. Coyotes could not run as fast as pronghorns, but since the pronghorn usually ran in wide circles, the members of a pack of coyotes would sometimes run in relays until the pronghorn became exhausted.

Rodents and lagomorphs constitute the other principal groups of mammals. Ground squirrels, pocket gophers, mice, and jack rabbits are common nearly everywhere and sometimes reach plague proportions. Prairie dogs form towns, some in former days large enough to cover several square kilometers, with each animal feeding on the grasses and herbs only in the

Fig. 23-4 Bison on the Great Plains (courtesy U.S. Forest Service).

Fig. 23-5 Pronghorn (courtesy U.S. Forest Service).

vicinity of his own burrow. Prairie dogs are most numerous on the Great Plains where the grasses are shorter, and a century ago they probably numbered in the billions. They as well as other species of rodents have numerous predatory enemies such as coyotes, badgers, ferrets, foxes, weasels, owls, and rattlesnakes (Fig. 23-6). They have good vision, however, and quickly plunge into their extensive underground burrows at the approach of danger (Koford 1958).

Jack rabbits, other small mammals, prairie chickens, as well as most of the small birds, grasshoppers, and other insects rely considerably on their well-developed protective coloration—they freeze in the deep grass to escape notice of predators. If the predator comes too close, they take to running, jumping, or flying with such a burst of activity as to startle the intruder momentarily and give them a head start in their flight.

The development of hopping locomotion among grassland and desert animals is of special interest. We see it expressed in jack rabbits, kangaroo rats, pocket mice, and grasshoppers in North America; all these species have short forelegs and long, strong, hind ones.

Hopping enables the animals to get above the level of the grass for locomotion and is of advantage also in allowing greater visibility.

There is a nearly complete absence of animals above ground in the winter. Bison formerly shifted their main populations from the northern to the southern portions of the grassland, pronghorns sought shelter in the piñon-juniper woodlands and petran bush in the valleys and foothills of the mountains, and most of the birds left for the south. Ground squirrels, prairie dogs, and jumping mice hibernate in underground burrows, reptiles from over large areas aggregate into deep holes or crevices in the sides of hills, and invertebrates pass the winter in a dormant or inactive condition in the soil humus. Many of these animals also become inactive or aestivate during dry seasons, especially in late summer and autumn, and sometimes aestivation proceeds directly into hibernation so that the animals are active for only a short time each year.

Grassland birds are direct, strong fliers and can withstand the fierce winds prevalent in open country. They are adroit in walking and running on the ground.

Fig. 23-6 Above, coyote; right, badger (courtesy U.S. Forest Service).

They also appear better able to tolerate the continuous direct rays of the sun, but nevertheless on hot sunny days they commonly seek the shade of the tall grasses, scattered bushes, or line up in the narrow shadow of a fence post or telephone pole. With the absence of trees, many birds make themselves conspicuous during the nesting season by developing loud flight songs which they give high in the air. Flight songs are much more characteristic of grassland than of forest birds. Contrary to the herd instinct in mammals, flocking is not particularly characteristic of grassland birds, although prairie chickens and grouse exhibit gregariousness in their mating performances. Birds are usually widely spaced, breeding densities are low, and nests well concealed in the grass. In spite of the apparent simplicity and uniformity of the plant community, different species coexist by differentiation of niche relationships (Kendeigh 1941a, Frinzel 1964, Cody 1968, Wiens 1969).

The more humid portion of the grassland is studded with small ponds or potholes. These ponds are surrounded with marsh vegetation. Here are found numerous ducks, grebes, herons, bitterns, coots, rails, terns, and other marsh birds as well as muskrats. Animal life is especially concentrated around these

water holes (Brehm 1896), although larger species roam widely through the upland for feeding purposes.

HUMAN RELATIONS

Tall grass prairie is highly productive agricultural land for corn, wheat, soybeans, and cereal crops in general. This is evident both in North America and Europe. Hogs are raised on the corn that is grown. The more arid short grasslands, such as are found on the Great Plains and the northern Great Basin of North America and on the Russian steppe, are more hazardous to cultivate, as crops, mostly wheat, often fail during the dry years of climatic cycles. Plowed or overgrazed ground, destitute of grass or crop cover, is whipped up by strong winds to produce the great dust storms that became so well known in the 1930s. Dry farming for wheat is a common practice in some areas, where land is cropped only every 2 or more years and left idle between times to accumulate ground moisture, or there may be crop production under irrigation (Weaver 1927). Man has not always used grassland intelligently (Shelford 1944), because he has plowed up land where the grass cover should have been left intact. Arid grassland in the Great Plains had best be used only for stock raising, especially of beef cattle and sheep, and this is done extensively on our western ranches (Harlan 1956). In the arid parts of Asia, many different peoples have a nomadic existence in a never-ending search for pasture for the cattle, sheep, goats, camels, and horses that serve their everyday needs (Hadlow 1953).

The history of early exploration and settlement on the North American grassland is of considerable interest (Malin 1947). Before the advent of white man, Indians were scarce over the grassland because of difficulties in transportation and in hunting large game animals. With the escape of horses from the early Spaniards and their rapid multiplication, the Indians soon learned to use them, and several tribes took to the prairies and plains.

The white man was, at first, somewhat reticent about invading the prairies and kept his settlements to the forested areas along the streams. The prairies that he first encountered along the forest-edge in the East were flat, very wet in the spring, and poorly drained. The grass roots made a tough sod difficult to plow with the primitive equipment then available. There were difficulties in obtaining drinking water. Prairie flies were a nuisance, and the prairie fires that frequently swept across the country were dangerous. Furthermore, he was accustomed to using timber for buildings, fences, and fuel, and the forest gave him protection from cold winter winds (Vestal 1939). In the course of time, however, and under the pressure of increasing populations, he learned to surmount these difficulties. Ditches were dug to drain the land and the streams were deepened. Steel-pointed plows and other improved farm equipment made cultivation easier. Construction of roads, bridges, and railroads brought building equip-

ment for homes, supplies, and other comforts of life. At the present time the tall grass prairie is the "bread basket" of our modern civilization.

SUMMARY

Grasslands occur on all continents and are marked by low amounts of precipitation and high rates of evaporation. Climax grasses are mostly perennial and are characteristically tall, medium, or short in stature. Since their leaves grow at the base, they tolerate grazing by animals, and it is in this biome that ungulate mammals and rodents attain large population densities.

The large herbivorous ungulates commonly go in herds, are fleet on foot, and have long-range vision. Some species of rodents form large colonies and dig extensive underground burrows. Locomotion by hopping occurs in several mammals and some insects. Protective coloration is well developed in many kinds of animals. Upland birds are strong fliers and commonly have flight songs. Scattered ponds surrounded by marsh have concentrations of many nesting species of birds and are the source of drinking water for the large mammals. Migration is well developed among birds and hibernation among small mammals, reptiles, amphibians, and invertebrates so that there is a nearly complete absence of animals above ground during the winter months. Only one biociation is recognized in North America and in each of the other continents.

The more humid grassland areas make very fertile and productive agricultural land for man, while the drier portions are best used for grazing of domestic animals.

Chapter 24

DESERT BIOME

Extreme desert is arid wasteland, with practically no vegetation. In the ecological sense, however, deserts also include arid regions which contain considerable vegetation in the form of bushes, shrubs, and trees especially adapted to tolerate hot, dry climates. Deserts are unique in possessing a large number of different life-forms among the plant dominants (Shreve and Wiggins 1964) and a unique animal life (Shelford 1963, Miller and Stebbins 1964).

Deserts, like any other biome, occur in belts at similar latitudes north and south of the equator around the world and they cover about one-fifth of its surface. Prominent deserts occur in southwestern United States and northwestern Mexico; in Sahara, southern Africa, Arabia, central Asia, Australia, and in a narrow strip along the west coast of South America (McGinnies et al. 1969). All these areas are at low elevations. In spite of its arid character, the desert, like other biomes, has a fascination and charm for one who becomes familiar with its inhabitants and their problems (Jaeger 1955, 1957).

CLIMATE

Average annual precipitation in the desert scrub of North America is usually not more than 13 cm (5 in).

and snowfall is slight (Jaeger 1957). Because of the high rate of evaporation and lack of penetration into the soil, Weather Bureau statistics are not indicative of how much moisture of precipitation is actually available to organisms. Rainfall is irregular and long drought periods are typical. The little rainfall that occurs is often in the form of short, violent storms or cloudbursts. The ground surface is generally baked hard, and most of the rain runs off; flash-floods are not infrequent (Lowdermilk 1953). Precipitation is slightly greater and evaporation less in the Great Basin. Where rainfall is so slight, dew formation assumes great significance. In the deserts of Israel, dew forms 120–240 nights of the year (Duvdevani 1953).

The yearly evaporation from a pan of water may greatly exceed the actual amount that falls on an area of similar size. Climates are considered subhumid when the precipitation-evaporation ratio ($\times$ 100) lies between 0 and -20, semiarid when the ratio is -20 to -40, and arid when it is below -40 (McGinnies et al. 1968). The high evaporation rate correlates with the low relative humidity, which at noon averages less than 25–30 per cent.

There is little cloudiness, and the actually received percentage of possible annual sunshine averages close to 90. Ultraviolet radiation reaches the ground in high concentrations. Winds are more or less continuous.

The mean annual temperature in the Great Basin is approximately 10°C (50°F), but it is over 20°C (68°F) in parts of the desert scrub. Daily maximum summer temperature in the desert is 40°C (104°F) or more. Differences between daily maxima and minima are greater than in any other biome. Frosts are limited to mid-winter; the frost-free period averages more than 280 days. Frost and snow are common, however, in the Great Basin (Kincer 1941); consequently, the Great Basin area is sometimes called the *cold desert* in contrast to the *hot desert* adjacent to the south.

The Great Basin is not a single large basin; rather, it is broken up into a number of small ones separated by low mountain ranges running in a north-south direction and seldom exceeding heights of 1800 m (6000 ft). These ranges support piñon-juniper woodland. Water drains from the surrounding hills into these small basins from which there are usually no outlets. High rates of evaporation make the basin lakes very salty or may dry them up to produce alkaline flats.

A characteristic topographic feature of the desert is the alluvial fan of erosion products washed down mountain slopes by the torrential rains. Broad flat basins occur between adjacent mountains. Extensive sand dunes occur in some portions of the desert, and sand and dust storms are spectacular features.

VEGETATION

Desert bushes and shrubs in North America are seldom more than 1–2 m high and are spaced 3–10 m apart (Fig. 24-3). Joshua-trees, paloverdes, and saguaros are, however, more conspicuous. The shrubs seldom form a canopy except along washes. The intervening ground between the shrubs is usually a wind-swept desert pavement of either fine texture soil, gravel, or rock, and always with very little humus. The shrubs and bushes have shallow, wide-spreading, many-branched root systems adapted quickly to absorb any surface moisture. There is very little moisture in the subsoil. Stems and branches often bear prickles or thorns and intertwine to form a dense tangle. A rich variety of thorny, succulent cacti occurs in the western hemisphere only, and is divisible into tree, cholla, and barrel types. Between the shrubs, a few short annual grasses may grow, but after rains, the ground often becomes thickly covered with a carpet of flowers and grasses that quickly mature seed and then disappear in the dry weather that follows.

Some desert plants have sclerophyllous leaves in adaptation to retard transpiration and survive long periods of drought. In a large number of them foliage is greatly reduced, even absent altogether, during long periods; stems contain the chlorophyll necessary to carry on photosynthesis. Cell walls are thick, highly ligneous, and have thick cuticles. The cell sap increases in osmotic pressure and hydrophilous colloids. Cacti store considerable water in their stems as a reserve for use when there is no water in the soil. Many other kinds of adaptations occur in these xerophytes (Weaver and Clements 1938, Zohary, in Cloudsley-Thompson 1954).

Biotic succession is not conspicuous in the desert because of the low rate of reactions by organisms upon the habitat (Shreve and Wiggins 1964). When the vegetation is disturbed it is usually replaced directly by the same type without intervening seral stages (Muller 1940). Physiographic succession is evident, however, depending upon distances from water channels, differences in elevation, leaching of salts out of the alkali flats, and in sand dune areas. There is considerable variation in the distribution of different species because of local differences in the chemical and physical nature of the soil, soil moisture, and so forth (Shantz and Piemeisel 1924).

PLANT ASSOCIATIONS IN NORTH AMERICA

Larrea-Franseria association (desert scrub): Creosote bush, *L. divaricata*, is generally distributed through all parts of the desert in North America and with bur sage, *Franseria*, a semi-shrub, sometimes forms a monotonous, uniform growth on the flat intermontane plains, relieved only by the larger acacias, saguaros, paloverdes, and mesquites along the washes. The richest variety of vegetation is on the outwash plains and lower mountain slopes, where there is greater penetration and retention of soil moisture. The desert scrub may, perhaps, be divided into several associations, although here we will consider only the three principal deserts as faciations (Fig. 24-1) (Shreve and Wiggins 1964, Axelrod

Fig. 24-1 Faciations of the desert and Great Basin associations in North America (after Axelrod 1950, Jaeger 1957).

Fig. 24-2 Joshua-trees interspersed through low shrubs of the Mohave desert in California (courtesy U.S. Forest Service).

Fig. 24-3 The Sonoran desert of Arizona: saguaro (tree cactus at left), paloverde trees (in middle distance), tree cholla (cactus at right center), organ pipe cactus (upper right), creosote bush (the taller bushes in the foreground), and bur sage (the smaller bushes in the foreground).

Fig. 24-4 Sagebrush with sparse grass in Nevada (courtesy U.S. Forest Service).

1950). The *Mohave desert* to the west is a rolling plain with a monotonous uniform cover of low shrubs, interspersed conspicuously with the curious Joshua-tree (Fig. 24-2). The *Sonoran desert*, sometimes subdivided into Colorado and Arizona deserts (Benson and Darrow 1944), or as many as six sub-units (Jaeger 1957), is much more diversified, with tall shrubs, trees, and succulent cacti of many forms, especially along the washes, and a few grasses (Fig. 24-3). The *Chihuahuan desert* to the east is almost completely separated by mountain ranges from the rest of the desert, lacks the aborescent cacti and legumes of the Sonoran desert, and has yuccas and agaves more prominent along with sotol (Jaeger 1957). Desert shrubs, particularly mesquite, have spread extensively into grassland as the result of seed distribution by domestic stock and protection of grassland from fires.

Atriplex-Artemisia spinescens association (shadscale association); *Artemisia tridentata-Agropyron* association (sagebrush association): Shadscale, *Atriplex confertifolia*, and bud sage, *Artemisia spinescens*, as well as other small (less then 1 m high), widely scattered, more or less spinescent, microphyll shrubs are dominant in the southern part of the Great Basin and have contact with desert scrub. Greasewood and a few grasses are important in some localities. Sagebrush, *Artemisia tridentata*, often occurs in nearly pure stands (Fig. 24-4), but where grazing is limited, several species of perennial grasses, especially wheat grass, *Agropyron spicatum*, become intermixed to form a continuum leading into the bunch grass prairie to the north. Only a few opuntia cacti occur. Sagebrush is widely distributed as a subclimax, because of overgrazing, in the bunch-grass prairie and short-grass plains. Various species of *Artemisia*

also extend into central and southern California and together with associated species have been called *coastal sagebrush*. It may be subclimax to chaparral. The original vegetation of these two associations has become greatly modified as result of overgrazing by domestic animals and increased erosion with reduction in grasses and edible shrubs and introduction of exotic species (Fautin 1946, Cottam 1947, Billings 1949).

A *tropical thorn woodland* or *scrub*, sometimes considered a separate plant formation or biome, occurs on the west coast of Mexico, in northern Venezuela, and in other scattered localities. It is made up of a dense scrubby growth of small, often thorny and leguminous trees. Cacti are common. Some authors distinguish a thorn forest and a short-tree forest, but there is considerable intergrading between the two (Gentry 1942) as well as with the tropical deciduous forest.

PALEOECOLOGY

During early Tertiary, the present desert regions in North America were largely dominated by tropical and warm-temperate forests. Following Eocene, rainfall gradually decreased, and forests were replaced first by grassland, then by desert. Desert began to appear during middle Tertiary on the lee side of high mountain ranges, but did not become extensive until middle and late Pliocene time. The present deserts in North America are therefore of comparatively recent origin.

Desert plants have apparently originated through gradual adaptation to arid climates of more hardy species belonging to all three Tertiary geofloras. *Artemisia, Atriplex, Eurotia,* and *Suaeda* of the two Great

Fig. 24-5 Changes in the relative proportions of different types of vegetation in the Mohave desert region since Miocene time (from Axlerod 1950).

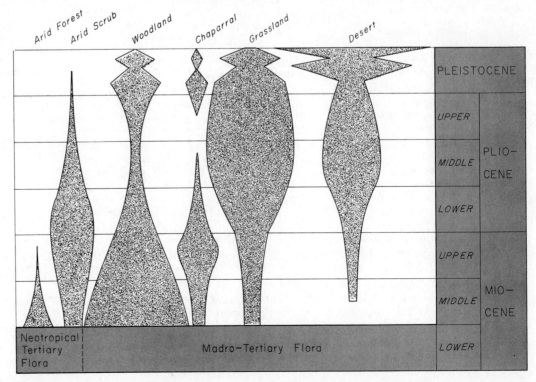

Basin associations are thought to have Arcto-Tertiary affinities. Related species in the same genera occur in the Eurasian deserts and to a more limited extent even in the Australian desert at the present time. Species of the desert scrub appear to be derived from the Neotropical and Madro-Tertiary Geofloras, and are for the most part unrelated to Eurasian forms. Related species and genera are found, however, in the arid regions of South America. The vegetation of the present Great Basin, Mohave, and Sonoran desert regions was largely woodland and chaparral through much of the Miocene and into the Pliocene (Fig. 24-5). Desert vegetation became differentiated with the increasing aridity of the mid-Pliocene. The Mohave and Sonoran deserts became distinct as cool winters in late Pliocene and Pleistocene restricted the less hardy succulent species to the Sonoran desert. It is probable that the origin and development of deserts elsewhere over the world has followed the same general pattern (Axelrod 1950, Clements 1936).

DESERT SCRUB BIOCIATION

Large mammals, such as the bison and pronghorn, are mostly absent from the desert scrub. The mule deer is present in small numbers both in the desert scrub and in the basin sagebrush, and the mountain lion, bobcat, and badger penetrate to some extent. The most common species of animals are the following (Dice 1939, Huey 1942), although additional species occur farther south in Mexico (Burt 1938, Van Rossem 1945, Baker 1956) (species with asterisks in this and following lists also found in the basin sagebrush biociation):

Mammals

Desert cottontail	*Canyon mouse
*Black-tailed jack rabbit	Cactus mouse
*White-tailed antelope squirrel	*Deer mouse
Spotted ground squirrel	Southern grasshopper mouse
Rock squirrel	White-throated wood rat
Round-tailed ground squirrel	*Desert wood rat
*Botta's pocket gopher	*Coyote
*Little pocket mouse	*Kit fox
Desert pocket mouse	Gray fox
Rock pocket mouse	*Western spotted skunk
Merriam's kangaroo rat	Collared peccary
Desert kangaroo rat	Mountain sheep

In the mesquite vegetation of New Mexico, the mouse and rat populations are highest in May with about 8.5 individuals per hectare (3.4 per acre) and with the kangaroo rats the most numerous species. Likewise in the Mohave Desert of California, Merriam's kangaroo rat, followed by the little pocket mouse, are the most common species. The pocket mouse is

dormant, however, most of the autumn and winter (Chew and Butterworth 1964). The energy relationships of small mammals in the desert ecosystem have been studied by Chew and Chew (1970). These small desert mammals tend to have larger home ranges than do comparable species in deciduous forest and grassland (Blair 1943). Several of these species extend their ranges well south through the Chihuahuan faciation (Dalquest 1953).

Birds

*Red-tailed hawk	Wied's crested flycatcher
Harris' hawk	Vermilion flycatcher
Caracara	Verdin
Gambel's quail	Cactus wren
Mourning dove	Bendire's thrasher
White-winged dove	Curve-billed thrasher
Ground dove	LeConte's thrasher
Roadrunner	*Crissal thrasher
*Great horned owl	Black-tailed gnatcatcher
Elf owl	Phainopepla
Lesser nighthawk	*Loggerhead shrike
Costa's hummingbird	Lucy's warbler
Gilded flicker	Abert's towhee
Gila woodpecker	*Black-throated sparrow
Ladder-backed woodpecker	

In addition to these, several species from the deciduous forest-edge, woodland, and chaparral penetrate into the desert. Bird populations are very low in the open desert (0–37 pairs per 40 hectares or 100 acres) but may reach 108 pairs per 40 hectares in washes or near water where there is a greater diversity of vegetation (Miller 1951, Hensley 1954, Dixon 1959, Balda 1967, Raitt and Maze 1968). A greater proportion of the avifauna is of North and South America origins than in the coniferous forest (Table 24-1).

Reptiles

Gopher turtle	*Whip-tailed lizard
Banded gecko	Western blind snake
Crested lizard	Boa snake
*Leopard lizard	Whip snake
Chuckwalla	Leaf-nosed snake
Zebra-tailed lizard	*Bull snake
Fringe-toed lizard	Common king snake
Spiny lizard	Western shovel-nosed snake
*Side-blotched uta	
Long-tailed uta	Mohave rattlesnake
Tree uta	Diamond rattlesnake
*Desert horned toads	Sidewinder rattlesnake
Desert gila monster	
Night lizard	

The list of reptiles is compiled from the studies of Mosauer (1935), Dice (1939), Huey (1942), and Johnson et al. (1948). Equivalent species occur in the

Table 24-1 Origin of avifauna in desert communities of southwestern North America. Percentages of species and breeding pairs of different ancestry (Hensley 1954, Balda 1967).

Community	Locality	Eurasia		North America		South America		Unanalyzed	
		Species	Pairs	Species	Pairs	Species	Pairs	Species	Pairs
Colorado desert	California	11	14	78	49	11	37	—	—
Arizona desert	Organ Pipe Cactus Nat. Mon., Ariz.	22	6	35	54	27	39	17	1
Arizona desert	Chiricahua Mts., Ariz.	20	29	44	40	24	27	12	4
Desert grassland	Chiricahua Mts., Ariz.	25	27	38	35	33	35	4	3
Oak and oak-juniper woodlands	Chiricahua Mts., Ariz.	24	26	31	40	33	28	12	6

deserts of Israel and Australia (Warburg 1965). Amphibians are not common, but the red-spotted toad occurs in small ponds. Little quantitative investigation has been made of the invertebrate populations of desert scrub, but grasshoppers and other orthopterans are especially conspicuous (Tinkham 1948), and the scorpion and tarantula spider are much in evidence. The spider population corresponds in number of species, density, and habits to those in the early stages of the sand dune sere in the deciduous forest biome (Chew 1961).

BASIN SAGEBRUSH BIOCIATION

This community inhabits both the shadscale and sagebrush associations. There is considerable overlap of species between the desert scrub and basin sagebrush biociations, as indicated by the species marked with an asterisk in the above lists, but there is a sufficient difference to warrant calling the two areas faunistically distinct. At the subspecies level, the contrast between the two animal communities is more striking. In addition to a strong intrusion of grassland species, the basin sagebrush has the following noteworthy species (Linsdale 1938, Hall 1946, Fautin 1946):

Mammals

Merriam's shrew
Pigmy rabbit
Nuttall's cottontail
Least chipmunk
Townsend's ground
    squirrel
Great Basin pocket mouse

Long-tailed pocket mouse
Dark kangaroo mouse
Ord's kangaroo rat
Chisel-toothed kangaroo rat
Northern grasshopper
    mouse
Sagebrush vole

Rodents, exclusive of ground squirrels and pocket gophers, average about 40 per hectare (16 per acre) in western Utah, with deer mice and kangaroo rats most numerous. Ground squirrels are widespread and numerous although sometimes locally restricted. Black-tailed jack rabbits are important constituents of the community and average numbers seen in different plant communities range from less than 0.1 to 0.5 per hectare. The pronghorn was once numerous in the Great Basin sagebrush, but not the bison (Fautin 1946).

Birds

There is greater contrast in the avifauna between the desert scrub and basin sagebrush biociations than in the mammalian fauna. However, more species enter this community from the grassland and forest-edge biociations, such as the Swainson's hawk, prairie falcon, burrowing owl, and horned lark, than venture into the desert scrub. Bird populations are low during the breeding season, averaging only about 25 pairs per 40 hectares (100 acres). The principal avian species in the basin sagebrush in addition to those listed above are (Fautin 1946, Miller 1951):

Sage grouse (Fig. 24-6)
Poor-will
Sage thrasher

Green-tailed towhee
Sage sparrow
Brewer's sparrow

Reptiles

Collared lizard
Sagebrush lizard
Striped whip snake

Long-nosed snake
Prairie rattlesnake

Lizards are numerous and conspicuous. Counts of only those seen above ground gave an average for the summer season of 6.5 per hectare (2.6 per acre). A few amphibians, particularly the western spadefoot toad and western toad, occur near bodies of water (Linsdale 1938, Fautin 1946).

Invertebrates

Actually, only two strata occur in this community, the shrub and ground, since herbs are few and scattered most of the year. In the shrub stratum, arachnids,

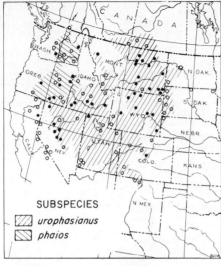

Fig. 24-6 Distribution of the sage grouse, *Centrocercus urophasianus,* in the basin sagebrush (Aldrich and Duvall 1955).

cicadellids, fulgorids, coccids, chrysomelids, and mirids are most numerous. Grasshoppers feed on the foliage and lay their eggs in open areas of the ground. Arachnids, tenebrionid beetles, and ants are the most conspicuous ground invertebrates. The harvester ant and honey ant build conspicuous mounds, the number of mounds of the former averaging over 15 per hectare (6.2 per acre) in the sagebrush community. Invertebrates are most numerous in sagebrush and greasewood and least abundant in shadscale. Maximum populations occur in May on most of the vegetation, after which they decline as temperature increases, but on the greasewoods, which retain their leaves and remain green, populations remain more constant throughout the summer (Fautin 1946, Allred *et al.* 1963).

OTHER BIOCIATIONS

The vegetation and animal life of extreme deserts around the world are impoverished, but in semideserts, similar to those in southwestern North America, ecologically equivalent species occur, although they show little taxonomic relationships with each other. These organisms are derived from adjacent, more humid floras and faunas and have many similarities in adjustments and adaptations. Rodents, for example, are generally numerous everywhere. In North American deserts the genera *Dipodomys* and *Perognathus* of the family Heteromyidae are especially important; in the Eurasian deserts, *Gerbillus*, *Meriones*, and *Dipodillus* belonging to the family Muridae are found; in South Africa *Pedetes*, family Pedetidae, occurs; and in the Australian desert the family Muridae is represented by *Notomys*. All are bipedal in locomotion and have elongated hind legs (Schmidt-Nielsen in Cloudsley-Thompson 1954). It is also of interest that tenebrionid beetles are represented by different subfamilies in different deserts of the world, but, contrary to the prevailing desert colors, these diurnal beetles are predominantly black (Brehm 1896, Buxton 1923, Haviland 1926, Kachkarov and Korovine 1942, Bodenheimer 1953, Dementiev 1958, McGinnies *et al.* 1968).

ANIMAL ADJUSTMENTS

The characteristic animals of the desert are the small herbivorous rodents and the reptiles. Large animals, including the carnivores, are relatively scarce, and population levels of the rodents appear determined more by the availability of food and water than by predation. Among birds on the Arizona desert, insectivorous species are most numerous, then seed-eaters, and lastly carnivores (Hensley 1954).

Most adjustments grassland species make to their environment continue to be expressed in the desert and additional ones become conspicuous (Buxton

1923, Sumner 1925, Heim de Balsac 1936, Linsdale 1938, Fautin 1946, Hensley 1954, Schmidt-Nielsen 1964, Cloudsley-Thompson 1954, Brown 1968). The two most critical environmental factors are the high temperatures, especially during mid-day, and the lack of water. Reptiles have some advantage in that their scaly skin is adapted to prevent rapid evaporation. Moist-skinned amphibians and snails are absent except in the immediate vicinity of springs or other bodies of water.

Animals tend to avoid extreme high temperatures rather than to tolerate them for any length of time. They do this in various ways. The small mammals, snakes, and even insects are largely nocturnal. Birds are active chiefly in the cooler hours of early morning and evening and tend to remain quiet and concealed during the middle of the day. Lizards are the most conspicuous animals during the day. Nearly all animals spend their time above ground in the shade cast by the scattered shrubs or rocks, and it is here that they have their burrows or nests. Bird nests occur most frequently on the east and northeast side of plants, where they are shaded from the hot afternoon sun. The intervening ground, fully exposed to the sun's rays, heats up much higher than the air temperature and may not cool down completely even at night, so that small mammals and other animals scurry quickly from the protection of one bush to another in their travels for food. Ground surface temperatures go well above the upper limit of tolerance of snakes, but some lizards can hold their bodies away from contact with the ground on their long thin legs. Even the grasshoppers come to rest in bushes to avoid the hot ground surface as much as possible. Grasshopper species confined to hot sandy areas have, like the lizards, long slender legs that hold their bodies away from the ground. Many mammals, reptiles, and insects (ants, crickets) burrow deeply into the ground and thereby avoid the surface heat (Fig. 24-7); for example, the burrows of kangaroo rats penetrate 50–65 cm below the surface near Tucson, Arizona (Sumner 1925). On one day when the maximum air temperature in the shade reached 42.5°C (108.5°F), and the temperature of the ground surface was 71.5°C (160.7°F), at a depth of 10 cm in the burrows the temperature was only 40.1°C (104.2°F), at 30 cm 29.8°C (85.6°F), and at 45 cm 27.9°C (82.2°F). The amount of moisture in the air inside these burrows is also more favorable, being 3 or 4 times higher than it is outside (Schmidt-Nielsen in Cloudsley-Thompson 1954). The percentage of mammal species that burrow increases from 6 in forest communities to 47 in short-grass plains, to 72 in deserts. This is in contrast to the decrease in percentage of mammals that are active on the ground level; from 68 to 53 to 28 (Bodenheimer 1957).

Many desert animals are adapted to go a long time without drinking water, but those species that depend on drinking water, which includes many of the larger

mammals, are restricted to the vicinity of springs, lakes, or ponds. Dew is often a source of water in the early morning. Much of the desert vegetation, particularly the cacti, is succulent and is a source of water to animals. The development of an armor on the plants in the form of thorns and prickles serves as a defense against excessive browsing by animals. The flowers and fruits of such plants as the saguaro are important sources of water to birds and other animals. The blood and body fluids of prey furnish ample water for carnivores. Metabolic water obtained with the oxidation of fats and carbohydrates in the food eaten is apparently sufficient for many species of small size. Even some of the larger game mammals of Africa find green pasture sufficient for satisfying their moisture needs if they can also obtain shade (Vesey-Fitzgerald 1960). Water is conserved in the bodies of birds, insects, and many desert reptiles by kidney wastes excreted as solid uric acid salts rather than as urine. The urine of mammals is more highly concentrated than in non-desert species, and feces are egested in a dry condition, the excess water having been reabsorbed in the large intestine.

After rains sufficient to soak the soil or to refill the shallow ponds, a rapid cycle of events occurs. Herbaceous plants become abundant and bloom. Snails come out of aestivation in the mud. Immature insects and crustaceans become abundant in the water. Termites and ants produce winged forms and mate, and other insects appear in large numbers. Frogs come out of their underground burrows and deposit their eggs; tadpoles hatch quickly, grow rapidly, and metamorphose into adults. Animals previously aggregated

Fig. 24-7 Burrow temperatures experienced by three scorpions, *Hadrurus arizonensis* (A, B, C) in southern Arizona at various depths over a 24-hour period. Soil temperatures at the surface and at −20 cm are shown also. Scorpion A, in a shallow burrow, died, but B at intermediate depth and C at about −40 cm both survived (Hadley 1970).

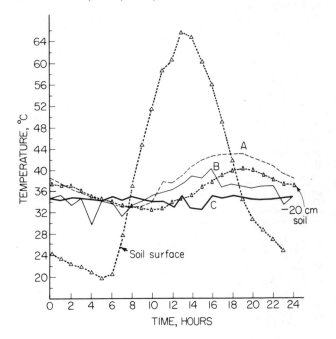

around water holes scatter widely over the surrounding country. In Australia, the coming of rains stimulates the initiation of reproductive activities in birds comparable to lengthening of photoperiods and rising temperatures in the spring for birds in cooler climates. However, as the rains stop and the ponds again dry up, populations contract sharply, and many species go back into aestivation, sometimes for years, until the next wet period occurs. Not only invertebrates but also birds and mammals largely confine their reproductive activities to the rainy season (Buxton 1923). Some rodents undergo seasonal aestivation to avoid unfavorable drought periods and even daily torpor to avoid excessive heat (Bartholomew and Dawson in Brown 1968).

In the northern cooler portions of the desert, hibernation over winter is necessary for many cold-blooded forms. Lizards and snakes may hibernate $\frac{1}{2}$ m or more in sand, under rocks, or in burrows of other animals (Cowles 1941). Rattlesnakes hibernate in natural cavities on hillsides or elsewhere.

The possession of poisons seems especially well developed among desert lizards, snakes, spiders, scorpions, and others. This permits predators to kill or immobilize scarce prey that may otherwise be too large or active to be taken. Likewise, activity of prey species in open spaces between scattered vegetation presents hazards that may be mitigated to some extent by being dangerous, unpleasant, or distasteful (Minton in Brown 1968).

Although much of the desert soil is a hard, gravelly pavement, loose soil and sand dunes are not uncommon. Many reptiles have special structural and behavior adaptations to cope with sandy habitats so that a good herpetofauna occurs in such places (Mosauer 1935). The texture and hardness of the soil, its depth, slope, and color influence the niche segregation of small mammals (Hardy 1945). The light gray, yellow, and brown tone of desert soils is reflected in the pale coloration of many desert birds and mammals. Desert animals are generally less heavily pigmented and are smaller in size than close relatives in humid regions (*Gloger's rule*). It is not certain how much of this is a response to the arid climate (Buxton 1923, Sumner 1925) or high light intensities (Meinertzhagen 1950), and how much is a response to the color of the soil. Deer mice and pocket mice occurring in the White Sands National Monument of New Mexico are very light in color, but a few kilometers away, in the Tularosa lava beds, their color is very dark (Dice and Blossom 1937). This blending coloration doubtless furnishes protection from the attacks of predators on moonlit nights. Even in humid climates, the dark coloration of animals may be an adaptation to the darker-colored vegetation and ground litter (Bowers 1960).

The fish of the desert present features of special interest. The small ponds and pools are widely isolated from each other and are without outlets to the sea.

The salt concentration in some of them is high as a result of centuries of continuous evaporation of water. Some spring-fed pools contain only a few dozen or a few hundred individuals, but because of their isolation these individuals have often evolved into distinct varieties or species found nowhere else. These various fish populations, however, show a relation to each other. In many cases, particularly in the Great Basin, the ponds are deep holes that persisted after the drying up of large shallow Pleistocene lakes, such as Lake Lahontan and Lake Bonneville (Fig. 24-8). The fish in these ponds are descendants, therefore, of populations formerly widespread throughout the Pleistocene lakes (Hubbs 1940a, Hubbs and Miller 1948).

HUMAN RELATIONS

In semi-deserts, there is production of cattle, sheep, goats, horses, and camels, but the carrying capacity of the land is low. Likewise, the biotope is very vulnerable to overgrazing, evident in the decline of palatable plants and spread of shrubs. Recovery from overgrazing is very slow because of the low productivity. In the Great Basin shadscale association, 5.8 jack rabbits consume as much of the same kinds of plants during the winter as does one sheep (Currie and Goodman 1966). The stock needs to have access to springs, ponds, or rivers, which are of course widely scattered. Nomadic primitive people roam the semi-deserts of Arabia, Africa, and Australia in search of pasture for their herds.

Since the scanty rainfall does not wash away the salts and nutrients, the soil is fertile where irrigation is possible. Vegetable and other crops and fruit can be raised advantageously. Desalted seawater may be an ample source for irrigation provided it can be made economically feasible. Lands that have been irrigated from streams for long periods of time may have harmful accumulations of salt resulting from evaporation of the water. Desalinization produces distilled water and avoids this difficulty (Young 1970). Where the groundwater table comes close to the surface locally, oases of vegetation occur, even in otherwise extreme desert, and may support small settlements. Air-conditioning may make indoors comfortable, but, on the whole, man does not find the desert an amenable habitat.

SUMMARY

Deserts, like grasslands, occur in all continents. They develop in areas with very low annual precipitation and generally high temperatures. The vegetation consists of often widely spaced shrubs with sclerophyllous or transitory leaves and commonly armed with spines, along with succulents, semi-shrubs, and herbs, including many species of annuals. Succession is not con-

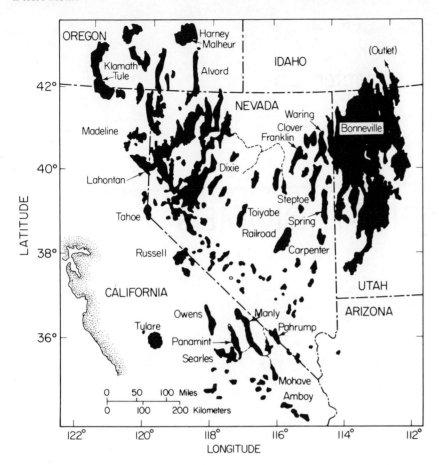

Fig. 24-8 Distribution and size of lakes in western United States formed by rainfall and the melting of glacial ice in late Wisconsin. Almost all of these lakes have dried up in the last 10,000 years (Engel 1969).

spicuous and the prevailing vegetation varies considerably with local soil and moisture conditions. Desert scrub and basin sagebrush biociations are distinguished in North America. These communities commenced to emerge as distinct biotic entities in mid-Pliocene, as organisms became adapted to the increasing aridity of the climate.

The most prominent animals of the desert are small herbivorous rodents and reptiles. Animals tend to avoid extremely high temperatures by becoming nocturnal, spending much of their time in shady places, or burrowing into the ground. Many desert animals are able to go a long time without drinking water, getting what they need from succulent foods and the oxidation of fats and carbohydrates. There is reabsorption of much water from urinary wastes, or uric acid is excreted instead of urea. After periods of rain, there is temporarily a rich expansion of plant and animal reproductive activities. Aestivation or dormancy is common during periods of drought. Many reptiles have structural and behavioral adaptations to cope with sandy habitats. There is a general tendency for desert animals to be lighter in color and smaller in size than close relatives in humid regions. Since ponds and lakes are widely isolated, speciation among fishes has developed extensively.

Chapter 25

TROPICAL BIOMES

Tropical communities and habitats vary from rain forest to desert. The largest continuous mass of tropical evergreen forest lies in the Amazon valley of South America and extends from lower Mexico across northern South America from the Pacific to the Atlantic Ocean. It is interrupted by tropical deciduous forest and savanna, as well as cloud forests in the Andes Mountains. Elsewhere tropical vegetation of various types covers extensive areas in central and western Africa and almost the whole of the Oriental Region. Tropical vegetation also occurs in Australia, New Guinea, and the Pacific Islands.

CLIMATE

The conspicuous features of tropical climate are high, even temperatures throughout the year; relatively uniform lengths of day and night; and seasonal variation in rainfall (Richards 1952). Mean monthly temperatures do not drop below 18°C (64°F) and may rise to 32°C (90°F) or higher. Lowest mean temperatures usually but not always occur during the wet season, but the difference between monthly means may be less than 1°C (1.8°F) and is seldom more than 13°C (23°F). There may be a greater range in temperature at different times of day and night than in mean monthly temperature throughout the year. Mean daily minimum

and maximum temperatures are seldom below 10°C (50°F) or above 43.3°C (110°F). As one ascends mountain slopes there is, of course, a drop in temperature.

On the equator, the lengths of day and night are approximately 12 hours each throughout the year. The seasonal variation increases away from the equator, both north and south, but in an opposite manner. The shortest daylength in the tropics is about 10.5 hours, the longest about 13.5 hours. In the rainy season, the actual amount of sunshine is low, averaging only 5 or 6 hours per day.

In contrast to the uniformity of temperature and the length of daylight is the considerable diversity in rainfall and humidity in different parts of the tropics. Deserts with insufficient moisture to support any vegetation occur at one end of a climatic gradient, while at the other end large areas exist where annual precipitation is between 250 cm (100 in.) and 400 cm (160 in.). Rainfall is largely convectional and results from the cooling of the air that rises from heated land surfaces. Sudden showers are often accompanied by lightning and thunder and may bring a sudden drop of as much as 4°C (7°F) in temperature. These storms commonly occur in the afternoon and may come regularly day after day. Because of this influence of the sun, the rainy season typically occurs when the sun is directly overhead.

Adjacent to the equator, there is considerable

rainfall every month in most areas. Between latitudes 3° and 10°–15°, North and South, the two periods of the year when the sun is at its zenith are far enough apart so that there are two rainy and two dry seasons each year (Fig. 25-1). At still higher latitudes there is only one wet and one dry season. What constitutes a dry season is arbitrary, but in the wetter parts of the tropics, it is considered the period when the rainfall is less than 10 cm (4 in.) per month. Under extreme conditions no rainfall may fall during the dry season, while in the rainy season some localities may receive over 100 cm (40 in.) in a single month. During the dry season the soil may become desiccated, while during the wet season it may become waterlogged. Grass fires frequently occur during the dry season. The periodic monsoons of India and southeastern Asia result from the outflow of dry winds from a high-pressure area that persists in central Asia during the winter and the inflow of moisture-laden winds from the surrounding oceans toward a continental low-pressure area during the summer.

In the wettest parts of the tropics, relative humidity is always very high. It seldom drops below 60 per cent of saturation during the hottest part of the day, and may average over 90 per cent for the entire day. On tropical mountains, mean relative humidity rises with increase in elevation until at 1000 m (3300 ft) in some localities there is almost continuous fog and drizzle.

Where rainfall is scant throughout the year (Fig. 18-18), there occurs desert scrub and tropical thorn forest belonging to the desert biome. Climax tropical savanna occurs where rainfall ranges from 90 to 150 cm (36–60 in.), but there is a dry season that lasts 4 or 5 months. Tropical deciduous forest replaces savanna where the dry season is shorter and less severe. Probably an annual total of 160 cm (64 in.) is the minimum that permits development of tropical broad-leaved evergreen forest. Occasional months may have as little as 6 cm (2.4 in.) but there is no true dry season. In the "cloud" forests on the mountains, rainfall may not be particularly high, but this is compensated for by almost continuous condensation of moisture on all the vegetation and the very low rate of evaporation.

VEGETATION

Tropical vegetation has been described in detail by Richards (1952). It is essentially a continuum from desert or thorn scrub to savanna or tropical deciduous forest to broad-leaved evergreen or rain forest (Fig. 25-2). There is no true climatic grassland in the tropics except in relatively small areas (Pendleton 1949). Treeless grassland is the result of excessive burning, cultivation, grazing, or unfavorable soil conditions for the growth of trees. In some parts of Africa, tree destruction by excessive populations of elephants, combined with fire set by natives, is converting woodlands, rain forests, and riparian or gallery forests into

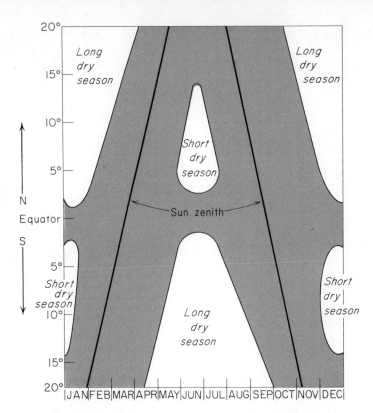

Fig. 25-1 Wet (indicated by the shading) and dry seasons in the tropics in relation to latitude (from Richards 1952, after E. de Martonne).

Fig. 25-2 Tropical rain forest on slopes of Mt. Aoyo, Republic of the Congo (courtesy S. Glidden Baldwin).

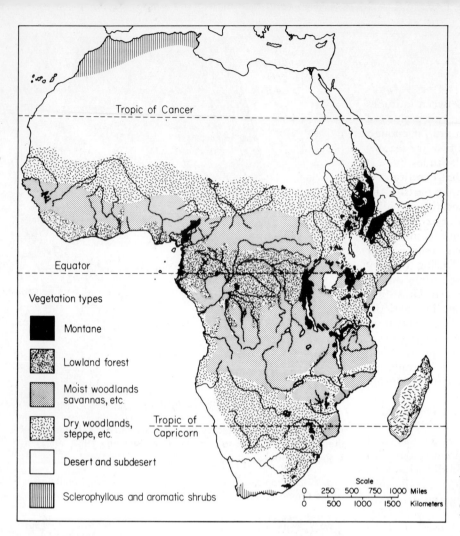

Vegetation types

- ■ Montane
- ▨ Lowland forest
- ▨ Moist woodlands savannas, etc.
- ▨ Dry woodlands, steppe, etc.
- □ Desert and subdesert
- ▥ Sclerophyllous and aromatic shrubs

Tropic of Cancer

Equator

Tropic of Capricorn

Scale
0   250 500 750 1000 Miles
0   500   1000   1500 Kilometers

Fig. 25-3 Distribution of the main vegetative types in Africa (from Moreau 1966, based on the map published by L'Association pour l'Etude de la Flore d'Afrique Tropicale).

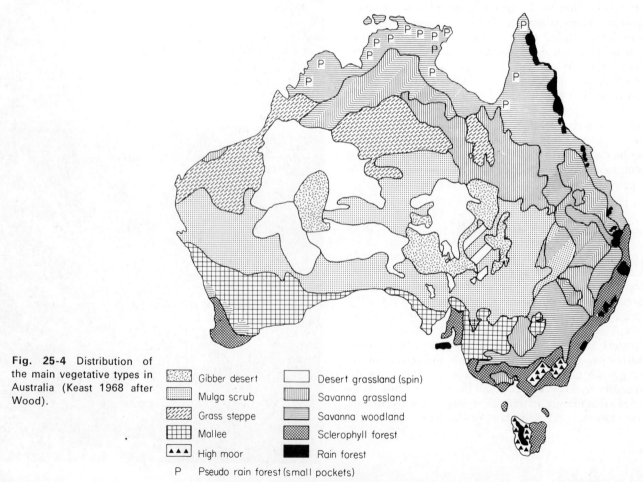

Fig. 25-4 Distribution of the main vegetative types in Australia (Keast 1968 after Wood).

- ▨ Gibber desert
- ▨ Mulga scrub
- ▨ Grass steppe
- ▦ Mallee
- ▲ High moor
- □ Desert grassland (spin)
- ▨ Savanna grassland
- ▤ Savanna woodland
- ▨ Sclerophyll forest
- ■ Rain forest
- P Pseudo rain forest (small pockets)

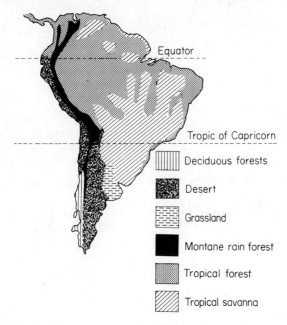

Fig. 25-5 Distribution of the main vegetative types in South America (after Küchler 1953, Smith and Johnston 1945).

grassland (Boughey 1963). Likewise, overgrazing and browsing by cattle, sheep, and goats, together with cultivation, are inducing an increase in erosion and a conversion of savanna and grassland into desert. The Sahara Desert is increasing in extent at an alarming rate (Cloudsley-Thompson 1969, Bourliere and Hadley 1970).

Tropical savanna is, however, extensive (Figs. 25-3, 25-4, 25-5). Much of the savanna in Africa probably represents a climatic climax, but in those areas with a high precipitation, the savanna may be subclimax due to fire, biotic, or edaphic conditions. This latter is the common situation in South America (Beard 1955). Savannas are communities having a *dominant stratum of more or less xeromorphic herbaceous plants, of which grasses and sedges are the principal components, and with scattered shrubs, trees or palms present* (Beard 1953). The grasses in the savanna may be tall, mid, or short and related taxonomically to those in temperate regions. The grasses and forbs die down in the dry season at the same time that the trees shed their leaves. Sedges are more common than grasses where there is more rainfall. The trees may form a dense narrow stand along rivers—the so-called *gallery forest*—or may be more or less uniformly scattered through the grassland to give the appearance of a park or orchard (Burtt 1942). The trees are often thorny or xerophilous, crooked in growth, and seldom over 20 m high. There is very little shrubby undergrowth; lianas and epiphytes are scarce. Probably savanna is increasing in extent at the present time because of human influence in destroying or opening up closed stands of the deciduous and evergreen forests.

Tropical deciduous forest, including monsoon for-

est, is more or less leafless during the dry season, is less lofty than the broad-leaved evergreen forest, but has a higher and more continuous canopy than savanna forest. The forest is rich in woody lianas and herbaceous epiphytes, but not in woody epiphytes. There are usually two tree strata; in the upper stratum the trees are scattered and strictly deciduous. A proportion of the trees in the lower stratum is evergreen, and the number of these broad-leaved evergreen trees increases in both strata as the climate becomes more humid and less seasonal. The deciduous forest is less susceptible to burning than is the savanna, while the evergreen forest is practically immune.

At its height, the tropical broad-leaved evergreen forest is nearly completely evergreen, hygrophilous, 36–55 m (100–180 ft) high, and rich in thick-stemmed woody lianas and in both woody and herbaceous epiphytes. It is commonly called a rain forest because of the continuous high humidity. Seasonal changes are minimal, the aspect being perpetually that of mid-summer. There are usually three strata of trees, one of shrubs and giant herbs, and one of low herbs and undershrubs.

The trees are extremely varied in size, but the dominants tend to be tall, slender, and unbranched except at their tops. The bark is thin, smooth, light-colored, and often covered with lichens. Tree bases are commonly provided with plank buttresses or stilts. Palms and tree ferns may be frequent. There is an extreme variety of species; for instance, there are seldom less than 15 and sometimes over 30 species of trees over 30 cm in diameter in a single hectare. The Indo-Malayan rain forest is richer in species than either tropical America or Africa; the African forest, the poorest and most uniform in flora.

The undergrowth is not a thick jungle as is popularly supposed. Because of the dense shade cast by the several tree strata, shrubs and herbs are scattered, there is little or no moss on the forest floor, and one can walk through this forest as easily as through one in a temperate climate. It is only where trees are blown down or the forest is undergoing secondary succession that increased light reaches the ground and jungle growth develops.

At Barro Colorado Island in the Panama Canal Zone, light intensity under the dense forest canopy is less than 1 per cent of the direct rays of the sun, but there are numerous sun-flecks where the intensity is greater. Full intensity of sunlight may reach 20,000 foot-candles. The temperature near the forest floor out of the sun-flecks is nearly constant. In the tree-tops, temperature rises rapidly from a nightly minimum not very different from that on the forest floor to a maximum in the early afternoon. Air movements within the forest are almost nil, and evaporation is only about one-half what it is in the open. Animals in the lower strata live in remarkably constant environments, although by climbing into the tree-tops they can enter

Fig. 25-6 Wildebeest in the tropical savanna of Tanganyika (courtesy S. Glidden Baldwin).

into environments comparable to those of open plains (Allee 1926, Moreau 1935a).

Tropical America has the richest variety of epiphytes; Africa, the poorest. These epiphytes include lichens, algae, mosses, liverworts, as well as vascular plants, such as bromeliads. The epiphytes are relatively small in size, have high light requirements, and tolerate a precarious water supply and lack of soil. Often they are more numerous than flowering herbs on the ground.

On the mountains, the tropical evergreen forest of the lowland gives way to a submontane and then to a montane rain forest. The trees are still evergreen but lower in stature, simpler in structure, and poorer in species. In addition to strictly tropical species, the forest may include many genera and species both of plants and animals that are of temperate origin (Miranda and Sharp 1950, Martin and Harrell 1957). Tree ferns are common. On exposed peaks and ridges, the trees become still more dwarfed and crooked but remain covered with many epiphytic ferns, mosses, and lichens; the whole is aptly described as elfin forest. Lianas are scarce in the montane rain forest as compared with the lowland rain forest, the shrub layer is dense, and there are only two tree strata.

Although plant ecologists are gradually working out a detailed classification of tropical vegetation (Beard 1955), we will base our recognition of major communities on the physiognomy of that vegetation which is of importance to animals. Tropical vegetation may be divided, then, into *desert* (including thorn scrub), *tropical savanna*, and *tropical forest biomes*. Deserts may be extreme with little or no vegetation present, or may be covered with scrub or thorn forest, as already discussed. Savannas represent a forest-edge community in which animals make use of both the trees and the intervening grassland to varying extent among different species. Animals may well distinguish by their segregation into

niches between park savanna and gallery forests and between grassland composed of tall or short bunchgrasses or sedge (Beard 1953). There appears to be no critical dividing line between tropical deciduous and rain forests, but together they give a continuous closed forest contrasting with the open forest or scattered groves of the savanna, and this is of importance to animals. Geographically the tropical forest is separable into American (Dansereau 1947, Leopold 1950), African, and Indo-Malayan units. The family predominance, for instance, of the Dipterocarpaceae in the Indo-Malayan rain forest, Leguminosae in the American rain forest, and the Meliaceae in some of the west African rain forests is noteworthy. This geographic distinction between the continents in plant life is reflected also in the animal life.

BIOCIATIONS

Brief attention has already been given to desert and grassland animals in tropical regions (Table 23-1). We may recognize one or more *tropical savanna biociations* in Africa to include both the more strictly grassland animals and those of the forest-edge. On the basis

Fig. 25-7 Zebras in Tanganyika (courtesy S. Glidden Baldwin).

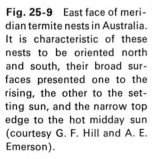

Fig. 25-8 Lions in the Amboseli Game Refuge, Kenya (courtesy S. Glidden Baldwin).

Fig. 25-9 East face of meridian termite nests in Australia. It is characteristic of these nests to be oriented north and south, their broad surfaces presented one to the rising, the other to the setting sun, and the narrow top edge to the hot midday sun (courtesy G. F. Hill and A. E. Emerson).

of differences in bird distribution, Moreau (1952) recognizes Sudanese arid, Somali arid, southwest arid, northern savanna, and southern savanna faunal areas. Among mammals, this community is especially characterized by extensive numbers of wildebeest, zebras, gazelles, antelopes, and lions (Figs. 25-6, 25-7, 25-8). that occur in open country; and the elephant, hippopotamus, rhinoceros, giraffe, warthog, and African buffalo that spend considerable time in thickets, swamps, and forests. Excellent accounts of African life

are those of Selous (1908), Chapman (1922), Moreau (1935, 1937, 1966), Darling (1960), Curry-Lindahl (1961), and Cloudsley-Thompson (1969). Termites are worldwide in distribution in tropical regions and their large mound nests are conspicuous in some parts of the savanna (Fig. 25-9). Another noteworthy insect of the savanna is the tsetse fly, the vector of a parasite which is a scourge to both man and beast (Buxton 1955). A different tropical savanna biociation may be distinguished in Australia, where a variety of marsupials are characteristic, and there are ecological equivalents for birds and other animals. However, the savannas of South America lack these herds of large mammals so conspicuous in Africa, and this is doubtless due to the savannas of South America being seral in nature and relatively recent in origin. We do not at present recognize a climax savanna biociation in South America. Forest and forest-edge communities have been distinguished for the tropical avifauna of Mexico (Edwards and Tashian 1959).

The animal inhabitants of the tropical forest biome differ so greatly in taxonomic composition on the different continents that they are placed in three different zoological regions. We may therefore designate the *Indo-Malayan*, *African*, and *American tropical forest biociations* (Chapter 18). Noteworthy studies of the animals of the American tropics are those of Bates (1864), Belt (1873), Beebe *et al.* (1917, 1947), Haviland (1926), Allee (1926), Strickland (1945), Goodnight and Goodnight (1956), and Shelford (1963).

PALEOECOLOGY

The tropical forest is of very great age and, along with associated habitats, may represent the center of origin for many modern groups of both plants and animals (Darlington 1957). In past geological ages both on the western and eastern hemispheres, the tropical forest has expanded over great areas when climates were moist and contracted into refugia when they became dry (Fig. 25-10). During periods in the Pleistocene,

Fig. 25-10 Presumed distribution of tropical forest (stippled), non-forest (white), and elevations above 1000 m (3280 ft) (black) in northern South America during warm-dry climatic periods of the Pleistocene. The horizontal hatched area is an Amazonian embayment [sea level raised about 50 m (164 ft)] (from Haffer 1969; copyright 1969 by the American Association for the Advancement of Science).

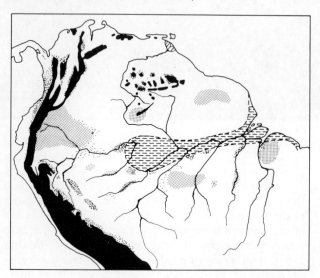

Table 25-1 Diversity of fauna at various latitudes (Dobzhansky 1950).

Birds		Snakes	
Region	Number of species	Region	Number of species
Labrador	81	Canada	22
New York	195	United States	126
Panama	1,100	Mexico	293
Colombia	1,395	Brazil	210

glaciers formed in the Andes Mountains of South America. Perhaps, if one were to work out a phylogenetic tree, all other biomes could be traced back in origin to the tropical biota of Mesozoic and Paleozoic times. Arctic communities have been derived from temperate ones, and temperate communities from tropical ones, as organisms became adapted to occupy climatic and environmental conditions that differed more and more from the primitive optimum of tropical regions.

In South America it is apparent that the fauna of the lowland rain forest (tropical zone) long antedates the elevations of the Andes mountains. After the mountains were formed, they were invaded by those species of plants and animals adapted to the lower temperatures and other differences in habitat moving gradually upward. The animal life of the submontane rain forest (subtropical zone) is derived almost entirely from the rain forest at lower elevations. The montane rain forest (temperate zone) is of more recent origin than the submontane forest and consequently derived part of its fauna from it. In addition, the montane forest contains many species of both animals and plants that have dispersed into it from higher latitudes both to the north and south, where these species occur in temperate climates at lower elevations. Some very high peaks extend above the tree-line to produce alpine meadow (paramo zone). The fauna and flora of this community are derived almost entirely by lateral dispersal of organisms from extreme southern South America (Chapman 1917).

During Miocene and Pliocene, much of northern Africa, now desert, was covered with forest. The tropical savanna began to form about 10,000,000 years ago but underwent contraction during pluvial periods of the Pleistocene. The present-day distribution of forest, savanna, and desert dates back only about 12,000 years (Moreau 1933, 1966).

ANIMAL ADJUSTMENTS

Like the flora, the fauna in the tropical rain forest is very rich in species (Table 25-1). Bates (1864) tells of finding 18 species of swallow-tailed butterflies within 10 minutes' walk of his house in tropical South

Fig. 25-11 Column of army ants moving across the forest floor, and, right, a bivouac of army ants, Barro Colorado Island, Panama Canal Zone (courtesy T. C. Schneirla).

America. This is to be compared with only about 20 species in all of North America north of Mexico. Reasons for high species diversity in the tropics have been given (pp. 256–258). Significantly more bird species are of irregular occurrence than in temperate regions, flocking is more prevalent, and breeding territories less firmly established (Karr 1971).

Although total populations of all species may be high in tropical regions, the density of individuals per species is often low. This is understandable, since the greater the number of species in an area, the greater becomes the competition among them for living space and resources. The great number of tree and plant species provides a variety of niches, but the range of resources to a given species is limited, and the realized niches are narrower than in the temperate zone.

In contrast with temperate regions, where animal adaptations are so largely concerned with the physical environment and getting food, there needs to be little or no adaptation in the tropical rain forest to winter cold, inclement weather, lack of food, or desiccation. Interspecies competition and struggle, however, is harsh and exacting, and evolutionary forces tend to perfect specializations that enable organisms to fit better into their niches or invade new ones and thus avoid much of the competition and predation (Allee 1931, Mertens 1948, Dobzhansky 1950, Janzen 1967).

One specialization in this connection is the ability to hang from trees—animals suspended in mid-air are almost inaccessible to attack by predators. Many birds, particularly flycatchers, orioles, and honeycreepers build pendant nests, as do certain solitary wasps. The cocoons of many moths as well as chalcid wasps are suspended by thin threads. Spiders in their webs sus-

pended off the ground are immune from attacks of army ants.

An unusual number of birds nest in holes in trees—trogons, motmots, parrots, toucans, woodpeckers—but this may be to protect them from the heavy showers as much as from marauders. Stinging ants inhabit the hollow stems of sapling cecropia trees and the swollen bases of leaf petioles of the leguminous tachigalia tree, and for this protection defend the trees against the attacks of the leaf-cutting ants.

Many other interesting coactions exist. Collared peccaries on Barro Colorado Island in the Canal Zone make narrow trails through the bushy tangles and dense undergrowth and proceed single-file from one place to another. These well-defined paths are utilized also by the coati, octodonts, and several kinds of marsupials which probably could not otherwise penetrate these areas (Enders 1935). The tapir is a trail-maker in the South American jungle, and the elephant in Africa ploughs its way through the forest by sheer strength. The trails it breaks are followed at later times by the hippopotamus, rhinoceros, buffalo, lion, leopard, hyena, pig, and baboon, which in turn make the trail more passable for lesser forms (Hesse, Allee, and Schmidt 1951).

In contrast with the sporadic occurrences of most species, ants and termites are abundant in the American tropics. The leaf-cutting ants and the insect-eating army ants (Fig. 25-11) are especially characteristic (Belt 1873). Termite nests occur in all strata, and the wood-eating habits of these insects hasten the destruction of woody materials. The Hymenoptera, Diptera, and Coleoptera are, in general, the best-represented groups among the insects (Briscoe 1952).

With temperatures high and uniform throughout the year, the developmental period of cold-blooded animals is shortened, and there is a general speeding up of the life cycle. Insects possessing only a single generation per year in the north temperate zone may complete their life cycle in 3–4 weeks in the tropics, and may have several generations during the year. On the other hand, high uniform temperatures are depressing for the metabolism and activities of warm-blooded animals, and the pace of their activities is comparatively slow.

Cold-blooded animals, particularly reptiles and arthropods, reach their largest adult sizes in the tropics. In the Amazon forest there is a spider large enough to catch and feed on small birds that are caught in its web (Bates 1864). Some moths have a wing-spread of 30 cm; a millipede reaches a length of 28 cm, and a slug 20 cm. Birds and mammals, however, are usually smaller than their relatives in temperate regions. Birds also lay fewer eggs in a clutch than they do in temperate regions.

Since there is little variation in the duration of light per day throughout the year, photoperiodism is largely absent. In desert regions, the gonads of birds may remain inactive and reproduction inhibited for a succession of seasons during a prolonged drought, but their sexual cycles respond quickly to rainfall, and nesting may begin within a few days after heavy precipitation (Serventy and Marshall 1957). Breeding is generally more common during the rainy season, although a few species nest only during the dry season. In regions where wet and dry seasons are not developed, bird species may breed throughout the year, although individual birds, after breeding, need a period of rest before they can breed again (Miller 1963). Continuous breeding is also characteristic of many other forms besides birds. Such continuous breeding tends to dampen annual fluctuations in abundance, reduces chances for the species to become extinct, and promotes species diversity (Inger and Greenberg 1966). In the evergreen rain forest of Borneo, where precipitation is extremely high, the onset of breeding among mammals comes mainly during the period of year when precipitation is minimum rather than maximum (Wade 1958).

There is no definite period of migration for the birds of the rain forest. Movements are largely localized and in quest of ripening fruit or other scattered food supplies. Away from the immediate vicinity of the equator and toward the periphery of the rain forest, where wet and dry seasons become important, the annual cycle of breeding, migration, and other activities becomes more pronounced (Baker *et al.* 1936, Davis 1945, Moreau 1950). The influx of large numbers of migrants from the Palaearctic during the northern winter into the dry woodlands and steppe of western Africa comes at a time when food is most plentiful. The resident species do not exhaust these energy resources, as the expansion of their populations is held in check by the more limited food available during the dry season (Morel and Bourliere 1962).

In dense rain forest on the equator, the daily rhythm of animal activities is striking. Many naturalists have commented on the great hush of the forest during the middle of the day. The forest appears empty of both birds and mammals. There is an occasional note of a bird, but birds do not have the varied and conspicuous songs that they do in temperate regions. They may be glimpsed in the tree-tops or searched out in the undergrowth moving through the forest in loosely formed groups, each group composed of a few individuals each of several species. These social groups occur at all seasons, although nesting birds must withdraw temporarily from them (Davis 1946). The cicada chorus is often loud and persistent; with the onset of darkness, other orthopteran insects burst into song to which tree frogs, night birds, monkeys, and others add their voices.

Nectar feeding is well developed in some tropical birds, and many flowers depend on birds for their pollination. Hummingbirds are numerous in the western hemisphere. Many insects are also nectar and pollen feeding, and these species are largely limited to the tropics. Fruits are an important food for many birds and mammals. Fruit-eating bats are confined to the tropics. Sloths and ant-eaters have powerful claws and long sticky tongues with which they open and plunder the nests of ants. Army ants moving in large numbers over the forest floor in their search for prey often attract a large and varied group of birds, but these birds are after the other insects that the ants stir up rather than the ants themselves (Johnson 1954).

As in most other terrestrial biomes, invertebrate animals occur in the greatest variety and numbers in the forest floor. Here they are much more numerous during the wet than the dry season. In the Panama Canal Zone during the wet season there are between 4000 and 10,000 animals per m^2, representing 294 species. Of this fauna, mites constitute 25 per cent, springtails 34 per cent, ants 25 per cent, and all others 16 per cent. Planaria-like flatworms, a leech, and a land crab are found, although in temperate climates they are usually limited to aquatic habitats. Many different kinds of millipedes are a characteristic feature of the fauna. Centipedes and snails are not abundant (Williams 1941).

Many animals of all sorts have developed an arboreal-living habit although their close relatives outside the tropics are ground-dwellers (Table 25-2). Arboreal mammals tend to be limited in size and possess opposable toes. The New-World monkeys have prehensile tails but not the Old-World monkeys and apes. Porcupines, climbing ant-eaters, the coatis, and the kinkajou also possess prehensile tails (Haviland 1926). Some sloths and lemurs hang upside down as they

Table 25-2 Stratal distribution of mammals in British Guiana; number of species shown in parentheses (Haviland 1926).

Order	Terrestrial Species		Arboreal Species
	Cursorial	Amphibious	
Marsupialia		*Chironectes* (1)	
Rodentia	Agoutis (2)	Capybara (1)	Porcupine (1)
	Paca (1)	Paca (1)	Squirrels (2)
	Spiny rats (2)		
	Rats and mice (8)		
Edentata	Armadillos (4)		Sloths (2)
	Anteater (1)		Anteaters (2)
Carnivora	Huntingdog (1)	Others (2)	Cats (5)
			Raccoon (1)
			Coati (1)
			Kinkajou (1)
			Tayra (1)
			Grison (1)
Ungulata	Peccaries (2)	Tapir (1)	
	Deer (2)		
Primates			Monkeys (6)
			Marmoset (1)

climb around through the branches of the trees. Tree-dwelling snakes and lizards are either long and whip-like or heavy-bodied and with prehensile tails. Parachutes, similar in function to those of the flying squirrels of temperate forests, have developed in such diverse forms as marsupials, lizards, and frogs. Some frogs are entirely arboreal and have sucking discs on their toes to aid in climbing. Some species of frogs lay their eggs in the trees in sacs made of leaves, others glue their eggs to leaves, still others carry them on their backs and the tadpole stage is passed through before hatching. Snails climb to the topmost branches of the trees. Some butterflies fly continuously about the tree-tops and appear never to alight on the ground. There is a group of tree-dwelling tiger-beetles, Odontocheilae. Termite nests located in trees are often connected to the ground by covered passages. Leeches climb into bushes to get onto the bodies of warm-blooded animals more easily. During the wet season, the mosquito species *Anopheles* and *Culex* are mostly crepuscular or nocturnal in activity while the sabethine group and certain *Aedes* are diurnal (Bates 1949).

Many of the epiphytes, especially the bromeliads, hold small quantities of water within the clump of leathery leaves high up in the trees. These small reservoirs contain algae, bacteria, and detritus which serve as a food base for a microcommunity of gastrotrichs, rotifers, flatworms, oligochaetes, small crustaceans, insect larvae, and frog tadpoles (Laessle 1961).

A half-dozen stratal societies of birds and mammals may be distinguished: (1) upper air with insectivorous and carnivorous birds and bats; (2) canopy with birds and mammals feeding largely on leaves and fruits; (3) middle zone with flying insectivorous birds and bats; (4) middle zone with scansorial mammals of mixed feeding habits ranging up and down the trunks; (5) large herbivorous ground animals; and (6) small ground mammals and birds on varied diets (Harrison 1962).

The true forest inhabitants keep well within the forest shade and, like the monkeys, are quite sensitive to direct exposure to the sun. Vertebrate animal life is less abundant, however, in the depths of the forest interior than it is on the forest margin. It follows that the fauna is richer both in species and in numbers in the tropical savanna than in the tropical rain forest itself.

HUMAN RELATIONS

The tropics are the native home of the black or Negroid races of man (p. 312). White man can seldom spend more than a few months at a time there without impairment of health and vigor. Relatively little effort is required by the native, living in the back country, to secure food, and the need for clothing is minimal. The biological environment, however, is harsh and exacting. He must guard against malarial plasmodia, hookworms, and a variety of skin parasites (Dobzhansky 1950).

Most of the food grown in the tropics is rich in carbohydrates and low in nitrogen. Fish culture, both marine and fresh water, is important in some areas but the large game mammals are potentially the best source of protein, especially in Africa. Savanna lands will support a substantially higher standing crop of mixed wild ungulates than of domestic livestock, and marginal lands in semi-arid regions, unsuitable for domestic animals, will support big game. Domestic stock is limited from large portions of grazing land by trypanosomiasis or nagana, transmitted by the tsetse fly, to which native species are immune. Domestic stock also has higher requirements for drinking water, which is often scarce, than do the native species. The high diversity of native species is adapted to use a wider range of non-duplicating plant diets. Per unit weight the wild ungulates are significantly more productive than domestic livestock from the standpoints of age of maturity, rate of reproduction, and rate of growth. Harvesting of wild game is a relatively new industry but has been demonstrated to be practical. There is also the possibility of domestication and game farming (Talbot 1966).

Natives in the tropical rain forest make their living by hunting and fishing. They live in rude huts made of branches and leaves. Their hunting is done with bows and poisoned arrows, blow pipes, and pits dug in the ground. African natives in tropical savanna are nomads and have herds of cattle, goats, and other animals which furnish them with milk, meat, and blood meals. In better developed equatorial lands, crops produced include manioc, from which tapioca and flour are obtained, cassava, sweet potatoes, yams, sugar cane, pineapples, bananas, and cocoa. Cocoanuts are important food in some places. Expanded research is required to increase food productivity, but tropical soils, compared with temperate ones, are essentially depleted of mineral reserves and humus because of rapid decomposition, rapid reabsorption of nutrients by the vegetation, and extensive runoff and leaching from heavy rains. The tropics are unlikely to become an agricultural panacea (Fosberg 1970).

Coffee is cultivated extensively in South America. Sap from which rubber is made was originally collected from scattered naturally growing trees in the forest, but rubber trees are now grown extensively in plantations. Rubber plantations have become important in southwest Asia.

The monsoon region of India supports a very large population. The land is divided into tiny plots, plowed by donkey, ox, or water buffalo, and cultivated by hand tools. Elephants do some of the heavier work. Tea leaves are harvested from bushes, rice is grown, and teak lumber obtained from the forest. Cattle and goats supply milk.

SUMMARY

Noteworthy of tropical climates is the uniformity of temperature and length of daylight throughout the year. Rainfall varies from a distinct seasonal distribution in some regions to constant and very heavy in others. Correlated with the rainfall gradient is a vegetation continuum from desert, to savanna, to tropical deciduous forest, to tropical rain and cloud forests. The biomes recognized are those of desert, tropical savanna, and tropical forest. Aside from deserts there are one or more tropical savanna biociations in Africa and in Australia and tropical forest biociations in Indo-Malaya, South Pacific islands, Africa, and America. The tropical forest flora and fauna are of great age and continuity, and it may well be that all other biomes can be traced back in origin to the tropical biota of mesozoic and paleozoic times.

The flora and fauna of the tropical forest are marked by the richness of their species compositions. Correlated with this is the harshness and severity of interspecific competition and predation. On the other hand, except for ants and termites, few species reach high levels of population density. Large herds of ungulate mammals occur, however, in the tropical savanna of Africa.

With uniform climate throughout the year, cold-blooded animals may have several generations per year and species of birds may breed during every month. There is no hibernation or period of dormancy, nor is there migration, except in regions of pronounced wet and dry seasons.

Cold-blooded forms, especially reptiles and arthropods, reach a large size in the tropics, but birds and mammals are generally smaller than their relatives in temperate regions. Although animals occur in greatest numbers and variety on the forest floor, many different groups have evolved members largely restricted to the arboreal stratum.

The tropics are the native home of the black races. Originally they made their living by hunting and fishing, grazing of domesticated animals, and a primitive form of agriculture. Intensive agriculture, comparable to that in temperate climates, is limited by inability of the soils to build up and maintain fertility. Exploitation of native big game may become an important source of protein for the exploding human population of Africa.

Part Seven

MARINE ECOLOGY

Chapter 26

OCEANOGRAPHY

The geographic distribution of organisms in the sea depends on their responses to currents, temperatures, and physical barriers; local distribution is affected by waves and tides, type of bottom, salinity, and depth. Marine ecology is concerned with environmental factors and problems of organismic adjustments quite different from those on land and also different in many respects from those in fresh water. Animals are relatively more conspicuous than plants. Succession is less evident, but such ecological processes as represented by chemical cycles, cooperation and disoperation, food chains, productivity, population dynamics, niche segregation, speciation, and dispersal are fully as important as on land.

Distinct self-contained community units are more difficult to recognize in the sea than on land because of the apparently greater interrelation of benthic species and the freer movement with circulating currents of plankton and nekton. Plankton is everywhere a basic link in food chains, but the general distribution and importance of plankton species in the sea is no more remarkable than that of soil organisms in terrestrial biomes. To consider the entire ocean community as a single biome, as has been suggested by some investigators, is stretching the concept beyond its usefulness. Since we identify biomes by differences in the life-forms and functional adjustments of the conspicuous dominant or predominant organisms, we may properly recognize biomes that occur in the open ocean, on eroding rocky shores, on muddy and sandy beaches, and composing the coral reefs and atolls. Each of these biomes may be subdivided by the taxonomic composition of the predominant organisms into secondary communities equivalent to the biociations that we have recognized on land. Much of the early literature on marine communities has been reviewed by Gislen (1930).

We can only hope in this chapter to present a brief summary of some of the more salient features of marine ecology. For more thorough treatments, the reader is referred to the publications of Sverdrup *et al.* (1942), Ekman (1953), Harvey (1955), Hedgpeth (1957), Zenkevitch (1963), Tait (1968), and Kinne (1970).

HABITAT

The marine biocycle is considered to have *benthic* (bottom) and *pelagic* (open water) divisions (Fig. 26-1). The *littoral zone* of the ocean shore extends between the limits of high and low tides. The *sublittoral zone* covers the continental shelf to a depth of about 200 m, the approximate depth at which maximum wave action produces any effect. The average depth of the ocean is about 3800 m, but oceanic trenches (*hadal zone*) extend much deeper; the Marianas Trench in the Pacific Ocean to approximately 11,600 m. The *neritic biochore* is above the continental shelf and is commonly 16–240 km (10–150 mi) wide. The *oceanic biochore* is subdivided vertically with the boundary between the epi- and mesopelagic zones, depending on the extent of effective light penetration.

Tides

The level of water in the ocean rises and falls usually twice each day or at an interval of 12 hours and 26 minutes. In some parts of the world the tides are less regular or there may be but one daily. *Flood-tide* is the period in which the level is rising and covering more and more of the shoreline; *ebb-tide* is the period in which the waters are receding. In the open sea the change in water level is less than a meter, but the change may be much more than this on the shore, depending on its configuration. The Bay of Fundy opens broadly to the sea and tapers to a narrow head landward, and tides may be 6 to 10 or even 15 m. On the other hand, when bodies of water have only a relatively narrow connection with the sea, as does the Gulf of Mexico with the Atlantic, the range in water level is less than 30 cm. Even lakes have seiches, but they are hardly perceptible except in the larger lakes, where it may amount to a few centimeters.

Tides are caused by the attraction of the moon and, to a lesser extent, the sun. When the sun's attraction is added to that of the moon, as occurs twice each month at times of full moon and new moon, the fluctuations of the tides are unusually high and unusually low. These are called *spring tides*. When the tidal influences of sun and moon are opposed, as happens twice each month, the tides have the least amount of flow and ebb and are called *neap tides*.

Tides have their greatest effect on animals on the seashore, because of the associated pounding of waves and the alternate submergence in water and exposure to the air. However, the organisms appear well adjusted to this rhythmic submergence and exposure (Flattely and Watson 1922, Korringa 1947). For instance, as the stones on which the chiton occurs become exposed, the animals react positively to gravity and negatively to strong light, and move downward. They travel at maximum speed while the stone is still moist and become aggregated on the damp lower sides of the stones. When the stone again becomes immersed by the returning tide, the animals lose their geotactic orientation, and, since illumination becomes more or less equal on all sides of the stones, they move about at random until they reach the upper surfaces again.

On the other hand, some ciliated and flagellated protozoans and diatoms in intertidal habitats are active only when the tide is out and become encysted or inactive and attached to surfaces when the tide is

Fig. 26-1 Subdivisions of the ocean biocycle (Hedgpeth 1957).

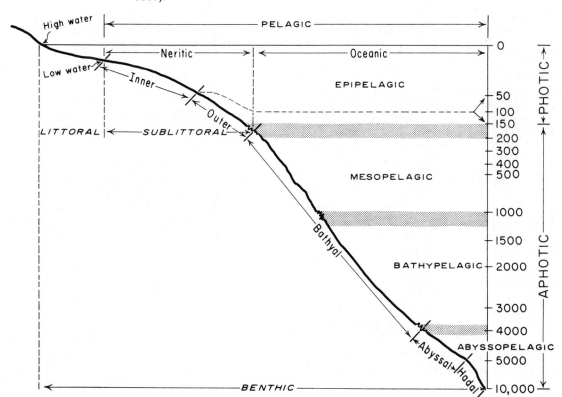

in (Fauré-Fremiet 1951). These rhythms in intertidal organisms may persist for days, even when the organisms are placed experimentally under constant conditions (Brown 1959).

Substratum

The pounding action of waves on rocky shore may have tremendous force, estimated in one instance at 15,000 kg/cm². Animals occupying exposed rocky shores in the surf belt must be strongly protected and firmly attached (Flattely and Walton 1922). The conical-shaped limpets present a minimum of surface to the waves. Barnacles are protected by heavy shells and grow fast to the rocks, snails and chitons hold themselves by powerful suction apparatus on their feet, mussels like *Mytilus* have a glandular byssus, while some species of sea urchins bore shallow craters into the rock. Advantage is taken of nooks, crannies, and spaces underneath stones and rocks (Glynne-Williams and Hobart 1952). Large depressions in the rock retain water at ebb tide to form tidal pools and thus may contain the more delicate species because of the protection they afford. The various sea weeds absorb some of the wave shock for the animals living with them. The shape, form, and size of corals, sponges, and other colonial types are affected by the amount of wave action to which the animals are exposed. A shell of the *Mytilus* mussel may weigh 58 g where the animal is exposed to a heavy surf but only 26.5 g in more quiet waters.

Sandy beaches occur only where the force of waves is reduced by being spread over a more gentle slope. Even here, especially during storms, the sand makes a very unstable substratum and not many animals except mollusks and some of the echinoderms can keep from being smothered or buried. Mud buttoms occur only in relatively quiet waters. Burrows made in mud hold their form better than in sand, so larger populations of animals can occur in mud.

The sea-floor at greater depths is covered with a variety of sediments. *Terrigenous* deposits of mineral and organic matter derived from the land and from the littoral and neritic biochores are relatively rich in nutrient substances and extend into the bathyal zone. All other deposits on the sea-bottom are *pelagic*, being derived, in part, from the skeletons of dead plankton and other organisms. In the long slow journey of these dead organisms to the bottom of the sea, much of the organic matter decomposes, releasing carbon dioxide, nitrates, phosphates, and the many other elements in the composition of the protoplasm. Even various amounts of skeletal material may dissolve, but enough of the organisms reach the bottom to create a substratum of loose flocculent ooze and to furnish food for the living animals that spend their lives in this habitat.

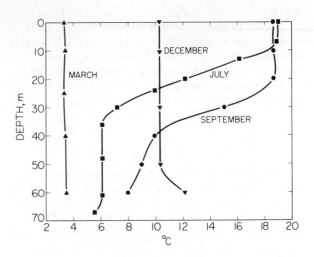

Fig. 26-2 Seasonal variation of temperature in neritic waters in the western Atlantic (Grice and Hart 1962).

Pelagic deposits that contain less than 30 per cent of organic remains are known as *red clay*. These deposits are the most widely spread of all, especially at the greater depths of the ocean. They are probably derived from wind-blown desert dust, terrestrial volcanic dust, and submarine eruptions. The very hard earbones of whales and teeth of sharks are regularly found in red clay. Animal life is scant, consisting only of shell-less holothurians and worms, probably because of the poor nutrient content and great depth.

Organic deposits are either calcareous or siliceous, the former being derived from the shells of foraminiferans, small pelagic mollusks, or flagellate coccolithophorids, and the latter from skeletal material of diatoms and radiolarian protozoans.

Pressure

There is an enormous increase in the pressure of water upon the bodies of animals at great depths, and this may affect the vertical distribution of such species as crabs and mussels (Menzies and Wilson 1961). This is not, however, an important limiting factor in the vertical distribution of animals in general, as internal pressures closely counterbalance external pressures and life is known to exist at the greatest depths. Adjustments of internal pressures are not so rapid, however, to prevent injury in many species that are dredged at great depths and quickly hauled to the surface. Furthermore, individual species have different limits of pressure tolerance.

Temperature and Currents

The temperatures of surface waters vary between the freezing point (−1.9°C) in polar regions and 25°–

30°C in the tropics. Seasonal variations are small in polar and tropical waters but somewhat greater in the temperate zones.

Temperature varies with depth, more so in the tropics than elsewhere. At 60°N latitude in the Atlantic Ocean, the mean temperature of the warmest and coldest months beneath the surface is about 10°C, while at a depth of 2000 m it is 3.5°C. On the equator the temperature beneath the surface is approximately 26°C, at 200 m 13°C, at 400 m 7.5°C, at 1000 m 4.5°C, and at 2000 m 3.3°C (Ekman 1953). In temperate regions, a seasonal thermocline develops near the surface during summer and is destroyed in autumn and winter when vertical mixing creates a layer of relatively uniform temperature in the upper 20 to 300 m (Fig. 26-2).

Currents moving toward the poles from the equator consist of warm water, and currents moving in the opposite direction of cold water. Surface currents make wide circular movements in opposite directions in the northern and southern hemisphere (Fig. 26-3).

Organisms living in the intertidal zone on the shore are ordinarily exposed to great variations of tempera-ture twice during each day, as they are alternately flooded by the tides and exposed to the air and direct solar radiation. Unusually severe cold spells during the winter have been known to produce extensive mortality of fish and invertebrates in shallow waters off the coasts of Texas and Florida (Gunter 1941). On the other hand, one of the characteristic features of the deep-sea habitat is its low and almost constant temperature.

Light

Much of the solar radiation is reflected from the surface or absorbed in the upper layers (p. 57). Even in the clearest waters and at maximum radiation, the red, orange, and ultraviolet are absorbed in the first 20 m. Green, yellow, and blue wavelengths penetrate farther, depending on the water color. When the sun is not at the zenith, light penetration is reduced, and the maximum penetration in the winter at high latitudes is much less than during the summer (Clarke 1939, Jerlov 1951).

Fig. 26-3 Surface currents of the oceans (after Huntington and Carlson 1934).

1. Labrador Current	6. South Equatorial Current	11. Indian Counter Current	16. California Current
2. Gulf Stream	7. Brazil Current	12. Equatorial Current	17. Peru Current
3. North Atlantic Drift	8. Brnguela Current	13. West Australian Current	18. North Equatorial Current
4. Canaries Current	9. Mozambique Current	14. East Australia Current	19. South Equatorial Current
5. North Equatorial Current	10. Monsoon Drift	15. Japan Current	20. West Wind Drift

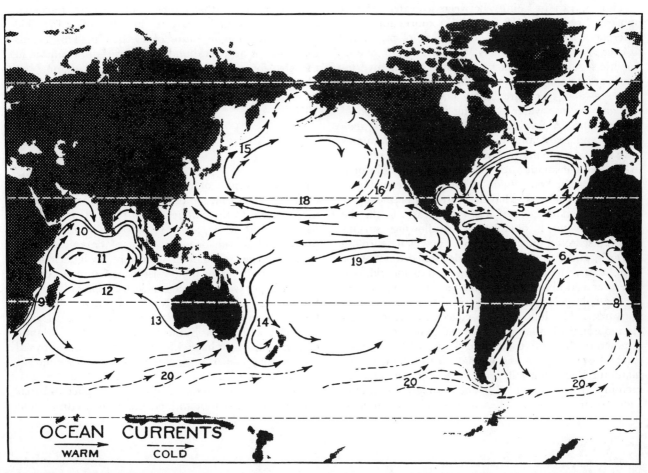

OCEAN CURRENTS
WARM   COLD

The *compensation point*, or the depth at which the amount of oxygen released in photosynthesis by algae just balances the oxygen needs of the plants for respiration over 24 hours, has been found to vary during the daytime between 1 and 100 m, according to the locality, turbidity, and season. The upper illuminated layer where photosynthesis exceeds respiration is often called the *photosynthetic zone* (Harvey 1955).

Salinity

The salinity of sea water varies from place to place, depending largely on the amount that it is diluted by the inflow of fresh water from rivers or melting glaciers or the amount that it is concentrated by evaporation. The Red Sea, for instance, has a salinity of 40‰ (40 g of dry salts in 1000 g of seawater) while in some polar seas the salinity is less than 30‰. The average salinity of the oceans as a whole is commonly given as 35‰, of which the chloride ion constitutes about 19‰ and the sodium ion a little over 10‰. The various major salts occur nearly everywhere in definite and constant proportions. As one would expect, the pH of seawater is high, averaging about 8. There is some similarity in relative proportions and concentrations of the various ions in seawater and in the blood or body fluids of many invertebrate organisms. This may indicate that the sea is the habitat in which living forms first evolved.

The contrast in salinity between seawater (35,000 ppm) and fresh water (15–660 ppm) requires important differences in physiological adjustment of organisms to occupy these two habitats. The problem is one of osmotic regulation (Black 1951). Most marine invertebrates are poikilosmotic in that they are nearly isotonic with seawater; they are highly permeable to water, and gain or lose water according to concentration of the medium. A few marine segmented worms, flatworms, and crabs and all marine fish and mammals have at least some internal osmotic regulation and tend to be homoiosmotic. All except the elasmobranch fishes maintain body fluids hypotonic to seawater in various ways. The skin has decreased its permeability to the free movement of water back and forth, the necessary water is obtained by swallowing, surplus salts are secreted outside the body, especially through the gills, and there is a general decreased function or atrophy of water-secreting organs such as the kidneys. The practical absence of insects and amphibians from the sea is largely due to their inability to secrete salts outwardly. The high osmotic concentration found in elasmobranchs is the result of huge quantities of urea retained in the body tissues and fluids.

Fresh-water organisms, in contrast to marine forms, maintain body fluids hypertonic to the surrounding medium by excretion of water through contractile vacuoles in lower organisms or highly functioning kidneys in higher ones, active absorption of salts from the surrounding water by special cells in the gills, and reabsorption of salts from the urine. There is no swallowing of water, as sufficient amounts are absorbed by osmosis through the gills and mouth surfaces and incidentally with feeding (Hutchinson 1967).

Probably the most extensively utilized of the dissolved substances in the sea are the nitrogen compounds (nitrates, nitrites, ammonia salts), phosphates, calcium salts, and silicates. Nitrates and phosphates are particularly important as nutrient material for phytoplankton. Calcium is required in large amounts for the shells of mollusks, the skeletons of corals, some protozoans and worms, certain algae, and other organisms and may be precipitated out of the water by bacteria. Silicon is required by sponges, some protozoans, and the phytoplankton diatoms.

These salts keep cycling through the ecosystem, but additions to the supply come continually from the land, being washed into the oceans by the rivers. Neritic waters are especially fertile and support a great mass and variety of animal life because of this land drainage and the pattern of water circulation on the continental shelf. Biological productivity decreases progressively from shallow waters over the continental shelf, to deeper waters, to the open ocean, but is also high over offshore banks and in areas of upwelling. Substantial amounts of nitrogen salts are also swept out of the air by precipitation, and there is nitrogen-fixation by bacteria.

It is of interest that atoms of phosphorus, nitrogen, and carbon occur in seawater in ratios of 1:15:1000 and in plankton in ratios of 1:16:106. This means that there is an overabundance of carbon available in the sea for absorption by the phytoplankton, but phosphorus and nitrogen may be limiting for further increases in the population of organisms (Redfield 1958).

Oxygen

The oxygen supply of seawater comes by diffusion from the air at the surface and from photosynthesis of green plants down to the compensation point. It is continuously used at all depths in respiration of animals and plants and in the decomposition of organic matter.

The oxygen content of seawater (Hedgpeth 1957) is seldom limiting for the occurrence of animals, except in the deeper waters of the brackish Black and Caspian Seas, where it is practically absent. Oxygen concentration is especially high on shores where there is splashing of waves. Surface waters of the Atlantic Ocean commonly have 4.5 to 7.5 cc/liter and abyssal regions may run over 5 cc/liter. Oxygen is somewhat less abundant in the Pacific and Indian Oceans. Oxygen may be reduced to lower concentrations between 100 and 1500 m, because of its use in animal respiration and in decomposition, than at lesser depths, where there is photosynthesis, or at greater depths, where the abundance of animals is greatly diminished.

Marine animals have a variety of mechanisms and adaptations for respiration (Flattely and Walton 1922). Greatest difficulties occur in shore animals at low tide when they are exposed to the air, but the need for oxygen at this time is decreased in many forms by curtailment of activity. Some crabs, barnacles, snails, and fish have become almost amphibious in being capable of respiring in air, although at reduced rates, as well as in water. Pure mud bottoms may present anaerobic conditions a short distance below the surface, but mud bottoms mixed with sand contain an abundant and diversified fauna.

PLANKTON

Composition

The plankton of the sea includes a great variety of forms, even more than in fresh water (Biglow 1926, Hardy 1956). Rotifers, however, are uncommon in marine plankton and cladocerans are much less important.

The nannoplankton consists mostly of protozoans, algae, bacteria, fungi, and viruses. The bacteria are largely *periphytic*, in that they are attached to the surfaces of floating plants, animals, and to particles of detritus. Very few occur freely suspended in the water (Harvey 1955). Bacteria, including nitrifying forms, occur at all depths but are especially abundant in or close to the bottom. They are generally more numerous in the winter than in the summer. Nitrogen-fixing organisms are chiefly blue-green algae and possibly yeasts (Oppenheimer 1962).

Fig. 26-4 Characteristic holoplankton. Protozoa: (a) forminifera *Globigerina*; (b) dinoflagellate *Gymnodinium*; (c) tintinnid *Stenosomella*; (d) tintinnid *Flavella*; (e) radiolarian *Protocystis*; (f) another radiolarian; (g) dinoflagellate *Noctiluca* (Sverdrup *et al.* 1942).

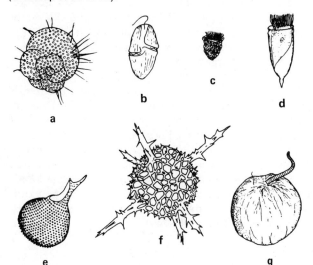

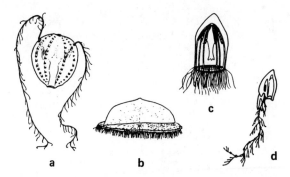

Fig. 26-5 Coelenterates and ctenophores: (a) comb-jelly *Pleurobrachia*; (b) siphonophore *Velella*; (c) jellyfish *Aglantha*; (d) siphonophore *Diphyes* (Sverdrup *et al.* 1942).

The green phytoplankton is composed primarily of diatoms, dinoflagellates, and small unarmored flagellates, but several other kinds of algae are present and occasionally important. The dinoflagellates *Noctiluca* and *Ceratium* are luminescent and in some regions may give a glow at night to the entire sea. Bioluminescence is not limited to these organisms, however, but occurs also in various forms of bacteria, radiolarians, sponges, coelenterates, ctenophores, nemertineans, worms, crustaceans, brittlestars, mollusks, balanoglossids, tunicates, and fish (Harvey 1952).

The most important groups of protozoan zooplankton, other than the green flagellates which are usually considered with the phytoplankton, are the rhizopod Foraminifera, the actinopod Radiolaria, and the ciliate tintinnids (Fig. 26-4). They may be enormously abundant at times.

Among the Coelenterata, many hydrozoans have medusae and larval floating stages in their life cycle, but only the siphonophores, the best known example of which is the Portuguese man-of-war, are pelagic throughout their life cycle (*holopelagic*). The true jellyfish of the class Scyphozoa are often conspicuous and ctenophores of the related phylum are often abundant (Fig. 26-5). Some of these forms are so large they are called *macroplankton*.

The various phyla of worms are represented in the plankton by only a few forms, of which the chaetognath *Sagitta* or arrow worm and the polychaete *Tomopteris* are often abundant (Fig. 26-6). Many benthic worms, however, produce larvae that are temporarily part of the plankton (*meropelagic*).

Many molluskan and echinoderm species are meropelagic and it is by means of their larvae that heavy, slow-moving benthic forms become widely dispersed. Two groups of snails are holopelagic: the heteropods that inhabit tropical and subtropical waters and the pteropods which occur in cold waters and are important food for whalebone whales.

Crustaceans form one of the principal groups of the net plankton, and of these the holopelagic copepods are by far the most abundant (Digby 1954). *Calanus finmarchicus* is one of the most noteworthy species and

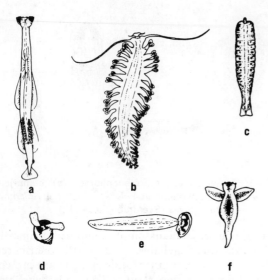

Fig. 26-6 Miscellaneous: (a) arrow worm *Sagitta*; (b) annelid *Tomopteris*; (c) nemertean *Nectonemertes*; (d) pteropod mollusk *Limacina*; (e) tunicate *Oikopleura*; (f) pteropod mollusk *Clione* (Sverdrup *et al.* 1942).

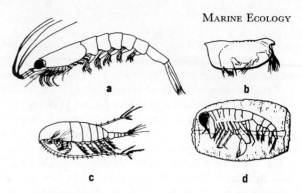

Fig. 26-7 Crustaceans: (a) euphausiid *Euphausia*; (b) ostracod *Conchoecia*; (c) copepod *Calanus*; (d) amphipod, *Phrononemia*, in empty mantle of the pelagic tunicate *Salpa* (Sverdrup *et al.* 1942).

is the principal food of the commercially important herring fish. Other important crustaceans that enter the plankton either as larvae or adults are ostracods, cumaceans, amphipods, mysidaceans, euphausiaceans, decapod shrimps and prawns, and stomatopods (Fig. 26-7). Many of these forms are also benthic or nektonic during a part of the life cycle.

Among the chordates are the remarkable and sometimes abundant tunicates. The eggs and immature stages of many fish are pelagic in that they absorb just enough water shortly after being spawned to have almost precisely the same density as the surrounding water. The eggs of skates and rays, some of the sharks, and some other fishes, such as the herring, however, sink to the bottom, where they remain until they hatch.

The species composition of zooplankton is not uniform throughout the ocean. Groups of species commonly occur together and are characteristic of particular water mass habitats (Fager and McGowan 1963).

Flotation Mechanisms

The specific gravity of seawater is 1.02 to 1.03, while that of naked cells or protoplasm varies from 1.02 to 1.08. The specific gravity of the entire organism may be considerably higher if it possesses a skeleton or shell.

Organisms have various devices to remain afloat, aside from swimming: absorption of large amounts of water to form jelly-like tissues or sap, storage of gas or air bubbles within the body, formation of lightweight fat deposits in the body or oil droplets within the cells, and increase of surface area in proportion to body mass, thereby increasing frictional resistance.

Increase in the relative amount of body surface is achieved by decrease in size, flattening, attenuation of body form, extensions of body parts as antennae, spines, tentacles, or cerci, surface hairs or tubercles of various sorts, surface sculpturing, or formation of colonies (Marshall 1954, Davis 1955). These devices result in many strange shapes among plankton organisms. When the organisms die, the protoplasm disintegrates, special flotation mechanisms are usually destroyed, there is a loss of swimming movements, and what is left of the organism sinks to the bottom (Fig. 26-8).

Abundance

The actual abundance of plankton varies greatly from place to place and from one season to the next. Smaller species tend to be more numerous than larger ones. The mean annual abundance of diatoms is commonly in the tens of thousands per liter and for shorter periods during the year algal blooms may increase the population to hundreds of thousands of cells per liter (Ricketts and Calvin 1948). Zooplankton is, however, much less abundant. Large populations of zooplankton generally follow large populations of phytoplankton (Fig. 26-9) and by their grazing maintain the standing crop of phytoplankton at the size suited to prevailing conditions (Nielsen 1958).

The total net zooplankton (cc per unit volume of water) is some 50 times more numerous in the neritic coastal waters off the Atlantic coast of North America than in the Sargasso Sea (Grice and Hart 1962). The abundance of plankton is generally higher in cold than in warm ocean waters, correlated with the greater amount of phosphate present in colder waters (Harvey 1955). Cold-water plankton tends to be of larger individual size. There is generally in most taxonomic groups, however, a lower variety of species in cold than in warm waters (Fig. 26-10). Annual productivity is also less in cold waters, correlated with fewer generations per year. Productivity of zooplankton varies between 14 and 58 per cent of the net primary productivity (Mullin 1969).

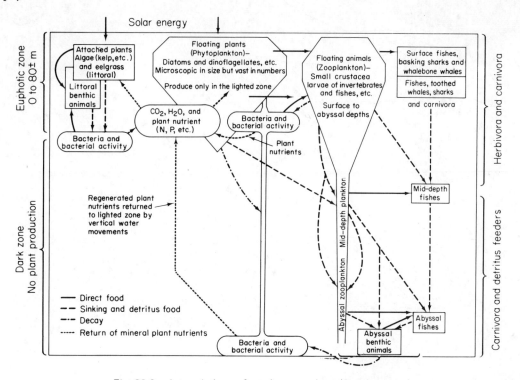

Fig. 26-8 Interrelations of marine organisms (Sverdrup *et al.* 1942).

Yearly Cycle

In general, winter is characterized by minimum levels of plankton. Nitrogen and phosphorus salts increase in surface waters because of the decomposition of organisms that have died during the preceding months, to the lesser absorption by phytoplankton, and to the greater mixing of waters from various depths accompanying the loss of thermal stratification. During the spring, as the result of increasing light, reduction of vertical turbulence of the water, and rising temperatures, the phytoplankton increases to a maximum for the year. In temperate and boreal regions the phytoplankton consists largely of diatoms. The zooplankton at this time abounds in immature stages.

By summer, nitrogen and phosphorus become diminished in the surface waters because of their use by phytoplankton and lack of replenishment from greater depths with the re-establishment of thermal stratification. Phytoplankton consequently declines rapidly, lacking nutrients and being consumed by the increasing zooplankton population. Zooplankton reaches its maximum during the summer but as it exhausts its phytoplankton food supply, it also declines. Decomposition of dead plankton in shallow waters and the destruction of the seasonal thermocline with vertical mixing of waters in the open ocean again return nitrogen and phosphorus to surface in the autumn, and this usually allows a second smaller maximum of plankton to develop in temperate climates.

High arctic seas usually have a single maximum

Fig. 26-9 Seasonal fluctuation of phyto- and zooplankton in neritic waters of the western Atlantic: (a) Curve at top is photosynthetic rate of *phytoplankton;* the second curve is the production rate; the bottom curve is the rate of change of the phytoplankton (Riley 1946). (b) Summation of *zooplankton* growth processes. (Riley 1947). Rates are in terms of grams of carbon per gram of plankton carbon per day

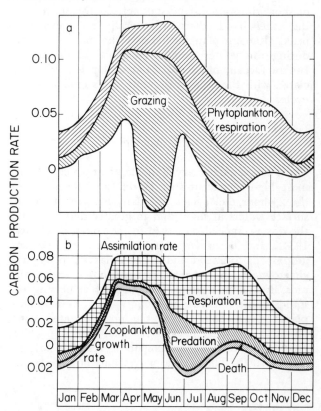

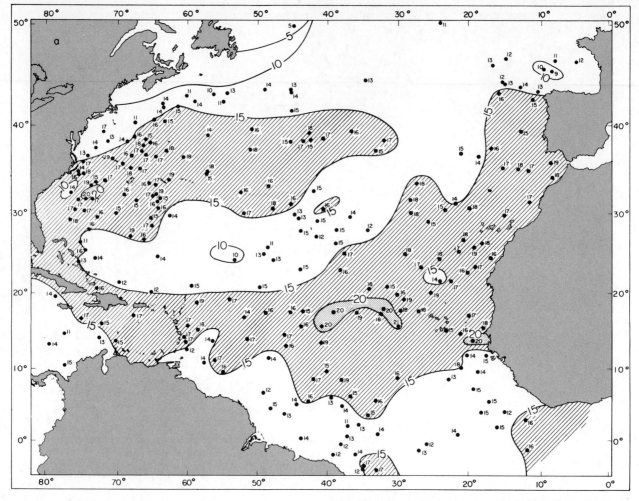

Fig. 26-10 Distribution of foraminiferan protozoans in the North Atlantic: (a) number of species per core sample, exclusive of rare forms; (b) species predominance, the number being the per cent of the predominant species in each core sample,

of short duration in the summer. In the tropics, on the other hand, there is generally no conspicuous peak, although plankton tends to be more abundant during the winter months (Bogorov 1958). In the Indian Ocean, physical oceanographic changes associated with the monsoons create a seasonal plankton cycle.

Diel Movements

Many zooplankters vary in the depth at which they are most highly concentrated at different times during the 24-hour diel cycle. The daily movements to the surface waters at night and to greater depths during the day is especially marked in the copepods (Clarke 1934), and occurs in euphausiaceans, mysidaceans, amphipods, ostracod and decapod larvae, pteropods, chaetognaths, polychaetes, siphonophores, and tintinnids. Even some nektonic animals, such as herring and squid, show vertical diel migrations. Most phytoplankton are confined to the upper lighted zone, although dinoflagellates have been shown to have short, vertical, daily movements in response to light (Hasle 1950). The vertical movements of these organ-

isms and of small fish are probably responsible for the shifts in the position of the deep *scattering layer* evident in the reverberation of high-frequency sound waves sent out form the surface (Eyring *et al.* 1948, Backus and Barnes 1957).

NEKTON

Mollusks, fishes, birds, and mammals make up the nekton of the sea. Mollusks are represented by the squids; fish, by the sharks, flying fish, herrings, mackerels, as well as many others, including numerous varieties of small species; and mammals, by the seals, porpoises, dolphins, and whales. The distribution of fish is irregular, but in general they occur more abundantly in neritic waters than in the open ocean. Likewise, they are much more numerous in the epipelagic than in lower strata. Most pelagic fish, except sharks, possess a swim-bladder useful for maintaining hydrostatic equilibrium at the depth where they occur; those fish that lack one are commonly bottom forms (Marshall 1954). In Arctic waters, fish are less abundant, and mammals relatively more important, than is the case farther south.

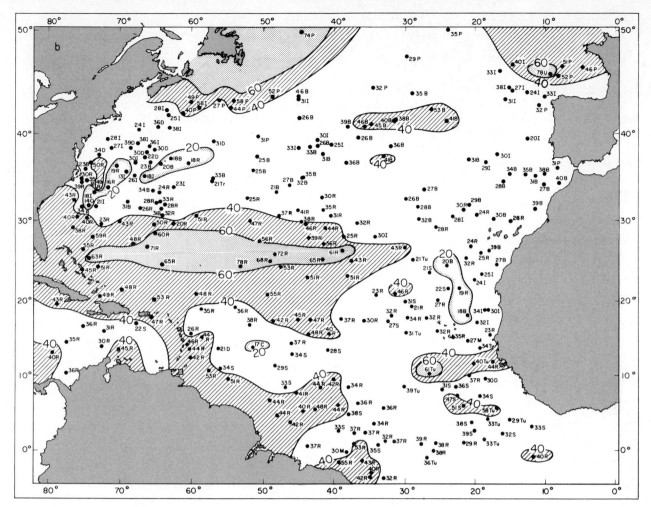

the letter referring to the species. Note that predominance is most pronounced in regions of low species variety and vice versa (Ruddiman 1969; copyright 1969 by the American Association for the Advancement of Science).

Birds, like many other marine animals, are more numerous in the neritic biochore than in the open ocean. In the oceans far from land occur only penguins, albatrosses, shearwaters, and petrels, and even these species become more common shoreward. Other marine species in neritic waters are tropic-birds, pelicans, gannets, boobies, cormorants, frigate-birds, ducks, gulls, terns, skimmers, auks, and murres. These marine birds may spend many days or weeks feeding and traveling over the water, but all must search out some shore, cliff, or isolated island on which to nest. Here they sometimes concentrate in enormous numbers during the nesting season because of the limited number of suitable nesting locations available.

BENTHOS

Benthos is of much greater variety in marine than in fresh-water habitats. These animals are very abundant in the littoral zone and decrease in numbers with depth until only scattered individuals are found in the deep ocean trenches (Saunders *et al.* 1965). Benthos consists of *sessile* forms, the sponges, barnacles, mussels, oysters, crinoids, corals, hydroids, bryozoans, and some worms;

creeping forms, such as crabs, lobsters, certain copepods, amphipods, other crustaceans, many protozoans, snails, echinoderms, some bivalves, and some fishes; and *burrowing* forms, including most clams, worms, and some crustaceans. Sessile and creeping forms are often grouped as *epifauna* and the burrowing forms as *infauna*. Epifauna in the littoral zone decreases in variety toward the poles since it is subjected to cold and ice erosion, but the species composition of infauna remains about the same.

In food habits, different species may be grouped as seston-eaters, mostly sessile or semi-sessile forms that capture suspended food particles; sluggish motile forms collecting detritus or food particles from the bottom surface; sluggish forms that extract food particles from bottom material that they swallow; carnivores, and scavengers (Sasilov 1957).

OCEANIC PLANKTON AND NEKTON BIOME

This biome is characterized by the predominance of organisms possessing life-forms adapted to keep them afloat. Plankton and nekton predominate, although the deep-sea benthos may also be considered as belong-

393

ing to this biome. Seasonal aspection may bring drastic changes in species composition, especially in plankton. Dominance, in the sense used for terrestrial communities, probably does not exist, except possibly in the Sargasso Sea, where the floating vegetation establishes the habitat. The ecosystem is self-contained, however, since energy is derived from the sun and nutrient material continues to recirculate with little or no dependence on terrestrial resources.

The *Sargassum* community of the Atlantic Ocean is of special interest. The floating *Sargassum* alga accumulates and is held within a limited area by circular ocean currents. This plant belongs to the intertidal zone of the Caribbean islands but is torn loose in large amounts along with attached animals during the hurricane season. It continues to grow thereafter, but does not reproduce. The fauna that it contains is a truly littoral one, rather than pelagic, but because the alga accumulates in fresh amounts as fast as old plants die, the animals reproduce and maintain a continued existence far from any shore.

Composition and Characteristics

The species composition of this biome varies consistently with depth so that a series of overlapping secondary communities may be recognized (Murray and Hjort 1912, Ekman 1953, Marshall 1954, Bruun in Hedgpeth 1957).

The *epipelagic community* or stratal society has the greatest abundance of plankton, nekton, and birds, as already described. The aquatic animals are generally colorless, transparent, or of a blue cast. Planktonic foraminiferans are of special interest in showing how changes occur regionally in species composition, richness of fauna, and abundance. Their abundant fossil remains, obtained from core samples to various depths in the ocean bottom, give evidence for the distribution of past climates when compared with present conditions. In the Atlantic Ocean, there is a general decline in species variety of living forms toward the North Pole, but this is interrupted by a paucity of species in the Sargasso Central Sea. Low species variety is correlated with certain species attaining high predominance; high species variety, with more equal densities between species (Ruddiman 1969).

In the *mesopelagic community* the fishes are usually small, laterally compressed, often silvery or grayish in color, with very large or telescopic eyes, and usually provided with luminescent organs (Fig. 26-11). Some velvety black or brown fishes also occur here. Invertebrates are reduced in number and variety and tend to be reddish in color. Since red rays do not penetrate to the depths where these animals live, they are essentially invisible.

The *bathy- and abyssopelagic communities* contain a highly diversified fauna, with every major marine taxon being represented, although often in small numbers. Most numerous benthic forms are foraminiferan protozoans followed by polychaete worms. They tend to be small in size and deposit feeders. Surface animals commonly have flat bodies, very long legs, or other means of distributing their weight over the loose, flocculent ooze. Many species rise above the ooze on stalks. The fragile glass sponges, long-stemmed crinoids, and long-legged crabs are possible only in very quiet waters that occur at great depths. Skeletons of all animals are fragile because of the difficulty of forming lime at low temperatures. Abundance decreases with depth, but even at 8300 m in the hadal zone, some 20 species have been found, chiefly holothurians, polychaetes, and sea anemones. Fishes are often long and slender. In the bathypelagic community, animals tend to be red or black in color; in the abyssopelagic community they are mostly pale or drab (Menzies 1965).

Bioluminescence occurs in a wide variety of invertebrates from protozoans to crustaceans, and it is exceptionally well developed in fish (Boden and Kampa 1964), among deep-sea forms. In some invertebrates, light-producing organs are scattered over the body. In other invertebrates and in pelagic fishes, there are special luminescent organs. It is estimated that two-thirds of the bathypelagic fish species and over 96 per cent of the individuals are luminous. Although several species of organisms occurring in surface waters are luminous, bioluminescence is more highly developed in the twilight zone, between 300 and 800 m, and occurs at still greater depths in the complete absence of natural light. The adaptive significance of bioluminescence is highly speculative. It may serve, in part, for attracting and seeing prey. Luminescent display may also serve for species and sex recognition,

Fig. 26-11 Representative deep-water fishes (Sverdrup *et al.* 1942).

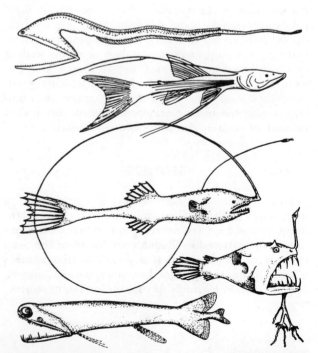

as does color in many surface animals. Joined with this bioluminescence is often the development of large eyes and special structures to permit vision at the very low light intensities that are produced. Perhaps as compensation for the difficulties of vision is the extensive development of antennae on some crustaceans and the very long rays in the fins of some fish which may serve for contact reception. In those fish where the eyes are small there is a reciprocally large development of olfactory organs.

The benthos and pelagic forms of the greater depths are doubtless derived from intermediate-depth forms, and these in turn from forms occurring on the continental shelf. Species have come to live in the deeper waters only as they became progressively adapted to this rigorous habitat. Relatively few forms have reached the hadal zone.

The deep-sea habitat has existed relatively unchanged since very early geological time except for the increasing deposition of bottom sediments and for some fluctuations in temperature. This uniform habitat has allowed some very ancient forms to persist to the present time. The coelacanth fish *Latimeria*, the mollusk *Neopilina*, and certain crustaceans belong to taxonomic groups that supposedly became extinct many millions of years ago. The examination of deep-sea bottom cores will doubtless give us information as to what kinds of animals were present in past ages. Determination of ratios of different oxygen isotopes and of different minerals in the composition of the fossil skeletons in these cores may make possible the determination of water temperatures and salinities at the time these fossil organisms were living (Ladd 1959).

Food Chains

As in fresh-water and terrestrial communities, bacteria in the sea are largely responsible for the final decomposition of excreta and dead bodies to make their essential nutrients available for reabsorption by the green phytoplankton (Ketchum 1947, Harvey 1955).

Nitrogen and phosphorus are least concentrated near the surface of the ocean, since this is the stratum in which they are most rapidly absorbed by the photoplankton. Excreta and dead organisms sink during the process of decomposition, so nitrogen regeneration is most evident at depths of 500 to 1500 m. The total non-living organic matter, either dissolved or in the form of particulate matter, is generally much larger than the biomass of living organisms at any one time. Organic matter makes up less than one-third of the total particulate matter, and all of it sinks slowly through the water, requiring months or even years to reach the bottom (Riley 1970). Much of the particulate organic matter is still undissolved and accumulates on the ocean bottom. Numerous species of

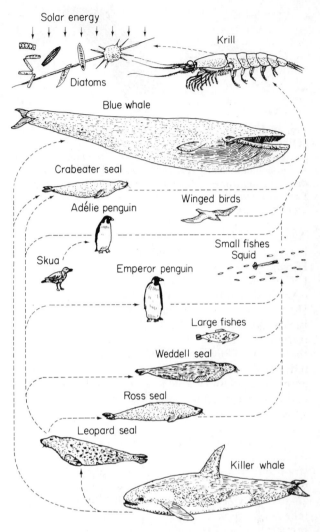

Fig. 26-12 Food cycle of marine animals in the Antarctic (Murphy 1962; from *Scientific American*).

invertebrates depend on it for food, especially on the bacteria that it contains (ZoBell 1952). The deep-sea fishes feed on these invertebrates or are carnivorous on other fish. Many of them have very wide mouths, distensible stomachs, and formidable teeth. In addition to these food coactions, it is also likely that many deep-sea fish and larger invertebrates undergo vertical migrations, so that they obtain food by preying on living organisms at more moderate depths. Much that is known about the life histories of these deep benthic species has been summarized by Marshall (1954).

Especially fertile regions of the open ocean occur when there is deep mixing of waters by *turbulence* and *upwelling*. Vertical water currents bring nutrients up to the surface from intermediate depths where they had accumulated. Prominent regions of more or less permanent upwelling occur around the Antarctic continent, off the coasts of California, Peru, and Somali, and off the west coasts of both north and south

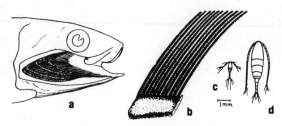

Fig. 26-13 Filter-feeding apparatus of the California sardine: (a) gill cover and gills removed to show one side of branchial sieve formed by gill-rakers; (b) enlarged drawing of a section of the branchial sieve; (c) small copepod, *Oithona plumifera*, drawn to the same scale as (b); (d) medium-sized copepod, *Calanus finmarchicus*, drawn to the same scale as (b) (Sverdrup *et al.* 1942).

Africa. In many areas, very small phytoplankton and bacteria form the first link in the food chain (Fig. 26-12), which goes next to the microzooplankton before reaching the carnivorous net zooplankton of copepods, euphasiids, and other crustacean forms (Ryther 1969). Most animals depending on small organisms and organic detritus for food have various filter-feeding mechanisms for straining the material out of the water. They do not actively search and catch individual items through directed actions. Invertebrate animals may also be able to absorb some essential salts and dissolved organic compounds to build their skeletal structures and for general metabolism, but there is considerable controversy on this point (Collier 1953).

The baleen or whalebone whales (Mysticeti) are toothless but possess large plates in their mouths that strain out the plankton (especially copepods, euphausiaceans, mysidaceans) that they use as food (Fig. 26-13). Only occasionally are small fish or other invertebrates ingested. Some whales reach tremendous proportions, and the differential in size between these animals and their food is one of the most remarkable in the animal kingdom. Much more common is the feeding on plankton by squids, the young stages of most fishes, and such adult fishes as sardine, anchovy, menhaden, herring, and mackerel. The menhaden is unique in having such fine-mesh gill-rakers that it can feed extensively on diatoms, which because of their smaller size cannot be readily secured by other large marine animals (Clarke 1954).

Small nektonic species are in turn preyed on by larger species. Sharks commonly hold the last link in the food chain. The pelagic birds are also fish-eaters or depend on floating carrion for their food.

Productivity

Most studies of plankton productivity have been conducted in the neritic zone (Raymont 1963). In the English Channel, the mean annual standing crops of phytoplankton, zooplankton, and pelagic fish in dry weight of organic matter are 0.4, 1.5, and 1.8 g/m².

This is unusual in giving a larger biomass of herbivores and carnivores than of photosynthetic plant material. However, the daily productivity of phytoplankton makes up for this because it is over 100 per cent, while that of zooplankton is only 10 per cent and that of fish 0.09 per cent. The productivity ratio of phytoplankton, zooplankton, and fish is approximately 280: 100: 1 (Harvey 1950).

The daily net productivity of phytoplankton in the upper 20 m of Block Island Sound near the eastern end of Long Island has been estimated at 26 per cent of the standing crop in excess of that consumed by zooplankton and bacteria in the surface layer. The zooplankton consumes not more than 4 per cent of the phytoplankton per day. Most of the excess daily production (19 per cent) in the surface waters settles downward and is used by animals and bacteria on or near the bottom, with the rest (7 per cent) becoming laterally dispersed into adjacent areas (Riley 1952). The daily productivity of zooplankton in this same area was calculated at 17 per cent of the standing crop (Deevey 1952).

Phytoplankton productivity varies with the time of the year. Near Kiel, Germany, in August there is a surplus of phytoplankton production over the amount consumed by animals; the productivity amounts to 350 mm³/m³/day, while animal consumption is 60 mm³/m³/day. During February, the productivity of plankton is only 10 mm³/m³/day, while the food requirements of animals is 18 mm³/m³/day. This deficiency in food production is correlated with the decrease at that time in the standing crop of both plants and animals (Sverdrup *et al.* 1942).

Productivity is especially high in those parts of the ocean where there is upwelling. It is in these areas that the yield of commercial fish of economic interest to man is the greatest.

BALANOID-GASTROPOD-THALLOPHYTE BIOME

This community extends from high- to below low-tide levels on rocky shores (Lewis 1964). Benthic animals and attached algal plants are conspicuous and important. The benthos is mostly epifauna, as the substratum is too hard to permit development of extensive infauna. When the tide is out, the organisms are subjected to drying, the occasional inflow of fresh water, higher temperatures, and greater light intensities. Organisms avoid desiccation when the tide is out by variously crawling under stones or thick algal growths, closing thick shells or operculae, retreating into crevices, or secreting a mucous seal. Most organisms are also faced with the pounding action of waves. Various holdfast or anchoring devices have developed, and many species protect their more delicate structures with a hard shell. The adaptations for life on the seashore are many and varied (Yonge 1949). The plankton and nekton

Table 26-1 Vertical zonation of mollusks on a rocky shore at Cape Ann, Mass. Numbers given represent the density of individuals per square meter (after Dexter 1945).

Zone	*Littorina saxatilis*	*Littorina littorea*	*Mytilus edulis*	*Mya arenaria* (seed)	*Littorina obtusata*	*Thais lapillus*	*Acmaea testudinalis*	*Anomia aculeata*	*Crepidula fornicata*
High-tide level	4	4							
105 cm lower	23	112							
115 cm lower	248	2,225	23	77					
131 cm lower	58	1,339	116	81	248	35			
140 cm lower	31	387	132	0	151	8			
156 cm lower		704	31	8	341	15			
174 cm lower		1,300		0	163	70	8		
184 cm lower		813		0	77	15	15	15	8
199 cm lower (near spring low-tide level)		387		15					45

associated with the benthos include many species not common to the oceanic biome.

Zonation

Vertical zonation of species on rocky shores is usually conspicuous (Table 26-1), although individual species may extend widely into adjacent areas (Hewatt 1937, Yonge 1949, Stephenson 1949, Southward 1958).

Beginning on the landward side there is a *supralittoral zone* mostly above the action of tides and inhabited as much by land as by marine animals (Fig. 26-14). This is followed seaward by a *supralittoral* or *Littorina fringe* which is wetted by the highest tides and by the splashing of waves. Because of the presence of either Myxophyceae or lichens, this zone is often discolored; commonly, black. The fringe is especially characterized by large numbers of small snails and sometimes isopods.

Next below this fringe is the *midlittoral* or *balanoid* zone. It is strictly intertidal, being covered and uncovered every day, and is occupied characteristically by acorn barnacles. This zone is often divided into subzones with the barnacles predominant in the upper portion, while polychaets, colonial hydroids, or other forms are relatively more important in the lower part. The subzonation of algae is often also well marked.

The lowest zone ever exposed, and then only at extreme low tides, is called the *infralittoral fringe*. It is a transition area. The entire area between extreme high and low tides, including the mid-littoral zone and its supralittoral and infralittoral fringes, when considered as a unit, may be referred to as the *littoral, eulittoral,* or *tidal zone* to distinguish it from the *infralittoral* or *sublittoral* zone that extends from the lowest of low tides to the edge of the continental shelf.

Zonation is brought about in large part by differences between species in tolerance to length of exposure and submergence. Animals get into the proper zones by one of several ways (McDougall 1943). Motile species move in and out of favorable areas in direct response to stimuli. In sessile forms, however, it is the motile larvae which are dispersed uniformly, but die off in unfavorable microhabitats. In some forms the larvae are attracted into certain areas by environmental factors, although the exact nature of the factors often remains obscure. Young periwinkles are transported by wave action to the lower margin of stony beaches, and further shoreward movement is mainly locomotor. They achieve their proper zonation by the end of the first year of life (Smith and Newell 1955). Superimposed on these responses to habitat factors are the coactions between species. The breeding population of a barnacle on the coast of Washington, for instance, is limited to a narrow zone near the high water mark as the result of predatory snails eliminating the barnacle larvae that settle at lower levels. In

Fig. 26-14 Terminology of zonation on rocky coasts (Stephenson 1949).

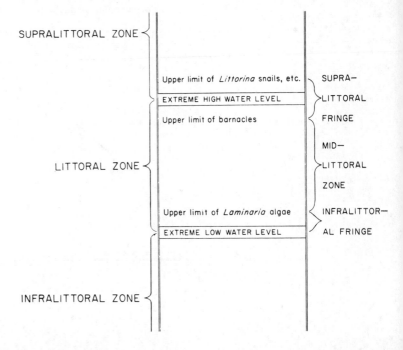

general, the upper limit of occurrence of intertidal organisms is determined by their tolerance to increasing severity of physical conditions, while the lower limit more often depends on competition for space or predation (Connell 1970).

Littoral Zone

Brown algae form thick masses and give protection to those animals that find shelter in or under them. A fauna of copepods, ostracods, water mites, and young littorinids inhabit these seaweeds. In England, the numbers of individuals per 100 g of seaweed vary from about 44 on brown algae to over 13,000 on lichens (Colman 1940).

The animal life on rocky shores is varied and luxuriant. Several species of acorn barnacles, snails, marine limpets, marine mussels, goose barnacles, sea anemones, chitons, sponges, hydroids, bryozoans, flatworms, annelids, amphipods, isopods, crabs, sea urchins, starfishes, tunicates, and insects are present. Total abundance of animals may run into tens of thousands of individuals per square meter (Allee 1923, Newcombe 1935, Dexter 1947, Yonge 1949, Stephenson 1950, 1952, 1954, 1961, Shelford *et al.* 1935, Hewatt 1937, Ricketts and Calvin 1948).

Sublittoral Zone

This community is not subjected to exposure by tides or to the pounding of surf, but is affected considerably by wave action and the complete circulation of water (Fig. 26-15). Animals move around somewhat more freely and there is less need for strong holdfast structures. Most organisms lack physiological tolerance for long exposure to the air and hence differ fundamentally in structure and mores from the community described above.

Laminarias or kelps are the largest of the brown algae and occur commonly in this community. They have root-like holdfasts attached to the bottom and their stalks, which are often several meters long, bear leaf-like branches that float at the surface in the larger species. A long list of animals find shelter and food in the kelp beds and especially in the protection of the holdfasts (Andrews 1945). Polychaete worms are particularly abundant in these holdfasts (Colman 1940). Filamentous red algae (Rhodophyceae) are also prominent.

Abundant characteristic animals on the Pacific coast are sea urchins, sea cucumbers, starfishes, snails, rock oyster, chitons, limpets, scallops, mussels, nudibranchs, barnacles, crabs, hermit crabs, hydroids, tunicates, shrimps, and various fish. Distribution of

Fig. 26-15 Schematic representation of vertical distribution of biological zones on the continental shelf and slope in the Okhotsk Sea of the north Pacific: I, zone of predominant development of the encrustation fauna (sessile seston-eaters) on stony bottom; II, zone of predominant development of filtering bivalves; III, zone of predominate development of large species of detritus-collecting bivalves; IV, zone of predominant development of detritus-collecting forms (excluding large bivalve species); V, zone of predominant development of encrustation fauna (sessile seston-eaters) on muddy-sand bottom; VI, subzone of predominant development of polychaetes (detritus-collecting forms); VII, zone of predominant development of deposit-swallowing forms (Savilov 1957).

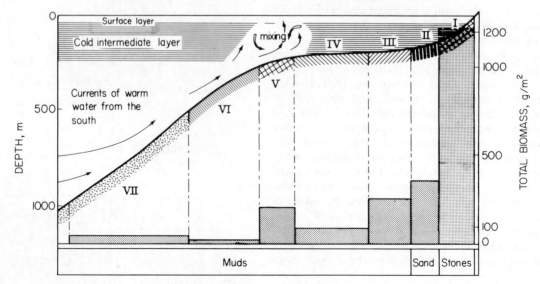

fish species correlates strongly with the type of bottom or benthos that is present (Popov 1931).

The variety of animals in this community is great, both in genera and species, but the density of any one benthic species is seldom greater than 10 per m^2 on the Pacific coast of North America (Shelford *et al.* 1935), in contrast with the littoral zone. Off the coast of California, the average fresh weight of the standing crop of plants decreases at depths of 1.5 to 22 m from 4667 to 606 g/m^2, while animals increase from 125 to 377 g/m^2 (Aleem 1956). At greater depths, the biomass of animals decreases.

Tidal Pools

Seawater is often retained in depressions or pools in the littoral zone and hence organisms here are never completely exposed to the air. They are, however, subject to high light intensity and increases in the temperature between tides (Klugh 1924). Tidal pools are usually rich in both plant and animal life, and some species are largely restricted to them. Red algae and kelps prefer the more shaded, cooler pools; the green algae and some of the smaller brown algae predominate in the well-insolated pools. Animals of both the sublittoral and littoral zones are found here.

Food Chains

The basic food elements in these rocky shore communities are the free-floating plankton and detritus in the water, the algae, and the organic debris adhering to the rock surfaces (Dexter 1947). Many animals have straining mechanisms that automatically collect food materials out of the large volume of water with which they have contact. Snails crawl over the rocks and seaweeds, scraping away at the algae and plant tissue and eroding the rock surfaces. Crabs and fish have a wealth of invertebrates upon which to feed, and are generally at the top of food chains along with birds that feed on the shore.

Marine shore animals have developed many devices to protect them from predators (Flattely and Walton 1922). Some crabs are concealed by seaweeds, hydroids, or other organisms growing on their carapaces. Hermit crabs take refuge in the shells of snails. Protective and warning coloration is common. Protective armor occurs in the form of shells, chitinous exoskeletons, spicules, spines, setae, bristles, and constructed tubes. Various forms of weapons have evolved, some of them poisonous, as the nematocysts of coelenterates, stylets of some gastropods, spines of the king crab, and chelae of crustaceans. Autotomy, or the ability to throw off an appendage grasped by an enemy, is highly developed in crabs, lobsters, and echinoderms. Well-developed powers of regeneration of lost parts occur in these forms, while in worms, sponges, hydroids, and other groups, regeneration of entire new bodies from small fragments is often possible.

Dominance and Succession

Dominance is exerted by those organisms that compete most successfully for the space that is available. When they become established, they largely control the presence of other species. This is true both with the seaweeds and with the more abundant and successful animals. In describing competition on wharf pilings at Beaufort, North Carolina, McDougall (1943: 367) states:

> So many barnacle larvae, for example, may settle on a small area of clean surface that only a fraction of one per cent of their number will ultimately find space to grow to full size. Incrusting bryozoans, such as Schizoporella, spread over and smother barnacles and other low-growing species in their vicinity. Colonial hydroids, sponges, and ascidians often form densely matted tangles which accumulate quantities of sediment and effectively smother barnacles, oysters, bryozoans, and other species less luxuriant than themselves.

The colonial hydroid *Tubularia crocea* dominates the pilings in April and May but during June, with water temperatures becoming higher, the animals die and slough off, taking with them many associated species. In the bare areas thus exposed, various other sessile species become established. By the end of August, two other colonial hydroids, *Pennaria tiarella* and *Eudendrium carneum*, and a colonial bryozoan, *Bugula neritina*, become dominant. In late October, as water temperatures fall, these summer species die and *Tubularia* again becomes active and reproduces abundantly. Low winter temperatures temporarily curtail its activities. It is apparent that true succession actually occurs but is passed through quickly and may be obscured by seasonal changes in the species composition of the biota (Redfield and Deevey 1952).

PELECYPOD-ANNELID BIOME

Habitat

This biome develops on depositing sand and mud bottoms in contrast to the biome just described that occurs on eroding rocky shores. There is still a good deal of wave action over sandy bottoms. Fine sand particles shift about almost continuously, and animals have difficulty in preventing their burrows from collapsing. In general, the water over muddy shores is shallower, quieter, and warmer. The mud forms a soft, compact bottom, but is also easily moved or shifted around by storms and wave action. Animal burrows in mud are more permanent. Species tend to segregate depending on the fineness of the soil particles and on the amount of organic matter present. Shores of high mud content may be low in oxygen because of decaying organic matter, so animal populations tend to be largest and most varied in a mixture of mud and sand. Tidal currents are weaker, and change in the level of water less pronounced on sand and muddy shores than on rocky ones.

Composition and Characteristics

Important plants in this biome are the marine eelgrass, which is a seed plant, and green algae, particularly the sea-lettuce, which grows in sheets either attached to the substratum or lying fragmented over large areas, and *Enteromorpha*, which grows in tufts or tangles. Occurring on eelgrass and sea-lettuce may be several kinds of epiphytic algae. These plants form extensive stands and are important to animals for attachment, shelter, and food. Eelgrass was almost eliminated from the Atlantic coast in 1931–32, possibly because of a protozoan disease. This disturbance had a profoundly deleterious effect on the abundance of many animals, including the brant, a bird that depended on it almost exclusively for food, and on scallop and other coastal fisheries. Twenty years later there was evidence that eelgrass was recovering much of its former abundance (Cottam and Munro 1954).

Predominant animals are pelecypods, polychaete worms, particularly *Arenicola* and *Nereis*; starfishes, brittle-stars, sea cucumbers, crabs, amphipods, and snails. Populations may run to several thousands of individuals per square meter. A variety of small fish occur here. Birds include sandpipers, plovers, and herons.

The biome is worldwide in distribution, but the characteristic life-forms are represented by different species locally. Thus a number of secondary communities (biociations) may be recognized (Petersen 1914, Jones 1950, Thorson in Hedgpeth 1957).

Many of the animal constituents in this biome are burrowing forms. The substratum of mud and sand holds considerable water and when the tide is out on exposed flats, pelecypods, worms, and other animal constituents retract their fleshy organs into their burrows or shells and remain in a water saturated environment (Hesse, Allee, and Schmidt 1951). They are thus not exposed to the atmosphere with changes in the tide. Furthermore, most forms are generally tolerant of low oxygen and high carbon dioxide concentrations. In order to maintain respiration when retracted in their underground burrows these animals have long siphons, sometimes longer than their bodies, or long tubes or canals that extend to the surface. Through these they maintain a circulation of water, often by means of special pumping organs.

Burrowing crustaceans have setose appendages, modified for digging, and small eyes. Those that remain near the surface have long antennae and robust bodies; those that live in deep burrows have short antennae and slender bodies. For the same reason, burrowing clams that remain near the surface have heavy shells while those that burrow deeper have more fragile shells. These clams either have wide, slimy feet and small shells and crawl with ease through the sand, or have a slender foot that can expand at the end to give enough anchorage so that the animal can pull itself downward into the sand or mud (Pearse *et al.* 1942).

The infauna also includes many microscopic forms, such as nematodes, flatworms, copepods, ostracods, foraminiferans, and other protozoans. These small organisms may be enormously abundant in number of individuals. In respect to the ciliated protozoans, some species are ubiquitous, but other species are characteristic of this habitat (Fauré-Fremiet 1950). Some species of ciliates occur in the intergranular spaces of the sand and mud, other species are associated with the surface film of diatoms (Webb 1956).

In contrast to the great variety of infauna in this biome, the epifauna is more restricted, although some species are abundant. The fiddler crab browses in great armies on beaches left exposed by the tide, but retreats into its burrow and plugs up the opening when the tide comes in. The characteristic ghost crab of sandy beaches of middle latitudes spends most of its time above high-tide level but must return occasionally to water to dampen its gill chamber.

Zonation

Zonation is less conspicuous than on rocky shores because of the prevalence of burrowing forms and of running and swimming species that move up and down with the tides (Brady 1943). Some evidence for zonation occurs on sandy beaches with crustaceans (Dahl 1953) and with pelecypods and annelids (Stephen 1953). A more pronounced change in species composition occurs in the sublittoral zone; in the North Pacific this change comes at a depth of 3 m (Shelford *et al.* 1935). Pelecypods and annelids still predominate (Holme 1953, Sanders 1956).

Food Coactions and Productivity

Bacteria are more abundant on mud bottoms than in the open sea (ZoBell 1946). On mud flats they may average 10 million cells/cc of mud, or with a biomass of nearly 40 g/m^3. Bacteria are, of course, vital for the decomposition of organic debris, dead organisms, and wastes, but may also be used directly as food. Assuming that the bacteria divide rapidly enough to increase their biomass 10 times per day, this gives a 24-hour production of 400 g/m^3, mostly concentrated in the 5 cm beneath the surface. Gross primary productivity on an intertidal sandflat in the Puget Sound region of the Pacific Coast varies from 998 to 1580 kcal/m²-yr, of which 68 per cent is used in respiration and the remainder is available to consumers or may be lost by emigration or drift (Pamatmat 1968).

Many protozoans and zooplankters feed on bacteria and detritus, and a large number of mud- and sand-dwelling invertebrates of larger size, such as pelecypods, annelids, and crustaceans, are deposit-feeders, or have special straining devices in the form of sieves, brushes, and hairy or mucous nets for collecting bacteria, organic particles, and small plankton organisms suspended in the water (Blegvad 1914, Yonge 1953). Deposit-feeders tend to predominate where the bottom is composed of fine sediments, while suspension-feeders predominate where the bottom is made of coarser materials (Sanders 1956). Snails, amphipods, and others feed on the plant tissues of eelgrass, sea-lettuce, and algae. Crabs, echinoderms, and fish feed partly on organic debris and partly on the smaller invertebrates. Fish feed on the invertebrates and larger fish feed on smaller ones. On sand and mud flats on the coast of California, only 5 per cent of all animal species are strictly carnivorous. Only a few species of animals feed directly on plants, but plant tissue becomes more available after its death and partial decomposition. The principal food chain is plants: detritus and bacteria: detritus- and bacteria-feeders: animal-feeders: birds (MacGinitie 1935).

At low tide, feeding and other activities are at a minimum as the bottom forms retract into their burrows and motile forms retreat seaward. However, some of the snails continue to feed on plants, and insects and birds come into the exposed area in search of debris and small animals for food. As the tide returns, the insects and birds retreat landward, but pelecypods extend their siphons, annelids rise from their burrows, shrimps, crabs, and fish move about over the surface, and the whole community becomes a scene of bustling activity.

Petersen (1918) and Jensen (1919) of Denmark were concerned with measuring the productivity of invertebrates in the sea, especially benthos, as a source of food for commercially important fish. More accurate calculation in the English Channel (Harvey 1950) give the ratio between the mean annual dry organic weight of bottom invertebrates and the bottom-dwelling fish as 15: 1. Annual productivity of the invertebrate fauna is twice as great as that of the fish, so the productivity ratio between the two groups is 30: 1.

Dominance

In contrast to rock-bottom communities, competition for space is not an important factor in mud- and sand-bottom communities. Actual evidence purporting dominance to exist is not convincing (MacGinitie 1939). Eelgrass and algae do not appreciably react on the physical characteristics of the habitat except to increase the supply of oxygen.

Starfishes and brittle-stars may exert control to a limited extent where they are numerous by feeding on and preventing pelecypods from becoming established in the community (Clements and Shelford 1939). It is possible that fishes may at times modify the species composition of an area, but for the most part it appears that the presence and distribution of species are controlled directly by the physical conditions of the habitat (Jones 1950) and that only population size is modified by predation and competition.

CORAL REEF BIOME

Coral reefs are formed by the accumulation of the calcareous skeletons of myriads of organisms. They extend from the sea bottom at depths of 46 m, or rarely 74 m, to slightly above low-tide level. The best formation of coral reefs is confined to warm waters above 18°C, although individual species may extend into colder regions (Vaughan 1919, Wells in Hedgpeth 1957).

Predominant organisms involved are commonly the anthozoan stony corals and organ corals and the hydrozoan milliporids. Some reefs, however, are formed principally by Foraminifera and still others by calcareous algae. All massive coral structures employ calcareous algae as cement. These algae not only thrive in the pounding surf on the windward side of the reef, but by their growth are able to repair damage to the reef caused by storms. Most typical reef-building animals are colonial and of shapes varying from closely compact, globose, or encrusting, to loosely branched or dendritic, depending in part on their exposure to wave action. Each polyp in a colony secretes its own calcareous skeleton and when it dies the next generation builds on top of the old so that the accumulation of a lime structure is fairly rapid (Yonge 1963).

The bright yellow or red colors of corals near enough to the water's surface for adequate light penetration are the result of algae, the zooxanthellae, which are either embedded in the body wall or free in the internal cavities. In addition, there are bands of green filamentous algae growing to a depth of 2 or 3 cm in the pores of the inert coral skeleton that may have a biomass 16 times that of the zooxanthellae. In their photosynthesis, these algae may absorb carbon dioxide and nutrients derived from animals of the coral reef and liberate oxygen of value to the animals (Odum and Odum 1955). Perhaps because of this symbiotic relationship, which requires solar radiation, living corals are largely confined to the upper, shallower waters. Coral animals actively ingest zooplankton, but apparently not phytoplankton, from the surrounding water (Hand 1956, Yonge 1958).

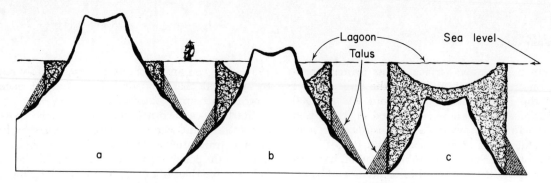

Fig. 26-16 Three types of coral reefs and their possible manner of origin, according to Charles Darwin: (a) fringing reef; (b) barrier reef; (c) atoll.

Reefs may be either fringing, barrier, or atoll (Fig. 26-16). Fringing reefs are in direct contact with the shore; barrier reefs are separated from the shore by a lagoon of varying width; atolls are annular or horseshoe-shaped, surrounding a lagoon that does not contain any central land mass. The Great Barrier Reef that extends for great distances off the east coast of Australia is a good example of the second type of reef, and atolls are numerous in the South Pacific. According to a theory first proposed by Darwin, barrier and atoll reefs form from fringing reefs either as the land subsides or the water level rises (Vaughan 1919). Atoll islands are formed either when the water level falls or when waves break off and pile up chunks of coral limestone to build the reefs a few feet above the high tide level. As the stone dissolves or becomes pulverized, it forms a soil on which plants can grow and terrestrial communities of animals invade. At times of low ocean levels, as during glacial periods, the tops of atolls may have extended high above the sea. Reefs are often not continuous because the organisms are intolerant of fresh water brought down by streams, and because they are very sensitive to smothering by mud or sand. On the exposed ocean side there is generally a zonation of different species from the shore outward (Odum and Odum 1955, Goreau 1959, Wells in Hedgpeth 1957).

The coral organisms, particularly the algae and the coelenterates, are true dominants in this biome since they build the substratum that makes possible the development of the community and the occurrence of other organisms. Competition for space, light, and protection from wave action is keen.

There is a great variety of secondary species associated with the corals. These include many kinds of alcyonarians; numerous brittle stars, crinoids, and holothurians; a great variety of chaetopod, echiurid, and sipunculid worms; crustaceans, including hermit crabs; mollusks; and large numbers of brilliantly colored, strikingly marked fish. The many crevices, holes, and cavities in and between the coral provide excellent hiding places and refuge from predators so that the impressive development of color among the fishes may be due, in part, to lack of predation pressure. The fishes have a variety of food habits and are represented in all consumer trophic levels (Hiatt and Strasburg 1960).

At the Eniwetok atoll in the South Pacific, the average dry weight biomass of the living photosynthetic plant material is estimated at 703 g/m², that of the herbivorous and carnivorous animals at 132 g and 11 g, respectively. These weights exclude the dead skeletal materials associated with the protoplasm. The ratio between plants and herbivores is 5.3:1, between herbivores and carnivores 11:1, or a composite ratio of 64:12:1. The total primary productivity per year as the result of photosynthesis was estimated at 12.5 times the biomass of the standing crop. This is sufficient to balance approximately the total plant and animal energy needs of the reef and thus render the coral reef a self-contained steady-state ecosystem (Odum and Odum 1955).

In recent years, coral reefs in the Pacific and Indian Oceans have been subjected to considerable destruction by attacks of the crown-of-thorns starfish, *Acanthaster planci*. Rare or unnoticed until about 1963, the species has since exploded in population and appears to threaten the very existence of this community. The starfish consumes the coral polyps from just below spring-tide level to the depth limit at which living corals occur. After the polyps disappear the coral structure becomes overgrown with algae and most of the fish leave. The rate of destruction is often rapid, 1 m² per individual starfish per month, and the starfish population over extensive areas may average 1 per m². There is no generally accepted explanation of how or why this coaction got started, and no effective control measure has yet been found (Chesher 1969). In some localities, however, there is evidence that the threat is subsiding, and the coral community is recovering on the basis of its own resiliency.

SUCCESSION TO LAND

The three great biocycles—ocean, fresh-water, and land—come into contact with each other around the margins of the seas. The change in the physical nature of the habitat from salt water to fresh water is a drastic one, but not more drastic than the change from salt water to land. The transition of animal and plant life is abrupt, and a zonation or physiographic succession of communities can be recognized. This transition from the ocean to fresh water and from the ocean to land as we see it today is of special interest since it parallels the probable evolution and dispersal of life in past ages.

Life is generally believed to have originated in the littoral region. Apparently no great groups [phyla] of animals originated except in the ocean. The routes by which animals probably left the ocean and reached fresh-water and land have been various. Some animals probably migrated directly across sea beaches; others probably ascended rivers, passed through marshes and swamps, or burrowed through soil. Some animals were transferred from the ocean by land elevations which isolated them in bodies of water which gradually became fresh. . . . Emigration from the sea did not take place at any one time. It has occurred many times in the past and is slowly progressing on many shores today. . . . The most successful animal colonizers of the land have been: (1) the arthropods, which have in many cases developed book-lungs or tracheae for breathing air; (2) the vertebrates, with lungs and dry skins; and (3) the snails, with slime and spirally coiled shells to prevent desiccation. . . . There are at present many examples of animals which are in the midst of their transformation from marine to fresh-water animals, or from marine or fresh-water into land animals. Not only have plants and animals emigrated from sea to land, but there are countless instances when migrations have taken, and are taking place in the opposite direction. Grasses, insects, reptiles, birds, and mammals have left the land for the sea. . . . Fishes began in fresh water, but now range through the ocean at all depths [Pearse 1950: pp. 9–10, 14].

On rocky shores and cliffs there is a splash or supralittoral zone above high tide level. Green algae occur here and scattered individuals of marine snails, acorn barnacles, limpets, amphipod sandfleas, and flatworms, as well as insects, especially Diptera and other forms that come from the land. Above the influence of splashing, the rocks may be covered with lichens and mosses, representing the initial stages in the terrestrial rock sere. However, salt spray is often blown inland a considerable distance to affect conspicuously the development of normal terrestrial vegetation and its accompanying animal life. Cliffs along

the ocean, as well as sandy beaches and islands, are favorite nesting places for large numbers of pelagic birds.

Above water action on sandy shores, the wind may blow the sand into dunes with the consequent development of the dune sere. On muddy flats there is typically a development of salt marshes, particularly in protected embayments or along the margins of outflowing rivers. The high marshes are flooded completely only during the spring tides, but the ground water is more or less continuously saline. As sediment accumulates the marsh eventually becomes dry land (Chapman 1960).

The seashore snails (*Littorina*), the marsh snail (*Melampus*), mussels (*Brachidontes*, *Mytilus*), crabs (*Carcinides*, *Cancer*), amphipods (*Gammarus*, *Orchestia*), and isopods (*Philoscia*) occur through the extensive salt marshes on the Atlantic Coast of North America and there are numerous insects (Table 26-2, Fig. 26-17). Killifishes are abundant and devour many mosquito larvae. Herons, plovers, sandpipers, ducks, rails, bitterns, redwinged blackbirds, marsh wrens, and sharp-tailed sparrows feed or nest. Muskrats and meadow voles, as well as other species of mammals, occur in salt marshes but are not particularly characteristic of them (McAtee 1939).

Salt marshes are among the most fertile areas of the world. Dry-matter productivity may be 2000–3000 g/m²-yr, compared with 200–400 on the continental shelf and 100–200 in the open ocean, and with 340–1400 for wheat and corn. However, a large fraction of wheat and corn production is directly usable by man as opposed to only a small fraction of the production of the salt marsh. Food chains in the salt marsh are based more on detritus feeding than directly on the vegetation; hence nutrients and energy must pass to higher consumer levels before they can be harvested for human use. The management problem here is one of utilization rather than production.

In tropical regions mangroves may develop instead of marshes on mucky, poorly aerated bottoms. The red mangrove has an extensive prop root system and grows in deep water not ordinarily exposed even at low tide. The mangroves protect the shore from erosion and aid

Table 26-2 Relative abundance of different orders of insects in successive plant stages in North Carolina salt marshes (Davis and Gray 1966).

Seral stage	Homoptera	Diptera	Hemiptera	Orthoptera	Coleoptera	Hymenoptera	Other orders
Spartina alterniflora, intertidal	90	7	2	1	+	+	+
Juncus roemerianus, reached by extra high tides	72	10	3	11	1	3	+
Distichlis spicata, inundated only by spring tides	57	19	19	1	2	1	+
Spartina patens, inundated only by hurricanes	30	44	9	3	4	9	1

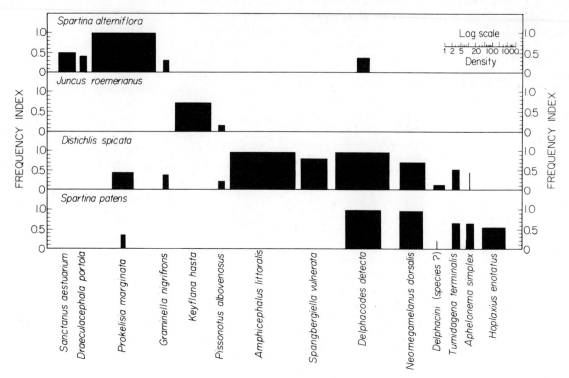

Fig. 26-17 Frequency-density diagram of the principal species of Homoptera from the herb layers of four seral stages of salt-marsh vegetation (Davis and Gray 1966).

in the accumulation of deposits of peat and mud that build up the shore and form islands. The black mangrove at higher levels usually produces erect roots that stick up through the mud and serve as pneumatophores. Mangroves are usually heavily populated beneath by snails, crabs, and other marine species (Golley *et al.* 1962, Macnae 1968).

SUCCESSION TO FRESH WATER

Where rivers flow into the ocean on low coastal plains and there are extensive embayments or estuaries, as along the Atlantic Coast, there is a very gradual change from salt water to brackish water (salinity:

Fig. 26-18 Relation of estuaries to salt and fresh water in respect to salinity (‰) and water circulation (after Cronin and Mansueti 1971).

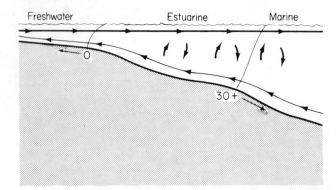

0.5–30‰) to entirely fresh water (Fig. 26-18). This habitat gradient fluctuates back and forth with the tides. Since fresh water is less dense and often warmer, it flows over the top of the salt water, with the result that strata with different physical characteristics are formed and these different strata are inhabited by different kinds of fish and other organisms. Influx of fresh water is one of the principal sources of dissolved nutrients, and the gradient of salinity is dependent upon continuous inflow of fresh water. Any reduction of inflow, as by damming of streams, will allow salt water to penetrate farther inland.

Species of marine organisms extend toward fresh water as far as permitted by their tolerance of reduced salinity (Fig. 26-19). Since this tolerance varies between species, the marine flora and fauna become impoverished as the fresh-water flora and fauna become enriched. There are, however, many more marine species than fresh-water species in estuaries. A few species find optimum environmental conditions in brackish waters and decrease in abundance both toward fresh water and toward the open sea. Economically important brackish and shallow water species on the Atlantic coast are the blue crab, lobster, American oyster, scallops, hard-shell clam, soft-shell clam, and such fishes as the Atlantic croaker, striped bass, American shad, scup, weakfish, and others.

Of special interest are fish that perform long migrations between fresh and salt water for spawning pur-

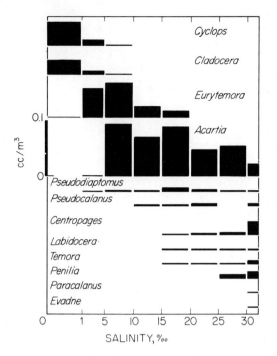

Fig. 26-19 Annual average volume (cc/m³) of the principal entomostracan species occurring in the salinity gradient of the Delaware estuary in the eastern United States (from Cronin *et al.* 1962).

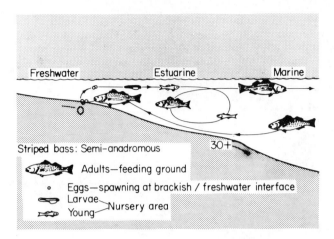

Fig. 26-20 Striped bass, one of the great estuarine fishes of the world, spawns close to or well into fresh water. Eggs and larvae drift downstream and mature within the system (Cronin and Mansueti 1971).

poses (Fig. 26-20). *Anadromous* fish, principally salmon, shad, striped bass, and some trout, come from the ocean into fresh-water streams; *catadromous* fishes, like the fresh-water eels, reverse the process. The chum salmon spends several years in the sea until it becomes sexually mature, then it ascends fresh-water streams to their cool, gravelly bottom headwaters to spawn, after which it dies. The Atlantic salmon, however, has a shorter developmental period and may spawn more than once. Tagged salmon and shad are commonly known to ascend the same streams in which they were hatched, and recent studies indicate that they recognize their home waters by chemical stimuli (smell), memory of which is retained from very young stages (Hasler 1954). Apparently salmon eggs must be laid in fresh water as otherwise the outer membranes do not harden and the number hatching is reduced (Black 1951). Furthermore, young salmon cannot tolerate sea water until they have developed chloride-secreting cells in the gills.

The adult female eel spends 5 to 20 years in the fresh-water streams that drain into the Atlantic Ocean in both the western hemisphere and in Europe. They often ascend these streams far into the interior of the continents. The male, however, remains in the brackish water of the bays and estuaries. It is here that mating takes place as the female returns to the sea to spawn. The journeys of the females have been difficult to follow but apparently both the American and European species spawn in the depths of the tropical sea north-

east of the West Indies. The females then die. It is here that the smallest immature forms occur in mixed populations in the open sea. It is still a mystery how the young of the two species become separated and get to their respective continents.

ZOOGEOGRAPHY

There have been several attempts to recognize taxonomic and geographic divisions in marine communities, beginning with Petersen (1914), but probably the best and most complete is that of Ekman (1953). Ekman divides marine life first of all into *faunas*, based partly on temperature and partly on geography, then into *regions* and *subregions* (Fig. 26-21). Subregions may be still further divided into *provinces* (Stephenson 1954). Families and genera that are endemic or restricted in their distribution have been most useful for distinguishing the major geographic divisions. These divisions are faunistic ones and each division may contain two or more of the biomes that we have just described.

Pelagic Biome

There is enough interchange of water between the different oceans to give considerable uniformity in the taxonomic composition of pelagic organisms. Ubiquitous species found in all oceans and in both equatorial and polar regions include species of siphonophores, ctenophores, polychaetes, copepods, chaetognaths, and amphipods.

The principal division of the epipelagic community is into a warm-water fauna lying between summer isotherms of 14°–15°C north and south of the equator and into Arctic and Antarctic faunas. A large number of species in the warm-water fauna are worldwide.

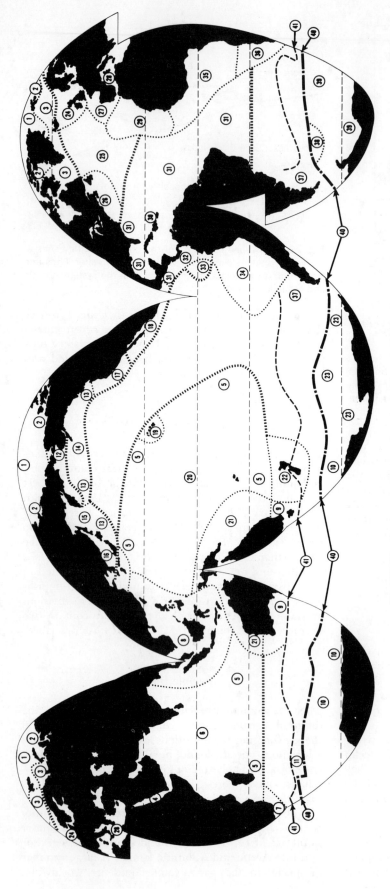

Fig. 26-21 Regions and subregions in the faunal distribution on the continental shelf. The anti-boreal and antarctic convergences represent areas where warm surface waters sink, because of their greater salt content, below the cold surface waters of the antarctics (prepared from information given by Ekman 1953).

1. Arctic
2. Polar-Arctic
3. Atlantic-Arctic
4. Red Sea
5. Indo-West Pacific
6. Indian Ocean
7. Cape

8. Indo-Malayan
9. South Australian
10. East Antarctic
11. Kerguelen
12. Northwest Pacific
13. North Pacific
14. Aleutian

15. Asiatic
16. Japanese
17. Californian
18. Gulf of California
19. Hawaiian
20. Central Pacific
21. Australian

22. New Zealand
23. Antarctic
24. European
25. North Atlantic
26. American
27. Lusitanian
28. Mediterranean

29. Mauritanian
30. West Indian
31. East Pacific-Atlantic
32. Panamic
33. Galapagos
34. Peruvian
35. Guinean

36. Namaquan
37. South American
38. High Antarctic
39. West Antarctic
40. Anti-Boreal Convergence
41. Antarctic Convergence

In general, the fauna of the Indian and Pacific Oceans is richer in species than that of the Atlantic Ocean. The arctic and antarctic faunas contain several characteristic and endemic species, some of which may at times become very numerous. The blue and fin whales of the antarctic have long been sought by man for their oil.

Inhabitants of the bathyal- and abyssal-benthos around the world show considerable uniformity with many genera of animals widely distributed. Doubtless this is due to the considerable uniformity of environmental conditions in the various ocean bottoms. Three major regions may be distinguished, however, the Atlantic, Pacific, and Antarctic (Vinogradov 1960).

Biomes of the Continental Shelf and Coral Reefs

The shelf fauna, made up of benthos and associated organisms, is divisible into warm-water, temperate, arctic, and antarctic regions. The poleward limits of the faunas are correlated with progressively lower temperatures and shorter growing seasons. Each fauna may in turn be subdivided into smaller units (Ekman 1953, Hall 1964, Salomonsen 1965).

The tropical subfauna is by far the richest in species and contains numerous endemic elements which do not penetrate extensively even into the subtropical zone. Coral reefs are found only in the tropics. The variety of forms making up each fauna becomes progressively less poleward. The tropics have been the center of origin, differentiation, and dispersal of these cold-blooded organisms, and species have invaded colder waters only as they have been able to acclimatize to them.

Many tropical genera and families are circum-tropical in distribution; that is, they are found in the Indian, Pacific, and Atlantic Oceans, although represented by different species in each area. There are, however, a few species that are also circumtropical, including the brittle-star (*Amphipholis squamata*), certain crabs (*Grapsus grapsus, Planes minutus, Plagusia depressa*), the hammerhead shark (*Zygaena malleus*), the porcupinefish (*Diodon hystrix*), and nearly all the marine turtles.

It is not possible here to describe the fauna found in the various regions and subregions or to analyze the interesting paleoecological history of each region. In general the *Indo-West Pacific region* and specifically the Indo-Malayan subregion have the greatest abundance, variety, and distinctness of animal life. This is expressed in many different taxonomic groups with significant percentages of families, genera, and species being exclusive to the region or subregion. This may represent the ancient profusion of forms that during the early Tertiary extended more or less around the world. This ancient fauna persisted here because the region was not subjected to the cooling of the climate

and waters that occurred elsewhere during late Tertiary and the Quaternary and which brought impoverishment of the fauna.

The West Indian subregion of the *Atlanto-East Pacific region* ranks next to the Indo-Malayan subregion in size and richness of fauna. The shelf fauna of the *North Pacific region* and adjacent Polar-Arctic, especially on the American side of the Pacific Ocean, is much richer than that of the *North Atlantic region*. Both the *Arctic* and *Antarctic regions* have a number of endemic forms, but in general the Antarctic fauna, especially of invertebrates, is much richer in species. There are a few species of crabs, *Cancer*, a starfish, *Ctendiscus crispatus*, and some other organisms that occur in both polar regions with continuous intermediate distribution. However, the species occur in shallow waters in the polar regions and only in the deeper cooler waters of the tropics. Various other species or related forms are found in the two opposite polar or temperate regions only, with presumably the interconnecting tropical linkage having become broken sometime during past geological time.

APPLIED ECOLOGY

Food Production

Although 71 per cent of the earth's surface is occupied by oceans and only 29 per cent by land, nearly all of the food and raw materials used by man is derived from the latter. This is in spite of the fact that agricultural soil is only a few inches thick and must be cultivated, protected from erosion, and fertilized, while the ocean with its chemical fertility, its photosynthetic production of basic plant food, and its great fisheries is an entirely natural resource ready to be harvested. However, fish harvesting in the ocean has definite limits determined by photoplankton primary productivity (Table 26-3). It is estimated that maximum sustained yield to man is of the order of 100–150 million metric tons per year, and in 1967, 60,000,000 tons were being harvested. The open ocean is relatively infertile, yet it covers most of the area; much more productive are the coastal areas, particularly the areas of upwelling, but these are of comparatively small size. A large part of the fish production is taken by predators other than man, and part of it must be left to maintain the breeding stock. The possibility of the ocean serving as an abundant future source of food for man is probably exaggerated. Seafood is important mostly as a supplementary source of protein. Man must look to other resources for his main energy requirements (Ryther 1969, Ricker 1969).

The plankton of the sea represents a possible food supply for man (Davis 1955). Its energy value is approximately 4 kcal/g dry weight, and it is more or less palatable (Clarke and Bishop 1948). However,

Table 26-3 Relative productivity of different parts of the ocean in metric tons. Note that transfer of nutrients and energy is calculated as passing through more links in the food chain before being assimilated by fish and the efficiency of transfer from one trophic level to another is calculated as lower in the open ocean than in the other two areas. These assumptions need verification before the absolute values of fish productivity can be fully accepted (Ryther 1969). (Copyright 1969 by the American Association for the Advancement of Science.)

Region	Per cent of total area	Primary productivity		Trophic levels in food chain	Ecological efficiency, I_{n+1}/I_n	Annual fish productivity (tons, fresh weight)
		g C/m²-yr	10⁹ tons C/yr			
Open ocean	90.0	50	16.3	5	10	1,600,000
Coastal waters	9.9	100	3.6	3	15	120,000,000
Upwelling areas	0.1	300	0.1	1.5	20	120,000,000

the expense involved in filtering and concentrating it is too great to be economical; poisonous species sometimes occur, it is not easily digested and assimilated, and consequently it has not as yet proved to be a feasible diet. The energy value of the plankton is used by man at the present time primarily as it is transferred into higher links of the food chain. Aside from the fishes, the chief marine organisms used as food are the oysters and other mollusks, shrimp, crabs, lobsters, and sea turtles. These are mostly animals of the continental shelf and estuaries.

There are a number of problems in the use and conservation of marine organisms. Natural beds of American oysters on the Atlantic coast have nearly all been exhausted through overfishing and pollution. Because of heavy erosion of the land, silt deposition has become excessive in most of the bays and estuaries and there is increasing difficulty for oyster spat to find clean hard surfaces on which to set. Surfaces that are loose or covered with silt are not suitable since the spat is very small and easily smothered. A common practice is to return to suitable areas all shells of mollusks removed or to introduce other suitable hard objects to furnish the necessary substratum for oyster setting. Control of erosion over the watershed would greatly alleviate the problem. The trend is increasing to lease suitable areas of water and to farm oysters in the manner of an agricultural crop (Korringa 1952).

In spite of their position at or near the top of the food chain, the greatest utilizable food resource of the sea is its fin fishes. The service of transferring the basic fertility of the sea through successive stages in the food cycle to fishes is performed by nature, and man needs only to harvest the final crop. The great bulk of commercial fish is in the families of herrings, codfish, salmons, flounders, and mackerels. Probably most kinds of fish are potentially useful, although some species that occur in coral reefs are inedible or poisonous. Tunas were not widely eaten in the United States until 1928; swordfish, once anathema to fishermen, are now as expensive as steak. Sharks were not used until a few years ago, but are now a major source of vitamins. Even rough fish or fish of small size may be ground into "fish flour," which has a high concentration of protein. The loss of elements from the ecosystem with the removal of fish is replaced by the continued inflow of nutrients from the land by way of the rivers. A fishery, temporarily exhausted, will usually become replenished by natural processes if left alone for a period of time. When one realizes that with the same expenditure of effort a man in a year's time can harvest two and one-half times as much edible fish as he can pork in pigs, it would appear that the ocean community is one that should be more extensively utilized (Taylor 1951, Walford 1958).

Ocean "farming" is largely limited to estuaries and salt marshes. Yields in the Orient occasionally reach or even surpass 1000 kg/hectare-yr (880 lb/acre-yr). Principal species are milkfish, gray mullets, eels, salmon, sea trout, tilapia, oysters, and shrimp. Conversion of marsh into ponds where water level, fertility, and cropping can be controlled has met with varying success in different parts of the world, but may be expected to expand as fish harvesting in the ocean itself approaches its yield limit (Pinchot 1970, Hickling 1970).

The large marine mammals will soon be lost for man's use unless drastic steps are soon taken. Whales have been pursued so vigorously for their oils that certain species are in danger of extinction. Seals have been taken extensively for their fur. International regulations have now been set up to limit the take of both groups of species.

Fouling of Ship Bottoms

The applied ecologist is also concerned in the fouling of ship bottoms by growth of organisms, particularly those belonging to the balanoid-gastropod-thallophyte biome. This is of major economic importance because of reduction of speed and the greater fuel consumption imposed on fouled ships. The problem has stimulated intensive studies of the behavior of the organisms concerned and the searching for chemicals or methods of treatment of ship bottoms to prevent their setting (Iselin 1952).

Oil Pollution

Oil production and consumption have greatly increased in recent years and with them pollution of large areas of the open ocean and adjacent shores as the result of oil leaks and wreckage of oil tankers. Oil must be transported over long distances of open ocean from places where it is obtained to places where it is consumed. Oil slicks may move with water currents far from the place of original contamination and may become concentrated in thick mats, killing vast numbers of plants and animals, including birds whose plumage becomes so badly soaked that they are unable to fly. The oil is subject to bacterial decomposition, but this requires time. Accelerated pollution may exceed the capacity for absorption into the ecosystem, in which case temporary depletion of marine life may become permanent destruction. If the sea is to take a larger share in food production for man, its productive capacity must not be allowed to deteriorate as the result of oil pollution or pollution from any other source (Cowell 1970, Nelson-Smith 1970). Estuaries and coastal waters are most exposed to pollution from other sources, since these wastes are derived from the land (Cronin 1971).

SUMMARY

Geographic distribution of marine organisms depends on their responses to current, temperature, and physical barriers; their local distribution is affected by waves and tides, type of bottom, salinity, and depth. Major divisions of the marine biocycle are pelagic (open water) and benthic (bottom). Major communities recognized are the oceanic plankton and nekton biome in the open sea, the balanoid-gastropod-thallophyte biome on rocky shores, the pelecypod-annelid biome on sand and mud bottoms, and the coral reef biome.

Organisms making up the oceanic biome are widely distributed around the world but may be divided into warm-water, arctic, and antarctic faunas. Coral reefs are found only in the tropics. The two biomes on the continental shelf subdivide into warm-water, temperate, arctic, and antarctic faunas and into more restricted regions and subregions. The warm-water faunas are richest in species, especially in the Indo-Malayan and West Indian subregions.

Marine plankton include a greater variety of forms than does fresh-water plankton, although rotifers are nearly absent; cladocerans, less important. They possess various unique mechanisms for flotation. Although abundance varies greatly from place to place and from season to season, plankton is generally much more numerous in neritic coastal waters than in the open sea. Diel movements between the surface at night and greater depths during the day are pronounced.

Mollusks (squids), fishes, birds, and mammals constitute the nekton. The taxonomic composition of the fish fauna varies with depth. Bioluminescence is exceptionally well developed among deep-sea nekton and benthos.

Benthos includes a great variety of sessile, creeping, and burrowing forms. It is very abundant in the littoral zone, and decreases in numbers with depth, although individuals are found in the deepest ocean trenches. There is considerable difference in the life-form and species composition of benthos occurring on rocky shores and on sand and muddy ones. Zonation of species is more prominent on rocky than on depositing shores. Succession and dominance occurs in some situations, but is less important than in terrestrial communities. Coral reefs have many special features, but are now exposed in the Pacific and Indian Oceans to damage by a species of starfish.

Food chains in the sea are similar to those in fresh water but the number of links and species composition varies in different communities. Productivity is especially high in regions where upwelling and turbulence bring nutrients from deeper levels up to the surface.

The three great biocycles of ocean, fresh-water, and land come into contact around the margins of the seas. The succession from the ocean to land is abrupt and from the ocean to fresh water through estuaries only slightly less so.

Effective use of the ocean resources as food for man is largely limited to the continental shelf, regions of upwelling, estuaries, and coastal marshes. Plankton is abundant and has high productivity in these regions but is likely to be useful to man only as its nutrient material and energy are transferred with considerable loss into the higher food chain links of mollusks, larger arthropods, fishes, and other vertebrates. Coastal regions need protection and management for food production and all waters kept clean from pollution of all sorts.

BIBLIOGRAPHY

Ackefors, Hans, Göran Löfroth, and Carl-Gustaf Rosén
  1970. A survey of the mercury pollution problem in Sweden with special reference to fish. Oceanogr. Mar. Biol. Ann. Rev. 8: 203–224.
Adams, Charles C.
  1905. The postglacial dispersal of the North American biota. Biol. Bull. 9: 53–71.
  1909. An ecological survey of Isle Royale, Lake Superior. Rep't B'd Geol. Surv. 1908, Lansing, Mich.: i-xiv, 1–468.
  1915. The variations and ecological distribution of the snails of the genus Io. Mem. Nat. Acad. Sci., Washington 12: 1–92.
Adamstone, Frank B.
  1924. The distribution and economic importance of the bottom fauna of Lake Nipigon. Univ. Toronto Stud., 24: 33–100.
Adamstone, Frank B. and W. J. K. Harkness
  1923. The bottom organisms of Lake Nipigon. Univ. Toronto Stud., 22: 123–170.
Aikman, J. M. and A. W. Smelser
  1938. The structure and environment of forest communities in central Iowa. Ecol. 19: 141–150.
Alabaster, J. S. and W. G. Hartley
  1962. The efficiency of a direct current electric fishing method in trout streams. J. An. Ecol. 31: 385–388.
Albrecht, William A.
  1944. Soil fertility and wildlife—cause and effect. Trans. 9th North Amer. Wildl. Conf.: 19–28.
  1956. Physical, chemical, and biochemical changes in the soil community. In W. L. Thomas, Jr.'s "Man's Role in Changing the Face of the Earth," University Chicago Press: 648–673.
Aldrich, John W.
  1943. Biological survey of the bogs and swamps of northeastern Ohio. Amer. Midl. Nat. 30: 346–402.
  1963. Geographic orientation of American Tetraonidae. J. Wildl. Man. 27: 529–545.
Aldrich, John W. and Allen J. Duvall
  1955. Distribution of American gallinacious game birds. U. S. Dept. Int., Fish Wildl. Serv., Circ. 34, Gov. Print. Off., Washington, D. C.: 1–23.
Aldrich, S. R.
  1972. Some effects of crop-production technology on environmental quality. Biosci. 22: 90–95.
Aleem, Anwar Abdel
  1956. Quantitative underwater study of benthic communities inhabiting kelp beds off California. Sci. 123: 183.
Alexander, Gordon
  1951. The occurrence of Orthoptera at high altitudes, with special reference to Colorado Acrididae. Ecol. 32: 104–112.
Alexander, Gordon and J. R. Hilliard, Jr.
  1969. Altitudinal and seasonal distribution of Orthoptera in the Rocky Mountains of northern Colorado. Ecol. Mono. 39: 385–431.
Alexander, Maurice M.
  1958. The place of aging in wildlife management. Amer. Sci. 46: 123–137.
Alexander, Richard D.
  1961. Aggressiveness, territoriality, and sexual behavior in field crickets (Orthoptera: Gryllidae). Behav. 17: 130–223.
  1962. The role of behavioral study in cricket classification. Syst. Zool. 11: 53–72.
Allard, Gilles O. and V. J. Hurst
  1969. Brazil-Gabon geologic link supports continental drift. Sci. 163 (3867): 528–532.
Allee, W. C.
  1912–13. An experimental analysis of the relation between physiolog-ical states and rheotaxis in Isopoda. J. Exp. Zool. 13: 269–344; 15: 257–295.
  1919. Note on animal distribution following a hard winter. Biol. Bull. 36: 96–104.
  1923. Studies in marine ecology. Biol. Bull. 44: 167–191, 205–253.
  1926. Distribution of animals in a tropical rainforest with relation to environment factors. Ecol. 7: 445–468.
  1931. Animal ecology of the tropical rain forest. Ohio St. Univ. Bull., 36: 95–102.
  1931. Animal aggregations. Univ. Chicago Press: ix + 431.
  1938. Cooperation among animals with human implications. Henry Schuman, New York: xiv + 233 (revised 1951).
  1940. Concerning the origin of sociality in animals. Scientia 67: 154–160.
  1951. Cooperation among animals. Henry Schuman, New York: 1–233.
Allee, W. C. and Edith S. Bowen
  1932. Studies in animal aggregations: mass protection against colloidal silver among goldfishes. J. Exp. Zool. 61: 185–207.
Allee, W. C., Alfred E. Emerson, Orlando Park, Thomas Park, and Karl P. Schmidt.
  1949. Principles of animal ecology. W. B. Saunders Co., Philadelphia: i-xii, 1–837.
Allee, W. C. and Peter Frank.
  1949. The utilization of minute food particles by goldfish. Physiol. Zool. 22: 346–358.
Allee, W. C. and Karl P. Schmidt.
  1951. Ecological animal geography. John Wiley & Sons, New York: i-xiii, 1–715.
Allee, W. C., and Janet Wilder
  1939. Group protection for Euplanaria dorotecephala from ultra-violet radiation. Physiol. Zool. 12: 110–135.
Allen, Durward L.
  1938. Ecological studies on the vertebrate fauna of a 500 acre farm in Kalamazoo County, Mich. Ecol. Mono. 8: 347–436.
Allen, Joel Asaph
  1892. The geographical distribution of North American mammals. Bull. Amer. Mus. Nat. Hist. 4: 199–243.
Allen, K. R.
  1935. The food and migration of the perch (Perca fluviatilis) in Windermere. J. An. Ecol. 4: 264–273.
Allred, Dorald M., D. E. Beck, and C. D. Jorgensen
  1963. Biotic communities of the Nevada test site. Brigham Young Univ. Sci. Bull., Biol. Ser. 2 (2): 1–52.
Altmann, Margaret
  1952. Social behavior of elk, Cervus canadensis Nelsoni, in the Jackson Hole area of Wyoming. Beh. 4: 116–143.
American Ornithologists' Union
  1957. Check-list of North American birds. Amer. Ornith. Union: i-xiii, 1–691.
American Public Health Association
  1955. Standard methods for the examination of water, sewage, and industrial wastes. New York: A.P.H.A., A.W.W.A., and F.S.I.W.A., 10th Ed.: 1–522.
Andersen, F. Søgaard
  1946. East Greenland lakes as habitats for chironomid larvae. Medd. Grønland 100 (10): 1–65.
Andersen, Johs
  1957. Studies in Danish hare population. 1. Population fluctuations. Danish Rev. Game Biol. 3: 85–131.
Anderson, Edgar
  1948. Hybridization of the habitat. Evol. 2: 1–9.

1949. Introgressive hybridization. John Wiley & Sons, New York: i–ix, 1–109.

1960. The evolution of domestication. In S. Tax's "The evolution of man", Univ. of Chicago Press: 67–84.

Anderson, N. H. and D. M. Lehmkuhl
1968. Catastrophic drift of insects in a woodland stream. Ecol. 49: 198–206.

Anderson, Richard O. and Frank F. Hooper
1956. Seasonal abundance and productivity of littoral bottom fauna in a southern Michigan Lake. Trans. Amer. Micro. Soc., 75: 259–270.

Anderson, Wallace L.
1950. Biology handbook. U. S. Dept. Agric. Soil Cons. Serv. (3rd ed.): 1–121.

Andrewartha, H. G.
1961. Introduction to the study of animal populations. Univ. Chicago Press: i–xvii, 1–281.

Andrewartha, H. G. and L. C. Birch
1954. The distribution and abundance of animals. Univ. Chicago Press, Chicago, Ill.: i–xv, 1–782.

Andrews, Harry L.
1945. The kelp beds of the Monterey region. Ecol. 26: 24–37.

Andrzejewska, Lucyna
1965. Stratification and its dynamics in meadow communities of *Auchenorrhyncha* (Homoptera). Ekol. Polska, A, 13 (31): 685–715.

Andrzejewska, L., A. Breymeyer, A. Kajak, and Z. Wojcik
1967. Experimental studies on trophic relationships of terrestrial invertebrates. In K. Petrusewicz's "Sec. Prod. Terr. Ecosystems," Inst. Ecol., Polish Acad. Sci., Warsaw: 477–495.

Andrzejewska, Lucyna and Zdzislawa Wojcik
1970. The influence of Acridoidea on the primary production of a meadow (field experiment). Ekol. Polska, A, 18(5): 89–109.

Antevs, Ernst
1929. Maps of the Pleistocene glaciations. Bull. Geol. Soc. Amer. 40: 631–720.

Anthrop, Donald F.
1969. Environment noise pollution: a new threat to sanity. Bull. Atomic Scient. 25: 11–16.

Ardrey, Robert
1967. The territorial imperative. Atheneum: New York. i–xii + 1–390.

Armelagos, George J. and John R. Dewey
1970. Evolutionary response to human infectious diseases. BioSci. 20: 271–275.

Ashby, W. Ross
1956. An introduction to cybernetics. John Wiley and Sons Inc., N. Y.: i–ix + 1–295.

Ashton, P. S.
1969. Speciation among tropical forest trees: some deductions in the light of recent evidence. J. An. Ecol. 38, 1969: 4p.

Atwood, Sanford S. (Chairman)
1970. Land use and wildlife resources. Nat. Acad. Sci., Committee Agr. Land Use and Wildl. Res., Washington, D. C.: i–vi + 1–262.

Auclair, Allan N. and F. Glenn Goff
1971. Diversity relations of upland forests in the western Great Lakes area. Amer. Nat. 105: 499–528.

Austin, Mary L.
1948. The killing action and rate of production of single particles of paramecin 51. Physiol. Zool. 21: 69–86.

Austin, Oliver L. and Oliver L. Austin, Jr.
1956. Some demographic aspects of the Cape Cod population of common terns (*Sterna hirundo*). Bird-Banding 27: 55–66.

Axelrod, Daniel I.
1950. Studies in late tertiary paleobotany. Carnegi Inst. Wash. Publ. 590: 1–322.

1952. A theory of angiosperm evolution. Evol. 6: 29–60.

1957. Late Tertiary floras and the Sierra Nevadan uplift. Bull. Geol. Soc. Amer. 68: 19–45.

1958. Evolution of the madro-tertiary geoflora. Bot Rev. 24: 433–509.

1960. The evolution of flowering plants. In S. Tax's "Evolution after Darwin." I: 227–305.

1970. Mesozoic paleogeography and early Angiosperm history. Bot. Rev. 36: 277–319.

Ayala, Francisco J.
1968. Environmental factors limiting the productivity and size of experimental populations of *Drosophila serrata* and *D. birchii*. Ecol. 49: 562–565.

1969. Experimental invalidation of the principle of competitive exclusion. Nat. 224: 1076–1079.

Ayres, Robert U. and Allen V. Kneese
1971. Economic and ecological effects of a stationary economy. An. Rev. Ecol. Syst. 2: 1–22.

Backus, R. H. & H. Barnes
1957. Television-echo sounder observations of mid-water sound scatters. Deep-Sea Res. 4 (2): 116–119.

Baer, Jean G.
1951. Ecology of animal parasites. Univ. Illinois Press, Urbana: i–x, 1–224.

Bagenal, T. B.
1951. A note on the papers of Elton and Williams on the generic relations of species in small ecological communities. J. An. Ecol. 20: 242–245.

Bailey, Alfred M.
1948. The birds of Arctic Alaska. Colorado Mus. Nat. Hist., Pop. Ser. 8: 1–317

Bailey, Irving W.
1920. Some relations between ants and fungi. Ecol. 1: 174–189.

Bailey, Reeve M.
1960. A list of common and scientific names of fishes from the United States and Canada. Waverly Press, Baltimore: 1–102.

Bajema, Carl Jay
1971. The genetic implications of population control. BioSci. 21: 71–75.

Baker, Frank Collins
1918. The productivity of invertebrate fish food on the bottom of Oneida Lake, with special reference to mollusks. New York St. Coll. For., Tech. Publ. 9: 1–264.

1919. The ecology of North American Lymnaeidae. Sci. 49: 519–521.

1920. The life of the Pleistocene or glacial period. Univ. Illinois Bull. 17 (41): i–xiv, 1–476.

Baker, John R., Ina Baker, Zita Baker, T. F. Bird and T. H. Harrison
1936. The seasons in a tropical rain forest (New Hebrides). J. Linn. Soc. London 39: 443–463, 507–519: 40: 123–161.

Baker, Rollin H.
1956. Mammals of Coahuila, Mexico. U. Kansas Publ., Mus. Nat. Hist. 9 (7): 125–335.

Balda, Russell P.
1967. Ecological relationships of the breeding birds of the Chiracahua Mountains, Arizona. Univ. Illinois, Ph.D. thesis.

1969. Foliage use by birds of the oak-juniper woodland and ponderosa pine forest in southeastern Arizona. Condor 71: 399–412.

Balduf, Walter Valentine
1935. The bionomics of entomophagous Coleoptera. Pt. I, John S. Swift Co., Chicago: 1–220.

1939. The bionomics of entomophagous insects. Pt. II, John. S. Swift Co., St. Louis: 1–384.

Baldwin, Paul H.
1953. Annual cycle, environment and evolution in the Hawaiian honeycreepers (Aves: Drepaniidae). Univ. California Publ. Zool. 52: 285–398.

Baldwin, S. Prentiss
1919. Bird-banding by means of systematic trapping. Abs. Proc. Linnean Soc. New York 31: 25–56.

Ball, Robert C.
1948. Relationship between available fish food, feeding habits of fish and total fish production in a Michigan lake. Michigan St. Coll. Agr. Exp. Sta. Tech. Bull. 206: 1–59.

Ball, Robert C. and Don W. Hayne
1952. Effects of the removal of the fish population on the fishfood organisms of a lake. Ecol. 33: 41–48.

Ballantine, Dorothy
1953. Comparison of the different methods of estimating nanno-plankton. J. Mar. Biol. Assoc. U. K. 32: 129–147.

Balogh, Janos
1958. Lebensgemeinschaften der Landtiere. Hungarian Acad. Sci., Budapest: 1–560.

Bamberg, Samuel A. and Jack Major
1968. Ecology of the vegetation and soils associated with calcareous parent materials in three alpine regions of Montana. Ecol. Mono. 38: 127–167.

Banfield, A. W. F.
1947. A study of the winter feeding habits of the shorteared owl (*Asio flammeus*) in the Toronto region. Canad. J. Res. D. 25: 45–65.

1951. Notes on the mammals of the Mackenzie District Northwest territories. Arctic 4: 113–121.

1963. The post glacial dispersal of American caribou. Proc. 16th Inter. Cong. Zool. 1: 206.

Bardach, John E.
  1968. Aquaculture. Sci. 161 (3846): 1098–1106.
Barick, F. B.
  1950. The edge effect of the lesser vegetation of certain Adirondack forest types with particular reference to deer and grouse. Roosevelt Wildl. Bull. 9: 1–146.
Barker, Roy J.
  1958. Notes on some ecological effects of DDT sprayed on elms. J. Wildl. Man. 22: 269–274.
Barlow, George W.
  1961. Causes and significance of morphological variation in fishes. Syst. Zool. 10: 105–117.
Barnes, H. and H. T. Powell
  1950. The development, general morphology and subsequent elimination of barnacle populations, *Balanus crenatus* and *B. balanoides*, after a heavy initial settlement. J. An. Ecol. 19: 175–179.
Bateman, A. J.
  1950. Is gene dispersion normal? Hered. 4: 353–363.
Bates, Henry Walter
  1864. The naturalist on the river Amazons. John Murray, London: i-xii, 1–466.
Bates, Marston
  1949. The natural history of mosquitoes. Macmillan Co., New York: i-xv, 1–379.
Bateson, William
  1894. Materials for the study of variation. Macmillan Co., New York: i-xv, 1–598.
Baum, Werner A. and James M. Havens
  1956. Recent climatic fluctuations in maritime provinces. Trans. 21st No. Amer. Wildl. Conf.: 436–453.
Baumgartner, Luther L.
  1938. Population studies of the fox squirrel in Ohio. Trans. 3rd No. Amer. Wildl. Conf.: 685–689.
Baumgras, Philip S.
  1943. Winter food productivity of agricultural land for seed-eating birds and mammals. J. Wildl. Man. 7: 13–18.
  1944. Experimental feeding of captive fox squirrels. J. Wildl. Man. 8: 296–300.
Baumhover, A. H., A. J. Graham, B. A. Bitter, D. E. Hopkins, W. D. New, F. H. Dudley, and R. C. Bushland
  1955. Screw-worm control through release of sterilized flies. J. Econ. Ent. 48: 462–466.
Beard, Elizabeth B.
  1953. The importance of beaver in waterfowl management at the Seney National Wildlife Refuge. J. Wildl. Man. 17: 398–436.
Beard, J. S.
  1953. The savanna vegetation of northern tropical America. Ecol. Mono. 23: 149–215.
  1955. A note on gallery forests. Ecol. 36: 339–340.
  1955a. The classification of tropical American vegetation types. Ecol. 36: 89–100.
Beatley, Janice C.
  1956. The winter-green herbaceous flowering plants of Ohio. Ohio J. Sci. 56: 349–377.
Beauchamp, R. S. A. and P. Ullyott
  1932. Competitive relationships between certain species of fresh-water triclads. J. Ecol. 20: 200–208.
Beaufort, L. F. de
  1951. Zoogeography of the land and inland waters. Sedgwick and Jackson, London: i-viii, 1–208.
Beaver, R. A.
  1967. The regulation of population density in the bark beetle *Scolytus scolytus* (F). J. An. Ecol. 36: 435–451.
Beck, Stanley D.
  1965. Resistance of plants to insects. Ann. Rev. Ent. 10: 207–232.
Beckel, D. K. Brown
  1957. Studies on seasonal changes in the temperature gradient of the active layer of soil at Fort Churchill, Manitoba. Arctic 10: 151–183.
Beckwith, Stephen L.
  1954. Ecological succession on abandoned farm lands and its relation to wildlife management. Ecol. Mono. 24: 349–376.
Bee, James W.
  1958. Birds found on the Arctic slope of northern Alaska. Univ. Kansas Publ. 10: 163–211.
Beebe, William and Jocelyn Crane
  1947. Ecology of Rancho Grande, a subtropical cloud forest in northern Venezuela. Zoologica 32: 43–60.

Beebe, William, G. Innes Hartley and Paul G. Howes
  1917. Tropical Wild Life in British Guiana. N. Y. Zool. Soc.: i-xx, 1–504.
Beecher, William J.
  1942. Nesting birds and the vegetation substrate. Chicago Ornith. Soc., Chicago Mus. Nat. Hist.: 1–69.
Beeson, Kenneth C., Victor A. Lazar and Stephen G. Boyce
  1955. Some plant accumulators of the micronutrient elements. Ecol. 36: 155–156.
Behney, W. H.
  1937. Food organisms of some New Hampshire trout streams. New Hampshire Fish and Game Dept., Surv. Rept. 2: 77–80.
Bellis, Edward C.
  1965. Home range and movements of the wood frog in a northern bog. Ecol. 46: 90–98.
Bellrose, Frank C. and Clair T. Rollings.
  1949. Wildlife and fishery values of bottomland lakes in Illinois. Biol. Notes, Illinois Nat. Hist. Surv. 21: 1-24.
Belt, Thomas
  1873. The naturalist in Nicaragua. J. M. Dent & Sons, London; i-xxxiv, 1–306.
Bendell, J. F.
  1955. Disease as a control of a population of blue grouse, *Dendragapus obscurus fuliginosus* (Ridgway). Can. J. Zool. 33: 195–223.
Bennett, George W.
  1962. Management of artificial lakes and ponds. Reinhold Pub. Corp., New York: i-xvii + 1–283.
Bennitt, Rudolf and Werner O. Nagel
  1937. A survey of the resident game and furbearers of Missouri. Univ. Missouri Stud. 12 (2): 1–215.
Benson, Lyman and Robert Darrow
  1944. A manual of southwestern desert trees and shrubs. Univ. Arizona Biol. Sci. Bull. 6: 1–411.
Berg, Kaj
  1948. Biological studies on the river Susaa. Folia Limn. Scand. 4: 1–203, 285–318.
  1951. On the respiration of some mollusks from running and stagnant water. Année Biol. 27: 561–567.
Berg, L. S.
  1950. Natural regions of the U.S.S.R. Macmillan Co., New York: i-xxxi, 1–436.
Bergmann, C.
  1847. Ueber die Verhältnisse der Wärmeökonomie der Thiere zu ihrer Grösse. Gottinger Studien 3: 595–708.
Berndt, Rudolf and Helmut Sternberg
  1968. Terms, studies and experiments on the problems of bird dispersion. Ibis 110: 256–269.
Berry, Edward W.
  1937. Tertiary floras of eastern North America. Bot. Rev. 3: 31–46.
Betts, Monica M.
  1954. Experiments with an artificial nestling. British Birds 47: 229–231.
  1955. The food of titmice in oak woodland. J. An. Ecol. 24: 282–323.
Beverton, R. J. H. and S. J. Holt
  1957. On the dynamics of exploited fish populations. Fish. Invest., Ser. II. 19: 1–533.
Beyers, Robert J.
  1963. The metabolism of twelve aquatic laboratory microecosystems. Ecol. Mono. 33: 281–306.
  1964. The microcosm approach to ecosystem biology. Amer. Biol. Teacher 26: 491–498.
Bigelow, Henry B.
  1926. Plankton of the offshore waters of the Gulf of Maine. Bull. Bur. Fish. 40, Pt. II, Doc. 968: 1–509.
Bigelow, N. K.
  1928. The ecological distribution of microscopic organisms in Lake Nipigon. Univ. Toronto Stud. 35: 57–74.
Billings, W. D.
  1949. The shadscale vegetation zone of Nevada and eastern California in relation to climate and soils. Amer. Midl. Nat. 42: 87–109.
  1951. Vegetational zonation in the great basin of western North America. C. R. Colloque: 101–122.
Birch, L. C.
  1948. The intrinsic rate of natural increase of an insect population. J. An. Ecol. 17: 15–26.
  1953. Experimental background to the study of the distribution and abundance of insects. III. Evol. 7: 136–144.
Birch, L. C. and D. P. Clark
  1953. Forest soil as an ecological community with special reference to the fauna. Quart. Rev. Biol. 28: 13–36.

Bird, Ralph D.
    1930. Biotic communities of the aspen parkland of central Canada. Ecol. 11: 356–442.
Birge, Edward A.
    1904. The thermocline and its biological significance. Trans. Amer. Micros. Soc. 25: 5–33.
Birge, Edward A. and Chancey Juday.
    1911. The dissolved gases of the water and their biological significance. Wisconsin Geol. Nat. Hist. Surv., 22: i–xx, 1–259.
    1929. Transmission of solar radiation by the waters of inland lakes. Trans. Wisconsin Acad. Sci. Arts. Lett., 24: 509–580.
    1934. Particulate and dissolved organic matter in inland lakes. Ecol. Mono. 4: 440–474.
Bishop, John E.
    1969. Light control of aquatic insect activity and drift. Ecol. 50: 371–380.
Black, Joe B.
    1963. Observations on the home range of stream-dwelling crawfishes. Ecol. 44: 592–595.
Black, Virginia Safford
    1951. Osmotic regulations in teleost fishes. Publ. Ontario Fish. Res. Lab., 71; Biol. Ser. 59: 53–89.
Blackman, F. F.
    1905. Optima and limiting factors. Ann. Bot. 19: 281–295.
Blair, W. Frank
    1939. Some observed effects of stream-valley flooding on mammalian populations in eastern Oklahoma. J. Mam. 20: 304–306.
    1943. Populations of the deer-mouse and associated small mammals in the mesquite association of southern New Mexico. Cont. Lab. Vert. Biol., 21: 1–40.
    1951. Interbreeding of natural populations of vertebrates. Amer. Nat. 85: 9–30.
    1951a. Population structure, social behavior, and environmental relations in a natural population of the beach mouse (Peromyscus polionotus leucocephalus). Cont. Lab. Vert. Biol., Ann Arbor, 48: 1–47.
    1953. Population dynamics of rodents and other small mammals. Adv. Gen. 5: 1–41.
    1954. Mammals of the mesquite plains biotic district in Texas and Oklahoma, and speciation in the central grasslands. Texas J. Sci. 6: 235–264.
    1955. Size difference as a possible isolation mechanism in Microhyla. Amer. Nat. 89: 297–300.
    1956. The mating calls of hybrid toads. Texas J. Sci. 8: 350–355.
    1957. Changes in vertebrate populations under conditions of drought. Cold Spring Harbor Sym. Quant. Biol. 22: 273–275.
    1958. Distributional patterns of vertebrates in the southern United States in relation to past and present environments. In C. L. Hubbs "Zoogeography", A.A.A.S.: 433–468.
    1962. Non-morphological data in anuran classification. Syst. Zool. 11: 72–84.
Blake, Irving Hill
    1926. A comparison of the animal communities of coniferous and deciduous forests. Illinois Biol. Mon. 10: 1–149.
    1945. An ecological reconnoissance in the Medicine Bow Mountains. Ecol. Mono., 15: 207–242.
Blatchley, W. S.
    1920. Orthoptera of northeastern America. Nature Publishing Co., Indianapolis: 1–784.
Blegvad, H.
    1914. Food and conditions of nourishment among the communities of invertebrate animals found on or in the sea bottom in Danish waters. Rep. Danish Biol. Sta. 22: 41–78.
Bliss, L. C.
    1956. A comparison of plant development in microenvironments of arctic and alpine tundras. Ecol. Mono. 26: 303–337.
    1962. Adaptations of arctic and alpine plants to environmental conditions. Arctic 15: 117–144.
    1963. Alpine plant communities of the presidential range, New Hampshire. Ecol. 44: 678–697.
Bloom, Arthur L. and Minze Stuiver
    1963. Submergence of the Connecticut Coast. Sci. 139 (3552): 332–334.
Blumenstock, David I. and C. Warren Thornthwaite
    1941. Climate and the world pattern. Climate and man. Yearb. Agri., U. S. Dept. Agri., Washington, D. C.: 98–127.
Bobek, Boguslaw
    1969. Survival, turnover and production of small rodents in a beech forest. Acta Theriol. 14: 191–210.

Bock, Walter J.
    1959. Preadaptation and multiple evolutionary pathways. Evol. 13: 194–211.
Boden, Brian P. and Elizabeth M. Kampa
    1964. Planktonic bioluminescence. Oceanogr. Mar. Biol. Ann. Rev. 2: 341–371.
Bodenheimer, F. S.
    1937. Population problems of social insects. Biol. Rev., 12: 393–430.
    1938. Problems of animal ecology. Oxford Univ. Press, London: i–vi, 1–183.
    1953. Problems of animal ecology and physiology in deserts. Desert Res. Proc., Jerusalem Post Press: 205–229.
    1957. The ecology of mammals in arid zones. Arid Zone Res. VIII, UNESCO, Paris, France: 100–137.
Boerma, Addeke H.
    1970. A world agricultural plan. Sci. Amer. 223 (2): 54–69.
Bogert, Charles M.
    1947. A field study of homing in the Carolina toad. Amer. Mus. Nov. 1355: 1–24.
Boggess, W. R.
    1956. Amount of throughfall and stemflow in a shortleaf pine plantation as related to rainfall in the open. Trans. Illinois Acad. Sci. 48: 55–61.
Boggs, S. W.
    1931. Seasonal variations in daylight, twilight and darkness. Geog. Rev. 21: 656–659.
Bogorov, B. G.
    1958. Perspectives in the study of seasonal changes of plankton and of the number of generations at different latitudes. In Buzzati-Traverso's "Perspectives in marine biology." Univ. California Press: 145–158.
Bond, Richard M.
    1945. Range rodents and plant succession. Trans. 10th No. Am. Wildl. Conf.: 229–234.
Bonner, James
    1962. The upper limit of crop yield. Sci. 137 (3523): 11–15.
Borchert, John R.
    1950. The climate of the central North American grassland. Ann. Assoc. Amer. Geog. 40: 1–39.
Bormann, F. H.
    1970. Subtraction by multiplication. Bull. Ecol. Soc. Amer. 51 (2): 3–10.
Bormann, F. H., G. E. Likens and J. S. Eaton
    1969. Biotic regulation of particulate and solution losses from a forest ecosystem. BioSci. 19: 600–610.
Bornebusch, C. H.
    1930. The fauna of forest soil. Nielsen and Lydiche, Copenhagen: 1–224.
Borutsky, E. V.
    1939. Dynamics of the total benthic biomass in the profundal of Lake Beloie. Proc. Kossino Limnol. Stat. Hydrometeor. Serv. U.S.S.R. 22: 196–218 (Engl. transl. Michael Ovchynnyk).
Botkin, Daniel B., G. M. Woodwell and N. Tempel
    1970. Forest productivity estimated from carbon dioxide uptake. Ecol. 51: 1057–1060.
Boughey, A. S.
    1963. Interaction between animals, vegetation and fire in southern Rhodesia. Ohio J. Sci. 63: 193–209.
    1968. Ecology of populations. The Macmillan Co., New York: i–viii + 1–135.
    1971. Man and the environment. The Macmillan Co., New York: i–viii + 1–472.
Bourlière, F.
    1946. Longévité moyenne et longévité maximum chez les Vertébrés. Ann. Biol. 3rd Ser., 22: 249–270.
    1959. Lifespans of mammalian and bird populations in nature. Ciba Found. Sym. "The Lifespan of Animals": 90–103.
Bourlière, Francois and Malcolm Hadley
    1970. The ecology of tropical savannas. An. Rev. Ecol. Syst. 1: 125–152.
Bovbjerg, Richard V.
    1952. Comparative ecology and physiology of the crayfish Orconectes propinquus and Cambarus fodiens. Phys. Zool. 25: 34–56.
Bowers, Darl E.
    1960. Correlation of variation in the wrentit with environmental gradients. Condor 62: 91–120.
    1964. Natural history of two beach hoppers of the genus Orchestoidea (Crustacea: Amphipoda) with reference to their complemental distribution. Ecol. 45: 677–696.

Bowman, Robert I.
  1961. Morphological differentiation and adaptation in the Galápagos finches. Univ. California Publ. Zool. 58: i–vii + 1–326.
Box, Thadis W., J. Powell, and D. L. Drawe
  1967. Influence of fire on south Texas chaparral communities. Ecol. 48: 955–961.
Boycott, A. E.
  1934. The habitats of land Mollusca in Britain. J. Ecol. 22: 1–38.
  1936. The habitats of fresh-water Mollusca in Britain. J. An. Ecol. 5: 116–186.
Boyden, Alan
  1954. The significance of asexual reproduction. Syst. Zool. 3: 26–37, 47.
Brady, F.
  1943. The distribution of the fauna of some intertidal sands and muds on the Northumberland Coast. J. An. Ecol. 12: 27–41.
Braestrup, F. W.
  1940. The periodic die-off in certain herbivorous mammals and birds. Sci. 92: 354–355.
  1941. A study on the Arctic fox in Greenland. Meddelelser om Grønland 131 (4): 1–101.
Bragg, Arthur N. and Charles Clinton Smith
  1943. Observations on the ecology and natural history of Anura IV. Ecol. 24: 285–309.
Bragg, Arthur N. *et al.*
  1950. Researches on the amphibia of Oklahoma. Univ. Okla. Press, Norman: 1–154.
Bramble, William C. and Roger H. Ashley
  1955. Natural revegetation of spoil banks in central Pennsylvania. Ecol. 36: 417–423.
Braun, E. Lucy
  1950. Deciduous forests of eastern North America, Blakiston Co., Philadelphia: i–xiv, 1–596.
Braun-Blanquet, J.
  1932. Plant sociology. McGraw-Hill Book Co., New York: i–xviii, 1–439 (English translation).
Bray, J. Roger
  1961. Measurement of leaf utilization as an index of minimum level of primary consumption. Oikos 12: 70–74.
  1964. Primary consumption in three forest canopies. Ecol. 45: 165–167.
Bray, J. R. and E. Gorham
  1964. Litter production in forests of the world. Adv. Ecol. Res. 2: 101–158.
Bray, R. J. and R. E. Loughhead
  1965. Sunspots. John Wiley & Sons Inc. i–xvi + 1–303.
Breder, C. M., Jr.
  1936. The reproductive habits of North American sunfishes (family Centrarchidae). Zoologica 21: 1–48.
Brehm, Alfred Edmund
  1896. From North Pole to Equator: studies of wild life and scenes in many lands. Blackie & Sons, London: i–xxxi, 1–592 (English translation).
Breidenbach, Andrew W. and R. W. Eldredge
  1969. Research and development for better solid waste management. BioSci. 19: 984–988.
Brereton, J. L.
  1957. The distribution of woodland isopods. Oikos 8: 85–106.
Brewer, Richard
  1960. A brief history of Ecology. Part 1—pre-nineteenth century to 1919. Occ. Papers of the C. C. Adams Center Ecol. Stud., 1: 1–18.
  1967. Bird populations of bogs. Wilson Bull. 79: 371–396.
Breymeyer, Alicja
  1966. Relations between wandering spiders and other epigeic predatory Anthropoda. Ekol. Polska, A, 14 (2): 27–71.
Brian, M. V.
  1952. The structure of a dense natural ant population. J. An. Ecol. 21: 12–24.
  1956. Exploitation and interference in interspecific competition. J. An. Ecol. 25: 339–347.
Briscoe, M. S.
  1952. The relation of insects and insect-borne diseases to the vegetation and environment in Liberia. Ecol. 33: 187–214.
Britton, Max E.
  1957. Vegetation of the Arctic tundra. 18th Biol. Colloq. Oregon St. College: 26–61.
Broadhead, Edward and Ian W. B. Thornton
  1955. An ecological study of three closely related psocid species. Oikos, 6: 1–50.

Brody, Samuel
  1945. Bioenergetics and growth. Reinhold Publishing Corp., New York (chap. 16).
  1952. Facts, fables, and fallacies on feeding the world population. Fed. Proc. 11: 681–693.
Broecker, Wallace S.
  1965. Isotope geochemistry and the Pleistocene climatic record. In Wright, H. E., Jr. and D. G. Frey's "The Quaternary of the United States", Princeton Univ. Press: 737–753.
  1970. Man's oxygen reserves. Sci. 168 (3939): 1537–1538.
Brooks, C. E. P.
  1949. Climate through the ages. Ernest Benn, London: 1–395.
  1951. Geological and historical aspects of climatic change. Compendium of Meteor., Boston, Mass: 1004–1018.
Brooks, John Langdon
  1946. Cyclomorphosis in *Daphnia*. Ecol. Mono. 16: 409–447.
Brooks, John L. and S. I. Dodson
  1965. Predation, body size, and composition of plankton. Sci. 150 (3692): 28–35.
Brooks, Maurice
  1940. The breeding warblers of the central Allegheny Mountain region. Wilson Bull. 52: 249–266.
  1943. Birds of the Cheat Mountains. Cardinal 6: 25–45.
  1955. An isolated population of the Virginia varying hare. J. Wildl. Man. 19: 54–61.
Brooks, Ronald J. and Edwin M. Banks
  1971. Radio-tracking study of lemming home range. Comm. Beh. Biol. 6: 1–5.
Brower, Lincoln Pierson
  1969. Ecological chemistry. Sci. Amer. 220 (2): 22–29.
Brown, C. J. D. and Robert C. Ball
  1942. An experiment in the use of derris root (rotenone) on the fish and fish-food organisms of Third Sister Lake. Trans. Amer. Fish Soc. 72: 267–284.
Brown, Charles R. and Melville H. Hatch
  1929. Orientation and "fright" reactions of whirligig beetles (Gyrinidae). J. Comp. Psych. 9: 159–189.
Brown, Frank A., J. W. Hastings, and J. D. Palmer
  1970. The biological clock—two views. Academic Press, N. Y.: i–viii + 1–194.
Brown, G. W., Jr.
  1968. Desert biology. Academic Press, N. Y.: i–xvii + 1–635.
Brown, Harold W. and Minna E. Jewell
  1926. Further studies on the fishes of an acid lake. Trans. Amer. Micros. Soc. 45: 20–34.
Brown, James H.
  1971. Mechanisms of competitive exclusion between two species of chipmunks. Ecol. 52: 305–311.
Brown, Jerram L.
  1969. The buffer effect and productivity in tit populations. Amer. Nat. 103: 347–354.
Brown, Jerram L. and Gordon H. Orians
  1970. Spacing patterns in mobile animals. An. Rev. Ecol. Syst. 1: 239–262.
Brown, Larry N.
  1965. Selection in a population of house mice containing mutant individuals. J. Mam., 46: 461–465.
Brown, Lester R.
  1967. The world outlook for conventional agriculture. Sci. 158: 604–611.
Brown, R. T. and J. T. Curtis
  1952. The upland conifer-hardwood forests of northern Wisconsin. Ecol. Mono. 22: 217–234.
Brown, Robert Zanes
  1953. Social behavior, reproduction and population changes in the house mouse (*Mus musculus* L.). Ecol. Mono. 23: 217–240.
Brown, William L., Jr.
  1960. Ants, acacias and browsing mammals. Ecology 41 (3): 587–592.
Brown, W. L., Jr. and E. O. Wilson
  1956. Character displacement. Syst. Zool. 5: 49–64.
Browne, Charles A.
  1942. Liebig and the law of the minimum. Liebig and after Liebig, Publ. AAAS 16: 71–82.
Brues, Charles T.
  1924. The specificity of food-plants in the evolution of phytophagous insects. Amer. Nat., 58: 127–144.
Bryson, Reid A.
  1966. Air masses, streamlines, and the boreal forest. Geog. Bull. 8: 228–269.

Bryson, Reid A., David A. Baerreis, and Wayne M. Wendland
1970. In W. Dort, Jr. and J. K. Jones, Jr.'s "Pleistocene and recent environments of the central Great Plains," Univ. Press of Kansas, Lawrence: 53–74.

Bryson, Reid A., W. N. Irving and J. A. Larsen
1965. Radiocarbon and soil evidence of former forest in the southern Canadian tundra. Sci. 147 (3653): 46–48.

Buchner, Paul
1953. Endosymbiose der Tiere mit pflanzlichen Mikroorganismen. Birkhäuser, Basel: 1–771.

Buck, D. Homer
1956. Effects of turbidity on fish and fishing. Trans. 21st No. Amer. Wildl. Conf.: 249–261.

Buck, D. Homer and C. F. Thoits III
1965. An evaluation of Petersen estimation procedures employing seins in 1-acre ponds. J. Wildl. Man. 29: 598–621.

Buckner, C. H.
1966. The role of vertebrate predators in the biological control of forest insects. Ann. Rev. Ent. 11: 449–470.

Buechner, Helmut K.
1960. The bighorn sheep in the United States, its past, present, and future. Wildl. Mono. 4: 1–174.

Buechner, H. K. and H. C. Dawkins
1961. Vegetation change induced by elephants and fire in Murchison Falls National Park, Uganda. Ecol. 42: 752–766.

Bullock, Theodore Holmes
1955. Compensation for temperature in the metabolism and activity of poikilötherms. Physiol. Rev. 30: 311–342.

Bump, Gardiner, Robert W. Darrow, Frank C. Edminster and Walter F. Crissey
1947. The ruffed grouse, life history, propagation, management. New York St. Cons. Dept., Albany: i–xxxvi, 1–915.

Burckhardt, Dieter
1953. Spielen die Raubvögel eine Rolle als "Gesundheitspolizei"? Ornith. Beob. 50: 149–152.

Burger, J. Wendell
1949. A review of experimental investigations of seasonal reproduction in birds. Wilson Bull. 61: 211–230.

Burger, W. Leslie
1950. Novel aspects of the life history of ambystomas. J. Tennessee Acad. Sci. 25: 252–257.

Burges, A. and F. Raw (editors)
1967. Soil biology. Academic Press, New York: i–x + 1–532.

Burkenroad, Martin D.
1946. Fluctuations in abundance of marine animals. Sci. 103: 684–686.

Burkholder, Paul R.
1952. Cooperation and conflict among primitive organisms. Amer. Sci. 40: 601–631.

Burnett, Thomas
1949. The effect of temperature on an insect host-parasite population. Ecol. 30: 113–134.

Burns, George W.
1958. Wisconsin age forests in western Ohio. II. Vegetation and burial conditions. Ohio J. Sci. 58: 220–230.

Burns, William
1969. Noise and man. J. B. Lippincott, Philadelphia: i–x + 1–336.

Burt, William Henry
1930. Adaptive modifications in the woodpeckers. Univ. California Publ. Zool. 32: 455–524.
1938. Faunal relationships and geographic distribution of mammals in Sonora, Mexico. Mus. Zool., Univ. Michigan, Misc. Publ. 39: 1–77.
1940. Territorial behavior and populations of some small mammals in southern Michigan. Univ. Michigan, Mus. Zool., Misc. Publ. 45: 1–58.
1943. Territoriality and home range concepts as applied to mammals. J. Mam. 24: 346–352.

Burton, George W. and Eugene P. Odum
1945. The distribution of stream fish in the vicinity of Mountain Lake, Virginia. Ecol. 26: 182–194.

Burtt, B. D.
1942. Some east African vegetation communities. J. Ecol. 30: 65–146.

Bustard, H. Robert
1970. The role of behavior in the natural regulation of numbers in the gekkonid lizard Gehyra variegata. Ecol. 51: 724–728.

Butler, L.
1953. The nature of cycles in populations of Canadian mammals. Can. J. Zool. 31: 242–262.

Butzer, Karl W.
1971. Environment and archeology, an ecological approach to prehistory. Aldine-Atherton, Inc., Chicago. i–xxvi + 1–703.

Buxton, Patrick A.
1923. Animal life in deserts. Edward Arnold & Co., London: i–xv, 1–176.
1955. The natural history of tsetse flies. H. K. Lewis & Co., London: i–viii, 1–816.

Buzas, Martin A. and Thomas G. Gibson
1969. Species diversity: benthonic Foraminifera in western North Atlantic. Sci. 163 (3862): 72–75.

Bylinsky, Gene
1970. The limited war on water pollution. Fortune, 1970: 102–107, 193–195, 197.

Cade, Tom J.
1955. Variation of the common rough-legged hawk in North America. Condor 57: 313–346.

Cade, Tom J., J. L. Lincer, C. M. White, D. G. Roseneau and L. G. Swartz
1971. DDE residues and eggshell changes in Alaskan falcons and hawks. Sci. 172 (3986): 955–957.

Cagle, Fred R.
1944. Home range, homing behavior, and migration in turtles. Misc. Publ. Mus. Zool., Univ. Michigan, 61: 1–34.

Cahalane, Victor H.
1942. Caching and recovery of food by the western fox squirrel. J. Wildl. Man. 6: 338–352.

Cahn, Alvin Robert
1927. An ecological study of southern Wisconsin fishes. Illinois Biol. Mono. 11: 1–151.
1929. The effect of carp on a small lake: the carp as a dominant. Ecol. 10: 271–274.

Cailliet, G. M., P. Y. Setzer, and M. S. Love
1971. Everyman's guide to ecological living. Macmillan Co., New York, N. Y.: i–viii + 1–119.

Cain, Stanley A.
1939. The climax and its complexities. Amer. Mid. Nat. 21: 146–181.
1944. Foundations of plant geography. Harper & Bros., New York: i–xiv, 1–556.
1947. Characteristics of natural areas and factors in their development. Ecol. Mono. 17: 185–200.

Cain, Stanley A. and G. M. de Oliveira Castro
1959. Manual of vegetation analysis. Harper & Brothers, New York: i–xvii + 1–325.

Caldwell, Lynton K.
1970. Authority and responsibility for environmental administration. Ann. Amer. Acad. Pol. Soc. Sci. 389: 107–115.

Calhoun, John B.
1944–46. Twenty-four hour periodicities in the animal kingdom. J. Tennessee Acad. Sci.: 19–21.
1956. Population dynamics of vertebrates. Compilations of research data. Rel. 5, Nat. Inst. Mental Health, Bethesda, Md.: 1–164.

Cantlon, John E.
1953. Vegetation and microclimates on north and south slopes of Cushetunk Mountain, New Jersey. Ecol. Mono. 23: 241–270.

Carlander, Kenneth D.
1955. The standing crop of fish in lakes. J. Fish. Res. Bd. Canada 12: 543–570.
1961. Handbook of freshwater fishery biology. Iowa State Univ. Press, Ames, Iowa, 1: 1–vi + 1–752.

Carlson, Clarence A.
1968. Summer bottom fauna of the Mississippi River, above Dam 19, Keokuk, Iowa. Ecol. 49: 162–169.

Carpenter, Charles C.
1952. Comparative ecology of the common garter snake (Thamnophis s. sirtalis), the ribbon snake (Thamnophis s. sauritus), and Butler's garter snake (Thamnophis butleri) in mixed populations. Ecol. Mono. 22: 235–258.

Carpenter, C. R.
1958. Territoriality: a review of concepts and problems. In Roe and Simpson's "Behavior and evolution," Yale Univ. Press: 224–250.

Carpenter, J. Richard
1940. The grassland biome. Ecol. Mono., 10: 617–684.
1940a. Insect outbreaks in Europe. J. An. Ecol. 9: 108–147.

Carpenter, Kathleen E.
1928. Life in inland waters with special reference to animals. Sidgwick & Jackson, London: i–xv, 1–267.

Carrick, Robert
1936. Experiments to test the efficiency of protective adaptations in insects. Trans. Roy. Ento. Soc. London (A) 85: 131–139.

Carson, Hampton L.
1968. The population flush and its genetic consequences. In R. C. Lewontin's "Population biology and evolution", Syracuse Univ. Press, Syracuse, New York: 123–137.

Carson, Rachel
1962. Silent spring. Houghton Mifflin Co., Boston: i-x + 1-368.

Caspari, Ernst
1951. On the biological basis of adaptedness. Amer. Sci. 39: 441–451.

Castenholz, Richard W.
1961. The effect of grazing on marine littoral diatom populations. Ecol. 42: 783–794.

Castri, Francesco di, R. Covarrubias and E. Hajek
1966. Soil ecosystems in sub-antarctic regions. In "Symposium on ecology of sub-arctic regions", UNESCO, Paris: 1–23.

Caughley, Graeme
1970. Eruption of ungulate populations, with emphasis on Himalayan thar in New Zealand. Ecol. 51: 53–72.

Caughley, Graeme and L. C. Birch
1971. Rate of increase. J. Wildl. Man. 35: 658–663.

Cavanaugh, William J. and Josephine E. Tilden
1930. Algal food feeding and case-building habits of the larva of the midgefly, Tanytarsus dissimilis. Ecol. 11: 281–287.

Chandler, David C.
1940. Plankton and certain physical-chemical data of the Bass Islands region, from September 1938 to November 1939. Ohio J. Sci. 40: 291–336.
1942. Light penetration and its relation to turbidity. Ecol. 23: 41–52.
1944. Relation of limnological and climatic factors to the phytoplankton of 1941. Trans. Amer. Micros. Soc. 63: 203–236.

Chandler, Robert F.
1941. The amount and mineral nutrient content of freshly fallen leaf litter in the hardwood forests of central New York. J. Amer. Soc. Agron. 33: 859–871.
1944. Amount and mineral nutrient content of freshly fallen needle litter of some northeastern conifers. Proc. (1943) Soil Sci. Sco. Amer. 8: 409–411.

Chaney, Ralph W.
1947. Tertiary centers and migration routes. Ecol. Mono. 17: 139–148.
1948. The bearing of the living Metasequoia on problems of Tertiary paleobotany. Proc. Nat. Acad. Sci. 34: 503–515.

Chapman, Abel
1922. Savage Sudan; its wild tribes, big-game, and bird life. G. P. Putnam's Sons, New York: i-xx, 1–452.

Chapman, Frank M.
1917. The distribution of bird life in Columbia; a contribution to a biological survey of South America. Bul. Amer. Mus. Nat. Hist. 36: i-x, 1–729.

Chapman, H. H.
1932. Is the longleaf type a climax? Ecology, 8 (4): 328–334.

Chapman, R. F. and J. H. P. Sankey
1955. The larger invertebrate fauna of three rabbit carcasses. J. An. Ecol. 24: 395–402.

Chapman, Royal N., C. E. Mickel, J. R. Parker, G. E. Miller, and E. G. Kelley
1926. Studies in the ecology of sand dune insects. Ecol. 7: 416–426.

Chapman, V. J.
1960. Salt marshes and salt deserts of the world. Interscience Publishers, Inc., New York: i-xvi + 1–392.

Cheatum, E. L. and C. W. Severinghaus
1950. Variations in fertility of white-tailed deer related to range conditions. Trans. 15th No. Amer. Wildl. Conf.: 170–190.

Cherkov, Yu I.
1961. A preliminary study of the faunal population of soils of the Arctic tundra of Yakutsk [In Russian with English Summary]. Zool. Zhur. 40: 326–333; Referat. Zhur. Biol. 1961, No. 4 Zh 423.

Chesher, Richard H.
1969. Destruction of Pacific corals by the sea star Acanthaster planci. Sci. 165 (3890): 280–283.

Chettleburgh, M. R.
1952. Observations on the collection and burial of acorns by jays in Hainault Forest. Brit. Birds 45: 359–364.

Chew, Robert M.
1951. The water exchanges of some small mammals. Ecol. Mono. 21: 215–225.
1961. Ecology of the spiders of a desert community. J. N. Y. Ent. Soc., 69: 5–41.

Chew, Robert M. and B. B. Butterworth
1964. Ecology of rodents in Indian Cove (Majave Desert), Joshua Tree National Monument, California. J. Mam. 45: 203–225.

Chew, Robert M. and Alice Eastlake Chew
1970. Energy relationships of the mammals of a desert shrub (Larrea tridentata) community. Ecol. Mono. 40: 1–21.

Chiang, H. C. and A. C. Hodson
1950. An analytical study of population growth in Drosophila melanogaster. Ecol. Mon. 20: 173–206.

Child, Ch. M.
1903. Studies on regulation. Arch. Entwick. Org. 15: 187–237, 355–420.

Childers, William and Hurst H. Shoemaker
1953. Time of feeding of the black crappie and the white crappie. Illinois Acad. Sci. Trans. 46: 227–230.

Chitty, Dennis
1954. Tuberculosis among wild voles: with a discussion of other pathological conditions among certain mammals and birds. Ecol. 35: 227–237.
1960. Population processes in the vole and their relevance to general theory. Canadian J. Zool. 38: 99–113.

Chitty, Helen
1950. Canadian Arctic wild life enquiry, 1945–49: with a summary of results since 1933. J. An. Ecol. 19: 180–193.

Christian, John J.
1956. Adrenal and reproductive responses to population size in mice from freely growing populations. Ecol. 37: 258–373.
1961. Phenomena associated with population density. Proc. Natl. Acad. Sci. 47: 428–449.
1971. Fighting, maturity, and population density in Microtus pennsylvanicus. J. Mam. 52: 566–567.

Christian, John J. and David E. Davis
1956. The relationship between adrenal weight and population status of urban Norway rats. J. Mam. 37: 475–486.
1964. Endocrines, behavior, and population. Sci. 146 (3651): 1550–1560.

Christy, Harlan R.
1952. Vertical temperature gradients in a beech forest in central Ohio. Ohio J. Sci. 52: 199–209.

Churchill, Ethan D.
1955. Phytosociological and environmental characteristics of some plant communities in the Umiat region of Alaska. Ecol. 36: 606–627.

Clapham, A. R.
1936. Over-dispersion in grassland communities and the use of statistical methods in plant ecology. J. Ecol. 24: 232–251.

Clark, Austin H.
1914. Nocturnal animals. J. Washington Acad. Sci. 4: 139–142.

Clark, John R.
1969. Thermal pollution and aquatic life. Sci. Amer. 220 (3): 18–27.

Clark, L. R., P. W. Geier, R. D. Hughes, and R. F. Morris
1967. The ecology of insect populations in theory and practice. Methuen, London: i-xiii + 1–232.

Clarke, Bryan and James Murray
1969. Ecological genetics and speciation in land snails of the genus Partula. Biol. J. Linn. Soc. 1: 31–42.

Clarke, C. H. D.
1940. A biological investigation of the Thelon Game Sanctuary. Nat. Mus. Canada Bull. 96, Biol. Ser. 25: i-iv, 1–135.

Clarke, George L.
1934. The role of copepods in the economy of the sea. Fifth Pacific Sci. Cong.: 2017–2021.
1939. The utilization of solar energy by aquatic organisms. Publ. A.A.A.S., 10: 27–38.
1940. Comparative richness of zooplankton in coastal and offshore areas of the Atlantic. Biol. Bull. 78: 226–255.
1946. Dynamics of production in a marine area. Ecol. Mono. 16: 321–335.
1954. Elements of ecology. John Wiley & Sons, New York: i-xiv, 1–534.

Clarke, George L. and David W. Bishop
1948. The nutritional value of marine zooplankton with a consideration of its use as an emergency food. Ecol. 29: 54–71.

Clarke, J. Lyell
1937. New and significant experiences in mosquito control in the Desplaines Valley Mosquito Abatement District. Proc. 24th Ann. Meeting New Jersey Mosquito Ext., Assn. 16 pp.

Clarke, J. R.
1953. The effect of fighting on the adrenals, thymus and spleen of the vole (Microtus agrestis). J. End. 9: 114–126.

Clausen, Ralph G.
1931. Orientation in fresh water fishes. Ecol. 12: 541–546.

1936. Oxygen consumption in fresh water fishes. Ecol. 17: 216–226.

Clements, Frederic E.
1916. Plant succession. Carnegie Inst. Washington Publ. 242: 1–512.
1936. The origin of the desert climax and climate. In "Essays in Geobotany." Univ. Calif. Press, Berkeley: 87–140.
1942. Cycles and climaxes. Chron. Bot. 7: 241–243.

Clements, Frederic E. and Victor E. Shelford
1939. Bio-ecology. John Wiley & Sons, New York: i-vi, 1–425.

Clements, Frederic E., John E. Weaver and Herbert C. Hanson.
1929. Plant competition, an analysis of community functions. Carnegie Inst. Washington, Publ. 398: i-xvi, 1–340.

Clench, Harry K.
1967. Temporal dissociation and population regulation in certain hesperine butterflies. Ecol. 48: 1000–1006.

Cleveland, L. R.
1924. The physiological and symbiotic relationships between the intestinal Protozoa of termites and their host, with special reference to Reticulitermes flavipes Kollar. Biol. Bull. 46: 178–227.
1934. The wood-feeding roach Crytocercus, its Protozoa, and the symbiosis between Protozoa and roach. Mem. Amer. Acad. Arts & Sci. 17: 185–342.

Cloudsley-Thompson, J. L.
1954. Biology of deserts. Hafner Publishing Co., New York: 1–224.
1969. The zoology of tropical Africa. W. W. Norton & Co. Inc., New York: i-xv + 1–356.

Cloudsley-Thompson, J. L. and M. J. Chadwick
1964. Life in deserts. Dufour Editions, Philadelphia: i-xvi + 1–218.

Cochran, William W. and Rexford D. Lord, Jr.
1963. A radio-tracking system for wild animals. J. Wildl. Man. 27: 9–24.

Cody, Martin L.
1968. On the methods of resource division in grassland bird communities. Amer. Nat. 102: 107–147.

Coe, Wesley R.
1955. Ecology of the bean clam Donax gouldi on the coast of southern California. Ecol. 36: 512–514.

Coker, R. E., A. F. Shira, H. Walton Clark, and A. D. Howard
1922. Natural history and propagation of fresh-water mussels. Bull. U. S. Bur. Fish. 37: 75–181.

Cole, Gerald A.
1953. Notes on copepod encystment. Ecol. 34: 208–211.
1955. An ecological study of the microbenthic fauna of two Minnesota lakes. Amer. Midl. Nat. 53: 213–230.

Cole, Lamont C.
1946. A study of the crypotozoa of an Illinois woodland. Ecol. Mono. 16: 49–86.
1946a. A theory for analyzing contagiously distributed populations. Ecol., 27: 329–341.
1951. Population cycles and random oscillations. J. Wildl. Man. 15: 233–252.
1954. Some features of random population cycles. J. Wildl. Man. 18: 2–24.
1954a. The population consequences of life history phenomena. Quart. Rev. Biol. 29: 103–137.
1956 (1958). Population fluctuations. Proc. 10th Intern. Cong. Ent. 2: 639–647.
1968. Can the world be saved? BioSci. 18: 679–684.
1969. Thermal pollution. BioSci. 19: 989–992; 20 (1970): 72.

Collias, Nicholas E.
1944. Aggressive behavior among vertebrate animals. Physiol. Zool. 17: 83–123.
1964. The evolution of external construction by animals. Amer. Zool. 4: 175–243.

Collias, Nicholas E. and Elsie C. Collias
1963. Evolutionary trends in nest building by the weaverbirds (Ploceidae). Proc. XIII Intern. Ornith. Cong. 518–530.

Collier, Albert
1953. The significance of organic compounds in sea water. Trans. 18th No. Amer. Wildl. Conf.: 463–472.

Colman, John
1940. On the faunas inhabiting intertidal seaweeds. J. Mar. Biol. Assoc. U. K. 24: 129–183.

Colquhoun, M. K. and Averil Morley.
1943. Vertical zonation in woodland bird communities J. An. Ecol. 12: 75–81.

Commoner, Barry
1971. The closing circle—nature, man; and technology. Alfred A. Knopf, Inc., New York, N. Y.: i-x + 1–326.

Conant, Roger
1958. A field guide to reptiles and amphibians. Houghton Mifflin Co., Boston: i-xv, 1–366.

Connell, Joseph H.
1961. The influence of interspecific competition and other factors on the distribution of the barnacle Chthamalus stellatus. Ecol. 42: 710–723.
1970. A predator-prey system in the marine intertidal region. 1. Balanus glandula and several predatory species of Thais. Ecol. Mono. 40: 49–78.

Connell, Joseph H. and Eduardo Orias
1964. The ecological regulation of species diversity. Amer. Nat. 98: 399–414.

Conway, Gordon R.
1971. Better methods of pest control. In W. W. Murdoch's "Environment, resources, pollution & society." Sinauer Associates Inc., Stamford, Conn.: 302–325.

Cook, A., O. S. Bamford, J. D. B. Freeman, and D. J. Teideman
1969. A study of the homing habit of the limpet. An. Behav. 17: 330–339.

Cook, R. E.
1969. Variation in species density of North American birds. Syst. Zool. 18: 63–84.

Cook, William C.
1929. A bioclimatic zonation for studying the economic distribution of injurious insects. Ecol. 10: 282–293.

Cooke, Lloyd M. (chairman)
1969. Cleaning our environment, the chemical basis for action. Rep't. Subcom. Env. Imp., Com. Chem. Publ. Aff., Amer. Chem. Soc., Washington, D. C.: i-ix + 1–249.

Cooke, May Thatcher
1942. Returns from banded birds: some longevity records of wild birds. Bird-Banding 13: 34–37, 70–74, 110–119, 176–181.

Coon, Carleton
1962. The origin of races. A. Knopf, New York: i-xli + 1–724.

Cooper, Edwin L., C. C. Wagner, and G. E. Krantz
1971. Bluegills dominate production in a mixed population of fishes. Ecol. 52: 280–290.

Cooper, J. G.
1859. On the distribution of the forests and trees of North America, with notes on its physical geography. Ann. Rep't. Bd. Reg. Smithsonian Inst.: 246–280.

Cooper, L. H. N.
1937. The nitrogen cycle in the sea. J. Mar. Biol. Assoc. 22: 183–204.

Coppel, H. C. and K. Leius
1955. History of the larch sawfly, with notes on origin and biology. Can. Ento. 87: 103–111.

Costello, David F.
1944. Natural revegetation of abandoned plowed land in the mixed prairie association of northeastern Colorado. Ecol. 25: 312–326.
1969. The prairie world. Thomas Y. Crowell Publ. Co., New York: i-xii + 1–242.

Cott, Hugh B.
1940. Adaptive coloration in animals. Methuen & Co., London: i-xxxii, 1–508.
1947. The edibility of birds. Proc. Zool. Soc. London, 116: 371–524.

Cottam, Clarence
1965. The ecologists' role in problems of pesticide pollution. BioSci. 15: 457–463.

Cottam, Clarence and David A. Munro
1954. Eelgrass status and environmental relations. J. Wildl. Man. 18: 449–460.

Cottam, Grant and J. T. Curtis
1956. The use of distance measures in phytosociological sampling. Ecol. 37: 451–460.

Cottam, Walter P.
1947. Is Utah Sahara bound? Bull. Univ. Utah, 37: 1–40.

Cowell, Eric B.
1970. Some biological effects of oil pollution. Your Environment 1: 84, 93–94.

Cowles, Henry C.
1899. The ecological relations of the vegetation on the sand dunes of Lake Michigan. Bot. Gaz. 27: 95–117, 167–202, 281–308, 361–391.

Cowles, Raymond B.
1941. Observations on the winter activities of desert reptiles. Ecol. 22: 125–140.

Cox, Clare Francis
1933. Alpine plant succession on James Peak, Colorado. Ecol. Mon. 3: 299–372.

Cox, George W.
1968. The role of competition in the evolution of migration. Evol. 22: 180–192.
Cox, Wm. T.
1936. Snowshoe rabbit migration, tick infestation, and weather cycles. J. Mam. 17: 216–221.
Craig, Wallace
1908. The voices of pigeons regarded as a means of social control. Amer. J. Sociol. 14: 86–100.
1908a. North Dakota life: plant, animal, and human. Bul. Am. Geo. Soc. 40: 321–332, 401–415.
Crawford, Bill T.
1950. Some specific relationships between soils and wildlife. J. Wildl. Man. 14: 115–123.
Creaser, Edwin P.
1931. Some cohabitants of burrowing crayfish. Ecol. 12: 243–244.
Crew, F. A. E. and L. Mirskaia
1931. The effects of density on an adult mouse population. Biol. Gen. 7: 239–250.
Criddle, Norman
1932. The correlation of sunspot periodicity with grasshopper fluctuation in Manitoba. Canad. Field-Nat. 46: 195–199.
Crisler, Lois
1956. Observations of wolves hunting caribou. J. Mam. 37: 337–346.
Critchley, Macdonald
1960. The evolution of man's capacity for language. In S. Tax's "The evolution of man", Univ. of Chicago Press: 289–308.
Crombie, A. C.
1946. Further experiments on insect competition. Proc. Roy. Soc. London (B) 133: 76–109.
1947: Interspecific competition. J. An. Ecol. 16: 44–73.
Cronin, L. E.
1971. Pollution prevention. Proc. Royal Soc. London, B, 177: 439–450.
Cronin, L. E., Joanne C. Daiber, and E. M. Hulbert
1962. Quantitative seasonal aspects of zooplankton in the Delaware River Estuary. Chesapeake Sci., 3: 63–93.
Cronin, L. E. and Alice J. Mansueti
1971. The biology of the estuary. In Philip A. Douglas and Richard H. Stroud's "A symposium on the biological significance of estuaries", Sport Fishing Inst., Washington, D. C.: 14–39.
Crook, J. H.
1965. The adaptive significance of avian social organizations. Symp. Zool. Soc. London 14: 181–218.
Crosby, Irving B.
1937. Methods of stream piracy. J. Geol. 45: 465–486.
Crossley, Jr., D. A. and Kurt K. Bohnsack
1960. Long-term ecological study in the Oak Ridge area. III. The oribatid mite fauna in pine litter. Ecol. 41: 628–647.
Crossley, D. A., Jr. and D. E. Reichle
1969. Analysis of transient behavior of radioisotopes in insect food chains. BioSci. 19: 341–343.
Crowcroft, Peter and F. P. Rowe
1957. The growth of confined colonies of the wild house-mouse (Mus musculus L.) Proc. Zool. Soc. London 129: 359–370.
Crowell, John C. and Lawrence A. Frakes
1970. Phanerozoic glaciation and the causes of ice ages. Amer. J. Sci. 268: 193–224.
Crowell, Kenneth L.
1962. Reduced interspecific competition among the birds of Bermuda. Ecol. 43: 75–88.
Culver, David C.
1970. Analysis of simple cave communities: niche separation and species packing. Ecol. 51: 949–958.
Cummins, Kenneth W.
1962. An evaluation of some techniques for the collection and analysis of benthic samples with special emphasis on lotic waters. Amer. Midl. Nat. 67: 477–504.
Cunha, A. Brito da, A. M. El-Tabey Shehata and W. de Oliveira
1957. A study of the diets and nutritional preferences of tropical species of Drosophila. Ecol. 38: 98–106.
Currie, Pat O. and D. L. Goodwin
1966. Consumption of forage by black-tailed jackrabbits on salt-desert ranges of Utah. J. Wildl. Man. 30: 304–311.
Curry-Lindahl, Kai
1961. Contribution a l'étude des vertébrés terrestres en Afrique tropicale. Inst. Parcs Nat. Congo et Ruanda-Urundi: 1–331.
Cushing, D. H.
1951. The vertical migration of planktonic Crustacea. Biol. Rev. 26: 158–192.

Cushing, Edward J.
1965. Problems in the Quaternary phytogeography of the Great Lakes region. In H. E. Wright, Jr. and D. G. Frey's "The Quaternary of the United States" Princeton Univ. Press: 403–416.
Cushing, John E., Jr.
1941. Non-genetic mating preference as a factor in evolution. Condor 43: 233–236.
1944. The relation of non-heritable food habits to evolution. Condor 46: 265–271.
Dachnowski, Alfred
1912. Peat deposits of Ohio. Geol. Surv. Ohio, 4th Ser., Bull. 16: i-viii, 1–424.
Dahl, Erik
1953. Some aspects of the ecology and zonation of the fauna on sandy beaches. Oikos 4: 1–27.
Daiber, Franklin C.
1952. The food and feeding relationships of the freshwater drum, Aplodinotus grunniens Rafinesque in western Lake Erie. Ohio J. Sci. 52: 35–46.
1956. A comparative analysis of the winter feeding habits of two benthic stream fishes. Copeia: 141–151.
Dale, M. B.
1970. Systems analysis and ecology. Ecol. 51: 2–16.
Dalke, Paul D.
1935. Dropping analyses as an indication of pheasant food habits. Trans. 21st Amer. Game Conf.: 387–391.
Dalquest, Walter W.
1953. Mammals of the Mexican state of San Luis Potosi. Louisiana St. Univ. Stud., Biol. Ser. 1: 1–229.
Damant, G. C. C.
1924. The adjustment of the buoyance of the larva of Corethra plumicornus. Jour. Physiol. 59: 343–456.
Dambach, Charles Arthur
1948. A study of the ecology and economic value of crop field borders. Ohio St. Univ., Grad. School Stud., Biol. Sci. Ser. 2: i-xi, 1–205.
Dammerman, K. W.
1948. The fauna of Krakatau. 1883–1933. K. Nederland. Akad. Wetensch., Afd. Natuurkunde, sect. 2, pt. 44: 1–594.
D'Ancona, Umberto
1954. The struggle for existence. Biblio. Bioth. 6: i-xi, 1–274 (English translation).
Daniels, Farrington
1949. Solar energy. Sci. 109 (2821): 51–57.
Dansereau, Pierre
1947. The distribution and structure of Brazilian forests. For. Chron. 23: 261–277.
1951. Description and recording of vegetation upon a structural basis. Ecol. 32: 172–229.
1960. A combined structural and floristic approach to the definition of forest ecosystems. Silva Fenn. 105: 16–21.
Dansereau, Pierre, P. F. Buell, and R. Dagon
1966. A universal system for recording vegetation. II. A methodological critique and an experiment. Sarracenia 10: 1–64.
Dansereau, Pierre and Fernando Segadas-Vianna
1952. Ecological study of the peat bogs of eastern North America. Can. J. Bot. 30: 490–520.
Darling, F. Fraser
1960. An ecological reconnaissance of the Maria Plains in Kenya Colony. Wildl. Mono. 5: 1–41.
Darlington, P. J., Jr.
1938. The origin of the fauna of the Greater Antilles with discussion of dispersal of animals over water and through air. Quart. Rev. Biol. 13: 274–300.
1957. Zoogeography: the geographical distribution of animals. John Wiley & Sons, New York: i-xi, 1–675.
Darnell, Rezneat M.
1964. Organic detritus in relation to secondary production in aquatic communities. Verh. Inter. Ver. Limn. 15: 462–470.
Daubenmire, R. F.
1943. Vegetational zonation in the Rocky Mts. Bot. Rev. 9: 325–393.
1947. Plants and environment. John Wiley & Sons, Inc., New York: i-xiii, 1–424.
1968. Ecology of fire in grasslands. Adv. Ecol. Res. 5, Academic Press: 209–266.
Davenport, C.B.
1908. Experimental morphology. Macmillan Co., New York: 1–509.
Davenport, Demorest
1955. Specificity and behavior in symbiosis. Quart. Rev. Biol. 30: 29–46.
Davidson, J. and H. G. Andrewartha
1948. The influence of rainfall, evaporation and atmospheric tem-

perature on fluctuations in the size of a natural population of *Thrips imaginis* (Thysanoptera). J. An. Ecol. 17: 200–222.

Davidson, Vera Smith
1930. The tree layer society of the maple-red oak climax forest. Ecol. 11: 601–606.
1932. The effect of seasonal variability upon animal species in total populations in a deciduous forest succession. Ecol. Mono. 2: 305–333.

Davis, Charles C.
1955. The marine and fresh-water plankton. Michigan St. Univ. Press, East Lansing: i–xi, 1–562.
1958. An approach to some problems of secondary production in the western Lake Erie region. Limnology & Ocean. 3: 15–28.

Davis, David E.
1946. A seasonal analysis of mixed flocks of birds in Brazil. Ecol. 27: 168–181.
1957. The existence of cycles. Ecol. 38: 163–164.

Davis, David E., John J. Christian, and Frank Bronson
1963. Effect of exploitation on birth, mortality, and movement rates in a woodchuck population. J. Wildl. Man. 28: 1–9.

Davis, John
1951. Distribution and variation of the brown towhees. Univ. California Publ. Zool. 52: 1–120.

Davis, John H., Jr.
1940. The natural features of southern Florida, especially the vegetation and the Everglades. Florida Geol-Survey. Bull. 25: 1–311.

Davis, Luckett V. and I. E. Gray
1966. Zonal and seasonal distribution of insects in North Carolina salt marshes. Ecol. Mono. 36: 275–295.

Davis, Margaret B.
1967. Late-glacial climate in northern United States: a comparison of New England and the Great Lakes area. In E. J. Cushing and H. E. Wright, Jr.'s "Quaternary paleoecology". Yale Univ. Press, New Haven, Conn.: 11–43.
1969. Climatic changes in southern Connecticut recorded by pollen deposition at Rogers Lake. Ecol. 50: 409–422.

Davis, Margaret B. and John C. Goodlett
1960. Comparison of the present vegetation with pollen-spectra in surface samples from Brownington Pond, Vermont. Ecol. 41: 346–357.

Davis, William B.
1938. Relation of size of pocket gophers to soil and altitude. J. Mam. 19: 338–342.

Davison, Verne E.
1940. A field method of analyzing game bird foods. J. Wildl. Man. 4: 105–116.

Day, Gordon M.
1953. The Indian as an ecological factor in the northeastern forest. Ecol. 34: 329–346.

Day, W. R.
1957. Sitka spruce in British Columbia, a study in forest relationships. For. Com. Bull. 28, Her Majesty's Stat. Off., London: 1–110.

Dayton, William A.
1953. Checklist of native and naturalized trees of the United States (including Alaska). Agr. Handb'k No. 41, For. Serv.: 1–472.

Deacon, James E., Wm. C. Bradley, and Karl M. Larsen
1964. Ecological distribution of the mammals of Clark Canyon, Charleston Mountains, Nevada. J. Mam., 45: 397–409.

DeBach, Paul
1949. Population studies of the long-tailed mealybug and its natural enemies on citrus trees in southern California, 1946. Ecol. 30: 14–25.

DeBach, Paul and Pelagija Sisojevic
1960. Some effects of temperature and competition on the distribution and relative abundance of *Aphytis lingnanensis* and *A. chrysomphali* (Hymenoptera: Aphelinidae). Ecol. 41: 153–160.

DeBach, Paul and H. S. Smith
1941. The effect of host density on the rate of reproduction of entomophagous parasites. J. Econ. Ento. 34: 741–745.
1947. Effects of parasite population density on rate of change of host and parasite populations. Ecol. 28: 290–298.

Deevey, Edward S., Jr.
1939. A contribution to regional limnology. Am. J. Sci. 238: 717–741.
1941. The quantity and composition of the bottom fauna of thirty-six Connecticut and New York lakes. Ecol. Mon. 11: 413–455.
1947. Life tables for natural populations of animals. Quart. Rev. Biol. 22: 283–314.
1949. Biogeography of the Pleistocene. Bull. Geol. Soc. America 60: 1315–1416.

1960. The human population. Sci. Amer. 203: 194–198, 200, 202–204.

Deevey, Edward S. and Richard Foster Flint
1957. Postglacial hypsithermal interval. Sci. 125: 182–184.

Deevey, Georgiana Baxter
1952. Quantity and composition of the zooplankton of Block Island Sound, 1949. Bull Bingham Oceanographic Coll. 13: 120–164.

Defant, Albert
1958. Ebb and flow; the tides of earth, air, and water. Univ. Michigan Press, Ann Arbor: 1–121.

DeLury, D. B.
1947. On the estimation of biological populations. Biom. 3: 147–167.
1958. The estimation of population size by a marking and recapture procedure. J. Fish. Res. Bd. Canada 15: 19–25.

Dementiev, G. P.
1958. La faune desertique du Turkestan. La Terre et la Vie 105: 3–44.

Dendy, Jack S.
1944. The fate of animals in stream drift when carried into lakes. Ecol. Mono. 14: 333–357.

Denham, Stacey C.
1938. A limnological investigation of the West Fork and Common Branch of White River. Invest. Indiana Lakes and Streams, No. 5: 17–71.

Dennis, John V.
1950. Bird dominance at the feeding station. Audubon Mag. 52: 349–400.

Dethier, Vincent G.
1947. Chemical insect attractants and repellants. Blakiston Co., Philadelphia: i–xv, 1–289.
1954. Evolution of feeding preferences in phytophagous insects. Evol. 8: 33–54.

Detwyler, Thomas R.
1971. Man's impact on environment. McGraw-Hill Book Co., New York, N. Y.: i–xiii + 1–731.

Dewey, Edward R.
1960. The case for exogenous rhythms. J. Cycle Res. 9: 131–176.
1965. The 9.6-year cycle. Found. Study of Cycles, Inc. 124 S. Highland Ave., Pittsburgh, Penn.: 3pp + tables.

DeWitt, Robert M.
1954. The intrinsic rate of natural increase in a pond snail (*Physa gyrina* Say). Amer. Nat., 88: 353–359.
1955. The ecology and life history of the pond snail *Physa gyrina*. Ecol. 36: 40–44.

Dexter, Ralph W.
1954. Zonation of the intertidal marine mollusks at Cape Ann, Massachusetts. Nautilus 58: 135–142.
1946. A demonstration of suspended animation. Turtox News, 24: 118–119.
1947. The marine communities of a tidal inlet at Cape Ann, Massachusetts: a study in bio-ecology. Ecol. Mono. 17: 261–294.

Dice, Lee R.
1938. Some census methods for mammals. J. Wildl. Man. 2: 119–130.
1939. The Sonoran biotic province. Ecol. 20: 118–129.
1940. Speciation in *Peromyscus*. Amer. Nat. 74: 289–298.
1942. Ecological distribution of *Peromyscus* and *Neotoma* in parts of southern New Mexico. Ecol. 23: 199–208.
1943. The biotic provinces of North America. Univ. Michigan Press, Ann Arbor: 1–78.
1945. Minimum intensities of illumination under which owls may find dead prey by sight. Amer. Nat. 79: 385–416.
1947. Effectiveness of selection by owls of deermice (*Peromyscus maniculatus*) which contrast in color with their background. Univ. Michigan, Cont. Lab. Vert. Biol. 34: 1–20.
1948. Relationship between frequency index and population density. Ecol. 29: 389–391.
1952. Measure of the spacing between individuals within a population. Cont. Lab. Vert. Biol., Univ. Michigan, 55: 1–23.
1952a. Natural communities. Univ. Michigan Press, Ann Arbor: i–x, 1–547.

Dice, Lee R. and Philip M. Blossom
1934. Studies of mammalian ecology in southwestern North America with special attention to the colors of desert mammals. Carnegie Inst. Washington Publ. 485: i–iv, 1–129.

Dice, Lee R. and Walter E. Howard
1951. Distance of dispersal by prairie deermice from birthplaces to breeding sites. Cont. Lab. Vert. Biol., Univ. Michigan, 50: 1–15.

Dickman, Mike
1968. Some indices of diversity. Ecol. 49: 1191–1193.

Dietz, Robert S. and John C. Holden
1970. The breakup of Pangea. Sci. Amer. 223 (4): 30–41.

Digby, P. S. B.
1954. The biology of the marine planktonic copepods of Scoresby Sound, East Greenland. J. An. Ecol. 23: 298–338.

Dilger, William C.
1956. Adaptive modifications and ecological isolating mechanisms in the thrush genera *Catharus* and *Hylocichla*. Wilson Bull. 68: 171–199.
1956. Hostile behavior and reproductive isolating mechanisms in the avian genera *Catharus* and *Hylocichla*. Auk 73: 313–353.
1962. Behavior and genetics. In E. L. Bliss' "Roots of behavior". Harper and Brothers: 35–47.
1962. Methods and objectives of ethology. The Living Bird. First Ann., Cornell Lab. Ornith.: 83–92.

Dillon, Lawrence S.
1956. Wisconsin climate and life zones in North America. Sci. 123: 167–176.

Dimond, John B.
1967. Evidence that drift of stream benthos is density related. Ecol. 48: 855–857.

Dirks-Edmunds, Jane C.
1947. A comparison of biotic communities of the cedar-hemlock and oak-hickory associations. Ecol. Mono. 17: 235–260.

Dix, Ralph L.
1964. A history of biotic and climatic changes within the North American grassland. In D. J. Crisp's "Grazing in terrestrial and marine environments," Blackwell Sci. Publ., Oxford, England: 71–89.

Dix, R. L. and J. M. A. Swan
1971. The roles of disturbance and succession in upland forest at Candle Lake, Saskatchewan. Can. J. Bot. 49: 657–676.

Dixon, A. F. G.
1970. Factors limiting the effectiveness of the coccinellid beetle, *Adalia bipunctata* (L.), as a predator of the sycamore aphid, *Drepanosiphum platanoides* (Schr.). J. An. Ecol. 39: 739–775.

Dixon, Keith L.
1959. Ecological and distributional relations of desert scrub birds of western Texas. Condor 61: 397–409.

Dobzhansky, Theodosius
1950. Evolution in the tropics. Amer. Sci. 38: 209–221.
1951. Genetics and the origin of species. Columbia Univ. Press, New York: 1–364.
1956. Genetics of natural populations. Evol. 10: 82–92.
1958. Evolution at work. Sci. 127: 1091–1098.

Dobzhansky, Th., D. M. Cooper, H. J. Phaff, E. P. Knapp, and H. L. Carson
1956. Studies on the ecology of *Drosophila* in the Yosemite region of California. Ecol. 37: 544–550.

Dodds, G. S. and Frederick L. Hisaw
1924–25. Ecological studies of aquatic insects. Ecol. 5: 137–148, 262–271; 6: 123–137.

Dole, Jim W.
1968. Homing in leopard frogs, *Rana pipiens*. Ecol. 49: 386–399.

Donn, William L. and M. Ewing
1966. A theory of ice ages III. Sci. 152 (3730): 1706–1712.

Dorf, Erling
1960. Climatic changes of the past and present. Amer. Sci. 48: 341–364.

Dorn, Harold F.
1962. World population growth: an international dilemma. Sci. 125: 283–290.

Dorst, J.
1946. Quel est le role des oiseaux dans la vie des fleurs? L'Ois. et Rev. Franc. Ornith. NS 16: 113–128.

Doudoroff, Peter and Max Katz
1953. Critical review of literature on the toxicity of industrial wastes and their components to fish. Sewage and Indust. Wastes 25: 802–839.

Douglass, A. E.
1928. Climatic cycles and tree growth. Carnegie Inst. Washington Publ. 289, 2: 1–166.

Dow, Robert L.
1969. Cyclic and geographic trends in seawater temperature and abundance of American lobster. Sci. 164 (3883): 1060–1063.

Dowdy, W. W.
1944. The influence of temperature on vertical migration of invertebrates inhabiting different soil types. Ecol. 25: 449–460.
1951. Further ecological studies on stratification of the Arthropods. Ecol. 32: 37–52.

Downs, Albert A. and William E. McQuilken
1944. Seed production of southern Appalachian oaks. J. For. 42: 913–920.

Dozier, Herbert L.
1953. Muskrat, production and management. U. S. Fish and Wildl. Serv. Circ. 18: 1–42.

Dresser, Peter van
1956. The coming solar age. Landscape 5: 30–32.

Drost-Hansen, Walter
1969. Allowable thermal pollution limits—a physico-chemical approach. Chesapeake Sci. 10: 281–288.

Drury, W. H., Jr.
1953. Birds of the Saint Elias quadrangle in southwestern Yukon Territory. Can. Field-Nat. 67: 103–128.

Duffey, Eric
1968. An ecological analysis of the spider fauna of sand dunes. J. An. Ecol. 37: 641–674.

Dumas, M.
1841. On the chemical statics of organized beings. London, Edinburgh and Dublin. Phil. Mag. and Jour. Sci., Ser. 3. 19: 337–347, 456–469.

Dunbar, Carl O.
1949. Historical geology. John Wiley & Sons, New York: i–xii, 1–573 (*2nd* edition, 1960).

Dunbar, M. J.
1968. Ecological development in polar regions: a study in evolution. Prentice-Hall, Englewood Cliffs, New Jersey: i–viii + 1–119.

Dunlavy, Joseph C.
1935. Studies on the phyto-vertical distribution of birds. Auk 52: 425–431.

Dunn, Emmett Reid
1931. The herpetological fauna of the Americas. Copeia: 106–119.

Dunnet, George M.
1955. The breeding of the starling, *Sturnus vulgaris*, in relaation to its food supply. Ibis, 97: 619–662.

Durango, S.
1949. The nesting associations of birds with social insects and with birds of different species. Ibis, 91: 140–143.

Dusi, Julian L.
1949. Methods for the determinination of food habits by plant microtechniques and histology and their application to cottontail rabbit food habits. J. Wildl. Man. 13: 295–298.

DuToit, Alex L.
1937. Our wandering continents: an hypothesis of continental drifting. Oliver and Boyd, Edinburgh: i–xiii, 1–366.

Duvdevani, S.
1953. Dew gradients in relation to climate, soil and topography. Desert Res. Proc., Jerusalem Post Press: 136–152.

Duvigneaud, P. and S. Denaeyer-De Smet
1970. Biological cycling of minerals in temperate deciduous forest. In David E. Reichle's "Ecol. Stud. I, Analysis of temperate forest ecosystems," Springer-Verlag, New York: 199–225.

Dyksterhuis, E. J. and E. M. Schmutz
1947. Natural mulches or "litter of grasslands: with kinds and amounts on a southern prairie." Ecol. 28: 163–179.

Dymond, J. R.
1947. Fluctuations in animal populations with special reference to those of Canada. Trans. Royal Soc. Canada, 41 (5): 1–34.

Eaton, Theodore H., Jr. and Robert F. Chandler, Jr.
1942. The fauna of the forest-humus layers in New York. Cornell Univ. Agr. Exp. Sta. Mem. 247: 1–26.

Eberhardt, L. L.
1969. Population estimates from recapture frequencies. J. Wildl. Man. 33: 28–39.

Ecke, Dean H.
1954. An invasion of Norway rats in southwest Georgia. J. Mam. 35: 521–525.

Eddy, Samuel
1928. Succession of Protozoa in cultures under controlled conditions. Trans. Amer. Micros. Soc. 47: 283–319.
1934. A study of fresh-water plankton communities. Illinois Biol. Mono. 12: 1–93.

Edelstam, Carl and Carina Palmer
1950. Homing behavior in gastropodes. Oikos 2: 259–270.

Edmondson, W. T.
1946. Factors in the dynamics of rotifer populations. Ecol. Mono. 16: 357–372.

Edney, E. B.
1954. Woodlice and the land habitat. Biol. Rev. 29: 185–219.

Edwards, Clive A.
1969. Soil pollutants and soil animals. Sci. Amer. 220 (4): 88–92, 97–99.

Edwards, C. A., D. E. Reichle, and D. A. Crossley, Jr.
1970. The role of soil invertebrates in turnover of organic matter and nutrients. In David E. Reichle's "Ecol. Stud. I, Analysis of temperate forest ecosystems", Springer-Verlag, New York: 147–172.

Edwards, Ernest P. and Richard E. Tashian
1959. Avifauna of the Catemaco Basin of southern Veracruz, Mexico. Condor 61: 325–337.

Edwards, R. W. and M. Owens
1962. The effect of plants on river conditions. IV. The oxygen balance of a chalk stream. J. Ecol. 50: 207–220.

Edwards, R. Y. and C. David Fowle
1955. The concept of carrying capacity. Trans. 20th N. Amer. Wildl. Conf: 589–602.

Eggleton, Frank E.
1931. A limnological study of the profundal bottom fauna of certain fresh-water lakes. Ecol. Mono. 1: 231–331.
1952. Dynamics of interdepression benthic communities. Trans. Amer. Microsc. Soc., 71: 189–228.

Egglishaw, Henry J.
1969. The distribution of benthic invertebrates on substrata in fast-flowing streams. J. An. Ecol. 38: 19–33.

Egler, Frank E.
1968. Herbicides. In F. B. Golley and H. K. Buechner's "A practical guide to the study of the productivity of large herbivores." I.B.P. Handbook No. 7, Blackwell Sci. Publ., Oxford: 252–255.

Ehrlich, Paul R. and Anne H. Ehrlich
1970. Population, resources, environment—issues in human ecology. W. H. Freeman and Co., San Francisco: 1–383.

Ehrlich, Paul R. and J. P. Holdren
1969. Population and panaceas—a technological perspective. BioSci. 19: 1065–1071.

Eidmann, H.
1931. Zur Kenntnis der Periodizitat der Insektenepidemien. Zeits. Angewandte Ent. 18: 537–567.

Eisenberg, Robert M.
1966. The regulation of density in a natural population of the pond snail, Lymnaea elodes. Ecol. 47: 889–906.
1970. The role of food in the regulation of the pond snail, Lymnaea elodes. Ecol. 51: 680–684.

Eisner, Thomas and J. Meinwald
1966. Defensive secretions of arthropods. Sci. 153 (3742): 1341–1350.

Ekman, S.
1953. Zoogeography of the sea. Sedgwick & Jackson, London: i–xii, 1–542.

Ellerman, J. R. and T. C. S. Morrison-Scott
1951. Checklist of palaearctic and Indian mammals—1758 to 1946. British Nat. Hist. Mus., London: 1–810.

Elliott, Frank R.
1930. An ecological study of the spiders of the beech-maple forest. Ohio J. Sci. 30: 1–22.

Elliott, J. M.
1967. The life histories and drifting of the Plecoptera and Ephemeroptera in a Dartmoor stream. J. An. Ecol. 36: 343–362.
1969. Diel periodicity in invertebrate drift and the effect of different sampling periods. Oikos 20: 524–528.
1971. Upstream movements of benthic invertebrates in a Lake District stream. J. An. Ecol. 4: 235–252.

Elliott, J. M. and G. W. Minshall
1968. The invertebrate drift in the River Duddon, English Lake District. Oikos 19: 39–52.

Ellis, M. M.
1936. Erosion silt as a factor in aquatic environments. Ecol. 17: 29–42.

Elton, Charles
1925. The dispersal of insects to Spitsbergen. Trans. Ento. Soc. London, 73: 289–299.
1927. Animal ecology. Macmillan Co., New York: i–xx, 1–207.
1930. Animal ecology and evolution. Clarendon Press, Oxford: 1–96.
1932. Territory among wood ants (Formica rufa L.) at Picket Hill. J. An. Ecol. 1: 69–76.
1939. On the nature of cover. J. Wildl. Man. 3: 332–338.
1942. Voles, mice and lemmings. Clarendon Press, Oxford: 1–496.
1946. Competition and the structure of ecological communities. J. An. Ecol. 15: 54–68.
1958. The ecology of invasions by animals and plants. Methuen & Co., London: 1–181.
1966. The pattern of animal communities. John Wiley & Sons, Inc., New York: 1–432.

Emerson, Alfred E.
1938. Termite nests—a study of the phylogeny of behavior. Ecol. Mono. 8: 247–284.
1952. The supraorganismic aspects of the society. Colloq. intern., Paris, 1950: 333–353.

Emiliani, Cesare
1955. Pleistocene temperatures. J. Geol. 63: 538–578.
1966a. Isotopic paleotemperatures. Sci. 154 (3751): 851–857.
1966b. Paleotemperature analysis of Caribbean cores P6304–8 and P6304–9 and a generalized temperature curve for the past 425,000 years. J. Geol. 74: 109–126.

Emlen, John T., Jr.
1940. Sex and age ratios in survival of the California quail. J. Wildl. Man.: 4: 92–99.
1966. Some quantitative representations of ecological distribution and faunal structure applied to African birds. Ostrich, Sup. 6: 271–283.
1967. A rapid method for measuring arboreal canopy cover. Ecol. 48: 158–160.

Enders, R. K.
1935. Mammalian life histories from Barro Colorado Island, Panama. Bull. Mus. Comp. Zool. 78: 385–502.

Engel, A. E. J.
1969. Time and the earth. Amer. Sci. 57: 458–483.

Engelmann, Manfred D.
1961. The role of soil arthropods in the energetics of an old field community. Ecol. Mono. 31: 221–238.

Enright, J. T. and W. M. Hamner
1967. Vertical diurnal migration and endogenous rhythmicity. Sci. 157 (3791): 937–941.

Epling, Carl
1944. Contributions to the genetics, taxonomy, and ecology of Drosophila pseudoobscura and its relatives. III. The historical background. Carnegie Inst. Washington, Publ. 554: 145–183.

Epstein, H.
1955. Domestication features in animals as functions of human society. Agr. History 29: 137–146.

Ericson, David B. and G. Wollin
1968. Pleistocene climates and chronology in deep-sea sediments. Sci. 162 (3859): 1227–1234.

Eriksen, C. H.
1968. Ecological significance or respiration and substrate for burrowing Ephemeroptera. Can. J. Zool. 46: 93–103.

Errington, Paul L.
1932. Technique of raptor food habits study. Condor 34: 75–86.
1937. Emergency values of some winter pheasant foods. Trans. Wisconsin Acad. Sci., Arts and Let. 30: 57–68.
1939. Reactions of muskrat populations to drought. Ecol. 20: 168–186.
1943. An analysis of mink predation upon muskrats in north-central United States. Iowa State College, Agr. Exp. Sta., Res. Bull. 320: 797–924.
1945. Some contributions of a fifteen-year local study of the northern bobwhite to a knowledge of population phenomena. Ecol. Mono. 15: 1–34.
1946. Predation and vertebrate populations. Quart. Rev. Biol.: 144–177; 221–245.
1951. Concerning fluctuations in populations of the prolific and widely distributed muskrat. Amer. Nat. 85: 273–292.
1957. Of population cycles and unknowns. Cold Spring Harbor Symp. Quant. Biol. 22: 287–300.

Errington, Paul L., Frances Hamerstrom, and F. N. Hamerstrom, Jr.
1940. The great horned owl and its prey in north-central United States. Iowa St. Coll., Agr. Exp. Sta., Res. Bull. 277: 759–847.

Errington, Paul L. and F. N. Hamerstrom, Jr.
1937. The evaluation of nesting losses and juvenile mortality of the ring-necked pheasant. J. Wildl. Man. 1: 3–20.

Evans, Francis C.
1942. Studies of a small mammal population in Bagley Wood, Berkshire. J. An. Ecol. 11: 182–197.
1950. Relative abundance of species and the pyramid of numbers. Ecol. 31: 631–632.

Evans, F. C. and W. W. Murdoch
1968. Taxonomic composition, trophic structure and seasonal occurrence in a grassland insect community. J. Ecol. 37: 259–273.

Evans, Francis C. and Frederick E. Smith
1952. The intrinsic rate of natural increase for the human louse, Pediculus humanus L. Amer. Nat. 86: 299–310.

Evans, Howard Ensign
1953. Comparative ethology and the systematics of spider wasps. Sys. Zool. 2: 155–172.

Evemar, Anders
1959. On the determination of the size and composition of a passerine bird population during the breeding season. Vår. Fågelvärld. Suppl. 2: 1–114.

Ewing, Maurice and William L. Donn
1956. A theory of ice ages. Sci. 123: 1061–1066; Sci. 127 (1958): 1159–1162.

Eyre, S. R.
1963. Vegetation and soils, a world picture. Aldine Publ. Co., Chicago: i–xvi + 1–324.

Eyring, Carl F., Ralph J. Christensen and Russell W. Raitt
1948. Reverberation in the sea. J. Acoust. Soc. Amer. 20: 462–475.

Fager, Edward W.
1968. The community of invertebrates in decaying oak wood. J. An. Ecol. 37: 121–142.

Fager, E. W. and J. A. McGowan
1963. Zooplankton species groups in the North Pacific. Sci. 140 (3566): 453–460.

Fairbridge, Rhodes W. (editor)
1961. Solar variations, climatic change, and related geophysical problems. Ann. New York Acad. Sci. 95: 1–739.

Farner, Donald S.
1945. The return of robins to their birthplaces. Bird-Banding 16: 81–99.
1945a. Age groups and longevity in the American robin. Wilson Bull. 57: 56–74.
1950. The annual stimulus for migration. Condor, 52: 104–122.
1955. Birdbanding in the study of population dynamics. In Alfred Wolfson's "Recent studies in avian biology", Univ. Illinois Press, Urbana: 397–449.

Fauré-Fremiet, E.
1951. The tidal rhythm of the diatom Hantzschia amphioxys. Biol. Bull. 100: 173–177.

Fautin, Reed W.
1946. Biotic communities of the northern desert shrub biome in western Utah. Ecol. Mono. 16: 251–310.

Fawver, Ben Junior
1950. An analysis of the ecological distribution of breeding bird populations in eastern North America. Univ. Illinois Ph.D. thesis.

Feeny, Paul
1970. Seasonal changes in oak leaf tannins and nutrients as a cause of spring feeding by winter moth caterpillars. Ecol. 51: 565–581.

Fenner, Frank and F. N. Ratcliffe
1965. Myxomatosis. Univ. Press, Cambridge: i–xiv + 1–379.

Fernald, Merritt Lyndon
1950. Gray's manual of botany. Amer. Book Co., New York: i–lxiv, 1–1632.

Fichter, Edson
1939. An ecological study of Wyoming spruce-fir forest arthropods with special reference to stratification. Ecol. Mono. 9: 183–215.
1954. An ecological study of invertebrates of grassland and deciduous shrub savanna in eastern Nebraska. Amer. Midl. Nat. 51: 321–439.

Findley, James S. and Sydney Anderson
1956. Zoogeography of the montane mammals of Colorado. J. Mam. 37: 80–82.

Finn, Frank
1895–97. Contributions to the theory of warning colors and mimicry. J. Asiatic Soc. Bengal. 64, Pt. II: 344–356; 65, Pt. II, 42–48; 66, Pt. II, 528–533; 613–668.

Finzel, Jean F.
1964. Avian populations of four herbaceous communities in southeastern Wyoming. Condor 66: 496–510.

Fischer, A. G.
1960. Latitudinal variations in organic diversity. Evol. 14: 64–81.

Fisher, James
1939. Birds as animals. William Heinemann, Toronto: i–xviii, 1–281.

Fisher, R. A.
1930. The genetical theory of natural selection. Clarendon Press, Oxford: 1–272.

Fisher, Stuart G. and G. E. Likens
1972. Stream ecosystem: organic energy budget. BioSci. 22: 33–35.

Fiske, W. F.
1910. Superparasitism: an important factor in the natural control of insects. J. Econ. Ento. 3: 88–97.

Fisler, George F.
1965. Adaptations and speciation in harvest mice of the marshes of San Francisco Bay. Univ. Calif. Publ. Zool. 77: 1–108.

Fitch, Henry S.
1947. Predation by owls in the Sierran foothills of California. Condor 49: 137–151.
1948. Ecology of the California ground squirrel on grazing lands. Amer. Midl. Nat. 39: 513–596.
1957. Aspects of reproduction and development in the prairie vole (Microtus ochrogaster). Univ. Kansas Publ., Mus. Nat. Hist., 10: 129–161.

Fitch, Henry S. and J. R. Bentley
1949. Use of California annual-plant forage by range rodents. Ecol. 30: 306–321.

Flattely, F. W. and C. L. Walton
1922. The biology of the sea-shore. Macmillan Co., New York: i–xvi, 1–336.

Flemer, David A.
1969. Chlorophyll analysis as a method of evaluating the standing crop phytoplankton and primary productivity. Chesapeake Sci. 10: 301–306.
1970. Primary productivity of the North Branch of the Raritan River, New Jersey. Hydrobiol. 35: 273–296.

Flerov, C. C. and M. A. Zablotski
1961. On the causative factors responsible for the change in the bison range. Bull. Moscow Soc. Nat. 66: 99–109 (In Russian, transl. by W. A. Fuller).

Fleschner, Charles A.
1959. Biological control of insect pests. Sci. 129: 537–544.

Flint, Richard Foster
1952. The ice age in the North American arctic. Arctic 5: 135–152.
1971. Glacial and quaternary geology. John Wiley & Sons, Inc., New York: i–ix + 1–892.

Fonda, R. W. and L. C. Bliss
1969. Forest vegetation of the montane and subalpine zones, Olympic Mountains, Washington. Ecol. Mono. 39: 271–301.

Fontaine, A. R. and Fu-Shiang Chia
1968. Echinoderms: an autoradiographic study of assimilation of dissolved organic molecules. Sci. 161 (3846): 1153–1155.

Forbes, Stephen A.
1887. The lake as a microcosm. (Reprinted) Bull. Illinois Nat. Hist. Surv., 15, 1925: 537–550.
1914. Fresh water fishes and their ecology. Illinois St. Lab. Nat. Hist.: 1–19.
1928. The biological survey of a river system—its objects, methods, and results. Bull. Illinois St. Nat. Hist. Surv., 17: 277–284.

Ford, E. B.
1964. Ecological genetics. John Wiley & Sons, Inc., New York: i–xv + 1–335.

Formosof, A. N.
1933. The crop of cedar nuts, invasions into Europe of the Siberian nutcracker (Nucifraga caryocatactes macrorhynchus Brehm) and fluctuations in numbers of the squirrel (Sciurus vulgaris L.) J. Ecol. 2: 70–81.

Fosberg, F. Raymond
1970. The tropical agriculture panacea. BioSci. 20: 793.

Foster, Mercedes S.
1969. Synchronized life cycles in the orange-crowned warbler and its mallophagan parasites. Ecol. 50: 315–323.

Foster, T. Dale
1937. Productivity of a land snail, Polygyra thyroidus (Say). Ecol. 18: 545–546.

Foster, William L. and J. Tate, Jr.
1966. The activities and coactions of animals at sapsucker trees. The Living Bird, 5th Ann., Cornell Lab. Ornith.: 87–113.

Fox, H. Munro, B. G. Simmonds and R. Washbourn
1935. Metabolic rates of ephemerid nymphs from swiftly flowing and from still water. J. Exp. Biol. 12: 179–184.

Fox, Jackson L., Theron O. Odlaug and Theodore A. Olson
1969. The ecology of periphyton in western Lake Superior. Part 1—Taxonomy & distribution. Water Resources Res. Center, Univ. Minnesota Grad. Sch., Bull. 14: i–ix + 1–127.

Fraenkel, Gottfried
1932. Die Wanderungen der Insekten. Ergeb. Biol. 9: 1–238.

Fraenkel, Gottfried and M. Blewett
1943. The basic food requirements of several insects. J. Exp. Biol., 20: 28–34.

Fraenkel, Gottfried S. and Donald L. Gunn
1940. The orientation of animals. Clarendon Press, Oxford: i–vi, 1–352.

Frank, Fritz
1953. Experiments concerning the crash of fieldmouse plagues (Microtus arvalis Pallas). Zool. Jb. (Systemmatik) 82 (English translation).

1957. The causality of microtine cycles in Germany. J. Wildl. Man., 21: 113–121.

Frank, Peter W.
1957. Coactions in laboratory populations of two species of *Daphnia*. Ecol. 38: 510–519.

Frank, Peter W., Catherine D. Boll and Robert W. Kelly
1957. Vital statistics of laboratory cultures of *Daphina pulex* DeGeer as related to density. Physiol. Zool. 30: 287–305.

Franz, H. and L. Leitenberger
1948. Biologisch-chemische Untersuchungen uber Humusbildung durch Bodentiere. Österreich. Zool. Zeits. 1: 498–518.

Franz, J. M.
1961. Biological control of pest insects in Europe. Ann. Rev. Ent. 6: 183–200.

Freeman, J. A.
1946. The distribution of spiders and mites up to 300 ft. in the air. J. An. Ecol. 15: 69–74.

French, N. R., C. D. Jorgensen, M. H. Smith and B. G. Maza
1971. Comparison of some IBP population estimate methods for small mammals. Special Report, Off. Chairman, USNC/IBP: 1–25.

Frenzel, Burkhard
1968. The Pleistocene vegetation of northern Eurasia. Sci. 161 (3842): 637–649.

Frenzel, B. and C. Troll
1952. Die Vegetationszonen des nordlichen Eurasiens während der letzten Eiszeit. Eiszeitalter u. Gegenwart, Öhringen 6: 154–167.

Freuchen, Peter and Finn Salomonsen
1958. The Arctic year. G. P. Putnam's Sons, New York: 1–438.

Frey, David G.
1955. A time revision of the Pleistocene pollen chronology of southeastern North Carolina. Ecol. 36: 762–763.
1963. Limnology in North America. Univ. Wisconsin Press, Madison: i-xvii + 1–734.

Frey, D. G. and J. B. Stahl
1958. Measurements of primary production of Southampton Island in the Canadian Arctic. Limn. and Ocean. 3: 215–221.

Friauf, James J.
1953. An ecological study of the Dermaptera and Orthoptera of the Welaka area in northern Florida. Ecol. Mono. 23: 79–126.

Friedmann, Herbert
1946. Ecological counterparts in birds. Sci. Month. 63: 395–398.

Frison, Theodore H.
1935. The stoneflies, or Plecoptera, of Illinois. Bull. Illinois Nat. Hist. Surv. 20: i-v, 281–471.

Fritsch, F. E.
1931. Some aspects of the ecology of fresh-water algae. J. Ecol. 19: 233–272.

Fritz, Sigmund
1957. Solar energy on clear and cloudy days. Sci. Month. 84: 55–65.

Frohne, W. C.
1956. The provendering role of the larger aquatic plants. Ecol. 37: 387–388.

Frost, Winifred E. and W. J. P. Smyly
1952. The brown trout of a moorland fishpond. J. An. Ecol. 21: 62–86.

Fry, F. E. J.
1937. The summer migration of the cisco, *Leucichthys artedi* (Le Sueur) in Lake Nipissing, Ontario. Univ. Toronto Stud., 44: 1–91.
1947. Effects of the environment on animal activity. Univ. Toronto Stud., Biol. Ser. 55: 1–62.

Frye, John C. and A. Bryan Leonard
1952. Pleistocene geology of Kansas. St. Geol. Surv. Kansas, Bull. 99: 1–230.

Fryer, G.
1957. The food of some freshwater cyclopoid copepods and its ecological significance. J. An. Ecol. 26: 263–286.

Fuller, W. A.
1967. Ecologie hivernale des lemmings et fluctuations de leur populations. La Terre et la Vie 114: 97–115.

Fulton, B. B.
1933. Inheritance of song in hybrids of two subspecies of *Nemobius fasciatus* (Orthoptera). Ann. Ento. Soc. Amer. 26: 368–376.

Gaarder, T. and H. H. Gran
1927. Investigations of the production of plankton in the Oslo Fjord. Rapp. et Proc.—Verb., Cons. Int. Explor. Mer. 42: 1–48.

Gale, William F.
1971. An experiment to determine substrate preference of the fingernail clam, *Sphaerium transversum* (Say). Ecol. 52: 367–370.

Gardner, Leon L.
1925. The adaptive modifications and the taxonomic value of the tongue in birds. Proc. U. S. Nat. Mus. 67(19): 1–49.

Gasdorf, Edgar C. and Clarence J. Goodnight
1963. Studies on the ecology of soil arachnids. Ecol. 44: 261–268.

Gashwiler, Jay S.
1970. Plant and mammal changes on a clearcut in west-central Oregon. Ecol. 51: 1018–1026.

Gates, David M.
1962. Energy exchange in the biosphere. Harper and Row, Publishers, New York: i-viii + 1–151.

Gates, Frank C.
1942. The bogs of northern lower Michigan. Ecol. Mono. 12: 214–254.

Gause, G. F.
1934. The struggle for existence. William & Wilkins Co., Baltimore: i-ix, 1–163.
1935. Experimental demonstration of Volterra's periodic oscillations in the numbers of animals. J. Exp. Biol. 12: 44–48.

Gaylor, Dona
1921. A study of the life history and productivity of *Hyalella knickerbokeri* Bate. Proc. Indiana Acad. Sci. 37: 239–250.

Geier, P. W. and L. R. Clark
1960. An ecological approach to pest control. Sym. 8th Tech. Meet. Int. Union Cons. Nat. and Nat. Res., Warsaw, E. J. Brill, Leiden: 10–18.

Geis, James W. and William R. Boggess
1968. The prairie peninsula: its origin and significance in the vegetational history of Central Illinois. In R. E. Bergstrom's "The Quaternary of Illinois", Univ. Illinois, Coll. Agr., Sp. Publ. 14: 89–95.

Gelting, Paul
1937. Studies on the food of the east Greenland ptarmigan especially in its relation to vegetation and snow cover. Meddelelser om Grønland, 116(3): 1–196.

Gentry, Howard Scott
1942. Rio Mayo plants. Carnegie Inst. Washington, Publ. 527: i-vii, 1–328.

Gerking, Shelby D.
1949. Characteristics of stream fish populations. Invest. Indiana Lakes and Streams, 3: 285–309.
1953. Evidence for the concepts of home range and territory in stream fishes. Ecol. 34: 347–365.
1954. The food turnover of a blue gill population. Ecol. 35: 490–498.
1957. A method of sampling the littoral macrofauna and its application. Ecol. 38: 219–226.
1962. Production and food utilization in a population of bluegill sunfish. Ecol. Mono. 32: 31–78.
1967. The biological basis of freshwater fish production (a symposium). John Wiley & Sons Inc.: i-xiv + 1–495.

Gerstell, Richard
1939. Certain mechanics of winter quail losses revealed by laboratory experimentation. Trans. 4th No. Amer. Wildl. Conf.: 462–467.

Getz, Lowell L.
1960. A population study of the vole, *Microtus pennsylvanicus*. Amer. Midl. Nat. 64: 392–405.

Ghilarov, M. S.
1964. Connection of insects with the soil in different climatic zones. Pedobiol. 4: 310–315.
1967. Abundance, biomass and vertical distribution of soil animals in different zones. In K. Petruzewicz's "Secondary productivity of terrestrial ecosystems", Warsaw, Poland: 611–629.
1971. Litter-destroying invertebrates and ways of increasing their useful activity. Soviet J. Ecol. (transl.) 1: 99–109.

Gibb, John
1954. Feeding ecology of tits, with notes on treecreeper and goldcrest. Ibis 96: 513–543.

Gilbert, John J.
1966. Rotifer ecology and embryological induction. Sci. 151 (3715): 1234–1237.

Gilbert, Perry W.
1944. The alga-egg relationship in *Ambystoma maculatum*, a case of symbiosis. Ecol. 25: 366–369.

Gillespie, Walter L.
1960. Breeding bird and small mammal populations in relation to the forest vegetation of the subarctic region of northern Manitoba. Univ. Illinois, Ph.D. thesis.

Gislen, Torsten
1930. Epibioses of the Gullmar Fjord. Kristinebergs Zool. Sta. 1877–1927, No. 4: 1–380.
1948. Aerial plankton and its conditions of life. Biol. Rev. 23: 109–126.

Gleason, Henry Allen
1922. The vegetational history of the Middle West. Ann. Assoc. Am. Geog. 12: 39–85.

1926. The individualistic concept of the plant association. Bull. Torrey Bot. Club, 53: 7–26.

Glenny, Fred H.
1954. Antarctica as a center of origin of birds. Ohio J. Sci. 54: 307–314.

Glick, P. A.
1939. The distribution of insects, spiders, and mites in the air. U. S. Dept. Agr., Tech. Bull, 673: 1–150.

Gliwicz, Z. M.
1969. The food sources of lake zooplankton. Ekol. Polska, B, 15: 205–223. (English Summary).

Glynne-Williams, J. and J. Hobart
1952. Studies on the crevice fauna of a selected shore in Anglesey. Proc. Zool. Soc. London 122: 797–824.

Godfrey, Gillian K.
1954. Tracing field voles (Microtus agrestis) with a Geiger-Müller counter. Ecol. 35: 5–10.

Goff, Carlos C.
1952. Floodplain animal communities. Amer. Midl. Nat. 47: 478–486.

Goldie, A. H. R.
1936. Some characteristics of the mean annual circulation over the British Isles. Quart. J. Roy. Met. Soc. 62: 81–102.

Goldman, Charles R. (ed.)
1965. Primary productivity in aquatic environments. Mem. 1st Ital. Idrobiol., 18 Suppl., Univ. California Press, Berkeley: 1–464.

Goldman, E. A.
1937. The Colorado River as a barrier in mammalian distribution. J. Mam. 18: 427–435.

Goldsby, Richard A.
1971. Race and races. The Macmillan Co., New York: i–xi, 1–132.

Goldschmidt, Richard
1945. Mimetic polymorphism. A controversial chapter of Darwinism. Quart. Rev. Biol. 20: 147–164, 205–230.

Goldthwait, Richard P.
1959. Scenes in Ohio during the last ice age. Ohio J. Sci. 59: 193–216.

Golley, Frank B.
1960. Energy dynamics of a food chain of an old-field community. Ecol. Mono. 30: 187–206.

Golley, Frank B. and John B. Gentry
1964. Bioenergetics of the southern harvester ant, Pogonomyrmex badius. Ecol. 45: 217–225.

Golley, F. B., J. B. Gentry, L. D. Caldwell, and L. B. Davenport, Jr.
1965. Number and variety of small mammals on the AEC Savannah River Plant. J. Mam. 46: 1–18.

Golley, Frank, Howard T. Odum, and Ronald F. Wilson
1962. The structure and metabolism of a Puerto Rican red mangrove forest in May. Ecol. 43: 9–19.

Goodhart, C. B.
1962. Variation in a colony of the snail Cepaea nemoralis (L.). J. An. Ecol. 31: 207–237.

Goodnight, Clarence J. and Marie L. Goodnight
1956. Some observations in a tropical rain forest in Chiapas, Mexico. Ecol. 37: 139–150.

Goodrich, Calvin
1937. Studies of the gastropod family, Pleuroceridae. Occ. Papers Mus. Zool., Univ. Michigan 347: 1–12.

Goreau, Thomas F.
1959. The ecology of Jamaican coral reefs. 1. Species composition and zonation. Ecol. 40: 67–90.

Graff, Fr. de.
1957. The microflora and fauna of a quaking bog in the nature reserve "Het Hol" near Kortenhoet in the Netherlands. Hydrobiol. 9: 210–317.

Graham, Alan and Charles Heimsch
1960. Pollen studies of some Texas peat deposits. Ecol. 41: 751–763.

Graham, Samuel A.
1925. The felled tree trunk as an ecological unit. Ecol. 6: 397–411.
1941. Climax forests of the Upper Peninsula of Michigan. Ecol. 22: 355–362.
1952. Forest entomology. McGraw-Hill Book Co., New York: i–xii, 1–351.
1954. Changes in northern Michigan forests from browsing by deer. Trans. 19th No. Amer. Wildl. Conf.: 526–533.

Green, R. G. and C. A. Evans
1940. Studies on a population cycle of snowshoe hares on the Lake Alexander area. J. Wildl. Man. 4: 220–238, 267–278, 347–358.

Green, R. G. and C. L. Larson
1938. A description of shock disease in the snowshoe rabbit. Amer. J. Hyg. 28: 190–212.

Greenbank, D. O.
1957. The role of climate and dispersal in the initiation of outbreaks of the spruce budworm in New Brunswick. Can. J. Zool. 35: 385–403.

Greenbank, John
1945. Limnological conditions in ice-covered lakes, especially as related to winter-kill of fish. Ecol. Mono. 15: 343–392.

Greene, Robert A. and Charles Reynard
1932. The influence of two burrowing rodents, Dipodomys spectabilis spectabilis (kangaroo rat), and Neotoma albigula albigula (pack rat) on desert soils in Arizona. Ecol. 13: 73–80.

Gressitt, J. Linsley
1956. Some distribution patterns of Pacific island faunae. Sys. Zool. 5: 11–32.

Grice, George D. and Arch D. Hart
1962. The abundance, seasonal occurrence and distribution of the epizooplankton between New York and Bermuda. Ecol. Mono. 32: 287–309.

Griffin, Donald R.
1953. Sensory physiology and the orientation of animals. Amer. Sci. 41: 209–244, 281.

Griggs, Robert F.
1936. The vegetation of the Katmai district. Ecol. 17: 380–417.
1946. The timberlines of northern America and their interpretation. Ecol. 27: 275–289.

Grinnell, Joseph
1904. The origin and distribution of the chestnut-backed chickadee. Auk 21: 364–382.
1917. Field tests of theories concerning distributional control. Amer. Nat. 51: 115–128.
1924. Geography and evolution. Ecol. 5: 225–229.
1928. Presence and absence of animals. Univ. California Chron. 30: 429–450.
1936. Up-hill planters. Condor, 38: 80–82.

Griscom, Ludlow
1950. Distribution and origin of the birds of Mexico. Bull. Mus. Comp. Zool., Cambridge, Mass, 103: 339–382.

Grodzinski, W., B. Bobek, A. Drozdz, A. Gorecki
1970. Energy flow through small rodent populations in a beech forest. In K. Petrusewicz and L. Ryszkowski's "Energy flow through small mammal populations", Polish Sci. Publ., Warsaw: 291–298.

Grodzinski, W. and A. Gorecki
1967. Daily energy budgets of small rodents. In K. Petrusewicz's "Secondary productivity of terrestrial ecosystems", Inst. Ecol., Polish Acad. Sci., Warsaw: 295–312.

Grodzinski, W., Z. Pucek, and L. Ryszkowski
1966. Estimation of rodent numbers by means of prebaiting and intensive removal. Acta Theriol. 11: 297–314.

Gross, Alfred O.
1947. Cyclic invasion of the snowy owl and the migration of 1945–1946. Auk 64: 584–601.

Gross, Jack E.
1969. Optimum yield in deer and elk populations. 34th No. Amer. Wildl. Conf.: 372–387.

Gudmundsson, Finnur
1951. The effects of the recent climatic changes on the bird life of Iceland. Proc. 10th Int. Ornith. Cong. Uppsala, 1950: 502–514.
1960. Some reflections on ptarmigan cycles in Iceland. Proc. XIIth Inter. Ornith. Cong.: 259–265.

Guilday, John E.
1963. Pleistocene zoogeography of the lemming, Dicrostonyx. Evol. 17: 194–197.

Gunter, Gordon
1941. Death of fishes due to cold on the Texas coast, January 1940. Ecol. 22: 203–208.

Haartman, Lars von
1949. Der Trauerfliegenschnäpper. I. Ortstreue und Rassenbildung. Act. Zool. Fenn. 56: 1–104.

Hadley, Neil F.
1970. Micrometeorology and energy exchange in two desert arthropods. Ecol. 51: 434–444.

Hadlow, Leonard
1953. Climate, vegetation and man. Philosophical Library, Inc. New York: 1–288.

Haeckel, Ernst
1866. Generelle Morphologie der Organismen. Georg Reimer, Berlin: 2 vols.

Haffer, Jürgen
1969. Speciation in Amazonian forest birds. Sci. 165 (3889): 131–137.

Hagmeier, Edwin M.
1966. A numerical analysis of the distributional patterns of North

American mammals. II. Re-evaluation of the Provinces. Syst. Zool. 15: 279–299.

Hairston, Nelson G.
1951. Interspecies competition and its probable influence upon the vertical distribution of Appalachian salamanders of the genus *Plethodon*. Ecol. 32: 266–274.
1964. Studies on the organization of animal communities. Brit. Ecol. Soc. Jubilee Symp., J. An. Ecol. 33 (Suppl.): 227–239.
1967. Studies on the limitation of a natural population of *Paramecium aurelia*. Ecol. 48: 904–910.

Hairston, Nelson G., Frederick E. Smith, and Lawrence B. Slobodkin
1960. Community structure, population control, and competition. Amer. Nat. 94: 421–425.

Haldane, J. B. S.
1948. The theory of a cline. J. Gen. 48: 277–284.

Halftorn, Svein
1958. Population changes, especially geographical changes, in the Norwegian avifauna during the last 100 years (English summary). Sterna, 3: 105–137.

Hall, Clarence A., Jr.
1964. Shallow-water marine climates and molluscan provinces. Ecol. 45: 226–234.

Hall, E. Raymond
1946. Mammals of Nevada. Univ. California Press, Berkeley: i-xi, 1–710.
1957. Vernacular names for North American mammals north of Mexico. Univ. Kansas, Mus. Nat. Hist., Misc. Publ. 14: 1–16.

Hall, F. G.
1937. Adaptations of mammals to high altitudes. J. Mam. 18: 468–472.

Halliday, W. E. D.
1937. A forest classification for Canada. Canada Dept. Mines & Resources, For. Serv. Bull. 89: 1–50.

Halliday, W. E. D. and A. W. A. Brown
1943. The distribution of some important forest trees in Canada. Ecol. 24: 353–373.

Hamilton, W. J., Jr.
1937. Activity and home range of the field mouse, *Microtus pennsylvanicus* (Ord). Ecol. 18: 255–263.
1937a. The biology of microtine cycles. J. Agr. Res. 54: 779–790.
1940. Life and habits of field mice. Sci. Month. 50: 425–434.
1940a. The feeding habits of larval newts with reference to availability and predilection of food items. Ecol. 21: 351–356.
1941. Reproduction of the field mouse *Microtus pennsylvanicus* (Ord). Cornell Univ., Agr. Exp. Sta., Mem. 237: 1–23.

Hamilton, W. J., Jr., and David B. Cook
1940. Small mammals and the forest. J. For. 38: 468–473.

Hamilton, William J. III
1973. Life's color code, McGraw-Hill Book Co., New York: i-x, 1–238.

Hammel, H. T.
1956. Infrared emissivities of some Arctic fauna. J. Mam. 37: 375–378.

Hammen, Carl S. and Paul J. Osborne
1959. Carbon dioxide fixation in marine invertebrates: a survey of major phyla. Sci. 130: 1409–1410.

Hammer, Marie
1937. A quantitative and qualitative investigation of the microfauna communities of the soil at Angmagsaalik and in Mikis Fjord. Medd. Grønland 108 (2): 1–53.

Hammond, Allen L.
1971. Mercury in the environment: natural and human factors. Sci. 171 (3973): 788–789.
1972. Ecosystem analysis: biome approach to environmental science. Sci. 175 (4017): 46–48.

Hammond, E. Cuyler
1938–39. Biological effects of population density in lower organisms. Quart. Rev. Biol. 13: 421–438; 14: 35–59.

Hanavan, Mitchell and Bernard Einar Skud
1954. Intertidal spawning of pink salmon. U. S. Fish & Wildl. Serv., Fish. Bull. 95: 167–176.

Hand, Cadet
1956. Are corals really herbivores? Ecol. 37: 384–385.

Hanes, Ted L.
1971. Succession after fire in the chaparral of southern California. Ecol. Mono. 41: 27–52.

Hansen, Henry P.
1947. Postglacial forest succession, climate, and chronology in the Pacific Northwest. Trans. Amer. Phil. Soc. 37: 1–130.

Hanson, Harold C. and Robert L. Jones
1968. Use of feather minerals as biological tracers to determine the breeding and molting grounds of wild geese. Illinois Nat. Hist. Surv., Biol. Notes 60: 1–8.

Hanson, Harold C., Murray Rogers and Edward S. Rogers
1949. Waterfowl of the forested portions of the Canadian Pre-Cambrian shield and the Paleozoic Basin. Can. Field-Nat., 63: 183–204.

Hanson, Herbert C.
1951. Characteristics of some grassland, marsh, and other plant communities in western Alaska. Ecol. Mono. 21: 317–378.
1953. Vegetation types in northwestern Alaska and comparisons with communities in other Arctic regions. Ecol. 34: 111–140.
1962. Dictionary of ecology. Philosophical Library Inc., New York: 1–382.

Hanson, Hugh
1948. Life-histories, populations, and distribution of millipeds. Univ. Illinois, Ph.D. thesis.

Hanson, William R.
1957. Density of wood rat houses in Arizona chaparral. Ecol. 38: 650.

Hardin, Garrett
1960. The competitive exclusion principle. Sci. 131: 1292–1297.
1968. The tragedy of the Commons. Sci. 162 (3859): 1243–1248.
1970. Parenthood: right or privilege? Sci. 169 (3944): 427.

Hardy, Alister C.
1956. The open sea. Houghton Mifflin Co., Boston: i-xv, 1–335.

Hardy, A. C. and R. Bainbridge
1954. Experimental observations on the vertical migrations of plankton animals. J. Mar. Biol. Ass. U. K. 33: 409–448.

Hardy, Ross
1945. The influence of types of soil upon the local distribution of some mammals in southwestern Utah. Ecol. Mono. 15: 71–108.

Hare, F. Kenneth
1950. Climate and zonal divisions of the boreal forest formation in eastern Canada. Geog. Rev. 40: 615–635.

Harlan, Jack R.
1956. Theory and dynamics of grassland agriculture. D. van Nostrand Co., Inc., Princeton: i-x + 1–281.

Harper, Francis
1953. Birds of the Nueltin Lake Expedition, Keewatin, 1947. Amer. Midl. Nat. 49: 1–116.
1955. The barren ground caribou of Keewatin. Univ. Kans., Mus. Nat. Hist., Misc. Publ. No. 6, 1–163.
1956. The mammals of Keewatin. Univ. Kans. Mus. Nat. Hist., Misc. Publ. 12: 1–94.
1964. Plant and animal associations in the interior of the Ungava Peninsula. Univ. Kansas Misc. Publ. 38: 1–58.

Harper, J. L.
1969. The role of predation in vegetational diversity. In Woodwell and Smith's "Diversity and stability in ecological systems", Brookhaven Sym. Biol., 22: 48–62.

Harrington, Robert W., Jr. and William L. Bidlingmayer
1958. Effects of dieldrin on fishes and invertebrates of a salt marsh. J. Wildl. Man. 22: 76–82.

Harris, David R.
1967. New light on plant domestication and the origins of agriculture: a review. Geog. Rev. 57: 90–107.

Harris, Van T.
1952. An experimental study of habitat selection by prairie and forest races of the deermouse, *Peromyscus maniculatus*. Cont. Lab. Vert. Biol., Univ. Michigan 56: 1–53.

Harrison, J. L.
1962. The distribution of feeding habits among animals in a tropical rain forest. J. An. Ecol. 31: 53–63.

Hart, Charles A.
1907. Zoological studies in the sand regions of the Illinois and Mississippi River valleys. Bull. Illinois St. Lab. Nat. Hist., 7: 195–272.

Hartley, P. H. T.
1948. The assessment of the food of birds. Ibis 90: 361–381.
1953. An ecological study of the feeding habits of the English titmice. J. An. Ecol. 22: 261–288.

Harvey, E. Newton
1952. Luminescent organisms. Amer. Sci. 40: 468–481.

Harvey, H. W.
1950. On the production of living matter in the sea off Plymouth. J. Mar. Biol. Assoc. U. K. 29: 97–137.
1955. The chemistry and fertility of sea waters. Univ. Press, Cambridge: i-viii, 1–224.

Hasle, Grethe Rytter
1950. Phototactic vertical migration in marine dinoflagellates. Oikos 2: 162–175.

Hasler, Arthur D.
1954. Odour perception and orientation in fishes. J. Fish. Res. Bd. Canada 11: 107–129.
1960. Guideposts of migrating fishes. Sci. 132: 785–792.

1969. Cultural eutrophication is reversible. BioSci. 19: 425–431.

Hasler, Arthur D., O. M. Brynildson and William T. Helm
1951. Improving conditions for fish in brown-water bog lakes by alkalization, J. Wildl. Man. 15: 347–350.

Hasler, Arthur D. and Wilhelm G. Einsele
1948. Fertilization for increasing productivity of natural inland waters. Trans. 13th No. Amer. Wildl. Conf.: 527–555.

Haviland, Maud D.
1926. Forest, steppe and tundra. Univ. Press, Cambridge: 1–218.

Hawbecker, Albert C.
1951. Small mammal relationships in an *Ephedra* community. J. Mam. 32: 50–60.

Haws, T. Glenn and Clarence J. Goodnight
1962. Some aspects of the hematology of two species of catfish in relation to their habitats. Phys. Zool. 35: 8–17.

Hayes, William P.
1927. Prairie insects. Ecol. 8: 238–250.

Hayne, Don W.
1949. Two methods for estimating population from trapping records. J. Mam. 30: 399–411.
1949a. An examination of the strip census method for estimating animal populations. J. Wildl. Man. 13: 145–157.

Hayne, Don W. and Robert C. Ball
1956. Benthic productivity as influenced by fish predation. Limn. and Ocean. 1: 162–175.

Hayne, Don W. and D. Q. Thompson
1965. Methods for estimating microtine abundance. Trans. 30th No. Amer. Wildl. and Nat. Res. Conf.: 393–400.

Hayward, C. Lynn
1945. Biotic communities of the southern Wasatch and Uinta Mountains, Utah. Great Basin Nat. 6: 1–124.
1948. Biotic communities of the Wasatch chapparral, Utah. Ecol. Mono. 18: 473–506.
1952. Alpine biotic communities of the Uinta Mountains, Utah. Ecol. Mono. 22: 93–120.

Healey, Michael C.
1967. Aggression and self-regulation of population size in deermice. Ecol. 48: 377–392.

Heape, Walter
1931. Emigration, migration and nomadism. W. Heffer & Sons, Cambridge: i-xii, 1–369.

Heath, James Edward
The origins of thermoregulation. In E. T. Drake's "Evolution and environment," Yale Univ. Press, New Haven, Connecticut: 259–278.

Heatwole, Harold and R. Levins
1972. Biogeography of the Puerto Rican bank: flotsam transport of terrestrial animals. Ecol. 53: 112–117.

Hedgpeth, Joel W.
1957. Treatise on marine ecology and paleoecology. Geo. Soc. Amer., Mem 67: i-viii, 1–1296.

Hefley, Harold M.
1937. Ecological studies on the Canadian River floodplain in Cleveland County, Oklahoma. Ecol. Mono. 7: 345–402.

Heilfurth, Fritz
1936. Beitrag zur Fortpflanzungsökologie der Hochgebirgsvögel. Beit. Fortpfl.-biol. Vögel 12: 98–105.

Heim de Balsac, Henri
1936. Biogeographie des mammiferes et des oiseaux de l'Afrique du Nord. Bull. biol. France et Belgique, Paris, suppl. 21: 1–446.

Heinis, Fritz
1910. Systematik und Biologie der moosbewohnen den Rhizopoden, Rotatorien und Tardigraden. Arch. f. Hydrobiol. 5: 89–166, 217–256.

Heller, H. Craig
1971. Altitudinal zonation of chipmunks (*Eutamias*): interspecific aggression. Ecol. 52: 312–319.

Heller, H. Craig and D. M. Gates
1971. Altitudinal zonation of chipmunks (*Eutamias*): energy budgets. Ecol. 52: 424–433.

Henderson, Junius
1927. The practical value of birds. The Macmillan Co., New York: i-xii, 1–342.

Hendricks, Sterling B.
1969. Food from the land. In Preston Cloud's "Resources and man", W. H. Freeman and Co., San Francisco: 65–85.

Henning, Daniel H.
1970. Comments on an interdisciplinary social science approach for conservation administration. BioSci. 20: 11–16.

Henrici, Arthur T.
1939. The distribution of bacteria in lakes, Publ. AAAS, 10 (Problems of Lake Biology): 39–64.

Hensley, M. Max
1954. Ecological relations of the breeding bird population of the desert biome of Arizona. Ecol. Mono. 24: 185–207.

Hensley, M. Max and James B. Cope
1951. Further data on removal and repopulation of the breeding birds in a spruce-fir forest community. Auk 68: 483–493.

Herman, Carlton M.
1969. The impact of disease on wildlife populations. BioSci. 19: 321–325, 330.

Herrick, Glenn W.
1926. The "ponderable" substance of aphids (Homop.) Ento. News 37: 207–210.

Herrington, Wm. C.
1947. The role of intraspecific competition and other factors in determining the population level of a major marine species. Ecol. Mono. 17: 317–323.

Hertz, David B.
1970. The technological imperative—social implications of professional technology. Ann. Amer. Acad. Pol. Soc. Sci. 389: 95–106.

Hess, A. D. and Albert Swartz
1941. The forage ratio and its use in determining the food grade of streams. Trans. 5th No. Amer. Wildl. Conf. 1940: 162–164.

Hesse, Richard, W. C. Allee, and Karl P. Schmidt
1951. Ecological animal geography. John Wiley & Sons, Inc., New York: i-xiii, 1–715.

Hewatt, W. G.
1937. Ecological studies on selected marine intertidal communities of Monterey Bay, California. Amer. Midl. Nat. 18: 161–206.

Hewes, Leslie
1950. Some features of early woodland and prairie settlement in a central Iowa County. Ann Assoc. Amer. Geog. 40: 40–57.
1951. The northern wet prairie of the United States: nature, sources of information, and extent. Ann. Assoc. Amer. Geog. 41: 307–323.

Heyward, Frank and A. N. Tissot
1936. Some changes in the soil fauna associated with forest fires in the longleaf pine region. Ecol. 17: 659–666.

Hiatt, Robert W. and Donald W. Strasburg
1960. Ecological relationships of the fish fauna on coral reefs of the Marshall Islands. Ecol. Mono. 30: 65–127.

Hickey, Joseph J.
1952. Survival studies of banded birds. U. S. Fish & Wildl. Serv. Washington, Special Sci. Rept: Wildl. No. 15: 1–177.
1954. Mean intervals in indices of wildlife population. J. Wildl. Man. 18: 90–106.
1955. Some American population research on gallinaceous birds. In Albert Wolfson's "Recent studies in avian biology." Univ. Illinois Press, Urbana: 326–396.

Hickling, C. F.
1970. Estuarine fish farming. Adv. Mar. Biol. 8: 119–213.

Hillbricht-Illkowska, Anna and Irena Spodniewska
1969. Comparison of the primary production of phytoplankton in three lakes of different trophic type. Ekol. Polska, A, 17 (14): 241–261.

Himmer, A.
1932. Die Temperaturverhältnisse bei den sozialen Hymenopteren. Biol. Rev. 7: 224–253.

Hinde, R. A.
1956. The biological significance of the territories of birds. Ibis 98: 340–369.

Hindwood, K. A.
1955. Bird/wasp nesting associations. Emu, 55: 263–274.

Hitchcock, A. S.
1951. Manual of the grasses of the United States. U. S. Dept. Agric., Misc. Publ. No. 200; 1–1051.

Hjort, Johan, Gunnar Jahn and Per Ottestad
1933. The optimum catch. Hvalradets Skrifter 7: 92–127.

Hobbs, William Herbert
1926. The glacial anticyclons. Macmillan Co., New York: i-xxiv, 1–198.
1950. The Pleistocene history of the Mississippi River. Sci. 111: 260–262.

Hoff, C. Clayton
1943. Seasonal changes in the ostracod fauna of temporary ponds. Ecol. 24: 116–118.
1957. A comparison of soil, climate, and biota of conifer and aspen

communities in the central Rocky Mts. Amer. Midl. Nat. 58: 115–140.

Hoffman, Robert S. and J. Knox Jones, Jr.
1970. Influence of late-glacial and post-glacial events on the distribution of recent mammals on the northern Great Plains. In W. Dort, Jr. and J. K. Jones, Jr.'s "Pleistocene and recent environments of the central Great Plains," Univ. Press of Kansas, Lawrence: 355–394.

Holdgate, M. W. (editor)
1970. Antarctic ecology. Academic Press, New York: 2 vols.

Holling, C. S.
1959. The components of predation as revealed by a study of small-mammal predation of the European pine sawfly. Canadian Ent. 91: 293–320.
1966. The functional response of invertebrate predators to prey density. Mem. Ent. Soc. Canada 48: 1–86.

Holm, L. G., L. W. Weldon, and R. D. Blackburn
1969. Aquatic weeds. Sci. 166: 699–709.

Holm, Theodore
1927. The vegetation of the alpine region of the Rocky Mountains in Colorado. Nat. Acad. Sci. Mem. 19 (3): 1–45.

Holm-Hansen, Osmund
1963. Algae: nitrogen fixation by Antarctic species. Sci. 139 (3539): 1059–1060.

Holme, N. A.
1953. The biomass of the bottom fauna in the English Channel off Plymouth. J. Mar. Biol. Assoc. U. K. 37: 1–49.

Holmes, Richard T.
1966. Breeding ecology and annual cycle adaptations of the red-backed sandpiper (Calidris alpina) in northern Alaska. Condor 68: 3–46.
1970. Differences in population density, territoriality, and food supply of dunlin on Arctic and Subarctic tundra. In Adam Watson's "Animal populations in relation to their food resources", Blackwell Sci. Publ., Oxford, England: 303–319.

Holt, Charles S. and Thomas F. Waters
1967. Effect of light intensity on the drift of stream invertebrates. Ecol. 48: 225–234.

Holttum, R. E.
1922. The vegetation of West Greenland. J. Ecol. 10: 87–108.

Hopkins, David M.
1967. The Bering land bridge. Stanford Univ. Press, Stanford, Clifornia: xiii + 495.

Hora, Sunder Lal
1930. Ecology, bionomics and evolution of the torrential fauna, with special reference to the organs of attachment. Phil. Trans. Roy. Soc. London, Ser. B, 218: 171–282.

Horn, Henry S.
1971. The adaptive geometry of trees. Mono. Pop. Biol., 3: i-xi + 1–144.

Horner, B. Elizabeth
1954. Arboreal adaptations of Peromyscus, with special reference to use of the tail. Cont. Lab. Vert. Biol., Univ. Michigan, 61: 1–84.

Hornocker, Maurice G.
1970. An analysis of mountain lion predation upon mule deer and elk in the Idaho Primitive Area. Wildl. Mono. 21: 6–39.

Horsfall, William R.
1955. Mosquitoes; their bionomics and relation to disease. Ronald Press Co., New York: i-viii, 1–723.

Horvath, Otto
1963. Contributions to nesting ecology of forest birds. Ph:D. thesis, Univ. British Columbia, 1–181.

Hough, Jack L.
1963. The prehistoric Great Lakes of North America. Amer. Sci. 51: 84–109.

House, H. L. and J. S. Barlow
1958. Vitamin requirements of the house fly, Musca domestica L. (Diptera: Muscidae). Ann. Ento. Soc. Amer. 51: 299–302.

Hovanitz, William
1941. Parallel ecogenotypical color variation in butterflies. Ecol. 22: 259–284.
1948. Differences in the field activity of two female color phases of Colias butterflies at various times of day. Cont. Lab. Vert. Biol., Univ. Michigan, 41: 1–37.
1949. Increased variability in populations following natural hybridization. In Jepsen, Mayr, and Simpson's "Genetics, paleontology, and evolution": 339–355.

Howard, David I., J. I. Frea, R. M. Pfister, P. R. Dugan
1970. Biological nitrogen fixation in Lake Erie. Sci. 169 (3940): 61–62.

Howard, H. Eliot
1920. Territory in bird life. Murray, London: 1–308.

Howard, L. O. and W. F. Fiske
1911. The importation into the United States of the parasites of the gipsy moth and the brown-tail moth. U. S. Dept. Agric. Bur. Ento. Bull. 91: 1–312.

Howard, Walter E.
1949. Dispersal, amount of inbreeding, and longevity in a local population of prairie deermice on the George Reserve, southern Michigan. Cont. Lab. Vert. Biol., Univ. Michigan, 43: 1–50.
1960. Innate and environmental dispersal of individual vertebrates. Amer. Midl. Nat. 63: 152–161.
1969. The population crisis is here now. BioSci. 19: 779–784.

Howell, Henry H.
1941. Bottom organisms in fertilized and unfertilized fish ponds in Alabama. Trans. Amer. Fish. Soc. 71: 165–179.

Hubbard, John P.
1969. The relationships and evolution of the Dendroica coronata complex. Auk 86: 393–432.

Hubbert, M. King
1969. Energy resources. In Preston Cloud's "Resources and man", W. H. Freeman & Co., San Francisco, California: 157–242.

Hubbs, Carl L.
1940. Fishes of the desert. Biol. 22: 61–69.
1940a. Speciation of fishes. Amer. Nat. 74: 198–211.

Hubbs, Carl L. and R. W. Eschmeyer
1938. The improvement of lakes for fishing. Bull. Inst. Fish. Res. 2: 1–233.

Hubbs, Carl L. and Robert R. Miller
1948. The Great Basin, with emphasis on glacial and postglacial times. The zoological evidence. Univ. Utah. Bull. 38, Biol. Ser. 10 (7): 17–166.

Hubbs, Clark and Kirk Strawn
1956. Interfertility between two sympatric fishes, Notropis lutrensis and Notropis venustus. Evol. 10: 341–344.

Huey, Laurence M.
1942. A vertebrate faunal survey of the Organ Pipe Cactus National Monument, Arizona. Trans. San Diego Soc. Nat. Hist. 9: 355–375.

Huffaker, C. B.
1957. Fundamentals of biological control of weeds. Hilgardia 27: 101–157.

Huffaker, C. B. and C. E. Kennett
1956. Experimental studies on predation: predation and cyclamen-mite populations on strawberries in California. Hilgardia 26: 191–222.

Hughes, D. A.
1966. On the dorsal light response in a mayfly nymph. An. Behav. 14: 13–16.
1970. Some factors affecting drift and upstream movements of Gammarus pulex. Ecol. 51: 301–305.

Hughes, Roger N.
1970. An energy budget for a tidal-flat population of the bivalve Scrobicularia plana (Da Costa). J. An. Ecol. 39: 357–379.

Huheey, James E.
1964. Studies of warning coloration and mimicry. Ecol. 45: 185–188.

Hulett, H. R.
1970. Optimum world population. BioSci. 20: 160–161.

Hultén, Eric
1937. Outline of the history of Arctic and boreal biota during the Quarternary Period. Bokförlags Aktiebolaget Thume, Stockholm: 1–168.

Hultin, L., B. Svensson, and S. Ulfstrand
1969. Upstream movements of insects in a South Swedish small stream. Oikos 20: 553–557.

Humphries, Carmel F.
1936. An investigation of the profundal and sublittoral fauna of Windermere. J. An. Ecol. 5: 28–52.

Hungate, R. E.
1939. Experiments on the nutrition of Zootermopsis. III. The anaerobic carbohydrate dissimilation by the intestinal Protozoa. Ecol. 20: 230–245.
1960. Microbial ecology of the rumen. Bact. Rev. 24: 353–364.

Hunt, Eldridge G. and Arthur I. Bischoff
1960. Inimical effects on wildlife of periodic DDD applications to Clear Lake. California Fish and Game 46: 91–106.

Huntington, C. C. and Fred A. Carlson
1934. The geographic basis of society. Prentice-Hall, Inc., New York: i-xxi, 1–626.

Huntington, Charles E.
1952. Hybridization in the purple grackle, *Quiscalus quiscula*. Syst. Zool., 1: 149–170.

Huntington, Ellsworth
1924. Civilization and Climate. Yale Univ. Press: i-xixx, 1–453.
1945. Mainsprings of civilization. John Wiley & Sons, Inc., New York: i-xii, 1–660.

Huntsman, A. G.
1948. Method in ecology-biapocrisis. Ecol. 29: 30–42.

Hurd, L. E., M. V. Mellinger, L. L. Wolf, and S. J. McNaughton
1971. Stability and diversity at three trophic levels in terrestrial successional ecosystems. Sci. 173: 1134–1136.

Hurley, D. E.
1959 Notes on the ecology and environmental adaptations of the terrestrial Amphipoda. Pacific Sci. 13: 107–129.

Hurley, Patrick M.
1968. The confirmation of continental drift. Sci. Amer. 218 (4): 52–64.

Hutchinson, G. Evelyn
1938. On the relation between the oxygen deficit and the productivity and typology of lakes. Int. Rev. ges. Hydrobiol. u. Hydrogr. 36: 336–355.
1944. Nitrogen in the biogeochemistry of the atmosphere. Amer. Sci. 32: 178–195.
1948. Circular causal systems in ecology. Ann. N. Y. Acad. Sci. 50: 221–246.
1950. The biogeochemistry of vertebrate excretion. Amer. Mus. Nat. Hist. 96: i-xviii, 1–554.
1951. Copepodology for the ornithologist. Ecol. 32: 571–577.
1957. A treatise on limnology, Vol. 1: Geography, Physics and Chemistry. John Wiley & Sons, Inc., New York: i-xiv, 1–1015.
1957. Concluding remarks. Cold Spring Harbor Sym. Quant. Biol. 22, Cold Spring Harbor, N. Y.: 415–427.
1959. Homage to Santa Rosalia or why are there so many kinds of animals? Amer. Nat., 93: 145–159.
1967. A treatise on limnology. Volume II. Introduction to lake biology and the limnoplankton. John Wiley & Sons, Inc., New York: i-ix + 1–1115.

Hutchinson, G. E. and E. S. Deevey, Jr.
1949. Ecological studies on animal populations. In Survey of Biological Progress, Vol. I. Academic Press. Inc., New York: 325–359.

Huxley, Julian
1943. Evolution, the modern synthesis. Harper & Bros., New York: 1–645.

Huxley, T. H.
1868. On the classification and distribution of the *Alectoromorphae* and *Heteromorphae*. Proc. Zool. Soc. London: 294–319.

Hynes, H. B. N.
1954. The ecology of *Gammarus duebeni* Lilljeborg and its occurrence in fresh water in western Britain. J. An. Ecol. 23: 38–84.
1960. The biology of polluted water. Liverpool Univ. Press: i-xiv + 1–202.
1970. The ecology of running waters. Univ. Toronto Press, Toronto, Canada: i-xxiv + 1–555.

Ide, F. P.
1935. The effect of temperature on the distribution of the mayfly fauna of a stream. Univ. Toronto Stud. Biol. Ser. 39: 1–76.

Inger, Robert F. and B. Greenberg
1966. Annual reproductive patterns of lizards from a Bornean rain forest. Ecol. 47: 1007–1021.

Ingles, Lloyd Glenn
1929. The seasonal and associational distribution of the fauna of the Upper Santa Ana River Wash. J. Ento. and Zool. 21: 1–48, 57–96.
1931. The succession of insects in tree trunks as shown by the collections from the various stages of decay. J. Ento. and Zool. 23: 57–59.
1950. Nesting birds of the willow-cottonwood community in California. Auk 67: 325–332.

Interstate Deer Herd Committee
1951. The Devils Garden deer herd. California Fish and Game 37: 233–272.

Irving, George W., Jr.
1970. Agricultural pest control and the environment. Sci. 168 (3938): 1419–1424.

Irving, Laurence
1957. The usefulness of Scholander's views on adaptive insulation of animals. Evol. 11: 257–259.

Iselin, C. O.
1952. Marine fouling and its prevention. U. S. Naval Inst., Annapolis, Md.: i-viii, 1–388.

Isely, F. B.
1938. Survival value of acridian protective coloration. Ecol. 19: 370–389.
1938. The relations of Texas Acrididae to plants and soils. Ecol. Mono. 8: 551–604.
1941. Researches concerning Texas Tettigoniidae. Ecol. Mono. 11: 457–475.

Istock, Conrad A.
1965. Distribution, coexistence and competition of whirligig beetles. Evol. 20: 211–234.

Ito, Yosiaki
1954. Sympatric occurrence of two species of aphids and their leaf preference, with special reference to the ecological significance (English summary). Bull. Nat. Inst. Agr. Sci., Ser. C, 4: 187–199.
1959. A comparative study on survivorship curves for natural insect populations. Jap. J. Ecol. 9: 107–115.

Ito, Y. and K. Miyashita
1965. Studies on the dispersal of leaf and planthoppers. III. An examination of the distance-dispersal rate curves. Jap. J. Ecol. 15: 85–89.

Ivlev, Victor S.
1939. Transformation of energy by aquatic animals. Coefficient of energy consumption by *Tubifex tubifex* (Oligochaeta). Int. Rev. ges. Hydrobiol. u. Hydrogr. 38: 449–458.
1945. The biological productivity of waters (English translation). Uspekhi Sovremennoi Biologii 19: 98–120.

Jackson, Alfred S.
1947. A bobwhite quail irruption in northwest Texas Lower Plains terminated by predation. Trans. 12th No. Amer. Wildl. Conf.: 511–519.

Jackson, C. H. N.
1933. On the true density of tsetse flies. J. An. Ecol. 2: 204–209.

Jackson, Hartley H. T.
1956. The return of the vanishing musk oxen. Audubon Mag. 58: 262–265, 289.

Jacobs, Merle E.
1955. Studies on territorialism and sexual selection in dragonflies. Ecol. 36: 566–586.

Jacot, Arthur Paul
1936. Soil structure and soil biology. Ecol. 17: 359–379.

Jaeger, Edmund C.
1955. The California deserts. Stanford Univ. Press, Stanford, California: i-x, 1–211.
1957. The North American deserts. Stanford Univ. Press, Stanford, California: i-vii, 1–308.

Jahn, Hermann
1942. Zur Oekologie und Biologie der Vögel Japans. J. Ornith. 90: 1–302.

Jameson, E. W., Jr.
1947. Natural history of the prairie vole (mammalian genus *Microtus*). Univ. Kansas Publ., Mus. Nat. Hist., 1: 125–151.

Jameson, E. W., Jr. and James M. Brennan
1957. An environmental analysis of some ectoparasites of small forest mammals in the Sierra Nevada, California. Ecol. Mono. 27: 45–54.

Janzen, Daniel H.
1967. Why mountain passes are higher in the tropics. Amer. Nat. 101: 233–249.
1969. Seed-eaters versus seed size, number, toxicity and dispersal. Evol. 23: 1–27.
1971. Herbivores and the number of tree species in tropical forests. Amer. Nat. 104: 501–528.

Jenkins, Dale W.
1950. Northern extension of the maple-basswood forest during the xerothermic period. Bull. Ecol. Soc. Amer. 31: 53–54.

Jennings, H. S.
1906. Behavior of the lower organisms. Columbia Univ. Press, New York: i-xiv, 1–366.

Jensen, P. Boysen
1919. Valuation of the Limfjord. I. Studies on the fish-food in the Limfjord 1909–1917, its quantity, variation and annual production. Rep't. Danish Biol. Sta. 26: 3–44.

Jerlov, N. G.
1951. Optical studies of ocean waters. Rep'ts. Swedish Deep Sea Expedition, 3. Physics and Chemistry, No. 1.: 1–59.

Jewell, Minna E.
1935. An ecological study of the fresh-water sponges of northeastern Wisconsin. Ecol. Mono. 5: 461–504.

Jewell, Minna E. and Harold W. Brown
1929. Studies on northern Michigan bog lakes. Ecol. 10: 427–475.

Johannes, R. E. and Kenneth L. Webb
1965. Release of dissolved amino acids by marine zooplankton. Sci. 150: 76–77.

Johansen, Frits
1922. The crustacean life of some arctic lagoons, lakes and ponds. Rep't. Can. Arctic Exp., 1913–1918, 7: 1–31.

Johansen, Hans
1956–58. Revision und entstehung der arktischen vogelfauna. Acta Arctica 8: 1–98, 9: 1–131.
1963. Zoogeographical aspects of the birds of the Subarctic. Proc. XIII Inter. Ornith. Cong.: 1117–1123.

Johnsgard, Paul A. and Irven O. Buss
1956. Waterfowl sex ratios during spring in Washington State and their interpretation. J. Wildl. Man. 20: 384–388.

Johnson, C. G.
1951. The study of wind-borne insect populations in relation to terrestrial ecology, flight periodicity and the estimation of aerial populations. Sci. Prog. 39: 41–62.
1957. The distribution of insects in the air and the empirical relation of density to height. J. An. Ecol. 26: 479–494.

Johnson, David H., Monroe D. Bryant and Alden H. Miller
1948. Vertebrate animals of the Providence Mountains area of California. Univ. California Publ. Zool. 48: 221–376.

Johnson, M. S. and Francis Munger
1930. Observations on excessive abundance of the midge Chironomus plumosus) at Lake Pepin. Ecol. 11: 110–126.

Johnson, R. A.
1954. The behavior of birds attending army ant raids on Barro Colorado Island, Panama Canal Zone. Proc. Linn. Soc. N. Y., Nos. 63–65: 41–70.

Johnston, David W. and Eugene P. Odum
1956. Breeding bird populations in relation to plant succession on the Piedmont of Georgia. Ecol. 37: 50–62.

Johnston, Richard F. and Robert K. Selander
1964. House sparrows: rapid evolution of races in North America. Sci. 144 (3618): 548–550.

Jones, Frank Morton
1932. Insect coloration and the relative acceptability of insects to birds. Ento. Soc. London, 80: 345–385.

Jones, N. S.
1950. Marine bottom communities. Biol. Rev. 25: 283–313.

Jordan, Carl. F.
1971. A world pattern in plant energetics. Amer. Sci. 59: 425–433.

Jordan, David Starr
1905. The origin of species through isolation. Sci. 22: 545–562.
1928. The distribution of fresh-water fishes. Ann. Rep't. Smiths. Inst. for 1927: 355–385.

Jordan, James S.
1971. Yield from an intensively hunted population of eastern fox squirrels. NE. Forest Exp. Sta., Upper Darby, Pa.: 1–8.

Jørgensen, C. Barker
1955. Quantitative aspects of filter feeding in vertebrates. Biol. Rev. 30: 391–454.

Juday, Chancey
1940. The annual energy budget of an inland lake. Ecol. 21: 438–450.
1942. The summer standing crop of plants and animals in four Wisconsin lakes. Trans. Wisconsin Acad. Sci., Arts., Let. 34: 103–135.

Kachkarov, D. N. and E. P. Korovine
1942. La vie dans les deserts (translated, revised and expanded by T. Monod), Payot, Paris: 1–360.

Kajak, Zdzislaw, Krzysztof Dusoge, and Andrzej Prejs
1968. Application of the flotation technique to assessment of absolute numbers of benthos. Ekol. Polska, A. 16: 607–620.

Kalabuchov, N. J.
1937. Some physiological adaptations of the mountain and plains forms of the wood-mouse (Apodemus sylvaticus) and of other species of mouse-like rodents. J. An. Ecol. 6: 254–272.

Kalabukhov, N. I.
1935. On the causes of fluctuations in number of mouse-like rodents. Review of the literature (English translation). Zoologicheskii Zhurnal, 14: 209–242.
1965. The structure and dynamics of natural foci of plague. J. Hyg., Epidem., Microbiol. and Imm. 9: 147–159.

Kalela, Olavi
1938. Über die regionale Verteilung der Brutvogelfauna im Flussgebiet des Kokemäenjoki. Ann. Zool. Soc. Zool.-Bot. Fenn. Vanamo. 5: 1–191.
1948. The occurrence of roe deer in Finland and changes in its distribution in the adjoining areas (English summary). Suomen Riista 3: 34–56.

1949. Changes in geographic ranges in the avifauna of northern and central Europe in relation to recent changes in climate. Bird-Banding 20: 77–103.
1955. Die neuzeitliche Ausbreitung des Kiebitzes Vanellus vanellus (L.), in Finnland. Ann. Zool. Soc. Vanamo 16: 1–80.

Kalela, Olavi and Terttu Kopenen
1971. Food consumption and movements of the Norwegian lemming in areas characterized by isolated fells. Ann. Zool. Fenn. 8: 80–84.

Kalmbach, E. R.
1934. Field observation in economic ornithology. Wilson Bull. 46: 73–90.
1939. Nesting success: its significance in waterfowl reproduction. Trans. 4th No. Amer. Wildl. Conf: 591–604.

Kapoor, Inder P., Robert L. Metcalf, Robert F. Nystrom, and Gurcharan K. Sangha
1970. Comparative metabolism of methoxychlor, methioclor, and DDT in mouse, insects, and in a model ecosystem. Agr. Food Chem. 18: 1145–1152.

Karplus, Martin
1949. Bird activity in the continuous daylight of Arctic summer. Bull. Ecol. Soc. Amer. 30: 66.

Karr, James R.
1968. Habitat and avian diversity of strip-mined land in east-central Illinois. Condor 70: 348–357.
1971. Structure of avian communities in selected Panama and Illinois habitats. Ecol. Mono. 41: 207–229.

Karr, James R. and Roland R. Roth
1971. Vegetation structure and avian diversity in several New World areas. Amer. Nat. 105: 423–435.

Kear, Janet
1962. Food selection in finches with special reference to interspecific differences. Proc. Zool. Soc. London, 138: 163–204.

Keast, Allen
1968. Australian mammals: zoogeography and evolution. Quart. Rev. Biol. 43: 373–408.
1969. Evolution of mammals on southern continents. Quart. Rev. Biol. 44: 121–167.
1970. Food specializations and bioenergetic interrelations in the fish faunas of some small Ontario waterways. In J. H. Steele's "Marine food chains", Oliver & Boyd, Edinburgh: 377–411.
1970. Adaptive evolution and shifts in niche occupation in island birds. Biotrop. 2: 61–75.

Keever, Catherine, H. J. Oosting, and L. E. Anderson
1951. Plant succession on exposed granite of Rocky Face Mountain, Alexander County, North Carolina. Bull. Torrey Bot. Club 78: 401–421.

Keith, L. B.
1963. Wildlife's ten-year cycle. Univ. of Wisconsin Press, Madison, i–xvi + 1–201.

Keller, Barry L. and Charles J. Krebs
1970. Microtus population biology; III. Reproductive changes in fluctuating populations of M. ochrogaster and M. pennsylvanicus in southern Indiana, 1965–67. Ecol. Mono. 40: 263–294.

Kellogg, Vernon Lyman
1913. Distribution and species-forming of ecto-parasites. Amer. Nat. 47: 129–158.

Kellogg, W. W., R. D. Cadle, E. R. Allen, A. L. Lazrus, and E. A. Martell
1972. The sulfur cycle. Sci. 175 (4022): 587–596.

Kemp, Gerald A. and L. B. Keith
1970. Dynamics and regulation of red squirrel (Tamiasciurus hudsonicus) populations. Ecol. 51: 763–779.

Kendeigh, S. Charles
1934. The role of environment in the life of birds. Ecol. Mono. 4: 299–417.
1941. Length of day and energy requirements for gonad development and egg-laying in birds. Ecol. 22: 237–248.
1941a. Birds of a prairie community. Condor 43: 165–174.
1941b. Territorial and mating behavior of the house wren. Illinois Biol. Mono. 18: 1–120.
1942. Analysis of losses in the nesting of birds. J. Wildl. Man. 6: 19–26.
1942a. Research areas in the national parks, January 1942. Ecol. 23: 236–238.
1944. Measurement of bird populations. Ecol. Mono. 14: 67–106.
1945. Community selection by birds on the Helderberg Plateau of New York. Auk 62: 418–436.
1946. Breeding birds of the beech-maple-hemlock community. Ecol. 27: 226–244.

1947. Bird population studies in the coniferous forest biome during a spruce budworm outbreak. Ontario Dept. Lands and Forests, Biol. Bull. 1: 1–100.

1948. Bird populations and biotic communities in northern Lower Michigan. Ecol. 29: 101–114.

1951. Nature sanctuaries in the United States and Canada—a preliminary inventory. Living Wilderness 15: 1–45.

1952. Parental care and its evolution in birds. Ill. Biol. Mono. 22: i-x, 1–356.

1954. History and evaluation of various concepts of plant and animal communities in North America. Ecol. 35: 152–171.

1956. A trail census of birds at Itasca State Park, Minnesota. Flicker 28: 90–104.

1969. Tolerance of cold and Bergmann's Rule. Auk 86: 13–25.

1973. Monthly variations in the energy budget of the house sparrow throughout the year. In S. C. Kendeigh and J. Pinowski's "Productivity, population dynamics, and systematics of granivorous birds", Inst. Ecol., Pol. Acad. Sci., Warsaw 1–410.

Kendeigh, S. Charles and S. Prentiss Baldwin
1937. Factors affecting yearly abundance of passerine birds. Ecol. Mono. 7: 91–124.

Kendeigh, S. Charles, George C. West and George W. Cox
1960. Annual stimulus for spring migration in birds. An. Beh. 8: 180–185.

Kenk, Roman
1949. The animal life of temporary and permanent ponds in southern Michigan. Misc. Publ. Mus. Zool., Univ. Michigan 71: 1–66.

1953. The fresh-water triclads (Turbellaria) of Alaska. Proc. U. S. Nat. Mus. 103: 163–186.

Kennedy, Clarence H.
1922. The ecological relationships of the dragonflies of the Bass Islands of Lake Erie. Ecol. 325–336.

1927. The metabolic gradient of the group or the ecological factors in zoogeographic distribution. Bull. Ecol. Soc. Amer. 8: 7.

Kennedy, J. S. (ed.)
1961. Insect polymorphism. London: Roy. Ent. Soc. (Sym. 1): 1–115.

Kennedy, Robert Vix
1958. Distribution of overwintering populations of insects in an oak-maple forest of east-central Illinois. Univ. of Illinois, M. S. thesis.

Kessel, Brina
1953. Distribution and migration of the European starling in North America. Condor 55: 49–67.

Ketchum, Bostwick H.
1947. The biochemical relations between marine organisms and their environment. Ecol. Mono. 17: 309–315.

Ketchum, Bostwick H., Lois Lillick and Alfred C. Redfield
1949. The growth and optimum yields of unicellular algae in moss culture. J. Cell. Comp. Physiol. 33: 267–279.

Kettlewell, H. B. D.
1965. Insect survival and selection for pattern. Sci. 148: 1290–1296.

Kevan, D. Keith McE. (editor)
1955. Soil zoology. Academic Press, Inc., New York: i-xiv, 1–512.

Kilgore, W. W. and R. L. Doutt (eds.)
1967. Pest control: biological, physical, and selected chemical methods. Academic Press, New York: i-xii + 1–477.

Kincer, Joseph B.
1923. The climate of the Great Plains as a factor in their utilization. Ann. Asso. Amer. Geog. 13: 67–80.

1941. Climate and weather data for the United States. Climate and Man. Yearbook of Agr., U. S. Dept. Agric.: 685–747.

King, Charles E.
1967. Food, age, and the dynamics of a laboratory population of rotifers. Ecol. 48: 111–128.

Kingsolver, John M. and Milton W. Sanderson
1967. A selected bibliography of insect-vascular plant associational studies. Agr. Res. Serv., U.S. Dept. Agr., 33–115: 3–33.

Kinne, Otto (ed.)
1970. Marine ecology. John Wiley & Sons Ltd, New York, 1: i-ix + 1–681.

Kinsey, Alfred C.
1930. The gall wasp genus *Cynips*. A study in the origin of species. Indiana Univ. Stud. 16: 84–86 (1929): 1–577.

1942. Isolating mechanism in gall wasps. Biol. Sym. 6: 251–269.

Kirby, Harold, Jr.
1937. Host-parasite relations in the distribution of Protozoa in termites. Univ. California Publ. Zool. 41: 189–211.

Kiritani, Keizi
1964. Natural control of populations of the southern green stink bug, *Nezara viridula*. Res. Pop. Ecol. 6: 88–98.

Kiritani, Keizi, N. Hokyo, and K. Kimura
1967. The study on the regulatory system of the population of the southern green stink bug, *Nezara viridula* L. (Heteroptera: Pentatomidae) under semi-natural conditions. Appl. Ent. Zool. 2: 39–50.

Kirk, Dudley
1967. Prospects for reducing natality in the underdeveloped world. Ann. Amer. Acad. Pol. Soc. Sci. 369: 48–60.

Kitazawa, Yuzo
1959. Bio-economic study of natural populations of animals. Jap. J. Zool. 12: 401–448.

Klaauw, C. J. Van Der
1948. Ecological morphology. Bibl. Bioth. D, 4: 27–111.

Klein, David R.
1970. Food selection by North American deer and their response to overutilization of preferred plant species. In Adam Watson's "Animal populations in relation to their food resources". Blackwell Sci. Publ., Oxford, England: 25–46.

Klekowski, R. Z., T. Prus, and H. Zyromska-Rudzka
1967. Elements of energy budget of *Tribolium castaneum* (Hbst.) in its developmental cycle. In K. Petrusewicz's "Secondary productivity of terrestrial ecosystems", Inst. Ecol., Pol. Acad. Sci., Warsaw: II 859–879.

Klomp, H.
1962. The influence of climate and weather on the mean density level, the fluctuations and the regulations of animal populations. Arch. Neerl. Zool. 15: 68–109.

Klopfer, Peter
1964. Parameters of imprinting. Amer. Nat. 48: 173–182.

Klopfer, P. H. and A. Jolly
1970. The stability of territorial boundaries in a lemur troop. Folia primat. 12: 199–208.

Klopfer, Peter H. and R. H. MacArthur
1961. On the causes of tropical species diversity: niche overlap. Amer. Nat. 95: 223–226.

Klugh, A. Brooker
1924. Factors controlling the biota of tidepools. Ecol. 5: 192–196.

1927. The ecology, food-relations and culture of fresh-water entomostraca. Trans. Roy. Canadian Inst., Toronto 16 (Pt. 1): 15–98.

Kluijver, H. N.
1951. The population ecology of the great tit, *Parus m. major* L. Ardea 39: 1–135.

Kluijver, H. N. and L. Tinbergen
1953. Territory and the regulation of density in titmice. Arch. Neerl. Zool. 10: 265–289.

Klute, Fritz
1928. Die Bedeutung der Depression der Schneegrenze fur eiszeitliche Probleme. Zeits. Gletscherkunde, 16: 70–93.

Knight, Henry G.
1937. Selenium and its relation to soils, plants, animals and public health. Sigma Xi Quart. 25: 1–9.

Knipling, E. F.
1963. A new era in pest control: the sterility principle. Agr. Sci. Rev. 1: 2–12.

Koeppe, Clarence Eugene
1931. The Canadian climate. McKnight & McKnight, Bloomington, Illinois: 1–280.

Kofoid, C. A.
1908. The plankton of the Illinois River. Bull. Illinois St. Lab. Nat. Hist. 8: 1–361.

Koford, Carl B.
1958. Prairie dogs, whitefaces, and blue grama. Wildl. Mono. No. 3: 1–78.

Korringa, P.
1947. Relation between the moon and periodicity in the breeding of marine animals. Ecol. Mono. 17: 347–381.

1952. Recent advances in oyster biology. Quart. Rev. Biol. 27: 266–308, 339–365.

Kozlovsky, Daniel G.
1969. A critical evaluation of the trophic level concept. I. Ecological efficiencies. Ecol. 49: 48–60.

Krafka, Jr., Joseph
1920. The effect of temperature upon facet number in the bar-eyed mutant of *Drosphila*. J. Gen. Physiol. 2: 409–464.

Krebs, Charles J.
1964. The lemming cycle at Baker Lake, Northwest Territories, during 1959–62. Arctic Inst. No. Amer., Tech. Pap. 15: 1–104.

1964. Cyclic variation in skull-body regressions of lemmings. Can. J. Zool. 42: 631–643.

1966. Demographic changes in fluctuating populations of *Microtus californicus*. Ecol. Mono. 36: 239–273.

1970. *Microtus* population biology: behavioral changes associated with the population cycle in *M. ochrogaster* and *M. pennsylvanicus*. Ecol. 51: 34–52.

Krebs, John R.
1970. Regulation of numbers in the great tit (Aves: Passeriformes). J. Zool., London 162: 317–333.

Krecker, F. H.
1919. The fauna of rock bottom ponds. Ohio. J. Sci. 19: 427–474.

Krecker, Frederick H. and L. Y. Lancaster
1933. Bottom shore fauna of western Lake Erie: a population study to a depth of six feet. Ecol. 14: 79–93.

Krefting, Laurits W., and Eugene I. Roe
1949. The role of some birds and mammals in seed germination. Ecol. Mono. 19: 269–286.

Krivolutskii, D. A.
1972. Current concepts of animal "life forms". Soviet J. Ecol. (transl.): 2: 202–207.

Krumbiegel, Ingo
1932. Untersuchungen über physiologische Rassenbildung. Zool. Jahrb., Abt. Syst., Ökol., Geog. Tiere, 63: 183–280.

Kucera, C. L.
1956. Grazing effects on composition of virgin prairie in north central Missouri. Ecol. 37: 389–391.

Küchler, A. W.
1953. Natural vegetation. Goode's World Atlas, Rand McNally Co., Chicago, Ill: 16–17, 52–53.

Kuehne, Robert A.
1962. A classification of streams, illustrated by fish distribution in an eastern Kentucky creek. Ecol. 43: 608–614.

Kuenzler, E. J.
1961. Structure and energy flow of a mussel population in a Georgia salt marsh. Limnol. Oceanogr. 6: 191–204.

Kühnelt, Wilhelm
1944. Uber Beziehungen zwischen Tier-und Pflanzen-Gesellschaften. Biol. Gen. 17: 566–593.
1961. Soil biology. Faber and Faber, London: 1–397.

Kulicke, H.
1960. Wintervermehrung von Rötelmaus (*Clethrionomys glareolus*), Erdmaus (*Microtus agrestis*) and Gelbhalsmaus *Apodemus flavicollis*. Z. Saugetierk. 25: 89–91.

Kulp, J. Laurence
1961. Geologic time scale. Sci. 133 (3459): 1105–1114.

Kurten, Björn
1958. A differentiation index, and a new measure of evolutionary rates. Evol. 12: 146–157.

Kusnezov, N.
1957. Numbers of species of ants in faunae of different latitudes. Evol. 11: 298–299.

Lack, David
1937. The psychological factor in bird distribution. Brit. Birds 31: 130–136.
1944. Ecological aspects of species-formation in passerine birds. Ibis 86: 260–286.
1946. Competition for food by birds of prey. J. An. Ecol. 15: 123–129.
1947. The significance of clutch-size. Ibis 47: 302–352; 48: (1948) 25–45.
1948. Natural selection and family size in the starling. Evol. 2: 95–110.
1951. Population ecology in birds. Proc. 10th Intern. Orinth. Congress, Uppsala, 1950. Uppsala-Stockholm: 409–448.
1952. Reproductive rate and population density in the great tit: Kluijver's study. Ibis 94: 167–173.
1954. The evolution of reproductive rates. In J. S. Huxley's "Evolution as a process." Allen & Unjvin, London: 143–156.
1954. The natural regulation of animal numbers. Clarendon Press, Oxford: i–viii, 1–343.
1965. Evolutionary ecology. J. Ecol. 53: 237–245.
1966. Population studies of birds. Clarendon Press, Oxford: i–v + 1–341.
1967. The significance of clutch-size in waterfowl. Wildfowl Trust 18th An. Rep.: 125–128.
1971. Ecological isolation in birds. Blackwell Sci. Publ., Oxford: i–xi + 1–404.

Lack, David, John Gibb, and D. F. Owen
1957. Survival in relation to brood-size in tits. Proc. Zool. Soc. London, 128: 313–326.

Lack, David and L. S. V. Venables
1939. The habitat distribution of British woodland birds. J. An. Ecol. 8: 39–71.

Ladd, Harry S.
1959. Ecology, paleontology and stratigraphy. Sci. 129: 69–78.

Laessle, Albert M.
1961. A micro-limnological study of Jamaican bromeliads. Ecol. 42: 499–517.

Lagler, Karl F.
1952. Freshwater fishery biology. Wm. C. Brown Co., Dubuque, Iowa: i–x, 1–360.

Lake, Charles T.
1936. The life history of the fan-tailed darter *Catonotus flabellaris flabellaris* (Rafinesque). Amer. Midl. Nat. 17: 816–130.

Lakhani, K. H. and J. E. Satchell
1970. Production by *Lumbricus terrestris* (L.). J. An. Ecol. 39: 473–492.

Lanciani, Carmine A.
1970. Resource partitioning in species of the water mite genus *Eylais*. Ecol. 51: 338–342.

Landsberg, Helmut E.
1970. Man-made climatic changes. Sci. 170 (3964): 1265–1274.

Langford, R. R.
1938. Diurnal and seasonal changes in the distribution of the limnetic crustacea of Lake Nipissing, Ontario. Univ. Toronto Stud. 45: 1–142.

Lanyon, Wesley E.
1957. The comparative biology of the meadowlarks (*Sturnella*) in Wisconsin. Publ. Nuttall Ornith. Club, 1: 1–67.

Lapage, Geoffrey
1951. Parasitic animals. Cambridge Univ. Press: i–xxi, 1–351.

Larimore, R. Weldon and Philip W. Smith
1963. The fishes of Champaign County, Illinois, as affected by 60 years of stream changes. Bull. Ill. Nat. Hist. Surv., 28: 295–382.

La Roi, George H.
1967. Ecological studies in the boreal spruce-fir forests of the North American taiga. I. Analysis of the vascular flora. Ecol. Mono. 37: 229–253.

Larson, Floyd
1940. The role of the bison in maintaining the short grass plains. Ecol. 21: 113–121.

Larson, Sten
1957. The suborder Charadrii in arctic and boreal areas during the Tertiary and Pleistocene, a zoogeographic study. Acta Vert. 1: 1–84.
1960. On the influence of the Arctic fox, *Alopex lagopus*, on the distribution of Arctic birds. Oikos 11: 276–305.

Lauckhart, J. Burton
1957. Animal cycles and food. J. Wildl. Man. 21: 230–234.

Laughton, A. S.
1971. South Labrador Sea and the evolution of the North Atlantic. Nat. 232: 612–617.

Laurence, B. R.
1954. The larval inhabitants of cow pats. J. An. Ecol. 23: 234–260.

Lawrence, Donald B.
1958. Glaciers and vegetation in south-eastern Alaska. Amer. Sci. 46: 89–122.

Lawrence, George E.
1966. Ecology of vertebrate animals in relation to chaparral fire in the Sierra Nevada foothills. Ecol. 47: 278–291.

Lawrence, R. F.
1953. The biology of the cryptic fauna of forests. A. A. Balkema, Amsterdam: 1–408.

Le Cren, E. D.
1965. A note on the history of mark-recapture population estimates. J. An. Ecol. 34: 453–454.

Lees, John
1948. Winter feeding hours of robins, blackbirds and blue tits. Brit. Birds 41: 71–76.

Lehmann, V. W.
1953. Bobwhite population fluctuations and vitamin A. Trans. 18th No. Amer. Wildl. Conf.: 199–246.

Lemon, Edgar, D. W. Stewart, and R. W. Shawcroft
1971. The sun's work in a cornfield. Sci. 174 (4007): 371–378.

Leopold, A. Carl and Robert Ardrey
1972. Toxic substances in plants and the food habits of early man. Sci. 176 (4034): 512–513.

Leopold, A. Starker
1950. Vegetation zones of Mexico. Ecol. 31: 507–518.

Leopold, Aldo
1933. Game management. Charles Scribner's Sons, New York: i–xxi, 1–481.

Leopold, Aldo and Sara Elizabeth Jones
  1947. A phenological record for Sauk and Dane Counties, Wisconsin, 1935–45. Ecol. Mono. 17: 81–122.
Leopold, Aldo, Lyle K. Sowls and David L. Spencer
  1947. A survey of over-populated deer ranges in the United States. J. Wildl. Man. 11: 162–177.
Lerner, I. Michael
  1954. Genetic homeostasis. Oliver and Boyd, London: i–vii, 1–134.
Lewis, J. R.
  1964. The ecology of rocky shores. The English Universities Press Ltd, London: i–xii + 1–323.
Lewis, William M. and S. Flickinger
  1967. Home range tendency of the largemouth bass (*Micropterus salmoides*). Ecol. 48: 1020–1023.
Li, Ching Chun
  1955. Population genetics. Chicago Univ. Press, Chicago: i–xi, 1–366.
Libby, W. F.
  1961. Radiocarbon dating. Sci. 133: 621–629.
Licht, Lawrence E.
  1967. Growth inhibition in crowded tadpoles: intraspecific and interspecific effects. Ecol. 48: 736–745.
Lidicker, William Z., Jr.
  1962. Emigration as a possible mechanism permitting the regulation of population density below carrying capacity. Amer. Nat. 96: 29–33.
  1965. Comparative study of density regulation in confined populations of four species of rodents. Res. Pop. Ecol. 7: 57–72.
Lieth, Helmut
  1971. Mathematical modelling for ecosystem analysis. In "Productivity of forest ecosystems", Proc. Brussels Sym. 1969 (Unesco): 567–575.
  1972. Uber die Primarproduktion der Pflanzendecke der Erde. Angew. Bot. 46: 1–37.
Lin, Norman
  1963. Territorial behavior in the cicada killer wasp, *Sphecius speciosus* (Drury) (Hymenoptera: Sphecidae). Behav. 20: 115–133.
Lincoln, Frederick C.
  1947. Manual for bird banders. U.S. Dept. Int., Washington, D. C.: 1–116.
  1950. Migration of birds. U. S. Fish & Wildl. Serv., Washington, D. C., Circ. 16: 1–102.
Lindauer, M.
  1955. The water economy and temperature regulation of the honeybee colony. Bee World 36: 62–72; 81–92: 105–111.
Lindeman, Raymond L.
  1941. Seasonal food cycle dynamics in a senescent lake. Amer. Midl. Nat. 26: 636–673.
  1942. The trophic-dynamic aspect of ecology. Ecol. 23: 399–418.
Lindroth, Carl Hildebrand
  1957. The faunal connections between Europe and North America. John Wiley & Sons, NY.: 1–344.
Lindsey, Alton A.
  1940. Recent advances in Antarctic bio-geography. Quart. Rev. Biol. 15: 456–465.
Linsdale, Jean M.
  1938. Environmental responses of vertebrates in the Great Basin. Amer. Midl. Nat. 19: 1–206.
Livingstone, D. A.
  1955. Some pollen profiles from Arctic Alaska. Ecol. 36: 587–600.
Livingstone, D. A., Kirk Bryan, Jr. and R. G. Leahy
  1958. Effects of an Arctic environment on the origin and development of freshwater lakes. Limn. and Ocean. 3: 192–214.
Lloyd, M. and R. J. Ghelardi
  1964. A table for calculating the 'equitability' component of species diversity. J. An. Ecol. 33: 217–225.
Lloyd, Monte, Jerrold H. Zar, and James R. Karr
  1968. On the calculation of information-theoretical measures of diversity. Amer. Midl. Nat. 79: 257–272.
Loeb, Jacques
  1918. Forced movements, tropisms, and animal conduct. Lippincott Co., Philadelphia: 1–209.
Longstaff, T. G.
  1932. An ecological reconnaissance in west Greenland. J. An. Ecol. 1: 119–142.
Loomis, W. F.
  1967. Skin-pigment regulation of vitamin-D biosynthesis in man. Sci. 157: 501–506.
Lorenz, Konrad Z.
  1935. Der Kumpan in der Umwelt des Vogels. J. Ornith. 83: 137–213 (Auk 54, 1937: 245–273).

Louch, Charles D.
  1956. Adrenocortical activity in relation to the density and dynamics of three confined populations of *Microtus pennsylvanicus*. Ecol. 37: 701–713.
  1958. Adrenocortical activity in two meadow vole populations. J. Mam. 39: 109–116.
Loucks, Orie L.
  1970. Evolution of diversity, efficiency, and community stability. Amer. Zool. 10: 17–25.
Love, Doris
  1959. The post-glacial development of the flora of Manitoba: a discussion. Can. J. Bot. 37: 547–585.
Low, Jessop B.
  1941. Nesting of the ruddy duck in Iowa. Auk 58: 506–517.
  1945. Ecology and management of the redhead, *Nyroca americana*, in Iowa. Ecol. Mono. 15: 35–69.
Low, Richard M.
  1971. Interspecific territoriality in a pomacentrid reef fish, *Pomacentrus flavicauda* Whitley. Ecol. 52: 648–654.
Lowdermilk, W. C.
  1953. Floods in deserts. Desert Res. Proc., Jerusalem Post Press: 365–377.
Lowe-McConnell, R. H.
  1969. Speciation in tropical freshwater fishes. Biol. J. Linn. Soc.: 51–75.
Lowrie, D. C.
  1948. The ecological succession of spiders of the Chicago area dunes. Ecol. 29. 334–351.
Lucas, C. E.
  1947. The ecological effects of external metabolites. Biol. Rev. 22: 270–295.
Luckiesh, Matthew
  1946. Applications of germicidal, erythemal and infrared energy. Van Nostrand Co., New York: i–viii, 1–463.
Luczak, Jadwiga
  1960. The distribution of spiders in the different strata of the pine wood. Ekol. Polska, B, 6: 39–50.
Lunt, H. A. and H. G. M. Jacobson
  1944. The chemical composition of earthworm casts. Soil Sci. 58: 367–375.
Lutz, H. J. and R. F. Chandler, Jr.
  1946. Forest Soils. John Wiley & Sons, Inc., New York: i–vi, 1–514.
Lyman, F. Earle
  1943. A pre-impoundment bottom-fauna study of Watts Bar Reservoir Area (Tennessee). Trans. Amer. Fish. Soc. 72: 52–62.
  1956. Environmental factors affecting distribution of mayfly nymphs in Douglas Lake, Michigan. Ecol. 37: 568–576.
Lyon, E. P.
  1905. On rheotropism. I. Rheotropism in fishes. Amer. J. Physiol. 12: 149–161.
Macan, T. T.
  1961. Factors that limit the range of freshwater animals. Biol. Rev. 36: 151–198.
  1963. Freshwater ecology. Longmans, Green and Co. Ltd. London: i–x + 1–338.
Macan, T. T. and E. B. Worthington
  1951. Life in lakes and rivers. Collins Co., London: i–xvi, 1–272.
MacArthur, Robert H.
  1955. Fluctuations of animal populations, and a measure of community stability. Ecol. 36: 533–536.
  1958. Population ecology of some warblers of northeastern coniferous forests. Ecol. 39: 599–619.
  1965. Patterns of species diversity. Biol. Rev. 40: 510–533.
MacArthur, Robert H. and Joseph H. Connell
  1966. The biology of populations. John Wiley & Sons, Inc., New York: i–xv + 1–200.
MacArthur, Robert H. and John W. MacArthur
  1961. On bird species diversity. Ecol. 42: 594–598.
MacArthur, R. H. and E. O. Wilson
  1967. The theory of island biogeography. Princeton Univ. Press, Princeton, New Jersey: i–xi + 1–203.
MacDonald, Gordon J. F.
  1971. Pollution, weather and climate. In W. W. Murdock's "Environment, resources, pollution and society", Sinauer Associates Inc., Stamford, Connecticut: 326–336.
Macdonald, W. W.
  1956. Observations on the biology of chaoborids and chironomids in Lake Victoria and on the feeding habits of the 'elephant-snout fish' (*Mormyrus kannume* Forsk.) J. An. Ecol. 25: 36–53.

Macfadyen, A.
1952. The small arthropods of a *Molinia* fen at Cothill. J. An. Ecol. 21: 87–117.
1954. The invertebrate fauna of Jan Mayen Island (East Greenland). Jour. An. Ecol. 23: 261–297.
1957. Animal ecology, aims and methods. Sir Isaac Pitman & Sons, London: i-xx, 1–264.
1961. Improved funnel-type extractors for soil arthropods. J. An. Ecol. 30: 171–184.

MacGinitie, G. E.
1932. Animal ecology defined? Ecol. 13: 212–213.
1935. Ecological aspects of a California marine estuary. Amer. Midl. Nat. 16: 629–765.
1939. Littoral marine communities. Amer. Midl. Nat. 21: 28–55.

Mackenthun, Kenneth M.
1958. The chemical control of aquatic nuisances. Committee on Water Pollution, Madison, Wis.: 1–64.

Mackenzie, J. M. D.
1951. Control of forest populations. Quart. J. For.: 1–8.
1952. Fluctuations in the numbers of British tetraonids. J. An. Ecol. 21: 128–153.

Mackie, Richard J.
1970. Range ecology and relations of mule deer, elk, and cattle in the Missouri River Breaks, Montana. Wildl. Mono. 20: 1–79.

MacLagan, D. Stewart
1932. The effect of population density upon rate of reproduction with special reference to insects. Proc. Roy. Soc. B, 111: 437–454.

MacLulich, D. A.
1937. Fluctuations in the numbers of the varying hare (*Lepus americanus*). Univ. Toronto Stud. Biol. Šerv. 43: 1–136.
1957. The place of chance in population processes. J. Wildl. Man. 21: 293–299.

Macnab, James A.
1958. Biotic aspection in the Coast Range Mountains of north-western Oregon. Ecol. Mono. 28: 21–54.

Macnae, William
1968. A general account of the fauna and flora of mangrove swamps and forests in the Indo-West-Pacific region. Adv. Mar. Biol. 6: 73–270.

Maddox, D. M., L. A. Andres, R. D. Hennessey, R. D. Blackburn, and N. R. Spencer
1971. Insects to control alligatorweed. BioSci. 21: 985–991.

Madison, D. M. and C. R. Shoop
1970. Homing behavior, orientation, and home range of salamanders tagged with tantalum-182. Sci. 168 (3938): 1484–1487.

Madsen, Bent Lauge
1969. Reactions of *Brachyptera risi* (Morton) (Plecoptera) nymphs to water current. Oikos 20: 95–100.

Magnuson, John J.
1962. An analysis of aggressive behavior, growth, and competition for food and space in medaka (*Oryzias latipes* [Pisces, Cyprinodontidae]). Can. J. Zool., 40: 313–362.

Maguire, Bassett, Jr.
1963. The exclusion of *Colpoda* (Ciliata) from superficially favorable habitats. Ecol. 44: 781–784.
1963. The passive dispersal of small aquatic organisms and their colonization of isolated bodies of water. Ecol. Mono. 33: 161–185.

Maher, William J.
1970. The pomarine jaeger as a brown lemming predator in Northern Alaska. Wilson Bull. 82: 130–157.

Malin, James C.
1947. The grassland of North America. Prolegomena to its history. Lawrence, Kansas: i-vii, 1–398.

Mani, M. S.
1968. Ecology and biogeography of high altitude insects. Ser. Ent. 4. D. W. Junk N. V. Publ., The Hague: i-xvi + 1–528.

Mann, K. H.
1965. Energy transformations by a population of fish in the river Thames. J. An. Ecol. 34: 253–275.
1969. The dynamics of aquatic ecosystems. Adv. Ecol. Res. 6: 1–81.

Manniche, A. L. V.
1910. The terrestrial mammals and birds of north-east Greenland. Medd. Grønland 45 (1): 1–200.

Manning, T. H.
1946. Bird and mammal notes from the east side of Hudson Bay. Can. Field-Nat. 60: 71–85.
1948. Notes on the country, birds and mammals west of Hudson Bay between Reindeer and Baker Lakes. Can. Field-Nat. 62: 1–28.
1952. Birds of the west James Bay and southern Hudson Bay coasts. Nat. Mus. Canada, Bull. 125: 1–114.

Manning, T. H., E. O. Höhn, and A. H. Macpherson.
1956. The birds of Banks Island. Nat. Mus. Canada, Bull. 143: i-iv, 1–144.

Manning, Winston M. and Richard E. Juday
1941. The chlorophyll content and productivity of some lakes in northeastern Wisconsin. Trans. Wisconsin Acad. Sci., Arts, Let. 33: 363–393.

Manville, Richard H.
1949. A study of small mammal populations in northern Michigan. Mus. of Zool., Univ. Michigan, Misc. Pub. 73: 1–83.

Margalef, D. Ramon
1957. Information theory in ecology. Mem. Real Acad. Cien. Artes Barcelona 23: 373–449 (transl. from Spanish: Gen. Systems 3, 1958: 36–71).
1963. On certain unifying principles in ecology. Amer. Nat. 97: 357–374.
1968. Perspectives in ecological theory. Univ. Chicago Press, Chicago, Ill: i-viii + 1-111.

Marsh, Frank L.
1937. Ecological observations upon the enemies of *Cecropia*, with particular reference to its hymenopterous parasites. Ecol. 18: 106–112.

Marshall, A. J.
1952. Non-breeding among Arctic birds. Ibis, 94: 310–333.

Marshall, Joe T., Jr.
1957. Birds of pine-oak woodland in southern Arizona and adjacent Mexico. Pacific Coast Avifauna, Berkeley, California, 32: 1–125.

Marshall, N. B.
1954. Aspects of deep sea biology. Hutchinson's Sci. and Tech. Publ., New York: 1–380.

Marshall, William H. and Murray F. Buell
1955. A study of the occurrence of amphibians in relation to a bog succession, Itasca State Park, Minnesota. Ecol. 36: 381–387.

Martin, Alexander C., Herbert S. Zim and Arnold L. Nelson
1951. American wildlife and plants. McGraw-Hill Co., Inc., New York: i-ix, 1–500.

Martin, Edwin P.
1956. A population study of the prairie vole (*Microtus ochrogaster*) in northeastern Kansas. Univ. Kansas. Publ., Mus. Nat. Hist. 8 (6): 361–416.

Martin, Norman Duncan
1960. An analysis of bird populations in relation to plant succession in Algonquin Park, Ontario. Ecol. 41: 126–140.

Martin, Paul S.
1963. The last 10,000 years. A fossil pollen record of the American Southwest. Univ. Arizona Press, Tucson: 1–87.

Martin, Paul S. and Byron E. Harrell
1957. The pleistocene history of temperate biotas in Mexico and eastern United States. Ecol. 38: 468–480.

Martin, Paul S. and Peter J. Mehringer, Jr.
1965. Pleistocene pollen analysis and biogeography of the Southwest. In H. E. Wright, Jr. and D. G. Frey's "The Quaternary of the United States". Princeton Univ. Press.: 433–451.

Martin, P. S. and H. E. Wright, Jr. (eds.)
1967. Pleistocene extinctions. Yale Univ. Press, New Haven, Conn.: i-x + 1–453.

Mason, Karl E.
1939. Relation of the vitamins to the sex glands. In Allen's "Sex and internal secretions": 1149–1212.

Masure, Ralph H. and W. C. Allee
1934. The social order in flocks of the common chicken and pigeon. Auk. 51: 306–327.

Matteson, Max R.
1948. Life history of *Elliptio complanatus* (Dillwyn, 1817). Amer. Midl. Nat. 40: 690–723.

Matthew, W. D.
1915. Climate and evolution. Ann. N. Y. Acad. Sci. 24: 171–318.

Maxwell, Arthur E., R. P. von Herzen, K. J. Hsu, J. E. Andrews, T. Saito, S. F. Percival, Jr., E. Deal Milow, R. E. Boyce
1969. Deep sea drilling in the South Atlantic. Sci. 168 (3935): 1047–1059.

Maxwell, John C.
1968. Continental drift and a dynamic earth. Amer. Sci. 56: 35–51.

Mayer, William V.
1953. A preliminary study of the Barrow ground squirrel, *Citellus parryi barrowensis*. J. Mam. 34: 334–345.

Mayr, Ernst
1939. The sex ratio in wild birds. Amer. Nat. 73: 156–179.
1944. Wallace's line in the light of recent zoogeographic studies. Quart. Rev. Biol. 19: 1–14.

1946. History of the North American bird fauna. Wilson Bull. 58: 3–41.

1947. Ecological factors in speciation. Evol. 1: 263–288.

1949. Speciation and systematics. In Jepson, Mayr, and Simpson's "Genetics, paleontology and evolution": 281–298.

1951. Speciation in birds. Proc. 10th Int. Ornith. Cong. 1950: 91–131.

1952. The problem of land connections across the South Atlantic with special reference to the Mesozoic. Bull. Amer. Mus. Nat. Hist. 99: 83–258.

1964. Inferences concerning the tertiary American bird faunas. Proc. Natl. Acad. Sci. 51: 280–288.

1969. Bird speciation in the tropics. Biol. J. Linn. Soc. 1: 1–17.

1970. Populations, species, and evolution. Harvard Univ. Press, Cambridge, Massachusetts: i–xv + 1–453.

Mayr, Ernst, E. Gorton Linsley, and Robert L. Usinger
1953. Methods and principles of systematic zoology. McGraw-Hill Book Co., Inc., New York: i–ix, 1–328.

McAtee, W. L.
1912. Methods of estimating the contents of bird stomachs. Auk 29: 449–464.

1922. Local suppression of agricultural pests by birds. Ann. Rep't. Smiths. Inst. 1920: 411–438.

1932. Effectiveness in nature of the so-called protective adaptations in the animal kingdom, chiefly as illustrated by the food habits of nearctic birds. Smithsonian Misc. Coll. 85 (7): 1–201.

1939. Wildlife of the Atlantic Coast salt marshes. U. S. Dept. Agric., Circ. 520: 1–28.

1947. Distribution of seeds by birds. Amer. Midl. Nat. 38: 214–223.

McCabe, Robert A.
1966. Vertebrates as pests: a point of view. In "Scientific Aspects of Pest Control", Nat. Acad. Sci., Nat. Res. Council, Washington, D. C., Publ. 1402: 115–134.

McCabe, Thomas T. and Barbara D. Blanchard
1950. Three species of Peromyscus. Rood Associates, Santa Barbara, Calif.: i–v, 1–136.

McClure, H. Elliott
1943. Aspection in the biotic communities of the Churchill Area, Manitoba. Ecol. Mono. 13: 1–35.

McDiarmid, A.
1969. Diseases in free-living wild animals. Sym. Zool. Soc. London, 24: i–xiv + 1–332.

McDiffett, Wayne F.
1970. The transformation of energy by a stream detritivore, Pteronarcys scotti (Plecoptera). Ecol. 51: 975–988.

McDougall, Kenneth Dougal
1943. Sessile marine invertebrates at Beaufort, North Carolina. Ecol. Mono. 13: 321–374.

McErlean, Andrew J. and Joseph A. Mihursky
1969. Species diversity. Species abundance of fish populations: an examination of various methods. Proc. 22nd Ann. Conf., SE Assoc. Game Fish Commissioners: 367–372.

McErlean, Andrew J., Joseph A. Mihursky, and Howard J. Brinkley
1969. Determination of upper temperature tolerance triangles for aquatic organisms. Chesapeake Sci. 10: 293–296.

McGauhey, P. H.
1968. Manmade contamination hazards. Ground Water 6: 10–13.

McGinnies, William G., Bram J. Goldman, and Patricia Paylore (eds.)
1968. Deserts of the world. Univ. Arizona Press, Tucson: i–xxviii + 1–788.

McIntosh, Robert P.
1963. Ecosystems, evolution and relational patterns of living organisms. Amer. Sci. 51: 246–267.

McKeever, Sturgis
1961. Relative populations of small mammals in three forest types of northeastern California. Ecol. 42: 399–402.

McNab, Brian K.
1971. The structure of tropical bat faunas. Ecol. 52: 352–358.

McNaughton, S. J. and L. L. Wolf
1970. Dominance and the niche in ecological systems. Sci. 167: 131–139.

McNeill, S.
1971. The energetics of a population of Leptopterna dolabrata (Heteroptera: Miridae). J. An. Ecol. 40: 127–140.

Meadows, Donella H., Dennis L. Meadows, Jørgen Randers, and William W. Behrens III
1972. The limits to growth. Universe Books, New York City, N. Y.: 1–205.

Meglitsch, Paul A.
1954. On the nature of the species. Syst. Zool. 3: 49–65.

Meinertzhagen, R.
1950. Some problems connected with Arabian birds. Ibis 92: 336–340.

Meinzer, Oscar E.
1922. Map of the Pleistocene lakes of the basin-and-range province and its significance. Bull. Geol. Soc. Amer. 33: 541–552.

Mengel, Robert M.
1964. The probable history of species formation in some northern wood warblers (Parulidae). The Living Bird, 3rd Ann., Lab. Ornith., Cornell Univ., Ithaca, N. Y.: 9–43.

1970. The North American central plains as an isolating agent in bird speciation. In W. Dort, Jr. and J. K. Jones, Jr.'s "Pleistocene and recent environments of the central Great Plains". Univ. Kansas Press, Lawrence: 279–340.

Menhinick, Edward F.
1962. Comparison of invertebrate populations of soil and litter of mowed grasslands in areas treated and untreated with pesticides. Ecol. 43: 556–561.

1963. Estimation of insect population density in herbaceous vegetation with emphasis on removal sweeping. Ecol. 44: 617–621.

Menzies, Robert J.
1965. Conditions for the existence of life in the abyssal sea floor. Oceanogr. Mar. Biol. Ann. Rev. 3: 195–210.

Menzies, Robert J. and J. B. Wilson
1961. Preliminary field experiments on the relative importance of pressure and temperature on the penetration of marine invertebrates into the deep sea. Oikos 12: 302–309.

Merriam, C. Hart
1890. Results of a biological survey of the San Francisco Mountain region and desert of the Little Colorado, Arizona. U. S. Dept. Agr., North Amer. Fauna No. 3: i–vii, 1–136.

Merriam, C. Hart, Vernon Bailey, E. W. Nelson, and E. A. Preble
1910. Zone map of North America. Biol. Surv., U. S. Dept. Agr. Washington, D. C.

Mertens, Robert
1948. Die Tierwelt des tropischen Regenwaldes. Kramer, Frankfurt: 1–144.

Miall, L. C.
1934. The natural history of aquatic insects. Macmillan Co., New York: i–xi, 1–395.

Migula, P., W. Grodzinski, A. Jasinski, and B. Musialek
1970. Vole and mouse plagues in south-eastern Poland in the years 1945–1967. Acta theriol. 15: 233–252.

Mihursky, J. A.
1967. On possible constructive uses of thermal additions to estuaries. BioSci. 17: 698–702.

Mihursky, J. A., A. J. McErlean, and V. S. Kennedy
1970. Thermal pollution, aquaculture, and pathobiology in aquatic systems. J. Wildl. Dis. 6: 347–355.

Miller, Alden H.
1951. An analysis of the distribution of the birds of California. Univ. California Publ. Zool. 50: 531–644.

1959. Response to experimental light increments by Andean sparrows from an equatorial area. Condor 61: 344–347.

1963. Seasonal activity and ecology of the avifauna of an American equatorial cloud forest. Univ. California Publ. Zool., 66 (1): 1–78.

Miller, Alden H. and Robert C. Stebbins
1964. The lives of desert animals in Joshua Tree National Monument. Univ. California Press, Berkeley: i–vi + 1–452.

Miller, Charles R.
1971. The socio-economic roots of environmental pollution. Biol. 53: 57–65.

Miller, Gerrit S. and Remington Kellogg
1955. List of North American recent mammals. Smithsonian Instit., Washington, D. C.: i–xii, 1–954.

Miller, Richard B.
1941. A contribution to the ecology of the Chironomidae of Costello Lake, Algonquin Park, Ontario. Univ. Toronto Stud. 49: 1–63.

Miller, Richard S.
1964. Ecology and distribution of pocket gophers (Geomyidae) in Colorado. Ecol. 45: 256–272.

1964. Larval competition in Drosophila melanogaster and D. simulans. Ecol. 45: 132–148.

1968. Conditions of competition between redwings and yellowheaded blackbirds. J. An. Ecol. 37: 43–62.

Milne, A.
1957. Theories of natural control of insect populations. Cold Spr. Harb. Symp. Quant. Biol. 22: 253–271.

Milum, V. G.
　1928. Temperature relations of honeybees in winter. Ann. Rept. Illinois Beekeepers Assoc. 28: 98–130.

Minckley, W. L.
　1963. The ecology of a spring stream, Doe Run, Meade County, Kentucky. Wildl. Mono. 11: 5–124.

Minshall, G. Wayne
　1967. Role of allochthonous detritus in the trophic structure of a woodland springbrook community. Ecol. 48: 139–149.

Miranda, F. and A. J. Sharp
　1950. Characteristics of the vegetation in certain temperate regions of eastern Mexico. Ecol. 31: 313–333.

Möbius, Karl
　1877. Die Auster und die Austernwintschaft. Berlin. (English translation) U. S. Commission Fish and Fisheries Report 1880: 683–751.

Moffett, James W.
　1936. A quantitative study of the bottom fauna in some Utah streams variously affected by erosion. Bull. Univ. Utah 26: 1–33.

Mohr, Carl O.
　1940. Comparative populations of game, fur and other mammals. Amer. Midl. Nat. 24: 581–584.
　1943. Cattle droppings as ecological units. Ecol. Mono. 13: 275–298.
　1947. Table of equivalent populations of North American small mammals. Amer. Midl. Nat. 37: 223–249.

Möller, Carl M., D. Müller and Jorgen Nielsen
　1954. Ein Diagramm der Stoffproduktion im Buchenwald. Ber. Schweig. bot. Ges. 64: 487–494.

Monk, Carl D.
　1967. Tree species diversity in the eastern deciduous forest with particular reference to north central Florida. Amer. Nat. 101: 173–187.

Monro, J.
　1967. The exploitation and conservation of resources by populations of insects. J. An. Ecol. 36: 531–547.

Moore, Emmeline, et al.
　1934. A problem in trout stream management. Trans. Amer. Fish. Soc. 64: 68–80.

Moore, Hilary B.
　1958. Marine ecology. John Wiley & Sons, New York: i–xi, 1–493.

Moore, John A.
　1946. Incipient intraspecific isolating mechanisms in Rana pipiens. Genet. 31: 304–326.
　1952. Competition between Drosophila melanogaster and Drosophila simulans. I. Population cage experiments. Evol. 6: 407–420.

Moore, Joseph C.
　1946. Mammals from Welaka, Putnam County, Florida. J. Mam. 27: 49–59.

Moore, Robert E.
　1965. Olfactory discrimination as an isolating mechanism between Peromyscus maniculatus and Peromyscus polionotus. Am. Midl. Nat. 73: 85–100.

Moore, Walter G. and A. Burn
　1968. Lethal oxygen thresholds for certain temporary pond invertebrates and their applicability to field situations. Ecol. 49: 349–351.

Moral, Roger Del and R. G. Cates
　1971. Allelopathic potential of the dominant vegetation of western Washington. Ecol. 52: 1030–1037.

Moreau, R. E.
　1933. Pleistocene climatic changes and the distribution of life in east Africa. J. Ecol. 21: 415–435.
　1935. A synecological study of Usambara, Tanganyika Territory, with particular reference to birds. J. Ecol. 23: 1–43.
　1935a. Some eco-climatic data for closed evergreen forest in tropical Africa. J. Linn. Soc. London (Zool.) 39: 285–293.
　1937. The avifauna of the mountains along the Rift Valley in North Central Tanganyika Territory (Mbulu District), Part 1. Ibis 14th Ser., 1, Pt. 2: 760–786.
　1944. Clutch-size: a comparative study, with special reference to African birds. Ibis 44: 286–347.
　1948. Ecological isolation in a rich tropical avifauna. J. An. Ecol. 17: 113–126.
　1950. The breeding seasons of African birds. Ibis 92: 223–267, 419–433.
　1952. Africa since the Mesozoic: with particular reference to certain biological problems. Proc. Zool. Soc. London 121, Pt. IV: 869–913.
　1955. Ecological changes in the Palaearctic region since the Pliocene. Proc. Zool. Soc. London 125: 253–295.
　1963. Vicissitudes of the African biomes in the late Pleistocene. Proc. Zool. Soc. London 141: 395–421.
　1966. The bird faunas of Africa and its islands. Academic Press, N. Y.: 1–424.

Morel, G. and F. Bourlière
　1962. Relations ecologiques des avifaunes sedentaire et migratrice dans une savane sahelienne du bas Senegal. La Terre et la Vie, 4: 371–393.

Morgans, J. F. C.
　1956. Notes on the analysis of shallow-water soft substrata. J. An. Ecol. 25: 367–387.

Morris, Desmond
　1967. The naked ape. McGraw-Hill Book Co., New York: 1–252.

Morris, R. F.
　1959. Single-factor analysis in population dynamics. Ecol. 40: 580–588.
　1960. Sampling insect populations. Ann. Rev. Ent., 5: 243–264.
　1967. Influence of parental food quality on the survival of Hyphantria cunea. Can. Ent. 99: 24–33.

Morris, R. F. and C. A. Miller
　1954. The development of life tables for the spruce budworm. Can. J. Zool. 32: 283–301.

Morris, William M.
　1947. The forest fire hazard in the sunspot cycle. Fire Control Notes (U. S. For. Serv.) 8 (4): 4–9.

Morrison, P. R.
　1966. Biometeorological problems in the ecology of animals of the Arctic. Int. J. Biometeor. 10: 273–292.

Morse, Marius
　1939. A local study of predation upon hares and grouse during the cyclic decimation. J. Wildl. Man. 3: 203–211.

Mosauer, Walter
　1932. Adaptive convergence in the sand reptiles of the Sahara and of California: a study in structure and behavior. Copeia: 72–78.
　1935. The reptiles of a sand dune area and its surroundings in the Colorado desert, California: a study in habitat preference. Ecol. 16: 13–27.

Mosby, Henry S.
　1960. Manual of game investigational techniques. Edwards Brothers, Ann Arbor, Michigan. 20 sections.

Moss, E. H.
　1944. The prairie and associated vegetation of southwestern Alberta. Can. J. Res., Sect. C, 22: 11–31.

Moyle, John B.
　1956. Relationships between the chemistry of Minnesota surface waters and wildlife management. J. Wildl. Man. 20: 303–320.

Moynihan, M.
　1968. Social mimicry; character convergence versus character displacement. Evol. 22: 315–331.

Mozley, Alan
　1932. A biological study of a temporary pond in western Canada. Amer. Nat. 66: 235–249.

Mucibabic, Smilja
　1957. The growth of mixed populations of Chilomonas paramecium and Tetrahymena pyriformis. J. Gen. Microbiol. 16: 561–571.

Muir, F.
　1914. Presidential address. Proc. Hawaii Ento. Soc. 3: 28–42.

Mullen, David A.
　1968. Reproduction in brown lemmings (Lemmus trimucronatus) and its relevance to their cycle of abundance. Univ. California Publ. Zool., 85: 1–24.

Muller, Cornelius H.
　1940. Plant succession in the Larrea-Flourensia climax. Ecol. 21: 206–212.
　1966. The role of chemical inhibition (allelopathy) in vegetational composition. Bull. Torrey Bot. Club, 93: 332–351.

Mullin, Michael M.
　1969. Production of zooplankton in the ocean: the present status and problems. Oceanogr. Mar. Biol. Ann. Rev. 7: 293–314.

Muma, Martin H. and Katherine E.
　1949. Studies on a population of prairie spiders. Ecol. 30: 485–503.

Mumford, Lewis
　1956. The natural history of urbanization. In W. L. Thomas' "Man's role in changing the face of the earth", Univ. Chicago Press, Chicago: 382–398.

Munroe, Eugene
　1956. Canada as an environment for insect life. Can. Ento. 88: 372–476.

Murie, Adolph
　1944. The wolves of Mount McKinley. U. S. Dept. Int., Nat. Park Serv., Fauna Ser. 5: i–xix, 1–238.

Murphy, Robert Cushman
   1938. Birds collected during the Whitney South Seas expedition. Amer. Mus. Nov. 977: 1–17.
   1962. The oceanic life of the Antarctic. Sci. Amer. (October): 187–210.
Murray, Jr., Bertram G.
   1971. The ecological consequences of interspecific territorial behavior in birds. Ecol. 52: 414–423.
Murray, J. and J. Hjort
   1912. The depths of the ocean. Macmillan & Co., London: i–xx, 1–82.
Murray, Keith F.
   1957. Pleistocene climate and the fauna of Burnet Cave, New Mexico. Ecol. 38: 129–132.
Musser, R. H.
   1948. Engineering handbook for farm planners. U. S. Dept. Agr., Soil Cons. Serv.: 1–196.
Mutch, Robert W.
   1970. Wildland fires and ecosystems—a hypothesis. Ecol. 51: 1046–1051.
Muttkowski, Richard Anthony
   1918. The fauna of Lake Mendota, a qualitative and quantitative survey with special reference to insects. Wisconsin Acad. Sci. Arts, Let., 19: 374–482.
Muttkowski, Richard A. and Gilbert M. Smith
   1929. The food of trout stream insects in Yellowstone National Park. Roosevelt Wildl. Annals 2: 241–263.
Myers, Everett Clark
   1927. Relation of density of population and certain other factors to survival and reproduction in different biotypes of *Paramecium caudatum*. J. Exp. Zool. 49: 1–43.
Myers, Judith H. and Charles J. Krebs
   1971. Genetic, behavioral, and reproductive attributes of dispersing field voles *Microtus pennsylvanicus* and *Microtus ochrogaster*. Ecol. Mono. 41: 53–78.
Mykytowycz, Romar
   1968. Territorial marking by rabbits. Sci. Amer. 218 (5): 116–119, 123–126.
Nash, Carroll Blue
   1950. Associations between fish species in tributaries and shore waters of western Lake Erie. Ecol. 31: 561–566.
Needham, James G.
   1949. The ditch at Archbold Biological Station and the dragonflies resident in it. Ecol. 30: 450–460.
Needham, James G., Jay R. Traver, and Hsu Yin-Chi
   1935. The biology of mayflies. Comstock Publ. Co., Ithaca: i–xiv, 1–759.
Needham, James G. and Minter J. Westfall, Jr.
   1955. A manual of the dragonflies of North America (Anisoptera). Univ. California Press, Berkeley: i–vii, 1–615.
Needham, Paul R.
   1928. A net for the capture of stream drift organisms. Ecol. 9: 339–342.
   1932. Bottom foods in trout streams. Field and Stream, 36: 40–44.
   1934. Quantitative studies of stream bottom foods. Trans. Amer. Fish. Soc. 64: 238–247.
   1938. Trout streams. Comstock Publ. Co., Ithaca, N. Y.: i–x, 1–233.
Neel, James V.
   1970. Lessons from a primitive people. Sci. 170: 815–822.
Negus, N. C., E. Gould, and R. K. Chipman
   1961. Ecology of the rice rat, *Oryzomys palustris* (Harlan), on Breton Island, Gulf of Mexico, with a critique of the social stress theory. Tulane Stud. Zool 8 (4): 93–123.
Neill, Wilfred T. and E. Ross Allen
   1954. Algae on turtles: some additional considerations. Ecol. 35: 581–584.
Neilson, M. M. and R. F. Morris
   1964. The regulation of European spruce sawfly numbers in the Maritime Provinces of Canada from 1937 to 1963. Can. Ent. 96: 773–784.
Neldner, Kenneth H. and Robert W. Pennak
   1955. Seasonal faunal variations in a Colorado alpine pond. Amer. Midl. Nat. 53: 419–430.
Nelson, Gid E., Jr.
   1952. The birds of Welaka. J. Florida Acad. Sci. 15: 21–39.
Nelson-Smith, A.
   1970. The problem of oil pollution of the sea. Adv. Mar. Biol. 8: 215–306.
Nestler, Ralph B.
   1949. Nutrition of bobwhite quail. J. Wildl. Man., 13: 342–358.
Netolitzky, Fritz
   1932. Zur Frage der zirkumpolaren Verbreitung der Tiere. Arch. Naturgesch. N. S. 1: 351–353.

Newbigin, Marion I.
   1950. Plant and animal geography. Methuen & Co., London: i–xv, 1–298.
Newcombe, Curtis L.
   1935. Certain environmental factors of a sand beach in the St. Andrews region, New Brunswick, with a preliminary designation of the intertidal communities. J. Ecol. 23: 334–355.
Newman, Murray A.
   1956. Social behavior and interspecific competition in two trout species. Physiol. Zool. 29: 64–81.
Newton, I.
   1967. The adaptive radiation and feeding ecology of some British finches. Ibis 109: 33–98.
   1970. Irruptions of crossbills in Europe. In Adam Watson's "Animal populations in relation to their food resources", Blackwell Sci. Publ., Oxford, England: 337–357.
Nice, Margaret Morse
   1937. Studies in the life history of the song sparrow I. Trans. Linn. Soc. New York 4: i–vi, 1–247.
   1941. The role of territory in bird life. Amer. Midl. Nat. 26: 441–487.
   1943. Studies in the life history of the song sparrow. II. Trans. Linn. Soc. New York, 6: i–viii + 1–328.
   1962. Development of behavior in precocial birds. Trans. Linn. Soc. N. Y., 7: i–xii + 1–126.
Nicholson, A. J.
   1933. The balance of animal populations. J. An. Ecol. 2: 132–178.
   1954. An outline of the dynamics of animal populations. Aust. J. Zool. 2: 9–65.
   1954a. Compensatory reactions of populations to stresses, and their evolutionary significance. Aust. J. Zool. 2: 1–8.
Nicholson, A. J. and V. A. Bailey
   1935. The balance of animal populations. Part I. Proc. Zool. Soc. London, Pt. 3: 551–598.
Nielsen, Anker
   1950. The torrential invertebrate fauna. Oikos 2: 176–196.
Nielsen, C. Overgaard
   1953. Studies on Enchytraeidae I. A technique for extracting Enchytraeidae from soil samples. Oikos 4: 187–196.
Nielsen, E. Steemann
   1958. The balance between phytoplankton and zooplankton in the sea. J. Cons. Exp. Mer. 23: 178–188.
   1964. Recent advances in measuring and understanding marine primary production. Jub. Sym. Supp. Brit. Ecol. Soc.: 119–130.
Niering, William A.
   1968. The effects of pesticides. BioSci. 18: 869–875.
Nikolski, G. V.
   1933. On the influence of the rate of flow on the fish fauna of the rivers of central Asia. J. An. Ecol. 2: 266–281.
Noble, G. Kingsley
   1931. The biology of the amphibia. McGraw-Hill Book Co., Inc. New York: i–xiii, 1–577.
Nørgaard, Edwin
   1951. On the ecology of two lycosid spiders *Pirata piraticus* and *Lycosa pullata* from a Danish sphagnum bog. Oikos 3: 1–21.
Northcote, Thomas G.
   1954. Observations on the comparative ecology of two species of fish, *Cottus asper* and *Cottus rhotheus*, in British Columbia. Copeia: 25–28.
Northcote, T. G. and P. A. Larkin
   1956. Indices of productivity in British Columbia lakes. J. Fish. Res. Bd. Canada, 13: 515–540.
Oberle, Mark
   1969. Forest fires: suppression policy has its ecological drawbacks. Sci. 165 (3893): 568–571.
O'Brien, R. D.
   1967. Insecticides, action and metabolism. Academic Press, N. Y.: i–xi + 1–332.
O'Connell, Jr., Timothy R. and Robert S. Campbell
   1953. The benthos of Black River and Clearwater Lake, Missouri. In "The Black River Studies," 8: Univ. Missouri Studies 26 (2): 25–41.
O'Connor, F. B.
   1957. An ecological study of the enchytraeid worm population of a coniferous forest soil. Oikos 8: 161–199.
Odum, Eugene P.
   1961. The role of tidal marshes in estuarine production. N. Y. St. Cons.: 63–66.
   1964. Primary and secondary energy flow in relation to ecosystem structure. Proc. 16th Inter. Cong. Zool. 4: 336–338.
   1969. The strategy of ecosystem development. Sci. 164(3877): 262–270.

1970. Optimum population and environment: a Georgian microcosm. Current History 58: 355–359, 365–366.

1971. Fundamentals of ecology. W. B. Saunders Co., Philadelphia: i–xiv + 1–574.

Odum, Eugene P. and Thomas D. Burleigh
1946. Southward invasion in Georgia. Auk 63: 388–401.

Odum, Eugene P., Clyde E. Connell, and Leslie B. Davenport
1962. Population energy flow of three primary consumer components of oldfield ecosystems. Ecol. 43: 88–96.

Odum, Eugene P. and Armando A. de la Cruz
1963. Detritus as a major component of ecosytems. AIBS Bull. 13 (3): 39–40.

Odum, Eugene P. and Frank B. Golley
1963. Radioactive tracers as an aid to the measurement of energy flow at the population level in nature. In "Radioecology." Proc. First Nat. Sym. Radioecology. Reinhold Publishing Corp., New York: 403–410.

Odum, E. P. and A. E. Smalley
1959. Comparison of population energy flow of a herbivorous and a deposit-feeding invertebrate in a salt marsh ecosystem. Proc. Nat. Acad. Sci., 45: 617–622.

Odum, Howard T.
1956. Primary production of flowing waters. Limn. and Ocean. 1: 102–117.

1957. Primary production measurements in eleven Florida springs and a marine turtle-grass community. Limn. and Ocean. 2: 85–97.

1957a. Trophic structure and productivity of Silver Springs, Florida. Ecol. Mono. 27: 55–112.

1971. Environment, power, and society. Wiley-Interscience, New York: i–ix + 1–331.

Odum, Howard T. and W. A. Allee
1954. A note on the stable point of populations showing both intraspecific cooperation and disoperation. Ecol. 35: 95–97.

Odum, Howard T. and Eugene P. Odum
1955. Trophic structure and productivity of a windward coral reef community on Eniwetok Atoll. Ecol. Mono. 25: 291–320.

Oesting, R. B. and W. C. Allee
1935. Further analysis of the protective value of biologically conditioned fresh water for the marine turbellarian, *Procerodes wheatlandi*. IV. The effect of calcium. Biol. Bull. 68: 314–326.

Old, Sylvia M.
1969. Microclimate, fire, and plant production in an Illinois prairie. Ecol. Mono., 39: 355–384.

Oliff, W. D.
1953. The mortality, fecundity and intrinsic rate of natural increase of the multimammate mouse, *Rattus* (*Mastomys*) *natalensis* (Smith) in the laboratory. J. An. Ecol. 22: 217–226.

Olmsted, Charles E.
1937. Vegetation of certain sand plains of Connecticut. Bot. Gaz. 99: 209–300.

Olson, Everett C.
1952. The evolution of a Permian vertebrate chronofauna. Evol. 6: 181–196.

1966. Community evolution and the origin of mammals. Ecol. 47: 291–302.

Olson, Jerry S.
1963. Energy storage and the balance of producers and decomposers in ecological systems. Ecol. 44: 322–331.

1970. Carbon cycles and temperate woodlands. Geographical index of world ecosystems. In D. E. Reichle's "Analysis of temperate forest ecosystems," Springer-Verlag, New York: 226–241, 297–304.

O'Neil, Ted
1949. The muskrat in the Louisiana coastal marshes. La. Dept. Wild. Life and Fish., New Orleans: i–xii, 1–152.

O'Neill, Robert V.
1967. Niche segregation in seven species of diplopods. Ecol. 48: 983.

1968. Population energetics of the millipede, *Narceus americanus* (Beauvois). Ecol. 49: 803–809.

Oosting, Henry J.
1942. An ecological analysis of the plant communities of Piedmont, North Carolina. Amer. Midl. Nat. 28: 1–126.

Oosting, Henry J. and Philippe F. Bourdeau
1955. Virgin hemlock forest segregates in the Joyce Kilmer Memorial Forest of western North Carolina. Bot. Gaz. 116: 340–359.

Oppheimer, Carl H. (ed.)
1962. Symposium on marine microbiology. Charles C. Thomas, Publ., Springfield, Illinois: i–xviii + 1–769.

Orians, Gordon H.
1962. Natural selection and ecological theory. Amer. Nat. 96: 257–263.

1969. The number of bird species in some tropical forests. Ecol. 50: 783–801.

Orians, Gordon H. and Gerald Collier
1963. Competition and blackbird social systems. Evol. 17: 449–459.

Orians, Gordon H. and M. F. Willson
1964. Interspecific territories of birds. Ecol. 45: 736–745.

Ortmann, A. E.
1920. Correlation of shape and station in fresh-water mussels (Naiades). Proc. Amer. Phil. Soc. 59: 269–312.

Osborn, Ben and Philip F. Allan
1949. Vegetation of an abandoned prairie-dog town in tall grass prairie. Ecol. 30: 322–332.

Osburn, Raymond C., Louis I. Dublin, H. W. Shimer and R. S. Lull
1903. Adaptation to aquatic, arboreal, fossorial, and cursorial habits in mammals. Amer. Nat. 37: 651–665; 731–736; 819–825; 38: 1904: 1–11.

Osterhaus, Sister M. Benitia
1962. Adaptive modifications in the leg structure of some North American warblers. Amer. Midl. Nat. 68: 474–486.

Ovington, J. D.
1957. Dry-matter production by *Pinus sylvestris*. Ann. Bot., N. S. 21: 287–314.

Ovington, J. D., Dale Heitkamp and Donald B. Lawrence
1963. Plant biomass and productivity of prairie, savanna, oakwood, and maize field ecosystems in central Minnesota. Ecol. 44: 52–63.

Owen, D. F.
1961. Industrial melanism in North American moths. Amer. Nat. 95: 227–233.

1963. Variation in North American screech owls and the subspecies concept. Syst. Zool. 12: 8–14.

Packard, Fred Mallery
1942. Wildlife and aspen in Rocky Mountain National Park, Colorado. Ecol. 23: 478–482.

Paine, Robert T.
1963. Trophic relationships of 8 sympatric predatory gastropods. Ecol. 44: 63–73.

1966. Food web complexity and species diversity. Amer. Nat. 100: 65–75.

Palmgren, Pontus
1930. Quantitative Untersuchungen über die Vogelfauna in den Wäldern Sudfinnlands. Acta. Zool. Fenn. 7: 1–218.

1932. Zur Biologie von *Regulus r. regulus* (L) und *Parus atricapillus borealis* Selys. Acta Zool. Fenn. 14: 1–113.

1936. Bemerkungen über die ökologische Bedeutung der biologischen Anatomie des Fusses bei einiger Kleinvogelarten. Ornis Fenn, 13: 53–58.

1949. Some remarks on the short-term fluctuations in the numbers of northern birds and mammals. Oikos 1: 114–121.

Pamatmat, Mario M.
1968. Ecology and metabolism of a benthic community on an intertidal sandflat. Int. Revue ges. Hydrobiol. Hydrogr. 53: 211–298.

Paris, Oscar H. and Frank A. Pitelka
1962. Population characteristics of the terrestrial isopod *Armadillidium vulgare* in California grassland. Ecol. 43: 229–248.

Park, Barry C.
1942. The yield and persistence of wildlife food plants. J. Wildl. Man. 6: 118–121.

Park, Orlando
1930. Studies in the ecology of forest Coleoptera. Ann. Ento. Soc. Amer. 23: 57–80.

1931. The measurement of daylight in the Chicago area and its ecological significance. Ecol. Mono. 1: 189–230.

1940. Nocturnalism—the development of a problem. Ecol. Mono. 10: 485–536.

Park, Orlando, Stanley Auerbach and Glenna Corley
1950. The tree-hole habitat with emphasis on the pselaphid beetle fauna. Bull. Chicago Acad. Sci. 9: 19–57.

Park, Thomas
1932. Studies in population physiology: the relation of numbers to initial population growth in the flour beetle *Tribolium confusum* Duval. Ecol. 13: 172–181.

1934. Studies in population physiology. III. The effect of conditioned flour upon the productivity and population decline of *Tribolium confusum*. J. Exp. Zool. 68: 167–182.

1938. Studies in population physiology VIII. The effect of larval population density on the post-embryonic development of the flour beetle, *Tribolium confusum* Duval. J. Exp. Zool. 79: 51–70.

1939. Analytical population studies in relation to general ecology. Amer. Midl. Nat. 21: 235–255.

1945. Life tables for the black flour beetle, *Tribolium madens* Charp. Amer. Nat. 79: 436–444.

1948. Experimental studies of interspecies competition. 1. Competition between populations of the flour beetles, *Tribolium confusum* Duval and *Tribolium castaneum* Herbst. Ecol. Mono. 18: 265–308.

1954. Experimental studies of interspecies competition. II. Temperature, humidity, and competition in two species of *Tribolium*. Physiol. Zool. 27: 177–238.

Park, Thomas and Nancy Woollcott
1937. Studies in population physiology VII. The relation of environmental conditioning to the decline of *Tribolium confusum* populations. Physiol. Zool. 10: 197–211.

Parker, G. H.
1902. The reactions of copepods to various stimuli and the bearing of this on daily depth-migrations. Bull. U. S. Fish. Comm. for 1901, 21: 103–123.

Parker, Richard A.
1958. Some effects of thinning on a population of fishes. Ecol. 39: 304–317.

Parsons, John Davis
1957. Literature pertaining to formation of acid-mine wastes and their effects on the chemistry and fauna of streams. Ill. St. Acad. Sci. Trans. 50: 49–59.

Patten, Bernard C.
1959. An introduction to the cybernetics of the ecosystem: the trophic-dynamic aspect. Ecol. 40: 221–231.

Patterson, J. T. and W. S. Stone
1952. Evolution in the genus *Drosophila*. Macmillan Co., New York: 1–610.

Payne, Jerra A.
1965. A summer carrion study of the baby pig *Sus scrofa* Linnaeus. Ecol. 46: 592–602.

Payne, N. M:
1934. The differential effect of environmental factors upon *Microbracon hebetor* Say (Hymenoptera: Braconidae) and its host, *Ephestia kühniella* Zeller (Lepidoptera: Pyralidae) II. Ecol. Mon. 4: 1–46.

Peakall, David B. and R. J. Lovett
1972. Mercury: its occurrence and effects on the ecosystem. BioSci. 22: 20–25.

Pearl, Raymond
1922. The biology of death. J. B. Lippincott Company, Philadelphia: 1–275.

1923. An introduction to medical biometry and statistics. W. B. Saunders Co., Philadelphia: 1–379.

1925. The biology of population growth. Alfred A. Knopf, Inc., New York: i-xiv, 1–260.

1927. The growth of populations. Quart. Rev. Biol. 2: 532–548.

1932. The influence of density of population upon egg production in *Drosophila melanogaster*. J. Exp. Zool. 63: 56–84.

Pearl, Raymond and John R. Miner
1935. Experimental studies on the duration of life. XIV. The comparative mortality of certain lower organisms. Quart. Rev. Biol. 10: 60–79.

Pearl, Raymond, John Rice Miner and Sylvia L. Parker
1927. Experimental studies on the duration of life. XI. Density of population and life duration in *Drosophila*. Amer. Nat. 61: 289–318.

Pearl, Raymond and Lowell J. Reed
1920. On the rate of growth of the population of the United States since 1790 and its mathematical representation. Proc. Nat. Acad. Sci. 6: 275–288.

Pearse, A. S.
1934. Ecology of lake fishes. Ecol. Mon. 4: 475–480.

1939. Animal ecology. McGraw-Hill Book Co., Inc., New York: i-xii, 1–642.

1946. Observations on the microfauna of the Duke Forest. Ecol. Mono. 16: 127–150.

1950. The emigrations of animals from the sea. Sherwood Press, Dryden, N. Y.: i-xii, 1–210.

Pearse, A. S., H. J. Humm, and G. W. Wharton
1942. Ecology of sand beaches at Beaufort, North Carolina. Ecol. Mono. 12: 135–190.

Pearson, Jay Frederick Wesley
1933. Studies on the ecological relations of bees in the Chicago region. Ecol. Mono. 3: 373–441.

Pearson, Oliver P.
1963. History of two local outbreaks of feral house mice. Ecol. 44: 540–549.

1964. Carnivore-mouse predation: an example of its intensity and bioenergetics. J. Mam. 45: 177–188.

1966. The prey of carnivores during one cycle of mouse abundance. J. An. Ecol. 35: 217–233.

1971. Additional measurements of the impact of carnivores on California voles (*Microtus californicus*). J. Mam. 52: 41–49.

Peiponen, V. A.
1970. Animal activity patterns under subarctic summer conditions. In "Ecology of the subarctic regions", Proc. Helsinki Sym., UNESCO: 281–287.

Pendleton, Robert C.
1949. The rain shadow effect on the plant formations of Guadalcanal. Ecol. Mono. 19: 75–93.

Penfound, Wm. T.
1956. Primary production of vascular aquatic plants. Limn. and Ocean. 1: 92–101.

Pennak, Robert W.
1940. Ecology of the microscopic Metazoa inhabiting the sandy beaches of some Wisconsin lakes. Ecol. Mono. 10: 537–615.

1944. Diurnal movements of zooplankton organisms in some Colorado mountain lakes. Ecol. 25: 387–403.

1946. The dynamics of freshwater plankton populations. Ecol. Mono. 16: 339–355.

1953. Fresh-water invertebrates of the United States. Ronald Press Co., New York: i-ix, 1–769.

1955. Comparative limnology of eight Colorado mountain lakes. Univ. Colorado Stud., Ser. Biol. 2: 1–75.

1957. Species composition of limnetic zooplankton communities. Limn. and Ocean. 2: 222–232.

Pennak, Robert W. and Ernest D. Van Gerpen
1947. Bottom fauna production and physical nature of the substrate in a northern Colorado trout stream. Ecol. 28: 42–48.

Pennington, Winifred
1941. The control of the numbers of fresh water phytoplankton by small invertebrate animals. J. Ecol. 29: 204–211.

Percival, E. and H. Whitehead
1929. A quantitative study of the fauna of some types of streambed. J. Ecol. 17: 282–314.

Perrins, C. M.
1965. Population fluctuations and clutchsize in the great tit, *Parus major* L. J. An. Ecol. 34: 601–647.

Petersen, C. G. Joh.
1914. The animal communities of the sea-bottom and their importance for marine zoogeography. Rep't. Danish Biol. Sta. 21: 1–44.

1918. The sea bottom and its production of fish food. Rep't. Danish Biol. Sta. 25: 1–62.

Peterson, Walburga
1926. Seasonal succession of animals in a chara-cattail pond. Ecol. 7: 371–377.

Peterson, Randolph L.
1953. Notes on the eastern distribution of *Eutamias minimus*. Cont. Roy. Ontario Mus. Zool. and Palaeontology, 37: 1–4.

1955. North American moose. Univ. Toronto Press, Toronto: i-xi, 1–280.

Petrides, George A.
1956. Big game densities and range carrying capacities in east Africa. Trans. 21st No. Amer. Wildl. Conf.: 525–537.

1961. The management of wild hoofed animals in the United States in relation to land use. La Terre et la Vie 2: 181–202.

Petrides, George A. and W. G. Swank
1965. Population densities and the range-carrying capacity for large mammals in Queen Elizabeth National Park, Uganda. Zool. Africana 1: 209–225.

Petrusewicz, Kazimierz
1963. Population growth induced by disturbance in the ecological structure of the population. Ekol. Polska, A, 11 (3): 87–125.

1967. Secondary productivity of terrestrial ecosystems (principles and methods). Inst. Ecol., Polish Acad. Sci., Warsaw: 2 vols.

Phelps, Chester F.
1954. Plan your farm for more wildlife. Virginia Widl. 15: 5–7.

Phillips, D. A., J. G. Torrey, and R. H. Burris
1971. Extending symbiotic nitrogen fixation to increase man's food supply. Sci. 174 (4005): 167–171.

Phillips, John
1934–35. Succession, development, the climax, and the complex organism: an analysis of concepts. J. Ecol. 22: 554–571, 23: 210–246, 488–508.

Phillipson, J.
1960. A contribution to the feeding biology of *Mitopus morio* (F) (Phalangida). J. An. Ecol. 29: 35–43.

Pianka, Eric R.
1966. Latitudinal gradients in species diversity: a review of concepts. Amer. Nat. 100: 33–46.
1966. Convexity, desert lizards, and spatial heterogeneity. Ecol. 47: 1055–1059.
1967. On lizard species diversity: North American flatland deserts. Ecol. 48: 333–351.

Picken, L. E. R.
1937. The structure of some protozoan communities. J. Ecol. 25: 368–384.

Pieczynska, Ewa, E. Pieczynski, T. Prus and K. Tarwid
1963. The biomass of the bottom fauna of 42 lakes in the Wegorzewo district. Ekol. Polska, A, 11 (19): 495–502.

Pielou, E. C.
1966. The measurement of diversity in different types of biological collections. J. Theoret. Biol. 13: 131–144.

Pimentel, David
1961. Species diversity and insect population outbreaks. Ann. Ent. Soc. Amer. 54: 76–86.
1961. Competition and the species-per-genus structure of communities. Ann. Ent. Soc. Amer. 54: 323–333.
1968. Population regulation and genetic feedback. Sci. 159 (3822): 1432–1437.

Pimlott, Douglas H.
1967. Wolf predation and ungulate populations. Amer. Zool. 7: 267–278.

Pinchot, Gifford B.
1970. Marine farming. Sci. Amer. 223 (6): 15–21.

Piper, Stanley E.
1928. The mouse infestation of Buena Vista Lake basin, Kern County, California, September 1926 to February 1927. Month. Bull., Dept. Agr., California: 17: 538–560.

Pitelka, Frank A.
1941. Distribution of birds in relation to major biotic communities. Amer. Midl. Nat. 25: 113–137.
1964. Predation in the lemming cycle at Barrow, Alaska 1951–63. Proc. 16th Inter. Cong. Zool. 1: 265.
1951a. Speciation and ecologic distribution in American jays of the genus *Aphelocoma*. Univ. California Publ. Zool. 50: 195–464.
1957. Some characteristics of microtine cycles in the Arctic. 18th Biology Colloq.: 73–88.

Pitelka, F. A., P. Quentin Tomich and George W. Treichel
1955. Ecological relations of jaegers and owls as lemming predators near Barrow, Alaska. Ecol. Mono. 25: 85–117.

Pleske, Theodore
1928. Birds of the Eurasian tundra. Mem. Boston Soc. Nat. Hist. 6: 111–485.

Polunin, Nicholas
1934. The vegetation of Akpatok Island. J. Ecol. 22: 337–395; 23 (1935): 161–209.
1948. Botany of the Canadian eastern Arctic. Part III. Vegetation and ecology. Nat. Mus. Canada, Bull. 104: i–iii, 1–304.

Popham, E. J.
1942. Further experimental studies of the selective action of predators. Proc. Zool. Soc. London A 112: 105–117.

Popov, A. M.
1931. Distribution of fishes in the Black Sea with reference to bottom conditions. Ecol. 12: 468–475.

Porsild, A. E.
1943. Birds of the Mackenzie Delta. Can. Field-Nat., 57: 19–35.

Portmann, Adolf
1959. Animal camouflage. Univ. Michigan Press, Ann Arbor: 1–111.

Potzger, J. E.
1951. The fossil record near the glacial border. Ohio J. Sci. 51: 126–133.
1956. Pollen profiles as indicators in the history of lake filling and bog formation. Ecol. 37: 476–483.

Poulton, E. B.
1887. The experimental proof of the protective value of color and markings in insects in reference to their vertebrate enemies. Proc. Zool. Soc. London: 191–274.

Pound, Roscoe and Frederic E. Clements
1900. The phytogeography of Nebraska. Univ. Nebraska

Pournelle, George H.
1950. Mammals of a north Florida swamp. J. Mam. 31: 310–319.

Pramer, David
1964. Nematode-trapping fungi. Sci. 144: 382–388.

Pratt, David M.
1943. Analysis of population development in *Daphnia* at different temperatures. Biol. Bull. 85: 116–140.

Prebble, M. L.
1954. Canada. Review of forest entomology, 1948–1953. Rep't. 6th Commonwealth Ento. Conf. London: 206–224.

Preble, Edward A.
1908. A biological investigation of the Athabaska-Mackenzie region. No. Amer. Fauna, U. S. Biol. Surv. No. 27: 1–574.

Prejs, Krystyna
1970. Some problems of the ecology of benthic nematodes (Nematoda) of Mikolajskie Lake. Ekol. Polska, A, 18 (9): 225–242.

Preston, F. W.
1948. The commonness, and rarity, of species. Ecol. 29: 254–283.
1960. Time and space and the variation of species. Ecol. 41: 611–627.

Price, Peter W.
1970. Characteristics permitting coexistence among parasitoids of a sawfly in Quebec. Ecol. 51: 445–454.

Price, Raymond and Robert B. Evans
1937. Climate of the west front of the Wasatch plateau in central Utah. Month. Weath. Rev. 65: 291–301.

Proctor, Vernon W.
1968. Long-distance dispersal of seeds by retention in digestive tract of birds. Sci. 160 (3825): 321–322.

Proctor, Vernon W. and C. R. Malone
1965. Further evidence of the passive dispersal of small aquatic organisms via the intestinal tract of birds. Ecol. 46: 728–729.

Prosser, C. Ladd
1950. Comparative animal physiology. W. B. Saunders Co., Philadelphia: i-ix, 1–888.
1955. Physiological variation in animals. Biol. Rev. 30: 229–262.
1957. The species problem from the viewpoint of a physiologist. Ernst Mayr's "The Species Problem," AAAS publ., Washington, D. C.: 339–369.

Protsch, Reiner and Rainer Berger
1973. Earliest radiocarbon dates for domesticated animals. Sci. 179 (4070): 235–239.

Pruitt, William O., Jr.
1957. Observations on the bioclimate of some taiga mammals. Arctic 10: 131–138.

Prychodko, W.
1958. Effect of aggregation of laboratory mice (*Mus musculus*) on food intake at different temperatures. Ecol. 39: 500–503.

Purchon, R. D.
1968. The biology of the Mollusca. Pergamon Press Ltd., New York: i-xix + 1–560.

Quarterman, Elsie and Catherine Keever
1962. Southern mixed hardwood forest: climax in the southeastern coastal plain, U.S.A. Ecol. Mono. 32: 167–185.

Rabinowitch, Eugene I.
1945-56. Photosynthesis and related processes. Interscience Publ., New York: 3 vols.

Raitt, Ralph J. and R. L. Maze
1968. Densities and species composition of breeding birds of a creosote-bush community in southern New Mexico. Condor 70: 193–205.

Ramaley, Francis
1940. The growth of a science. Univ. Colorado Stud. 26: 3–14.

Ramensky, L. G.
1926. Die Grundgesetzmässigkeiten im aufbau der Vegetationsdecke (German abstract). Bot. Centbl., N. S. 7: 453–455.

Rand, Austin L.
1945. Mammal investigations on the Canol Road, Yukon and Northwest Territories, 1944. Nat. Mus. Canada, Bull. 99: 1–52.
1948. Glaciation, an isolated factor in speciation. Evol. 2: 314–321.
1954. The ice age and mammal speciation in North America. Arctic 7: 31–35.

Randall, John E.
1961. Overgrazing of algae by herbivorous marine fishes. Ecol. 42: 812.

Rasmussen, D. Irvin
1941. Biotic communities of Kaibab Plateau, Arizona. Ecol. Mono. 11: 229–275.

Rasmussen, D. I. and Everett R. Doman
1943. Census methods and their application in the management of mule deer. Trans. 8th No. Amer. Wildl. Conf.: 369–380.

Rasool, S. I. and S. H. Schneider
1971. Atmospheric carbon dioxide and aerosols: effects of large increases on global climate. Sci. 173 (3992): 138–141.

Ratcliffe, D. A.
  1970. Changes attributable to pesticides in egg breakage frequency and eggshell thickness in some British birds. J. Appl. Ecol. 7: 67–115.
Raunkiaer, C.
  1934. Life forms of plants and statistical plant geography. Clarendon Press, Oxford: i-xvi + 1-632.
Raup, Hugh M.
  1941. Botanical problems in boreal America. Bot. Rev. 7: 147–248.
  1947. Some natural floristic areas in boreal America. Ecol. Mono. 17: 221–234.
  1951. Vegetation and cryoplanation. Ohio J. Sci. 51: 105–116.
Rausch, Robert
  1953. On the status of some arctic mammals. Arctic 6: 91–148.
Raw, F.
  1960. Earthworm population studies: a comparison of sampling methods. Nat. 187: 257.
Rawson, Donald S.
  1930. The bottom fauna of Lake Simcoe and its role in the ecology of the lake. Univ. Toronto Stud. 40: 1-183.
  1953. The bottom fauna of Great Slave Lake. J. Fish. Res. Bd. Canada 10: 486–520.
  1955. Morphometry as a dominant factor in the productivity of large lakes. Proc. Int. Assoc. theoret. appl. Limn. 12: 164–175.
  1956. The net plankton of Great Slave Lake. J. Fish. Res. Bd. Canada 13: 53–127.
Ray, Carleton
  1960. The application of Bergmann's and Allen's rules to poikilotherms. J. Morph. 106: 85–108.
Ray, Louis L.
  1951. Permafrost. Arctic 4: 196–203.
Raymont, John E. G.
  1963. Plankton and productivity in the oceans. The Macmillan Co., N. Y.: i-viii + 1-660.
Redfield, Alfred C.
  1958. The biological control of chemical factors in the environment. Amer. Sci. 46: 205–221.
Redfield, A. C. and E. S. Deevey
  1952. Temporal sequences and biotic successions. In "Marine fouling and its prevention." U. S. Naval Inst., Annapolis: 42–47.
Reed, Guilford and A. Brooker Klugh
  1924. Correlation between hydrogen iron concentration and biota of granite and limestone pools. Ecol. 5: 272–275.
Reeves, Cora D.
  1907. The breeding habits of the rainbow darter (Etheostoma coeruleum Storer), a study in sexual selection. Biol. Bull., 14: 23–59.
Regier, Henry A. and Douglas S. Robson
  1967. Estimating population number and mortality rates. In S. D. Gerking's "The biological basis of freshwater fish production", John Wiley & Sons, Inc.: 31–66.
Reichle, David E.
  1967. Radioisotope turnover and energy flow in terrestrial isopod populations. Ecol. 48: 351–366.
  1969. Distribution and abundance of bog-inhabiting pselaphid beetles. Trans. Ill. St. Acad. Sci. 62: 233–264.
Reichle, David E. and D. A. Crossley, Jr.
  1965. Radiocesium dispersion in a cryptozoan food web. Health Physics, Pergamon Press 11: 1375–1384.
Reid, R. W.
  1955. The bark beetle complex associated with lodgepole pine slash in Alberta. Can. Ento. 87: 311–323.
Reighard, Jacob
  1908. Methods of studying the habits of fishes with an account of the breeding habits of the horned dace. Bull. U. S. Bur. Fish, 28, Pt. 2: 1111–1136.
  1920. The breeding behavior of the suckers and minnows. Biol. Bull. 38: 1–32.
  1943. The breeding habits of the river chub, Nocomis micropogon (Cope). Papers Michigan Acad Sci., Arts, Letters, 28: 387–423.
Reimers, Norman and Bobby D. Combs
  1956. Method of evaluating temperature in lakes with description of thermal characteristics of Convict Lake, California. U. S. Fish and Wildl. Serv., Fish Bull. 105: 535–553.
Reiners, W. A.
  1972. Structure and energetics of three Minnesota forests. Ecol. Mono. 42: 71–94.
Reiners, William A., I. A. Worley, and D. B. Lawrence
  1970. Plant diversity in a chronosequence at Glacier Bay, Alaska. Ecol. 52: 55–69.

Reiser, Oliver L.
  1937. Cosmecology: a theory of evolution. J. Hered. 28: 367–371.
Remane, Adolf
  1952. Die Besiedelling des Sandbodens im Meere und die Bedeutung der Lebensformtypen für die Ökologie. Zool. Anz. 16, Suppl: 327–359.
Renn, Charles E.
  1941. The food economy of Anopheles quadrimaculatus and A. crucians larvae. A symposium on hydrobiology, Univ. Wisconsin Press, Madison: 329–342.
Renner, F. G.
  1938. A selected bibliography on management of western ranges, livestock, and wildlife. U. S. Dept. Agr., Misc. Publ. 281: i-ii, 1–468.
Retzlaff, Elmer G.
  1939. Studies in population physiology with the albino mouse. Biol. Gen. 14: 238–265.
Reynoldson, T. B. and R. W. Davies
  1970. Food niche and co-existence in lake-dwelling triclads. J. An. Ecol. 39: 599–617.
Rhoades, Rendell
  1962. The evolution of crayfishes of the genus Orconectes section Limosus (Crustacea: Decopoda). Ohio J. Sci. 62: 65–96.
Rhodda, Michael
  1967. Noise and society. Oliver and Boyd, London: i-vii + 1 − −113.
Rice, Elroy L.
  1964. Inhibition of nitrogen-fixing and nitrifying bacteria by seed plants. Ecol. 45: 824–837.
Rice, Lucille A.
  1954. Observations on the biology of ten notonectoid species found in the Douglas Lake, Michigan region. Amer. Midl. Nat. 51: 105–132.
Rich, Earl R.
  1956. Egg cannibalism and fecundity in Tribolium. Ecol. 37: 109–120.
Richards, P. W.
  1952. The tropical rain forest. An ecological study. Univ. Press, Cambridge: i-xviii, 1–450.
Richardson, Frank
  1942. Adaptive modifications for tree-trunk foraging in birds. Univ. California Publ. Zool. 46: 317–368.
Richman, Sumner
  1958. The transformation of energy by Daphnia pulex. Ecol. Mono. 28: 273–291.
Richter, Curt P.
  1927. Animal behavior and internal drives. Quart. Rev. Biol. 2: 307–343.
  1942. Physiological psychology. Annual Rev. Physiol 4: 561–574.
Ricker, William E.
  1934. An ecological classification of certain Ontario streams. Univ. Toronto Stud., Biol. Ser. 37: 1–114.
  1938. "Residual" and kokanee salmon in Cultus Lake. J. Fish. Res. Bd. Canada 4: 192–218.
  1946. Production and utilization of fish populations. Ecol. Mono. 16: 373–391.
  1948. Methods of estimating vital statistics of fish populations. Indiana Univ. Publ., Sci. 15: 1–101.
  1952. The benthos of Cultus Lake. J. Fish. Res. Bd. Canada 9: 204–212.
  1954. Stock and recruitment. J. Fish. Res. Bd. Canada 11: 559–623.
  1958. Handbook of computations for biological statistics of fish populations. Fish. Res. Bd. Canada, Bull. 119: 1–300.
  1969. Food from the sea. In Preston Cloud's "Resources and Man", W. H. Freeman and Co., San Francisco: 86–108.
Ricketts, Edward F. and Jack Galvin
  1948. Between Pacific tides. Stanford Univ. Press, Stanford, Calif.: i-xxvii, 1–365.
Ricklefs, Robert E.
  1967. A graphical method of fitting equations to growth curves. Ecol. 48: 978–983.
Riley, Charles V.
  1960. The ecology of water areas associated with coal strip-mined lands in Ohio. Ohio J. Sci. 60: 106–121.
Riley, Gordon A.
  1940. Limnological studies in Connecticut. Part III. The plankton of Linsley Pond. Ecol. Mono. 10: 279–306.
  1946. Factors controlling phytoplankton populations on Georges Bank. J. Mar. Res. 6: 54–73.
  1947. A theoretical analysis of the zooplankton population of Georges Bank. J. Mar. Res. 6: 104–113.

1970. Particulate organic matter in sea water. Adv. Mar. Biol. 8: 1–118.

1952. Phytoplankton of Block Island Sound, 1949. Bull. Bingham Ocean. Coll. 13: 40–64.

1956. Oceanography of Long Island Sound, 1952–1954. Production and utilization of organic matter. Bull. Bingham Ocean. Coll. 15: 324–344.

Riley, Gordon A., Henry Stommel and Dean F. Bumpus
1949. Quantitative ecology of the plankton of the western North Atlantic. Bull. Bingham Ocean. Coll. 22: 1–169.

Robert, Adrien
1955. Les associations de gyrins dans les étangs et les lacs du Parc du Mont Tremblant. Can. Ento. 87: 67–78.

Robertson, Charles
1927. Flowers and insects. XXIV. Ecol. 8: 113–132.

Robertson, Forbes W. and James H. Sang
1945. The ecological determinants of population growth in a *Drosophila* culture. Proc. Roy. Soc. London (B) 132: 258–291.

Robertson, O. H.
1954. A method for securing stomach contents of live fish. Ecol. 26: 95–96.

Robertson, William Beckwith
1945. An analysis of breeding-bird populations of tropical Florida in relation to the vegetation. Univ. Illinois, Ph. D. thesis.

Robins, C. Richard and Ronald W. Crawford
1954. A short accurate method for estimating the volume of stream flow. J. Wildl. Man. 18: 366–369.

Robinson, Thane S.
1963. Illumination preferenda of bobwhites. Occ. Papers, C. C. Adams Center Ecol. Stud., Western Michigan Univ., 8: 1–10.

Robson, G. C. and O. W. Richards
1936. The variation of animals in nature. Longman, Green & Co., New York: i–xvi, 1–425.

Rodin, L. E. and N. I. Bazilevich
1965. Production and mineral cycling in terrestrial vegetation. Transl. from Russian, 1967, Oliver and Boyd, London: 1–288.

Roe, Frank Gilbert
1951. The North American buffalo. Univ. Toronto Press, Toronto: i–viii, 1–957.

Rogers, J. Speed
1933. The ecological distribution of the crane-flies of northern Florida. Ecol. Mono. 3: 1–74.

Rohlich, Gerard A. (Chairman)
1969. Eutrophication: causes, consequences, correctives. Nat. Acad. Sci., Washington, D. C. (Sym.): i–vii + 1–661.

Romell, L. G.
1935. Ecological problems of the humus layer in the forest. Cornell Univ., Agr. Exp. Sta., Mem. 170: 1–28.

Root, Richard B.
1960. An estimate of the intrinsic rate of natural increase in the planarian, *Dugesia tigrina*. Ecol. 41: 369–372.

Rosato, Peter and Denzel E. Ferguson
1968. The toxicity of endrin-resistant mosquitofish to eleven species of vertebrates. BioSci. 18: 783–784.

Rose, S. Meryl
1960. A feedback mechanism of growth control in tadpoles. Ecol. 41: 188–199.

Rosenzweig, Michael L. and Jerald Winakur
1969. Population ecology of desert rodent communities: habitats and environmental complexity. Ecol. 50: 558–572.

Ross, Herbert H.
1944. The caddis flies, or Trichoptera, of Illinois. Bull. Illinois Nat. Hist. Surv. 23: 1–326.

1951. The origin and dispersal of primitive caddisflies. Evol. 5: 102–115.

1953. On the origin and composition of the Nearctic insect fauna. Evol. 7: 145–158.

1957. Principles of natural coexistence indicated by leafhopper populations. Evol. 11: 113–129.

1962. A synthesis of evolutionary theory. Prentice-Hall, Inc., Englewood Cliffs, N. J.: i–xiii + 1–387.

1970. The ecological history of the Great Plains: evidence from grassland insects. In W. Dort, Jr. and J. D. Jones, Jr.'s "Pleistocene and recent environments of the central Great Plains," Univ. Press of Kansas, Lawrence: 226–240.

Roth, Roland R.
1967. An analysis of avian succession on upland sites in Illinois. M. S. thesis, University of Illinois.

Rowan, Wm. and L. B. Keith
1956. Reproductive potential and sex ratios of snowshoe hares in northern Alberta. Canadian J. Zool. 34: 273–281.

Rowley, Ian
1967. Sympatry in Australian ravens. Proc. Ecol. Soc. Aust. 2: 107–115.

Royama, T.
1970. Factors governing the hunting behaviour and selection of food by the great tit (*Parus major* L.) J. An. Ecol. 39: 619–667.

Ruddiman, William F.
1969. Recent planktonic Foraminifera: dominance and diversity in North Atlantic surface waters. Sci. 164 (3884): 1164–1167.

Rudebeck, Gustaf
1950. The choice of prey and modes of hunting of predatory birds with special reference to their selective effect. Oikos 2: 65–88; 3 (1951): 200–231.

Ruiter, L. de
1955. Countershading in caterpillars. Arch. Neerlandaises Zool. 11: 1–56.

Rukeyser, William Simon
1972. Fact and foam in the row over phosphates. Fortune 85: 71–73, 166, 168, 170.

Runcorn, S. K. (ed.)
1962. Continental drift. Academic Press, New York: i–xii + 1–338.

Russell, Carl Parcher
1932. Seasonal migration of mule deer. Ecol. Mono. 2: 1–46.

Russell, E. S.
1931. Some theoretical considerations of the "overfishing" problem. J. Cons. Perm. Int. Explor. Mer, 6: 3–20.

Russell, Richard J.
1957. Instability of sea level. Amer. Sci. 45: 414–430.

Ruttner, Franz
1953. Fundamentals of limnology (English translation). Univ. Toronto Press, Toronto: i–xi, 1–242.

Rydberg, P. A.
1954. Flora of the Rocky Mountains and adjacent Plains. Hafner Publ. Co., New York: i–xii, 1–1144.

Ryder, John P.
1970. A possible factor in the evolution of clutch size in Ross' goose. Wilson Bull. 82: 5–13.

Ryszkowski, Lech
1970. Estimates of consumption of rodent populations in different pine forest ecosystems. In Petrusewicz and Ryszkowski's "Energy flow through small mammal populations," Polish Sci. Publ., Warsaw: 283–289.

Ryther, John H.
1956. The measurement of primary production. Limn. and Ocean. 1: 72–84.

1969. Photosynthesis and fish production in the sea. Sci. 166 (3901): 72–76; 168 (3930), 1970: 503–505.

1970. Is the world's oxygen supply threatened? Nat. 227: 374–375.

Ryther, John H. and W. M. Dunstan
1971. Nitrogen, phosphorus, and eutrophication in the coastal marine environment. Sci. 171 (3975): 1008–1013.

Ryther, J. H., W. M. Dunstan, K. R. Tenore, and J. E. Huguenin
1972. Controlled eutrophication—increasing food production from the sea by recycling human wastes. BioSci. 22: 144–152.

Ryther, J. H. and C. S. Yentsch
1957. The estimation of phytoplankton production in the ocean from chlorophyll and light data. Limn. and Ocean. 2: 281–286.

Ryther, J. H., C. S. Yentsch, and G. H. Lauff
1959. Sources of limnological and oceanographic apparatus and supplies. Limn. & Ocean. 4: 357–365.

Saila, Saul B.
1956. Estimates of the minimum size-limit for maximum yield and production of chain pickerel, *Esox niger* Le Sueur, in Rhode Island. Limn. and Ocean. 1: 195–201.

Salomonsen, Finn
1950–51. The birds of Greenland. Part I. Einar Munksgaard, Copenhagen, Denmark: 1–604.

1965. The geographical variation of the fulmar (*Fulmarus glacialis*) and the zones of marine environment in the North Atlantic. Auk 82: 327–355.

Salt, George W.
1957. An analysis of avifaunas in the Teton Mountains and Jackson Hole, Wyoming. Condor 59: 373–393.

1967. Predation in an experimental protozoan population (*Woodruffia-Paramecium*). Ecol. Mono. 37: 113–144.

Salt, George., F. S. J. Hollick, F. Raw and M. V. Brian
  1948. The arthropod population of pasture soil. J. An. Ecol. 17: 139–150.
Sanders, Howard L.
  1956. The biology of marine bottom communities. Bull. Bingham Ocean. Coll. 15: 345–414.
  1960. Benthic studies in Buzzards Bay. III. The structure of the soft-bottom community. Limn. Ocean. 5: 138–153.
  1968. Marine benthic diversity: a comparative study. Amer. Nat. 102: 243–282.
Sanders, Howard L. and Robert R. Hessler
  1969. Ecology of the deep-sea benthos. Sci. 163 (3874): 1419–1424.
Sanders, H. L., R. R. Hessler, and G. R. Hampson
  1965. An introduction to the study of deep-sea benthic faunal assemblages along the Gay Head-Bermuda transect. Deep-Sea Res. 12: 845–867.
Sanders, William T. and Barbara J. Price
  1968. Mesoamerica. The evolution of a civilization. Random House, New York: i–xix + 1–264.
Sandner, Henryk
  1962. Investigations of the role and mechanism of the action of the density factor in populations of the common bean weevil (Acanthoscelides obsoletus Say). Ekol. Polska, B, 8: 179–186.
Sandon, H.
  1927. The composition and distribution of the protozoan fauna of the soil. Edinburgh, Oliver & Boyd: i–xv, 1–237.
Saunders, Aretas A.
  1936. Ecology of the birds of Quaker Run Valley, Allegany State Park, New York. New York St. Mus. Handbook 16: 1–174.
Saunders, George W.
  1964. Studies of primary productivity in the Great Lakes. Great Lakes Res. Div., Univ. Michigan, 11: 122–129.
Savage, Donald E.
  1958. Evidence from fossil land mammals on the origin and affinities of the western Nearctic fauna. In C. L. Hubbs' "Zoogeography," AAAS Publ. 51.
Savely, Harvey E.
  1939. Ecological relations of certain animals in dead pine and oak logs. Ecol. Mono. 9: 321–385.
Savilov, A. I.
  1957. Biological aspect of the bottom fauna groupings of the North Okhotsk Sea. Marine Biology, Trans. Inst. Oceanology 20, translation Amer. Inst. Biol. Sci., Washington 6, D. C. 1959: 67–136.
Scalon, W. N.
  1937. Les Oiseaux de Sud des Taimir. IV. Les Remarques Ecologiques. Le Gerfaut, 27: 181–195.
Schackleford, Martha W.
  1939. New methods of reporting ecological collections of prairie arthropods. Amer. Mid. Nat., 22: 676–683.
Schäfer, Ernst
  1938. Ornithologische Ergebnisse zweier Forschungsreisen nach Tibet. J. Ornith. 86 (Sonderheft): 1–349.
Schaller, Friedrick
  1968. Soil animals. Univ. Michigan Press, Cambridge, Mass.: 1–144.
Scheffer, Victor B.
  1955. Body size with relation to population density in mammals. J. Mam. 36: 493–515.
Schjelderup-Ebbe, Thorleif
  1922. Beiträge zur Sozialpsychologie des Hanshuhns. Zeit. Tierpsch. 88: 225–252.
Schmalhausen, I. I.
  1949. Factors of evolution. Blakiston Co., Philadelphia: i–xiv, 1–327.
Schmidt, Karl P.
  1938. Herpetological evidence for the postglacial eastward extension of the steppe in North America. Ecol. 19: 396–407.
  1946. On the zoogeography of the holarctic region. Copeia: 144–152.
  1953. A check list of North American amphibians and reptiles. Univ. Chicago Press, Chicago: i–viii, 1–280.
Schmidt-Nielsen, Knut
  1964. Desert animals. Physiological problems of heat and water. Clarendon Press, Oxford: i–xv + 1–277.
Schoener, Thomas W.
  1965. The evolution of bill size differences among sympatric congeneric species of birds. Evol. 19: 189–213.
  1968. The anolis lizards of Bimini: resource partitioning in a complex fauna. Ecol. 49: 704–726.
  1969. Models of optimal size for solitary predators. Amer. Nat. 103: 277–313.

Scholander, P. F.
  1955. Evolution of climatic adaptation in homeotherms. Evol. 9: 15–26.
Scholander, P. F., Walter Flagg, Vladimir Walters, and Laurence Irving
  1953. Climatic adaptation in arctic and tropical poikilotherms. Physiol. Zool. 26: 67–92.
Scholander, P. F., Raymond Hock, Vladimer Walters, Laurence Irving and Fred Johnson
  1950. Body insulation, heat regulation, and adaptation to cold in arctic and tropical mammals and birds. Biol. Bull. 99: 225–271.
Schomer, Harold A.
  1934. Photosynthesis of water plants at various depths in the lakes of northeastern Wisconsin. Ecol. 15: 217–218.
Schouw, Joakim Frederik
  1823. Pflanzengeographicher Atlas. Berlin.
Schuchert, Charles
  1955. Atlas of paleogeographic maps of North America. John Wiley & Sons, Inc., New York: i–xi, 1–177.
Schultz, Arnold M.
  1964. The nutrient-recovery hypothesis for Arctic microtine cycles. II. Ecosystem variables in relation to Arctic microtine cycles. In D. J. Crisp's "Grazing in terrestrial and marine environments," Blackwell Sci. Publ., Oxford: 57–68.
Schulz, Willy
  1931. Die Orientierung des Rückenschwimmers zum Licht und zur Strömung. Zeit. vergl. Physiol. 14: 392–404.
Schwarzbach, Martin
  1963. Climates of the past. Van Nostrand Co. Ltd., New York: i–xii + 1–328.
Sclater, Philip Lutley
  1858. On the general geographical distribution of the members of the class Aves. J. Proc. Linn. Soc. (Zool.) 2 (for 1857): 130–145.
Scott, Donald C.
  1949. A study of a stream population of rock bass, Ambloplites rupestris. Invest. Indiana Lakes and Streams 3: 169–234.
Scott, Robert F.
  1954. Population growth and game management. Trans. 19th No. Amer. Wildl. Conf.: 480–504.
Scott, Thomas G. and Willard D. Klimstra
  1955. Red foxes and a declining prey population. So. Illinois Univ., Mono. Ser. 1: 1–123.
Scotter, George W.
  1967. Effects of fire on barren-ground caribou and their forest habitat in northern Canada. Trans. 32nd No. Amer. Wildl. Nat. Res. Conf.: 246–259.
Sears, Paul B.
  1942. Xerothermic theory. Bot. Rev. 8: 708–736.
  1948. Forest sequence and climatic change in northeastern North America since early Wisconsin time. Ecol. 29: 326–333.
  1958. The inexorable problem of space. Sci. 127: 9–16.
Selander, Robert K.
  1965. Avian speciation in the Quaternary. The Quaternary of the U. S., Princeton Univ. Press: 527–542.
  1966. Sexual dimorphism and differential niche utilization in birds. Condor 68: 113–151.
Selous, Frederick Courteney
  1908. African nature notes and reminiscences. Macmillan & Co., London,; i–xxx, 1–356.
Selye, Hans
  1955. Stress and disease. Sci. 122: 625–631.
Serventy, D. L. and A. J. Marshall
  1957. Breeding periodicity in western Australian birds: with an account of unseasonal nestings in 1953 and 1955. Emu 57: 99–126.
Seton, Ernest Thompson
  1909. Life histories of northern animals. Charles Scribner's Sons, New York: 2 vols.
  1912. The arctic prairies. Constable & Co., London: i–xvi, 1–415.
  1925–28. Lives of game animals. Doubleday, Doran Co., New York: 4 vols.
Severtzoff, S. A.
  1934. On the dynamics of populations of vertebrates. Quart. Rev. Biol. 9: 409–437.
Shannon, Claude E. and Warren Weaver
  1949. The mathematical theory of communication. Univ. Illinois Press: 1–125.
Shantz, H. L.
  1923. The natural vegetation of the Great Plains region. Ann. Asso. Amer. Geog. 13: 81–107.

Shantz, H. L. and R. L. Piemeisel
  1924. Indicator significance of the natural vegetation of the south-
    western desert region. J. Agr. Res. 28: 721–801.
Shantz, H. L. and Raphael Zon
  1924. Atlas of American agriculture. Natural vegetation. U. S. Dept.
    Agr., Bur. Agr. Econ., Adv. Sheets 6: 1–29.
Sharp, A. J.
  1953. Notes on the flora of Mexico: world distribution of the woody
    dicotyledonous families and the origin of the modern vegetation.
    J. Ecol. 41: 374–380.
Shaw, Gretchen
  1932. The effect of biologically conditioned water upon rate of growth
    in fishes and amphibia. Ecol. 13: 263–278.
Sheaffer, John R.
  1970. Reviving the Great Lakes. Sat. Rev., Nov. 7: 62–65.
Sheldon, Andrew L.
  1968. Species diversity and longitudinal succession in stream fishes.
    Ecol. 49: 193–198.
  1969. Equitability indices: dependence on the species count. Ecol. 50:
    466–467.
Shelford, Victor E.
  1908. Life-histories and larval habits of the tiger beetles (Cicindelidae).
    Linnean Soc. Jour. Zool. 30: 157–184.
  1911. Ecological succession. Biol. Bull. 21: 127–151, 22: 1–38.
  1911a. Physiological animal geography. J. Morph. 22: 551–618.
  1913. Animal communities in temperate America. Univ. Chicago
    Press, Chicago: i-xiii, 1–368.
  1914. A comparison of the responses of sessile and motile plants and
    animals. Amer. Nat. 48: 641–674.
  1914. An experimental study of the behavior agreement among the
    animals of an animal community. Biol. Bull. 26: 294–315.
  1915. Principles and problems of ecology as illustrated by animals.
    J. Ecol. 3: 1–23.
  1926. Naturalist's Guide to the Americas. Williams & Wilkins Co.,
    Baltimore: i-xv, 1–761.
  1929. Laboratory and field ecology. Williams & Wilkins Co., Balti-
    more: i-xii, 1–608.
  1932. Basic principles of the classification of communities and habitats
    and the use of terms. Ecol. 13: 105–120.
  1942. Biological control of rodents and predators. Sci. Month. 15:
    331–341.
  1943. The abundance of the collared lemming (Dicrostonyx groenlandicus
    (Tr.) var. richardsoni Mer.) in the Churchill area, 1929 to 1940.
    Ecol. 24: 472–484.
  1944. Deciduous forest man and the grassland fauna. Sci. 100: 135–
    140, 160–162.
  1945. The relation of snowy owl migration to the abundance of the
    collared lemming. Auk 62: 592–596.
  1951. Fluctuation of non-forest animal populations in the upper
    Mississippi basin. Ecol. Mono. 21: 149–181.
  1951a. Fluctuation of forest animal populations in east central Illinois.
    Ecol. Mono. 21: 183–214.
  1952. Paired factors and master factors in environmental relations.
    Illinois Acad. Sci. Trans. 45: 155–160.
  1954a. The antelope population and solar radiation. J. Mam. 35:
    533–538.
  1954b. Some lower Mississippi Valley flood plain biotic communities;
    their age and elevation. Ecol. 35: 125–142.
  1963. The ecology of North America. Univ. Illinois Press, Urbana:
    i-xxii + 1–610.
Shelford, V. E. and M. W. Boesel
  1942. Bottom animal communities of the island area of western Lake
    Erie in the summer of 1937. Ohio J. Sci. 42: 179–190.
Shelford, V. E. and W. P. Flint
  1943. Populations of the chinch bug in the upper Mississippi Valley
    from 1823 to 1940. Ecol. 24: 435–455.
Shelford, V. E. and Sigurd Olson
  1935. Sere, climax and influent animals with special reference to the
    transcontinental coniferous forest of North America. Ecol. 16:
    375–402.
Shelford, V. E. and A. C. Twomey
  1941. Tundra animal communities in the vicinity of Churchill, Mani-
    toba. Ecol. 22: 47–69.
Shelford, V. E., A. O. Weese, Lucile A. Rice, D. I. Rassmussen, and
    Archie MacLean
  1935. Some marine biotic communities of the Pacific coast of North
    America. Ecol. Mono. 5: 249–354.
Sheppard, P. M.
  1954. Evolution in bisexually reproducing organisms. In Huxley,
    Hardy and Ford's "Evolution as a process," George Allen &
    Unwin, London: 201–218.
  1958. Natural selection and heredity. Hutchinson, London: 1–212.
Shimada, Bell M.
  1958. Diurnal fluctuation in photosynthetic rate and chlorophyll "a"
    content of phytoplankton from eastern Pacific waters. Limn. &
    Ocean. 3: 336–339.
Shimek, B.
  1930. Land snails as indicators of ecological conditions. Ecol. 11:
    673–686.
Shorrocks, B.
  1970. Population fluctuations in the fruit fly (Drosophila melanogaster)
    maintained in the laboratory. J. An. Ecol. 39: 229–253.
Short, Lester L., Jr.
  1965. Hybridization in the flickers (Colaptes) of North America. Amer.
    Mus. Nat. Hist. Bull. 129: 311–428.
Shreve, Forrest and Ira L. Wiggins
  1964. Vegetation and flora of the Sonoran Desert. Stanford Univ.
    Press, Stanford, California, 2 vols.
Sibley, Charles G.
  1954. Hybridization in the red-eyed towhees of Mexico. Evol. 8:
    252–290.
  1957. The evolutionary and taxonomic significance of sexual dimor-
    phism and hybridization in birds. Condor 59: 166–191.
Sigafoos, R. S.
  1951. Soil instability in tundra vegetation. Ohio J. Sci. 51: 281–
    298.
Siivonen, Lauri
  1954. Some essential features of short-term population fluctuation.
    J. Wildl. Man. 18: 38–45.
  1956. The correlation between the fluctuations of partridge and Euro-
    pean hare populations and the climatic conditions of winters in
    south-west Finland during the last thirty years. Pepers on Game
    Res., Helsinki, 17: 1–30.
  1957. The problem of the short-term fluctuations in numbers of tetra-
    onids in Europe. Papers on Game Res. (Finland) 19: 1–
    44.
  1962. Die schneemenge als überwinterungsökologischer factor. Sitz.
    Finn. Akad. Wissensch.: 111–125.
Silliman, Ralph P.
  1969. Population models and test populations as research tools. BioSci.
    19: 524–528.
Silliman, Ralph P. and James S. Gutsell
  1957. Response of laboratory fish populations to fishing rates. Trans.
    22nd No. Amer. Wildl. Conf.: 464–471.
  1958. Experimental exploitation of fish populations. U. S. Fish Wildl.
    Serv., Fish. Bull. 58: 215–252.
Simberloff, D. S. and E. O. Wilson
  1969. Experimental zoogeography of islands: the colonization of
    empty islands. Ecol. 50: 278–296.
Simonson, Row W.
  1957. What soils are. Soils, Yearbook of Agriculture, Dept. Agric.:
    17–31.
  1962. Soil classification in the United States. Sci. 137 (3535): 1027–
    1034.
Simpson, George Gaylord
  1940. Mammals and land bridges. J. Washington Acad. Sci. 30:
    137–163.
  1947. Holarctic mammalian faunas and continental relationships
    during the Cenozoic. Bull. Geol. Soc. Amer. 58: 613–687.
  1949. Rates of evolution in animals. In Jepson, Mayr, and Simpson's
    "Genetics, paleontology and evolution": 205–228.
  1953. The major features of evolution. Columbia Univ. Press, New
    York: i-xx, 1–434.
  1956. Zoogeography of West Indian land mammals. Amer. Mus.
    Nov. 1759: 1–28.
  1964. Species density of North American Recent mammals. Syst.
    Zool., 13: 57–73.
Singer, S. Fred
  1970. Human energy production as a process in the biosphere. Sci.
    Amer. 223 (3): 175–190.
Sirkin, Leslie A.
  1967. Late-Pleistocene pollen stratigraphy of western Long Island and
    eastern Staten Island, New York. In E. J. Cushing and H. E.
    Wright, Jr.'s "Quaternary paleoecology," Yale Univ. Press,
    New Haven, Conn.: 249–274.
Sladen, B. K. and F. B. Bang
  1969. Biology of populations. American Elsevier Publ. Co. Inc., New
    York: i-xxii + 1–449.

Slobodkin, L. Basil
  1954. Population dynamics in *Daphnia obtusa* Kurz. Ecol. Mono. 24: 69–88.
  1960. Ecological energy relationships at the population level. Amer. Nat. 94: 213–236.
  1961. Growth and regulation of animal populations. Holt, Rinehart and Winston, New York: i–viii + 1–184.
  1962. Energy in animal ecology. Adv. Ecol. Res. 1: 69–101.
  1963. Trophic chains; experimental approaches to marine biology. Marine Biology I, Amer. Inst. Biol. Sci., Washington 6, D. C.: 235–239.
  1964. Experimental populations of hydrida. Jub. Supp. Sym. Brit. Ecol. Soc.: 131–148.
Slobodkin, L. Basil and Sumner Richman
  1956. The effect of removal of fixed percentages of the newborn on size and variability in populations of *Daphnia pulicaria* (Forbes). Limn. and Ocean. 1: 209–237.
Slobodkin, L. B., F. E. Smith, and N. G. Hairston
  1967. Regulation in terrestrial ecosystems and the implied balance of nature. Amer. Nat. 101: 109–124.
Smalley, Alfred E.
  1960. Energy flow of a salt marsh grasshopper population. Ecol. 41: 672–677.
Smirnov, Eugen and W. Polejaeff
  1934. Density of population and sterility of the females in the coccid *Lepidosaphes ulmi* L. J. An. Ecol. 3: 29–40.
Smith, A. C. and I. M. Johnston
  1945. A phytogeographic sketch of Latin America. In Verdoorn's "Plants and plant science in Latin America," Chronica Bot. Co., Waltham, Mass.: 11–18.
Smith, Charles Clinton
  1940. Biotic and physiographic succession on abandoned eroded farmland. Ecol. Mono. 10: 421–484.
  1940. The effect of overgrazing and erosion upon the biota of the mixed-grass prairie of Oklahoma. Ecol. 21: 381–397.
Smith, Christopher C.
  1970. The coevolution of pine squirrels (*Tamiasciurus*) and conifers. Ecol. Mono. 40: 349–371.
Smith, Frank
  1925. Variation in the maximum depth at which fish can live during the summer in a moderately deep lake with a thermocline. Bull. U. S. Bur. Fish 41: 1–7.
  1930. Records of spring migration of birds at Urbana, Illinois, 1903–1922. Bull. Illinois St. Nat. Hist. Surv. 19: 105–117.
Smith, Frederick E.
  1963. Density-dependence. Ecol. 44: 220.
Smith, Harry S.
  1935. The role of biotic factors in the determination of population densities. J. Eco. Ent. 28: 873–898.
Smith, J. E. and G. E. Newell
  1955. The dynamics of the zonation of the common periwinkle (*Littorina littorea* (L.)) on a stony beach. J. An. Ecol. 24: 35–56.
Smith, Lloyd L., Jr. and John B. Moyle
  1944. A biological survey and fishery management plan for the streams of the Lake Superior north shore watershed. Minnesota Dept. Cons., Tech. Bull. 1: 1–228.
Smith, Philip W.
  1957. An analysis of post-Wisconsin biogeography of the prairie peninsula region based on distribution phenomena among terrestrial vertebrate populations. Ecol. 38: 205–219.
  1965. Recent adjustments in animal ranges. In H. E. Wright, Jr. and D. G. Frey's "The Quaternary of the United States," Princeton Univ. Press: 633–642.
Smith, Robert Lee
  1972. The ecology of man: an ecosystem approach. Harper & Row, Publ., New York: 112–126, 270–278.
Smith, Stanford H.
  1968. Species succession and fishery exploitation in the Great Lakes. J. Fish. Res. Bd. Canada 25: 667–693.
  1972. Factors of ecological succession in oligotrophic fish communities of the Laurentian Great Lakes. J. Fish. Res. Bd. Canada 29: 717–730.
Smith, Stanley G.
  1954. A partial breakdown of temporal and ecological isolation between *Choristoneura* species (Lepidoptera: Tortricidae). Evol. 8: 206–224.
Smith, V. G.
  1928. Animal communities of a deciduous forest succession. Ecol. 9: 479–500.

Snow, D. W.
  1952. The winter avifauna of Arctic Lapland. Ibis 94: 133–143.
  1971. Evolutionary aspects of fruit-eating by birds. Ibis 113: 194–202.
Snow, W. E.
  1955. Feeding activities of some blood-sucking Diptera with reference to vertical distribution in bottomland forest. Ann. Ento. Soc. Amer. 48: 512–521.
Snyder, Dana Paul
  1950. Bird communities in the coniferous forest biome. Condor 52: 17–27.
Snyder, L. L.
  1935. A study of the sharp-tailed grouse. Univ. Toronto Stud., Biol. Ser. 40: 1–66.
Sokal, R. R. and T. J. Crovello
  1970. The biological species concept: a critical evaluation. Amer. Nat. 104: 127–153.
Solomon, M. E.
  1949. The natural control of animal populations. J. An. Ecol., 18: 1–35.
  1953. The population dynamics of storage pests. Trans. 9th Int. Ento. Cong. 2: 235–248.
  1957. Dynamics of insect populations. Ann. Rev. Ento. 2: 121–142.
Sonneborn, T. M.
  1957. Breeding systems, reproductive methods, and species problems in Protozoa. AAAS Symp. 50: 155–324.
Soper, J. Dewey
  1944. The mammals of southern Baffin Island, Northwest Territories, Canada. J. Mam. 25: 221–254.
  1946. Ornithological results of the Baffin Island expeditions of 1928–1929 and 1930–1931, together with more recent records. Auk 63: 1–24.
Sorensen, Thorvald
  1941. Temperature relations and phenology of the northeast Greenland flowering plants. Medd. Grønland 125 (9): 1–305.
Southern, H. N.
  1959. Mortality and population control. Ibis 101: 429–436.
Southward, A. J.
  1958. The zonation of plants and animals on rocky sea shores. Biol. Rev. 33: 137–177.
Southwick, Charles H.
  1955. The population dynamics of confined house mice supplied with unlimited food. Ecol. 36: 212–225.
Southwood, T. R. E.
  1966. Ecological methods. Methuen & Co. Ltd, London: i–xviii + 1–391.
Soveri, Jorma
  1940. Die Vogelfauna von Lammi, ihre regionale Verbreitung und Abhängigkeit von den ökologischen Faktoren. Acta Zool. Fennica 27: 1–176.
Spärck, R.
  1935. On the importance of quantitative investigation of the bottom fauna in marine biology. J. Conseil, 10: 3–19.
Spector, William S.
  1956. Handbook of biological data. ASTIA Doc. Serv. Center, Knott Bldg., Dayton, Ohio: i–xxxvi, 1–584.
Speirs, J. Murray
  1939. Fluctuations in numbers of birds in the Toronto region. Auk 56: 411–419.
Spieth, Herman T.
  1952. Mating behavior within the genus *Drosophila* (Diptera). Bull. Amer. Mus. Nat. Hist. 99: 395–474.
  1958. Behavior and isolating mechanisms. In Roe and Simpson's "Behavior and evolution," Yale Univ. Press, New Haven, Conn.: 363–389.
Sprules, William M.
  1940. The effect of a beaver dam on the insect fauna of a trout stream. Trans. Amer. Fish. Soc. 70: 236–248.
  1947. An ecological investigation of stream insects in Algonquin Park, Ontario. Univ. Toronto Stud., Biol. Ser. 56: 1–81.
Stanchinsky, V. V.
  1931. On the importance of the mass of *Artensubstanz* in the dynamic equilibrium of the biocoenose. (In Russian, German summary). J. Ecol. and Biocenology 1: 88–98 (Abstracted in English by Richard Brewer).
Starrett, William C.
  1951. Some factors affecting the abundance of minnows in the Des Moines River, Iowa. Ecol. 32: 13–27.

Stebbins, Robert C.
1949. Speciation in salamanders of the plethodontid genus *Ensatina*. Univ. California Publ. Zool. 48: 377–526.

Steel, Paul E., Paul D. Dalke and Elwood G. Bizeau
1956. Duck production at Gray's Lake, Idaho 1949–1951. J. Wildl. Man. 20: 279–285.
1957. Canada goose production at Gray's Lake, Idaho, 1949–1951. J. Wild. Man. 21: 38–41.

Steele, J. H., A. D. McIntyre, R. R. C. Edwards, and Ann Trevallion
1970. Interrelations of a young plaice population with its invertebrate food supply. In Adam Watson's "Animal populations in relation to their food resources," Blackwell Sci. Publ., Oxford, England: 375–388.

Steere, J. B.
1894. On the distribution of genera and species of non-migratory land-birds in the Philippines. Ibis, 6th Ser., 6: 411–420.

Stegman, LeRoy C.
1960. A preliminary survey of earthworms of the Tully Forest in central New York. Ecol. 41: 779–782.

Stegmann, B.
1932. Die Herkunft der paläaktischer Taiga-Vögel. Arch. Naturges. N. F. 1: 355–398.
1938. Grundzuge der Ornithogeographischen Gliederung des paläarktischen Gebietes. Faune de l'URSS (N. S. 19) Oiseaux 1 (2): 77–156.

Stehli, Francis G., Robert G. Douglas, and Norman D. Newell
1969. Generation and maintenance of gradients in taxonomic diversity. Sci., 164 (3882): 947–949.

Stehr, Wm. C. and J. Wendell Branson
1938. An ecological study of an intermittent stream. Ecol. 19: 294–310.

Stein, Robert C.
1963. Isolating mechanisms between populations of Traill's flycatchers. Proc. Amer. Philos. Soc., 107: 21–50.

Stephen, A. C.
1935. Life on some sandy shores. Essays in Mar. Biol. Oliver & Boyd, London: 50–72.

Stephenson, T. A. and Anne Stephenson
1949. The universal features of zonation between tide-marks on rocky coasts. J. Ecol. 37: 289–305.
1950. Life between tide-marks in North America. I. The Florida Keys. J. Ecol. 38: 354–402.
1952. Life between tide-marks in North America. II. Northern Florida and the Carolinas. J. Ecol. 40: 1–49.
1954. Life between tide-marks in North America. III. Nova Scotia and Prince Edward Island. J. Ecol. 42: 14–70.
1961. Life between tide-marks in North America. IV. Vancouver Island. J. Ecol. 49: 1–29, 227–243.

Stern, Vernon M., Ray F. Smith, Robert van den Bosch, and Kenneth S. Hagen
1959. The integrated control concept. Hilgardia 29: 81–101.

Stetson, H. T.
1947. Sunspots in action. Ronal Press, New York: i–xiv + 1–252.

Stevanovic, Dragica
1956. *Collembola* populations in the forest associations on Kopaonik (English summary). Inst. D'Ecologie et De Biogeographie 7 (4): 1–16.

Stewart, Omer C.
1956. Fire as the first great force employed by man. In W. L. Thomas' "Man's role in changing the face of the earth," Univ. Chicago Press, Chicago: 115–133.

Stewart, Robert E. and John W. Aldrich
1952. Ecological studies of breeding bird populations in northern Maine. Ecology 33: 226–238.

Stewart, W. D. P.
1967. Nitrogen-fixing plants. Sci. 158 (3807): 1426–1432.

Stickel, Lucille F.
1946. Experimental analysis of methods for measuring small mammal populations. J. Wildl. Man. 10: 150–159.
1948. Observations on the effect of flood on animals. Ecol. 29: 505–507.
1950. Populations and home range relationships of the box turtle, *Terrapene c. carolina* (Linnaeus). Ecol. Mono. 20: 351–378.

Stimson, John
1970. Territorial behavior of the owl limpet *Lottia gigantea*. Ecol. 51: 113–118.

Stirton, R. A.
1947. Observations on evolutionary rates in hypsodonty. Evol. 1: 32–41.

Stockstad, D. S., Melvin S. Morris and Earl C. Lory
1953. Chemical characteristics of natural licks used by big game animals in western Montana. Trans. 18th No. Amer. Wild. Conf.: 247–258.

Stoddard, Herbert L.
1931. The bobwhite quail. Charles Scribner's Sons, New York: i–xxix, 1–559.

Stoddart, Lawrence A. and D. I. Rasmussen
1945. Big game-range livestock competition on western ranges. Trans. 10th No. Amer. Wildl. Conf.: 251–256.

Stoddart, Lawrence A. and Arthur D. Smith
1943. Range management. McGraw-Hill Book Co., Inc., New York: i–xii, 1–547.

Stone, Edward C.
1965. Preserving vegetation in parks and wilderness. Sci. 150 (3701): 1261–1267.

Storer, Tracy I., Francis C. Evans, and Fletcher G. Palmer
1944. Some rodent populations in the Sierra Nevada of California. Ecol. Mono. 14: 165–192.

Strahler, Arthur N.
1951. Physical geography. John Wiley & Sons, Inc. New York: Chap. 11.

Strandine, Eldon J.
1941. Quantitative study of a snail population. Ecol. 22: 86–91.

Strecker, Robert L.
1954. Regulatory mechanisms in house-mouse populations: the effect of limited food supply on an unconfined population. Ecol. 35: 249–253.

Strecker, Robert L. and John T. Emlen, Jr.
1953. Regulatory mechanisms in house-mouse populations: the effect of limited food supply on a confined population. Ecol. 34: 375–385.

Strickland, A. H.
1945. A survey of the arthropod soil and litter fauna of some forest reserves and cacao estates in Trinidad, British West Indies. J. An. Ecol. 14: 1–11.

Strohecker, H. F.
1937. An ecological study of some Orthoptera of the Chicago area. Ecol. 18: 231–250.

Stross, Raymond G. and J. R. Stottlemyer
1965. Primary production in the Patuxent River. Chesapeake Sci., 6: 125–140.

Sukachev, V. and N. Dylis
1964. Fundamentals of forest biogeocoenology. Oliver and Boyd, London: i–viii + 1–672 (Transl. from Russian).

Summerhayes, U. S.
1941. The effect of voles (*Microtus agrestis*) on vegetation. J. Ecol. 29: 14–48.

Sumner, F. B.
1924. The stability of subspecific characters under changed conditions of environment. Amer. Nat. 58: 481–505.
1925. Some biological problems of our southwestern deserts. Ecol. 6: 352–371.
1935. Studies of protective color change. III. Experiments with fishes both as predators and prey. Proc. Nat. Acad. Sci. 21: 345–353.

Surber, Eugene W.
1937. Rainbow trout and bottom fauna production in one mile of stream. Trans. Amer. Fish. Soc. 66: 193–202.

Svärdson, Gunnar
1949. Competition and habitat selection in birds. Oikos 1: 157–174.

Sverdrup, H. U., Martin W. Johnson and Richard H. Fleming
1942. The oceans. Prentice-Hall, Inc., New York: i–x, 1–1087.

Swaine, J. M. and F. C. Craighead
1924. Studies on the spruce budworm (*Cacoecia fumiferana* Clem.) Dom. Canada, Dept. Agric., Bull. 37: 1–27.

Swan, Lawrence W.
1961. The ecology of the high Himalayas. Sci. Amer. 205: 68–78.

Swingle, Homer S. and E. V. Smith
1941. The management of ponds for the production of game and pan fish. A symposium in hydrobiology. Univ. Wisconsin Press: 218–226.
1942. Management of farm fish ponds. Alabama Poly. Inst. Agr. Exp. Sta.: 1–23.

Taber, Richard D.
1956. Uses of marking animals in ecological studies: marking of mammals; standard methods and new developments. Ecol. 37: 681–685.

Taber, Richard D. and Raymond F. Dasmann
1957. The dynamics of three natural populations of the deer *Odocoileus hemionus columbianus*. Ecol. 38: 233–246.

Tait, R. V.
1968. Elements of marine ecology. Plenum Press, New York: i–vii + 1–304.

Talbot, Lee M.
1966. Wild animals as a source of food. Bur. Sport Fish. Wildl., Sp. Sci. Rep't.-Wildl. 98: i–iv + 1–16.

Talbot, Mary
1934. Distribution of ant species in the Chicago region with reference to ecological factors and physiological toleration. Ecol. 15: 416–439.
1953. Ants of an old field community on the Edwin S. George Reserve, Livingston County, Michigan. Univ. Michigan Cont. Lab. Vert. Biol. 63: 1–13.

Tanner, James T.
1966. Effects of population density on growth rates of animal populations. Ecol. 47: 733–745.

Tanner, V.
1944. Newfoundland-Labrador: The plant life, the animal life. University Press, Cambridge: 330–436.

Tansley, A. G.
1935. The use and abuse of vegetational concepts and terms. Ecol. 16: 284–307.

Tappa, Donald W.
1965. The dynamics of the association of six limnetic species of *Daphnia* in Aziscoos Lake, Maine. Ecol. Mono. 35: 395–423.

Tarzwell, C. M.
1965. Biological problems in water pollution. U. S. Dept. Health, Education and Welfare: i–ix + 1–424.

Tatschl, John L.
1967. Breeding birds of the Sandia Mountains and their ecological distributions. Condor, 69: 479–490.

Tatton, J. O'G. and J. H. A. Ruzicka
1967. Organochlorine pesticides in Antarctica. Nat. 215: 346–348.

Taverner, P. A.
1934. Birds of the eastern Arctic. Canada's Eastern Arctic, Dept. Int.: 113–128.

Taverner, Percy A. and George Miksch Sutton.
1934. The birds of Churchill, Manitoba. Ann. Carnegie Mus. 23: 1–83.

Tax, Sol (ed.)
1960. Evolution after Darwin. Vol. 1. The evolution of life. Univ. Chicago Press: i–viii + 1–629.

Taylor, A. W.
1967. Phosphorus and water pollution. J. Soil Water Cons. 22: 228–231.

Taylor, Clyde C., Henry B. Bigelow and Herbert W. Graham
1957. Climatic trends and the distribution of marine animals in New England. U. S. Fish and Wildl. Serv., Fish. Bull. 115: 293–345.

Taylor, Harden F.
1951. Survey of marine fisheries of North Carolina. Univ. of North Carolina Press, Chapel Hill: i–xii, 1–555.

Taylor, L. R.
1960. The distribution of insects at low levels in the air. J. An. Ecol. 29: 45–63.
1970. The optimum population for Britain. Academic Press, New York: i–xxiii + 1–182.

Taylor, L. R. and H. Kalmus
1954. Dawn and dusk flight of *Drosophila subobscura* Collin. Nat. 174: 221–222.

Taylor, Walter P.
1930. Methods of determining rodent pressure on the range. Ecol. 11: 523–542.
1934. Significance of extreme or intermittent conditions in distribution of species and management of natural resources, with a restatement of Liebig's law of the minimum. Ecol. 15: 374–379.
1935. Some animal relations to soils. Ecol. 16: 127–136.
1936. Some effects of animals on plants. Sci. Month. 43: 262–271.

Teal, John M.
1957. Community metabolism in a temperate cold spring. Ecol. Mono. 27: 283–302.
1962. Energy flow in the salt marsh ecosystem of Georgia. Ecol. 43: 614–624.

Tebo, Leonidas B.
1955. Bottom fauna of a shallow eutrophic lake, Lizard Lake, Pocahontas County, Iowa. Amer. Midl. Nat. 54: 89–103.

Thayer, Gerald H.
1910. Concealing-coloration in the animal kingdom. Macmillan & Co., London; i–xix, 1–260.

Thienemann, August
1926. Der Nahrungekreislauf im Wasser. Verh. deutsch. Zool. Ges. 31: 29–79.
1950. Verbreitungsgeschichte der Süsswassertierwelt Europas. Die Binnengewässer, E. Schweizerbart'sche Verlagsbuchhandlung (Erwin Nageln), Stuttgart, 18: i–xvi, 1–809.

Thomas, Eberhard
1969. Die Drift von *Asellus coxalis septentrionalis* Herbet (Isopoda). Oikos 20: 231–247.

Thomas, Edward S.
1951. Distribution of Ohio animals. Ohio J. Sci. 51: 153–167.

Thomas, Lyell J.
1944. Researches in life histories of parasites of wildlife. Trans. Illinois Acad. Sci. 37: 7–24.

Thomas, William A.
1968. Decomposition of loblolly pine needles with and without addition of dogwood leaves. Ecol. 49: 568–571.

Thomas, William L. (ed.)
1956. Man's role in changing the face of the earth. Univ. Chicago Press, Chicago: i–xxxviii + 1–1193.

Thomasson, K.
1956. Reflections on arctic and alpine lakes. Oikos 7: 117–143.

Thompson, Daniel Q.
1955. The 1953 lemming emigration at Point Barrow, Alaska. Arctic, 8: 37–45.
1955. The role of food and cover in population fluctuations of the brown lemming at Point Barrow, Alaska. Trans. 20th No. Amer. Wildl. Conf.: 166–176.

Thompson, David H.
1933. The migration of Illinois fishes. Illinois Nat. Hist. Surv., Biol. Notes 1 (mimeo): 1–25.
1941. The fish production of inland streams and lakes. A symposium on hydrobiology, Univ. Wisconsin Press: 206–217.

Thompson, Donald R.
1958. Field techniques for sexing and aging game animals. Wisconsin Cons. Dept., Madison, Sp. Wildl. Rep't. 1: 1–44.

Thompson, W. R.
1928. A contribution to the study of biological control and parasite introduction in continental areas. Parasit. 20: 90–112.
1929. On the relative value of parasites and predators in the biological control of insect pests. Bull. Ent. Res. 19: 343–350.
1939. Biological control and the theories of the interactions of populations. Parasit. 31: 299–388.

Thornthwaite, C. Warren
1940. Atmospheric moisture in relation to ecological problems. Ecol. 21: 17–28.

Thorp, James
1949. Effects of certain animals that live in soil. Sci. Month. 68: 180–191.

Thorpe, W. H.
1939. Further studies on pre-imaginal olfactory conditioning in insects. Proc. Royal Soc., London 127: 424–433.
1940. Ecology and the future of systematics. In Huxley's "The new systematics", Clarendon Press, Oxford: 341–364.
1945. The evolutionary significance of habitat selection. J. An. Ecol. 14: 67–70.
1950. Plastron respiration in aquatic insects. Biol. Rev. 25: 344–390.
1951. The definition of some terms used in animal behavior studies. Bull. An. Beh. 9: 1–7.
1956. Learning and instinct in animals. Methuen and Co., Ltd., London: i–viii, 1–493.

Tiger, Lionel
1970. Dominance in human societies. An. Rev. Ecol. Syst. 1: 287–306.

Tiger, Lionel and Robin Fox
1971. The imperial animal. Holt, Rinehart and Winston, New York: i–xi + 1–308.

Tilly, Laurence J.
1968. The structure and dynamics of Cone Spring. Ecol. Mono. 38: 169–197.

Timofeeff-Ressovsky, N. W.
1933. Uber die relative Vitalität von *Drosophila melanogaster* Meigen und *D. funebris* (Diptera, Muscidae acalypteratae) unter verschiedenen Zuchtbedingungen, in Zusammenhang mit der Verbreitungsarealen dieser Arten. Frabricius. Arch. Naturgesch (N. S.) 2: 285–290.
1940. Mutations and geographical variation. In Huxley's "The new systematics", Clarendon Press, Oxford: 73–136.

Tinbergen, Lukas
1946. De Sperwer als Roofvijand van Zangvogels (English summary). E. J. Brill, Leiden: 1–213.

1960. Factors influencing the intensity of predation by songbirds. Arch. Neerl. Zool., 13: 265–336.

Tinbergen, N.
1933. Die ernährungsökologischen Beziehungen zwischen *Asio otus otus* L. und ihren Beutetieren, inbesondre, den *Microtus—* Arten. Ecol. Mon. 3: 443–492.
1951. The study of instinct. Clarendon Press, Oxford: i–xii, 1–228.
1968. On war and peace in animals and man. Sci. 160: 1411–1418.

Tinkham, Ernest R.
1948. Faunistic and ecological studies on the Orthoptera of the Big Bend region of Trans-Pecos Texas. Amer. Midl. Nat. 40: 521–663.

Tinkle, Donald W., Don McGregor, and Sumner Dana
1962. Home range ecology of *Uta stansburiana* Stejnegeri. Ecol. 43: 223–229.

Tischler, Wolfgang
1951. Zur Synthese Biozönotischer Forschung. Acta Bioth. 9: 135–162.

Tobins, Jerome D.
1971. Differential niche utilization in a grassland sparrow. Ecol. 52: 1065–1070.

Todd, Valerie
1949. The habits and ecology of the British harvestmen (Arachnida, Opiliones), with special reference to those of the Oxford district. J. An. Ecol. 18: 209–229.

Tompkins, Grace
1933. Individuality and territoriality as displayed in winter by three passerine species. Condor 35: 98–106.

Tothill, John D.
1922. The natural control of the fall webworm (*Hyphantria cunea* Drury) in Canada. Dom. Canada Dept. Agr. Bull. 3, N. S. Tech.: 1–107.

Tramer, Elliot J.
1969. Bird species diversity: components of Shannon's formula. Ecol. 50: 927–929.

Transeau, Edgar Nelson
1935. The prairie peninsula. Ecol. 16: 423–437.

Trautman, Milton B.
1942. Fish distribution and abundance correlated with stream gradients as a consideration in stocking programs. Trans. 7th No. Amer. Wildl. Conf.: 211–224.

Trippensee, Reuben Edwin
1948. Wildlife management. McGraw-Hill Book., Inc., New York: 2 vols (Vol. 2, 1953).

Trojan, Przemyslaw
1970. Energy flow through a population of *Microtus arvalis* (Pall.) in an agrocenosis during a period of mass occurrence. An ecological model of the costs of maintenance of *Microtus arvalis* (Pall.). In K. Petrusewicz and L. Ryszkowski's "Energy flow through small mammal populations," Polish Sci.Publ., Warsaw: 113–122, 267–279.

Turček, Frantisek J.
1951. On the stratification of the avian population of the Querceto-Carpinetum forest communities in southern Slovakia (English summary). Sylvia 13: 71–86.
1952. Ecological analysis of the bird and mammalian population of a primeval forest on the Pol'ana-mountain (Slovakia) (English extract). Rozpravy II tridy Ceske akademi, 62: 1–51.
1955. Bird populations of three types of forest communities in Slovakia (English summary). Biologia 10: 293–308.
1956. Zur Frage der Dominanz in Vogelpopulationen. Waldhygiene 8: 249–257.

Turk, Amos, J. Turk, and J. T. Wittes
1972. Ecology, pollution, environment. W. B. Saunders Co., Philadelphia: 1–217.

Turner, Frederick B.
1970. The ecological efficiency of consumer populations. Ecol. 51: 741–742.

Turner, Frederick B., Robert I. Jennrich, and Joel D. Weintraub
1969. Home ranges and body size of lizards. Ecol. 50: 1076–1081.

Twenhofel, W. H.
1939. Principles of sedimentation. McGraw-Hill Book Co., Inc., New York: Chap. VI.

Udvardy, Miklos D. F.
1951. The significance of interspecific competition in bird life. Oikos 3: 98–123.
1958. Ecological and distributional analysis of North American birds. Condor 60: 50–66.
1959. Notes on the ecological concepts of habitat, biotope and niche. Ecol. 40: 725–728.

1963. Bird faunas of North America. Proc. XIII Inter. Ornith. Cong.: 1147–1167.
1969. Dynamic zoogeography. Van Nostrand Reinhold Co., New York: i–xviii + 1–445.

Ullyett, G. C.
1950. Competition for food and allied phenomena in sheep-blowfly populations. Phil. Trans. Roy. Soc. London, B 234: 77–174.

Umbgrove, J. H. F.
1946. Evolution of reef corals in East Indies since Miocene time. Bull. Amer. Assoc. Petrol. Geol. 30: 23–31.

Uramoto, Masanori
1961. Ecological study of the bird community of the broad-leaved deciduous forest of Japan. Yamashina's Inst. Ornith. Zool., Misc. Reports 3: 1–32.

Urquhart, F. A.
1957. Changes in the fauna of Ontario. Univ. Toronto Press, Toronto: i–iv, 1–75.
1957. A discussion of Batesian mimicry as applied to the monarch and viceroy butterflies. University Toronto Press, Canada: 1–27.
1958. A discussion of the use of the word "migration" as it relates to a proposed classification for animal movements. Roy. Ont. Mus. Zool. and Paleo., Contr. 50: 1–11.

Utida, Syunro
1953. Interspecific competition between two species of bean weevil. Ecol. 34: 301–307.

Uvarov, B. P.
1928. Locusts and grasshoppers. Imp. Bur. Ento., London: i–xiii, 1–352.
1931. Insects and climate. Trans. Ento. Soc. London 79: 1–247.

Valentine, James W.
1968. Climatic regulation of species diversification and extinction. Bull. Geol. Soc. Amer. 79: 273–275.

Vallentyne, J. R.
1965. Net primary productivity and photosynthetic efficiency in the biosphere. In C. R. Goldman's "Primary productivity in aquatic environments." Univ. California Press, Berkeley: 311 (abstract).

Valverde, Jose A.
1964. Remarques sur la structure et l'evolution des communautes de vertebres terrestres. I. Structure d'une communaute. II. Rapports entre predateurs et proies. La Terre et la Vie, 111: 121–154.

Van der Drift, Joseph
1950. Analysis of the animal community in a beech forest floor. Ponsen and Looijen, Wageningen: 1–168.

Vandersal, William R.
1937. The dependence of spoils on animal life. Trans. 2nd. No. Amer. Wildl. Conf.: 458–467.

Van der Schalie, Henry
1941. Zoogeography of naiades in the Grand and Muskegon rivers of Michigan as related to glacial history. Michigan Acad. Sci., Arts, Letters, Paper 26: 297–310.

Van Deusen, R. D.
1954. Maryland freshwater stream classification by watersheds. Chesapeake Biol. Lab., Cont. 106: 1–30.

Van Deventer, William Carl
1937. Studies on the biology of the crayfish, *Cambarus propinquus* Girard. Illinois Biol. Mono. 15: 1–67.

Van Dyne, George M. (ed.)
1969. The ecosystem concept in natural resource management. Academic Press, Inc., N. Y.: i–xii + 1–383.

Van Hook, Jr., Robert I.
1971. Energy and nutrient dynamics of spider and orthopteran populations in a grassland ecosystem. Ecol. Mono. 41: 1–26.

Van Rossem, A. J.
1945. A distributional survey of the birds of Sonora, Mexico. Mus. Zool., Louisiana St. Univ., Occup. Papers 21: 1–379.

Van Vleck, David B.
1969. Standardization of *Microtus* home range calculation. J. Mam. 50: 69–80.

Varley, George C.
1947. The natural control of population balance in the knapweed gall-fly (*Urophora jaceana*). J. An. Ecol. 16: 139–187.
1959. The biological control of agricultural pests. J. Roy. Soc. Arts, 107: 475–490.
1970. The concept of energy flow applied to a woodland community. In Adam Watson's "Animal populations in relation to their food resources," Blackwell Sci. Publ., Oxford, England: 389–405.

Vaughan, Terry A.
1954. Mammals of the San Gabriel Mountains of California. Univ. Kans. Publ., Mus. Nat. Hist. 7 (9): 513–582.

Vaughan, Thomas Wayland
    1919. Corals and the formation of coral reefs. Smithsonian Report (1917): 189–276.
Verduin, Jacob
    1956. Energy fixation and utilization by natural communities in western Lake Erie. Ecol. 37: 40–50.
    1957. Daytime variations in phytoplankton photosynthesis. Limn. and Ocean. 2: 333–336.
    1969. Man's influence on Lake Erie. Ohio J. Sci. 69: 65–70.
Vesey-FitzGerald, Desmond Foster
    1960. Grazing succession among East African game animals. J. Mam., 41: 161–172.
Vestal, Arthur G.
    1913. An associational study of Illinois sand prairie. Bull. Illinois State Lab., 10: 1–96.
    1939. Why the Illinois settlers chose forest lands. Trans. Illinois St. Acad. Sci., 32: 85–87.
    1949. Minimum areas for different vegetations. Their determination from species-area curves. Illinois Biol. Mono. 20: 1–129.
Vinogradova, N. G.
    1960. The geographic distribution of deep water benthal fauna of the Antarctic. Ref. Zhur., Biol., 1961, No. 2 D653 (Translation).
Visher, S. S.
    1916. The biogeography of the northern Great Plains. Geog. Rev. 2: 89–115.
Vogtman, Donald B.
    1945. Flushing tube for determining food of game birds. J. Wildl. Man. 9: 255–257.
Volpe, E. Peter
    1952. Physiological evidence for natural hybridization of Bufo americanus and Bufo fowleri. Evol. 6: 393–406.
Volterra, Vito
    1931. Variations and fluctuations of the number of individuals in animal species living together. In R. N. Chapman's "Animal ecology," McGraw-Hill Book Co., Inc., New York: 409–448.
Voous, K. H.
    1963. The concept of faunal elements or faunal types. Proc. XIII Inter. Ornith. Cong.: 1104–1108.
Vos, A. de and Randolph L. Peterson
    1951. A review of the status of woodland caribou (Rangifer caribou) in Ontario. J. Mam. 32: 329–337.
Voûte, A. D.
    1946. Regulation of the density of the insect-populations in virgin forests and cultivated woods. Arch. Neerl. Zool. 7: 435–470.
Vuilleumier, Beryl Simpson
    1971. Pleistocene changes in the fauna and flora of South America. Sci. 173 (3999): 771–780.
Vuilleumier, Francois
    1970. Insular biogeography in continental regions. 1. The northern Andes of South America. Amer. Nat. 104: 373–388.
Waddington, C. H.
    1957. The strategy of the genes. George Allen & Unwin, Ltd., London: i–ix, 1–262.
Wade, Phyllis
    1958. Breeding season among mammals in the lowland rainforest of North Borneo. J. Mam. 39: 429–433.
Wadsworth, R. M. (ed.)
    1969. The measurement of environmental factors in terrestrial ecology (a symposium). Blackwell Sci. Publ., Oxford: i–x + 1–314.
Wagar, J. Alan
    1970. Growth versus the quality of life. Sci. 168 (3936): 1179–1184.
Wagner, Richard H.
    1971. Environment and man. W. W. Norton & Co., Inc., New York: i–xiii + 1–491.
Waksman, Selman A.
    1952. Soil microbiology. John Wiley & Sons, Inc., New York: i–viii, 1–356.
Walcheck, Kenneth C.
    1970. Nesting bird ecology of four plant communities in the Missouri River Breaks, Montana. Wilson Bull. 82: 370–382.
Waldichuk, Michael
    1956. Basic productivity of Trevor Channel and Alberni Inlet from chemical measurements. J. Fish. Res. Bd. Canada 13: 7–20.
Waldron, Ingrid
    1964. Courtship sound production in two sympatric sibling Drosophila species. Sci. 144: 191–192.
Walford, Lionel A.
    1958. Living resources of the sea. Ronald Press Co., New York: i–xv, 1–321.

Walker, Ilse
    1967. Effect of population density on the viability and fecundity in Nasonia vitripennis Walker (Hymenoptera, Pteromalidae). Ecol. 48: 294–301.
Wallace, Alfred R.
    1876. The geographical distribution of animals. Macmillan and Co., London: 2 vols.
Wallen, I. E.
    1955. Some limnological considerations in the productivity of Oklahoma farm ponds. J. Wildl. Man. 19: 450–462.
Walls, G. L.
    1942. The vertebrate eye and its adaptive radiation. Cranbrook Press, Bloomfield Hills, Michigan, Bull. 19: 1–785.
Wallwork, John A.
    1959. The distribution and dynamics of some forest soil mites. Ecol. 40: 557–563.
    1970. Ecology of soil animals. McGraw-Hill, New York: 1–283.
Walter, Herbert E.
    1906. The behavior of the pond snail, Lymnaeus elodes Say. Cold Spring Harbor Mono. 6: 1–35.
Wang, Chi-Wu
    1961. The forests of China with a survey of grassland and desert vegetation. Maria M. Cabot Foundation, Harvard Univ., Cambridge, Mass., Publ. 5: i–xiv + 1–313.
Wang, Chia Chi
    1928. Ecological studies on the seasonal distribution of Protozoa in a fresh-water pond. J. Morph., 46: 431–478.
Warburg, M. R.
    1965. The evolutionary significance of the ecological niche. Oikos 16: 205–213.
Warden, Carl J., Thomas N. Jenkins, and Lucien H. Warner.
    1935–36–40. Comparative Psychology. Ronald Press, New York: 3 vols.
Warming, Eug.
    1895. Plantesamfund (Revised for English edition as "Oecology of plants", 1909), Clarendon Press, Oxford: 1–xi, 1–422.
Warner, Richard W.
    1971. Distribution of biota in a stream polluted by acid mine-drainage. Ohio J. Sci. 71: 202–215.
Warnock, John E.
    1965. The effects of crowding on the survival of meadow voles (Microtus pennsylvanicus) deprived of cover and water. Ecol. 46: 649–664.
Warren, Charles E.
    1971. Biology and water pollution control. W. B. Saunders Co., Philadelphia, Pa.: i–xvi + 1–434.
Wasilewska, Lucyna
    1970. Nematodes of the sand dunes in the Kampinos forest. I. Species structure. Ekol. Polska, A, 18 (20): 429–443.
Waterhouse, F. L.
    1955. Microclimatological profiles in grass cover in relation to biological problems. Quart. J. Roy. Met. Soc. 81: 63–71.
Waters, Thomas F.
    1965. Interpretation of invertebrate drift in streams. Ecol. 46: 327–334.
    1966. Production rate, population density, and drift of a stream invertebrate. Ecol. 47: 595–604.
Watkins, Julian F, II; Frederick R. Gehlbach; and James C. Kroll
    1969. Attractant-repellent secretions of blind snakes (Leptotyphlops dulcis) and their army ant prey (Neivamyrmex nigrescens). Ecol. 50: 1098–1102.
Watmough, R. H. P.
    1968. Population studies on two species of Psyllidae (Homoptera, Sternorhyncha) on broom (Sarothamnus scoparius [L.] Wimmer). J. An. Ecol. 37: 283–314.
Watson, Adam
    1956. Ecological notes on the lemmings Lemmus trimucronatus and Dicrostonyx groenlandicus in Baffin Island. J. An. Ecol. 25: 289–302.
    1963. Bird numbers on tundra in Baffin Island. Arctic 16: 101–108.
Watson, Adam and David Jenkins
    1968. Experiments on population control by territorial behaviour in red grouse. J. An. Ecol. 37: 595–614.
Watson, Adam and Robert Moss
    1970. Dominance, spacing behaviour and aggression in relation to population limitation in vertebrates. In Adam Watson's "Animal populations in relation to their food resources," Blackwell Sci. Publ., Oxford, England: 167–220.
Watt, Kenneth E. F.
    1955. Studies on population productivity. I. Three approaches to the optimum yield problem in populations of Tribolium confusum. Ecol. Mono. 25: 269–290.

1966. Systems analysis in ecology. Academic Press, N. Y.: i–viii + 1–276.

1968. Ecology and resource management. McGraw-Hill Book Co., New York: i–xii + 1–450.

Wayne, William J.
1967. Periglacial features and climatic gradient in Illinois, Indiana, and western Ohio, east-central United States. In E. J. Cushing and H. E. Wright, Jr.'s "Quaternary paleoecology," Yale Univ. Press, New Haven, Connecticut: 393–414.

Weaver, C. Richard
1943. Observations on the life cycle of the fairy shrimp *Eubrancipus vernalis*. Ecol. 24: 500–502.

Weaver, J. E.
1927. Some ecological aspects of agriculture in the prairie. Ecol. 8: 1–17.

1944. North American prairie. Amer. Scholar 13: 329–339.

1954. North American prairie. Johnsen Publ. Co., Lincoln, Neb.: i–xi, 1–348.

Weaver, J. E. and F. W. Albertson
1956. Grasslands of the Great Plains. Their nature and use. Johnsen Publ. Co., Lincoln, Neb.: i–ix, 1–395.

Weaver, John E. and Frederic E. Clements
1938. Plant ecology. McGraw-Hill Book Co., Inc., New York: i–xx, 1–520.

Weaver, J. E. and T. J. Fitzpatrick
1934. The prairie. Ecol. Mono. 4: 111–295.

Weaver, J. E. and Evan L. Flory
1934. Stability of climax prairie and some environmental changes resulting from breaking. Ecol. 15: 333–347.

Weaver, J. E. and G. W. Tomanek
1951. Ecological studies in a midwestern range: the vegetation and effects of cattle on its composition and distribution. Nebraska Cons. Bull. 31: 1–82.

Webb, Glenn R.
1947. Studies of the sex-organs of mating polygyrid landsnails. Illinois Acad. Sci. Trans. 40: 218–227.

Webb, Majorie G.
1956. An ecological study of brackish water ciliates. J. An. Ecol. 25: 148–175.

1961. The effects of thermal stratification on the distribution of benthic protozoa in Esthwaite Water. J. An. Ecol. 30: 137–151.

Webb, William L.
1957. Interpretation of overbrowsing in northeastern forests. J. Wildl. Man. 21: 101–103.

Weber, Neal A.
1957. Fungus-growing ants and their fungi: *Cyphomyrmex costatus*. Ecol. 38: 480–494.

Wecker, Stanley C.
1963. The role of early experience in habitat selection by the prairie deer mouse, *Peromyscus maniculatus bairdi*. Ecol. Mono. 33: 307–325.

Weese, Asa Orrin
1924. Animal ecology of an Illinois elm-maple forest. Illinois Biol. Mon. 9: 1–93.

Wegener, Alfred
1924. The origin of continents and oceans (English trans.). Methuen & Co., London: i–xx, 1–212.

Weinberg, Alvin M. and R. Philip Hammond
1970. Limits to the use of energy. Amer. Sci. 58: 412–418.

Welch, Paul S.
1948. Limnological methods. McGraw-Hill Book Co., Inc. Blakiston Div., Philadelphia: i–xviii, 1–381.

1952. Limnology. McGraw-Hill Book Co., Inc., New York: i–xi, 1–538.

Weller, Milton W.
1959. Parasitic egg laying in the redhead (*Aythya americana*) and other North American Anatidae. Ecol. Mono. 29: 333–365.

Wellington, W. G.
1964. Qualitative changes in populations in unstable environments. Can. Ent. 96: 436–451.

Wells, John W.
1954. Recent corals of the Marshall Islands. U. S. Geol. Surv., Profess. Paper No. 260–I: 385–486.

Wells, M. M.
1918. The reactions and resistance of fishes to carbon dioxide and carbon monoxide. Illinois St. Lab. Nat. Hist., 11: 557–569.

Wells, Philip V.
1965. Scarp woodlands, transported grassland soils, and concept of grassland climate in the Great Plains region. Sci. 148: 246–249.

1966. Late Pleistocene vegetation and degree of pluvial climatic change in the Chihuahuan Desert. Sci. 153 (3739): 970–975.

1970. Postglacial vegetational history of the Great Plains. Sci. 167: 1574–1582.

Went, F. W. and N. Stark
1968. Mycorrhiza. BioSci. 18: 1035–1039.

Westlake, D. F.
1965. Some basic data for investigations of the productivity of aquatic macrophytes. In C. R. Goldman's, "Primary productivity in aquatic environments," Univ. California Press, Berkeley: 231–267.

Westveld, Marinus
1956. Natural forest vegetation zones of New England. J. For. 54: 332–338.

Wetmore, Alexander
1954. The sleeping habit of the willow ptarmigan. Auk 62: 638.

Wetzel, Ralph Martin
1949. Analysis of small mammal populations in the deciduous forest biome. University of Illinois, Ph. D. thesis.

1958. Mammalian succession on midwestern floodplains. Ecol. 39: 262–271.

Wheeler, William Norton
1923. Social life among the insects. Harcourt, Brace & Co., New York: i–vii, 1–375.

Whipple, George C.
1927. The microscopy of drinking water. John Wiley & Sons, Inc., New York: i–xix, 1–586.

White, Joseph James
1968. Bioenergetics of the woodlouse *Tracheoniscus rathkei* Brandt in relation to litter decomposition in a deciduous forest. Ecol. 49: 694–704.

White, Leslie A.
1959. The concept of culture. Amer. Anthro. 61: 227–251.

White, Ray J. and Oscar M. Brynildson
1967. Guidelines for management of trout stream habitat in Wisconsin. Wisconsin Dept. Nat. Res., Div. Cons., Tech. Bull. 39: 1–65.

Whitehead, Donald R.
1965. Palynology and Pleistocene phytogeography of unglaciated eastern North America. In H. E. Wright, Jr. and D. G. Frey's "The Quaternary of the United States," Princeton Univ. Press: 417–432.

Whiteside, Melbourne C. and Rodney V. Harmsworth
1967. Species diversity in chydorid (Cladocera) communities. Ecol. 48: 664–667.

Whitmore, Frank C., Jr. and R. H. Stewart
1965. Miocene mammals and central American seaways. Sci. 148 (3667): 180–185.

Whitney, R. J.
1939. The thermal resistance of mayfly nymphs from ponds and streams. J. Exp. Biol. 16: 374–385.

Whittaker, Robert H.
1951. A criticism of the plant association and climatic climax concept. Northwest Sci. 25: 17–31.

1952. A study of summer foliage insect communities in the Great Smoky Mountains. Ecol. Mono. 22: 1–44.

1953. A consideration of climax theory: the climax as a population and pattern. Ecol. Mono. 23: 41–78.

1956. Vegetation of the Great Smoky Mountains. Ecol. Mono. 26: 1–80.

1957. Recent evolution of ecological concepts in relation to the eastern forests of North America. Amer. J. Bot. 44: 197–206.

1961. Experiments with radiophosphorus tracer in aquarium microcosms. Ecol. Mono. 31: 157–188.

1962. The classification of natural communities. Bot. Rev. 28: 2–239.

1965. Dominance and diversity in land plant communities. Sci. 147 (3655): 250–260.

1967. Gradient analysis of vegetation. Biol. Rev. 42: 207–264.

1969. Evolution of diversity in plant communities. In G. M. Woodwell and H. H. Smith's "Diversity and stability in ecological systems," Brookhaven Sym. Biol. 22, Brookhaven Nat. Lab., Upton, N. Y.: 178–195.

1970. Communities and ecosystems. Macmillan Co., New York: i–xi + 1-162.

Whittaker, R. H. and P. P. Feeny
1971. Allelochemics: chemical interactions between species. Sci. 171 (3973): 757–770.

Whittaker, R. H. and G. M. Woodwell
1969. Structure, production and diversity of the oak-pine forest at Brookhaven, New York. J. Ecol. 57: 155–174.

1972. Evolution of natural communities. In J. A. Wiens' "Ecosystem structure and function," Proc. 31st Ann. Biol. Coll., Oregon St. Univ. Press: 137–159.

Wickler, Wolfgang
1968. Mimicry in plants and animals. McGraw Hill Book Co., New York: 1–255.

Wiebe, A. H.
1931. Notes on the exposure of several species of pond fishes to sudden changes in pH. Trans. Amer. Micros. Soc. 50: 380–393.

Wiegert, Richard G.
1964. Population energetics of meadow spittlebugs as affected by migration and habitat. Ecol. Mono. 34: 217–241.
1965. Energy dynamics of the grasshopper populations in old field and alfalfa field ecosystems. Oikos 16: 161–176.

Wiegert, R. G., D. C. Coleman, and E. P. Odum
1971. Energetics of the litter-soil subsystem. Proc. Sym., "Methods of study in soil ecology," IBP-UNESCO, Paris: 93–98.

Wiens, J. A.
1966. On group selection and Wynne-Edwards' hypothesis. Amer. Sci. 54: 273–287.
1969. An approach to the study of ecological relationships among grassland birds. Ornith. Mono. No. 8: 1–93.

Wilde, S. A.
1966. A new systematic terminology of forest humus layers. Soil Sci. 101: 403–407.

Wilhm, Jerry L.
1967. Comparison of some diversity indices applied to populations of benthic macroinvertebrates in a stream receiving organic wastes. J. Water Pollution Control Fed., Washington, D. C. (Pt. 1): 1673–1683.

Wilhm, Jerry L. and T. C. Dorris
1968. Biological parameters for water quality criteria. BioSci. 18: 477–481.

Williams, Arthur B.
1936. The composition and dynamics of a beech-maple climax community. Ecol. Mono. 6: 318–408.

Williams, C. B.
1947. The generic relations of species in small ecological communities. J. Ecol. 16: 11–18.
1954. The statistical outlook in relation to ecology. J. Ecol. 42: 1–13.
1958. Insect migration. Macmillan Co., New York: i–xiii, 1–235.
1964. Patterns in the balances of nature and related problems in quantitative ecology. Academic Press, New York: i–vii + 1–324.

Williams, Carroll W.
1967. Third-generation pesticides. Sci. Amer. 217 (1): 13-17.

Williams, Cecil S. and William H. Marshall
1938. Duck nesting studies, Bear River Migratory Bird Refuge, Utah, 1937. J. Wildl. Man. 2: 29–48.

Williams, Eliot C., Jr.
1941. An ecological study of the floor fauna of the Panama rain forest. Bull. Chicago Acad. Sci. 6: 63–124.

Williams, G. R.
1954. Population fluctuations in some northern hemisphere game birds (Tetraonidae). J. An. Ecol. 23: 1–34.
1963. A four-year population cycle in California quail, Lophortyx californicus (Shaw) in the South Island of New Zealand. J. An. Ecol., 32: 441–459.

Williamson, Penelope
1971. Feeding ecology of the red-eyed vireo (Vireo olivaceus) and associated foliage-gleaning birds. Ecol. Mono., 41: 130–152.

Willson, Mary F.
1971. Life history consequences of death rates. Biol. 53: 49–56.

Wilson, Carroll L. and William H. Matthews
1970. Man's impact on the global environment. Rep't. Study Crit. Env. Probl., M.I.T. Press, Cambridge, Massachusetts: i–xxii + 1–319.

Wilson, Edward O.
1959. Some ecological characteristics of ants in New Guinea rain forests. Ecol. 40: 437–447.

Wilson, Edward O. and William H. Bossert
1971. A primer of population biology. Sinauer Associates, Inc. Publishers, Stamford, Connecticut: 1–192.

Wilson, E. O. and W. L. Brown, Jr.
1953. The subspecies concept and its taxonomic application. Syst. Zool. 2: 97–111.

Wilson, Larry
1952. In search of lost forests. Forest & Outdoors 48: 8–9, 24–25.

Wing, Leonard William
1935. Wildlife cycles in relation to the sun. Trans. 21st Amer. Game Conf.: 345–363.

1943. Spread of the starling and English sparrow. Auk 60: 74–87.
1954. Cycles of lynx abundance. J. Cycle Res. 2: 28–51.
1961. The 3.864-year lemming cycle and latitudinal passage in temperature. J. Cycle Res. 10 (2): 59–70.

Wingfield, C. A.
1939. The function of the gills of mayfly nymphs from different habitats. J. Exp. Biol. 16: 363–373.

Winn, Howard E.
1958. Comparative reproductive behavior and ecology of fourteen species of darters (Pisces-Percidae). Ecol. Mono. 28: 155–191.

Witkamp, Martin
1971. Soils as components of ecosystems. An. Rev. Ecol. Syst. 2: 85–110.

Wolcott, George N.
1937. An animal census of two pastures and a meadow in northern New York. Ecol. Mono. 7: 1–90.

Wolfe, John N.
1951. The possible role of microclimate. Ohio J. Sci. 51: 134–138.

Wolfe, John N., Richard T. Wareham and Herbert T. Scofield
1949. Microclimates and macroclimate of Neotoma, a small valley in central Ohio. Ohio. Biol. Surv. Bull. 41: 8: 1–267.

Wolfenbarger, D. O.
1946. Dispersion of small organisms. Distance dispersion rates of bacteria, spores, seeds, pollen, and insects; incidence rates of diseases and injuries. Amer. Midl. Nat. 35: 1–152.

Wolfson, Albert
1948. Bird migration and the concept of continental drift. Sci. 108: 23–30.

Wolvekamp, H. P.
1955. Die physikalische Kieme der Wasserinsekten. Exper. 11: 294–301.

Womble, William H.
1951. Differential systematics. Sci. 114: 315–332.

Wood, Kenneth G.
1956. Ecology of Chaoborus (Diptera: Culicidae) in an Ontario lake. Ecol. 37: 639–643.

Woodbury, Angus M.
1933. Biotic relationships of Zion Canyon, Utah, with special reference to succession. Eco. Mono. 3: 147–245.
1936. Animal relationships of Great Salt Lake. Ecol. 17: 1–8.
1947. Distribution of pigmy conifers in Utah and northeastern Arizona. Ecol. 28: 113–126.

Woodbury, Richard B.
1958–59. Pre-spanish human ecology in the southwestern deserts. Arid Lands Colloquia (Univ. Arizona): 82–92.

Woodin, Howard E. and Alton A. Lindsey
1954. Juniper-pinyon east of the Continental Divide, as analyzed by the line-strip method. Ecol. 35: 473–489.

Wood-Jones, F.
1912. Coral and atolls. Lovell Reeve & Co., Ltd.: i–xxiii, 1–392.

Woodruff, Lorande Loss
1912. Observations on the origin and sequence of the protozoan fauna of hay infusions. J. Exp. Zool. 12: 205–264.
1913. The effect of excretion products of Infusoria on the same and in different species, with special reference to the protozoan sequence in infusions. J. Exp. Zool. 14: 575–582.

Woods, Frank W. and Royal E. Shanks
1959. Natural replacement of chestnut by other species in the Great Smoky Mountains National Park. Ecology 40: 349–361.

Woodwell, G. M.
1970. Effects of pollution on the structure and physiology of ecosystems. Sci. 168 (3930): 429–433.

Woodwell, G. M. and T. G. Marples
1968. The influence of chronic gamma irradiation on production and decay of litter and humus in an oak-pine forest. Ecol. 49: 456–465.

Woodwell, G. M. and H. H. Smith
1969. Diversity and stability in ecological systems. Brookhaven Sym. Biol., Upton, N. Y., 22: i–vii + 1–264.

Woodwell, George M. and Robert H. Whittaker
1968. Primary production in terrestrial ecosystems. Amer. Zool. 8: 19–30.

Wooster, L. D.
1939. An ecological evaluation of predatees of a mixed prairie area in western Kansas. Trans. Kansas Acad. Sci. 42: 515–517.

Wright, Anna Allen and Albert Hazen Wright
1933. Handbook of frogs and toads. Comstock Publ. Co., Ithaca, New York: i–xi, 1–231.

Wright, Bruce S.
1960. Predation on big game in East Africa. J. Wildl. Man. 24: 1–15.

Wright, H. E., Jr.
1968. History of the prairie peninsula. In R. E. Bergstrom's "The Quaternary of Illinois," Univ. Illinois, Coll. Agric., Spec. Publ. 14: 78–88.
1970. Vegetational history of the central plains. In W. Dort, Jr. and J. K. Jones, Jr.'s "Pleistocene and recent environments of the central Great Plains," Univ. Press of Kansas, Lawrence: 157–172.

Wright, John C.
1958. The limnology of Canyon Ferry Reservoir. I. Phytoplankton-zooplankton relationships in the euphotic zone during September and October, 1956. Limn. and Ocean 3: 150–159.

Wright, Sewall
1931. Evolution in Mendelian populations. Genet. 16: 97–159.

Wu, Yi Fang
1931. A contribution to the biology of *Simulium* (Diptera). Papers Michigan Acad. Sci., Arts, Letters 13: 543–599.

Wulff, E. V.
1943. An introduction to historical plant geography (English translation). Chron. Bot. Co., Waltham, Mass.: i–xv, 1–223.

Wynne-Edwards, V. C.
1962. Animal dispersion in relation to social behaviour. Oliver and Boyd, Edinburgh and London: i–xi + 1–653.

Yeager, Lee E. and Harry G. Anderson
1944. Some effects of flooding and waterfowl concentrations on mammals of a refuge area in central Illinois. Amer. Midl. Nat. 31: 159–178.

Yeatter, Ralph E. and David H. Thompson
1952. Tularemia, weather and rabbit populations. Illinois Nat. Hist. Surv. Bull. 25: 351–382.

Yonge, C. M.
1949. The sea shore. Collins, London: i–xiv, 1–311.
1953. Aspects of life on muddy shores. Essays Mar. Biol., Oliver & Boyd, London: 29–49.
1958. Ecology and physiology of reef-building corals. In Buzzati-Traverso's "Perspectives in marine biology". Univ. California Press: 117–135.
1963. The biology of coral reefs. Adv. Mar. Biol. 1: 209–260.

Young, Gale
1970. Dry lands and desalted water. Sci. 167 (3917): 339–343.

Young, Vernon A.
1936. Edaphic and vegetational changes associated with injury of a white pine plantation by roosting birds. J. For. 34: 512–523.

Yount, James L.
1956. Factors that control species numbers in Silver Springs, Florida. Limn. and Ocean. 1: 286–295.

Zenkevitch, L.
1963. Biology of the seas of the USSR. (transl. from the Russian). George Allen & Unwin, Ltd., London: 1–955.

Zeuner, Frederick E.
1945. The Pleistocene period. Bernard Quaritch, Ltd., London: i–xii, 1–322.

Zimmerman, Elwood C.
1948. Insects of Hawaii. I. Introduction. Univ. Hawaii Press, Honolulu: i–xx, 1–206.

Zimmerman, John L.
1971. The territory and its density dependent effect on *Spiza americana*. Auk 88: 591–612.

Zimmermann, Rud.
1932. Ueber quantitative Bestandsaufnahmen in der Vogelwelt. Mit. Ver. Sachs. Ornith. 3: 253–267.

Zippin, Calvin
1958. The removal method of population estimation. J. Wildl. Man. 22: 82–90.

ZoBell, Claude E.
1946. Marine microbiology. Chron. Bot. Co., Waltham, Mass.: 1–240.
1952. Bacterial life at the bottom of the Philippine Trench. Sci. 115: 507–508.

ZoBell, Claude E. and Catherine B. Feltham
1942. The bacterial flora of a marine mud flat as an ecological factor. Ecol. 23: 69–78.

SUBJECT INDEX

SPECIES INDEX